DATE			

FLORA

OF THE

PRAIRIES AND PLAINS

OF

CENTRAL NORTH

AMERICA

———

BY

PER AXEL RYDBERG, Ph.D.

Late Curator of the Herbarium of
The New York Botanical Garden

IN TWO VOLUMES

Volume
Two

DOVER PUBLICATIONS, INC.

NEW YORK

Published in Canada by General Publishing Company, Ltd., 30 Lesmill Road, Don Mills, Toronto, Ontario.

Published in the United Kingdom by Constable and Company, Ltd., 10 Orange Street, London WC 2.

This Dover edition, first published in 1971, is an unabridged and corrected republication of the work originally published in one volume by The New York Botanical Garden in 1932.

Library of Congress Catalog Card Number: 79-166434
International Standard Book Number: 0-486-22585-2

Manufactured in the United States of America
Dover Publications, Inc.
180 Varick Street
New York, N.Y. 10014

CONTENTS

[For this Dover reprint edition, the text has been divided into two volumes. Volume One consists of pages 1 through 503 (Family 87: Oxalidaceae); Volume Two begins with Family 88: Linaceae at the bottom of page 503 and continues through page 969.]

ABBREVIATIONS

The well-known abbreviations of the states of the United States are here omitted.

Adv. = adventive
Alp. = Alpine Zone
Alta. = Alberta*
Am. = America or American
Ap = April
Arctic = Arctic Zone.
Au = August
Auth. = Authors†
[B] = Britton's Manual
[BB] = Britton & Brown's Illustrated Flora
B. C. = British Columbia
Boreal = Boreal Zone
c = central
C. Am. = Central America
cm. = centimeter
D = December
dm. = decimeter
e = eastern
Eur. = Europe
Eurasia = Europe and northern Asia
F = February
(Fl. Colo.) = The Author's Flora of Colorado
(Fl. Mont.) = The Author's Catalogue of the Flora of Montana and Yellowstone Park
[G] = Gray's New Manual
Greenl. = Greenland
Ja = January
Je = June
Jl = July
Labr. = Labrador
L. Calif. = Lower California
L. Son. = Lower Sonoran Zone
m. = meter
Mack. = Mackenzie Territory
Man. = Manitoba

Mex. = Mexico‡
mm. = millimeter
Mont. = Montane Zone
Mont. = Montana
Mr = March
My = May
n = northern
N = November
N. Am. = North America
Nat. = Naturalized
N. B. = New Brunswick
ne = northeastern
Newf. = Newfoundland
N. S. = Nova Scotia
nw = northwestern
O = October
Ont. = Ontario
P. E. I. = Prince Edward's Island
Plain = Subboreal Plains Zone
Que. = Quebec
[R] = The Author's Flora of the Rocky Mountains
s = southern
S = September
S. Am. = South America
Sask. = Saskatchewan
se = southeastern
Son. = Sonoran Zone
St.Plains = Staked Plains Zone
Subalp. = Subalpine Zone
Subarctic = Subarctic Zone
Subboreal = Subboreal Zone
Submont. = Submontane Zone
sw = southwestern
Trop. = Tropics
w = western
W. Ind. = West Indies

Signs

— (short dash) between figures or words means that the two figures or two words denote the extremes of variation

§ Subgenus or section of a genus.

— (long dash) between the names of two or more states denotes the extent of distribution.

× denotes a hybrid between the two species mentioned.

Measurements

1 mm. = 1/25 inch
3 mm. = ⅛ inch
1 cm. = ⅖ inches

5 cm. = 2 inches
1 dm. = 4 inches
1 m. = 40 inches or 3⅓ feet

300 m. = 1000 feet

---

* Many people use the abbreviation "Alb.", but as far as the author has been able to ascertain, the official one is "Alta."
† Used in cases of misapplications of names where the author first using it in such a sense has not been ascertained.
‡ Observe the difference between N. M. (New Mexico) and n Mex. (northern Mexico).

FLORA

OF THE

PRAIRIES AND PLAINS

OF

CENTRAL NORTH

AMERICA

—

Family 88. **LINACEAE.** FLAX FAMILY.

Herbs with alternate leaves, with or without stipules. Flowers perfect, regular, racemose or paniculate. Sepals and petals 5, rarely 4 or 6. Stamens as many as the sepals, monadelphous. Gynoecium of 5, sometimes 2 or 3, united carpels. Fruit a capsule, opening by twice as many valves as there are carpels.

Stigmas introrse and more or less elongate; sepals glandless; flowers in ours blue, rarely white.   1. LINUM.
Stigmas terminal and capitate; sepals, at least the inner ones, with marginal glands; flowers in ours yellow.   2. CATHARTOLINUM.

1. LÌNUM L. FLAX, BLUE FLAX.

Annual or perennial glabrous plants, sometimes woody at the base. Leaves alternate, without stipules or stipular glands, narrow, entire. Sepals 5, persistent. Petals 5, in ours blue, or rarely white, unappendaged and entire at the base. Stamens 5; filaments dilated and united at the base, each sinus with a short staminodium. Gynoecium 5-carpellary, not cartilaginous at the base; styles 5, elongate, distinct or united; stigmas elongate, introrse. Capsules 5-celled, the carpels with incomplete false partitions. Seeds flat, elongate-lenticular.

Inner sepals ciliate; stigmas much elongate; introduced annual.   1. *L. usitatissimum*
Sepals not ciliate; stigmas rather short; native perennials.
    Sepals over 5 mm. long at maturity, more than one-half
      as long as the capsule.   2. *L. Lewisii*.
    Sepalsl less than 5 mm. long, less than one-half as long as
      the capsule.   3. *L. pratense*.

1. L. usitatíssimum L. Stem 2–8 dm. high, glabrous; leaves narrowly linear-lanceolate, 3-nerved, sessile; sepals acuminate, the outer elliptic or ellip-tic-lanceolate, the inner elliptic-ovate or ovate, ciliate, 7–9 mm. long, all 3-nerved at the base; petals blue, 1–1.5 cm. long; capsule 6–8 mm. long. Waste places and old fields; occasionally escaped from cultivation, native of Eur. My–S.

2. L. Lewísii Pursh. *Fig. 338*. Stem 2–7 dm. high, often branched at the base, obscurely striate; leaves erect, linear or nearly so, 1–2 cm. long; sepals 5.5–7 mm. long, the outer ovate, short-acuminate, the inner broader, mucronate; petals blue or rarely white, 1.5–2 cm. long. Plains and hills: Man.—Neb.—Tex.—Calif.—Alaska; n Mex. *W. Temp.—Plain—Mont.* My–Au.

f. 338.

3. L. praténse (Norton) Small. Stem 1–6 dm. high, striate in age, commonly branched at the base; leaves commonly numerous and crowded towards the base, narrowly linear or subulate, 0.5–1 cm. long, rather succulent; bracts subulate; sepals mostly 4–5 mm. long, ovate to oblong-ovate, the outer acute or short-acuminate, the inner broader, mucronulate; petals blue, 1–1.5 cm. long. Dry plains: Man.—Tex.—Ariz.—Nev.—Wyo.—Sask. *Plain—Submont.* My–S.

2. CATHARTOLÍNUM Reichenb. YELLOW FLAX.

Annual or perennial herbs. Leaves alternate or occasionally opposite, without stipules, but often with stipular glands, mostly narrow, entire or serrulate. Sepals 5, persistent or deciduous, at least the inner ones with gland-tipped teeth. Petals 5, yellow or white. Stamens 5; filaments united at the base, monadelphous, the free portion more or less dilated at the base, sometimes with short staminodia in the sinuses. Gynoecium 5-carpellary, rarely 2-carpellary. Styles filiform, distinct or united; stigmas terminal, capitate. Capsule ovoid or depressed, 5-celled, rarely 2-celled, or completely or incompletely 10-celled by false septa. Seeds flattened, often lunate.

Sepals persistent; capsule without thickenings at the base.
   Styles distinct; lower leaves spatulate or oblanceolate, the upper lanceolate.
                                                  1. *C. virginianum*.
   Styles more or less united; leaves all linear or linear-lance-
     olate.   2. *C. sulcatum*.
Sepals deciduous; capsule with cartilaginous thickenings at the
   base.
    Stem, branches, and pedicels densely puberulent.   3. *C. puberulum*.
    Stem, branches, and pedicels not densely puberulent, usually
     glabrous.
      Petals 5–9 mm. long.   4. *C. compactum*.
      Petals 10–20 mm. long.
        False septa of the capsule slightly thickened; sepals
          becoming 6–8 mm. long; petals 11–15 mm. long.   5. *C. rigidum*.
        False septa thickened half their width; sepals becom-
         ing 9–11 mm. long; petals 17–20 mm. long.   6. *C. Berlandieri*.

1. **C. virginiànum** (L.) Reichenb. Glabrous perennial; stem 2-7 dm. high, erect, corymbosely branched above; lower leaves opposite, the upper alternate, 1-2.5 cm. long, entire; outer sepals lanceolate, 3-3.5 mm. long, acuminate, entire, the inner ones shorter and broader, entire or erose, with marginal glands; petals yellow; capsule depressed-globose, 2.5 mm. broad. *L. virginianum* L. [G, B]. Shaded places: Me.—Iowa—Ala.—Ga. *Canad.- Allegh.* Je-Au.

2. **C. sulcàtum** (Riddell) Small. Glabrous, pale green annual; stem simple below, 0.5-4 dm. high, sulcate or wing-angled; leaves linear or linear-lanceolate, early deciduous, 1-2 cm. long, entire; outer sepals lanceolate, 5-6 mm. long, finely glandular-toothed at the middle; inner sepals slightly shorter, more toothed, acuminate. *Linum sulcatum* Riddell [G, B]. Dry sandy soil: Ont.—Man.—N.D.—Tex.—Ga. *E Temp.*

3. **C. pubérulum** (Engelm.) Small. Pale green, often glabrous, perennial, stem 0.5-3 dm. high, usually branched throughout; leaves numerous and crowded at the base, erect or ascending, thick, the lower linear or linear-spatulate, entire, the upper linear or subulate, often gland-toothed below the middle; sepals narrowly lanceolate, becoming 7-9 mm. long, acuminate, the inner ones shorter than the outer and more coarsely toothed towards the apex; petals light yellow, 12-17 mm. long *L. rigidum puberulum* Engelm. Dry plains and hills: S.D.—Kans.- Tex.—Ariz.—Nev. *Plain—Submont.* Ap-Au.

4. **C. compáctum** (A. Nels.) Small. Dull green perennial; stem 1-2 dm. high, branched at the base; leaves numerous towards the base, but early deciduous, the upper more persistent, erect or ascending, entire; sepals lanceolate, becoming 6-8 mm. long, acuminate, evenly glandular-toothed, the inner much shorter than the outer; petals yellow, 7-9 mm. long; capsule ovoid, 4 mm. long. *L. compactum* A. Nels. Dry plains: N.D.—Mo. Kans—Wyo.—Mont. *Plain —Submont.* Je-Au.

5. **C. rígidum** (Pursh) Small. Glabrous, bright green perennial; stem 1-5 dm. high, simple at the base, corymbosely branched above, prominently striate-angled; leaves relatively few, erect, early deciduous, linear to linear-lanceolate, 0.6-3 cm. long, acute or acuminate, the upper ones glandular ciliate or glandular-serrulate; sepals lanceolate, becoming 6-8 mm. long, acuminate, evenly glandular-toothed, the inner ones shorter and broader than the outer ones; petals yellow; capsule ovoid, 4-5 mm. long. *L. rigidum* Pursh [G, B]. Plains and hills: Man.—Tex.—N.M.—Alta. *Plain—Submont.* Je-O.

6. **C. Berlandièri** (Hook.) Small. Bright green glaucescent perennial; stem 0.5-3.5 dm. high, simple at the base, branched and angled above; leaves rather few, spreading or ascending, linear to linear-lanceolate, mostly 1.5-4 cm. long, acuminate, entire, the lateral veins marginal; sepals lanceolate, becoming 9-11 mm. long, evenly glandular-toothed, acuminate and minutely spinulose-tipped, the inner ones shorter; petals light yellow, 17-20 mm. long; capsule ovoid, 4-4.5 mm. long. *L. Berlandieri* Hook. *L. arkansanum* Osterhout. Sandy soil: S.D.—Neb.—Tex.—Colo. *Plain—Son.* Ap-Jl.

Family 89. **BALSAMINACEAE.** JEWEL-WEED FAMILY.

Herbs, with more or less succulent leaves and swollen nodes. Flowers perfect, irregular, zygomorphic. Sepals 3, rarely 5, the posterior one petaloid, strongly saccate or spurred at the base. Petals 5, or by union of two and two only 3. Stamens 5, alternate with the petals; anthers more or less united around the stigma. Gynoecium of 5 united carpels; styles obsolete. Pod a 5-celled capsule, elastically dehiscent.

1. IMPÀTIENS L. Touch-me-not, Jewelweed.

All our species annual. Sepals in ours 3. Lateral petals each united with the adjacent posterior one. Capsule elastically dehiscent. Seeds usually many, pendulous above each other in a single row.

Flowers racemose; pod splitting from the base into spirally twisted valves; corolla yellow.
    Spur porrect, *i. e.*, bent at a right angle to the sack, one-fourth as long as the same or less; sack as broad as long, pale yellow, unspotted or minutely so.
                                                                   1. *I. pallida.*
    Spur strongly incurved, one third as long as the sack or more; sack longer than broad.
        Sack at least two thirds as broad as long, abruptly contracted into the spur.
                                                      2. *I. biflora.*
        Sack one half as broad as long or less, gradually tapering into the spur.
                                                      3. *I. Nortonii.*
Flowers solitary on axillary pedicels; pod splitting from the apex into incurved valves; corolla purplish.
                                                      4. *I. Balsamina.*

1. **I. pállida** Nutt. Stem usually light green, 1–2 m. high; leaf-blades thin, bright green, oval or ovate, 3–15 cm. long, crenate-dentate; lateral sepals broadly ovate, acuminate, light green, 5–7 mm. long; posterior sepal light sulphur-yellow, usually unspotted, broadly conic, about 15 mm. long; spur 3–8 mm. long; petals of the same color as the spur or lighter, usually dotted, the anterior one broadly obovate, emarginate, 7–8 mm. long. River banks: N.S.—Ga.—Kans.—Sask. *E. Temp.—Plain.*

2. **I. biflòra** Walt. *Fig. 339.* Stem 1–1.5 m. high, often tinged with red; leaf-blades oval or ovate, 2–10 cm. long, green or purplish; lateral sepals light green or purplish, obovate, short-acuminate, 5–6 mm. long; posterior sepal orange, or rarely pink, usually copiously spotted with red or purple, sometimes wholly unspotted, elongate-helmet-shaped, 15–18 mm. long; spur about 1 cm. long; petals of the same color as the posterior sepal, the anterior one nearly orbicularly obcordate, 6–7 mm. long. *I. maculata* Muhl. *I. fulva* Nutt. River banks: Newf.—Fla.—Ala.—Ida.—Mack. *E. Temp.—Plain—Submont.*

3. **I. Nortònii** Rydb. Stem light green, about 1 m. high; leaf-blades slightly paler beneath, 4–10 cm. long; lateral sepals ovate, acuminate, 7–10 mm. long; posterior sepal orange, spotted with brown and purple, 2.5–3 cm. long; spur abruptly recurved, slightly clavate; petals spotted with purple, the anterior one obovate, 1 cm. long, fully as wide. Wet woods and river banks: Mo.—Kans. *Ozark.*

4. **I. Balsamìna** L. Annual; stem 3–10 dm. high, pubescent, at least when young; leaves oblanceolate, 5–15 cm. long, sharply serrate, sparingly pubescent or glabrate; flowers solitary or two in each axil; lateral sepals ovate, 2–3 mm. long; posterior sepal rose or purple, broadly conic, 5–8 mm. long, 10–15 mm. wide, with an arched cylindric spur; anterior petals obovate, 8–10 mm. long, lateral and posterior petals obovate, 2–2.5 mm. long. Around dwellings: Pa.—Kans.—La.—Fla.; escaped from cult.; nat. of Asia.

Family 90. LIMNANTHACEAE. False Mermaid Family.

Herbs of wet places, with pinnately dissected leaves. Flowers perfect, regular. Hypanthium present, but small, saucer-shaped. Sepals 3–5, persistent, valvate or nearly so. Petals 3–5, marcescent. Stamens 6–10. Gynoecium of 2–5 carpels; ovaries distinct, but the styles united to near the top, arising between the ovaries. Fruit of semi-drupaceous nutlets.

1. FLOÈRKEA Willd. FALSE MERMAID.

Low slender annuals. Leaves simply pinnately dissected. Flowers inconspicuous on axillary peduncles. Sepals 3, slightly imbricate, spreading in fruit. Petals 3, white, about half as long as the sepals. Stamens 6; filaments subulate, distinct, those opposite the petals shorter; anthers didymous, introrse, opening longitudinally. Ovaries 2 or 3; style 2- or 3-cleft at the apex; stigmas capitate.

1. **F. proserpinacoïdes** Willd. *Fig. 340.* Stem slender, fleshy, decumbent, 1–4 dm. long; leaves bipinnately divided into linear, elliptic, or oblance-olate segments; sepals ovate-lanceolate, 3 mm. long; petals erect, 1.5 mm. long, oblong-obovate; carpels subglobose, 2 2.5 mm. broad. Meadows and wet places. Que.—N.D.—Tenn.—Del. *Canad—Allegh.* Ap–Je.

f. 340.

Family 91. ZYGOPHVLLACEAE. CALTROP FAMILY.

Perennial herbs, shrubs, or trees, with articulate stems. Leaves alternate or opposite, digitate or evenly pinnate, with inequilateral punctate leaflets. Flowers perfect, regular or nearly so. Sepals 5, rarely 4 or 6, usually imbricate. Petals as many, hypogynous. Stamens twice as many. Gynoecium of 2–5 united carpels; ovary 2–5-celled, rarely 10 12-celled, angled or winged; styles wholly united. Fruit a capsule or splitting into as many or twice as many nutlets as there are carpels.

Fruit dividing into 5 (by abortion sometimes fewer) carpels, each with 3–5 one-
    seeded compartments, spiny.   1. TRIBULUS.
Fruit dividing into 8–12 (twice as many as the original carpels) one-seeded nutlets,
    merely tubercled.   2. KALLSTROEMIA.

1. TRÍBULUS L. BUR-NUT, CALTROP.

Diffuse herbs, with weak, often prostrate, stems and branches. Leaves abruptly pinnate, opposite. Stipules lanceolate or subulate, membranaceous. Flowers solitary on axillary peduncles. Sepals 5, lanceolate, herbaceous, soon caducous. Petals 5, obovate, yellow, orange, or rarely white, spreading, deciduous. Stamens 10, hypogynous; filaments filiform, naked, the 5 inner alternate with the petals, shorter than the outer, subtended by a small gland. Ovary sessile, 5-celled, surrounded at the base by an urceolate 10-lobed disk. Styles united into a short stout column; stigmas 5, more or less connate. Fruit depressed, 5-angled, tuberculate or spinose, separating at maturity into 5 bony carpels. Seeds oblong-obovate. Endosperm wanting.

f. 341.

1. **T. terréstris** L. *Fig. 341.* A diffuse trailing annual herb; stems hirsute and slightly swollen at the nodes; leaflets about 5–7 pairs, somewhat oblique, oblong or elliptic, 3–13 mm. long; petals obovate, about 4 mm. long, light yellow; carpels 5, bony, each with 2 divergent stout spines and 2 or more smaller ones, sparingly hispid. Waste places and sandy soil: S.C.—S.D.—Utah—Ariz.; Mex. and W.Ind.; nat. from Eur.

2. KALLSTROÈMIA Scop.

Diffuse and branching herbs. Stems procumbent. Leaves opposite, abruptly pinnate, the terminal pair larger. Flowers solitary, orange or yellow. Peduncles axillary. Sepals 5 or 6, mostly persisting. Petals 4–6, obovate or obcordate, spreading, caducous. Stamens 10 or 12, hypogynous; filaments filiform, those opposite the petals adnate to the petals, the alternate ones smaller and subtended externally by a small gland. Ovary sessile, 8–12-celled, without transverse septa; styles united, columnar, or subulate from a conic base; stigma capitate; fruit roughened or tuberculate, separating at maturity into 8–12 bony, indehiscent, 1- rarely 2-seeded tuberculate nutlets, leaving a more or less persistent thick styliferous central axis.

Beak of the fruit shorter than the body, decidedly conic at the base.
                                             1. *K. hirsutissima.*
Beak of the fruit longer than the body, scarcely thickened at
  the base.   2. *K. intermedia.*

1. **K. hirsutíssima** Vail. *Fig. 342.* A much branched annual; branches stout, trailing, 4–7 dm. long, appressed cinereous-pubescent and hirsute with longer spreading hairs; leaflets 3–4 pairs, elliptic, 8–20 mm. long, shaggy-hirsute, at least beneath; sepals linear-lanceolate, 4–6 mm. long; petals obovate-oblong, 5–6 mm. long, yellow or orange, fading whitish, retuse; fruit canescent-strigose; beak pubescent, conic at the base, 3–4 mm. long; carpels 10. Sandy soil: Tex.—Kans.—Colo.—Mex. *St. Plain—Son.*

2. **K. intermèdia** Rydb. A profusely branched annual; branches 3–7 dm. long, ascending or decumbent, hirsute as well as pilose, striate; leaflets 4 or 5 pairs, elliptic, 8–20 mm. long, glabrate above, hirsute beneath; sepals 5–6 mm. long, subulate; petals obovate, bright yellow, 7–10 mm. long; beak 5–6 mm. long. *K. maxima* A. Gray [G, B]; not T. & G. Sandy soil: Ill.—Kans.—N.M.—Tex.; Mex. *Austral. Ap–S.*

Family 92. RUTACEAE. Rue Family.

Aromatic trees or shrubs, with secreting glands in the foliage. Leaves mostly digitately or pinnately compound; leaflets usually inequilateral. Flowers perfect, polygamous or dioecious. Calyx with 3–5 lobes or sepals, rarely wanting. Petals 3–5, imbricate. Stamens as many as the petals or twice as many, sometimes more, inserted on a hypogynous disk. Gynoecium of 1–5, or more, free or united carpels. Fruit various, of follicles, a capsule, samara, or drupe.

Fruit of 1–5, 2-valved follicles; stem usually prickly; leaves in ours pinnate.
                                           1. ZANTHOXYLUM.
Fruit a samara; stem unarmed; leaves 3-foliolate.   2. PTELEA.

1. ZANTHÓXYLUM [Catesby] L. Prickly Ash.

Shrubs and trees, often prickly, with aromatic bark. Leaves alternate, usually pinnate, often with stipular prickles. Flowers mostly cymose or paniculate, perfect, dioecious, or polygamous. Sepals hypogynous, 3–5, sometimes wanting, usually more or less united. Petals 3–10. Stamens 3–5, if isom-

erous, alternate with the petals. Pistils 1–5, on a
fleshy gynophore, rudimentary in the staminate flowers. Follicles 1-celled, 2-ovuled, usually 1-seeded.
Xanthoxylum F. G. Gmelin.

1. Z. americànum Mill. *Fig. 343.* Shrub or
small tree, armed with sharp stout stipular prickles;
leaves odd-pinnate; petiole and rachis slightly
grooved, prickly; leaflets 3–11, oblong-ovate to oval,
1.5–8 cm. long, from emarginate to short-acuminate,
oblique at the base, entire or crenate; flowers dioecious, in small axillary cymes, appearing before
the leaves; sepals wanting; petals 4 or 5, oblong or
oval, 2.5–3 mm. long, fringed, greenish yellow; stamens 4 or 5; pistils 2–5; follicles ellipsoid, 4–6 mm.
long, glandular. Rocky woods and river banks:
Que. Minn.—Okla.—Ga. *Allegh.* Ap–My.

f. 343.

2. PTÈLEA L. HOP-TREE, SHRUBBY TREFOIL.

Unarmed shrubs or trees. Leaves alternate, rarely opposite, 3-foliolate,
rarely 4- or 5-foliolate; leaflets entire or toothed, pellucid-punctate. Flowers
polygamous, in corymbose or paniculate cymes.
Sepals 4 or 5, rarely 6. Petals as many, greenish or
yellowish white. Stamens 4 or 5, rarely 6, hypogynous, abortive in the pistillate flowers. Gynoecium
of 2 or 3 united carpels; ovules 2 in each cavity.
Fruit a 2-celled or rarely 3-celled samara with reticulate wings, surrounding the body.

1. P. trifoliàta L. *Fig. 344.* A shrub or small
tree, up to 8 m. high; leaflets 3–10 cm. long, 1.5–5.5
cm. wide, dark green and shining above, pale beneath, entire or undulate, the lateral ones oblique,
elliptic to obovate, the terminal one larger, rhombic-
obovate; fruit suborbicular, 1.5–2.8 cm. broad,
finely pubescent, rounded or acute at the apex.
Rocky places and waste grounds: N.Y.—Minn.—
Neb.—Tex.—Fla. *Allegh.—Austral.* Je.

f.344.

Family 93. SIMARUBACEAE. QUASSIA FAMILY.

Shrubs or trees, with oil-sacs in the bark. Leaves alternate, simple or
pinnate. Flowers perfect, monoecious, dioecious, or polygamous, in
clustered or paniculate cymes. Sepals 3–7, distinct or partly united.
Petals as many, imbricate or valvate, rarely wanting. Stamens as many
or twice as many, rarely numerous. Gynoecium of 2–5 distinct or united
carpels. Fruit a drupe, berry, capsule, or samara.

1. AILÁNTHUS Desf. TREE-OF-HEAVEN.

Trees with pinnate leaves, the leaflets numerous. Flowers polygamo-
dioecious, in large panicles. Sepals 5 or 6, imbricate. Petals 5 or 6, valvate.
Disk flat, 10-lobed. Stamens in the staminate flowers 10, inserted at the base
of the disk, in the perfect flowers 2 or 3, and in the pistillate flowers wanting.
Gynoecium of 2–5 united carpels; ovary 2–5-lobed, flattened; styles united.
Ovules solitary in each cavity. Fruit an elongate samara or 2–5 together, the
wings membranous, veiny, 1-seeded in the middle.

1. A. glandulòsa Desf. A large tree; leaves 3–9 dm. long; leaflets 11–41, oblong-lanceolate, entire or few-toothed near the base, 7–18 cm. long, acuminate or acute; petals greenish or yellowish, 2.5–4 mm. long; samaras oblong, 3–5 cm. long, twisted. Around dwellings: Ont.—Iowa—Kans.—Tex. —Ala.; cult. from China and escaped.

Family 94. POLYGALACEAE. MILKWORT FAMILY.

Herbs, or rarely shrubs or trees, with alternate, opposite, or whorled leaves; stipules wanting. Flowers irregular. Sepals 5, free, imbricate, the two inner petaloid, called wings. Petals 5 or 3, the lower one concave, often beaked or crested, called keel, the others more or less united. Stamens usually 8; filaments usually united into a tube, cleft on the back; anthers becoming 1-celled, opening by terminal pores or cracks. Gynoecium usually of 2, rarely 5, united carpels. Fruit usually a 2-celled capsule.

1. POLÝGALA (Tourn.) L. MILKWORT.
Filaments united. Capsule compressed laterally, 2-celled. Seeds caruncled.

Annuals.
  Lower leaves verticillate.
    Spike elongate, slender, the bracts deciduous; corolla greenish-white, or barely rose-tinged. 1. *P. verticillata.*
    Spike short and thick, the bracts persistent; corolla rose- or greenish-purple. 2. *P. cruciata.*
  Lower leaves alternate.
    Petals not united into an elongate tube, which is not longer than the sepals. 3. *P. viridescens.*
    Petals united into an elongate tube, longer than the sepals. 4. *P. incarnata.*
Perennials.
  Flowers white; cleistogamous flowers not present.
    Leaves linear or linear-oblanceolate. 5. *P. alba.*
    Leaves lanceolate to ovate. 6. *P. Senega.*
  Flowers rose-purple, showy; inconspicuous cleistogamous flowers present on the subterranean shoots.
    Flowers in terminal racemes; capsule 2–5 mm. long. 7. *P. polygama.*
    Flowers axillary, few; capsule 5–8 mm. long. 8. *P. paucifolia.*

1. P. verticillàta L. Stem 0.5–3 dm. high, angled, with numerous spreading branches; leaves in 4's or 5's, 1–3 cm. long; racemes spike-like, narrowly conic; flowers greenish or greenish-white; wings ovate, about 1 mm. long; keel crested with several thick processes; capsule about 2 mm. long. Dry soil: Que.—Fla.—Mex.—Wyo.—N.D.—Man. *E. Temp.—Plain—Submont.* Je–S.

2. P. cruciàta L. Stem 0.5–4 dm. high, 4-angled; leaves in whorls of 4's or 5's, spatulate to linear, 1–3 cm. long; spikes 1–5 cm. long, 1–2 cm. thick, short-peduncled or sessile; wings 3–6 mm. long, deltoid, caudate-acuminate; keel crested with few thick processes. Bogs and wet places: Me.—Minn.— Neb.—La.—Fla. *E. Temp.* Jl–S.

3. P. viridéscens L. Stem 1–5 dm. high, angled, simple or branched above; leaves linear or linear-oblong, 1–3 dm. long, acute to cuspidate; raceme ovoid or subglobose, 8–12 mm. thick; flowers reddish-purple, green, or white; wings broadly ovate, 3–6 mm. long, mucronate; keel crested with few thick processes; capsule suborbicular, 2 mm. broad. *P. sanguinea* L. [G]. Sandy moist soil: N.S.—Minn.—Okla.—N.C. *E. Temp.* Je–S.

4. P. incarnàta L. Stem 2–7 dm. high, simple or branched above, striate; leaves linear-subulate, 0.5–1.5 cm. long, fleshy; racemes cylindric, 8–10 mm. thick, dense; flowers pink and white; wings lanceolate, 4 mm. long; petals 6–8 mm. long; keel crested, with entire or cleft processes; capsule 4 mm. long, grooved. Sandy soil: N.J.—Iowa—Neb.—Okla.—Mex.—Fla. *Allegh.—Austral.* Je–Au.

5. P. álba Nutt. *Fig. 345.* Stems several from the base, erect or ascending, 1–5 dm. high, angled, simple; leaves 1–2.5 cm. long; racemes spike-like, 4–6 mm. thick; wings oval or obovate, 2–3 mm. long; corolla white; keel with a fimbriate crest; capsule ovoid, 2–3 mm. long. Dry plains: Minn.—Kans. —Tex.—Ariz.—Colo.—Mont.; Mex. *Prairie—Plain.* Ap–O.

f. 345.

6. P. Sénega L. Stems several from a thick root, erect, 1–5 dm. high, simple; leaves numerous, lanceolate to lance-elliptic or (in var. *latifolia*) ovate-lanceolate or ovate, scabrous on the margin; racemes spike-like, 2–6 cm. long; wings orbicular-ovate, 2–3 mm. long; corolla greenish white, 3–4 mm. long; keel crested with thick processes; capsule flat, broader than long. Dry soil: N.B.—N.C.—Ark.— S.D.—Alta. *Canad.—Allegh.—Submont.* My–Jl.

7. P. polýgama Walt. Stem branched at the base, 1–5 dm. high, striate; leaves spatulate to linear, 1–4 cm. long, apiculate; racemes loosely flowered, 12–15 cm. long; flowers rose-purple to pink, the lower ones drooping; wings obovate, 4–6 mm. long; keel crested with several large branched processes; capsule oblong, 2 mm. long. Dry sandy soil: N.S.—Man.—Tex.—Fla. *E. Temp.* Jl.

8. P. paucifólia Willd. Stem simple or branched below; branches 5–20 cm. high, pubescent above; leaves 3–6, near the ends of the branches, elliptic to oval or ovate, 1–3 cm. long, acute or mucronate, somewhat pubescent above; flowers 3 or 4, rose or purple, rarely white, 1.5–2 cm. long; wings obovate, 1–1.5 cm. long, narrowed at the base; keel with a conspicuous imbricate crest; stamens 6; capsule subglobose, 5–8 mm. broad, retuse at the apex. Moist woods: Que.—(Sask.)—Man.—Minn.—Ill.—Ga. *Canad.—Allegh.* My–Je.

Family 95. EUPHORBIACEAE. Spurge Family.

Monoecious or dioecious herbs (all ours), or shrubs or trees, often with milky sap, and alternate, opposite, or verticillate simple leaves. Flowers with or without petals, sometimes much reduced and subtended by a calyx-like involucre. Stamens few or many; filaments united or distinct. Ovary usually 3-celled. Styles as many as the cells of the ovary, simple, divided, or many-cleft. Capsule separating at maturity into 3 2-valved carpels.

Flowers not in involucres ; calyx of several sepals.
  Corolla present in either the staminate or pistillate flowers, or in both.
    Stamens 5–12 ; filaments distinct.
      Capsule 1-celled, indehiscent.   1. CROTONOPSIS.
      Capsule 3-celled, dehiscent.   2. CROTON.
    Stamens 10 ; filaments monadelphous.   3. DITAXIS.
  Corolla wanting.
    Stamens 8–16 ; staminate flowers in spikes or racemes ; pistillate flowers axillary.   4. ACALYPHA.
    Stamens 6 ; staminate flowers spicate.   2. CROTON.
    Stamens 1–5.
      Flowers axillary, one staminate and one pistillate flower in each axil; ovules and seeds 2 in each cavity.   5. PHYLLANTHUS.
      Flowers in racemes ; ovules and seeds solitary in each cavity.
        Pistillate flowers and capsules pedicelled.   6. TRAGIA.
        Pistillate flowers and capsules sessile.   7. STILLINGIA.

Flowers in involucres; calyx represented by a minute scale on the filament-like pedicels.

Glands of the involucres with petal-like appendages, these, however, sometimes much reduced.
  Leaves all opposite.
    Leaf-blades oblique at the base, inequilateral; glands 4.
                                      8. CHAMAESYCE.
    Leaf-blades equilateral, not oblique at the base; glands 5.
                                10. ZYGOPHYLLIDIUM.
  Leaves alternate or scattered, at least below the inflorescence.
    Annuals or biennials; stipules present but narrow; bracts petal-like.
                                 9. LEPADENA.
    Perennials; stipules wanting; bracts not petal-like. 12. TITHYMALOPSIS.

Glands of the involucres without petal-like appendages, naked, sometimes with crescent-shaped horns.
  Stem topped by an umbel; stipules none; involucres in open cymes, each with 4 glands and entire or toothed lobes.   11. GALARRHOEUS.
  Stem not topped by an umbel; stipules gland-like; involucres in conglomerate cymes, each with a single gland, or rarely 4 glands and fimbriate lobes.
                                  13. POINSETTIA.

1. CROTONÓPSIS Michx.

Annual silvery scurfy monoecious plants. Leaves mostly alternate, entire. Staminate flowers uppermost in the inflorescence, with 5 sepals, 5 petals, and 5 inflexed stamens. Pistillate flowers with 3–5 sepals, no petals, 5 petal-like glands opposite the sepals, and a 1-celled ovary. Fruit a small scaly or spiny indehiscent capsule.

  1. **C. elliptica** Willd. Stem 1–4 dm. high, branched; leaves few; blades oblong to linear-lanceolate, 1–2.5 cm. long, entire; fruit ovoid, scaly, the scales tubercled but not spine-bearing on the back. Sandy soil: N.J.—Kans.—Tex.—Fla. *Austral.* Jl–S.

2. CRÒTON L. CROTON.

Herbs (ours) or shrubs, monoecious or rarely dioecious, heavy-scented, stellate-pubescent or scaly, sometimes more or less glandular. Leaves mostly alternate, entire, toothed, or lobed. Flowers in axillary or terminal clusters. Staminate flowers uppermost. Sepals 4–6, usually 5. Petals usually present, but often small, alternating with the glands. Stamens 5 or more, inflexed in bud. Pistillate flowers below the staminate ones. Sepals 5–10. Petals usually wanting. Ovary 3-celled; ovules solitary in each cell; stigmas once, twice, or thrice 2-cleft. Capsule mostly splitting into 3 2-valved carpels. Seeds smooth or minutely pitted.

Petals present in the staminate flowers.
  Leaf-blades toothed.   1. *C. glandulosus.*
  Leaf-blades entire.
    Pistillate flowers clustered below the staminate ones, not on reflexed pedicels.   2. *C. capitatus.*
    Pistillate flowers on long reflexed pedicels.
      Capsule 3-celled and 3-seeded.   3. *C. Lindheimerianus.*
      Capsule 2-celled but 1-seeded.   4. *C. monanthogynus.*
Petals wanting in all the flowers.   5. *C. texensis.*

  1. **C. glandulòsus** L. Annual; stem 3–6 dm. high, umbellately branched; leaves in the form represented within the area (var. *septentrionalis* Muell. Arg.), coarsely stellate, mainly oblong, serrate, 1.5–7.5 cm. long, the base with a saucer-shaped gland on each side; staminate flowers in spikes, with 4 petals, a 4-rayed disk, and 8 stamens; pistillate flowers capitately clustered, with rudimentary petals; capsule 5 mm. long, subglobose. Waste places: Va.—Iowa —Kans.—Tex. *Austral.* Mr–D.

  2. **C. capitàtus** Michx. Annual; stem 3–8 dm. high; lower leaf-blades oval, the upper oblong, mucronate, 2–5 cm. long, entire, mostly rounded at the base, velvety above, densely stellate-tomentose beneath; racemes 1–3 cm. long; staminate flowers with 5 sepals, 5 spatulate ciliate petals, and 7–12 stamens;

pistillate flowers with 6–8 sepals, and no petals; capsule globose, 7–9 mm. thick. Dry soil: N.J.—Iowa—Kans.—Tex.—Ga. *Austral.* My–O.

3. C. Lindheimeriànus Scheele. Annual, greenish-pubescent; stem 3–5 dm. high, 2–4-dichotomous; leaf-blades oval to rhombic-ovate, 1–3 cm. long, rounded or retuse at the apex, undulate; racemes few-flowered; staminate flowers with 5 sepals, 5 spatulate ciliate petals, and 7-12 stamens; pistillate flowers 1–3 together, with 5 sepals and no petals; capsule oblong, 4–5 mm. long. Sandy prairies: Kans.—Tex.—Mex. *Texan.*

4. C. monanthógynus Mich. Annual; stem 3–5 dm. high, 2–4-dichotomous; lower leaf-blades suborbicular, the upper ovate or oblong, 1–4 cm. long, entire; staminate flowers with 3–5 ovate sepals, 3–5 spatulate petals, and 5–10 stamens; pistillate flowers 1–4 together, with 5 oblong sepals and no petals; capsule ovoid, 3–4 mm. long. Rocky soil: Ind. Iowa—Kans.—Tex.—Ga.; Mex. *Austral.*

5. C. texénsis (Klotzsch) Muell. Arg. *Fig. 346.* Lepidote, dioecious annual; stem 4–16 dm. high, di- or tri-chotomously branched; the staminate plant usually more slender than the pistillate one and with narrower leaves; leaf-blades linear, lanceolate, or oblong, 4–12 cm. long, entire; staminate racemes 1–3 cm. long; sepals oblong to ovate-oblong; stamens 8–12; pistillate flowers 2–4 together or solitary; sepals triangular; capsule globose or oval, 4–6 mm. long, warty. Plains and prairies: Ill.—Ala. Tex.—Ariz.—Wyo.—S.D.; Mon. *Prairie—Plain.* My–Au.

f. 346.

3. DITÁXIS Vahl.

Monoecious or rarely dioecious herbs or shrubs, often with rootstocks. Leaves alternate, silky or pilose, entire or rarely toothed, often strongly veined. Flowers in axillary and terminal clusters, usually bracted. Staminate flowers usually crowded at the ends of the racemes. Sepals 4 or 5, valvate. Petals 4 or 5. Stamens of the same number, or twice or thrice as many; filaments united into a column. Pistillate flowers with imbricate sepals and rudimentary petals. Ovary 3-celled, each cell with a solitary ovule. Styles 3, 2-cleft. Capsule 3-lobed, depressed, separating into three 2-valved carpels. Seeds subglobose, wrinkled or crested. *Argythamnia* P. Br.

Pistillate flowers with normal petals; flowers in small axillary clusters.
                    1. *D. humilis.*
Pistillate flowers with rudimentary petals; flowers in elongate racemes.
                    2. *D. mercurialina.*

1. D. hùmilis (Engelm. & Gray) Pax. Perennial, with a woody root and caudex; stems much branched, pubescent, spreading, 1–3 dm. long; leaf-blades ovate, oblong, or oblanceolate, 1–3 cm. long, entire; capsule short-pedicelled, depressed, 4–6 mm. in diameter; seeds oval-globose, 2 mm. long, muricate. Prairies: Kans.—La.—Tex.—N.M. *Texan—Son.* My–O.

2. D. mercurialina (Nutt.) Coulter. Perennial; stem slender, 1–6 dm. high, silky; leaves ovate to narrowly lanceolate or oblanceolate, 2–5 cm. long, undulate, 3-ribbed, glabrate; staminate flowers with lanceolate ciliate sepals and spatulate-oblong undulate petals; pistillate flowers with much longer sepals and rudimentary petals; capsule depressed, 6–9 mm. broad, silky, 3-lobed. *A. mercurialina* Muell. Arg. [G]. Dry soil: Ark.—Kans.—Tex. *Texan.* Je–Au.

4. ACALÝPHA L. Three-seeded Mercury.

Annual or perennial monoecious herbs or shrubs. Leaves alternate, petioled, with stipules. Flowers in terminal and axillary spikes or racemes, apetalous. Staminate clusters peduncled; each flower with 4 sepals and 8–16 stamens, the filaments united at the base. Pistillate flowers subtended by foliaceous bracts, with 3–5 sepals. Ovary 3-celled. Capsule of 3 two-valved carpels, each 1-seeded.

Staminate and pistillate flowers in the same spike.
  Foliage not glandular; bracts usually longer than the spikes.   1 *A. virginica.*
  Foliage glandular; bracts shorter than the spikes.   2. *A. gracilens.*
Staminate and pistillate flowers in different spikes.   3. *A. ostryaefolia.*

f.347.

1. **A. virgínica** L. *Fig. 347.* A somewhat pubescent annual; stem 1–7 dm. high; leaves ovate or elliptic, 2–10 cm. long, coarsely serrate; staminate and pistillate flowers in the same axillary spikes, the pistillate ones 1–3 at the base; bracts large, palmately lobed, usually enclosing the spike; capsule subglobose, 3-lobed, 3 mm. broad. Woods and thickets: Ont.—Minn.—Neb.—Tex.—Fla. *Allegh. Ozark.* Je–O.

2. **A. grácilens** A. Gray. A pale green annual; stem 1–8 dm. high, with slender branches; leaves lanceolate to linear-oblong, 1–5 cm. long, rather firm, serrate; staminate spike very slender; capsule subglobose, 3 mm. broad. Dry woods: Mass.—Kans. —Tex.—Fla. *Allegh.—Ozark.* Je–S.

3. **A. ostryaefòlia** Riddell. A dark green puberulent annual; stem stout, 3–10 dm. high; leaves thin, ovate, 5–10 cm. long, serrate, obtuse to cordate at the base; staminate and pistillate flowers in separate spikes, the bractlets of the former minute, those of the latter conspicuous, lobed; capsule depressed, spiny, 3–4 mm. broad. *A. caroliniana* Ell. Thickets and waste places: N.J.—Kans.—Mex.—Fla. *Ozark—Texan.* Je–N.

5. PHYLLÁNTHUS L.

Annual or perennial herbs, or in the tropics shrubs or trees. Leaves alternate, sometimes so arranged as to resemble leaflets of a compound leaf, entire, usually sessile. Flowers monoecious, apetalous, one staminate and one pistillate in an axil, sometimes on the edges of leaf-like branches. Sepals 5 or 6, imbricate. Stamens usually 3; filaments more or less united. Ovary 3-celled; styles 3, each 2-cleft. Capsule globose or depressed, each carpel 2-seeded.

1. **P. caroliniénsis** Walt. A dark green glabrous annual; stem slender, 1–5 dm. high; leaves numerous, obovate or oblong, 5–20 mm. long, obtuse; flowers nearly sessile in the axils, inconspicuous; sepals narrowly cuneate; capsule 2 mm. broad; seeds with minute papillae. Sandy soil and moist woods: Pa.—Ill.—Kans.—Tex.—Fla.; Mex. and Cent. Am. *Austral.—Trop.* My–O.

6. TRÀGIA (Plum.) L.

Perennial monoecious herbs, or shrubs, usually armed with stiff, stinging hairs. Leaves alternate, entire, toothed, or lobed, mostly cordate, petioled. Flowers in racemes or spicate racemes, bracteolate, apetalous. Staminate flowers with 3–5 sepals, and mostly 1–3 stamens. Pistillate flowers with 3–8 sepals. Ovary 3-celled; ovules solitary. Styles 3, often united to above the middle. Capsule 3-lobed, separating into three 2-valved carpels. Seeds subglobose.

Staminate flowers with 3 sepals and stamens.   1. *T. nepetaefolia.*
Staminate flowers with 4–6 sepals and stamens.   2. *T. ramosa.*

1. **T. nepetaefòlia** Cav. *Fig. 348.* Plant with stinging hairs; stem slender, 1–4 dm. high; leaf-blades triangular-ovate or lanceolate, 1–5 cm. long, dentate-serrate, cordate at the base; racemes many-flowered, 1–3.5 cm. long; pistillate flowers with 5 sepals; capsule depressed, 5–6 mm. broad, hirsute. Sandy soil: Kans.—N.M.—Mex. *Texan—Son.* Ap–Au.

2. **T. ramòsa** Torr. Light green perennial, with a woody caudex, hispid; stem 0.5–3 dm. high, usually much branched; leaf-blades lanceolate to triangular-lanceolate, 1–5 cm. long, coarsely and sharply serrate; staminate flowers with 4–5 sepals and 5–6 stamens; pistillate flowers solitary, with 5 sepals subtended by a 3-lobed bract; capsule depressed, 6–8 mm. thick, orange, more or less variegated. Dry soil: Mo.—Tex.—Ariz.—Colo. *Texan.* —*Son.*—*Submont.* Ap–Au.

f.348.

7. **STILLÍNGIA** L. Queen's Delight. Queen's-root.

Monoecious herbs or shrubs. Leaves glabrous, mostly alternate, with 2 glands at the base, entire or toothed. Staminate flowers several together in the axils of the bractlet; calyx 2- or 3-lobed; stamens 2 or 3. Pistillate flowers solitary in the lower axils; calyx 3-lobed. Capsule 2- or 3-lobed; styles stout, united at the base.

Stem-leaves lanceolate or elliptic, serrulate.   1. *S. salicifolia.*
Stem-leaves obovate, oval, or oblong, crenulate.   2. *S. sylvatica.*

1. **S. salicifòlia** (Torr.) Raf. *Fig. 349.* Herb; stems 3–10 dm. high, woody at the base, sometimes umbellately branched above; leaf-blades 4–10 cm. long, acute; spike 4–9 cm. long; capsule 12–15 mm. broad; seeds 7 8 mm. long. *S. lanceolata* Nutt. *S. Smallii* Woot. & Standl. Sandy soil: Ark.—Kans. —Tex. *Texan.*

2. **S. sylvàtica** L. Herb; stems 4–12 dm. high, stout, topped with an umbel; leaf-blades 3–10 cm. long, obtuse; spikes 5–12 cm. long; capsule 12–15 mm. broad; seeds 7–9 mm. long. Sandy soil: Va.— Kans.—Tex.—Fla. *Austral.* Mr–O.

8. **CHAMAESÝCE** S. F. Gray.
Spurge, Carpet-weed.

Annual or perennial herbs, or in the tropics shrubs. Stems often radially branched at the base, ascending or prostrate, forking. Leaves opposite, entire or toothed, more or less oblique at the base; stipules delicate, entire or fringed. Involucres solitary in the axils or in axillary cymes. Glands 4, sessile or stalked, naked or usually with an appendage; one sinus of each involucre usually glandless. Capsule 3-lobed, smooth, with sharp or rounded angles. Seeds angled, the faces smooth or transversely wrinkled. *Euphorbia* L., in part.

f.349.

Leaf-blades entire.
  Annuals or biennials.
    Seeds 1.5 mm. long or more.   1. *C. polygonifolia.*
    Seeds less than 1 mm. long.
      Plants prostrate; leaves oblong to orbicular.
        Leaf-blades manifestly longer than broad, usually more than twice as long.   2. *C. Geyeri.*

Leaf-blades as broad as long or nearly so.   3. *C. serpens.*
Plants more or less ascending or erect; leaf-blades
    linear or linear-lanceolate; stipules white
    scarious.
     Appendages of the glands longer than broad;
     Appendages of the glands broader than long.   4. *C. Nuttallii.*
      seeds 4-angled.   5. *C. petaloidea.*
Perennials.
  Leaves glabrous.
   Appendages crescent-shaped, much narrower than
    the gland; leaves broadly deltoid-ovate; seeds
    transversely ridged.   6. *C. Greenei.*
   Appendages conspicuous, semi-orbicular, fully as
    broad as the glands; leaves ovate or ovate-
    lanceolate; seed irregularly pitted.   7. *C. Fendleri.*
  Leaves pubescent.   8. *C. lata.*
Leaves toothed, but sometimes only at the apex; stems and
  branches prostrate; annuals.
  Capsules glabrous.
   Seeds strongly transversely wrinkled.
    Leaves broadly obliquely obovate.
     Plant erect or ascending.   9. *C. hyssopifolia.*
     Plant prostrate or spreading.   10. *C. aequata.*
    Leaves oblong or linear-oblong; stem mostly ascend-
     ing or erect.   11. *C. glyptosperma.*
   Seeds pitted or irregularly and faintly wrinkled.
    Leaves oblong; seeds usually with a white bloom.   12. *C. serpyllifolia.*
    Leaves linear; seeds brownish, usually without a
     bloom.   13. *C. albicaulis.*
  Capsule pubescent.
   Involucre deeply split on one side.   14. *C. humistrata.*
   Involucre not split on one side.
    Leaf-blades and involucres glabrous or sparingly
     pubescent.   15. *C. maculata.*
    Leaf-blades and involucres copiously hirsute.   16. *C. stictospora.*

1. **C. polygonifòlia** (L.) Small. A pale green, glabrous annual, branched at the base; branches spreading, prostrate, 1–2.5 dm. long; leaf-blades narrowly oblong to linear-lanceolate, 6–20 mm. long, fleshy, apiculate or acute; involucre turbinate-campanulate, 2 mm. high; glands 4, columnar; appendages obsolete; capsule globose-ovoid, 3–4 mm. long, wrinkled in age; seeds oblong-ovoid, pale gray, minutely pitted. *E. polygonifolia* L. [G, B]. Sand, mostly along the coast and Great Lakes, rarely inland: N.H.—Wis.—Iowa—Fla. *E. Temp.* Jl–S.

2. **C. Géyeri** (Engelm.) Small. Plant green; stem branched at the base; branches 0.5–4 dm. long; leaf-blades oblong to ovate, 4–12 mm. long, obtuse, usually mucronulate, truncate or subcordate at the base; involucres turbinate, 1.5 mm. high; glands 0.3 mm. broad; appendages inconspicuous, white or red; seeds narrowly ovoid, 1.5 mm. long, nearly terete, ash-colored. *Euphorbia Geyeri* Engelm. [G, B]. On plains and prairies: Ill.—Mo.—Tex.—Neb.—Minn. *Prairie—Plain.* Je–S.

3. **C. sérpens** (H. B. K.) Small. Plant pale green; stems branched at the base, 5–30 cm. long; leaf-blades orbicular to oval, 2–6 mm. long, obtuse or emarginate, rounded or subcordate at the base; involucres nearly 1 mm. high; appendages minute, irregular, crenate; capsules nodding, depressed-globose, 2 mm. broad; seeds 1 mm. long, smooth, obtusely 4-angled, light gray. *E. serpens* H. B. K. [G, B]. On prairies and plains: Ont.—Fla.—Ariz.—S.D.; Mex. *Plain—Son.* Je–O.

4. **C. Nuttállii** (Engelm.) Small. A bright green, glabrous annual; stems ascending or erect, branched, 1–6 dm. high; leaf-blades linear, 1–2.5 mm. long, often involute, truncate at the apex; involucres solitary in the axils, campanulate, scarcely 2 mm. long; glands saucer-shaped; appendages white, entire, oblong; capsule 2 mm. long; seeds ovoid, 1.5 mm. long, gray, 4-angled. *E. petaloidea Nuttallii* Engelm. *E. Nuttallii* Small [B]. *E. zygophylloides* Boiss. [G]. Dry prairies: Kans.—Tex.—Mex. *Texan.* Je–S.

5. C. petaloìdea (Engelm.) Small. Plant pale green; stem 1–6 dm. high; leaf-blades linear, oblong or linear-lanceolate, 1–2.5 cm. long, obtuse; involucres oblong-campanulate, 2 mm. long; glands about as long as the lobes; appendages white, reniform, ovate or suborbicular, entire or undulate; capsules globose-reniform, fully 2 mm. long; seeds oblong-ovoid, nearly 2 mm. long, ash-colored, minutely pitted, nearly tereto. *E. petaloidea* Engelm. [G, B]. Plains, in sandy soil: Minn.—Iowa—Mont.—Tex. *Prairie—Plain.* Jl–S.

6. C. Greènei (Millsp.) Rydb. *Fig. 350.* Perennial from a woody root; stems numerous, prostrate, diffuse, branched, 1–2 cm. long; leaves 4–7 mm. long and nearly as broad, glabrous, very oblique; involucres 1.5–2.5 mm. long; glands thick, purple; appendages thick, minutely 2-lobed; capsules 2.5–3 mm. thick; seeds ovoid, white, angled. *E. Greenei* Millsp. Confused with *C. Fendleri* [B]. Dry or sandy plains: S.D.—Okla.—Utah—Ida. *Plain Submont.* Je–Au.

f. 350.

7. C. Féndleri (T. & G.) Small. Plant cespitose, usually pale green; stems erect or ascending, about 1 dm. high; leaf-blades ovate to lance-ovate, 4–6 mm. long, 2–5 mm. wide; involucres 1.5–2.5 mm. high; appendages yellowish or purplish, often crenate-toothed; capsule 3 mm. in diameter; seeds ovoid, 1.5 mm. long. *E. Fendleri* T. & G. Plains. Tex. Ariz. Utah. *St. Plain—Son.* Ap–O.

8. C. làta (Engelm.) Small. Stems branched from a woody root, spreading or ascending, 5–15 cm. long, pubescent; leaf-blades ovate to lanceolate, 5–10 mm. long, canescent-pubescent, revolute-margined; involucre 1 mm. long; appendages narrowly crescent-shaped, undulate; capsule rounded, 3-angled, 2 mm. thick; seeds oblong, 1.5 mm. long, 4-angled and transversely wrinkled. *E. lata* Englem. [B]. Plains: Kans.—Tex.—N.M.—Colo. *St. Plain.*

9. C. hyssopifòlia (L.) Small. Stem branched above, 1–6 dm. high; leaf-blades oblong to ovate or obovate, sometimes falcate, oblique, 3-nerved, often with a red blotch; involucre obovoid, 1 mm. long; gland-appendages orbicular or reniform, entire, white or red; capsule glabrous, fully 2 mm. long; seeds ovoid-oblong, 1.5 mm. long, 4-angled, brown or black, with a grayish coating, transversely ridged. *E. nutans* Lag. [B]. *E. Preslii* Guss. [G]. Fields and thickets; Mass.—Minn.—Neb. *E. Temp.* My–O.

10. C. aequàta Lunell. Plant dark green; stem prostrate, much branched, 1–2 dm. long, terete, red; leaf-blades obliquely obovate or spatulate, 5–17 mm. long, serrate, rounded or obtuse at the apex, often red along the midrib; capsule glabrous; seeds dark, 4-angled, strongly rugose. Railroad banks, etc.: N.D. *Plain.* Jl.

11. C. glyptospérma (Engelm.) Small. Plant pale green, glabrous; stem 0.5–4 dm. long; leaf-blades oblong, linear-oblong, or rarely ovate, 2–12 mm. long, obtuse or subcordate at the base, serrulate towards the apex; involucres 1 mm. long, with 4 dark ribs; appendages narrow, crescent-shaped, often crenulate; capsule less than 2 mm. thick; seeds oblong, 1 mm. long, ash-colored, strongly 4-angled. *E. glyptosperma* Engelm. [G, B]. *C. erecta* Lunell. Sandy soil: Ont.—Mo.—Tex.—Mex.—B.C. *Temp. Plain—Submont.* My–S.

12. C. serpyllifòlia (Pers.) Small. Plant dark green or reddish; stems 1–4 dm. long; leaf-blades oblong or spatulate, 3–12 mm. long, obtuse or retuse, serrulate above the middle or nearly entire; involucres more than 1 mm. long;

appendages narrow, crescent-shaped, lobed; capsules 2 mm. broad; seeds ovoid, hardly 1 mm. long, obtusely 4-angled. *E. serpyllifolia* Pers. [G, B]. Plains and in dry soil: Mich.—Tex.—Calif.—B.C.—Alta.; Mex. *Temp.—Plain—Submont.* Ap–O.

13. C. albicaùlis Rydb. Plant pale green; stems straw-colored or light greenish yellow, shining; leaf-blades linear, minutely callous-toothed towards the apex, 1–1.5 cm. long, 2–3 mm. wide; involucres about 1 mm. long, turbinate; appendages white, crescent-shaped, crenulate; pod 2 mm. thick; seeds light olive-brown, oblong, acutely 4-angled, 1.3–1.5 mm. long, less than 0.5 mm. broad. *E. albicaulis* Rydb. [B]. Cultivated or sandy soil: Neb.—Mont.—N.M. *Plain.* Jl–Au.

14. C. humistràta (Engelm.) Small. Stem branched from the base, the branches prostrate, spreading, 1–3 dm. long; leaf-blades ovate-oblong, 4–10 mm. long, oblique, obtuse or subcordate at the base; involucre 1 mm. long; glands 4, disk-like; appendages narrow, irregular, red or white; capsule depressed-globose, less than 2 mm. broad, its lobes keeled; seeds oblong, 1 mm. long, papillose, ash-colored, obscurely wrinkled. *E. humistrata* Engelm. [G, B]. Sandy soil: Que.—Minn.—Kans.—Miss. *Allegh.—Ozark.* Au–O.

15. C. maculàta (L.) Small. Plant dark green; stems 5–40 cm. long, often dark red, villous; leaf-blades oblong or ovate-oblong, 4–16 mm. long, more or less serrate, usually blotched, oblique, semi-cordate at base; involucres 1 mm. long; appendages semi-orbicular, white, crenulate; capsules about 2 mm. thick, pubescent; seeds ovoid-oblong, less than 1 mm. long, obtusely angled, ash-colored, minutely pitted and inconspicuously transversely wrinkled. *E. maculata* L. [G, B]. Dry ground and waste places: Ont.—Fla.—Tex.—Wyo.; introduced in Calif. *E. Temp.—Plain.* Mr–O.

16. C. stictóspora (Engelm.) Small. Plant yellowish green; stems 5–30 cm. long, villous; leaf-blades oblong to suborbicular, densely villous, 4–6 mm. long, obtuse, dentate-serrate above the middle; involucres campanulate, 1 mm. high; appendages narrow, crescent-shaped; capsules 1.5–2 mm. in diameter; seeds narrowly ovoid, 1.2–1.5 mm. long, pointed, pitted, ash-colored, sharply angled. *E. stictospora* Engelm. [G, B]. Dry soil: S.D.—Kans.—Tex.—Ariz.— Mex. *Austral.—Son.* Ap–N.

9. LEPADÈNA Raf. SNOW-ON-THE-MOUNTAIN.

Annual herbs, with erect stems. Leaves scattered below the umbel-like inflorescence, often very showy and petal-like in the inflorescence. Stipules fugacious. Involucres campanulate; lobes fimbriate. Glands 5, peltate, with pink or white petal-like appendages. Capsule 3-lobed, pubescent, round-lobed. Seeds narrowed upwards, reticulate. *Dichrophyllum* Kl. & Garcke.

1. L. marginàta (Pursh) Nieuwl. *Fig. 351.* Stout annual; stem 3–9 dm. tall, usually pilose; leaves ovate or obovate, 2–9 cm. long, sessile; bracts large, white-margined; involucres campanulate, 4 mm. long, usually pubescent; appendages reniform, white; capsule depressed-globose, 6 mm. in diameter, usually pubescent; seeds ovoid-globose, terete, 4 mm. long, dark ash-colored, reticulate-tuberculate. *Euphorbia marginata* Pursh. *Dichrophyllum marginatum* (Pursh) Kl. & Garcke [G, B]. Prairies, plains, and river valleys: Minn.—Tex.—Colo.—Mont.; introduced eastward. *Plain—Submont.* Jl–S.

f. 351.

10. ZYGOPHYLLÍDIUM Small. SPURGE.

Annual herbs, with erect, forking stems. Leaves mostly opposite, narrow, equilateral, not oblique at the base. Stipules gland-like, often obsolete. Involucres short-peduncled in the upper forks. Glands 5, broader than long, with petal-like appendages. Capsule long-pediceled, 3-lobed. Seeds terete, narrowed upwards, more or less papillose.

1. **Z. hexagònum** (Nutt.) Small. Plant yellowish-green; stem slender, 1-5 dm. high, with ascending almost filiform branches; leaves linear, oblong, or lanceolate, short-petioled; involucres 2 3 mm. long, ciliate; appendages green or whitish; capsule 4 mm. thick; seeds ovoid or oblong, 3 mm. long, terete, papillose. *Euphorbia hexagona* Nutt. [G, B]. On sandy prairies and river valleys: Minn.—Iowa—Tex.—N.M.—Mont. *Prairie—Plain.* Jl–S.

11. GALARRHOÈUS Haw. SPURGE.

Annual or perennial herbs, rarely shrubby, with simple or branched stems. Leaves below the inflorescence scattered, without stipules, entire or toothed, often broadened upwards. Bracts of the umbel-like inflorescence different from the stem-leaves. Involucres sessile or peduncled; lobes often toothed. Glands 4, transversely oblong, reniform, or crescent-shaped, with the horn-like appendages pointing upwards. Capsule 3-lobed; lobes rounded, acute or keeled. Seeds pitted. *Tithymalus* (Tourn.) Adans.

Leaves not serrulate; glands of the involucre crescent shaped.
  Perennials with a woody caudex.
    Seeds smooth.
      Stem-leaves 4–12 mm. broad; capsule smooth.
        Leaf-like bracts subtending the umbel lanceolate or oblanceolate.
                                              1. *G. Esula.*
        Leaf-like bracts subtending the umbel oval or obovate.   2. *G. lucidus.*
      Stem-leaves 1–4 mm. broad; capsule rough.   3. *G. Cyparissias.*
    Seeds pitted.   4. *G. robustus.*
  Annuals, or perennials by means of sobols; seeds pitted.
    Seeds finely pitted; lobes of the capsule keeled.   5. *G. Peplus.*
    Seeds with large pits in longitudinal rows; lobes of the capsule rounded.   6. *G. commutatus.*
Leaves serrulate; glands of the involucre oblong to orbicular; annuals or biennials.
  Seeds smooth or nearly so; capsule warty.
    Gland stalked; warts on the capsule long; seeds faintly reticulate.   7. *G. obtusatus.*
    Glands sessile; warts on the capsule depressed; seeds smooth.   8. *G. platyphyllus.*
  Seeds strongly reticulate.
    Glands nearly sessile; capsule warty.
      Upper stem-leaves merely sessile; bracts about as broad as long.   9. *G. arkansanus.*
      Upper stem-leaves with small auricles at the base; bracts manifestly longer than broad.   10. *G. missouriensis.*
    Glands stalked; capsule smooth.   11. *G. Helioscopia.*

1. **G. Ésula** (L.) Rydb. Bright green cæspitose perennial; stems 2 6 dm. high; leaves few, scattered, 1.5–4 cm. long, those subtending the umbel lanceolate or oblanceolate; bracts reniform, mucronate; involucre campanulate, 2.5–3 mm. high; capsule nodding, smooth; seeds oblong. *Euphorbia Esula* L. [G, B]. *Tithymalus Esula* Hill. Waste places: Mass.—Man.—N.D.—Colo.—N.J.; nat. from Eur. Je–S.

2. **G. lùcidus** (Waldst. & Kit.) Rydb. Stout, glabrous perennial; stem 2–5 dm. high, erect; leaves linear-oblong or oblong-lanceolate, 4–10 cm. long, entire; bracts broadly cordate, mucronate; involucre campanulate, 3 mm. long; glands 4, oblong, the appendages crescent-shaped, short-horned; capsule finely wrinkled, globose-ovoid, 4 mm. broad; seeds oblong, 2.5–3 mm. long. *E. lucida* L. [B]. *T. lucidus* Kl. & Garcke. *E. nicaeensis* (BB). Around dwellings and fields: N.Y.—Pa.—Iowa—Man.; nat. from Europe. Jl–S.

3. **G. Cyparíssias** (L.) Small. Bright green, cespitose perennial; stems erect, usually simple, 2–3 dm. high, very leafy; leaves 1.5–3 cm. long, 0.5–3 mm. broad; bracts reniform or nearly orbicular; involucres 2 mm. long; glands crescent-shaped; appendages none; capsule subglobose, 3 mm. thick; seeds oblong, 2 mm. long. *E. Cyparissias* L. [G]. *T. Cyparissias* Lam. [R]. Waste places and around dwellings: Mass.—Va.—Colo.—Mont.; escaped and nat. from Eu. My–O.

4. **G. robústus** (Engelm.) Rydb. *Fig. 352.*
Robust cespitose perennial; stems 1–3 dm. high; usually glabrous; stem-leaves oblong to broadly ovate, 1–2 cm. long, often nearly as broad; involucres 2.5 mm. high, 1.5 mm. wide, hirsute inside; glands often dentate between the short horns, 1–2 mm. wide; capsule depressed-ovoid, 4–4.5 mm. wide; seeds ovoid, 2.3 mm. long, with shallow pits. *E. montana robusta* Engelm. *T. robustus* Small [R]. Mountains: S.D.—Mont.—Ariz.—N.M.—Neb. *Plain —Submont.* My–Jl.

f. 352.

5. **G. Péplus** (L.) Haw. Bright-green, glabrous annual; stem 1–3 dm. high; leaves alternate, oblong or obovate, 1–4 cm. long, obtuse or retuse, more or less crisp, slender-petioled; bracts opposite, ovate, apiculate, sessile; involucre campanulate, 1–1.5 mm. high; glands with subulate horns; capsule globose-ovoid, 2–3 mm. broad, nodding; seeds oblong, 1.5 mm. long, whitish. *E. Peplus* L. [G, B]. Waste places: Que.—N.Y.—Ala.—Man.; Alaska—Calif.; nat. from Eur. Ja–S.

6. **G. commutàtus** (Engelm.) Small. Annual or perennial by means of shoots from the decumbent stems, 1.5–3 dm. high, branched at the base; stem-leaves obovate, obtuse or retuse, 0.5–3 cm. long; those of the sterile branches oblanceolate; bracts triangular-reniform; glands 1–1.5 mm. wide, yellow or brown when old; horns white, twice as long; capsule 3 mm. thick; seeds subglobose. *E. commutata* Engelm. [G, B]. *T. commutatus* Kl. & Garcke. [R]. River valleys and sandy soil: Pa.—N.C.—Tex.—Wis.—Mont. *Canad.—Allegh. —Plain.* My–Jl.

7. **G. obtusàtus** (Pursh) Small. Yellowish-green, glabrous annual; stem 3–6 dm. high; leaves alternate, spatulate-oblong, obtuse, serrulate below the middle; bracts ovate-cordate, 1–2.5 cm. long; involucre 1 mm. long; capsule subglobose, 4 mm. thick; seeds oblong, lenticular, nearly 2 mm. long, dark brown. *E. obtusata* Pursh [G, B]. *E. dictyosperma* Auth.; not F. & M. *T. obtusatus* Kl. & Garcke. Dry soil: Pa.—Iowa—Kans.—Tex.—S.C. *Allegh.— Carol.—Ozark.* Je–S.

8. **G. platyphýllus** (L.) Haw. Bright green annual; stem 1–5 dm. high, often reddish; leaves scattered, oblong or oblong-spatulate, 2–3 cm. long, acute; bracts deltoid to reniform, mucronate; involucre campanulate, 2 mm. high; capsule subglobose, 4 mm. thick; seeds oblong or oval, brown. *E. platyphylla* L. [G, B]. Shores: Que.—N.Y.—Man.; nat. from Eur. Je–S.

9. **G. arkansànus** (Engelm. & Gray) Small. Annual or biennial; stem slender, 2–5 dm. high, branched; leaves few, cuneate or spatulate, acute, serrulate, 1–3 cm. long, 5–10 mm. wide; bracts broadly triangular-ovate to oblong, mucronate, serrulate, the larger 1 cm. wide, 1–2 cm. long; involucres 1 mm. high; glands transversely elliptic, yellow, less than 1 mm. wide; capsule depressed globose, verrucose, 2.5–3 mm. wide; seeds ovoid-lenticular, brown or purplish, wrinkled, reticulate, nearly 1.5 mm. long. *E. arkansana* Engelm. & Gray [B]. *T. arkansanus* Kl. & Garcke. [R]. Plains and sandy places: N.D.— Mo.—Ala.—Tex.—Ariz.—Wyo.; Mex. *Son.—Plain—Prairie.* Ap–Je.

10. G. missouriénsis (Norton) Rydb. Annual or biennial, olive-green; stem 3–6 dm. high, stout; stem-leaves spatulate, 2–3.5 cm. long, obtuse, serrate to below the middle; bracts ovate, inequilateral, acute, serrate; involucres less than 2 mm. high; glands transversely oblong; capsules 3–3.5 mm. wide, bearing elongated warts; seeds ovoid, distinctly and regularly reticulate, purplish-brown. *E. arkansana missouriensis* Norton [B]. *T. missouriensis* Small. Dry plains: Minn.—Kans.—N.M.—Wash. *Prairie—Plain.* My–Jl.

11. G. Helioscòpia (L.) Haw. Stem ascending, 1.5–3.5 dm. high; leaves obovate, rounded or retuse at the end, finely serrate; umbels 5-rayed, the branches 3-branched; glands orbicular, stalked; pod globose-ovoid, 4 mm. broad, smooth; seeds reddish-brown, strongly reticulate. WARTWEED. Waste places: Newf.—Pa.—Iowa—Man.; nat. from Eur. Je–O.

12. TITHYMALÓPSIS Kl. & Garcke. SPURGE.

Perennial herbs with rootstocks. Leaves alternate below, opposite or whorled in the inflorescence, entire, often revolute on the margins. Involucres with toothed or fimbriate lobes. Glands with white, pink, or rose petal-like appendages. Capsule broader than high, with 3 rounded lobes. Seeds narrowed upwards, punctate, without caruncle.

1. T. corollàta (L.) Small. Rootstocks stout; stem 2–9 dm. high, often spotted; leaves linear, oblong, or spatulate, 2–4 cm. long, those subtending the umbels whorled; bracts ovate to linear, green; involucre less than 2 mm. high; glands green, oblong, with white appendages; capsule 3–4 mm. broad. *E. corollata* L. [G, D]. Dry soil: Ont.—Minn.—Kans.—Tex.—Fla. *E. Temp.* Ap–O.

13. POINSÉTTIA Graham. SPURGE.

Annual or perennial herbs, or shrubby plants, with green or partially colored foliage. Leaves alternate below, opposite above. Stipules gland-like. Involucres in axillary or terminal cymes, or solitary; lobes fimbriate. Glands fleshy, solitary, or rarely 3 or 4, without appendages. Capsule 3-lobed, lobes rounded. Seeds narrowed upwards, tuberculate.

Gland or glands of the involucre stalked; bracts and upper leaves slightly if at all discolored.
  Seeds not prominently tubercled; glands of the involucre 3–4; leaf-blades linear or linear-lanceolate. 1. *P. cuphusperma.*
  Seeds prominently tubercled; gland of the involucre solitary; leaf-blades ovate to lanceolate (linear-lanceolate only in one variety). 2. *P. dentata.*
Gland of the involucre sessile or nearly so; bracts and upper leaves discolored at the base; leaf-blades very variable, the upper usually fiddle-shaped. 3. *P. heterophylla.*

f. 353.

1. P. cuphuspérma (Boiss.) Small. Simple annual; stem 2–4 dm. high, more or less pubescent, especially upwards, erect; leaf-blades 2–8.5 cm. long, entire or denticulate; involucres crowded at the ends of the branches, almost 4 mm. long; capsule 5 mm. in diameter; seeds narrowly ovoid, 3 mm. long, irregularly 4-angled, ridged and slightly tuberculate. *E. cuphusperma* Boiss. [B]. Plains and prairies: S.D.—Colo.—Ariz.—Tex.—Mex. *Plain—Son.—Submont.* Au–S.

2. P. dentàta (Michx.) Small. *Fig. 353.* Pubescent annual; stem erect or ascending, 2–4 dm. high, somewhat woody at the base, branched; leaf-blades from ovate to nearly linear or orbicular-oblong, 1–9 cm. long, coarsely dentate; involucres about 3 mm. long; capsule glabrous, 4–5 mm. in diameter; seeds ovoid-globose, ash-colored, inconspicuously 4-angled. *E. dentata* Michx. [G,B]. Dry soil: S.D.—Pa.—La.—Mex.—Utah. *Plain—Submont.* Je–S.

3. P. heterophýlla (L.) Small. Bright green annual or biennial; pubescent or nearly glabrous; stem erect, 3–10 dm. high, woody below; leaf-blades linear to orbicular, the lower ones often entire, the upper undulate, sinuate, or dentate, the uppermost often fiddle-shaped and like the bracts blotched with red; involucres 3 mm. long, with 5 ovate or oblong laciniate lobes, one sinus bearing a sessile gland; capsule glabrous or minutely pubescent, 6 mm. broad; seeds oblong-ovoid, 3–4 mm. long, transversely wrinkled and tuberculate. *E. heterophylla* L. [G,B]. Sandy soil: Ill.—Fla.—Tex.—Mont.; Mex. and trop. Am. *E. Temp.—Trop.—Plain.* Ja–D.

Family 96. CALLITRICHACEAE. WATER STARWORT FAMILY.

Small aquatic caulescent annuals or perennials, with opposite entire leaves, often crowded towards the end of the branches. Flowers inconspicuous, solitary in the axils of the leaves, polygamous. Calyx wanting. Corolla none. Stamens solitary; anthers 2-celled; cells sometimes confluent. Pistils solitary; ovary 4-celled; styles united in pairs. Fruit leathery, indehiscent, 4-lobed, 4-seeded. Endosperm fleshy, embryo straight, axile, terete.

1. CALLÍTRICHE L. WATER STARWORT.
Characters of the family.

Upper floating leaves obovate or spatulate, 3-nerved; flowers subtended by a pair of bracts.
  Fruit oval, longer than the styles, sharply keeled.   1. *C. palustris.*
  Fruit obovoid, not longer than the styles, blunt-angled.   2. *C. heterophylla.*
All leaves linear, 1-nerved; flowers bractless.   3. *C. autumnalis.*

1. C. palústris L. *Fig. 354.* Aquatic perennial; stems usually floating, 2–30 cm. long; submerged leaves sessile, linear, 1-nerved, 1–1.5 cm. long; floating leaves petioled; blades 5–10 mm. long; fruit oval, 1.5 mm. long, slightly notched, grooved and winged, especially above. *C. verna* L. Shallow water: Newf.—Fla.—Calif.—Alaska; Mex., S. Am., and Eurasia. *Temp.—Plain—Mont.* Je–Au.

f. 354

2. C. heterophýlla Pursh. Perennial, growing in water or mud; stem 2–10 cm. long; floating leaves crowded at the ends, broadly spatulate, often retuse, abruptly narrowed into the petiole; fruit 1 mm. long, deeply emarginate. Still water or mud: Newf.—Md.—La.—Iowa.—Man. *E. Temp.* Jl–S.

3. C. autumnàlis L. Submerged aquatic perennial; stem slender, 1–4 dm. long; leaves all linear, retuse or bifid at the apex, 0.5–2 cm. long; fruit orbicular or rounded ellipsoid, 1–2 mm. in diameter, grooved and winged. *C. bifida* (L.) Morong. In water: Que.—N.Y.—Colo.—Utah—Ore.—Man.; Eur. *Temp.—Plain—Mont.* Jl–S.

Family 97. CELASTRACEAE. STAFF-TREE FAMILY.

Shrubs, trees, or twining vines, sometimes spiny, with opposite, whorled, or alternate, simple leaves, with or without stipules. Inflorescence normally cymose. Flowers perfect, polygamous, or dioecious.

Sepals 4 or 5, imbricate. Petals 4 or 5, usually inserted under the fleshy disk, imbricate. Stamens 4 or 5 under the disk, on the margin thereof, or on top of it. Gynoecium of a compound pistil. Ovary 2–5-celled; style short or wanting; stigma 2–5-lobed. Fruit a capsule, drupe, or berry. Seeds often surrounded by a brightly colored aril.

Shrubby vines with alternate leaves ; ovary free from the disk.   1. CELASTRUS.
Shrubs with opposite leaves ; ovary immersed in the disk.   2. EUONYMUS.

1. CELASTRUS L. WAXWORK, SHRUBBY BITTERSWEET, STAFF-TREE.

Usually twining vines or shrubs. Leaves alternate, entire or toothed; stipules minute. Flowers in axillary or terminal racemes, or panicles. Sepals 5. Hypanthium urn-shaped. Petals 5, inserted under the disk, which is cup-shaped, 5-lobed. Stamens 5, inserted in the sinuses of the disk. Gynoecium of a compound pistil; ovary 2–4-celled, sometimes incompletely so; style short and stout; stigma 2–4-lobed. Ovules 2 in each cavity, erect. Capsule globose or ellipsoid, leathery, 2–4-celled, loculicidal. Seeds 1 or 2 in each cavity, surrounded by a scarlet aril.

1. C. scándens L. *Fig. 357.* A climbing shrub, usually with twining branches; leaf-blades elliptic, oval, oblong, or ovate, 6–10 cm. long, acuminate, crenate, glabrous or nearly so; petals oblong or obovate, greenish, erose; capsule subglobose, 1 cm. in diameter, finely wrinkled, orange, opening by 3 valves. Along streams and in thickets; Que.—Fla.—Mont. Man. *E. Temp.—Plain.* My–Je.

f.357.

2. EUÓNYMUS (Tourn.) L. BURNING BUSH, WAHOO.

Shrubs or trees, commonly with 4-angled branches. Leaves opposite, entire or toothed, deciduous; stipules caducous. Flowers solitary or cymose. Sepals 4 or 5. Petals as many, inserted under the disk. Stamens 4 or 5, inserted on the disk; filaments very short; anthers 2-celled. Disk flat, 4- or 5-lobed. Gynoecium of a compound pistil; ovary 3–5-celled; stigma with as many inconspicuous lobes; ovules 2 in each cavity, ascending. Capsule 3–5-lobed, angled or winged, loculicidal. Seeds 2 in each cavity, surrounded by an orange or scarlet aril.

1. E. atropurpùreus Jacq. *Fig. 358.* A shrub or small tree, 1–8 m. high; leaf-blades elliptic to ovate or obovate, 5–16 cm. long, acuminate, puberulent, crenate-serrate; flowers trichotomously cymose; corolla dark purple; petals 4, suborbicular, 2–3 mm. long; capsule 4-lobed, pendulous, not warty, depressed, the lobes wing-like; seeds 8–10 mm. long, with a scarlet aril. River banks: N.Y.—Fla.—Okla.—Mont. *E. Temp.—Plain.* Je.

f.358.

Family 98. AQUIFOLIACEAE. HOLLY FAMILY.

Shrubs or trees, with alternate, often evergreen leaves, without stipules. Flowers perfect, dioecious, or polygamous, usually cymose. Sepals 4–6, imbricate, persistent. Petals as many, deciduous, imbricate. Stamens alternate with the petals, inserted at the base of the corolla;

anthers introrse, opening lengthwise. Gynoecium compound, of 2–6 carpels; ovary with as many cells; stigmas usually sessile. Ovules 1 or 2 in each cavity, pendulous, anatropous. Fruit a drupe, with horny or crustaceous nutlets.

Stamens adnate to the base of the corolla ; petals slightly united, oval or obovate.
                                             1. ILEX.
Stamens free ; petals free, linear.   2. NEMOPANTHES.

1. ÌLEX L. HOLLY. WINTERBERRY.

Usually glabrous shrubs or trees. Flowers sometimes inclined to be dioecious, the staminate in axillary clusters, the pistillate solitary. Sepals, petals, and stamens 4–6; petals united at the base. Drupe subglobose, with 4–6, rarely 7 or 8, nutlets. Seeds without an aril.

Nutlets ribbed ; flowers solitary or several from the axil, but without common peduncle.   1. *I. decidua.*
Nutlets smooth ; peduncles with several cymose flowers.   2. *I. verticillata.*

1. **I. decídua** Walt. *Fig. 356.* Shrub or small tree, 1–10 m. high; leaves thickish, oblanceolate to elliptic, 2–6 cm. long, obtuse or retuse, crenate-serrate, dark green and glabrous above, paler and usually pubescent beneath; pedicels 5–15 mm. long; corolla white, 4–6 mm. broad; lobes 4, oblong; drupe 7–9 mm. broad, orange to scarlet. Wet places: Va. —Ill.—Kans.—Tex.—Fla. *Allegh.—Austral.* My.

2. **I. verticillàta** (L.) A. Gray. Shrub or small tree, up to 6 m. high, with glabrous twigs; leaves thin, elliptic, oval, or obovate, 2–8 cm. long, acute or acuminate, glabrous or slightly pubescent above; tomentose beneath, reticulate, serrate; corolla white, 6–7 mm. broad; lobes obtuse; drupe 6–8 mm. in diameter, red. Swamps or wet ground: N.S.—Minn. —Mo.—Fla. *E. Temp.* Je–Jl.

f. 356.

2. NEMOPÁNTHES Raf. MOUNTAIN HOLLY, CATBERRY.

Shrubs with ashy gray bark. Leaves alternate, deciduous, glabrous, slender-petioled. Flowers axillary, solitary or a few together, long-pedicelled, polygamo-dioecious. Calyx of the staminate flowers minute, 4- or 5-toothed, that of the fertile flowers none. Petals 4 or 5, linear, spreading. Stamens 4 or 5. Fruit a drupe with 4 or 5 bony nutlets. *Ilicioides* Dum.-Cours.

1. **N. mucronàta** (L.) Trel. Shrub erect, 1–5 m. high; leaves obovate, 3–5 cm. long, mucronate, subentire; flowers solitary or the staminate ones 2–4 together; pedicels 3–4 cm. long; drupe red, 7–8 mm. broad. *I. mucronata* (L.) Britt. [B]. Swamps: Newf.—Va.—Ind.—Minn. *Canad.* My.

Family 99. EMPETRACEAE. CROWBERRY FAMILY.

Low evergreen undershrubs. Leaves small, jointed to short pulvini, channeled. Flowers monoecious or dioecious or polygamous. Calyx of 3 sepals or none. Petals 2 or 3 or none. Stamens 2–4, the anthers 2-celled, longitudinally dehiscent. Pistil 2–several-celled, the styles united.

the stigma with as many lobes as the cells. Ovules solitary. Fruit drupe-like, with 1-seeded nutlets.

1. ÉMPETRUM L. CROWBERRY.
Depressed shrubs with densely leafy branches. Flowers polygamous in the axils of the leaves. Sepals 3, spreading, petaloid. Petals 3. Stamens 3. Ovary 6–9-celled, the stigma with 6–9 toothed lobes. Fruit pulpy, with 6–9 stones.

1. E. nigrum L. *Fig. 355.* Leaves linear-oblong, glabrous or puberulent, 4–7 mm. long, 1 mm. wide, rough on the revolute margins; flowers purplish; fruit 4–6 mm. thick, globose, black. Bogs: Greenl.—N.Y.—Minn.—Alta.—Alaska; also Eurasia. *Arctic–Subarctic.*

f.355.

Family 100. **ANACARDIACEAE.** SUMACH FAMILY.

Polygamous, monoecious, or dioecious, or rarely hermaphrodite shrubs or trees, with alternate, simple or pinnate leaves, without stipules. Inflorescence paniculate or spicate or racemose. Flowers regular. Sepals 3–5, distinct. Petals 3–5, usually inserted on a hypogynous disk, imbricate or rarely valvate. Stamens 3–5, rarely more, alternate with the petals. Gynoecium of 1, or 4 or 5 united or nearly distinct carpels; styles united, sometimes distinct; stigmas entire. Fruit a drupe or a berry; seeds solitary.

Drupe with pubescent exocarp; stone smooth.   1. RHUS.
Drupe with glabrous, smooth exocarp; stone ribbed.   2. TOXICODENDRON.

1. RHUS (Tourn.) L. SUMACH, SKUNK-BUSH.

Shrubs or trees, not poisonous. Leaves alternate, odd-pinnate, rarely unifoliolate; leaflets more or less toothed. Flowers in terminal panicles, polygamous or dioecious. Sepals commonly 5. Petals as many, imbricate. Ovary 1-celled; ovules pendulous. Drupe red, covered with acid-secreting hairs. Stone smooth. Stamens 5, inserted under the edge or between the lobes of the flattened disk. Flowers greenish white or yellowish.

Leaflets 5–31; inflorescences terminating leafy shoots.
  Leaf-rachis winged.   1. *R. copallina.*
  Leaf-rachis not winged.
    Twigs and leaf-rachis densely long-pubescent; branches of inflorescence and fruit long-hairy.   2. *R. hirta.*
    Twigs and leaf-rachis glabrous, or finely or sparingly pubescent; fruit short-hairy.
      Branches of the inflorescence hirsute.   3. *R. Sandbergii.*
      Branches of inflorescence puberulent or glabrous.
        Leaflets 13–31, dark green above, shining.   4. *R. glabra.*
        Leaflets 11–17, light or yellow-green, dull.   5. *R. cismontana.*
Leaflets 3, rarely 5; inflorescences axillary.
  Leaflets many-lobed above the middle, the terminal one mostly rhombic, the lateral ones obliquely ovate or oval.
    Leaves rather thin, sparingly pubescent, mostly on the veins; twigs glabrate.   6. *R. crenata.*
    Leaves rather firm, densely hirsute; twigs densely short-pubescent.   7. *R. Nortonii.*
  Leaflets 3–5-, rarely 7-lobed at the apex, all cuneate.
    Leaves densely villous.   8. *R. Osterhoutii.*
    Leaves puberulent or glabrous.
      Branchlets and leaves puberulent, at least when young.   9. *R. trilobata.*
      Branchlets and leaves glabrous, except for a few scattered cilia on the petioles and veins.   10. *R. oxyacanthoides.*

1. R. copallìna L. Shrub or small tree, 1–10 m. high; leaflets 9–21, oblong or lanceolate, 3–10 cm. long, acute or acuminate, undulate, glossy above, pale and dull beneath; panicle broad, 0.5–3 dm. long; flowers greenish; petals oblong, 2.5 mm. long; fruit globose, 4 mm. broad, bright red, finely pubescent. Dry soil: Me.—Minn.—Tex.—Fla. *E. Temp.* Je–Au.

2. R. hírta (L.) Sudw. Shrub or small tree, up to 12 m. high; leaflets 11–31, oblong to lanceolate, 4–15 cm. long, acuminate, coarsely serrate, glabrate above, pubescent beneath; panicle 1–2 dm. long; petals oblong-lanceolate, 3 mm. long; fruit bristly-pubescent, 4 mm. broad, bright red. *R. typhina* L. [G]. Dry rocky soil: N.B.—Minn.—Miss.—Ga. Je–Jl.

3. R. Sandbérgii (Vasey & Holz.) Greene. A low shrub, 2–10 dm. high; branchlets rusty-tomentulose and sparingly hirsute or glabrate; leaflets 5–15, subsessile, lanceolate, 4–8 cm. long, acuminate, coarsely serrate, dark green above, paler beneath, glabrous; panicle 3–7 cm. long, ovoid, its branches tomentulose and hirsute; flowers greenish; petals oblong, 2.5 mm. long; fruit globose, 4–5 mm. long. Rocky hills: n Minn. May be a hybrid between *R. hirta* and *R. cismontana. Boreal.* Jl.

4. R. glàbra L. Shrub, sometimes 4 dm. high; leaflets oblong or lanceolate, acuminate, serrate, dark green above, glaucous beneath, glabrous; flowers bright green; petals ovate, 2 mm. long, hooded; fruit bright red, 5 mm. broad, viscid-pubescent. *R. angustiarum* Lunell. *R. Hapemanii* Lunell (?). Dry soil: N.S.—Man.—N.D.—Kans.—Minn.—Fla. Je–Au.

5. R. cismontàna Greene. *Fig. 359.* Shrub, 1–2 m. high; branches glabrous, often reddish, somewhat glaucous; leaves 1.5–2 dm. long; leaflets 11–17, usually pallid green, often lighter along the veins, glaucescent beneath, 5–10 cm. long, lanceolate, abruptly acuminate, sharply serrate, thin; panicle about 1 dm. long; branches usually spreading, finely pubescent; calyx-lobes 1 mm. long; petals greenish yellow, 2–2.5 mm. long. *R. sambucina, R. tessellata, R. albida,* and *R. asplenifolia* Greene. Along streams: Minn.—Mo.—N.M.—Ariz.—Wyo. *Prairie—Plain—Submont.* My–Jl.

f.359.

6. R. crenàta (Mill.) Shrub, 1–10 m. high, aromatic; leaflets 3, thin-leathery, serrate, 2–6 cm. long, glabrous or slightly pubescent, the terminal one rhombic, cuneate at the base, the lateral ones obliquely ovate, 2–6 cm. long; flowers yellow, in dense spike-like racemes; petals oblong, 2.5 mm. long; fruit globose, 7–8 mm. broad, bristly-pubescent. *R. canadensis* Marsh. [G]. *R. aromatica* Ait. [B]. Sandy soil and rocky woods: Vt.—Minn.(?)—se Kans.—La.—Fla. *E. Temp.* Mr–Ap.

7. R. Nortònii (Greene) Rydb. Shrub, 1–3 m. high; twigs densely short-pubescent or puberulent, but glabrate the second year; leaflets firm, dark green above, paler beneath, densely hirsute on both sides; terminal leaflet rhombic, with cuneate base, 3–5 cm. long, 1.5–3 cm. wide, 7–15-lobed above the middle, the lateral ones 2–4 cm. long; flowers yellowish; petals oblong, 2 mm. long; fruit globose, 5–6 mm. broad, densely hairy, the hairs fully 1 mm. long. *Schmaltzia Nortonii* Greene. *S. lasiocarpa* Greene. Dry hills: Ill.—Kans.—Okla.—Mo. *Ozark.*

8. R. Osterhoùtii Rydb. Shrub, about 1 m. high; leaves trifoliolate; leaflets sessile, 1–2.5 cm. long, cuneate, 3–5-lobed and some of the lobes with rounded teeth, dark green above, pale beneath; otherwise like the following species. *Schmaltzia pubescens* Osterh. Hills: Neb.—S.D.—Colo. *Submont.* Je.

9. **R. trilobâta** Nutt. Shrub, 1–2 m. high, blooming before the leaves; leaves trifoliolate, with unpleasant odor; leaflets 1–3 cm. long, dark green above, paler and minutely pubescent beneath, the lateral ones elliptic or cuneate-obovate, the terminal one cuneate-spatulate, usually more or less 3-lobed and crenate; petals obovate, 2 mm. long; fruit globose, 5–6 mm. thick, more or less densely short pubescent. *Schmaltzia trilobata, S. Bakeri, S. cognata, S. glomerata, S. subpinnata*, and *S. glabrata* Greene. Hills and plains: Sask.—Mo.—Tex.—Calif.—Wash. *Plain—Submont.* Ap Je

10. **R. oxyacanthoïdes** (Greene) Rydb. Low shrub, about 1 m high; leaves 3-foliolate; leaflets obovate-cuneate, or the terminal one rhombic-obovate, rather thin, rounded-crenate, the terminal one 2–4 cm. long, and often 3-lobed, the lateral ones 1.5–3 cm. long; fruit bright red, puberulent. *Schmaltzia oxyacanthoides* Greene. Hills and cañons: S.D.—w Colo.—Utah—(? Mont.) *Submont.* Ap–My.

2. **TOXICODÉNDRON** (Tourn.) Mill. POISON IVY, POISON OAK, POISON SUMACH.

Shrubs, trees, or vines, with a resinous sap, poisonous to touch. Leaves alternate, pinnately 3–several-foliolate, with coarsely toothed or entire leaflets. Flowers polygamous, paniculate. Sepals 4–6, persistent. Petals 4–6, imbricate, yellowish or greenish. Disk annular. Stamens 4–6 or 10. Ovary 1-celled; ovules pendulous. Drupe whitish, smooth, shining, glabrous. Stone striately ribbed.

Leaves 7–11-foliolate.	1. *T. Vernix.*
Leaves 3-foliolate	
Leaflets undulate or rarely few-toothed.	
Plant climbing with aerial rootlets; fruit depressed-globose, 5–6 mm. broad.	2. *T. radicans.*
Plant with subterranean shoots, only the short erect floral branches above ground; fruit globose, 4–5 mm. broad.	3. *T. desertorum.*
Leaflets coarsely toothed.	
Leaves glabrous or nearly so, fruit 5–6 mm. broad.	4. *T. Rydbergii.*
Leaves distinctly pubescent, at least on the veins beneath.	
Leaflets rhombic-ovate in outline, distinctly acuminate; plant often climbing.	5. *T. Negundo.*
Leaflets very thin, large, suborbicular, acute or very shortly abruptly acuminate; stems mostly subterranean, only the erect branches above ground, not climbing.	6. *T. fothergilloides.*

1. **T. Vérnix** (L.) Kuntze. Shrub or small tree, up to 8 m. high; leaflets oblong to oval, or the terminal one obovate, 4–15 cm. long, acuminate at each end, undulate, glabrous above, pubescent beneath; panicles axillary, loose; flowers greenish; petals linear-oblong, curved, 2 mm. long; fruit about 5 mm. broad, white. POISON SUMACH. *Rhus Vernix* L. [G, B]. *R. venenata* DC. Swamps: R.I.—Minn.—La.—Fla. *E. Temp.* Je.

2. **T. radìcans** (L.) Kuntze. Vine, usually high-climbing; leaflets ovate-lanceolate, 4–20 cm. long, entire, undulate, rarely few-toothed, acuminate, deep green, sparingly pubescent beneath; panicles loose; flowers greenish; petals oblong or obovate. POISON IVY. *R. radicans* L. [B]. *R. Toxicodendron radicans* Torr. [G]. Thickets, woods, and along fences: N.S.—Man.—Kans.—Ark.—Fla. *E. Temp.* My–Je.

3. **T. desertòrum** Lunell. Stem trailing, often subterranean; leaflets 3–7 cm. long, 2.5–5 cm. wide, broadly ovate, acuminate, entire, shining, pilose on the veins beneath, entire or sinuate above; fruit globose, 4–5 mm. broad. Sand hills: N.D. Jl.

4. T. Rydbérgii (Small) Greene. *Fig. 360.*
A single-stemmed shrub, usually less than 1 m. high;
leaves pinnately 3-foliolate; leaflets 3–10 cm. long,
rhombic-ovate, thick, bright green, strongly veined
beneath, glabrous, usually coarsely and broadly
toothed; flowers in dense axillary panicles; petals
whitish yellow, with greenish veins; fruit depressed-
globose, 5–6 mm. in diameter. *R. Rydbergii* Small
[B]. *T. hesperinum* Greene, a large form. *T.
macrocarpum* Greene. Hillsides and open woods:
Mich. — Iowa — Kans. — N.M. — Ariz. — Ore. — B.C.
Prairie—Plain—Submont. Je–Jl.

5. T. Negúndo Greene. Stem erect or climbing over bushes; twigs and petioles pubescent; leaflets broadly rhombic-ovate in outline, 8–15 cm. long,
sparingly hirsutulous above, hirsute especially on the
veins beneath, 7–15 cm. long, coarsely toothed or lobed, acuminate; panicles
rather loose; flowers greenish; petals oval, 2 mm. long, with darker veins; fruit
globose, 4–5 mm. broad, shining. Low woods and thickets: Iowa—Kans.—
Tex.—Mo. *Ozark.* My–Je.

6. T. fothergilloìdes Lunell. Stems horizontal, subterranean; floral
branches erect, 2–3 dm. high; leaves pinnately 3-foliolate; petals 1.5–3 dm.
long; leaflets 10–12 cm. long and wide, suborbicular, coarsely few-toothed, acute
or abruptly short-acuminate, very thin, pilose on the veins beneath; flowers in
axillary panicles, 6–10 cm. long; fruit yellowish white, globose, 4–5 mm. in
diameter. Dense woodlands: Devils Lake, N.D. Je–Jl.

Family 101. STAPHYLEACEAE. Bladdernut Family.

Shrubs or trees, with alternate or opposite, compound leaves, without
stipules. Flowers perfect or polygamous, in racemes or panicles. Sepals
5. Disc present, cup-shaped, with a free edge. Petals 5, imbricate, inserted under the edge of the disc. Stamens 5; filaments distinct; anthers
introrse. Gynoecium of 2 or 3 partly united
carpels, styles 2 or 3, distinct or partly united.
Ovules in 1 or 2 rows on the ventral suture.
Fruit an inflated membranous capsule or a berry.
Seeds few or many.

1. STAPHYLÈA L. Bladdernut.

Shrubs. Leaves 3-foliolate or odd-pinnate.
Flowers perfect, regular, in axillary racemes or panicles. Sepals imbricate, deciduous. Carpels united
at the base; stigmas capitate. Fruit a membranous
capsule, 2- or 3-celled, 2- or 3-lobed. Seeds solitary
in each cavity.

1. S. trifoliàta L. *Fig. 361.* Shrub, 1–5 m.
high; leaves trifoliolate; leaflets elliptic to ovate or
obovate, 5–10 cm. long, acuminate, serrate; raceme
drooping, 5–10 cm. long; sepals lanceolate, 7–10 mm. long; petals slightly
longer, pubescent at the base; capsule bladder-like, 3-lobed, 4–6 cm. long.
Thickets: Que.—Minn.—Mo.—Ga. *Canad.—Allegh.* Ap–My.

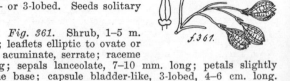

Family 102. HIPPOCASTANACEAE. Buckeye Family.

Shrubs or trees, with palmately compound alternate leaves, without
stipules. Flowers often polygamous, irregular, in racemes or panicles.

Sepals mostly 5, more or less united. Disk present, annular or one-sided.
Petals 4 or 5, unequal, clawed. Stamens 5–8, inserted inside the disk;
anthers introrse. Gynoecium of 3 united carpels, ovary 3-celled; styles
elongate, united. Ovules 2 in each cell. Fruit a loculicidal, 3-valved
capsule.

1. AÉSCULUS L. BUCKEYE. HORSE-CHESTNUT.

Characters of the family.

Capsule smooth ; calyx oblong-campanulate.	1. *A. octandra.*
Capsule spiny ; calyx campanulate.	
Shrubs ; leaflets 6–9, incised above the middle as well as serrate.	2. *A. arguta.*
Trees, leaflets usually 5, merely serrate.	3. *A. glabra.*

1. A. octándra Marsh. Shrub or tree, up to 4 m. high; leaflets mostly
5, oblanceolate or elliptic, 1–2.5 dm. long, acuminate, finely serrate; flowers 2–3
cm. long; calyx 1–1.5 cm. long, finely glandular; lobes broadly ovate, obtuse;
corolla yellow or purplish; upper petals with claws equaling the oval blades,
the lower ones with claws several times as long as the suborbicular blades;
stamens short-exserted; capsule 4–6 cm. thick. Woods: Pa.—Iowa—Tex.—Ga.
Allegh.—Austral. Ap–My.

2. A. argùta Buckl. Shrub, 1–3 m. high, with
smooth bark; leaflets narrowly elliptic, 6–12 cm.
long, acuminate, shining above, dull beneath; flowers 2.5 cm. long, calyx 4–5 mm. long, minutely
glandular; lobes rounded; corolla pale yellow, long-
clawed; blades oblong, those of the upper petals
broader; capsule subglobose, 3–4.5 cm. thick. *A.
glabra arguta* Robinson [G], Woods: Iowa—Mo.—
Tex.—Neb. *Ozark.* Mr–Ap.

3. A. glàbra Willd. *Fig. 362.* Tree, up to 25
m. high; leaflets elliptic to oblanceolate, short-acuminate, glabrate, 7–18 cm. long; flowers 12–18 mm.
long, finely pubescent; lobes broadly ovate, rounded;
corolla greenish yellow; upper petals with linear-
spatulate blades, the blades of the lower petals oval
or oblong; capsule obovoid, 3–7 cm. thick. Woods:
Iowa—Ala.—Kans.—Tex. *Austral.* Mr–Ap.

f.362.

Family 103. ACERACEAE. MAPLE FAMILY.

Shrubs or trees, with opposite, simple or compound leaves, without
stipules. Flowers perfect or polygamous, rarely dioecious or monoecious,
regular. Sepals 4 or 5, rarely 6–9, imbricate, often colored. Petals of the
same number, inserted on the margins of an indistinct disk, or very often
wanting. Stamens 4–9, often 8. Gynoecium of 2 united carpels; styles
united. Fruit of 2 winged carpels, united below (samaras).

Leaves palmately lobed or rarely palmately compound ; flowers polygamous or perfect.	
	1. ACER.
Leaves pinnately compound ; flowers dioecious.	2. NEGUNDO.

1. ÀCER (Tourn.) L. MAPLE.

Trees or rarely shrubs. Leaves opposite, petioled, more or less palmately
lobed or cleft, rarely palmately compound. Flowers polygamous or rarely
perfect, regular, in axillary or terminal racemes or corymbs. Sepals 4 or 5,
rarely 6–9. Disk cup-shaped, lobed, or more commonly obsolete or wanting.
Stamens 4–9. Petals as many as the sepals or wanting. Samaras 2, with reticulate wings.

Flowers in racemes or corymbs; disk well developed; lobes of the leaves toothed.
 Flowers in long racemes, appearing after the leaves; leaves long-acuminate, with
  small lateral lobes; petals longer than the sepals.   1. *A. spicatum.*
 Flowers corymbose, appearing with the leaves; leaves not
  long-acuminate, deeply 3–5-cleft; terminal lobe only
  slightly larger.
  Flowers long-pedicelled, drooping; petals wanting.
   Leaves pale and glabrous beneath; basal sinus open. 2. *A. Saccharum.*
   Leaves green and pubescent beneath; basal sinus
    narrow.   3. *A. nigrum.*
  Flowers short-pedicelled, erect; petals present, greenish,
   nearly equaling the sepals.   4. *A. glabrum.*
Flowers in umbels from separate lateral buds, appearing before
 the leaves.
 Petals wanting; fruit tomentose when young; wings diver-
  gent.   5. *A. saccharinum.*
 Petals present, linear-oblong; fruit glabrous; wings incurved. 6. *A. rubrum.*

1. A. spicàtum Lam. Shrub or small tree, sometimes 10 m. high, with
thin smooth bark; twigs somewhat velutinous-puberulent or glabrate in age;
leaf-blades longer than broad, cordate, 3-lobed, glabrate above, paler and pubescent beneath; petals linear or spatulate; samaras about 2 cm. long; wings
ascending-spreading at about 90°; backs nearly straight. Rocky woods and
mountain slopes: Newf.—Ga.—Iowa—Minn.—Sask. *Canad.—Allegh.* My–Je.

2. A. Sáccharum Marsh. Tree, up to 40 m. high; leaf-blades firm, 8–15
cm. broad, usually 5-lobed and sharply toothed, green above, paler beneath;
calyx 5 mm. long, hairy at the apex; sepals obtuse; fruit with slightly spreading wings, 3.5–4 cm. long. *A. saccharinum* Wang.; not L. Sugar Maple.
Hard Maple. Woods: Newf.—Man.—Tex.—Ga. *E. Temp.* Ap–My.

3. A. nìgrum Michx. f. Tree, up to 40 m. high, with gray bark; leaf-
blades 15–20 cm. broad, often broader than long, green on both sides, with
3–5 entire or undulate lobes; calyx campanulate, 5 mm. long; sepals rounded;
fruit with slightly spreading wings, 3–4 cm. long. Black Maple. *A. Saccharum nigrum* Britt. [G]. Woods: Ont.—Minn.—La.—Ga. *E. Temp.* Ap–My.

4. A. glàbrum Torr. *Fig. 363.* Usually a
small tree, 5–15 m. high; twigs glabrous, reddish;
older branches gray; leaf-blades broadly cordate or
rounded-reniform in outline, 4–8 cm. long, 4–10 cm.
wide, glabrous, 5–7-lobed, or usually on young shoots
and vigorous branches or sometimes on the whole
tree 3-divided to the base; samaras 2.5–3 cm. long,
usually straight on the back; sinus usually acute.
A. neomexicanum Greene. Mountains, usually along
streams: S.D.—Neb.—N.M.—Ariz.—Utah—Wyo.
Submont.—Mont. My–Je.

f. 363.

5. A. saccharinum L. Tree, rarely 35 m. high,
with gray flaky bark; leaf-blades 10–15 cm. long
and broad, with 3–5 incised lobes, silky when young,
glabrate and green above, glaucous or silvery white
beneath, truncate or cordate at the base; calyx
greenish or yellowish; fruit drooping, with widely spreading wings, 5–6 cm.
long. *A. dasycarpum* Ehrh. Silver Maple. Soft Maple. Woods: N.B.—
Ont.—N.D.—Okla.—Fla. *E. Temp.* F–Ap.

6. A. rùbrum L. Tree, up to 35 m. high, with pale bark; leaf-blades as
long as broad or longer, 7–10 cm. long, deep green above, pale or glaucous beneath, shallowly 3–5-lobed and evenly serrate; flowers red or yellowish, the
pistillate ones with longer pedicels; calyx deeply cleft; fruit with incurved
wings, 2–2.5 cm. long. Red Maple. Low ground: N.B.—Man.—Tex.—Fla.
E. Temp. Mr–Ap.

2. NEGÚNDO (Ray) Ludwig. BOX-ELDER, ASH-LEAVED MAPLE.

Trees, with light green twigs, widely branching. Leaves opposite, odd-pinnate, usually with toothed leaflets. Flowers dioecious, appearing before the leaves, the staminate in drooping small clusters, the pistillate racemose. Sepals 4 or 5, very small. Petals wanting. Stamens 4 or 5. Disk obsolete. Samaras 2, with reticulate wings. *Rulac* Adans.

Branchlets glabrous, usually with a bloom ; anthers acute, tapering into a tip 0.25–
   0.5 mm. long.   **1. *N. Nuttallii.***
Branchlets pubescent ; anthers obtuse, merely mucronate.   **2. *N. interius.***

f. 364.

1. N. Nuttállii (Niouw⁻¹) Rydb. *Fig. 364.*
Tree up to 15 m. high, with spreading branches; bark of the twigs and younger branches light green; leaflets usually 3, lanceolate or ovate, long-acuminate, usually rounded at the base, coarsely toothed or often more or less lobed, sparingly hairy or in age glabrate above, more densely so beneath, and often with tufts of hairs in the axils of the veins; fruit glabrous, gradually tapering below, wing scarcely decurrent. *Acer Negundo* Am. auth., in part [B,G]. River valleys: Mich.—Mo.—Kans.—Colo.—Mont.; often cultivated. *Prairie—Plain.* Ap–My.

2. N. intèrius (Britton) Rydb. Tree, up to 12 m. high; leaflets lanceolate or ovate, or the terminal rhombic, 5–12 cm. long, long-acuminate, usually lobed, coarsely toothed, otherwise as in the preceding; fruit as in the preceding. *A. interior* Britton. Along streams and in cañons, mostly in the foothills and mountains: Sask.—Man.—Mo.—Neb.—N.M.—Ariz.—Mont. *Plain—Submont.* My–Jo.

Family 104. SAPINDACEAE. SOAPBERRY FAMILY.

Shrubs, trees, or vines. Leaves opposite or alternate, usually with stipules, simple or pinnately compound. Flowers dioecious or polygamous, rarely perfect, in racemes or panicles. Sepals 4 or 5, imbricate. Disc present, entire or lobed. Petals 4 or 5, or wanting. Stamens 5 or 8–15, inserted on the disc. Gynoecium of 2–4 united carpels; ovary with as many cells; styles partly united; stigmas capitate or lobed. Fruit a capsule or berry-like.

Vines with tendrils ; fruit inflated.   **1. CARDIOSPERMUM.**
Shrubs or trees ; fruit baccate.   **2. SAPINDUS.**

1. CARDIOSPÉRMUM L. BALLOON VINE. HEART-SEED.

Herbaceous or shrubby vines, with tendrils. Leaves alternate, biternate or decompound, with toothed leaflets. Flowers polygamo-dioecious, irregular, in axillary racemes or corymbs, with tendril-bearing peduncles. Sepals 4, in 2 series, the outer smaller. Disk 1-sided, outside the stamens, with 2 glands. Petals 4, unequal, the two larger with a scale at the base, the two smaller crested. Stamens 8, filaments unequal. Ovary 3-celled, 3-angled; styles united at the base. Capsule inflated, 3-angled, membranous, veiny. Seeds solitary in each cavity.

1. C. Halicácabum L. An annual or biennial climbing herb; stem 1–5 m. long; leaflets 3, ovate or ovate-lanceolate, coarsely serrate, incised; outer sepals broader than long, the inner longer than broad; petals obovate; filaments

pubescent; capsule 3–3.5 cm long, pubescent. Thickets: N.J.—Kans.—Tex.—
Fla.; Trop. Am. *Allegh.—Austral.—Trop.* Je–S.

2. SAPÍNDUS L. SOAPBERRY.

Shrubs or trees. Leaves alternate, pinnate, with
entire or toothed leaflets. Flowers regular, polygamous, in racemes or panicles. Sepals 4 or 5, imbricate in 2 series. Petals 4 or 5, inserted under the
disk. Stamens 8–10, inserted on the disk; filaments
distinct; anthers versatile. Ovary 2–4-celled; styles
united or distinct. Fruit berry-like; seeds 1 in each
cavity.

1. S. Drummóndii H. & A. *Fig. 365.* Small
tree, with spreading branches; leaflets 9–19, narrowly lanceolate, acuminate, 4–8 cm. long, falcate,
glabrous above, soft-pubescent beneath; sepals ovate,
obtuse; petals twice as long as the sepals, rhombic-
lanceolate, lacerate at the apex, fruit globose, 1.5
cm. broad, yellow, drying black. Hillsides: Ark.—
Kans.—-Ariz.—La. *Austral—Son.*

Family 105. RHAMNACEAE. BUCKTHORN FAMILY.

Shrubs or trees, with alternate, rarely opposite, simple, usually several-ribbed leaves and small stipules. Flowers greenish, perfect or polygamous, rarely dioecious. Sepals 4 or 5, valvate. Hypanthium somewhat
developed, lined or filled with a thickened disk. Petals 4 or 5. Stamens
of the same number, inserted on the disk, opposite the petals. Gynoecium
of 2 or 3 united carpels; ovary 2- or 3-celled, partly immersed in the
disk. Styles and stigmas more or less united. Fruit a capsule, a drupe,
or rarely a samara.

Fruit pulpy; petals small, clawless, or wanting; stigmas usually 2. 1. RHAMNUS.
Fruit dry; petals hooded and long-clawed; stigmas 3. 2. CEANOTHUS.

1. RHÁMNUS (Tourn.) L. BUCKTHORN.

Shrubs or trees, unarmed, or with spinose branchlets. Leaves alternate,
entire or toothed, several-ribbed; stipules deciduous. Flowers perfect or polygamo-dioecious, axillary, often clustered, in racemes, cymes or umbels. Sepals
4 or 5, keeled within. Disk cup-shaped, lining the hypanthium. Petals 4 or 5,
or wanting, clawless, inserted on the margin of the disk. Stamens of the same
number, on the edge of the disk; filaments very short. Ovary 2–4-celled; styles
2–4, united at the base. Drupe berry-like, with 3–4 nutlets. Endosperm fleshy.

Flowers dioecious or polygamous; nutlets grooved.
  Shrub spinose; introduced species. 1. *R. cathartica.*
  Shrub not spinose; native species.
    Flowers 4-merous; petals present; seeds 2, grooved;
      leaves not strongly veiny. 2. *R. lanceolata.*
    Flowers 5-merous, apetalous; seeds 3, scarcely grooved;
      leaves strongly veiny. 3. *R. alnifolia.*
Flowers perfect; nutlets smooth. 4. *R. caroliniana.*

1. R. cathártica L. Shrub or low tree, 2–7 m. high, with rough bark;
leaves petioled; blades glabrous or nearly so, broadly ovate or oval, 3–6 cm.
long, crenate; flowers dioecious, greenish, 4-merous; petals lanceolate; drupe
globose, about 8 mm. in diameter; seeds 3–4, grooved. Around dwellings: Ont.
—Va.—Colo.—Minn.; cult. and occasionally escaped; native of Eur. Je.

2. R. lanceolàta Pursh. Unarmed shrub or small tree; leaves short-petioled; blades lanceolate or ovate-lanceolate, acuminate, with a blunt apex, minutely crenate, glabrous above, puberulent beneath, 3–8 cm. long; flowers 2–3 in the axils of the leaves, greenish, dioecious; petals lanceolate, shorter than the sepals; drupe black, 6 mm. in diameter. Wet ground: Pa.—Ala.—Tex.—Neb. *Allegh.— Austral—Plain.* My–Je.

3. R. alnifòlia L'Hér. *Fig. 366.* Unarmed shrub, 1–2 m. high; branches gray, finely pubescent; leaves short-petioled; blades oval or elliptic, from obtuse to somewhat acuminate, coarsely crenate-serrate, finely pubescent when young, soon glabrate, 3–10 cm. long, strongly veined, flowers solitary or 2 or 3 and umbellate; sepals spreading; drupe black, 7–8 mm. in diameter. Swamps: Me.—N.J. –Wyo.—n Calif.—B.C. *Temp,— Submont* My–Je.

4. R. caroliniàna Walt. Shrub or small tree, up to 10 m. high, with ashy-gray bark; leaves firm, oblong or elliptic, 5–12 cm. long, acute or short-acuminate, undulate or serrulate, short-petioled, densely tomentose or becoming glabrate; flower-clusters umbel like, few-flowered; sepals and petals 5, the former triangular-ovate, acute, the latter notched at the apex, one third as long; fruit subglobose, about 1 cm. thick, black, 3-seeded. River banks and hillsides: Va.—Kans.—Tex.—Fla. *Austral.* My–Je.

2. CEANÒTHUS L. New Jersey Tea, Mountain Laurel. Snow Brush, Deer Brush

Shrubs or small trees, often with spinulose branches. Leaves alternate, commonly 3-ribbed; stipules caducous. Flowers perfect, crowded in terminal panicles, cymes, or umbels. Hypanthium urn-shaped, filled by the disk. Sepals 5, deciduous, white. Petals 5, inserted under the disk, long-clawed and strongly hooded. Stamens 5; filaments filiform, exserted. Ovary immersed in the disk, 3-celled. Styles short, united below; stigmas 3. Fruit dry, 3-lobed, separating into 3 nutlets. Seeds flattened; endosperm fleshy.

Leaf-blades rounded-oval, often cordate at the base, closely glandular-dentate or crenate.
  Peduncles often naked from lateral buds; leaves sparingly pubescent beneath, soon glabrate.   1. *C. sanguineus.*
  Peduncles on leafy branches of the present season; leaves velutinous beneath.   2. *C. velutinus.*
Leaf-blades oblong to elliptic or ovate, glandular-serrate to sub-entire.
  Umbels panicled, mostly terminal; leaves dull beneath, glabrate or villous.
    Leaves acute, of an ovate type; common peduncle elongate.   3. *C. americanus.*
    Leaves obtuse, of an elliptic type; common peduncle short.
      Leaves thin, glossy above, sparingly hairy when young, glabrate in age; peduncles finely viscid-puberulent.   4. *C. ovatus.*
      Leaves thick, veiny, dull, as well as the peduncles more or less permanently villous.   5. *C. pubescens.*
  Umbels mostly axillary; leaves silky beneath, obsoletely denticulate or entire; branches often ending in spines.   6. *C. Fendleri.*

1. C. sanguíneus Pursh. Tall shrub, 1–3 m. high; young twigs greenish, those of preceding season red or purple; leaf-blades oval, rounded at the apex, rounded or cordate at the base, glabrate; petals white, spatulate; fruit obovoid, 3-lobed above, not crested, 4 mm. thick. Hillsides and copses: Mich.—S.D.—Calif.—B.C. *Submont.* Ap–Jl.

2. **C. velùtinus** Dougl. Shrub, 1–3 m. high, with olive twigs and brown branches; leaf-blades thick, broadly oval, often subcordate at base, glabrous and shining above, pale and velutinous beneath, 4–8 cm. long; inflorescence paniculate, ample; petals white, broadly spatulate; fruit obovoid, 5 mm. in diameter, deeply 3-lobed on the top. Hillsides: Mont.—S.D.—Colo.—Utah—B.C. *Submont.—Mont.* My–Jl.

3. **C. americànus** L. Shrub, 2–10 dm. high, branched, finely pubescent, becoming glabrate; leaves 3–10 cm. long, thin, 3-ribbed, serrate; panicle 1–4 cm. long; petals dipper-like, 1–1.5 mm. long; fruit 5–6 mm. broad, 3-lobed, the lobes crested on the back. Woods and hillsides: Ont.—Man.—Kans.—Tex.—Fla. *E. Temp.* My–Jl.

4. **C. ovàtus** Desf. Shrub, 2–10 dm. high; branches brownish; leaf-blades narrowly oblong to oval or ovate, 1.5–4 cm. long, serrate with gland-tipped teeth; petals spatulate, dipper-shaped, 1.5–2 cm. long; fruit 4–4.5 mm. in diameter, depressed-globose, 3-lobed. *C. ovalis* Bigel. Sandy soil: Vt.—Ga.—Tex.—Colo.—S.D.—Man. *E. Temp.—Plain—Submont.* Ap–Je.

5. **C. pubéscens** (T. & G.) Rydb. Shrub, 5–10 dm. high; leaf-blades from oblong to lanceolate or oval, 2–5 cm. long, glabrate except on veins above, villous beneath, glandular-serrate; petals white, 1.5 mm. long, spatulate; fruit subglobose, 5–6 mm. in diameter, slightly 3-lobed. *C. ovatus pubescens* T. & G. [G, B]. Sandy soil: Mich.—Mo.—Tex.—Colo.—S.D. *Prairie—Plain—Submont.* My–Jl.

6. **C. Féndleri** A. Gray. Low spinescent shrub, 1–5 dm. high, with canescent twigs; leaf-blades elliptic, 3-ribbed, entire-margined, 7–30 mm. long, green and strigose or glabrate above, pale and silky-canescent beneath; petals white, 1.5 mm. long, spatulate; fruit 4 mm. wide, subglobose, somewhat 3-lobed. Woods and hillsides: S.D.—N.M.—Ariz.—Utah—Wyo. *Submont.—Mont.* Je–Au.

Family 106. VITACEAE. Grape Family.

Woody vines, climbing by means of tendrils. Flowers perfect, polygamous, or dioecious, in axillary clusters. Sepals and petals 4 or 5, the latter valvate. Stamens as many as the petals and opposite them. Gynoecium of 2, or rarely 3–6, united carpels; styles united; stigmas capitate or peltate. Fruit a berry; seeds 1 or 2 in each cavity.

Flowers with a hypogynous disk ; leaves simple or 3-foliolate.
  Petals deciduous, cohering into a cap.
    Pith interrupted by a diaphragm at the nodes ; tendrils forked ; bark shreddy.
                                                  1. Vitis.
    Pith continuous at the nodes ; tendrils simple ; bark not shreddy.
                                                2. Muscadinia.
  Petals distinct, spreading.
    Flowers 4-merous ; disk 4-lobed.   3. Cissus.
    Flowers 5-merous ; disk entire or nearly so.   4. Ampelopsis.
Flowers without a hypogynous disk ; leaves palmately 5–7-foliolate. 5. Psedera.

1. VÌTIS (Tourn.) L. Grape.

Climbing or trailing vines, with shreddy bark and branched tendrils. Leaves alternate, simple, palmately lobed, petioled; stipules small and caducous. Flowers dioecious, polygamo-dioecious, or rarely perfect. Disk hypogynous. Calyx minute. Petals caducous, coherent at the apex as a cap. Stamens exserted. Ovary 2-celled, rarely 5- or 4-celled. Berry pulpy. Seeds few, pear-shaped.

Leaves green and glabrous beneath, at maturity.
  Leaves merely toothed, long-pointed.   1. *V. cordifolia.*
  Leaves lobed as well as toothed, acute or short-acuminate.
    Leaves and twigs arachnoid-floccose when young.   2. *V. Longii.*
    Leaves and twigs without tomentum when young.
      Berry with a bloom ; leaves with acute lobes and
        sinuses.   3. *V. vulpina.*

Berry without a bloom; leaves with acuminate lobes and obtuse sinuses.   4. *V. palmata.*
Leaves pale or rusty-brown beneath.
  Leaves glaucous or in age glabrate beneath.   5. *V. bicolor.*
  Leaves pubescent beneath.
    Leaves whitish beneath; berries small; no tendril or inflorescence opposite every third leaf.
     Berries with a bloom; branches terete.   6. *V. aestivalis.*
     Berries without a bloom; branches angular.   7. *V. cinerea.*
    Leaves rusty beneath; berries large; a tendril or inflorescence opposite every leaf.   8. *V. Labrusca.*

1. V. cordifòlia Michx. A high-climbing vine; leaf-blades thin, 7.5–10 cm. long, longer than broad, ovate in outline, rarely slightly lobed at the apex, irregularly toothed, deeply cordate at the base, sparingly pubescent beneath or glabrous, panicles 1–3 dm. long, drooping; berries globose, 8–10 mm. in diameter, usually ripening after frost, black and shining. Woods and thickets: N.Y.—Neb.—Tex.—Fla. *E. Temp.* My–Je.

2. V. Lóngii Prince. Rather shrubby, usually not profusely climbing; twigs and leaves floccose when young; leaf-blades broadly cordate, 8–10 cm., rarely 15 cm. wide, 7–12 cm. long, slightly 3-lobed, short-acuminate, the teeth broadly deltoid, the basal sinus rounded, hairy on both sides, in age dark green above, paler beneath; tendrils short, once or twice forked; panicle 5–7 cm. long, rather simple and compact; berries 8–12 mm. broad, subglobose, black, with a bloom. *V. Solonis* Planch. Sandy soil: Kans.—se Colo.—N.M.—Tex. *St. Plains.*

3. V. vulpìna L. *Fig. 367.* High-climbing vine; leaf-blades thin, broadly cordate, with a broad basal sinus, often somewhat 3-lobed, deeply and irregularly serrate, 5–12 cm. long, 6–15 cm. wide; panicles 5–12 cm. long; berries 1 cm. or less in diameter, bluish or purplish black, with a bloom, very sour. Woods and river banks: N.B.—W.Va.—Tex.—N.M. —Wyo.—Man. *E. Temp.—Plain—Mont.* Ap–Je.

4. V. palmàta Vahl. A slender, trailing vine; leaves thin, deeply 3–5-lobed, coarsely toothed, cordate at the base, glabrous or on the lower surface sparingly pubescent on the veins and glaucescent; panicles 5–12 cm. long; berries 7–10 mm. long, black, without bloom. River banks and rocky places: Ill.—Iowa—Tex.—La. *Ozark.* Je–Jl.

f. 367.

5. V. bícolor LeConte. A high-climbing vine; leaf-blades thin, longer than broad, 3–5-lobed, 1–3 dm. broad, glabrous above, glaucous and glabrous, or slightly pubescent on the veins beneath; panicles 5–10 cm. long; berries globose, 10–14 mm. broad, black with a bloom. Woods and river banks: N.H.—Minn.—Kans.- Tex.—N.C. *Allegh.*

6. V. aestivàlis Michx. A high-climbing vine; leaf-blades 3–5-lobed, 1–3 dm. broad, shallowly toothed, dull green and glabrate above, tomentose with brown hairs beneath, cordate at the base; panicles 1–2.5 dm. long; berries globose, 8–10 mm. broad, black, with a bloom, with a tough skin. Thickets and rocky places: N.H.—Iowa—Kans.—Tex.—Fla. *E. Temp.* My–Je.

7. V. cinèrea Engelm. A high vine; leaf-blades entire or 3-lobed, white or grayish beneath, 1–1.5 dm. broad, often longer than broad; panicles 1–2 dm. long; berries 6–8 mm. broad, black, without a bloom. Thickets: Ill.—Neb. —Tex. *Ozark—Prairie.*

8. V. Labrúsca L. A high-climbing vine; leaf-blades thickish, orbicular or round-ovate, mostly longer than broad, shallowly toothed or sinuate, entire or 3-lobed at the apex, sometimes deeply lobed, with rounded sinuses, 1–1.5 dm. broad, at last glabrate above, densely tomentose beneath; panicles 5–12 cm. long; berries globose, 15–20 mm. broad. Thickets: Mass.—Minn. (?)— Miss.—Ga. *Allegh.* My–Je.

2. **MUSCADÍNIA** (Planch.) Small. BULLACE GRAPE. MUSCADINE GRAPE.

Trailing or climbing vines, with close bark and simple tendrils. Leaves alternate, with simple, angled or coarsely toothed blades, not densely pubescent. Flowers dioecious or polygamo-dioecious, in racemes or panicles. Calyx minute. Corolla deciduous, the petals adherent at the top, and falling off as a cap. Disk present, hypogynous. Stamens alternate with the lobes of the disk. Ovary 2-celled; ovules 2 in each cell. Seeds semi-ellipsoid.

1. **M. rotundifòlia** (Michx.) Small. A vine, sometimes 3 m. high, the stem often with aerial roots; leaf-blades firm, suborbicular or round-ovate, 4–9 cm. broad, acuminate, cordate at the base, glabrous except in the axils of the veins beneath; panicle 2–3 cm. long, the staminate ones longer than the fertile ones; berries subglobose, 15–20 mm. broad, dull purple, without a bloom, with a tough skin. *Vitis rotundifolia* Michx. [G, B]. Sandy banks, swamps, and thickets: Del.—Kans.—Tex.—Fla. *Austral.* My.

3. CÍSSUS L.

Climbing vines, with simple, 3-parted, or 3-foliolate leaves, the leaves or leaflets entire or toothed. Flowers perfect or sometimes polygamous, in small cymes. Sepals and petals 4, the latter distinct. Disk present, cup-like, 4-lobed. Stamens 4, on the margins of the disk. Ovary 2-celled; ovules 2 in each cell. Berries small; pulp scant. Seeds triangled.

1. **C. incìsa** (Nutt.) DesMoul. *Fig. 368.* A vine, 1–10 m. high, with warty bark; tendrils forking; leaf-blades 3-foliolate, pale green, fleshy, the leaflets 3–10 cm. long, coarsely toothed, the terminal one sometimes 3-lobed, the lateral ones 2-lobed; cymes trichotomous, umbel-like; berries round-obovoid, 10–12 mm. long, black; seeds 1 or 2, obovoid. Sandy shores: Mo.—Kans.—Tex.—Fla. *Austral.*

f.368.

4. AMPELÓPSIS Michx.

Vines with few tendrils. Leaves alternate, simple to bipinnate. Flowers mostly perfect, cymose. Sepals and petals 5, the latter distinct, expanding. Disk present, entire or undulate. Ovary 2-celled; ovules 2 in each cell; style slender. Berry nearly dry. Seeds 2–4.

1. **A. cordàta** Michx. A high-climbing vine, with forking tendrils and warty bark; leaves simple, ovate or triangular-ovate, 4–12 cm. long, acuminate, acutely serrate, truncate or subcordate at the base, glabrous or sparingly pubescent beneath; cymes slender-peduncled, 3–8 cm. broad; berries subglobose, 6–8 mm. broad, bluish or greenish. *Cissus Ampelopsis* Pers. [G]. River banks: Va.—se Neb.—Kans.—Tex.—Fla. *Austral.* My–Je.

5. PSÉDERA Necker. VIRGINIA CREEPER, AMERICAN IVY, WOODBINE.

Climbing or trailing vines, with forking tendrils, their branches often with adhesive disks. Leaves alternate, palmately 5–7-foliolate. Flowers perfect or polygamo-dioecious, in compound cymes. Sepals and petals 5. Disk obsolete or wanting. Stamens 5. Ovary 2-celled, sessile. Berries with scant pulp, inedible. Seeds more or less 3-angled. *Parthenocissus* Planch.

Foliage pubescent, usually densely so.	1. *P. hirsuta.*
Foliage glabrous or nearly so.	
Aerial rootlets present; tendrils with disks.	2. *P. quinquefolia.*
Aerial rootlets lacking; tendrils without disks.	3. *P. vitacea.*

1. **P. hirsùta** (Donn) Greene. A trailing vine; leaflets 5, oval or elliptic, acute or short-acuminate, coarsely toothed above the middle; corymbs 8–12 cm. broad; berries subglobose, 8–9 mm. broad, dark blue, with a slight bloom; tendrils slender, usually with disks, and aerial rootlets often present. *Ampelopsis hirsuta* Donn. *Parthenocissus hirsuta* Small. Rocky places: Vt.—Minn.—Tex.— Ga.; Mex. E. *Temp.—Trop.* Jl.

2. **P. quinquefòlia** (I.) Greene. *Fig. 369.* Tall vine, climbing; branches warty; leaves usually 5-foliolate; leaflets ovate or obovate to oblong-lanceolate, acute or short-acuminate, serrate above the middle, teeth directed forward; corymbs 8–12 cm. broad; berries subglobose, 8–9 mm. in diameter, blue, with a scant bloom. *Parthenocissus quinquefolia* Planch. [B]. Woods and banks: Que.—Fla.—Tex.—S.D.—Man. E. *Temp.—Plain—Submont.* Je–Jl.

f 369.

3. **P. vitàcea** (Knerr) Greene. Straggling vine, with long tendrils and smoother bark; leaflets 5–6, thin, 4–10 cm. long, lanceolate or oval, acuminate, serrate, with large, often flaring, lanceolate teeth; corymb about 5 cm. broad; berries 6–7 mm. in diameter, bluish black. *Parthenocissus quinquefolia laciniata* Planch. [B]. Woods and banks: Mich.— Ohio— N.M.—Ariz.—Wyo. *Prairie—Plain—Submont.* Je–Jl.

Family 107. **TILIACEAE.** LINDEN FAMILY.

Trees or shrubs, with alternate or rarely opposite simple leaves and free stipules. Flowers perfect, regular, racemose or cymose. Sepals 4 or 5, deciduous, valvate. Petals 4 or 5, imbricate or convolute, clawed. Stamens usually many; filaments distinct, often in groups opposite the petals; anthers 2-celled. Gynoecium of 2–10 united carpels; styles united; stigmas capitate or lobed. Ovules few to many, in two rows in each cavity. Fruit a capsule, often nut-like or berry-like.

1. **TÍLIA** L. LINDEN, BASSWOOD, LIME-TREE.

Trees with tough bast. Leaves alternate, petioled, with oblique blades. Flowers in axillary or terminal cymes, with the conspicuous bract partly adnate to the peduncle. Sepals 5, thick. Petals 5, naked or with a scale at the base, imbricate. Stamens many; filaments branched, in 5 groups opposite the petals, one at the base of each scale; anthers extrorse. Ovary 5-celled; stigma 5-toothed. Ovules 2 in each cell. Fruit a nut-like berry. Seeds 1 or 2.

1. **T. americàna** L. *Fig. 370.* Tree, up to 40 m. high; leaf-blades orbicular-ovate, 8–15 cm. long, abruptly acuminate, serrate with gland-tipped teeth, oblique, cordate or subcordate at the base, glabrous or minutely pubescent in the axils of the veins beneath; bract decurrent to near the base of the peduncles; petals 9–11 mm. long, pale yellow; staminoidal scale spatulate, nearly as long as the petals; fruit oval, tomentose. *T. glabra* Vent. Woods: N.B.—Man.—N.D.—Tex.—Ga. E. *Temp.*

f. 370.

Family 108. **MALVACEAE.** MALLOW FAMILY.

Herbs (all ours), or rarely shrubs or trees, with alternate, palmately ribbed and usually lobed leaves, often with stellate or branched pubescence. Flowers perfect, regular. Calyx of 5, more or less united, valvate sepals. Petals 5, convolute. Stamens many, monadelphous. Gynoecium of several, usually united carpels. Fruit a several-celled capsule, or the carpels separating at maturity.

Fruit of several carpels separating at maturity.
  Style-branches filiform, stigmatose longitudinally on the outer side.
    Flowers perfect.
      Petals notched at the apex; carpels beakless, without internal processes.
        Involucral bracts 2–3.   1. MALVA.
        Involucral bracts 6–9.   2. ALTHAEA.
      Petals not notched at the apex; carpels beaked, with an internal process
        above the seed.   3. CALLIRRHOË.
    Flowers dioecious.   4. NAPAEA.
  Style-branches terminated by capitate or truncate stigmas.
    Carpels strongly reticulate on the lower part of the sides facing the adjacent
      carpels, this portion enclosing the seed, the upper part smooth and empty.
                              7. SPHAERALCEA.
    Carpels not differentiated into an upper and lower portion.
      Lower seed at least from an ascending ovule; calyx more or less bracte-
        olate.
        Perennials, with maple-like leaves; carpels hirsute, 2- or 3-seeded;
          calyx not accrescent.   6. PHYMOSIA.
        Annuals, with narrow leaves; carpels puberulent, 1-seeded; calyx
          accrescent.   5. SIDOPSIS.
      Lower seed at least resupinate-pendulous; calyx without involucel, or
        this represented by 1–3 subulate-setaceous bractlets.
        Carpels 1-ovuled, the cell filled with the seed.
          Bractlets present; plant stellate-lepidote, canescent.
                              8. DISELLA.
          Bractlets wanting; plant not stellate-lepidote. 9. SIDA.
        Carpels 3–9-ovuled, several-seeded, dehiscent apically and dorsally;
          bractlets wanting.   10. ABUTILON.
Fruit with the carpels permanently united into a loculicidal capsule.
                              11. HIBISCUS.

1. **MÁLVA** (Tourn.) L. MALLOW, CHEESES (fruit).

Annual or perennial herbs. Leaves alternate, with pubescent, lobed or dissected, reniform or suborbicular blades. Flowers solitary or clustered in the axils of the leaves, or rarely in terminal racemes or spikes, subtended by 2 or 3 bractlets, perfect and regular. Calyx of 5 partly united sepals. Petals 5, emarginate. Carpels many, 1-celled, reniform, indehiscent, beakless, arranged around a central axis. Seeds ascending; embryo circular.

Flowers in axillary fascicles surpassed by the petioles of the subtending leaves;
  leaves merely lobed.
    Corolla less than three times as long as the calyx.
      Corolla scarcely exceeding the calyx; carpels rugose-reticulate on the back,
        and with acute or winged margins.
        Calyx becoming much enlarged, and spreading under the fruit, nervose-
          reticulate; angles of the carpels margined and denticulate.
                            1. *M. parviflora.*
        Calyx not much enlarged in fruit, mostly erect, not
          reticulate; carpels merely acute on the margins. 2. *M. pusilla.*
      Corolla about twice as long as the calyx; carpels not ru-
        gose on the back, round-margined.
        Plant decumbent; leaves scarcely crisp; carpels
          smooth.   3. *M. rotundifolia.*
        Plant erect; leaves very crisp; carpels veiny-reticulate. 4. *M. crispa.*
    Corolla 3–4 times as long as the calyx; plant erect; carpels
      flat on the back, rugose-reticulate.
      Lower leaves with rounded lobes; petals obcordate, al-
        most crimson, shallowly notched.   5. *M. mauritania.*
      Lower leaves with triangular lobes; petals narrowly
        cuneate, deeply emarginate, reddish-violet with darker
        veins.   6. *M. sylvestris.*
Flowers in terminal leafy racemes or panicles, longer than the
  subtending petioles; cauline leaves deeply dissected. 7. *M. moschata.*

1. M. parviflòra L. Annual; stem erect or ascending, branched, glabrous or sparingly hairy; leaf-blades reniform in outline, with about 7 rounded crenate lobes, 2-10 cm. in diameter; calyx-lobes rounded, mucronate; corolla lilac, about 5 mm. long; carpels glabrous or pubescent. Waste places: N.D.—Tex.—Calif.; also on the Atlantic Coast; Mex.; adv. or nat. from Eur. F-O.

2. M. pusilla Smith. Annual; stem erect, branched, 1-3 dm. high; leaf-blades reniform, 2-5 cm. broad, obscurely round-lobed and crenate; calyx-lobes ovate, in fruit mostly erect; corolla white; carpels pubescent or glabrate. *M. borealis* Wallm. *M. vulgaris* Lunell; not L. Waste places: Mont.—Calif.; adv. from Eur. Jl-Au.

3. M. rotundifòlia L. Annual; stems procumbent; leaf-blades rounded-reniform, more or less distinctly round-lobed and crenate, usually more or less pubescent, calyx not much accrescent, its lobes ovate; corolla pale lilac or whitish, about 1 cm. long; carpels puberulent. *M. vulgaris* L. Waste places: Mass.—N.C.—Calif.—Wash.—Man.; nat. from Eur. My-S.

4. M. crispa L. Annual; stem erect, 5-20 dm. high; leaf-blades reniform in outline, 5-20 cm. in diameter, distinctly lobed and double-crenate; flowers nearly sessile; calyx accrescent, in fruit becoming membranous and velny, but not spreading; corolla purplish or white, about 1 cm. long; carpels not at all rugose. Waste places and around dwellings: N.S.—N.J.—Colo.—S.D.; escaped from cultivation. Jl-S.

5. M. mauritània L. Annual or biennial; stem 3-5 dm. high, glabrous or nearly so; petioles finely pilose on the upper side; leaf-blades rounded-reniform, 3-10 cm. broad; lobes of the lower leaves 5-7, rounded or crenate, those of the upper ones 3-5, more ovate; calyx 5 mm. long; corolla 2-2.5 cm. long; fruit about 8 mm. broad. Around dwellings: N.Y.—N.D.; escaped from cult.; nat. of the Mediterranean region. Ap-O.

6. M. sylvéstris L. Biennial; stem erect, 3-4 dm. high, hirsute; petioles long, hirsute; leaf-blades orbicular to reniform, 4-12 cm. wide; lobes 3-7, triangular or ovate; pedicels villous; calyx 5 mm. long, lobes triangular; corolla 2-2.5 cm. long; fruit 7-10 mm. broad. Waste places: Que.—N.D.—Colo.—Mex.—Fla.; nat. from Eur. My-O.

7. M. moschàta L. Perennial; stem 3-6 dm. high, erect, lower leaves long-petioled; blades orbicular, 7-10 cm. wide, with broad rounded dentate lobes; upper leaf-blades deeply divided into linear to cuneate divisions, which are cleft or divided; corolla 5-8 times as long as the calyx; petals obcordate; carpels 15-20, hairy, rounded on the back. Waste places: N.S.—N.J.—Neb.—Minn.; B.C.—Ore.; escaped from cult.; native of Eur.

2. ALTHAÈA L. HOLLYHOCK. MARSH MALLOW.

Tall herbs, with lobed or divided, palmately ribbed leaves. Flowers perfect, racemose or solitary, subtended by an involucre of 6-9 united bracts. Calyx 5-cleft. Petals 5, large. Carpels numerous, the styles as many as the 1-ovuled cavities, stigmatic on the inner side. Carpels indehiscent, 1-seeded. Seeds ascending.

1. A. ròsea Cav. Stem stout, pubescent, 1-2 m. high; leaf-blades rounded-cordate, lobed or sinuate, and crenate, 1-2 dm. broad; flowers in long leafy spike-like raceme; petals rose-colored or white, about 5 cm. long. Escaped from cult.; nat. of China.

3. CALLÍRRHOÈ Nutt. POPPY MALLOW.

Perennial herbs, with thick, farinaceous roots. Leaves alternate, with lobed or cleft blades or the upper stem-leaves palmately or pedately dissected. Flowers solitary, axillary, or sometimes in terminal racemes, subtended by 1-3 bractlets. Sepals 5, united below. Petals 5, purple to white, cuneate at the base, or fan-shaped. Carpels 10–20, 1-celled, 1-seeded, beaked. Seeds ascending; embryo curved.

Bractlets present.
  Peduncles several-flowered; leaf-blades triangular or hastate; carpels not rugose.
                                                               1. *C. triangulata.*
  Peduncles 1-flowered; leaf-blades orbicular in outline.
    Involucre contiguous to the calyx; carpels rugose.   2. *C. involucrata.*
    Involucre separated from the calyx; carpels tuberculate
      on the back.   3. *C. Papaver.*
Bractlets wanting; carpels coarsely rugose.
  Lower petioles strigose; carpels very pubescent; corolla 2–3
    cm. broad.   4. *C. alceoides.*
  Lower petioles hirsute; carpels scarcely pubescent; corolla
    3–5 cm. broad.   5. *C. digitata.*

1. **C. triangulàta** (Leavenw.) A. Gray. Perennial, with a thick fusiform root; stem 3–10 dm. high; leaves mostly basal, with long-pubescent petioles; blades triangular-ovate or hastate, 3–5 cm. long, crenate or lobed; upper leaves 3–5-cleft or parted; bractlets several, spatulate; calyx-lobes deltoid, acute, ciliate; petals deep purple, 2–2.5 cm. long, undulate at the apex; carpels pubescent, short-beaked. Prairies: Ill.—Minn.(?)—Tex.—N.C. *Prairie—Austral.* My–Au.

2. **C. involucràta** (T. & G.) A. Gray. *Fig. 371.* Cespitose perennial, with a napiform thick root; stems procumbent, more or less hirsute; leaf-blades rounded in outline, palmately or pedately 5–7-parted, the cuneate divisions deeply cleft into lanceolate or oblong lobes; peduncles surpassing the leaves; calyx subtended by 3 linear or oblong bractlets half as long as the lanceolate calyx-lobes; petals crimson or purple, 2–3 cm. long. Plains: Minn.—Mo.—Tex.—Utah—Wyo. *Prairie—Plain.* Je–Au.

f.371.

3. **C. Papàver** (Cass.) A. Gray. Perennial, with a thick root; stem decumbent or ascending, 1–8 dm. long, branched; leaf-blades palmately or pedately 5–7-parted, 5–10 cm. wide; segments cuneate, entire, toothed, or parted; pedicels longer than the leaves; bractlets linear to oblong; calyx-lobes lanceolate, twice as long as the bractlets; petals crimson to cherry-red, 2.5 cm. long, erose at the apex; carpels pubescent. Sandy soil: Ga.—Kans.—Tex.—Fla. *Austral.*

4. **C. alceoìdes** (Michx.) A. Gray. Perennial, with a thick root; stem erect, 2–5 dm. high, branched at the base; basal leaves long-petioled; blades ovate or triangular in outline, palmately lobed or incised; stem-leaves short-petioled, palmately cleft or parted into linear or linear-cuneate divisions; petals pink or rose, 1–1.5 cm. long, erose-fimbriate at the apex; flowers racemose-corymbose; calyx-lobes triangular. Dry soil: Ky.—Neb.—Tex.—Tenn.; introduced in Ida. *Prairie—Austral.* My–Au.

5. **C. digitàta** Nutt. Perennial, with a thick root; stem erect or decumbent, 3–8 dm. long, villous-hirsute below; basal leaves long-petioled; blades palmately lobed or parted or those of the earlier leaves crenate; stem-leaves short-petioled, palmately divided into linear divisions; peduncles slender; calyx-lobes lanceolate; petals red-purple, violet, or white, 1.5–2 cm. long, erose-fimbriate. Dry soil: Mo.—Kans.—Tex. *Ozark.* Ap–Jl.

4. **NAPAÈA** (Clayton) L. GLADE MALLOW.

Perennial herbs. Leaf-blades palmately lobed. Flowers dioecious, in corymbiform panicles. Bractlets none. Calyx 5-lobed. Petals 5, distinct, obcordate, white. Stamens in the staminate flowers 15–20, monadelphous, the anthers borne at the summit of the column, in the pistillate flowers the column present but the anthers wanting. Pistils 8–10, in the staminate flowers rudimentary; styles in the pistillate flowers stigmatic on their inner side. Carpels 8–10, imperfectly 2-valved. Seeds solitary in each carpel.

1. N. dioìca L. Stem simple, 1–3 dm. high; lower leaves long-petioled; blades orbicular, 7–11-parted; divisions acute, dentate or lobed; petals in the staminate flowers 2–3 times as long as the calyx, in the pistillate flowers somewhat smaller; carpels rugose-reticulate. Moist ground: Pa.—Minn.—Iowa—Va. *Allegh.* Jl.

5. SIDÓPSIS Rydb.

Annual, strigose herbs. Leaf-blades lanceolate or oblong, denticulate. Flowers perfect, solitary or glomerate, subsessile in the axils of the leaves. Bractlets present, setaceous. Calyx angular, accrescent in age, with 5 triangular lobes. Corolla yellow. Carpels 5 or 6, thin-chartaceous, puberulent, smooth, at length 2-valved. Seeds reniform, filling the whole cavity of the carpel.

1. S. hispida (Ell.) Rydb. Stem erect, 1–3 dm. high; leaf blades oblong-lanceolate to linear-oblong, 1.5–4 cm. long, remotely serrate; petals slightly exceeding the calyx-lobes; carpels reniform. *Sida hispida* Ell. *Malvastrum anqustum* A. Gray [G, B]. Dry soil: Tenn.—Iowa—Kans. *Ozark.* Jl–Au.

6. PHYMÒSIA Desv. Wild Hollyhock, Maple-leaved Mallow.

Tall leafy perennials, with rootstocks. Leaves alternate, palmately ribbed and lobed, with acute lobes, and maple-like in appearance. Flowers large, clustered in the upper axils or interruptedly spicate or corymbose at the summit, subtended by bractlets. Sepals united at the base. Petals rose-colored, purplish, or white, obcordate. Stamens monadelphous in single series. Carpels oblong, hirsute, thin, not reticulate, 2- or 3-seeded. *Iliamna* Greene.

1. P. rivulàris (Dougl.) Rydb. *Fig. 373.* Stem 6–20 dm. high, sparingly stellate; leaf-blades cordate or reniform in outline, 5–7-lobed, 5–15 cm. long and about as wide; lobes triangular or broadly lanceolate, usually coarsely toothed; pedicels and calyx stellate; petals pink or white, 2–2.5 cm. long. *Sphaeralcea rivularis* Torr. *I. rivularis* and *I. angulata* Greene. *Sphaeralcea acerifolia* A. Gray, in part [B]. Along streams: Alta.—S.D.—Colo.—Nev.—B.C. *Submont.—Mont.* Jl–S.

f.373.

7. SPHAERÁLCEA St.-Hil. Globe Mallow, Scarlet Mallow.

Mostly perennial herbs, often cinereous, with stellate or branched hairs. Leaves alternate, with palmately lobed or dissected blades. Flowers mostly in terminal spike-like racemes, subtended by 1–3 bractlets. Sepals 5, partially united. Petals 5, golden, pink, or scarlet, usually notched. Stamens monadelphous in a single series. Stigmas truncate or capitate. Carpels 5 or more; the lower portion of each enclosing 1 or 2 seeds, reticulate on the sides facing the adjacent carpels, the upper portion of the carpels smooth and empty. *Malvastrum* A. Gray.

Leaves lanceolate; fruit ellipsoid to ovate.
  Carpels with a cusp.   1. *S. cuspidata.*
  Carpels rounded at the apex.   2. *S. angustifolia.*
Leaves rounded or rhomboid to reniform in outline; fruit depressed-globose to broadly ellipsoid.   3. *S. coccinea.*

1. **S. cuspidàta** Torr. Perennial, with a woody base; stem 3–10 dm. high, stellate-pubescent; leaf-blades lanceolate or linear-lanceolate, crenulate, 5–10 cm. long, sometimes hastately lobed at the base; flowers axillary; sepals ovate, acute; petals pink, 6–8 mm. long; carpels 5–6 mm. long, strongly rugose at the base. Plains: Tex.—Kans.—Colo.—Ariz.; n Mex. *St. Plains—Son.* Ap–Au.

2. **S. angustifòlia** (Cav.) Don. Perennial, with a somewhat woody base; stem 5–15 dm. high, leafy, subcanescent; leaf-blades 5–12 cm. long, lanceolate or linear-oblong, crenate, the lower sometimes hastately lobed; flowers clustered in the axils; sepals lanceolate or ovate-lanceolate, acute or short-acuminate; petals pink, 8–10 mm. long; carpels 5–6 mm. long, only the lower fourth or less rugose. Plains and dry places: Tex.—Colo.—Utah—Ariz.; Mex. *St. Plains—Son.—Submont.* Ap–Au.

3. **S. coccínea** (Nutt.) Rydb. *Fig. 372.* Perennial, with a woody base; stems several, 1–3 dm. high; leaf-blades 3-cleft to the base and the lateral divisions again deeply 2-parted; divisions rhombic-cuneate to spatulate, usually 3-cleft at the apex, with oval or oblong lobes; inflorescence dense, short, raceme-like; sepals ovate, about 3 mm. long, villous-stellate; petals usually brick-red, 8–12 mm. long; carpels reniform, 3 mm. high. *Malvastrum coccineum* (Nutt.) A. Gray [G, B]. Dry plains and sandy valleys: Man.—Iowa—Tex.—Utah—Ore. *Prairie—Plain—Mont.* My–Au.

8. **DISÉLLA** Greene.

Canescent, stellate perennials. Leaves alternate, pubescent, reniform, or hastate. Flowers perfect, solitary or clustered in the axils of the leaves. Bractlets 1–3. Sepals 5, united into a usually angular base. Petals 5. Styles filiform; stigmas capitate. Carpels 1-seeded, five to many, 1-celled, indehiscent or partially 2-valved, obtuse or short acuminate. Seeds pendulous, 3-angled.

1. **D. hederàcea** (Dougl.) Greene. Perennial, with a cespitose caudex or rootstock; stems decumbent or spreading, branched, 1–4 dm. long, scurfy-canescent; leaf-blades reniform or suborbicular, dentate, scurfy-canescent, 1–5 cm. wide; petals pale yellow or white, about 1 cm. long; fruit short-conic, tomentulose. *Sida hederacea* (Dougl.) Torr. [B]. Low banks and salt-flats: Wash.—Utah—Ariz.—Tex.—se Kans.; Mex. *St. Plains—Son.—Submont.* Je–S.

9. **SÌDA** L.

Annual or perennial herbs. Leaves various, usually narrow. Flowers perfect, solitary or clustered in the axils of the leaves. Bractlets wanting. Calyx 5-lobed. Petals 5, delicate, of various colors. Styles filiform; stigmas capitate. Carpels 5 to many, indehiscent or partly 2-valved, 1-seeded, the seed filling the cavity.

1. **S. spinòsa** L. Minutely pubescent annual, stem erect, 1–6 dm. high; leaf-blades oblong to ovate or linear-lanceolate, 1–5 cm. long, serrate, cordate or truncate at the base; flowers axillary, the pedicels shorter than the petioles; calyx slightly accrescent, the lobes triangular; corolla pale yellow, 1–1.5 cm. broad; carpels 5, slightly wrinkled, 2-beaked, 4 mm. long. Waste places and fields: N.Y.—Neb.—Tex.—Fla.; Trop. Am. *Allegh.—Austral.—Trop.*

10. **ABÙTILON** (Tourn.) Mill. INDIAN MALLOW, VELVET LEAF.

Herbs (ours), shrubs, or trees. Leaves alternate, softly pubescent, entire, toothed, or lobed, usually cordate at the base. Flowers perfect, mostly axillary. Involucel wanting. Sepals 5, united below. Petals 5, distinct, usually yellow. Styles filiform or club-shaped, with terminal stigmas. Carpels 1-celled, leathery or parchment-like, more or less prominently beaked, 2-valved at the apex and on the back, with 1–6 reniform seeds.

1. **A. Theophrásti** Medic. Velvety-pubescent annual; stem erect, 3–20 dm. high; leaf-blades suborbicular, 1–3 dm. wide, abruptly acuminate at the apex, cordate at the base, dentate or crenate, petioled; calyx accrescent, the lobes ovate, abruptly pointed; petals yellow, 1–1.5 cm. long, obovate with a cuneate base, truncate or retuse at the apex; fruit 2–2.5 cm. thick, 1.5–2 cm. high; carpels villous, subulate-tipped. *A. Abutilon* (L.) Rusby [B]. *A. Avicennae* Gaertn. Waste places: Me.—N.D.—Tex.—Fla.; nat. from Asia. Au–O.

11. HIBÍSCUS L. Rose Mallow. Marsh Mallow.

Herbs, shrubs, or trees. Leaves alternate, with stipules, the blades entire, lobed, or parted. Flowers perfect, regular, peduncled, axillary. Bractlets 3–5, or more, distinct or united. Sepals 5, more or less united. Petals 5, showy. Staminoidal column antiferous on its upper part, but not at the 5-lobed or truncate summit. Ovary 5-celled; styles 5; stigmas capitate or peltate. Fruit a loculicidal, 5-celled capsule.

Calyx closely enclosing the capsule, not bladdery. 1. *H. militaris.*
Calyx bladdery, loosely enclosing the capsule. 2. *H. Trionum.*

1. **H. militáris** Cav. *Fig. 374.* Glabrous perennial; stem 1–2 m. high; leaf-blades ovate or broadly lanceolate, 5–15 cm. long, often hastately 3–5 lobed and serrate, truncate or cordate at the base, the lobes acute or acuminate; bractlets several, linear, 1.5–2 cm. long; calyx somewhat accrescent; lobes ovate; petals pink, with a purple blotch at the base, 5–8 cm. long; capsule 1.5–2.5 cm. long, pointed; seeds silky. River banks: Pa.—Minn. –Kans.—La.— Fla. Au–S.

f. 374

2. **H. Triónum** L. Hispid annual; stem branching, 1–4 dm. high; leaf-blades ovate to suborbicular in outline, pedately 3–7-lobed or parted, the lobes toothed or incised; bractlets linear, 1 cm. long; calyx prominently nerved, inflated in fruit, the lobes triangular; petals pale yellow or whitish, with a purplish or brown eye, 2.5–4 cm. long, spatulate; capsule globose-ovoid, hairy, about 1.5 cm. high; seeds rough. Waste places: N.S.—N.D.—Kans.—Fla.; also Utah and Ore.; nat. or adv. from Eur. Au–S.

Family 109. HYPERICACEAE. St. John's-wort Family.

Herbs or shrubby plants, with opposite, rarely whorled, entire or nearly so, pellucid-punctate leaves, without stipules. Flowers cymose, perfect, regular or nearly so. Sepals 4 or 5, herbaceous, equal or unequal. Petals 4 or 5, yellow, pink, or flesh-colored. Stamens few or many, usually 3- or 5-delphous. Gynoecium of several united carpels. Ovary 1-celled, with parietal placentae, or 3–7-celled, with central placentae. Styles distinct or united; stigmas often capitate. Fruit a septicidal capsule.

Sepals and petals 4; outer sepals much longer than the inner. 1. Ascyrum.
Sepals and petals 5; those of the same series subequal.
  Petals yellow; flowers without glands.
    Leaves with distinct, more or less spreading, flat blades. 2. Hypericum.

Leaves scale-like or linear-subulate, appressed or nearly erect, without distinct blades. 3. SAROTHRA.
Petals pink, often tinged with green or purple; flowers with three orange glands alternating with the three bundles of stamens. 4. TRIADENUM.

1. ÁSCYRUM L. ST. ANDREW'S CROSS. ST. PETER'S-WORT.

Low suffruticose leafy plants. Leaves opposite, black-dotted. Flowers perfect, solitary, terminal, on 2-bracted pedicels. Sepals 4, very unequal, the two outer much larger. Petals 4, yellow, oblique, convolute. Stamens distinct or the filaments slightly united at the base; anthers dehiscent lengthwise. Ovary 1-celled, with 2–4 parietal placentae; style distinct or nearly so. Capsule septicidal, narrowed upward.

1. **A. hypericoìdes** L. Stem 1–9 dm. high, decumbent at the base; leaves obovate to linear, usually oblong, 6–35 mm. long, 2–8 mm. wide, biglandular at the base; bractlets close under the flowers; outer sepals ovate or cordate, 6–9 mm. long; inner sepals smaller and narrower, somewhat petaloid; petals oblong, about equaling the outer sepals. *A. Crux-Andreae* L. ST. ANDREW'S CROSS. Banks and thickets: Mass.—Neb.—Tex.—Fla.; W.Ind.; Mex. and Cent. Am. *E. Temp.—Trop.* Jl–Au.

2. HYPÉRICUM (Tourn.) L. ST. JOHN'S-WORT.

Herbs or shrubs. Leaves opposite, rather thick, usually sessile, entire or nearly so, more or less punctate. Flowers cymose. Sepals 5, slightly unequal. Petals 5, yellow, convolute. Stamens numerous, more or less distinctly 3- or 5-delphous. Ovary 1-celled, with 3 or 5 parietal placentae, or 3- or 5-celled. Styles 3–5, sometimes coherent. Seeds numerous.

Stamens numerous.
  Styles 5; petals 2–5 cm. long; stamens in 5 bundles. 1. *H. Ascyron.*
  Styles 3.
    Tall shrubs; stamens distinct; stigmas elongate; capsule completely 3-celled. 2. *H. prolificum.*
    Perennial herbs, rarely shrubby at the base.
      Stamens distinct; stigmas elongate; capsule 1-celled, with 3 parietal placentae.
        Leaves linear; seeds transversely rugose, deeply pitted. 3. *H. cistifolium.*
        Leaves oblong; seeds minutely striate. 4. *H. ellipticum.*
      Stamens with the filaments united into 3 or 5 bundles; stigmas capitate; capsule 3-celled; leaves sessile or clasping.
        Leaves oblong; petals black-dotted only along the margins. 5. *H. perforatum.*
        Leaves elliptic to ovate; petals black-dotted in several lines.
          Petals 15–20 mm. long; leaves narrowed at the apex. 6. *H. pseudomaculatum.*
          Petals 8–10 mm. long; leaves rounded at the apex.
            Leaf-blades sessile or clasping. 7. *H. punctatum.*
            Leaf-blades short-petioled. 8. *H. subpetiolatum.*
Stamens 5–20, in 3 bundles; annual herbs.
  Bracts all broad and foliaceous. 9. *H. boreale.*
  Bracts, except the lowest ones, subulate.
    Leaves ovate, oval or oblong, somewhat clasping.
      Sepals obtuse or acutish; leaves obtuse. 10. *H. mutilum.*
      Sepals acuminate; leaves acute. 11. *H. gymnanthum.*
    Leaves linear or lanceolate, sessile.
      Leaves lanceolate, 5–7-ribbed at the base. 12. *H. majus.*
      Leaves linear, 1–3-ribbed. 13. *H. canadense.*

1. **H. Áscyron** L. Shrub, 5–15 dm. high; leaves ovate-lanceolate, clasping, acute, 5–15 cm. long, about 2.5 cm. wide, with elongate dots; sepals ovate-lanceolate; petals bright yellow, obovate; capsule conic. Banks: Que.—Man.—Kans.—N.J. *Canad.—Allegh.* Jl–Au.

2. H. prolíficum L. Shrub, 3–10 dm. high; leaves linear-lanceolate to oblong, obtuse or mucronate, 2.5–8 cm. long, 6–18 mm. wide, acute, short-petioled; flowers in small, axillary cymes; sepals somewhat unequal, lanceolate or ovate; corolla deep yellow, 1.5–2 cm. broad; capsule ovoid. Sandy soil: N.J.— Minn.—Ark.—Ga. *Allegh.* Jl–S.

3. H. cistifólium Lam. Stem 3–10 dm. high, 4-angled; leaves linear-oblong, obtuse, 5–8 cm. long, 6–12 mm. wide; more or less revolute, sessile; flowers in dichotomous cymes; sepals ovate or lanceolate, small; corolla 1–1.5 cm. broad; petals obliquely cuneate; capsule ovoid-globose. *H. sphaerocarpum* Michx. [B]. Banks: Ohio—Iowa—Kans.—Ark.—Ala. *Allegh.– Ozark.* Jl–S.

4. H. ellípticum Hook. Stem 2.5–5 dm. high, 4-angled; leaves elliptic-oblong, 1–4 cm. long, 6–10 mm. wide, with translucent veins; sepals oblong or oblanceolate, foliaceous; corolla 8–12 mm. broad, sometimes 4-merous; petals pale yellow; capsule ovoid. Moist ground: N.S.—Man.—N.D.—Md. *Canad.* Jl–Au.

f.375:

5. H. perforàtum L. *Fig. 375.* Stem much branched, 3–7 dm. high; leaves linear to oblong, obtuse, 1–3 cm. long, 2–10 mm. wide; flowers cymose; sepals 3–4 mm. long, glandular-punctate, acute or acuminate; corolla 2–2.5 cm. broad; petals obovate, 10–12 mm. long, toothed above the middle on one side; capsule ovoid. Waste places and fields: N.S.—Man.—Colo.—Va.; B.C.– Calif.; nat. from Eur. Je–S.

6. H. pseudomaculàtum Bush. Stem 4–8 dm. high, corymbosely branched; leaves ovate or lanceolate, 1.5–4 cm. long, often strongly revolute, black dotted; flowers cymose; sepals lanceolate, 4–5 mm. long, acute; corolla copper-yellow, 2–3 cm. broad; capsule ovoid. Woods and dry soil: Ill.—Kans.— Tex.—Ga. *Austral*

7. H. punctàtum Lam. Stem 2–10 dm. high, branched above; leaves leathery, ovate to oblong-ovate, 1–6 cm. long, obtuse, punctate; sepals lanceolate, 3–4 mm. long, corolla 8–13 mm. broad, pale yellow, with black lines and dots. *H. maculatum* Walt. [B]. Hillsides and dry ground: Ont.—Minn.— Kans.—Tex.—Fla. *E. Temp.* Jl–S.

8. H. subpetiolàtum Bickn. Stem 2–8 dm. high, branched above; leaves thin, oblong or elliptic, obtuse or retuse at the apex, usually narrowed at the base, 3–6 cm. long; sepals oblong or lanceolate, 1.5–2 mm. long; corolla dull yellow, 5–8 mm. broad; petals oblong or elliptic, finely spotted; capsule ovoid, 4–6 mm. long. Thickets: Que.—Iowa—Minn.—Ga. *Allegh.* Je–S.

9. H. boreàle (Britton) Bickn. Perennial; stem erect, from a decumbent base, 5–30 cm. high, subterete; leaves elliptic or oval, sessile, 5–20 mm. long, 3–5-ribbed; sepals linear-oblong, obtuse; petals elliptic, slightly shorter than the sepals; capsule oblong, obtuse, 4–5 mm. long. Bogs: Newf.—Minn.— Ohio—N.J. *Boreal.* Jl–Au.

10. H. mùtilum L. Annual or perennial; stem 1–8 dm. high, narrowly 4-winged; leaves oblong-ovate or ovate, 0.5–2.5 cm. long, obtuse, 5-ribbed, minutely punctate, glaucescent beneath; cymes open; sepals oblong, 3–3.5 mm. long, the veins pellucid; corolla 7–9 mm. broad; petals oblong, obtuse; capsule ovoid, acute. Low places: N.S.—Man.—Kans.—Tex.– Fla. *E. Temp.* Jl–S.

11. H. gymnánthum Engelm. & Gray. Stem 2–9 dm. high; leaves firm, ovate, 0.5–2 cm. long, acute or acuminate, 5–7-ribbed; cymes flat-topped; sepals lanceolate, 3–4 mm. long, acuminate; corolla 4–5 mm. broad; capsule slender, conic. Low ground: N.J.—Minn. (?)—Tex.—N.C. *Allegh.* Jl–S.

12. H. màjus (A. Gray) Britton. Erect annual, simple below; stem 2–9 dm. high; leaves lanceolate or oblong-lanceolate, 2–6 cm. long, 5–7-ribbed; sepals lanceolate, acuminate; petals oval or obovate, 2–3 mm. long; stamens 5–10; capsule narrowly conic. *H. canadense majus* A. Gray. Wet places: Me.—N.J.—Colo.—Wash.—B.C. *Boreal.—Plain—Submont.* Jl–Au.

13. H. canadénse L. Annual; stem 1–6 dm. high, angled; leaves linear, obtuse, tapering at the base, 1–5 cm. long; sepals lanceolate, 3–5 mm. long, acute; corolla 7–9 mm. broad; capsule slender, conic. Wet sandy soil: Newf.—Man.—S.D.—Ky.—Ga. *Canad.—Allegh.* Jl–S.

3. SARÒTHRA L. PINE-WEED. ORANGE-GRASS.

Annual herbs. Leaves scale-like or subulate, appressed or nearly erect, opposite, bladeless. Flowers perfect, regular, scattered along the branches. Sepals 5, equal. Petals 5, yellow. Stamens 5–20. Ovary 1-celled, sessile; styles 3, distinct; ovules several, on 3 parietal placentae. Capsule 1-celled, septicidal.

Flowers pedicelled; sepals equaling the capsule; leaves subulate.   1. *S. Drummondii.*
Flowers subsessile; sepals much shorter than the capsule; leaves
   short, scale-like.   2. *S. gentianoides.*

1. S. Drummóndii Grev. & Hook. Stem 1–6 dm. high, wing-angled, the branches alternate; leaves linear-subulate, nearly erect, 8–25 mm. long; cymes elongate; sepals linear-lanceolate, 3–4 mm. long; corolla 10–12 mm. broad; stamens 10–20; capsule ovoid. *Hypericum Drummondii* T. & G. [G, B]. Dry soil: Va.—Iowa.—Kans.—Tex.—Fla. *Allegh.—Austral.* Jl–S.

2. S. gentianoìdes L. Stem 1–5 dm. high, branched, the branches opposite; leaves 2 mm. long, appressed; sepals linear or linear-lanceolate, 2–3 mm. long, acute or acuminate; corolla 4–8 mm. broad; stamens 5–10; capsule conic, 4–5 mm. long. *Hypericum nudicaule* Walt. *H. gentianoides* (L.) E. S. P. [G]. *H. Sarothra* Michx. Sandy soil: Me.—Minn.—Tex.—Fla. Je–O.

4. TRIADÈNUM Raf. MARSH ST. JOHN'S-WORT.

Perennial herbs, with rootstocks. Leaves opposite, thickish, the veins curving along the margins. Flowers perfect, regular, cymose. Sepals 5, erect. Petals 5, pink, often tinged with greenish or purple, imbricate, oblique, deciduous. Stamens 9; filaments united into 3 bundles, alternating with 3 scale-like, orange glands. Ovary 3-celled, sessile; styles 3, distinct. Ovules numerous. Fruit a 3-celled, septicidal capsule, valvate at the apex.

1. T. virgínicum (L.) Raf. *Fig. 376.* Stem 3–6 dm. high; leaves oblong or elliptic or ovate, 2–10 cm. long, rounded or emarginate at the apex, black-punctate beneath, clasping; cymes few-flowered; sepals lanceolate or oblong, 5–6 mm. long, pale-margined; petals obovate, 8–10 mm. long; capsule oblong. *Hypericum virginicum* L. [G]. Swamps: Lab.—Man.—Neb.—La.—Fla. *E. Temp.* Jl–S.

f. 376.

Family 110. ELATINACEAE. WATERWORT FAMILY.

Herbs or shrubby plants, ours low water herbs, with opposite or whorled leaves, with stipules. Flowers inconspicuous, perfect, regular, solitary or clustered in the axils of the leaves.   Sepals 2–5, imbricate. Petals as many, hypogynous.   Stamens 2–5, or sometimes 10.   Gynoecium of 2–5 united carpels.   Ovary 2–5-celled, with central placentae; stigmas 2–5, distinct.   Fruit a septicidal capsule.

Flowers 2–4-merous ; sepals membranous, obtuse ; capsule membranous ; glabrous
    water plants.   1. ELATINE.
Flowers 5-merous ; sepals with thickened midrib, acute or pointed ; capsule firm ;
    pubescent land plants.   2. BERGIA.

1. ELATÍNE L.   WATERWORT, MUD-PURSLANE.

Low water plants.   Leaves opposite or whorled, entire.   Flowers usually solitary, axillary.   Sepals 2–4, membranous, obtuse, nerveless.   Petals and stamens as many as the sepals or the latter twice as many.   Ovary and fruit 2–4-celled.   Styles 2–4.   Capsule membranous, 2–4 valved.

Leaves oblanceolate ; flowers usually 3-merous.   1. E. triandra.
Leaves obovate.
  Flowers 3-merous ; seed with 20–30 acute cross-ribs and irregu-
    lar longitudinal ribs.   2. E americana.
  Flowers 2-merous ; seeds with 15–18 obtuse cross-ribs and regu-
    lar longitudinal ribs.   3. E. minima.

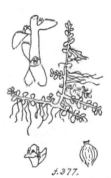

1. **E. triándra** Schkuhr.   Immersed water plant, flaccid; stems 5–10 cm. long; leaves opposite, oblong or oblanceolate; flowers sessile; sepals usually 3; seeds slightly curved, little sculptured.   Shallow water: Ill.—Colo.—Wash.; Eur.   *Prairie– Plain— Submont.*

2. **E. americàna** (Pursh) Arn.   *Fig. 377.* Diffuse-rooting annual; stems 1–4 cm. long; leaves obovate, very obtuse, 2–6 mm. long; petals in the terrestrial form rose-colored; pod globose, 1 mm. in diameter; seeds slightly curved, sculptured with 9–10 longitudinal and 20–30 cross-bars.   Mud and shallow water: Que.—Va.—Tex.—Calif.—B.C.   *Plain —Submont.*   Jl–S.

f.377.

3. **E. mínima** (Nutt.) Fisch. & Mey.   Creeping annuals, forming mats up to 1 dm. broad; leaves cuneate-obovate to oblong, sessile or subsessile, rounded at the apex, 0.7–5 mm. long.   Sandy shores or shallow water: Newf.—Va.—Minn.   *Canad.* Au–S.

2. BÉRGIA L.

Herbs or shrubby plants.   Leaves opposite, pubescent, thickish.   Flowers perfect, regular, solitary or clustered in the axils.   Sepals 5, acute or pointed, with a thick midrib.   Petals 5.   Stamens usually 5.   Ovary 5-celled; styles 5. Fruit a crustaceous, septicidal, 5-celled capsule.

1. **B. texàna** (Hook.) Seubert.   Stem branched at the base, prostrate or ascending, 1–3 dm. long; leaves elliptic or spatulate, 1–2.5 mm. long, serrulate, short-petioled, strongly nerved beneath; flowers solitary or 2 or 3 together in the axils; sepals ovate, 2.5–3 mm. long, acuminate, denticulate; petals shorter; capsule 2 mm. broad.   Sandy or alluvial soil: Ill.—S.D.—Kans. —Tex.—Ark.; Calif.—Nev.—Wash.   *Prairie—Austral.*   Je–Au.

Family 111. TAMARICACEAE.   TAMARIX FAMILY.

Shrubs or trees, with alternate, small or scale-like, often imbricate leaves.   Flowers perfect; inflorescence racemose, spicate, or paniculate.

Sepals and petals 4–6, mostly 5, imbricate. Stamens 5 to many; filaments free. Gynoecium 2–5-carpellary; ovary 1-celled with 3–5 basal placentae. Fruit a capsule. Seeds with a hairtuft.

1. TÁMARIX L. TAMARIX.

Shrubs or trees with scale-like leaves. Flowers in dense panicles. Petals inserted under a 10-lobed disk. Stamens 5–10; filaments adnate to the corolla. Seeds numerous. Endosperm wanting.

1. T. gállica. L. *Fig. 378.* Shrub or low tree, with spreading branches; leaves scale-like, imbricate, acute, 1 mm. long; flowers paniculate; sepals deltoid, 0.5 mm. long, petals white or pinkish. Roadsides, waste places and river bottoms: N.C.—Fla.—Tex.—Colo.; nat. from Eur.

f. 378.

Family 112. CISTACEAE. ROCK-ROSE FAMILY.

Shrubs or undershrubs. Leaves alternate or opposite. Flowers nearly regular, usually perfect, solitary, racemose, clustered, or paniculate. Sepals 3–5, persistent. Petals 5 or 3, sometimes wanting, fugaceous. Stamens 8, hypogynous. Gynoecium of several united carpels; ovary sessile, 1–several-celled; ovules orthotropous; styles united; stigma entire or 3-lobed. Fruit a capsule. Seeds several or numerous. Embryo slender; endosperm starchy or fleshy.

Petals 5, yellow, fugaceous or wanting ; ovary strictly 1-celled.
  Styles slender, elongate ; leaves scale-like, sometimes subulate.
                                                             1. HUDSONIA.
  Styles obsolete or short ; leaves with distinct flat blades.   2. CROCANTHEMUM.
Petals 3, persistent, marcescent, not yellow ; style short or none ; ovary partly
  3-celled by the valvate placentae.   3. LECHEA.

1. HUDSÒNIA L. BEACH HEATHER.

Low shrubs, tufted and much branched. Leaves small, scale-like or subulate, persistent. Flowers regular, perfect, small, terminating the branches. Sepals and petals 5, the latter yellow, oblong. Stamens 8. Gynoecium of 3 united carpels; style 1, filiform; stigma minute. Fruit a 3-valved, 1-celled capsule. Seeds 1 or 2 at the base of each of the 3 nerve-like parietal placentae.

1. H. tomentòsa Nutt. Stem densely tufted, 1–2 dm. high; leaves 2 mm. long, oval or oblong, densely imbricate, scale-like, hoary; flowers sessile; sepals obtuse; capsule 1-seeded. Sandy shores and pine-lands: N.B.—Mack.—N.D.—N.C. My–Au.

2. CROCÁNTHEMUM Spach. FROST-WEED.

Undershrubs. Leaves more or less coriaceous, entire, flat or revolute-margined. Flowers of two kinds, viz., some with large fugaceous petals and many stamens, the others cleistogamous, apetalous or with small petals, and 3–10 stamens. Styles obsolete or short; stigmas capitate or 3-lobed. Fruit a capsule. Embryo curved.

Petaliferous flowers 5–12, racemose, in fruit not overtopped by the short lateral
  branches.   1. C. Bicknellii.
Petaliferous flowers 1 or 2, in fruit overtopped by the long lateral
  branches.   2. C. majus.

1. C. Bicknéllii (Fern.) Britt. *Fig. 379.*
Hoary canescent herbs, slightly woody at the base,
3–6 dm. high; leaves oblanceolate or oblong-lanceolate, 1.5–3.5 cm. long, stellate on both sides, canescent beneath, greener above; petaliferous flowers
5–12, in terminal cymes; sepals densely canescent,
the outer nearly as long as the inner, petals yellow,
7–9 mm. long, oval; apetalous flowers clustered in
the axils of the leaves, nearly sessile. *Helianthe-
mum majus* Bickn. [G, B]; not B. S. P. *H. canadense Walkerae* W. H. Evans. Dry soil: Me.—Va.
—Tex.—Colo —S.D. *Plain—Submont.* Jl–Au.

f 379

2. C. màjus (L.) Britt. Like the preceding;
stem 1–6 dm. high, at first simple; leaves linear-
oblong or lanceolate, subsessile, 1.2–3 cm. long,
rough and dark green above, canescent beneath,
revolute margined; sepals of the petaliferous flowers pilose; the outer shorter
than the inner; petals 1-2 cm. long. *H. canadense* Michx. [G, B]. *C. canadense* (L.) Britton. *H. majus* (L.) B.S.P. Banks or sandy soil: Me.—Man.
—Iowa—Miss.—N.C. *Canad.—Allegh.* My-Jl.

3. **LÉCHEA** (Kalm) L. PINWEED.

Perennial branching herbs, developing leafy shoots from the base in the
fall. Leaves entire, narrow. Flowers perfect. Sepals 5, the outer two smaller
and narrower. Petals 3, ovate to linear, greenish or purplish, persistent. Stamens 3–12. Gynoecium of 3 united carpels; stigmas 3, nearly sessile, laciniate-
plumose. Fruit a globose or obovoid, 3-valved, 3-celled or sometimes 1-celled
capsule.

Leaves of the basal shoots elliptic or oblong ; pubescence spreading.	1. *L. villosa.*
Leaves of the basal shoots narrowly lanceolate or linear ; pubescence appressed,	
Inner sepals 1-nerved, usually shorter than the outer.	
Inner sepals 3-nerved, equaling or longer than the outer.	2 *L. tenuifolia.*
Stem and young parts strigose canescent ; pod globose, 1–2 mm. broad.	3. *L. stricta.*
Stem and young parts not canescent ; pod depressed-globose, 2–3 mm. broad.	4. *L. intermedia.*

1. L. villòsa Ell. Stem erect, 3–7 dm. high, stout, simple; leaves of
the stem elliptic, 1.5–2.5 cm. long, 6–12 mm. wide; leaves of the basal shoots
6–8 mm. long, 4–6 mm. wide; flowers secund-scorpioid; petals greenish purple;
pod depressed-globose. Dry poor soil: N.H.—Neb.—Tex.—Fla. *E. Temp.*
Jl-Au.

2 L. tenuifòlia Michx. Stems densely tufted, 1–3 dm. high, divaricately
branched above; leaves of the stem 4–15 mm. long, 1 mm. wide or less, narrowly linear; those of the basal shoots 6–8 mm. long; petals red-purple; pod
globose-oval, 1 mm. thick. Dry places: Mass.—Neb.—Tex.—Fla. *Allegh.—
Austral.* Jl-Au.

3. L. stricta Leggett. Stem erect, 3–4 dm. high, densely and fastigi-
ately branched, strigose, the branches ascending; leaves of the stem 1–2.5 cm.
long, 1-2 mm. wide, linear; those of the basal shoots linear-oblong, 4–6 mm.
long; pedicels slender; capsule globose, 1–2 mm. thick. Dry places: N.Y.—
Man.—Neb. *Canad.—Allegh.* Jl-Au.

4. L. intermèdia Leggett. Stem erect, strict, 3–4 dm. high, sparingly
pubescent or glabrate; leaves of the stem linear-lanceolate or -oblanceolate,
1–2.5 cm. long; those of the basal shoots lanceolate, 3–6 mm. long; petals
brown-purple. Open places: N.S.—Minn. -Pa.—N.J. *Canad.* Jl-Au.

Family 113. **VIOLACEAE.** Violet Family.

Low herbs, or in the tropics woody. Leaves simple, alternate or basal, with stipules. Flowers perfect, irregular. Sepals and petals 5; the latter imbricate in bud, the lowermost spurred or saccate at the base. Stamens 5; anthers united or connivent. Gynoecium of 3 united carpels. Ovary 1-celled, with 3 parietal placentae. Capsule loculicidal; seeds anatropous.

Sepals more or less auricled at the base; corolla spurred.   1. Viola.
Sepals not auricled at the base; corolla merely gibbous at the base.
  Petals nearly equal in length; stamens syngenecious.   2. Cubelium.
  Petals very unequal in length; anthers only connivent.   3. Calceolaria.

1. **VÌOLA** (Tourn.) L.   Violet, Heart's-ease, Pansy.

Usually perennial herbs, either bearing leaves and 1-flowered scapes from the crown of the rootstock, or stemmed, with manifest internodes between the leaves, and with axillary 1-flowered peduncles. Flowers usually of two kinds, those of spring with showy petals and those of summer with petals rudimentary or lacking, the latter never opening, but self-fertilized within the closed calyx; petaliferous flowers nodding, pentamerous and irregular as to calyx, corolla and stamens. Sepals 5, persistent in fruit, auricled at the base. Petals 5, the lowest one spurred. Stamens distinct, but more or less coherent, the two lower furnished with nectar-bearing appendages projecting into the spur. Capsule ovoid to cylindric, 3-valved, bearing 20–60 obovate seeds 1–3 mm. long.

Plant stemless; leaves and scapes directly from the rootstock; petals violet or white.
  Rootstock thick, without stolons.
    Style club-shaped, beakless, oblique at the apex, with the stigma near the center of the cavity; cleistogamous flowers wanting; petals beardless; leaves pedately dissected.   1. Pedatae.
    Style capitate at the apex, with a conical beak on the lower side, which ends in the stigma; cleistogamous flowers present; lateral petals of the petaliferous flowers usually bearded.
      Blades, at least of the lower leaves, palmately dissected to near the base.   2. Pedatifidae.
      Blades merely toothed or lobed, but not to the base.
        Leaf-blades reniform, cordate, or broadly deltoid in outline.   3. Cucullatae.
        Leaf-blades ovate or lanceolate in outline, the later ones usually more or less hastately lobed or toothed at the base.   4. Sagittatae.
  Rootstock slender; plants usually with stolons.   5. Palustres.
Plant leafy-stemmed, with axillary flowers.
  Perennials; style not strongly thickened above, without a hollow summit.
    Petals blue, purple, or white, yellow, if at all, only at the base.
      Stigma not capitate; stipules herbaceous, bristle-toothed; spur elongate.   6. Caninae.
      Stigma capitate, beakless, bearded at the summit; stipules entire, the lower scarious: spur short.   7. Canadenses.
    Petals yellow.   8. Pubescentes.
  Annuals; style much enlarged at the hollowed summit; stipules leaf-like, pectinately lobed towards the base.   9. Tricolores.

1. Pedatae.
One species.   1. V. pedata.

2. Pedatifidae.
One species.   2. V. pedatifida.

3. Cucullatae.
Later leaves more or less lobed or cleft, the earlier ones merely toothed.
  Cleistogamous flowers ovoid, on short prostrate peduncles, their capsules purplish; leaf-blade cordate, 5–11-lobed.   3. V. palmata.
  Cleistogamous flowers on erect peduncles, their capsules green.   4. V. viarum.
Later leaves, like the earlier ones, merely toothed.
  Cleistogamous flowers on short prostrate peduncles, or in V. septentrionalis ascending and more elongate in fruit; capsule purplish; leaf-blades cordate in outline.
    Leaves glabrous or nearly so.
      Flowers pale; leaf-blades, except the earliest ones, mostly longer than broad, deltoid-cordate.   5. V. missouriensis.

Flowers usually deeply blue-violet; leaf-blades usually as broad as long, reniform or broadly cordate.

Petals 14–18 mm. long; sepals without a gland or callosity at the apex; later leaves 5–10 cm. long, broadly cordate, with a deep broad sinus.   6. *V. papilionacea.*

Petals about 12 mm. long; sepals with a small gland or callosity at the apex; later leaves reniform, rounded or acute at the apex, 2–4 cm. long, with a rather shallow sinus.   7. *V. Lunellii.*

Leaves pubescent, especially beneath and on the petioles.

Leaves villous; spurred petal glabrous.   8. *V. sororia.*

Leaves hirsutulous; spurred petal villous.

Sepals and their auricles ciliolate.   9. *V. septentrionalis.*

Sepals and their auricles not ciliolate.   10. *V. Novae-Angliae.*

Cleistogamous flowers on erect peduncles, their capsules green.

Spurred petal glabrous, rounded at the apex.

Cleistogamous flowers ovoid.   11. *V. pratincola.*

Cleistogamous flowers long and slender.   12. *V. cucullata.*

Spurred petal more or less bearded, retuse or emarginate.

Mature leaves rounded or blunt at the apex; sepals obtuse.   13. *V. nephrophylla.*

Mature leaves abruptly acuminate at the apex; sepals acute.   14. *V. retusa.*

4. SAGITTATAE.

Leaves finely pubescent, ovate-oblong in outline.   15. *V. fimbriatula.*

Leaves glabrous, oblong-lanceolate in outline.   16. *V. sagittata.*

5. PALUSTRES.

Petals lilac or pale blue.

Leaves minutely hairy above; spur 5–8 mm. long.   17. *V. Selkirkii.*

Leaves glabrous on both sides; spur short, 2 mm. long.   18. *V. palustris.*

Petals white, the three lower ones with purple veins.

Cleistogamous capsules ovoid, purplish; woodland species.

Leaf-blades reniform; stolons short.   19. *V. renifolia.*

Leaf-blades ovate, acute or acuminate.

Lateral petals bearded; seeds obtuse at the base.   20. *V. incognita.*

Lateral petals beardless; seeds acute at the base.   21. *V. blanda.*

Cleistogamous capsules ellipsoid, green; peduncles erect; bog and wet meadow plants.

Leaf-blades broadly ovate or orbicular, cordate at the base, obtuse at the apex.   22. *V. pallens.*

Leaf-blades narrower.

Leaf-blades ovate, acute, the base tapering or merely subcordate.   23. *V. primulifolia.*

Leaf-blades lanceolate to elliptic-lanceolate.   24. *V. lanceolata.*

6. CANINAE.

Petals blue or violet.

Leaves glabrous or nearly so.

Leaf-blades suborbicular.   25. *V. conspersa.*

Leaf-blades ovate.   26. *V. adunca.*

Leaves puberulent; blades ovate.   27. *V. subvestita.*

Petals white or cream-colored.   28. *V. striata.*

7. CANADENSES.

Lower leaves with reniform blades, hirsutulous beneath.   29. *V. rugulosa.*

Lower leaves with broadly obovate acuminate blades, subglabrous.   30. *V. canadensis.*

8. PUBESCENTES.

Stem at first short; first flowers and leaves from near the base.

Basal leaves narrowly ovate or elliptic, obtuse.   31. *V. vallicola.*

Basal leaves lanceolate, acute or subacute.   32. *V. Nuttallii.*

Stem naked below; peduncles from the axils of the upper leaves only.

Plant sparingly pubescent or glabrate; basal leaves usually 1–3.   33. *V. eriocarpa.*

Plant conspicuously pubescent; basal leaves usually wanting.   34. *V. pubescens.*

9. TRICOLORES.

One species.   35. *V. Rafinesquii.*

1. V. pedàta L. Nearly glabrous, from a short erect rootstock; leaf-blades pedately divided to near the base, the segments 2–4-cleft; earlier leaves smaller and less divided; corolla 2–4 cm. broad, lilac-purple or the upper two petals dark violet. Dry fields: Mass.—Man.—La.—Fla. *E. Temp.* Ap–Je.

2. V. pedatífida G. Don. Leaves 3-parted, the divisions variously cleft and incised into linear lobes, usually truncate or cuneate at the base, the margins and midrib hirsutulous; peduncles of the petaliferous flowers taller than the leaves, those of the apetalous flowers shorter, but erect. *V. delphinifolia* Nutt. Prairies and valleys: Alta.—Ohio—N.M.—Ariz. *Prairie—Plain—Mont.* Ap–Je.

V. nephrophylla × pedatifida. Resembling *V. nephrophylla,* but the leaf-blades cut only half way to the base into 9–13 oblong lobes. S.D.

V. papilionacea × pedatifida. Leaves approaching those of *V. papilionacea,* but deeply toothed or cleft one-third to the base. Minn.

V. pedatifida × sororia. Intermediate between the two parents, characterized by the broad leaves, broader than long, cordate at the base, and more or less lobed at the apex. *V. Bernardi* Greene. Ill.—S.D.—Mo.

V. pedatifida × sagittata. Leaves intermediate between those of the two parents, shorter and broader than in *V. sagittata* and cleft half way to the base into lanceolate lobes. Ill.—Iowa.

3. V. palmàta L. Pubescent, with a thick oblique rootstock; leaf-blades ovate or cordate in outline, palmately lobed or cleft; segments 5–11, toothed, the terminal one much larger than the rest, villous on the petioles and the veins beneath; corolla violet-purple, 2–4 cm. broad. Wooded hills and dry soil: Mass. —Iowa—Fla. *E. Temp.* Ap–My.

4. V. viàrum Pollard. Glabrous; leaf-blades deltoid or cordate in outline, the earlier unlobed, the later pedately 3–7-lobed and serrate, the middle lobe broad, the lateral ones more or less lunate; corolla deep violet; petals narrow, the lowest and the upper two emarginate. Waysides and banks: Mo. —Kans.—Ark. *Ozark.* Ap–My.

5. V. missouriénsis Greene. Glabrous; with a short rootstock; leaf-blades deltoid, with a cordate base, 6–10 cm. wide, coarsely crenate-serrate; sepals lanceolate, obtuse, narrowly white-margined; corolla pale violet, with a darker band above the white center; capsule of the cleistogamous flowers ellipsoid, finely dotted. River bottoms: Mo.—Wis.—S.D.—Okla.—Tex.—La. *Prairie— Ozark.—Texan.* Ap–My.

6. V. papilionàcea Pursh. Plants robust and hardy from a stout branching rootstock, usually glabrous, but petioles sometimes sparsely pubescent; blades when fully grown often 12 cm. wide, broadly ovate-cordate, acute or abruptly pointed; outer sepals ovate-lanceolate; upper and lateral petals broadly obovate, 8–10 mm. wide. *V. obliqua* Britt. [BB]; not Mill. Fields and groves: Mass.—Ga.—Okla.—Minn.; apparently introduced in Denver, Colo., and vicinity. *Canad.—Allegh.*

7. V. Lunéllii Greene. Low, glabrous plant; leaf-blades reniform, 2–4 cm. long, 3–4 cm. wide, crenate, or the later ones 4 cm. long and 5 cm. wide, with a broad but rather shallow sinus; sepals lance-oblong, obtuse; corolla deep violet; petals obovate, spur short and rounded; cleistogamous flowers on horizontal pedicels, ovoid, 5 mm. long. Moist ground: N.D. *Prairie.* My.

8. V. soròria Willd. Villous-pubescent, with a thick rootstock; leaf-blades ovate to orbicular, or even reniform, cordate at the base, crenate-serrate, often 1 dm. wide; corolla violet or lavender, rarely white; outer sepals ovate-oblong, obtuse. *V. cuspidata* Greene [B]. Moist meadows: Que.—S.D.—Okla. —N.C. Ap–My.

V. nephrophylla × sororia. Early leaves pubescent on the petioles and lower surface as in *V. sororia,* not glabrous; the late leaves broadly deltoid-cordate, with open sinuses as in *V. nephrophylla,* not rounded-cordate as in *V. sororia.* Minn.

V. missouriensis × **sororia.** Intermediate between the parents, with the broadly deltoid, crenate-serrate later leaves of the former, but distinctly pubescent on the lower surface and petioles. Kans.—Miss.

V. papilionacea × **sororia.** Intermediate between the parents, differing from *V. sororia* in the longer petioles, thinner blades and the sparser pubescence, and from *V. papilionacea* in being pubescent. N.Y.—Minn.—Kans. —La.

9. **V. septentrionàlis** Greene. Basal stipules often 1 cm. long, bristly ciliolate and more or less glandular; leaves cordate-ovate, somewhat acuminate, the apex obtuse; flowers large; sepals obtuse, finely ciliate; cleistogamous flowers sagittate, on ascending peduncles; their capsules subglobose, usually purple. Woodlands: P.E.I.—Conn.—Pa. -Wash. –B.C. *Boreal.*

10. **V. Nòvae-Ángliae** House. Earlier leaf-blades narrowly cordate-triangular, 2–4 cm. long, the later often as broad as long; sepals and auricles not ciliolate; otherwise resembling the preceding. Gravelly beaches and rocky banks: Me.—Minn.

11. **V. pratíncola** Greene. Plant in every way smaller than the preceding; leaves deeply cordate-ovate, tapering gradually to a subacute apex, 2–3 cm. wide at petaliferous flowering, 5–8 cm. wide at maturity; petals spatulate, 4–6 mm. wide; peduncles taller than the leaves. *V. Sandbergii* Greene. Hills and prairies: Colo. –Wyo.—N.D.—Minn. *Plain— Submont.*

12. **V. cucullàta** Ait. Glabrous perennial; leaf-blades, except the earliest ones, ovate to reniform, acute, finely crenate-serrate, up to 8 cm. wide; peduncles much exceeding the leaves, corolla blue-violet, with a dark throat, rarely white; lateral petals bearded, the lowest one glabrous, slightly shorter; sepals narrowly lanceolate; cleistogamous flowers on long erect peduncles, very narrow; capsule green, 10–15 mm. long. Wet places: Que.—Man.—Iowa—Ark.—Ga. *Allegh.* Ap–Je.

f.380.

13. **V. nephrophýlla** Greene. *Fig. 380.* Leaves broadly cordate-ovate to reniform, obtuse or bluntly short-pointed; petals large, the upper pair often sparsely pubescent; sepals ovate to lanceolate, obtuse; cleistogamous flowers on erect peduncles; capsules green, short-ellipsoid; seeds olive-brown, 2 mm. long. *V. peramoena* and *V. cognata* Greene. Bogs and borders of cold streams: Newf.—Conn.—Minn.—N.M.—Ariz.—Wash.—B.C. *Boreal.—Plain—Mont.* My–Jl.

V. nephrophylla × **papilionacea.** Intermediate between the parents, glabrous throughout, the leaf-blades cordate-ovate as in *V. papilionacea* to reniform as in *V. nephrophylla,* the lower petal from densely bearded to beardless. Wis.—Iowa.

14. **V. retùsa** Greene. Glabrous throughout; leaves at flowering time broadly cordate-deltoid, finely serrate; later leaves much dilated, with a shallow sinus, more or less decurrent on the petiole, often abruptly acuminate; cleistogamous flowers on erect peduncles; capsules ellipsoid, green; seeds brown, 2 mm. long. Borders of streams: Kans.—Colo. *Plain.* My–Je.

15. **V. fimbriátula** Smith. More or less pubescent; leaf-blades of the earlier leaves ovate, the later ones oblong-ovate, finely pubescent, especially beneath, crenulate, but with an incised often somewhat hastate base; corolla violet-purple; cleistogamous flowers on erect peduncles; capsule green, ovoid, 7–10 mm. long. *V. ovata* Nutt. Fields and hillsides: N.S.—Minn.—Iowa— Ga. *Canad.—Allegh.* My–Je.

16. V. sagittàta Ait. Glabrous or finely puberulent; petioles usually longer than the blades; blades lanceolate or oblong-lanceolate, the later hastately or sagittately lobed or toothed at the base, ciliate, the earlier ones merely crenate; corolla violet-purple. Moist banks: Mass.—Minn.—La.—Tex. *E. Temp.*

V. sagittata × sororia. Leaf-blades usually more elongate than in *V. sororia*, distinctly serrate and often more or less lobed at the base. Wis.—Minn.

17. V. Selkírkii Pursh. Leaves broadly cordate-ovate, the basal lobes converging or overlapping, the margins crenate-serrate, the upper surface hirtellous; petals pale violet, all beardless; spur 5–8 mm. long, with an enlarged rounded end; capsule subglobose; seeds small, buff. Woods: N.B.—Pa.—Minn. —B.C.—Colo. *Boreal.—Mont.*

18. V. palústris L. Scapes and leaves glabrous throughout; arising from the ends of scaly creeping rootstocks; blades cordate-ovate to orbicular, remotely and lightly crenate-serrate; petals pale lilac to white, the lateral slightly bearded, the spur short and thick; seeds dark brown, 1.5 mm. long. *V. cyclophylla* and *V. eucycla* Greene. Cold bogs and wet borders of rivulets: Lab.—Vt.—S.D.—Colo.—Utah—Alaska. *Boreal.—Mont.—Subalp.* My–Au.

19. V. renifòlia Gray. Pubescent throughout, or often subglabrous, especially on the upper surface of the leaf; blades reniform, those of summer often ending in a short blunt tip; petals white, all beardless; capsule ellipsoid; cleistogamous flowers on horizontal pedicels, their capsules purple. Cold forests: Newf.—Pa.—Minn.—Colo.—Mack. *Boreal.—Mont.* Ap–My.

20. V. incógnita Brainerd. Peduncles, petioles, and lower surface of the leaves with long white hairs; leaf-blades orbicular or reniform, 2–4 cm. wide, abruptly pointed at the apex; summer leaves broadly ovate, rugose; corolla white, the upper two petals obovate; plant producing numerous runners. Mountains and woodlands: Newf.—N.D.—Tenn. *Canad.—Allegh.* Ap–My.

21. V. blánda Willd. Glabrous, except the upper surface of the leaves; leaf-blades ovate-cordate, with narrow sinuses, acute, 6 cm. wide or less, with scattered short hairs above; upper petals long and narrow, reflexed, sometimes twisted. *V. amoena* Le Conte. *V. alsophila* Greene [B]. *V. blanda palustriformis* A. Gray. Ravines and shaded slopes: Que.—Man.—Ga. *E. Temp.* Ap–My.

22. V. pállens (Banks) Brainerd. Petioles and scapes often red-dotted, somewhat hirsutulous; leaf-blades glabrous, broadly ovate-cordate or orbicular-cordate, 1–6 cm. wide, crenate, rounded at the apex; upper petals broadly obovate; stolons slender. *V. blanda* Le Conte [B]; not Willd. Springy ground: Labr.—Man.—N.D.—Tenn.—S.C. *Canad.—Allegh.* Ap–My.

23. V. primulifòlia L. Leaves obscurely crenate-serrate, glabrous; peduncles longer than the leaves; flowers like those of *V. lanceolata*, the lower three petals purple-veined, the lateral ones slightly bearded or beardless; cleistogamous flowers on short erect peduncles. Open ground: N.B.—Fla.— La.—Minn. *E. Temp.*

24. V. lanceolàta L. Glabrous, profusely stoloniferous; leaf-blades lanceolate or elliptic, 5–15 cm. long, 1–2 cm. wide, tapering at the base, obscurely crenulate; petals all beardless, the lower three purple-veined. Open bogs: N.S.—Minn.—Neb.—Miss.—Fla. *E. Temp.*

25. V. conspérsa Reichenb. Stem 8–15 cm. high; lower leaf-blades orbicular, cordate at the base, crenate, 2–4 cm. wide, the upper ones smaller, acute; stipules ovate-lanceolate, serrately ciliate; corolla pale violet, rarely white. *V. Muhlenbergii* Torr. *V. labradorica* Britt. [B]; not Schrank. Low shaded ground: Que.—Minn.—Ga. *Canad.—Allegh.* Ap–My.

26. **V. adúnca** Smith. Stems several, sparingly pubescent, 3–25 cm. long; leaf-blades ovate, obtuse, subcordate, finely crenate; upper leaf-blades narrower, short-petioled; stipules linear, spinulose-serrate; petals violet or purple, 10–15 mm. long; spur long, curved upward. *V. inamoena* Greene. *V. achlydophylla* Greene. Hillsides: N.B.—Alaska—Calif.—Colo. *Boreal.—Mont.* My–Jl.

27. **V. subvestita** Greene. Stems many, spreading, 5–15 cm. long, finely puberulent; stipules linear-lanceolate, incised, with bristly teeth; leaf-blades ovate, 1–2.5 cm. long, crenate, subcordate at the base; corolla violet; spur about 6 mm. long. *V. arenaria* Am. auth. [B, G]; not DC. *V. cardaminifolia* Greene. *V. anisopetala* Greene was described as glabrous, but Lunell's specimens are decidedly puberulent. Sandy soil: Me.—Man.—S.D. *Canad.* My–Jl.

28. **V. striàta** Ait. Stems several, angled, 1.5–6 dm. high, decumbent; leaf-blades glabrous, orbicular to ovate, subcordate at the base, 2.5–4 cm. wide, crenate; stipules large, oblong-lanceolate, fimbriate; sepals linear-lanceolate, ciliolate; spur thick, 4 mm. long. Low shady ground: N.Y.—Minn. (?) —Mo.—Ga. *Canad.—Allegh.* Ap–My.

29. **V. rugulòsa** Greene. Stems 2–6 dm. high; root-leaves cordate-reniform, abruptly short-pointed, often 10 cm. wide, densely hirsutulous beneath; lower stem-leaves similar, the upper successively smaller and shorter petioled, becoming ovate-acuminate; capsules ovoid to subglobose, often puberulent, 6–10 mm. long; seeds brown, 2 mm. long. *V. Rydbergii* Greene. Woods: Man.—Iowa—Colo.—Alaska. *Boreal.—Plain—Mont.* My–Jl.

30. **V. canadénsis** L. Stems several, 15–35 cm. high; leaves broadly cordate-ovate, abruptly acuminate, serrate with incurved teeth, nearly glabrous, or the smaller upper leaves muriculate to hirtellous, especially on the veins underneath; petals often purplish with age; spur short and rounded; capsule ovoid, usually glabrous; seeds brown, 2 mm. long. *V. neomexicana* and *V. geminiflora* Greene. Moist, wooded, and steep mountain slopes: N.D. S.C.—N.M.—Ariz.—B.C. *Temp.—Submont.—Mont.* My–Jl.

31. **V. vallícola** A. Nels. Glabrous or pubescent; leaves nearly entire, ovate to lanceolate, obtuse, the base often obliquely rounded; petals yellow, the two upper often tinged with purple; cleistogamous flowers borne on the upper part of the stem; capsule short-ellipsoid; seeds 2 mm. long. *V. physalodes* Greene. Moist valleys in the mountains: Colo.—N.D.—Sask.—Wash.—B.C. *Plain—Mont.* My–Je.

32. **V. Nuttállii** Pursh. More or less pubescent; stems commonly numerous, in moist situations 15–25 cm. high; leaves except the few earliest lanceolate to linear-lanceolate, tapering at the base into margined petioles, the apex acute or subacute, obscurely and remotely denticulate; seeds brown, 3 mm. long. On the foothills and plains: Man.—Mo.—Ariz.—Mont. *Prairie—Plain —Submont.* Ap–Je.

33. **V. eriocárpa** Schw. Glabrous, except the upper part of the stem and the leaf-veins beneath; stems usually 2–4 from one rootstock; basal leaves long-petioled; blades ovate to reniform, mostly pointed; cauline leaves only on the upper part of the stem, short-petioled or subsessile; capsule woolly or glabrous; seeds 2 mm. long. *V. scabriuscula* Schwein. [G, B]. Open woods: N.S.—Man.—Neb.—Tex.—Va. *Canad.—Allegh.* Ap–My.

34. **V. pubéscens** Ait. Stem often solitary; root-leaves few or none; stem-leaves short-petioled; blades reniform to ovate, cordate or truncate, decurrent, the apex slightly blunt-pointed; stipules large, obliquely ovate; capsules often white-woolly; seeds 3 mm. long. Rich woodlands: N.S.—Va.—Neb.—S.D. —Man. *Canad.—Plain—Submont.*

35. **V. Rafinésquii** Greene. A subglabrous annual, 10–25 cm. high, often branching from the base; lower leaves round, upper ovate to spatulate, attenu-

ate at the base; stipules foliaceous, pectinate, the terminal segment elongate, entire; petals cream-colored, about twice the length of the sepals. Fields and waste places: N.Y.—Ga.—Tex.—Colo.—Mich. *E. Temp.—Plain.* Ap–My.

2. CUBÈLIUM Raf. GREEN VIOLET.

Leafy perennials. Leaves alternate, entire or toothed. Flowers perfect, axillary, somewhat irregular. Sepals 5, subequal, not auricled. Petals 5, greenish, imbricate, nearly equal in length, the lowest one larger and gibbous at the base. Stamens 5, syngenecious; anthers almost sessile. Style hooked at the apex. Pod 3-valved, obscurely angled.

1. **C. cóncolor** (T. F. Forst.) Raf. Stem 3–9 dm. high; leaf-blades oblong or elliptic, acute or acuminate, entire; flowers 8 mm. long, on recurved, short pedicels, 1–3 in each axil; pod 2 cm. long. *Hybanthus concolor* Spreng. [G]. Rich woods: Ont.—Mich.—Kans.—Ga. Ap–Je.

3. CALCEOLÀRIA Loefl. GREEN VIOLET.

Perennial herbs or rarely shrubs, with alternate or opposite leaves. Flowers axillary and solitary or racemose, perfect. Sepals equal, not auricled. Petals unequal, the lowest largest and gibbous at the base; stamens 5; anthers connivent but distinct; the two lower filaments glandular at the base. Capsule 3-valved, opening elastically. *Ionidium* Vent.

1. **C. verticillàta** (Ortega) Kuntze. Low perennial herb; stems clustered, 1–4 dm. high, slightly pubescent; leaves alternate, but often fascicled; stipules subulate or wanting; blades linear, 2–4 cm. long; flowers solitary, nodding; petals white, 4–6 mm. long. *Ionidium lineare* Torr. Dry ground: Kans.—Tex.—Ariz.—Colo.; Mex. *Son.*

Family 114. LOASACEAE. LOASA FAMILY.

Usually scabrous-pubescent herbs, rarely shrubs, with opposite or alternate leaves, without stipules. Flowers perfect, regular, cymose. Hypanthium well developed, turbinate to cylindric, enclosing the ovary. Sepals 4 or 5, imbricate or convolute. Petals as many, yellow or red. Stamens numerous; filaments filiform or dilated, sometimes passing into staminodia closely resembling the petals. Gynoecium of 2–5 united carpels, enclosed in the hypanthium. Ovary 1-celled, with parietal placentae.

Placentae with horizontal lamellae between the seeds; these in two rows; seeds
    very flat, more or less winged.   1. NUTTALLIA.
Placentae without lamellae; seeds usually prismatic.
    Placentae filiform; ovules in one row, 10–40; seeds minutely muricate, not
      striate; filaments free or nearly so.   2. ACROLASIA.
    Placentae broad, band-like; ovules in 1 or 2 rows, few; seeds distinctly striate,
      often rugose; filaments at the base united with the petals into a ring.
                                                            3. MENTZELIA.

1. NUTTÁLLIA Raf. SAND LILY, BLAZING STAR.

Scabrous, pubescent, biennial or perennial herbs, with barbed or hooked hairs, mostly stout. Leaves alternate, toothed or pinnatifid, rarely entire. Sepals 5, persistent. Petals 5, or apparently 10, on account of the petaloid staminodia, yellow to straw-colored, usually conspicuous. Stamens numerous; filaments filiform, or the outer ones dilated. Hypanthium oblong-cylindric; ovules in two rows on each of the three parietal placentae. Capsule stout; seeds numerous, horizontal, flattened, winged or margined, separated by horizontal lamellae. *Touterea* Eat. & Wright. *Hesperaster* Cockerell.

Outer filaments dilated; petals 2–3 cm. long.
  Plant branched below; flowers subtended by solitary entire bracts.
                                 1. *N. nuda.*
  Plant simple below; flowers subtended by several toothed
    bracts.   2. *N. stricta.*
Filaments all filiform; petals 4–5 cm. long.   3. *N. decapetala.*

1. N. nùda (Pursh) Greene. Stout perennial
or biennial; leaves oblanceolate, 5–15 cm.
long, sinu-
ately lobed, with oblong or ovate lobes, less hispid
above; petals spatulate, acute, about 2 cm. long;
staminodia nearly as large, similar; capsule about 3
cm. long; seeds winged. *Bartonia nuda* Pursh.
Mentzelia nuda T. & G. [B]. Plains and hillsides:
w Neb.—Colo.—Wyo. *Plain–Submont.* Jl–Au.

2. N. stricta (Osterhout) Greene. Strict,
rather simple perennial; stem 5–10 dm. high,
branched above; leaves linear-lanceolate, 5–10 cm.
long, sinuately toothed, with triangular teeth; petals
oblanceolate, acute, 2.5–3.5 mm. long; petaloid
staminodia similar; capsule 2–3 cm. long; seeds
wing-margined. *Hesperaster strictus* Osterhout.
Plains and foothills: w Neb.—Tex.—Colo.— Wyo.
Plain– Submont. Je–O.

3. N. decapétala (Pursh) Greene. *Fig. 381.* Stout biennial; stem 5–10
dm. high, leafy, very scabrous; lower leaves oblanceolate, 1–3 dm. long, the
upper lanceolate, pinnately lobed, with lanceolate lobes; flowers subtended by
several pinnatifid bracts; petals 4–5 cm. long, oblanceolate, acute; petaloid
staminodia almost as large; capsule about 4 cm. long; seeds with a narrow
wing. *Bartonia decapetala* Pursh. *M. ornata* T. & G. *M. decapetala* (Pursh)
Urban & Gilg. [G, B]. Cañons: Man.— Iowa—Tex.—Nev. *Alta. Plain—Submont.* Jl–S.

2. ACROLÀSIA Presl.

Scabrous pubescent annuals. Leaves alternate, rarely some opposite, narrow, from entire to pinnatifid. Sepals 5, persistent. Petals 5, yellow, rather
inconspicuous. Stamens many, free, or nearly so; filaments filiform. Hypanthium cylindric; placentae 3, filiform; ovules 10–40, in a single row on each
placenta. Capsule slender, linear-cylindric or linear-clavate. Seeds rather few,
pendulous, more or less prismatic, muricate, not striate, not separated from
each other by horizontal lamellae.

Seeds rather strongly muricate; upper leaves narrow; sepals lanceolate-subulate, half
  as long as the petals or longer.   1. *A. albicaulis.*
Seeds minutely muricate (tubercles seen only under strong magni-
  fication); leaves broadly ovate.   2. *A. compacta.*

1. A. albicaùlis (Dougl.) Rydb. *Fig. 382.*
Stem white, glabrous, erect, or decumbent at the
base, 1–4 dm. high; lower leaves linear-oblanceolate,
usually dentate or entire; middle leaves with 1–5
pairs of linear lobes, the upper linear, usually entire; sepals lanceolate-subulate, 2–2.5 mm. long;
petals obovate, 3–4 mm. long, golden yellow, rather
prominently veined; capsule linear-cylindric, 10–15
mm. long, 2 mm. thick. *M. albicaulis* Dougl. [B].
Sandy soil: Mont.—Neb.—Tex.—N.M.—Ariz.- B.C.
Plain—Submont. Mr.–Au.

2. A. compácta (A. Nels.) Rydb. Stem whitish,
pilose; leaves numerous, 2–3 cm. long, rough-
hirsute; petals obovate, 3–4 mm. long; capsule
linear-clavate, 10–13 mm. long, 10–12-seeded. *M.
compacta* A. Nels. Plains and foothills: S.D.—
Colo.—Ida.—Wash. *Son.—Submont.* Je–Au.

f. 382.

3. MENTZÈLIA (Plum.) L. STICK-LEAF.

Annual, biennial, or perennial herbs, with white bark, scabrous, with barbed hairs. Leaves alternate, relatively broad, sinuate or lobed. Flowers perfect, cymose. Sepals 5, persistent. Petals also 5, deciduous. Stamens numerous; filaments filiform, united at the base with each other and with the petals into a ring. Hypanthium short, tapering at the base. Placentae 3, broad and band-like, bearing the ovules in 1 or 2 rows. Capsule 3-valved at the top. Seeds rather few, ellipsoid, angled, striate, neither winged nor separated by horizontal lamellae.

1. M. oligospérma Nutt. Perennial, with a fusiform root; stem straw-colored, rough-hirsute; leaves ovate or ovate-lanceolate, or sometimes rhombic, coarsely dentate and often somewhat lobed, 1–6 cm. long; sepals 5–8 mm. long, linear-subulate; petals 10–15 mm. long, obovate-cuneate, cuspidate, golden yellow; stamens about 20; capsule about 8 mm. long, 2 mm. thick, about 3-seeded. *M. Nelsonii* Greene, a small-flowered form. Rocky places and hillsides: Ill.—La.—Tex.—N.M.—N.D.; Mex. *Prairie—Plain—Submont.* Je–Au.

Family 115. PASSIFLORACEAE. PASSION-FLOWER FAMILY.

Perennial herbs, vines, shrubs, often climbing by means of tendrils from the axils of the leaves. Leaves alternate, simple or compound, usually with stipules. Flowers usually perfect, regular. Calyx of 4 or 5, partly united sepals. Petals 4 or 5, below a filamentous crown, or obsolete. Stamens 4 or 5; filaments monadelphous, adherent to the stipe of the ovary. Gynoecium of several united carpels; ovary 1-celled, with parietal placentae, stalked; styles as many as the placentae, clavate. Fruit a berry. Seeds numerous.

1. PASSIFLÒRA L. PASSION-FLOWER.

Vines, climbing by means of tendrils. Leaves simple, entire, lobed or parted. Flowers showy. Crown of the corolla of filiform spreading filaments. Styles 3, mostly recurved or spreading; stigmas capitate. Seeds flattened.

Flowers subtended by a conspicuous involucre of 3 bracts; leaf-blades with 2 glands at the base; petals lavender.   1. *P. incarnata.*
Flowers without an involucre; leaf-blades glandless; petals greenish-yellow.   2. *P. lutea.*

1. P. incarnàta L. *Fig. 383.* Stem 3–20 dm. long; leaf-blades palmately 3–5-lobed, serrate, 8–12 cm. long, minutely pubescent; bracts ovate, 3–5 mm. long, glandular; sepals linear-oblong, boat-shaped, with a horn at the apex, 2.5–3 cm. long, green without, lavender within; petals linear-oblong, obtuse; filaments of the crown lavender, striped with purple and white at the middle. Thickets: Va.—Kans.—Okla.—Fla. *Austral.* My–Jl.

2. P. lùtea L. Stem 3–10 dm. long; leaf-blades 3-lobed, 3–15 cm. long, the lobes rounded or ovate, entire; sepals oblong or linear, 8–12 mm. long, green; petals oblong, shorter than the sepals; crown of 3 series of yellow filaments. Woods and thickets: Pa.—Kans.—Tex.—Fla. *Allegh.—Austral.* My–Jl.

f. 383.

Family 116. CACTACEAE. CACTUS FAMILY.

Succulent shrubby plants, usually spiny, the spines arising from small hairy or bristly cushions (areolae); leaves usually wanting, if present, in

ours small and deciduous. Flowers perfect, in ours regular, solitary.
Hypanthium well developed, enclosing the ovary. Sepals usually many,
imbricate, the inner often petal-like. Petals many, in two or more series.
Stamens numerous, in several series. Gynoecium of several united carpels; ovary inferior, 1-celled, with several parietal placentae. Fruit a
fleshy or rather dry berry.

Leaves evident, but small and deciduous; flowers without definite tube; seeds with a
  bony aril; stems jointed.   **1. OPUNTIA.**
Leaves wanting; flowers with a more or less definite tube; seeds without aril; stems
  not conspicuously jointed.
  Flowers borne near the spine-bearing areolae.
    Plant-body more or less cylindric; spines arranged on definite ribs.
                                                 2. ECHINOCEREUS.
    Plant-body usually globular or ellipsoid; stem tubercled.   **3. PEDIOCACTUS.**
  Flowers borne remote from the spines, at the base of the tubercles of the stem.
                                               4. NEOMAMILLARIA.

1. OPÚNTIA Hill. PRICKLY PEAR, CHOLLA, INDIA FIG, TREE CACTUS.

Fleshy plants with conspicuously jointed stems, the joints flat or terete.
Leaves scale-like, caducous, spirally arranged. Areolae axillary, usually spine-
bearing. Flowers lateral, arising from the upper part of some spine-bearing
areolae. Hypanthium bearing areolae, not produced beyond the ovary. Sepals
spreading, in several rows. Style cylindric; stigma 5–8-lobed. Berry more or
less pear-shaped, fleshy or rather dry. Seeds disk-like and flattened. Embryo
curved around the endosperm.

Internodes cylindrical; spines covered with a delicate sheath; plants tall (1–8
  meters high); flowers red.   **1. O. imbricata.**
Internodes flat and broad; spines not covered by a sheath.
  Fruit dry when mature, usually very spiny
    Internodes very fleshy, often terete in section, the ter-
      minal one easily breaking loose.   **2. O fragilis.**
    Internodes never terete in section, but always much
      broader than thick.
      Petals normally yellow.   **3. O. polyacantha.**
      Petals red.   **4. O. rutila.**
  Fruit juicy, usually naked or nearly so.
    Roots fusiform or tuberous.   **5. O. macrorrhiza.**
    Roots fibrous, not fusiform.
      Internodes often naked or sometimes bearing long,
        stout spines.   **6. O. humifusa.**
      Internodes generally bearing several spines from each
        areole.
        Spines twisted.   **7. O. tortispina.**
        Spines not evidently twisted.   **8. O. camanchica.**

1. O. imbricáta (Haw.) Engelm. Erect, arborescent, 1.5–8 m. high, with
verticillate branches; internodes cylindric, 0.5–1.5 dm. long, 2 cm. thick; areolae oblong, 15–30 mm. long; pulvini short-woolly, but scarcely bristly; spines
8–30, terete, spreading, the inner often 2.5 cm. long; flowers purple, 6–7.5 cm.
broad; fruit subglobose, prominently tuberculate, unarmed. *O. arborescens*
Engelm. TREE CACTUS. Foothills and plains: Tex.—Colo.—Utah—Sonora. *St.
Plain—Son.—Submont.*

2. O. frágilis (Nutt.) Haw. Decumbent; internodes 3.5–5 cm. long,
ovate, only slightly compressed or subterete; pulvini large, white-woolly and
with few bristles; spines 1–4, divaricate, the uppermost stout, angular, 12–20
mm. long; flowers pale yellow, about 5 cm. broad; fruit ovate, almost naked,
about 2.5 cm. long. Plains and hills: B.C.—Man.—Wis.—Kans.—Utah. *Plain
—Submont.*

3. O. polyacántha Haw. Stem prostrate; internodes broadly obovate or
orbicular, pale green, 5–15 cm. long, about 12 mm. thick; pulvini densely
bristly; spines 5–15, rather stout, 1–5 cm. long, variegated, the 3–5 inner stout,
reddish-brown, 3–5 cm. long, some deflexed, some spreading; flowers yellow;
fruit ovoid, spiny, 2.5 cm. long. *O. missouriensis* DC. Plains and prairies:
Sask.—Man.—Wis.—Mo.—N.M. –Utah—B.C. *Prairie—Plain.* My–Je.

4. O. rùtila Nutt. *Fig. 384.* Stem erect; internodes obovate or more oblong, 7–12 cm. long, 5–10 cm. wide; pulvini with reddish-brown bristles; spines 2–4, soon becoming gray, up to 3 cm. long, the larger flattened, porrect or erect, the lower deflexed; flowers carmine; style rose-red; stigma green; fruit obovate, bristly, soon unarmed. *O. rhodantha* K. Schum. *O. utahensis* Purpus. Table-lands: Colo.— w Neb.—Wyo. *Plain—Submont.* Je.

f. 384.

5. O. macrorrhìza Engelm. Stem prostrate or ascending; internodes obovate or oblong-obovate, 5–8 cm. long; pulvini remote, with yellowish brown bristles; spines 1–3, white or variegated, spreading or reflexed; flowers pale yellow, red at the base, 5–7.5 cm. broad; fruit obovoid, green or pale purple, 3–5 cm. long. *Prairie.* Plains: Minn. (?) —Ark.—Kans.—Ariz.—Tex.

6. O. humifùsa Raf. Stem diffuse; internodes obovate or suborbicular, 7.5–12 cm. long; pulvini with slender reddish-brown bristles, mostly unarmed; spines when present few, only marginal, stout, straight, white, usually reddish at the base and apex, the longest 1.5–2.5 cm. long; flowers sulphur-yellow, 6–8.5 cm. broad; fruit clavate, naked, 3.5–5 cm. long, with purplish pulp. *O. Rafinesquii* Engelm. Sandy soil: Tex.—Colo.—N.D.—Minn. *Plain.*

7. O. tortispìna Engelm. Stem prostrate; internodes orbicular-obovate, 1.5–2 dm. long; pulvini with yellowish bristles, armed with 3–5 yellowish, angled, usually twisted spines, the longer ones 3–6 cm. long, with 2–4 slender ones below; flowers sulphur-yellow, 6–7.5 cm. broad; fruit ovate, 4.5–5 cm. long. Plains: Miss.—Neb.—Tex.—Colo. *Texan—Plain—Son.*

8. O. camánchica Engelm. Stem prostrate; internodes obovate-orbicular; pulvini with few greenish or yellowish-brown bristles, armed with 1–3 (or the marginal ones 3–6) spines, which are reddish brown or blackish brown, with paler tips, 3.5–7.5 cm. long, the upper erect, the rest deflexed; fruit oval, 3.5–5 cm. long, deep red, sweet. Plains and hillsides: Tex.—Kans. (?)—Colo.—Ariz. *Plain—Submont.*

2. ECHINOCÈREUS Engelm. HEDGEHOG CEREUS.

Plants with cylindric or rarely oblong stems, ribbed, or if tubercled, the tubercles more or less connected in vertical rows. Leaves none. Spine-bearing areolae on the ribs or the tubercles. Flowers arising close above the spine-bearing, fully developed areolae, hence appearing lateral on the stem. Hypanthium prolonged beyond the ovary, scaly. Style filiform. Fruit fleshy. Seeds tubercled, endosperm scanty; embryo straight.

Corolla greenish.   1. *E. viridiflorus.*
Corolla red or purple.   2. *E. caespitosus.*

1. E. viridiflòrus Engelm. Stem globose or sometimes oblong-cylindric, 2.5–7.5 cm. high; ribs 13, acute; central spines solitary, stout, straight or curved, 12–14 mm. long, variegated with purple and white; radial spines 12–18, strict and radiating, 2–6 mm. long; flowers 2.5 cm. wide, greenish brown without, yellowish green within; fruit elliptic, 10–12 mm. long. Plains and hills: Tex.—Kans.—Wyo.—N.M. *St. Plains—Son.—Submont.*

2. E. caespitòsus Engelm. & Gray. Stem globose to cylindric or ovoid, 2.5–5 cm. or rarely up to 15 cm. high, 1–12 together; ribs 12 or 13, straight; areolae close together; central spines 1 or 2, short, or more often wanting; radial spines 20–30, straight or slightly recurved, the lateral ones 4–8 mm. long, the upper and lower shorter; flowers rose-purple, 5–7.5 cm. long and nearly as broad; fruit ovoid, green, 12–20 mm. long. Plains: Kans.—Tex.—Mex. *St. Plains—Son.*

3. PEDIOCÁCTUS Britton & Rose.

Stem globose, leafless, tubercled; the tubercles arranged in spiral rows, nipple-shaped. Flowers borne on the tubercles, near the areolae. Hypanthium funnelform, bearing a few scales. Petals numerous, pink. Stamens numerous. Fruit globose, irregularly bursting, nearly or quite scaleless. Seed tubercled, with a basal hilum.

1. P. Simpsoni (Engelm.) Britton & Rose. Subglobose or turbinate at the base, sometimes clustered, 7.5–12.5 cm. in diameter; ribs 8–13, with prominent nipple-shaped tubercles, which are 12–16 mm. long; radial spines 20–30, slender, straight, 8–12 mm. long; central spines 8–10, stouter, yellowish to black, 10–14 mm. long; flowers 16–20 mm. long and nearly as broad, yellowish-green or purple; fruit green, 6–7 mm. long, suborbicular. *Echinocactus Simpsoni* Engelm. Table-lands and plains. Nev.—Utah,—Colo.—w Kans. *Plain—Submont.—Subalp.*

4. NEOMAMILLÀRIA Britton & Rose. BALL CACTUS, NIPPLE CACTUS.

Fleshy plants, with globular or oval, solitary or clustered stems, covered by spirally arranged tubercles with spine-bearing areoles at the end. Leaves none. Flowers borne near woolly aerolae in the axils or near the base of the tubercles. Hypanthium produced beyond the ovary, campanulate or funnelform, naked. Style filiform. Berry fleshy. *Mamillaria* Haw.

Flowers yellowish or greenish, merely tinged with red; central spines 1; berry scarlet, globose.	
Stem usually simple or nearly so; central spines stout; porrect; flowers about 2.5 cm. long.	1 *N. missouriensis.*
Stems tufted; central spines often wanting or small; flowers 3–5 cm. long.	2. *N. similis.*
Flowers purple; central spines several; berry green, ellipsoid.	
Stems tufted, depressed-globose; central spines 3 or 4.	3. *N. vivipara.*
Stem usually simple, ellipsoid or oblong; central spines 4–12, rarely 3.	4. *N. radiosa.*

1. N. missouriénsis (Sweet) Britton & Rose. Stem mostly simple, globose, 3–5 cm.; tubercles 12–15 mm. long, in about 8 spiral rows; spines gray, 10–20 together; central spine 10–12 mm. long; corolla greenish-yellow or tinged with reddish; berry globose, 6–8 mm. in diameter, ripening the next spring; seeds black, 1 mm. long. *M. missouriensis* Sweet [G]. *M. Notesteinii* Britton. *Cactus missouriensis* Kuntze [B]. Plains and hills: Man.—S.D.—Mont.—Colo.—Kans. *Plain.* My.

2. N. símilis (Engelm.) Britton & Rose. Stems clustered, forming masses 3 dm. broad; tubercles 1.5–2 cm. long, in 8 spiral rows; radial spines gray, 12–15, puberulent, the central one often lacking, not larger than the rest; corolla yellowish, 3–5 cm. long; berry globose, 8–10 mm.; seeds 1.5–2 mm. long. *M. similis* Engelm. *Cactus missouriensis similis* Coulter [B]. Plains: Kans. (?)—Colo.—Tex. *St. Plains.*

3. N. vivípara (Nutt.) Britton & Rose. *Fig. 385.* Stems usually tufted, 3–5 cm. high; tubercles terete, ellipsoid, slightly grooved; central spines 3 or 4, slender, reddish-brown, 8–12 mm. long, one of them deflexed, the others ascending; radial spines 12–20, white, often dark-tipped, 6–8 mm. long; flowers about 3.5 cm. long, bright purple; fruit pale green,

f. 385.

12–18 mm. long; seeds yellowish-brown, obliquely pear-shaped, 1.5 mm. long. *Cactus viviparus* Nutt. [B]. *M. vivipara* Haw. [G]. Plains: Man.—Alta.—Colo.—Kans. *Plain.*

4. **N. radiòsa** (Engelm.) Rydb. Stems mostly simple, ellipsoid or cylindric, 5–12 cm. high and 5 cm. thick; tubercles 8–12 mm. long; central spines 4–12, tawny or purplish, 8–12 mm. long, the upper the longer, the lowest small and porrect; radial spines 20–30, white, 6–8 mm. long; flowers 3.5–5 cm. long, dark purple; seeds 2 mm. long. *M. radiosa* Engelm. *Cactus radiosus* Coulter. Plains: Kans.—s Utah—Sonora—Tex. *St. Plains—Son.*

Family 117. **THYMELAEACEAE.** Mezereum Family.

Shrubs or trees, with acrid juice and tough bark. Leaves alternate or opposite, without stipules. Flowers perfect or polygamous. Calyx of 4 or 5 united sepals, often bearing 4 or 5 scales within. Hypanthium more or less well developed. Corolla wanting. Stamens as many or twice as many as the sepals; filaments distinct. Pistil solitary; ovary 1-celled; style usually excentric. Ovules solitary, anatropous, pendulous.

1. **DÍRCA** L. Leatherwood, Moosewood, Wicopy.

Shrubs, with alternate, deciduous leaves. Flowers appearing before the leaves. Hypanthium present. Calyx funnelform, corolla-like, with a wavy, slightly 4-lobed limb. Stamens 8, filaments exserted, unequal in length. Ovary 1-celled, 1-ovuled; style filiform. Fruit a drupe. Endosperm wanting.

1. **D. palústris** L. *Fig. 386.* Shrub, 5–20 dm. high, slender; bark smooth; leaves ovate, 4–8 cm. long, obtuse at both ends, entire, short-petioled; flowers usually 3 in each axil, racemose; hypanthium and calyx together 7–8 mm. long, lemon-yellow; drupe 7–9 mm. long, red. Woods: N.B.—Minn.—Mo.—Tenn.—Fla. *E. Temp.* Ap–My.

f. 386.

Family 118. **ELAEAGNACEAE.** Oleaster Family.

Shrubs or trees, with silvery, scaly, or stellate pubescence. Leaves entire, alternate or opposite. Flowers in axillary clusters, perfect, polygamous or dioecious. Hypanthium in the pistillate flowers tubular or urn-shaped, adnate to and enclosing the ovary. Sepals 4, deciduous. Corolla none. Stamens 4 or 8. Disk present, annular or lobed. Ovary 1-celled, 1-ovuled. Fruit drupe-like.

Stamens 4; flowers perfect or polygamous; leaves alternate.   1. Elaeagnus.
Stamens 8; flowers dioecious; leaves opposite.   2. Shepherdia.

1. **ELAEÁGNUS** (Tourn.) L. Silver-berry, Silver-bush.

Silvery-scaly shrubs or trees. Leaves alternate, petioled. Flowers solitary or 2–4 together, axillary, perfect or polygamous. Hypanthium tubular, constricted over the ovary. Sepals 4, deciduous, valvate. Stamens 4. Fruit drupe-like.

Branchlets with brown scales.   1. *E. commutata.*
Branchlets silvery, without brown scales.   2. *E. angustifolia.*

1. E. commutàta Bernh. *Fig. 387.* Shrub or small tree, 2–5 m. high, with brown-scurfy twigs; leaf-blades oblong or elliptic, densely silvery-scurfy on both sides, 2–10 cm. long; flowers 1–3 in the axils, fragrant, 12–16 mm. long; perianth silvery without, yellowish within; sepals ovate, 2 mm. long; fruit ellipsoid, silvery, 8–12 mm. long; stone 8-striate. *E. argentea* Pursh [G, D]. Banks and hillsides: Que.—Minn.—S.D.— Utah—Yukon. *Plain —Submont.* Je–Jl.

f 387

2. E. angustifòlia L. Shrub or small tree, up to 6 m. high, sometimes spiny; leaves lanceolate, entire, 5–7 cm. long, light green; perianth campanulate, yellow within; fruit oval, yellow, with silvery scales. *E. hortensis* Bieb. Around dwellings: N.H.—S.D.—N.D.; escaped from cult.; native of Eur.

2. SHEPHÉRDIA Nutt. Buffalo-berry, Bull-berry.

Shrubs, with silvery or brown, scaly or stellate pubescence. Leaves opposite, petioled. Flowers small, dioecious, in small clusters at the nodes of preceding season. Hypanthium of the pistillate flowers urn-shaped or ovoid, bearing an 8-lobed disk at its mouth. Stamens 8, alternate with the lobes of the disk. Fruit drupe-like. *Lepargyrea* Raf.

Leaves green above; shrub not thorny.   1. *S. canadensis.*
Leaves silvery white on both sides; plant usually thorny.   2. *S. argentea.*

1. S. canadénsis (L.) Nutt. A thornless shrub, 1–3 m. high, with brown scurfy branches; leaves ovate or oval, silvery stellate and brown-scurfy spotted beneath; flowers brown without, greenish-yellow within; fruit rounded-ellipsoid, red or yellowish, 4–6 mm. long, insipid. *L. canadensis* (L.) Greene [B]. Woods and banks· Newf.—N.Y.—Mich.—Colo.—Utah—Ore.—Alaska. *Boreal.—Submont.—Subalp.* Ap–Je.

2. S. argéntea Nutt. A shrub or small tree, 2–7 m. high, with whitish, more or less thorny branches; leaves oblong, 2–5 cm. long, rounded at the apex, acute at the base; flowers brown; fruit rounded-ellipsoid, sour, scarlet, red, or golden yellow, 4–6 mm. long. *L. argentea* (Nutt.) Greene [B]. River banks: Sask.—Man.—Iowa—Kans.—N.M.—Nev.—Alta. *Plain—Submont.* Ap–My.

Family 119. LYTHRACEAE. Loosestrife Family.

Herbs, rarely shrubs or trees, with opposite or alternate, mostly entire leaves. Flowers perfect, solitary or in axillary clusters or cymes. Hypanthium from globose or campanulate to cylindric, enclosing but free from the ovary. Sepals 4 or 5, often accompanied by as many accessory teeth. Petals 4 or 5, or wanting. Stamens few or many, in one or several series. Gynoecium of several united carpels; ovary 2–6-celled, rarely 1-celled; styles united. Fruit a capsule, rarely indehiscent.

Hypanthium campanulate or turbinate, becoming hemispheric or globose.
  Low herbs; flowers inconspicuous.
    Petals wanting; capsule indehiscent.   1. Didiplis.
    Petals usually present, 4; capsule dehiscent.
      Capsule bursting irregularly.   2. Ammannia.
      Capsule septicidal.   4. Rotala.
  Shrubs; flowers conspicuous.   3. Decodon.
Hypanthium cylindric or tubular.
  Flowers regular; hypanthium symmetrical.   5. Lythrum.
  Flowers irregular; hypanthium oblique.   6. Cuphea.

1. DÍDIPLIS Raf. WATER-PURSLANE.

Aquatic herbs, resembling *Callitriche,* with 4-angled stems. Leaves opposite, narrow, entire. Flowers solitary in the axils. Hypanthium campanulate, 4-angled. Sepals 4. Petals wanting. Stamens 2–4, mostly 4; filaments very short. Ovary 2-celled; style short; stigma 2-lobed. Capsule subglobose, 2-celled.

1. D. diándra (Nutt.) Wood. Stem submerged or creeping in the mud, 1–3 dm. long, glabrous; leaves of two kinds: the submerged ones thin, linear to lance-linear, acute or acuminate, the emersed ones thicker, linear or spatulate, often obtuse, 1–2 cm. long; hypanthium about 1 mm. long; capsule 1 mm. thick. *D. linearis* Raf. Shallow water and wet shores: N.C.—Wis.—Minn.—Tex.—Fla. *E. Temp.* Je–Au.

2. AMMÁNNIA (Houst.) L.

Annual herbs, with 4-angled stems. Leaves opposite, entire, usually auricled at the base. Flowers solitary or cymosely clustered in the axils. Hypanthium campanulate, 4-angled, becoming subglobular. Sepals 4, usually with as many small teeth in the sinuses between them. Petals 4, early deciduous. Stamens 4–8. Ovary subglobose, usually 2–4-celled; styles filiform; stigmas capitate. Capsule membranous, bursting irregularly. Seeds angular, with coriaceous coat.

Flowers and capsule sessile.   1. *A. coccinea.*
Flowers and capsule pedicelled in peduncled cymes.   2. *A. auriculata.*

1. A. coccínea Rottb. Annual; stem erect, glabrous, branched below, 1.5–5 dm. high, glabrous; leaves linear or linear-lanceolate, acutish at the apex, auriculate at the base, entire, 3–7 cm. long; flowers 1–5 in each axil, sessile or nearly so; petals purple, early deciduous; style elongate, very slender, usually more than half as long as the capsule. *A. latifolia* T. & G.; not L. (?) *A. alcalina* Blankinship. Swamps and wet places: Ind.—Fla.—Tex.—Wyo.—Mont.; Mex.; Brazil. *Trop.—E. Temp.—-Plain.* Jl–S.

2. A. auriculàta Willd. Stem slender, glabrous, 5–25 cm. high; leaves oblong to lance-linear, 1–3 cm. long, acute or acutish, entire, clasping and auricled at the base; petals white or pink, broadly spatulate; capsule 2–2.5 mm. thick. Ponds and swamps: S.D.—N.M.—Tex.; Mex., C.Am., S.Am. *Trop. —Plain.* My–Je.

3. DÉCODON J. F. Gmel. SWAMP LOOSE-
STRIFE, WILD OLEANDER.

Aquatic shrubs, with virgate branches. Leaves opposite or whorled, entire. Flowers in axillary spikes, trimorphous. Hypanthium campanulate, ribbed. Sepals 5–7, with an accessory tooth in each sinus. Petals as many, purple. Stamens 10, or rarely 8, alternately longer and shorter. Ovary 3–6-celled, globose; style filiform; stigma capitate. Capsule loculicidally 3–6-celled.

1. D. verticillàtus (L.) Ell. *Fig. 388.* Stem 1–3 m. high; leaves lanceolate, 3–20 cm. long, acuminate, undulate, short-petioled; hypanthium 4–5 mm. high, glabrous; sepals triangular, acuminate; petals lanceolate to ovate, 7–9 mm. long, purple, acuminate at each end. Swamps: Mass.—Minn.—La.—Kans.—Fla. *E. Temp.* Jl–S.

f. 388.

4. ROTÀLA L.

Annual or biennial swamp herbs, with 4-angled stems. Leaves opposite, narrow, entire. Flowers perfect, usually solitary in the axils. Hypanthium campanulate, becoming subglobose in fruit. Sepals 4, very small. Petals usually 4. Stamens 4; filaments short. Ovary subglobose, 4-celled; styles very short; stigma entire or nearly so. Capsule 4-celled, subglobose, septicidally 4-valved.

1. **R. ramòsior** (L.) Koehne. A glabrous annual, 5–25 cm. high; leaves oblong or linear-oblong, 1–3 cm. long, obtuse, acute at the base; flowers usually solitary in the axils; calyx 3–4 mm. long; lobes acuminate; petals minute; capsule ellipsoid. Wet places: Mass.—Fla.—Tex.—Neb.—Calif.—Ida.—Wash.; W. Ind. and Mex. Jl–S.

5. LÝTHRUM L. LOOSESTRIFE.

Herbs or shrubs, with angled stems; ours perennial herbs. Flowers solitary in the axils, or in spikes or racemes. Hypanthium cylindric, ribbed or grooved. Sepals 4–6, accompanied with as many accessory teeth. Petals 4–6, obovate or oblanceolate. Stamens 8–12, in one series; filament filiform. Ovary 2-celled; stigma capitate. Capsule membranous, 2-celled, or becoming 1-celled, septicidally 2 valved or bursting irregularly.

Leaves mostly alternate, lanceolate, rounded or cordate at the base.   1. *L. alatum.*
Leaves opposite, linear, tapering at the base.   2. *L. lineare.*

1. **L. alàtum** Pursh. Perennial; stem glabrous, 3–12 dm. high, 4-angled or -winged; leaves sessile, alternate or the lowest opposite, lanceolate or oblong, rounded at the base, acute at the apex, 2–3 cm. long; flowers solitary in the upper axils; petals deep purple, ascending; disk fleshy; calyx 5–6 mm. long in fruit, ribbed. Low ground: (?) Ont.—Mass.—D.C.—Tex.—Wyo.—B.C. *Temp.—Plain.* Je–Au.

2. **L. lineàre** L. *Fig. 389.* Perennial; stem 3–10 dm. high, 4-angled, branched above; leaves 1–4 cm. long, leathery, slightly revolute; hypanthium 1 mm. long; petals cuneate, rounded at the apex, pale purple or pink; capsule narrowly clavate, barely 4 mm. long. Salt meadows: N.J.—Fla.—Tex.—Kans. —Iowa. *Allegh.—Austral.* Jl–S.

f 389

6. CÙPHEA P. Br. BLUE WAXWEED, WAXBUSH.

Herbs or shrubs, usually clammy. Leaves opposite or whorled, entire. Hypanthium elongate, saccate or spurred at the base, oblique at the mouth, many-ribbed. Sepals 6, with or without a tooth in each sinus. Petals 6, unequal, clawed. Stamens 11–12; filaments short. Ovary 2-celled; style filiform; stigma 2-lobed. Capsule oblong, thin-membranous. *Parsonsia* R. Br.

1. **C. petiolàta** (L.) Koehne. Annual; stem 1–7 dm. high, clammy-pubescent; leaf-blades lanceolate to ovate-lanceolate, 1–4.5 cm. long, ciliate, oblique at the base, petioled; hypanthium 10–15 mm. long, spurred; petals violet-purple, oval or obovate, rounded or retuse at the apex. *Parsonsia petiolata* Rusby [B]. Hillsides or dry soil: R.I.—Iowa—Kans.—La.—Ga. *Allegh.* Jl–O.

Family 120. **MELASTOMATACEAE.** MEADOW-BEAUTY FAMILY.

Perennial herbs, shrubs, or trees, with the nodes of the stem and branches often enlarged. Leaves alternate, without stipules, usually with three or more strong ribs. Flowers perfect, usually regular or nearly so. Hypanthium well developed, campanulate to tubular. Sepals 3–6, valvate. Petals 3–6, contorted in aestivation, often oblique. Stamens twice as many as the sepals, or rarely of the same number, fertile, or those opposite the petals smaller and sterile. Anthers appendaged, opening by pores. Gynoecium of 2 to many (mostly 4) united carpels. Ovary 2- to many-celled, usually 4-celled; styles and stigmas united. Fruit berry-like or a valvate capsule, many-seeded.

1. RHÉXIA L. MEADOW-BEAUTY.

Perennial herbs, with rootstocks, often tuber-bearing. Leaves opposite, 3–5-ribbed, usually toothed. Hypanthium urn-shaped, produced above the ovary. Sepals 4. Petals 4, oblique, deciduous. Stamens 8, anthers 1-celled, opening by a terminal pore. Capsule 4-celled, 4-valved, many-seeded.

Anthers short, oblong, not spurred at the base ; hypanthium glabrous.
                  1. *R. petiolata.*
Anthers elongate, linear, spurred at the base ; hypanthium glandu-lar-hispid.
                  2. *R. virginica.*

1. **R. petiolàta** Walt. Stem 2–6 dm. high, 4-angled, glabrous; leaf-blades 1–2 cm. long, broadly ovate to ovate-lanceolate, acute, bristly serrate; sepals triangular; petals 9–15 mm. long, violet-purple or pink, rounded or apiculate at the apex. *R. ciliosa* Michx. Sandy swamps: Md.—Iowa—La. —Fla. *Austral.* Je–Au.

2. **R. virgínica** L. *Fig. 390.* Stem 2–10 dm. high, bristly pubescent or glabrate, sharply 4-angled or 4-winged; leaf-blades oblong-elliptic to oval or lanceolate, 2–10 cm. long, bristly serrulate, 5-ribbed; sepals lanceolate; petals 11–17 mm. long, bright purple, rounded or retuse at the apex. Damp soil: Me.—Iowa—La.—Fla. *E. Temp.* Jl–S.

f 390

Family 121. ONAGRACEAE. EVENING-PRIMROSE FAMILY.

Herbs, or rarely shrubs, with simple, alternate or opposite leaves. Flowers perfect, axillary or in terminal racemes. Hypanthium often elongate, enclosing and adnate to the ovary. Sepals 2–6, usually 4. Petals 2–9, usually 4, convolute in the bud, rarely wanting. Stamens as many or twice as many as the sepals. Gynoecium of 1–6, usually 4, united carpels. Ovary 1–6-celled, inferior; styles united. Fruit capsular or nut-like.

Flowers 4-merous.
 Fruit a many-seeded capsule, opening by valves.
  Seeds with a tuft of silky hairs.
   Hypanthium not prolonged beyond the ovary ; flowers large.
                1. CHAMAENERION.
   Hypanthium somewhat prolonged beyond the ovary. 2. EPILOBIUM.
  Seeds without a tuft of silky hairs, naked or tuberculate.
   Hypanthium not produced beyond the ovary.
    Stamens 8–12, in two series.
     Sepals deciduous ; capsule linear, terete ; petals at least partly purple. 3. GAYOPHYTUM.
     Sepals persistent ; capsule angled or ribbed ; petals white or yellow.
                4. JUSSIAEA.
    Stamens 4, in a single series ; sepals persistent.
     Leaves opposite ; stem prostrate or floating. 5. ISNARDIA.
     Leaves alternate ; stem erect or ascending. 6. LUDWIGIA.
   Hypanthium prolonged beyond the ovary into a cylindric or funnelform tube.
    Stigma divided into 4 linear lobes.
     Stamens equal in length ; capsule terete or round-angled.

Ovules and seeds inserted in 2 or rarely more rows; petals
yellow. 7. OENOTHERA.
Ovules and seeds in one row; buds drooping; petals white or
pink. 8. ANOGRA.
Stamens unequal in length, the alternate longer; capsule crested or
winged.
Ovules and seeds numerous, with slender funiculi.
Corolla yellow; capsule depressed at the apex.
Stigma lobes linear; capsule winged. 9. KNEIFFIA.
Stigma-lobes very short, almost obsolete; capsule merely
ridged. 10. PENIOPHYLLUM.
Corolla red, purple, or white; capsule pointed.
11. HARTMANNIA.
Ovules and seeds in 1 or 2 rows, sessile, few.
Capsules with more or less distinct double crests on the an-
gles; seed furrowed along the raphe. 12. PACHYLOPHUS.
Capsules winged or at least sharply ridged on the angles.
Seeds in a double row without a crown or ridge at the
upper end; the capsule obpyramidal.
Plant acaulescent with the leaves crowded on the
crown; capsule winged above. 13. LAVAUXIA.
Plant caulescent, diffuse, stem wiry; capsule merely
sharp-angled. 14. GAURELLA.
Seeds in a single row with a membranous crown or a
mere ridge at the top, enclosing the base of the seed
above; body of the pod cylindric or fusiform, with
distinct wings at least at the middle.
15. MEGAPTERIUM.
Stigma discoid or capitate.
Stigma discoid; hypanthium-tube funnelform above.
Hypanthium-tube longer than the ovary; stigma entire.
16. GALPINSIA.
Hypanthium-tube shorter than the ovary; stigma 4-toothed.
17. MERIOLIX.
Stigma capitate, or nearly so. 10. PENIOPHYLLUM.
Fruit indehiscent nut-like.
Hypanthium-tube filiform; filaments unappendaged; ovary 1-celled.
18. STENOSIPHON.
Hypanthium-tube obconic; filaments with scales at the base; ovary 4-celled.
19. GAURA.
Flowers 2-merous; fruit indehiscent, obovoid, and bristly with hooked hairs.
20. CIRCAEA.

1. CHAMAENÈRION (Gesn.) Ludw. FIRE-WEED.

Perennial herbs, somewhat woody at the base. Leaves alternate, narrow,
entire. Flowers perfect, somewhat irregular, showy, in terminal racemes.
Hypanthium not prolonged beyond the ovary. Sepals 4, deciduous. Petals
4, purple, forming an oblique cross, entire. Stamens 8, with declined filaments.
Stigma 4-lobed. Capsule elongate, nearly linear, obtusely 4-angled, loculicidal.
Seeds with a tuft of hairs (coma) at the upper end.

Style pubescent at the base; leaves lanceolate or linear-lanceolate, with the lateral
veins confluent in marginal loops. 1. C. spicatum.
Style glabrous; lateral veins of the leaves obsolete, not looped. 2. C. latifolium.

1. **C. spicàtum** (Lam.) S. F. Gray. *Fig. 391.*
Stem 0.5–2.5 m. high, glabrate below, puberulent
above; leaves lanceolate or linear-lanceolate, 5–15
cm. long, paler beneath; petals purple, rose-colored,
or rarely white; style exceeding the stamens, hairy
at the base; capsule 5–7.5 cm. long. *Epilobium angustifolium β* L. *E. angustifolium* Auth. [G]; not
L. *C. angustifolium* Scop., in part [B]. Edges of
woods and copses, burnt-over ground, etc.: Greenl.
—N.C.—Ill.—N.M.,—Calif.—Alaska; Eurasia. *Temp.
—Arct.—Submont.—Mont.* Je S.

2. **C. latifòlium** (L.) Sweet. Perennial, with
a cespitose rootstock; stem 1–4 dm. high, glabrous
below; leaves 2–5 cm. long, ovate or ovate-lanceolate,
pale; inflorescence usually short; petals 1.5–3 cm.
long, rose-colored, pale purple, or white, purple-

f 391

veined; style shorter than the stamens, glabrous; capsule 5–8 cm. long. *Epilobium latifolium* L. Wet places, especially along mountain streams: Greenl.—Que.—S.D.—Colo.—Wash.—Alaska; Eurasia. *Subalp.—Alp.—Arct.* Je–Au.

2. EPILÒBIUM (Gesn.) L. WILLOW-HERB, COTTON-WEED.

Caulescent herbs, rarely shrubby. Leaves alternate or opposite, with often toothed blades. Flowers perfect, racemose or spicate, or rarely solitary. Hypanthium prolonged beyond the ovary into an obconic short tube. Sepals and petals 4, the latter often notched, purplish, pink, or white, rarely yellow. Stamen 8, not declined. Stigma club-shaped, subentire, slightly notched, or rarely 4-lobed. Capsule elongate, subcylindric, slightly fusiform or clavate, 4-celled, 4-sided, loculicidal. Seeds with a tuft of hairs (coma) at the upper end.

Perennials; stigma entire or merely notched.
  Leaves oblong, oval, ovate, or lanceolate.
    Plants with rosettes or turions; leaves ovate or lanceolate, usually broadest
      below the middle, and distinctly denticulate or dentate.
      Flowers 7–8 mm. long; petals purple or dark pink; leaves ovate-lance-
        olate.   **1. *E. occidentale.***
      Flowers 3–5 mm. long.
        Leaves all, except the uppermost, short-peti-
          oled.
          Leaves narrowly lanceolate, pale green;
            corolla white.   **2. *E. americanum.***
          Leaves lanceolate, not pale; corolla usu-
            ally pink.
            Seeds abruptly contracted above, with-
              out a neck; coma cinnamon-colored
              when mature; leaves elongate-lance-
              olate, closely serrulate.   **3. *E. coloratum.***
            Seeds gradually tapering into a neck;
              coma white; leaves oblong-lance-
              olate to ovate, more distantly ser-
              rate.   **4. *E. adenocaulon.***
        Leaves all sessile or only the very earliest
          sometimes short-petioled.
          Leaf-blades ample, ovate or broadly lance-
            olate; plant 1–2 dm. high.   **5. *E. saximontanum.***
          Leaf-blades narrowly lanceolate, almost
            erect; plant slender, 2–4 dm. high.   **6. *E. Drummondii.***
    Plants with stolons or soboles. low, 1–2 (seldom 3)
      dm. high; leaf-blades oval or oblong, indis-
      tinctly denticulate.
      Petals white.   **7. *E. alpinum.***
      Petals purple or pink.   **8. *E. Hornemannii***
  Leaves narrow, linear or linear-oblong, entire or mi-
    nutely denticulate; capsule pubescent.
    Stem and pods densely short-pilose, with straight
      hairs.   **9. *E. strictum.***
    Stem and pods cinereous, strigose or crisp-hairy, or
      glabrate.
      Leaves and lower part of the stem glabrous.
        Leaves thick, oblanceolate, obtuse.   **10. *E. oliganthum.***
        Leaves thin, linear or lanceolate, acute.   **11. *E. wyomingense.***
      Leaves and stem crisp-hairy.   **12. *E. lineare.***
Annuals, with more or less sheddy, straw-colored bark;
  stigma 4-cleft.
  Capsules and pedicels glabrous or sparingly puberulent.   **13. *E. paniculatum.***
  Capsules and pedicels glandular-pubescent; pedicels very
    short.   **14. *E. adenocladon.***

1. E. occidentàle (Trelease) Rydb. Stem 3–10 dm. high, somewhat angled, glabrous below, crisp and glandular-pubescent above; lower leaves short-petioled, the upper subsessile; blades ovate to triangular-lanceolate, 3–7 cm. long; petals purple or rose, 5–6 mm. long; capsule 4–6 cm. long, more or less pubescent; seeds beakless; coma white. *E. adenocaulon occidentale* Trelease. Wet places: Alta.—S.D.—Colo.—Calif.—B.C. *W. Temp.—Submont.—Mont.* Jl–Au.

2. E. americànum Haussk. Stem 3–6 dm. high, slightly angled, glabrous below, slightly crisp-hairy above; leaves short-petioled, 2–4 cm. long, glabrous and pale, denticulate or subentire; petals about 4 mm. long, white; capsules glabrous or sparingly pilose when young, 3–5 cm. long; coma white. *E. adeno-caulon perplexans* Rydb.; scarcely Trelease. Wet places: Que.—Sask.—N.D.—Wyo.—Ida.—B.C. *Boreal.* Je–Au.

3. **E. colorátum** Muhl. Stem 3–10 dm. high, much branched, glabrous below, crisp-hairy in lines above; leaves 5–15 cm. long, short-petioled; petals pink, 3–5 mm. long; capsule finely pubescent; seeds minutely papillose. Low ground: Me.—Man.—S.D.—Kans.—S.C. *Canad.--Allegh.* Jl–S.

4. **E. adenocaúlon** Haussk. *Fig. 392.* Stem erect, 3–10 dm. long, somewhat angled, glabrous below, crisp and glandular in the inflorescence; leaves glabrous or the upper puberulent, ovate-lanceolate or elliptic-lanceolate, 3–6 cm. long; petals about 4 mm. long; capsules 4–5 cm. long, glabrate in age; coma white or nearly so. Wet places: N.B.--Pa.—Colo.—Nev.—Yukon. *Boreal.* -*Plain—Mont.* Je–Au.

f 392.

5. **E. saximontànum** Haussk. Perennial, with turions; stem simple, ascending or suberect, 1–2 dm. (rarely 3 dm.) high; leaves pale green, glabrous, 1–3 cm. long, the lower ovate, entire, the upper ovate-lanceolate, denticulate; petals 3–5 mm. long; capsules glabrous, about 4 cm. long, short-pedicelled; seeds acute at each end. Near springs: Alta.—S.D.—Wyo.—B.C. *Submont.—Subalp.* Jl–Au.

6. **E. Drummóndii** Haussk. Perennial, with turions; stem strict, erect, 2–4 dm. high, pulv, glabrous; leaves pale, 2–4 cm. long, repand-dentate; petals 4 mm. long; capsule 3–5 cm. long, slender-pedicelled; seeds acute at each end, minutely apiculate; coma dingy. Wet places, Sask.—S.D.—Colo.—Utah—B.C. *Submont.—Mont.* Je–Au.

7. **E. alpinum** L. Stem ascending or decumbent at the base, 1–3 dm. high, glabrous; leaves elliptic, rather obtuse, 2–5 cm. long, glabrous, the upper often alternate; petals about 3 mm. long; capsules slender, about 5 cm. long, with long pedicels; seeds smooth, beaked. *E. lactiflorum* Haussk. *E. canadense* Lev., a somewhat puberulent form. Wet places in the mountains and in arctic regions: Greenl.—N.H.—S.D.—Colo.--Calif.—Alaska; n Eur. and ne Asia. *Boreal.—Mont.—Alp.*

8. **E. Hornemánnii** Reichenb. Stem 1–3 dm. high, ascending, somewhat crisp-hairy on the decurrent lines and in the inflorescence; leaves 2–4 cm. long, elliptic or oval, mostly obtuse; flowers few, erect; petals 5–7 mm. long; capsule 5 cm. long, on slender pedicels; seeds usually rough, short-apiculate; coma dingy. Wet places on the mountain sides and in arctic regions: Greenl.—N.H. S.D.—Colo.—Calif.—Alaska; n Eurasia. *Boreal.—Mont—Subalp.* Je–Au.

9. **E. stríctum** Muhl. Stem erect, 2–15 cm. high, simple or branched above, with erect branches, velvety; leaves narrowly lanceolate to linear, with revolute margins; petals pink, 7–8 mm. long; capsule 5–7.5 cm. long, short-pedicelled; seeds minutely papillose; coma dingy. *E. molle* Torr. [G]; not Lam. Bogs and wet meadows: Que.—Man.—Kans.--Va. *Canad.—Allegh.* Jl–S.

10. **B. oligánthum** Michx. Stem 1–5 dm. high, mostly simple, minutely pubescent above; leaves thick, linear-oblanceolate, obtuse, 1–3 cm. long, ascending; flowers few; petals white or pinkish, 5–6 mm. long; capsule 2.5–5 cm. long, slightly pubescent or glabrous; coma white. *E. palustre monticolu* Haussk. [G]. Bogs: Newf.—Man.—Wis.—Pa.

11. **E. wyomingénse** A. Nels. Stem slender, erect, branched, glabrous below, minutely puberulent above; leaves thin, 3–5 cm. long; midrib evident,

but lateral veins obscure; petals 3–4 mm. long; capsules 4–7 cm. long, minutely cinereous; seeds fusiform, scarcely beaked; coma white. *E. palustre albiflorum* Lehm. In bogs and along streams: Sask.—S.D.—Colo.—Utah—Yukon. *Plain* —*Submont.* Jl–Au.

12. E. lineàre Muhl. Stem erect, branched, 2–4 dm. high; leaves linear, short-pubescent, with crisp hairs, revolute on the margins, 2–5 cm. long; flowers numerous; petals about 4 mm. long; capsule 4–5 cm. long, canescent, often slender-pedicelled; seeds fusiform, short-apiculate; coma dingy. *? E. densum* Raf. [G]. Swamps: N.B.—Del.—Colo.—Ida.—B.C. *Plain—Submont.* Jl–S.

13. E. paniculàtum Nutt. Stem branched, 4–8 dm. high, glabrous, straw-colored; leaves linear or linear-lanceolate, 3–5 cm. long; pedicels longer than the bracts; tube of hypanthium 2–3 mm. long; calyx-lobes about as long; petals lilac or rose-colored, 5–7 mm. long; pods clavate, about 2 cm. long; seeds brown, 2 mm. long; coma tawny. Sandy soil: Man.—S.D.—Colo.—Calif.— B.C. *Plain—Mont.* Jl–Au.

14. E. adenócladon (Haussk.) Rydb. Stem rigid, 4–8 dm. high, glabrous and straw-colored below, the smaller branches and pedicels densely glandular; leaves linear or linear-lanceolate, mostly entire, 2–4 cm. long; tube of the hypanthium about 3 mm. (rarely 4 mm.) long; calyx-lobes 2–3 mm. long; petals purple or rose, 5–7 mm. long; capsule clavate, 1.5–2.5 cm. long; seeds blackish-brown, obovoid, 2 mm. long; coma tawny. *E. paniculatum adenocladon* Haussk. Sandy soil: S.D.—Colo.—Utah—Mont.; Que. *Plain—Submont.*

3. GAYÓPHYTUM A. Juss. BABY'S BREATH.

Caulescent annuals. Leaves very narrow, linear, entire. Flowers axillary, minute, perfect. Hypanthium not produced above the ovary. Sepals 4, reflexed, deciduous. Petals 4, small, obovate or spatulate, white or rose-colored. Stamens 8, those opposite the petals shorter and with sterile anthers. Stigmas capitate or clavate. Capsule clavate or linear, 2-celled, 4-valved. Seeds without a coma.

Capsules torulose, less than three times as long as the pedicels, usually more or less clavate.
  Capsules decidedly clavate, rounded at the apex, seldom longer than the pedicels, spreading or reflexed.   1. *G. ramosissimum.*
  Capsules only slightly if at all clavate, narrow, usually longer than the pedicels, and erect.   2. *G. Nuttallii.*
Capsules neither torulose nor clavate; pedicel very short.   3. *G. racemosum.*

1. G. ramosíssimum T. & G. Stem diffusely branched, 3–6 dm. high; leaves linear, 2–4 cm. long; petals white or rose-colored; capsule 3–8 mm. long, decidedly torulose, spreading or usually reflexed. Sandy places and on hillsides: Mont.—S.D.—Colo.—Ariz.—Calif.—B.C. *Submont.—Mont.* My–Au.

2. G. Nuttállii T. & G. Stem 2–5 dm. high, glabrous, branched; leaves 1–3 cm. long, acute; capsule linear, 7–15 mm. long, somewhat torulose; pedicels 2–4 mm. long. Sandy or dry soil, hillsides, etc.: S.D.—Colo.—Ariz.—Calif.— Wash. *Submont.* Jl–S.

3. G. racemòsum T. & G. Stem 1–3 dm. high, more or less branched, glabrous or slightly strigose above; leaves linear, 2–4 cm. long, usually glabrous; sepals and petals mostly less than 1 mm. long; capsule narrowly linear, 1–1.5 cm. long, glabrous, erect, almost sessile. Sandy soil: S.D.—Colo.—Calif.— Wash. *Submont.—Subalp.* Je–Au.

4. JUSSIAÈA L. PRIMROSE-WILLOW.

Perennial herbs or shrubs, with mostly alternate entire leaves. Flowers perfect, regular, axillary. Hypanthium not produced beyond the ovary. Sepals 4–6, acute, persistent. Petals 4–6, white or yellow. Stamens 8–12, in two series. Ovary 4–6-celled; capsule linear, oblong, or clavate, narrowed at the base, angled or ribbed, septicidal or irregularly splitting.

1. J. diffúsa Forsk. Stem creeping or floating, glabrous, 3–10 dm. long; leaves oblong to obovate, 2.5–10 cm. long, slender-petioled; flowers long-peduncled; sepals and petals 5; corolla yellow, 2–2.5 cm. broad; stamens 10; capsule woody, 2.5–4 cm. long, cylindraceous; seeds quadrate. Shallow water and muddy banks: Ky.—Kans.—Tex.—Fla.; Trop. Am. and Asia. *Austral—Trop.* Je-Au.

5. ISNÁRDIA L. MARSH PURSLANE.

Succulent caulescent herbs, with creeping or floating stems. Leaves few, petioled, opposite. Flowers perfect, axillary, sessile. Hypanthium not produced beyond the ovary. Sepals 4, erect and persistent. Petals 4, small, reddish, or wanting. Stamens 4, with short filaments. Capsule short, truncate at the apex, mostly obpyramidal or obovoid. Seeds without a coma.

1. I. palústris L. Stem branching, 1–5 dm. long; leaf-blades oval, ovate, or spatulate, 1–2.5 cm. long, slender-petioled; flowers solitary; bractlets none; sepals triangular, acute; capsule 4-sided, slightly longer than broad, 3–4 mm. long. *Ludwigia palustris* Ell. [G]. Muddy places and shallow water: N.S.—Fla.—Calif.—Ore.; Mex. and the Old World. *Temp.—Trop.*

6. LUDWÍGIA L. FALSE LOOSESTRIFE.

Perennial (ours) or annual herbs, with alternate, sessile or subsessile, entire leaves. Flowers axillary, perfect, regular. Hypanthium not produced above the ovary, tapering at the base. Sepals 4, persistent. Petals 4, or wanting. Ovary 4- or 5-celled. Capsule short, terete, ribbed or winged, septicidal or opening by terminal pores.

Flowers peduncled; petals ample, yellow, equaling or longer than the sepals; capsule opening by terminal pores.   1. *L. alternifolia.*
Flowers sessile; petals small, greenish, or wanting; capsule septicidal, the valves separating from the disk-like base of the style.
    Capsule campanulate or slightly turbinate, nearly as broad as long; bractlets linear, nearly equaling the calyx.   2. *L. polycarpa.*
    Capsule short-cylindric, twice as long as broad; bractlets minute or wanting.   3. *L. glandulosa.*

f 393.

1. L. alternifòlia L. *Fig. 393.* Stem glabrous or puberulent, 5–12 dm. high, branching; leaves lanceolate to lance-linear, 5–12 cm. long, acute at each end, short-petioled; flowers short-peduncled; petals lanceolate, foliaceous; capsule cubic above, rounded at the base, wing-angled, 5–7 mm. high. Swamps: N.H.—Minn.—Neb.—Tex.—Fla. *E. Temp.* Je-S. '

2. L. polycárpa Short & Peter. Stem 3–10 dm. high, glabrous, more or less winged; leaves narrowly lanceolate, acute at each end, 5–10 cm. long, sessile; bractlets linear-subulate, adnate to the hypanthium; sepals triangular-lanceolate, often serrulate; capsule glabrous, 4-sided, 5 mm. high, longer than the sepals. Swamps: Ont.—Minn.—Kans.—Tenn. *Allegh.* Jl–O.

3. L. glandulòsa Walt. Stem glabrous, 3–10 dm. high; leaves alternate, linear-lanceolate, acute at each end, 5–10 cm. long, short-petioled or subsessile; flowers 1 or 2 in each axil; sepals deltoid, acute; petals wanting; capsule 4-grooved, 6–8 mm. long, glabrous, several times as long as the sepals. Swamps: Va.—Ill.—Kans.—Tex.—Fla. *Austral.* Jl–S.

7. OENOTHÈRA L. EVENING PRIMROSE.

Biennial or annual stout herbs, with taproots. Leaves alternate, undulate or toothed. Flowers in terminal spikes, nocturnal, from erect buds. Hypanthium-tube above the ovary, funnelform. Sepals 4, reflexed. Petals 4, usually obcordate, yellow, or in age turning purplish. Stigmas with 4 linear lobes. Stamens 8. Capsule usually tapering upwards, 4-celled, loculicidal, 4-valved. Seeds in two or more rows, horizontal. *Onagra* Adans.

Pods lance-cylindric or lance-prismatic, tapering from the thick base; seeds horizontal.
  Free tips of the sepals 3 mm. long or longer.
    Petals obovate.
      Spike not conspicuously leafy-bracted; bracts lance-linear, none except the very lowest ones exceeding the hypanthium-tube or the fruit in length.

Petals 3–4 cm. long, usually turning pinkish in drying.	1. *O. Hookeri.*
Petals 1.5–2.5 cm. long, not turning pinkish.	2. *O. biennis.*

      Spike conspicuously leafy-bracted; bract, even in the upper ones, exceeding the hypanthium-tube or the fruit in length.

Petals linear.	3. *O. muricata.*
	5. *O. cruciata.*
Free tips of the sepals about 2 mm. long.	4. *O. strigosa.*

Pods linear-cylindric, of uniform diameter throughout; seeds ascending.
  Flowers in a terminal spike, longer than the reduced subtending leaves; petals rhombic-ovate.

	6. *O. rhombipetala.*

  Flowers axillary, shorter than the subtending leaves which are not much reduced; petals obovate.

Petals 5–12 mm. long.	7. *O. laciniata.*
Petals 25–50 mm. long.	8. *O. grandis.*

1. O. Hoòkeri T. & G. Biennial, with a taproot; stem 5–10 dm. high, erect, angled, more or less hirsute and canescent; leaves lanceolate, repand-dentate, hirsute and canescent, 5–20 cm. long; free tube of the hypanthium 3.5–4 cm. long; sepals about 3 cm. long, more or less hirsute; petals 3–4 cm. long; capsules 4–5 cm. long, hirsute and more or less canescent. *Onagra Macbridei* A. Nels. Valleys: Mont.—Kans.—N.M.—Calif.—B.C.; n Mex. *W. Temp.— Plain—Submont.—Mont.* Je–S.

2. O. biénnis L. Biennial, with a taproot; stem 3–25 dm. high, more or less hirsute; leaves oblong to lanceolate, 2.5–15 cm. long, acute or acuminate, repand; sepals longer than the hypanthium; petals 1.5–2.5 cm. long; capsule pubescent, 2–2.5 cm. long. *Onagra biennis* (L.) Scop. [B]. Dry soil: Labr.—Man.—Kans.—La.—Fla.; naturalized in the Old World. *E. Temp.* Je–O.

3. O. muricàta L. Biennial, with a taproot; stem erect, 2–8 dm. high, very leafy, puberulent and usually hirsute, the hairs with reddish pustulate bases; leaves lanceolate, ascending, entire or sinuately denticulate, more or less pubescent; inflorescence dense; sepals 15–18 mm. long; petals light yellow, 12–15 mm., rarely 2 cm. long; capsule 2.5–3 cm. long, more or less hirsute. Sandy places: Newf.—N.Y.—Colo.—Mont. *Boreal.—Plain—Submont.* Jl–S.

4. O. strigòsa (Rydb.) Mack. & Bush. *Fig. 394.* Biennial or annual; stem erect, 3–10 dm. high, grayish-strigose and hirsute; lower leaves spatulate, the others lanceolate, 5–10 cm. long, more or less repand-dentate; inflorescence leafy; sepals 1.5–2 cm. long, grayish-hirsute; capsule 2.5–3 cm. long, grayish-hirsute. *O. biennis canescens* T. & G. *O. muricata canescens* B. L. Robinson [G]. *Onagra strigosa* Rydb. [B]. Valleys and sandy places: Minn.—Kans.—Utah—Wash.—B.C. *W. Temp.— Plain—Submont.* Je–Au.

f 394

5. O. cruciàta Nutt. Stem 3–8 dm. high, reddish, somewhat strigose or glabrous; leaves lanceolate, remotely denticulate; appendages of the sepals 3 mm. long, inserted below the tip of the sepals; petals linear, 5–12 mm. long, 1–3 mm. wide, light yellow; capsule 3 mm. long, villous. Sandy soil: Me.—N.Y.—Minn. *Canad.* Jl–O.

6. O. rhombipétala Nutt. Biennial; stem stout, erect, 5–12 dm. high, finely and densely pubescent; leaves lanceolate or the lower oblanceolate, 5–10 cm. long, the lower petioled, the upper sessile, acuminate, remotely denticulate; bracts linear-lanceolate, about equaling the hypanthium; sepals closely strigose, about 2 cm. long; petals 2–3 cm. long; capsule curved, 12–16 mm. long, closely pubescent. *Raimannia rhombipetala* Rose [B]. Sandy prairies: Ind.—S.D.—Tex. *Prairie—Plain.* Jc–Jl.

7. O. laciniàta Hill. Annual or more enduring, at first simple, later branched near the base; branches decumbent, 1–5 dm. long, sparingly hirsute; leaves oval or lanceolate, or the lower oblanceolate or spatulate, 2.5–5 cm. long, acute or obtuse, sinuately dentate to pinnatifid; bracts similar to the leaves; hypanthium 3–4 cm. long; sepals reflexed, petals yellow, turning rose; capsule 2.5–3 cm. long, pubescent. *O. sinuata* L. *R. laciniata* Rose [BB]. Dry sandy soil: Vt.—S.D. Tex.—N.J.; Mex. *E. Temp.—Trop.* My Je.

8. O. grándis (Britton) Rydb. Biennial, similar to the preceding in habit; hypanthium 5–7 cm. long; sepals 2–3.5 cm. long; capsule 2–3 cm. long. *O. laciniata grandis* Britton [B], v. *grandiflora* (B. Wats.) B. L. Robinson [G]. *R. grandis* Rose [BB]. Sandy soil: Mo.—Kans.—Tex. *Ozark.—Texan.* My Jl

8. ÁNOGRA Spach. WHITE EVENING PRIMROSE.

Annual or perennial caulescent herbs, often with white flaky bark. Leaves alternate, usually pinnatifid or toothed. Flowers axillary, usually diurnal, from drooping buds. Hypanthium produced beyond the ovary into a cylindric tube. Sepals 4, narrow, reflexed. Petals white, turning pink, obcordate, showy. Stigma with 4 linear lobes. Capsule cylindric, slightly angled, loculicidal, 4-valved, many-seeded. Seeds terete in one row in each cavity of the ovary.

Calyx in bud merely acutish; tips not free.
  Pubescence, at least in part, consisting of long silky hairs.
    Branches erect; petals 1.5–2 cm. long.   1. *A. perplexa.*
    Branches decumbent or merely ascending; petals 2.0–4 cm. long.   2. *A. albicaulis.*
  Pubescence wholly cinereous-strigose.   3. *A. Bradburiana.*
Calyx in bud acuminate or acute; tips free.
  Capsule linear-cylindric; throat of the corolla glabrous.
    Calyx sparingly long-hairy, glandular-puberulent, or glabrous, not strigose.
      Capsule strongly ascending, straight; leaves linear or lance-linear, entire or nearly so, strigose beneath.   4. *A. Nuttallii.*
      Capsules divergent, usually curved upwards; leaves usually denticulate or sometimes lobed, pubescent on both sides.   5. *A. cinerea.*
    Calyx and hypanthium densely grayish-strigose; leaves cinereous.   6. *A. latifolia.*
  Capsules oblong; throat of the corolla hairy; leaves deeply pinnatifid.   7. *A. coronopifolia.*

1. A. perpléxa Rydb. Biennial; stem 1–2 dm. high, more or less cinereous and with scattered hairs, branched at the base, with erect branches; leaves 2–6 dm. long, oblanceolate in outline, hirsute and somewhat canescent, the basal ones dentate, the cauline ones strongly ascending, deeply pinnatifid, with narrowly linear, more or less ascending lobes; tube of the hypanthium 1.5–2 cm. long, hirsute; calyx-lobes 1.5–1.75 cm. long, hirsute; capsule about 3 cm. long, and fully 3 mm. thick, erect. Plains: Neb.—Kans.—Colo.—Wyo. *Plain—Submont.* Je–Jl.

2. A. albicaùlis (Pursh) Britton. Annual or biennial; stem 1–3 dm. long, branched at the base; basal leaves spatulate, obtuse, merely toothed; stem-leaves 5–12 cm. long, deeply pinnatifid, with lanceolate or linear, rather divergent divisions; tube of hypanthium about 3 cm. long, sparingly pubescent;

calyx-lobes about 3 cm. long, sparingly silky-hirsute or sometimes glabrous; pod 3.5–5 cm. long, about 3 mm. thick. *Oenothera pinnatifida* Nutt. *O. albicaulis* Pursh. Hillsides and in sandy soil: Man.—S.D.—Tex.—Ariz.—Mont.; n Mex. *Plain—Submont.* My–Jl.

3. **A. Bradburiàna** (Nutt.) Rydb. Annual or biennial; stem erect, 1–2 dm. high, finely cinereous-strigose; leaves 3–5 cm. long, oblanceolate in outline; the basal ones subentire or dentate, the cauline ones deeply pinnatifid, cinereous-strigose; tube of hypanthium and calyx-lobes cinereous, each nearly 2 cm. long; petals fully 2 cm. long. *O. Bradburiana* Nutt. Sandy places: Wyo.— S.D.—Utah. *Plain.* Jl.

4. **A. Nuttállii** (Sweet) A. Nels. Perennial; stem 3–10 dm. high; leaves 5–10 cm. long; tube of hypanthium glabrous or sparingly glandular-puberulent, 3–4 cm. long; calyx-lobes 1.5–2 cm. long, glabrous or sparingly hirsute; free tips 3 mm. long; capsule 3 cm. long; petals 2–2.5 cm. long. *A. pallida* Britton [B]. *O. pallida* Robins. & Fern. [G]; not Lindl. *Oenothera albicaulis* Nutt. In sandy soil: Man.—Wis.—Colo.—Ida.—B.C. *Prairie—Plain—Mont.* Je–Au.

5. **A. cinèrea** Rydb. Branched perennial; stem 3–4 dm. high, cinereous-strigose when young; leaves lanceolate to ovate-lanceolate, mostly subsessile, 3–5 cm. long, sinuate-dentate or denticulate, cinereous; tips of the calyx free and rather long; petals 15–18 mm. long; pods 3–3.5 cm. long, 3 mm. thick, almost straight. Dry soil: S.D.—w Neb.—Ida.—Colo. *Plain —Submont.* Je–Jl.

6. **A. latifòlia** Rydb. *Fig. 395.* Perennial, whole plants strigose-canescent; stem divaricately branched, 2–6 dm. high; leaves ovate, oblong or lanceolate, sinuate-dentate, 2–6 cm. long; tube of hypanthium and calyx-lobes about 2 cm. long; free tips 1.2–2 mm. long; petals nearly 2 cm. long; capsule divaricate, usually curved, 3 cm. long, 3 mm. thick. Sandy soil; Neb. — Kans. — Utah — Ida. *Prairie—Plain—Mont.*

f.395.

7. **A. coronopifòlia** (T. & G.) Britton. Perennial, with a rootstock; stem 1–3 dm. high, strigose; leaves 1–6 cm. long, pinnatifid, with linear divisions, or the basal ones spatulate and merely dentate; tube of hypanthium 1–2 cm. long, strigose; calyx-lobes 1.5–2 cm. long; free tips less than 1 mm. long; petals 1–1.5 cm. long; capsule oblong, 1–1.5 cm. long, 4 mm. thick. *O. coronopifolia* T. & G. Prairies and plains: S.D.—Kans.—N.M.—Ariz.—Ida. *Prairie—Plain—Mont.* Je–Au.

9. KNEÌFFIA Spach. SUNDROPS.

Annual or perennial, caulescent herbs, often shrubby at the base. Leaves alternate, entire or toothed. Buds erect. Flowers perfect, regular, in terminal spikes. Sepals 4, narrow, reflexed. Petals 4, yellow, obovate. Stigma with 4 linear lobes. Ovary club-shaped, 4-angled. Ovules numerous, in many rows, slender-stalked. Capsule more or less club-shaped, or oblong with a tapering base, 4-winged. Seeds not angled, without a tubercle.

Capsule linear-clavate, subsessile; petals 20–35 mm. long.	1. *K. pratensis.*
Capsule more or less obovoid-clavate, distinctly stalked; petals 6–15 mm. long.	
Capsule crisp-hairy, glandless; petals of the earlier flowers 10–15 mm. long.	2. *K. Spachiana.*
Capsule sparingly glandular-hirsute; petals of the earlier flowers 5–10 mm. long.	3. *K. perennis.*

1. K. praténsis Small. Stem erect, 5–10 dm. high, hirsute; blades narrowly elliptic or lance-elliptic, 3–10 cm. long, undulate or sinuate, hirsute; sepals 17–22 mm. long, the free tips 2.5–4 mm. long; petals 2–3.5 cm. long; capsule 18–21 mm. long, hirsute. *Oenothera pratensis* (Small) B. L. Robinson [G]. Low ground: Ohio—Iowa—Ark. *Prairie* —*Ozark*.

f 396

2. K. Spachiàna (T. & G.) Small. Stem minutely pubescent or somewhat hirsute, erect, 1–4 dm. high, branched; basal leaves spatulate or oblong-spatulate; stem-leaves linear, 2–6 cm. long, undulate, petioled; sepals about 1 cm. long; petals 9–15 mm. long; capsule-body 1–1.5 cm. long. *Oenothera Spachiana* T. & G. Sandy soil: Ark.—Kans.— Tex.—La. *Austral.*

3. K. perénnis (L.) Pennell. *Fig. 396.* Stem finely puberulent or glabrate, erect, 2–6 dm. high; basal leaves oblanceolate or spatulate; stem-leaves elliptic, oblong, or linear, 2–8 cm. long, entire or very nearly so; sepals 5–7 mm. long; petals 6–9 mm. long; capsule-body 6–12 mm. long, glabrate. *O. perennis* L. *O. pumila* L. [G]. *K. pumila* Spach. [B]. Dry soil: N.S.— Man.—Kans.—Ga. *Canad.—Allegh.* Je–Au.

10. PENIOPHÝLLUM. Pennell.

Slender annual herbs. Leaves alternate. Flowers perfect, in terminal spikes. Hypanthium produced beyond the ovary. Sepals 4, partly cohering, in two groups, reflexed. Petals triangular obovate. Stamens 8; filaments unequal in length. Stigma broad, slightly 4-lobed. Capsule ellipsoid, 4-angled, not stalked. Seeds angled.

1. P. linifòlium (Nutt.) Pennell. Stem slender, 1.5–4.5 dm. high, finely pubescent; basal leaves petioled, ovate or spatulate, the blades 2–5 cm. long; stem-leaves filiform, 1–3 cm. long; bracts deltoid-ovate; petals 4 mm. long; capsule 4–6 mm. long. *Oenothera linifolia* Nutt. [G]. *Kneiffia linifolia* Spach. [B]. Dry soil: Ill.—Kans.—Tex.—La. *Ozark—Austral.* My–Jl.

11. HARTMÁNNIA Spach. PRIMROSE.

Annual or perennial herbs, with leafy stems. Leaves alternate, usually lyrate or pinnatifid. Flowers perfect, regular, diurnal, in terminal spikes or racemes, drooping in bud. Hypanthium funnelform, produced beyond the ovary. Sepals 4, reflexed. Petals 4, white, red, or purple. Ovary elongate, the ovules numerous, on slender stalks. Stigma with 4 linear lobes. Capsule short, clavate, oblong or ellipsoid, 4-winged, often narrowed at the base. Seeds tubercled.

1. H. speciòsa (Nutt.) Small. Stem stout, canescent-strigose, ascending or decumbent, 2–7 dm. high; leaves lanceolate or oblong, rarely linear, 2–12 cm. long, distantly toothed or pinnatifid; spike lax; sepals 2–4 cm. long; petals white, turning pink, 2.5–5 cm. long, retuse at the apex; capsule oblong, 1–2 cm. long, with 4 stout wings. *Oenothera speciosa* Nutt. [G]. Plains and prairies: Ill.—Neb.—N.M.—Tex.—La.; Mex. *Prairie—Tex.—Trop.* My–Je.

12. PACHÝLOPHUS Spach. ROCK ROSE, MOUNTAIN PRIMROSE.

Acaulescent or short-stemmed perennials. Leaves mostly crowded, toothed, lobed, or pinnatifid. Flowers axillary, perfect, night-blooming. Hypanthium much elongated beyond the ovary into a narrowly funnelform tube. Sepals 4, elongate, reflexed. Petals white, turning pink, showy. Stamens 8. Stigma with 4 linear lobes. Capsule lance ovoid or ovoid-oblong, large, 4-angled, with more or less wavy-crested angles, woody, tardily loculicidal. Seeds many, in 1 or 2 rows in each cell.

Plant glabrous, or nearly so.   1. *P. caespitosus.*
Plant hairy.
  Plant not canescent-strigose.
    Hypanthium, calyx, and fruit glabrous ; slightly strigose
      or with a few scattered long hairs ; pubescence short
      and usually appressed.   2. *P. montanus.*
    Hypanthium, calyx, and fruit densely hirsute ; plant more
      or less caulescent ; ridges of the fruit with lobed, more
      or less foliaceous crests.   3. *P. eximius.*
  Plant densely canescent-strigose throughout.   4. *P. canescens.*

    1. P. caespitòsus (Nutt.) Raim. Cespitose
acaulescent perennial; leaves with winged petioles,
1–2 dm. long; blades oblanceolate, glabrous, sinuate-
dentate, with triangular teeth; tube of the hypanthium 5–10 cm. long, glabrous; petals 3–4 cm. long;
fruit lance-ovoid, 3 cm. long, with low rounded tubercles on the angles. *Oenothera caespitosa* Nutt.
P. glabra A. Nels., a depauperate form. Dry hills:
Man.—Neb.—N.M.—Utah—Mont. *Plain—Mont.*—
Je–Au.

    2. P. montànus (Nutt.) A. Nels. *Fig. 397.*
Cespitose, acaulescent perennial; leaves short-petioled, 5–10 cm. long; leaf-blades oblanceolate,
coarsely sinuate-toothed, canescent-pubescent on the
margins, otherwise glabrous or sparingly strigose;
petals 2–3 cm. long; capsule sessile, ovoid, obtusely
tubercled on the blunt angles, 2 cm. long. *O. montana* Nutt. Dry hills: Sask.
—Neb.—Colo.—Nev.—Ida. *Plain—Submont.* My–Au.

    3. P. exímius (A. Gray) Woot. & Standl. Caulescent perennial; stem
stout, 1–3 dm. high, leafy; leaves oblanceolate or spatulate in outline, 1–2 dm.
long, more or less hirsute, especially on the margins and veins, sinuate-dentate
or lobed; tube of hypanthium 1–1.5 dm. long, hirsute; petals 2.5–4 cm. long;
capsule 3–4 cm. long, lance-ovoid. *O. eximia* A. Gray. *P. exiguus* Rydb. (Fl.
Colo., misprint). Hills and banks: S.D.—N.M. *Plain—Submont.* Je–Au.

    4. P. canéscens Piper. Acaulescent cespitose perennial, canescent with
fine appressed pubescence; leaves pale green, oblanceolate, sinuate-dentate or
subentire; tube of hypanthium 5–8 cm. long, canescent; petals 2–4 cm. long;
capsule 2–3 cm. long, lance-ovoid, canescent, strongly tubercled. Dry hills and
bench-lands: Mont.—w Neb.—Calif.—Wash. *Son.—Submont.* Je–Jl.

<h3 align="center">13. LAVAÙXIA Spach.</h3>

    Perennial or annual, mostly acaulescent herbs. Leaves mostly basal, numer-
ous, pinnatifid or lobed. Flowers few, axillary. Sepals narrow, reflexed. Petals
white, pink, or yellow. Stamens 8. Hypanthium produced beyond the ovary
into a slightly funnelform tube. Stigmas with 4 linear lobes. Capsule short,
sharply 4-angled or 4-winged above, few-seeded.

Annuals ; petals less than 1 cm. long ; flowers scarcely opening,
  fertilized in the buds.   1. *L. Watsonii.*
Biennials or perennials ; petals more than 1 cm. long ; flowers expanding normally.
  Petals yellow, turning pink, entire.   2. *L. flava.*
  Petals white, turning rose-colored, often 3-lobed.   3. *L. triloba.*

    1. L. Watsònii (Britt.) Small. Small acaulescent annual; leaves thin,
oblanceolate, sinuately toothed or lobed, rarely pinnatifid, 5–10 cm. long; flowers rarely fully opening, less than 5 cm. long; capsule often forming dense
clumps, 1–1.5 cm. long, about as broad as long, winged above. *O. triloba parviflora* S. Wats. [G]. *L. triloba Watsonii* Britt. [B]. Plains: Neb.—Okla.
Plain.

2. **L. flàva** A. Nels. *Fig. 398.* Acaulescent perennial; leaves oblong-lanceolate in outline, deeply runcinate-pinnatifid, 15–25 cm. long, glabrous or a little pubescent on the margins; tube of hypanthium 4–7 cm. long; petals yellow, turning pink, 12–18 mm. long; capsule 2–3 cm. long, narrowly winged. *Oenothera triloba ecristata* M. E. Jones. Valleys: Man. (?)—Sask.—Neb.—Colo.—Calif.—Wash. *Submont.—Mont.* Je–Au.

3. **L. tríloba** (Nutt.) Spach. Acaulescent perennial; leaves oblong-lanceolate in outline, 5–30 cm. long, runcinate-pinnatifid or sinuate, glabrous or ciliate on the margins; hypanthium 5–20 cm. long; calyx-tips free; petals 1.5–4 cm. long; capsule 2–3 cm. long, 4-winged above. *O. triloba* Nutt. [G]. Dry soil: Ky.—Kans.—Tex. Miss.; Mex. *Austral.* My–Jl.

f.398

14. GAURÉLLA Small. SPOTTED PRIMROSE.

Low caulescent perennials, with branching stems. Leaves alternate, entire or distantly toothed. Flowers perfect, axillary, sessile. Hypanthium produced beyond the ovary into a slightly funnelform tube. Sepals elongate, reflexed. Petals 4, obovate, white or pink, spotted or striped with red. Style short, slightly clavate; stigma with 4 linear lobes. Capsule short, 4-angled, with a short curved beak. Seeds few, angled, striate.

1. **G. canéscens** (Torr. & Frém.) A. Nels. Cespitose perennial; stems 1–5 dm. long, strongly branched, decumbent or ascending, strigose; leaves lanceolate or oblong, 8–16 mm. long, repand-denticulate or entire; petals white, turning pink, spotted, about 1 cm. long; capsule 8–10 mm. long, angled. *Oenothera canescens* Torr. & Frém. *O. guttulata* Geyer. *G. gultulata* Small [B]. Prairies: Neb.—Tex.—N.M.—Colo. *Plain.*

15. MEGAPTÈRIUM Spach. PRIMROSE.

Perennial leafy-stemmed herbs. Leaves alternate, entire or toothed, rarely pinnatifid. Flowers perfect, regular, showy, axillary. Sepals 4, elongate, reflexed. Hypanthium produced beyond the ovary. Petals 4, large, yellow. Stigma with 4 linear lobes; style filiform; ovules few, sessile. Body of capsule ellipsoid or cylindric, broadly 4-winged. Seeds crested, in a single row.

Body of the pod tapering at each end; wings narrower than the body; usually
    acaulescent plants with lobed or toothed leaves.   1. *M. brachycarpum.*
Body of the pod truncate or retuse at the apex; wings broader
    than the body; usually leafy-stemmed plants with en-
    tire leaves.
    Petals 2–4 cm. long; capsule less than 5 cm. long, oblong
      or oval in outline.
      Stem and leaves closely pubescent; capsule less than
        2 cm. wide.   2. *M. Fremontii.*
      Stem and leaves glabrous; capsule more than 2 cm.
        wide.   3. *M. oklahomense.*
    Petals 3.5–7 cm. long; capsule more than 5 cm. long.
      Stem 1.5–6 dm. high; leaves grayish, narrowly oblance-
        olate; hypanthium 5–15 cm. long; petals 5–7 cm.
        long; pod broadly oval.   4. *M. missouriense.*
      Stem 1–2 dm. high; leaves lanceolate to spatulate; hy-
        panthium 8–10 cm. long; petals 3.5–4 cm. long; pod
        orbicular in outline.   5. *M. argophyllum.*

1. **M. brachycárpum** (A. Gray) Rydb. Cespitose acaulescent perennial, canescent-strigose throughout; leaves oblanceolate or spatulate, sinuate-pinnatifid to entire, 5–15 cm. long; tube of the hypanthium about 1 dm. long; petals yellow, turning rose-colored, 4–5 cm. long; capsule ovoid, 2 cm. long, canescent.

Oenothera brachycarpa A. Gray. *Lavauxia brachycarpa* (A. Gray) Britton [B]. Plains and hills: Kans.—Tex.—Colo. *Plain—Submont.* My–Jl.

2. M. Fremóntii (S. Wats.) Britton. Stem 5–15 cm. high, appressed-pubescent; leaves lanceolate to oblanceolate or linear, 3–8 cm. long, acuminate, entire or nearly so, silvery-canescent; hypanthium 8–10 cm. long; capsule oblong in outline, 2–2.5 cm. long, rounded at the apex. *Oenothera Fremontii* S. Wats. Plains and prairies: Neb.—Kans.—Tex. *Plain—St. Plains.*

3. M. oklahoménse Norton. Stem 1.5–3 dm. high, glabrous, branched below; leaves lanceolate or oblong, 4–9 cm. long, remotely toothed or undulate, glabrous; hypanthium 8–10 cm. long; sepals nearly 2 cm. long; petals 1.2–2.5 cm. long; capsule oblong or oval, 2.5–3.5 cm. long. Hillsides and plains: Kans.—Tex. *Plains.*

4. M. missouriénse (Sims) Spach. *Fig. 399.* Stem densely canescent, 1.5–6 cm. long; leaves thick, linear-lanceolate to oblong-lanceolate, 5–15 cm. long, acuminate or acute at the apex, entire or remotely denticulate; hypanthium 5–15 cm. long; sepals 4.5–5.5 cm. long; petals 5–7 cm. long; capsule in outline broadly obovate, 5–8 cm. long, 4–6 cm. wide, often deeply retuse. *O. missouriensis* Sims [G]. Rocky or sandy places: Mo.—Neb.—Colo.—Tex. *Ozark—Plain.* My–Jl.

f. 399.

5. M. argophýllum R. R. Gates. Stem usually less than 2 dm. high; leaves oblanceolate, lanceolate, or spatulate, 8–10 cm. long, 1.5–2 cm. wide, distantly denticulate, densely and uniformly silvery on both sides; calyx silky-canescent, the lobes about 5 cm. long, the free tips 6–7 mm. long; corolla 3.5–4 cm. long, broadly obcordate; pod orbicular, 5–6 cm. long, scarcely retuse at the apex. *Oenothera missouriensis incana* Torr. Gravelly dry hills: Kans. —Tex. *Plain.* My–Je.

16. GALPÍNSIA Britton.

Suffruticose perennials, with branched stems. Leaves alternate, narrow. Flowers perfect, axillary to foliaceous bracts. Hypanthium produced beyond the ovary into a more or less funnelform tube. Sepals 4, reflexed. Petals 4, yellow. Stamens 8. Stigma disk-like, entire. Capsule cylindric or somewhat fusiform, narrowed at the base, more or less curved. Seeds often tuberculate.

Leaves green, glabrous or puberulent.
  Hypanthium about 1 cm. wide at the mouth; leaves linear.   1. *G. Hartwegi.*
  Hypanthium about 1.5 cm. wide at the mouth; leaves oblong or lanceolate.   2. *G. Fendleri.*
Leaves canescent.
  Whole plant strigose-cinereous.   3. *G. lavandulaefolia.*
  Stem and hypanthium more or less hirsute.   4. *G. interior.*

1. G. Hartwégi (Benth.) Britton. Perennial; stem shrubby and branched at the base, 2–4 dm. high; leaves linear, 1–3.5 cm. long, entire or repand-denticulate; hypanthium gradually dilated, its tube 3–4 cm. long; sepals 2 cm. long; petals 2–3 cm. long; capsule narrowly cylindric, 1.5–2.5 cm. long. *O. Hartwegi* Benth. Plains: Neb.—Ariz.—Tex.; Mex. *Plain—Son.* My–S.

2. G. Féndleri (A. Gray) Heller. Stems several, from a branched caudex, 2–4 dm. high, stout, glabrous or nearly so; leaves oblong to linear-lanceolate, 2–4 cm. long, entire or sinuate-dentate; hypanthium gradually dilated, 4–5 cm. long; sepals 2–2.5 cm. long; petals 2.5–3.5 cm. long; capsule 2–2.5 cm. long. *O. Fendleri* A. Gray. Rocky soil: Kans.—N.M.—Tex. *Texan—St. Plains.* My–Au.

3. G. lavandulaefòlia (T. & G.) Small. *Fig. 400.*
Cespitose perennial, with a woody base; stems 1–2
dm. high, decumbent or ascending, canescent; leaves
numerous, crowded, linear or linear-spatulate, entire
or nearly so, finely pubescent; hypanthium gradually
dilated; tube 3–4 cm. long; petals 13–19 mm. long;
capsule 2–2.5 cm. long. *Oenothera lavandulaefolia*
T. & G. *G. Hartwegi lavandulaefolia* T. & G. Plains
and prairies: S.D. – Kans. — Tex. — Ariz. — Wyo.
Plain—Son. My–Jl.

f. 400.

4. G. intèrior Small. Cespitose perennial, with
a woody base; stems 2–3 dm. high, hirsute as well as
puberulent; leaves oblong, 1–2.5 cm. long, entire,
partly clasping; hypanthium elongate, gradually dilated, 2–3.5 cm. long; petals 1.5–2.5 cm. long; capsule 1.5–2 cm. long. Plains and prairies: Neb. –
Kans.—Tex. *Plain.* My–Au.

17. MERÌOLIX Raf.

Biennial or perennial herbs, with branched stems, sometimes suffruticose at
the base. Leaves alternate, narrow or sharply serrate. Flowers perfect, regular, axillary. Hypanthium produced beyond the ovary, funnelform, the free
portion rather short. Sepals 4, keeled on the back. Petals yellow, often spotted
at the base. Stamens 8. Stigma disk-like, 4-toothed. Capsule linear, 4-angled,
sessile. Seeds numerous, longitudinally grooved.

Throat of the hypanthium dark purple within, 5–6 mm. broad at the mouth.
    1. *M. melanoglottis.*
Throat of the hypanthium orange within, sometimes with a
    darker ring at the base of the stamens, 3–4 mm. broad
    at the mouth.
  Petals more than 2 cm. long; mouth of the hypanthium
    6–10 mm. wide.   2. *M. intermedia.*
  Petals less than 1.5 cm. long; mouth of the hypanthium
    3–4 mm. wide.
    Leaf-blades linear-oblanceolate or narrowly oblance-
    olate, acute.   3. *M. serrulata.*
    Leaf-blades broadly oblanceolate or spatulate, coarsely
    dentate, mostly obtuse.   4. *M. oblanceolata.*

1. M. melanoglóttis Rydb. Perennial, shrubby and branched at the
base; stems 2.5–4 dm. high, the upper portion as well as the leaves grayish-
strigose; leaf-blades spatulate to almost linear, 2.5–6 cm. long, shallowly serrate; petals 10–12 mm. long; capsule straight, 2 cm. long, copiously pubescent.
Dry soil: Tex.—Kans.—Colo. *Plain.*

2. M. intermèdia Rydb. Perennial, woody and branched at the base;
stems 3–5 dm. high, grayish- or whitish-pubescent; leaves canescent,
spatulate to linear, 4–7 cm. long, more strongly serrate than in the next; sepals
7–8 mm. long; petals 2–2.5 cm. long; capsule 2–2.5 cm. long, pubescent. It
has been confused with *M. spinulosa* (T. & G.) Heller, which is glabrous or
nearly so. Dry soil: Man.—Colo.—Tex.—Iowa. *Prairie –Plain.* My–Jl.

3. M. serrulàta (Nutt.) Walp. Perennial, woody at the base; stem 1–5
dm. high, branched and leafy, more or less canescent; leaf-blades spatulate to
linear-oblong or linear, 2–5 cm. long, serrate, canescent-strigose; petals obovate,
8–12 mm. long, crenulate; capsule 1.5–2 cm. long. *Oenothera serrulata* Nutt.
[G]. Dry soil, plains and hills: Man.—Mo.—Tex.—Ariz.—Alta. *Prairie—
Plain—Submont.* Je–Au.

4. M. oblanceolàta Rydb. Low, densely branched, 1–2 dm. high, densely
strigose; leaves short-petioled, the blades broadly oblanceolate or spatulate,
coarsely serrate, 1–2 cm. long, 3–5 mm. wide, strigose-canescent on both sides;

hypanthium 9–12 mm. long, about 3 mm. wide at the mouth; sepals 4 mm. long, broadly ovate, mucronate; petals bright yellow, broadly obovate, 7–8 mm. long. Limestone: w Kansas. *Plain.* Jl.

18. STENOSÌPHON Spach.

Perennial herbs, with erect leafy stem. Leaves alternate, sessile, narrow. Flowers perfect, in elongate terminal spikes. Hypanthium much produced beyond the ovary into a filiform tube. Sepals 4, reflexed. Petals white, clawed. Stamens 8, declined, without scales at the base. Ovary 1-celled; stigma 4-lobed, surrounded by a cup-like border. Fruit 8-ribbed, indehiscent, 1-celled, 1-seeded.

1. **S. linifòlius** (Nutt.) Britton. Stem slender, glabrous, 6–15 dm. high; leaves lanceolate to linear, ₤.5–5 cm. long, entire, acuminate; spikes elongate, sometimes 3 dm. long; tube of the hypanthium about 1 cm. long; petals white, long-clawed, about 5 mm. long; fruit ovoid, hirsutulous, rugose, 3–4 mm. long. Dry plains and prairies: Neb.—Ark.—Tex.—Colo.; Mex. *Plain—Son.* Jl–S.

19. GAÙRA L. BUTTERFLY WEED.

Annual, biennial, or perennial caulescent herbs, with branching stems. Leaves alternate, usually narrow. Flowers perfect, in terminal racemes or panicles, or spikes. Hypanthium prolonged beyond the ovary into a slender but short tube. Sepals 4, rarely 3. Petals clawed, unequal. Stamens usually 8, declined, with a small scale at the base of the filament. Ovary 4-celled; style declined; stigma 4-lobed, with a cup-like border below. Fruit nut-like, indehiscent, ribbed or angled, 4-celled, but some of the cells usually empty. Seeds 1–4.

Anthers oval, attached near the middle; fruit fusiform, sessile, almost equally ribbed. 1. *G. parviflora.*
Anthers linear, or nearly so, attached near the base; fruit strongly 4-angled or 4-winged, at least above.
  Fruit not contracted below into a pedicel-like or thick and seedless base.
    Stem hirsute; leaves thin. 2. *G. biennis.*
    Stem puberulent or canescent; leaves comparatively thick. 3. *G. Pitcheri.*
  Fruit contracted below into a pedicel-like or thick and seedless base.
    Base of the fruit stout and short, angled or ribbed.
      Plants more or less pubescent.
        Leaves canescent, the lower usually oblong and sinuately toothed. 4. *G. coccinea.*
        Leaves sparingly strigose, all linear or nearly so and entire. 5. *G. parvifolia.*
      Plant glabrous or nearly so, except the strigose hypanthium. 6. *G. glabra.*
    Base of the fruit slender.
      Body of the fruit 3–4 mm. long, ovoid, pubescent. 7. *G. Michauxii.*
      Body of the fruit 5–8 mm. long, oblong or fusiform, glabrous or nearly so.
        Fruit fusiform, merely angled; leaves inconspicuously pubescent. 8. *G. sinuata.*
        Fruit oblong, wing-angled; leaves densely pubescent. 9. *G. villosa.*

1. **G. parviflòra** Dougl. Biennial; stem 6–15 dm. high, silky-pilose, with long spreading hairs; leaves lanceolate, ovate-lanceolate, or elliptic, 3.5–10 cm. long, acute or acuminate, repand-dentate, softly pilose; spike very long; petals 2.5–4 mm. long; fruit 6–8 mm. long, glabrous or nearly so. Plains, valleys, and prairies: S.D.—Mo.—La.—Ariz.—Wash.; Mex. *Prairie—Plain—Submont.* My–O.

2. **G. biénnis** L. Stem 5–15 dm. high, villous-hirsute; leaves lanceolate or elliptic, 3–10 cm. long, acute or acuminate at each end, denticulate or undulate; spike slender; petals white, sometimes becoming pink, 7–11 mm. long; fruit elliptic, acute at each end, 6–10 mm. long, 4-ribbed, hirsute. Dry soil: Que.—Minn.—Kans.—Miss.—Ga. *Canad.—Allegh.* Jl–S.

3. G. Pítcheri (T. & G.) Small. Stem 10–15 dm. high, puberulent, sometimes canescent; leaves elliptic to linear-lanceolate, 4–11 cm. long, remotely denticulate or entire; spike elongate; petals pink, 6–9 mm. long; fruit elliptic, 6–7 mm. long, puberulent. Prairies: Iowa—Neb.—Kans.—Ark. *Ozark— Prairie.* Jl–S.

4. G. coccínea Nutt. Stems several, branched, strigose or puberulent and more or less hirsute, 1–5 dm. high; leaves numerous, sessile, oblong or lanceolate, the upper linear-lanceolate, repand-dentate or entire, 1–3.5 cm. long, canescent; tube of hypanthium 6–8 mm. long; petals from pink or white to scarlet, 4–6 mm. long; capsule 5–7 mm. long, canescent. *G. marginata* Lehm. Plains and prairies: Man.—Tex.—Ariz.—Mont.; Mex. *Plain—Submont.* Ap-Jl.

5. G. parvifòlia Torr. Stem 3–4 dm. high, with erect branches, sparingly strigose or in age glabrate; leaves 1–2.5 cm. long; tube of hypanthium 5 mm. long; petals 4–5 mm. long, scarlet; capsule 6–7 mm. long. Dry hills and plains: Kans.—Tex.—Colo. —N.M. *St. Plains—Son.—Submont.* Jl–Au.

6. G. glàbra Lehm. *Fig. 401.* Stem 2–4 dm. high, in age straw-colored and shreddy; leaves oblong or lanceolate, or those of the branches linear-lanceolate; tube of hypanthium 6–7 mm. long; petals pink or brick-red, rarely white; fruit 5–6 mm. long. Dry plains and prairies: S.D.—Colo. Ariz.—Mont. *Plain–Submont.* Je–Au.

†401.

7. G. Michaûnii Spach. Perennial; stem 6–8 cm. long, puberulent; leaves linear or linear-oblong, acute or acuminate at each end, sinuate or remotely dentate, 3.5–8 cm. long; raceme elongate; petals white, turning reddish, 4–6 mm. long; fruit-body 3–4 mm. long, puberulent, sharply 4-angled above; base slender, as long as the body. *G. filipes* Spach [G]. Dry fields: Va.—Kans.—La.—Fla. *Austral.* Jl–Au.

8. G. sinuàta Nutt. Perennial; stem 3–9 dm. high, glabrous, branched at the base; leaves spatulate to lanceolate or linear, sinuately toothed to pinnatifid, 2.5–8 cm. long, acute or acuminate; petals pinkish, 8–10 mm. long; fruit-body 4-ridged above, fusiform, 5–7 mm. long, longer than the pedicel-like base. Dry soil: Kans.—Tex.—Ark.; Mex. My–Jl.

9. G. villòsa Torr. Stem erect, 3–10 dm. high, canescent or villous; leaves spatulate to lanceolate or linear, 2.5–7.5 cm. long, sinuately toothed or pinnatifid, acute or acuminate; petals 10–14 mm. long; fruit-body oblong, wing-angled, 8 mm. long; pedicel-like base shorter and slender. Plains and prairies: Kans.—Tex.—Ark. *Texan.* Je–S.

20. CIRCAÈA L. ENCHANTER'S NIGHTSHADE.

Low perennial herbs, with succulent but slender stems. Leaves opposite, petioled. Flowers small, in racemes. Hypanthium slightly produced beyond the ovary into a slender short tube. Sepals 2. Petals 2, notched, white. Stamens 2, alternate with the petals. Fruit 1–2-celled, 1–2-seeded, obovoid, indehiscent, usually covered with hooked hairs.

Fruit 2-celled; leaves of an ovate type.
  Hairs of the fruit copious, stiff.   1. *C. lutetiana.*
  Hairs of the fruit few, weak.   2. *C. intermedia.*
Fruit 1-celled; leaves of a cordate type.   3. *C. alpina.*

1. **C. lutetiàna** L. Stem pubescent, at least above, 3–6 dm. high; leaves ovate, 5–10 cm. long, acuminate, remotely denticulate; pedicels 4–8 mm. long; bracts none or caducous; fruit rounded-pyriform. Open woods: N.S.—Man.—S.D.—Kans.—Ga.; Eur. and Asia. *Canad.—Allegh.* Je–Au.

2. **C. intermèdia** Ehrh. Stem 2–4 dm. high, sparingly pubescent above; leaves thin, ovate, the upper cordate at the base, with salient teeth; bracts usually present, small; fruit obovoid. Shady places: Labr.—Iowa—Tenn.; Eur. *Canad.* Je–Au.

3. **C. alpìna** L. *Fig. 402.* Stem 0.5–2 dm. high, glabrous or pubescent above; leaf-blades cordate, 2.5–5 cm. long; pedicels 3–4 mm. long, reflexed in fruit; fruit narrowly obovoid, about 2 mm. long. Cold woods: Lab.—Ga.—S.D.—Colo.—Alaska; Eurasia. *Boreal.—Mont.* Jl–S.

f. 402.

Family 122. **HALORAGIDACEAE.** Water Milfoil Family.

Perennial or rarely annual, caulescent herbs, aquatic or amphibious, with alternate, opposite, or whorled leaves, the submerged ones often finely divided. Flowers perfect or monoecious, solitary or clustered, either axillary or in terminal spikes. Sepals 2–4. Petals usually wanting or small, 2–4. Stamens 1–8. Gynoecium of 1–4 somewhat united carpels. Ovary 1–4-celled, inferior, angled or winged; style wanting. Fruit a nutlet or drupe.

Flowers monoecious or polygamous, 4-merous; stamens 4 or 8; submerged leaves finely pinnatifid.
    Fruit at maturity splitting into 2 or 4, angled nutlets; flowers usually monoecious or polygamo-monoecious.   1. Myriophyllum.
    Fruit 3- or 4-angled, 3- or 4-celled, the carpels permanently united, flowers perfect.
        2. Proserpinaca.
Flowers perfect, without sepals and petals; stamen 1; leaves all entire.
    3. Hippuris.

1. **MYRIOPHÝLLUM** (Vaill.) L. Water Milfoil.

Aquatic herbs, with slender stems, usually floating. Leaves alternate or whorled, the emersed ones entire or pectinately lobed, the submerged ones finely pectinately dissected into filiform divisions. Flowers monoecious or polygamous, axillary or in terminal spikes, the upper ones usually staminate, with short hypanthium, 2–4 sepals, 2–4 small petals and 4–8 stamens; the intermediate ones often perfect; the lower ones pistillate, with 4 minute sepals, 4 small petals, or none, and a 2–4-celled ovary; stigmas 2–4, plumose. Fruit bony, splitting into 2 or 4, angled nutlets. Seeds solitary in each nutlet.

Carpels smooth; stamens 8.
    Floral leaves ovate, entire or dentate, usually shorter than the flowers.
        1. *M. spicatum.*
    Floral leaves pinnatifid, much longer than the flowers.   2. *M. verticillatum.*
Carpels 2-keeled and roughened on the back; floral leaves longer than the flowers; stamens 4.
    Floral leaves ovate or lanceolate, serrate.   3. *M. heterophyllum.*
    Floral leaves linear, pectinate.   4. *M. pinnatum.*

1. **M. spicàtum** L. Stem submerged, 3–20 dm. long; leaves verticillate in 4's or 5's, 1–3 cm. long, pinnatifid into filiform divisions; flowers verticillate in an interrupted spike; stigmas rounded, sessile, not elongate; sepals of the staminate flowers usually deep purple. Still water and slow streams: Newf.—Conn.—Kans.—N.M.—Calif.—Alaska; Eurasia. *Temp.—Plain—Submont.* Je–Au.

2. M. verticillàtum L. *Fig. 403.* Stem 3–20
dm. long; leaves flaccid, verticillate in 3's or 4's,
1.5–4 cm. long, pinnatifid, the rachis usually broader
than the filiform divisions; floral leaves or bracts
similar, but shorter and firmer, 6–20 mm. long;
stigmas elongate, recurved; sepals of the staminate
flowers pale green or pinkish. Still water and slow
streams: Que.—N.Y.—Neb.—Ida.—B.C. *Boreal.—
Plain—Submont.* Je–Au.

3. M. heterophýllum Michx. Stem 2–10 dm.
high; submerged leaves crowded, verticillate in 4's
or 5's, or scattered, 1.5–5 cm. long, pectinate-pinnatifid, with 6–10 pairs of filiform segments; floral
leaves in whorls of 3's–6's, sharply serrate; petals
of the staminate flowers oblong, pale, fruit 2–2.5
mm. long. Ponds and slow streams: Que.—S.D.—
Tex.—Fla.; Mex. *E. Temp.–-Trop.*

f403.

4. M. pinnàtum (Walt.) B.S.P. Stem usually 1–2 dm. high, sometimes
longer; leaves in whorls of 3's–6's, the submerged ones pectinate, with 3–5
pairs of filiform segments, the floral ones linear, pectinate-serrate to pectinate,
shorter; staminate flowers with purple petals; fruit 2 mm. long. *M. scabratum*
Michx. [G]. Ponds, ditches and muddy shores: R.I.--N.D.—Tex.—Fla.; Mex.
and C.Am. *E. Temp.—Trop.*

2. PROSERPINÀCA L. MERMAID-WEED.

Low aquatic perennial herbs, with the stems creeping at the base. Leaves
alternate, toothed to pectinate-pinnatifid, the lower more dissected than the
upper. Flowers perfect, axillary, minute, greenish. Hypanthium present,
adnate to the ovary, 3- or 4-angled. Sepals 3 or 4. Petals wanting. Stamens
3 or 4. Stigmas 3 or 4, cylindric to conic-subulate. Fruit nut-like, 3- or
4-angled, 3- or 4-celled. Ovules and seeds solitary in each cell.

1. P. palústris L. Stem 2–7 dm. long; leaves linear, usually of two
kinds, the floral ones emersed and merely serrate, the rest mostly submerged,
pectinate, with entire or serrulate segments, but occasionally there are found
pectinate leaves above the floral ones; sepals triangular; fruit pyramidal or
ovoid-pyramidal, 3 mm. long, acute-angled. Ponds and slow streams: N.B.—
Minn.—Tex.—Fla.; Mex., C.Am. and W.Ind. *E. Temp.—Trop.*

3. HIPPÙRIS L. MARE'S-TAIL.

Aquatic perennial herbs, with simple stems.
Leaves narrow, entire, vorticillate. Flowers perfect,
axillary. Hypanthium adherent to the ovary, with a
minute entire limb. Sepals and petals none. Stamen solitary. Pistil solitary; style filiform, stigmatic its whole length, placed in a groove of the
anther. Fruit 1-celled, 4-seeded.

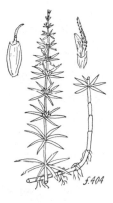

Leaves linear or lanceolate, in verticils of 5–12.
                                     1. *H. vulgaris.*
Leaves oblong, obovate or oblanceo-
   late, in verticils of 4–6.   2. *H. tetraphylla.*

1. H. vulgàris L. *Fig. 404.* Stem simple, 2–6
dm. high; leaves in whorls of 5's–12's, linear, acute,
1–3 cm. long; stamen with a short thick filament
and large anther; fruit nearly 2 mm. long. Swamps
and ponds: Greenl.—N.Y.—Neb.—N.M.—Calif.—
Alaska; Eurasia. *Boreal.—Plain—Mont.* Je–Au.

f.404.

2. H. tetraphýlla L. f. Stem 1–4 dm. high; leaves short and fleshy, 8–15 mm. long; fruit obovoid, 1 mm. long. Wet places mostly along the coast: Labr.—Que.—Man.—Alaska; Eurasia. *Arctic—Subarctic.* Jl.

Family 123. ARALIACEAE. Ginseng Family.

Aromatic herbs, shrubs, or trees. Leaves alternate or whorled, simple or compound. Flowers inconspicuous, variously disposed, the ultimate divisions of the inflorescence usually umbellate, perfect or polygamous, regular. Sepals 5, often minute. Petals 5 or 10, valvate, or imbricate, inserted in the margin of an epigynous disk. Stamens 5 or 10, alternate with the petals; filaments short, distinct; anthers introrse. Gynoecium of 2–5 united carpels; ovary 2–5-celled; styles usually as many; stigmas simple. Ovules solitary in each cavity, anatropous. Fruit a berry or a drupe. Endosperm copious, fleshy. Embryo straight.

Leaves pinnately decompound, alternate; plant mostly leafy-stemmed. 1. Aralia.
Leaves digitately compound, the stem-leaves whorled; plant subscapose. 2. Panax.

1. ARÀLIA (Tourn.) L. Sarsaparilla, Spikenard, Wild Ginseng.

Perennial herbs (ours), or shrubs or trees. Leaves alternate, petioled, pinnately or ternately compound, with toothed leaflets. Flowers perfect or polygamous, in racemose or paniculate umbels, or in compound umbels. Sepals 5, often obsolete. Petals 5, imbricate, with obtuse or inflexed tips. Stamens 5; filaments incurved. Ovary 5-celled; styles 5, spreading. Drupe 5-lobed, 5-celled, black, fleshy. Seeds flattened.

Umbels numerous, in large compound racemes. 1. *A. racemosa.*
Umbels few (2–7), corymbose.
  Plant leafy-stemmed, bristly. 2. *A. hispida.*
  Plant subscapose, the peduncles and leaves rising from the
    nodes of the rootstock. 3. *A. nudicaulis.*

1. A. racemòsa L. Stem spreading, somewhat woody below, 1–2 m. high, purple or spotted; leaves 3–5 times pinnate; leaflets thin, ovate to orbicular, 4–20 cm. long, acuminate at the apex, cordate at the base, doubly serrate; bractlets subulate; sepals minute, hooded; corolla greenish, 2 mm. wide; berries subglobose, 5 mm. thick, dark red or purple. Woods: N.B.—Man.—S.D.—Kans.—Ga. *E. Temp.* Jl–Au.

f.405.

2. A. híspida Vent. Stem from a horizontal creeping rootstock, 2–10 dm. high, bristly; leaves bipinnate, the lower long-petioled; leaflets oblong to oval or ovate, 1.5–5 cm. long, acute, irregularly serrate, acute or rounded at the base, glabrous, or pubescent on the veins beneath; bractlets subulate; corolla white, 3–4 mm. wide; berries 6–8 mm. thick, dark purple. Open woods: Labr.—Man.—Ind.—N.C. *E. Temp.* Je–Jl.

3. A. nudicaùlis L. *Fig. 405.* Acaulescent perennial herb, with a long rootstock; leaves ternate, long-petioled; primary divisions pinnately 3–5-foliolate; leaflets oval or ovate, acuminate, finely serrate, 5–13 cm. long; peduncles 2–3 dm. long; flowers greenish; fruit globose, purplish black. Woods: Newf.—N.C.—Colo.—Ida.—B.C. *Temp.—Submont.—Mont.* My–Je.

2. PÀNAX L. Ginseng.

Perennial scapose herbs, with rootstocks. Stem-leaves in a single whorl, petioled, digitately compound. Umbels simple, terminal, solitary. Flowers polygamous. Sepals minute or obsolete. Petals 5, valvate, spreading, white or greenish. Stamens 5, the filaments short. Gynoecium of 2 or 3, rarely 5, united carpels; styles short, distinct. Fruit a drupe-like berry, usually 2- or 3 celled, red or yellow.

Rootstock globular; leaflets sessile, obtuse; berry yellow. 1. *P. trifolium.*
Rootstock fusiform; leaflets petioluled, acuminate; berry crimson. 2. *P. quinquefolium.*

1. P. trifòlium L. Stem 1–2 dm. high, simple; leaves 3, whorled, 3–5-foliolate; leaflets oblong or elliptic to oblanceolate, 1–8 cm. long, doubly serrate; petals 1 mm. long or less; berry 2- or 3-angled, 4–5 mm. thick. Open woods: N.S.—Minn.—Neb.—Ga. Ap–Je.

2. P. quinquefòlium L. Stem 2–4 dm. high, leaves 3 or 4, whorled, 5- (rarely 6- or 7-) foliolate; leaflets obovate or oval, or suborbicular, 2–15 cm. long, acuminate, doubly serrate, acute to subcordate at the base; petals 1–1.5 mm. long, yellowish-green; berry usually didymous, 8–12 mm. broad. Rich woods: Que.—Man.—Neb.—Fla. *E. Temp.* Jl–Au.

Family 124. AMMIACEAE. CARROT FAMILY.

Herbs, with usually hollow stems. Leaves alternate, usually compound or decompound, rarely simple, with dilated sheathing leaf-bases. Flowers perfect or polygamous, in simple or compound umbels, or rarely in heads or head-like clusters, the umbels commonly subtended by involucres and involucels, consisting of several bracts or bractlets. Hypanthium adnate to the ovary. Sepals 5, usually small. Petals 5, usually with an inflexed tip, often emarginate. Stamens 5, inserted on an epigynous disk; anthers versatile. Gynoecium of 2 united carpels, each 1-ovuled. Styles 2, distinct, borne on a more or less developed thickened base (*stylopodium*). Fruit of two 1-seeded carpels, separating at maturity; each carpel usually with 5 principal ribs, sometimes with 4 secondary ribs; the faces where the two carpels meet are called the commissures, and the sinuses between the ribs, intervals; the pericarp usually containing oil-tubes in the intervals and on the commissural side; some or all the ribs often winged.

Fruit with the secondary ribs most prominent, in ours ellipsoid, armed with prickles; oil-tubes under the secondary ribs; leaves pinnately compound.
                              Tribe 1. DAUCEAE.
Fruit with primary ribs only or ribs wanting; oil-tubes (rarely lacking) in the intervals between the ribs.
  Fruit scaly or spiny.
    Fruit ovoid, covered with hyaline scales or tubercles; leaves coriaceous, spinulose toothed or divided; flowers in dense heads.
                              Tribe 2. ERYNGIEAE.
    Fruit with hooked spines, subglobose; flowers in simple or compound, few-rayed umbels; leaves not spinose, palmately (in ours) or pinnately divided.
                              Tribe 3. SANICULEAE.
  Fruit not spiny, only bristly in the ribs in *Osmorrhiza.*
    Fruit linear-oblanceolate or linear-lanceolate in outline.
                              Tribe 4. OSMORRHIZEAE.
    Fruit oblong to orbicular in outline.
      Fruit not compressed dorsally, terete in cross-section or somewhat obcompressed laterally; wings of the lateral ribs (if present) rarely much broader than those of the dorsal ribs.
        Fruit strongly flattened laterally; low plants with creeping rootstocks or stems; leaf-blades simple, peltate or reniform.
                         Tribe 5. HYDROCOTYLEAE.
        Fruit not strongly flattened laterally; leaf-blades compound or dissected.
          Ribs not conspicuously winged.
            Ribs, at least the dorsal ones, filiform, not corky.

Ribs nearly all alike.
    Carpels bearing on the commissure a corky projection,
        which connects with the gynophore; scapose plants
        with corm-like roots.   Tribe 6. ERIGENIEAE.
    Carpels sessile on the gynophore; leafy-stemmed plants
        or, if scapose, not with corm-like roots.
                         Tribe 7. CAREAE.
    Ribs unlike, the lateral ones more or less prominent, corky.
                       Tribe 8. CYNOSCIADIEAE.
      Ribs all corky.   Tribe 9. CICUTEAE.
    Ribs conspicuously winged.   Tribe 10. THASPIEAE.
Fruit flattened dorsally, with the lateral ribs more or less prominent.
                       Tribe 11. PEUCEDANEAE.

TRIBE 1. DAUCEAE.

Stylopodium obsolete; calyx-teeth obsolete; fruit flattened dorsally.
                         1. DAUCUS.
Stylopodium conical; calyx-teeth prominent; fruit flattened laterally.
                         2. TORILIS.

TRIBE 2. ERYNGIEAE.

One genus.   3. ERYNGIUM.

TRIBE 3. SANICULEAE.

One genus.   4. SANICULA.

TRIBE 4. OSMORRHIZEAE.

Oil-tubes absent in the mature fruit; fruit bristly on the ribs, attentuate at the base.
                         5. OSMORRHIZA.
Oil-tubes present; fruit not bristly.
    Fruit fusiform; oil-tubes present both in the intervals and under the ribs.
                       6. CRYPTOTAENIA.
    Fruit cylindric; oil-tubes present in the intervals only, solitary.
                       7. FALCARIA.

TRIBE 5. HYDROCOTYLEAE.

One genus.   9. HYDROCOTYLE.

TRIBE 6. ERIGENIEAE.

One genus.   8. ERIGENIA.

TRIBE 7. CAREAE.

Flowers white or pinkish.
    Plants with fascicled tuberous roots and few narrow, mostly entire leaf-segments.
      Seed-face concave with a longitudinal ridge; oil-tubes several in the intervals.
                       10. EULOPHUS.
      Seed-face plane; oil-tubes solitary in the intervals.   11. ATENIA.
    Plants with taproots or rhizomes.
      Fruit round, with subglobose carpels and very slender inconspicuous ribs;
        leaves pinnate; oil-tubes several.   12. BERULA.
      Fruit ovate or oblong, with more prominent ribs.
        Fruit oblong, neither tubercled nor spiny.
          Seed-face flat.
            Fruit not constricted on the commissure; bracts linear.
                       13. CARUM.
            Fruit constricted on the commissure, subdidymous; bracts 3-cleft.
                       14. AMMI.
         Seed-face concave to deeply sulcate.
            Involucre present; fruit oblong, acute; calyx-teeth obsolete.
                       15. CHAEROPHYLLUM.
            Involucre wanting; fruit subglobose; calyx-teeth ovate.
                       16. CORIANDRUM.
                           17. SPERMOLEPIS.
      Fruit ovoid, tubercled or spiny.
Flowers yellowish, sometimes white.
    Stylopodium present, prominent.
      Involucres and involucels wanting; fruit nearly terete; stylopodium stout.
                       18. FOENICULUM.
      Involucres and involucels present; fruit laterally compressed; stylopodium
      low.   19. APIUM.
    Stylopodium flat or wanting.
      Leaves simple; oil tubes wanting or continuous around the seed cavity.
                       20. BUPLEURUM.
    Leaves compound, or only the basal ones simple.
      Oil-tubes present in the fruit; ribs of the fruit not wavy.
        Caulescent perennials, with taproots.

Oil tubes solitary in the intervals; root neither deep-seated nor conspicuously thickened.   21. ZIZIA.

Oil-tubes several in the intervals.

  Calyx-teeth obsolete; stylopodium wanting; plants not with thick deep-seated roots.   22. TAENIDIA.

  Calyx-teeth prominent; stylopodium present; plants with thick deep-seated roots.   23. MUSINEON.

Acaulescent cespitose perennials, with short branched caudices; stylopodium wanting.   24. DAUCOPHYLLUM.

Oil-tubes wanting; ribs of the fruit prominent, wavy.

25. CONIUM.

TRIBE 8. CYNOSCIADIEAE.

Fruit oblong; seeds flattened dorsally, with plane face.   26. CYNOSCIADIUM.

Fruit ovoid or ellipsoid; seeds subterete.   27. PTILIMNIUM.

TRIBE 9. CICUTEAE.

Ribs equal and prominent.

  Annuals; fruit ovoid, terete, with acute ribs, oil-tubes strictly one in each interval; bractlets conspicuous, reflexed; leaves finely dissected.

28. AMMI

  Perennials; fruit ellipsoid with blunt ribs; oil-tubes 1–3 in each interval; bractlets small, narrow; leaves, at least the emersed ones, simply pinnate, with broad leaflets.   29. SIUM.

Ribs unequal, the dorsal ones low and broad; the lateral ones prominent and thick.

  Fruit smooth; dorsal and intermediate ribs blunt, the lateral ones acute; calyx-teeth prominent, seeds terete.   30. CICUTA.

  Fruit scabrous or tubercled on the acute dorsal and intermediate ribs; lateral ribs flat, very thick; calyx-teeth obsolete; seeds with a plane face.

31. AMMOSELINUM.

TRIBE 10. THASPIEAE.

Plant pseudo-scapose, from a deep-seated fleshy root; leaves fleshy, with obtuse segments.   32. PHELLOPTERUS.

Plants leafy-stemmed, from a rootstock.   03 THASPIUM.

TRIBE 11. PEUCEDANEAE.

Stylopodium wanting.

  Dorsal and intermediate ribs, or some of them, winged.

    Wings thickened and corky towards the margin; plants pseudo-scapose from a thick deep-seated root.   04. CYMOPTERUS.

    Wings not thickened towards the margin; leafy-stemmed plant with rootstock.

33. THASPIUM.

  Dorsal ribs filiform.

    Lateral wings thin, not corky.

      Oil-tubes more than one in each interval; plants mostly small, perennial, with a thick root.   35. COGSWELLIA.

      Oil-tubes solitary in each interval; plants tall annuals.

36. ANETHUM.

37. PLEIOTAENIA.

    Lateral wings thick and corky; plants large.   37. PLEIOTAENIA.

Stylopodium present.

  Stylopodium depressed (except in *Angelica*).

    Dorsal ribs more or less winged.

      Plant acaulescent or low-stemmed, with narrow leaf-segments; sepals evident; petals yellow.   38. CYNOMARATHRUM.

      Plant caulescent, with broad leaf-segments; sepals obsolete; petals white.

        Wings of the fruit thin; bractlets setaceous or none.

39. ANGELICA.

        Wings of the fruit thick; bractlets several, connate at the base.

40. LEVISTICUM.

    Dorsal ribs filiform; plant caulescent; sepals obsolete; petals yellow.

42. PASTINACA.

  Stylopodium conical.

    Plant villous, at least in the inflorescence; sepals obsolete; leaves ternate, with rounded-cordate leaflets; oil-tubes prominent, inversely clavate, only in the upper half of the carpels.   41. HERACLEUM.

    Plant glabrous, with linear or lanceolate leaf-segments.

      Sepals obsolete; leaves in ours three or four times compound; oil-tubes several in the lateral intervals.   43. CONIOSELINUM.

      Sepals evident; leaves in our species simply pinnate; oil-tubes solitary in all intervals.   44. OXYPOLIS.

1. DAUCUS (Tourn.) L. CARROT.

Bristly annuals or biennials. Leaves pinnately decompound. Flowers white or rarely pinkish, in compound umbels. Bracts foliaceous and pinnately cleft;

bractlets entire or toothed. Calyx-teeth obsolete. Fruit oblong, flattened dorsally. Primary ribs 5, slender, bristly; secondary ribs 4, strong, each bearing a row of strong, barbed prickles. Stylopodium depressed or wanting. Oil-tubes solitary under the secondary ribs, 2 on the commissure. Seeds flattened dorsally; face more or less concave.

Carpels broadest at the middle; each wing with 12 or more bristle-like prickles.
1. *D. Carota.*
Carpels broadest below the middle; each wing with 1–8 flat prickles.
2. *D. pusillus.*

1. D. Caròta L. *Fig. 406.* Stem erect, 4–12 dm. high, channeled, bristly-hispid; leaves pinnately decompound, their segments acute or acuminate, cleft or toothed; umbel compound, the rays numerous; flowers white or pink, the central of each umbel larger and often purple. Fields and waste places: Vt.—Va.—Calif.—Wash.; escaped from cultivation or nat. from Eur. Jl–S.

2. D. pusíllus Michx. Annual; stem repeatedly pinnately dissected into short linear divisions; bracts bipinnate, with linear divisions; bractlets linear-filiform; fruit ovoid-oblong, 3–5 mm. long. Dry or sandy soil and waste places: S.C.—Fla.—Calif.—B.C.—Ida.—Mex.—S.Am. *Austral.—Son.—Trop.* Mr–Au.

f. 406.

2. TÓRILIS Adans. HEDGE PARSLEY.
Annual leafy-stemmed herbs. Leaves pinnately decompound, pubescent. Umbels compound, head-like. Bracts few or wanting. Bractlets several, narrow. Sepals prominent, acute. Petals unequal, white. Stylopodium conic. Fruit laterally flattened, spiny; carpels with 5 primary and 4 secondary winged ribs. Oil-tubes solitary under the secondary ribs, 2 on the commissural side. Seed-face sulcate.

Umbels open, loose, long-peduncled.
1. *T. Anthriscus.*
Umbels dense, subcapitate, short-peduncled or subsessile.
2. *T. nodosa.*

1. T. Anthríscus (L.) Bernh. Stem 6–10 dm. high; leaves bipinnate; segments lanceolate or oblong, dentate; rays 3–8, 1–1.5 cm. long; pedicels 2–4 mm. long; fruit ovoid-oblong, densely bristly, 3–4 mm. long. Waste places: N.J.—D.C.—Kans.; nat. from Eur. Jl–S.

2. T. nodòsa (L.) Gaertn. Stem 1–4 dm. high, branched, flexuose; leaves 2–6 cm. long, once or twice pinnate; leaflets pinnatifid; umbels sessile or short-peduncled, opposite the leaves; fruits ovoid, 3 mm. long, the outer ones armed with barbed spines, the inner ones sharply tuberculate. Waste places: Pa.—Iowa—Tex.; Calif.; W.Ind., Mex., C.Am., and S.Am.; adv. from Eur. My–Au.

3. ERÝNGIUM (Tourn.) L. RATTLESNAKE-MASTER, ERYNGO.
Glabrous caulescent perennials. Leaves usually rigid, coriaceous, spinosely toothed or divided. Flowers white or blue, in dense, bracted heads. Bractlets intermixed with the flowers. Sepals very prominent, rigid, persistent. Stylopodium wanting. Fruit ovoid or oblong, laterally flattened, covered with hyaline scales or tubercles; ribs obsolete. Oil-tubes usually 5, of which 3 are on the dorsal and 2 on the commissural side. Seed-face plane.

Leaves not parallel-veined, the upper ones palmately parted or cleft.
1. *E. Leavenworthii.*
Leaves parallel-veined, linear or oblance-linear, entire.
2. *E. yuccifolium.*

1. E. Leavenwórthii T. & G. A glabrous
perennial; stem 3–10 dm. high, branched above;
lower leaves oblanceolate, spiny-toothed, 4–10 cm.
long, the upper ones palmately cleft or parted, their
segments incised-pinnatifid; head cylindric, 2.5–4
cm. long; bracts incised-pinnatifid, spinose; bractlets
3–7-toothed, or the upper ones like the bracts; fruit
2 mm. long. Dry prairies: Kans.—Tex. *Texan*
Jl–O.

2. E. yuccifòlium Michx. *Fig. 407.* A glaucous perennial; stem 3–18 dm. high, corymbose
above; leaves 1–9 dm. long, remotely bristly on the
margins; heads ovoid or subglobose, 1–2 cm. thick;
bracts lanceolate, entire; bractlets similar but
smaller; fruit 3–4 mm. long, scaly. *E. aquaticum*
L., in part [B]. Meadows: Conn. –S.D,—Tex.—
Fla. *E. Temp.* Je–S.

f 407.

4. SANÍCULA (Tourn.) L. SNAKE-ROOT, SANICLE.

Smooth perennials, with rootstocks and few-leaved stems. Leaves palmate
or rarely pinnate, with incised or pinnatifid divisions. Flowers greenish-yellow
or purple, in irregularly compound, few-rayed umbels. Involucres and involucels
present. Calyx-teeth foliaceous, persistent. Fruit globose, densely covered
with hooked bristles; ribs none. Stylopodium wanting. Oil-tubes from 3 to
many, usually 5, of these 3 dorsal and 2 commissural. Seed-face plane or concave.

Styles longer than the bristles of the fruit; staminate flowers often in separate um-
    bels; perennials.
  Petals and anthers greenish white; sepals linear-subulate; fruit 6 mm. long.
                                                                        1. *S. marilandica.*
  Petals and anthers yellow; sepals ovate; fruit 3 mm. long.   2. *S. gregaria.*
Styles shorter than the bristles of the fruit; staminate flowers
    always intermixed with the fertile ones; biennials.
  Leaves 3–5-divided; pedicels of the staminate flowers 1-2
    times as long as the hypanthium; fruit less than 4 mm.
    long; carpels with 5 oil-tubes.   3. *S. canadensis.*
  Leaves trifoliate; pedicels of the staminate flowers 3–4 times
    as long as the hypanthium; fruit 6 mm. long; carpels
    with 2 or 3 large oil-tubes.   4. *S. trifoliata.*

1. S. marilándica L. *Fig. 408.* Stem 4–12 dm.
high; basal leaves long-petioled; blades 3–5-divided
to the base and the lateral divisions 2-cleft; divisions oblanceolate or obovate, sharply cut and serrate, 5–10 cm. long; stem-leaves short-petioled
or the upper sessile; involucres of few more or less
leaf-like bracts; involucels of small bractlets; fruit
sessile, 6–7 mm. long, including the bristles. Rich
woods: Newf.—Ga.—Colo.—Wash.—B.C. *Temp.—
Plain—Submont.* Je–S.

2. S. gregària Bickn. Stem slender, 3–10 dm.
high; basal leaves bright green, long-petioled; blades
3–5-divided; divisions 8 cm. long, obovate-cuneate
to lanceolate, doubly serrate; stem-leaves 1 or 2, or
wanting, short-petioled; involucre of large 3-parted
bracts; involucels of foliaceous bractlets; fruit stipitate, broadly obovoid. Woods: Vt.—Minn.—Kans.—N.C.
My–Je.

f.408.

Canad. —Allegh.

3. S. canadénsis L. Stem more or less branched, 3–12 dm. high; basal leaves long-petioled, palmately 3-divided to the base; lateral divisions 2-parted; divisions incised; the upper leaves short-petioled or subsessile; umbels irregular, few-rayed; involucres and involucels of few small bracts and bractlets; fruit 3–6 mm. long, including the bristles. Rich woods: Vt.—Fla.—Tex.—Wyo. *E. Temp.—Plain—Submont.* Je–S.

4. S. trifoliàta Bickn. Stem slender, 3–8 dm. high; basal leaves long-petioled, ample; divisions ovate or rhombic, acute, 6–9 dm. long, coarsely serrate or sometimes cleft; bracts foliaceous; staminate flowers few; fruit ellipsoid. Rich woods: Vt.—Minn.—Tenn.—Pa. *Canad.—Allegh.* Je–Jl.

5. OSMORRHÌZA Raf. SWEET CICELY.

Glabrous or hirsute perennials, with thick aromatic roots and more or less leafy stems. Leaves ternately decompound, with broad, lanceolate or ovate, toothed leaflets. Flowers white or purplish, in few-rayed umbels. Involucres and involucels few-leaved or wanting. Calyx-teeth obsolete. Fruit linear-clavate, attenuate at the base, bristly on the equal ribs. Stylopodium conic or depressed. Oil-tubes obsolete in the mature fruit. Seed-face concave to deeply grooved. *Washingtonia* Raf.

Involucels of several bractlets.
  Stem, petioles, and leaf-rachises villous.
    Stylopodium and style about 1 mm. long.   1. *O. Claytoni.*
    Stylopodium and style 2–3 mm. long.   2. *O. villicaulis.*
  Stem (except at the nodes), petioles and leaf-rachises glabrous
    or nearly so; stylopodium and style 2–3 mm. long.   3. *O. longistylis.*
Involucels lacking or of a single small bractlet.
  Fruit obtuse at the apex, without a neck.   4. *O. obtusa.*
  Fruit more or less constricted at the apex into a short neck.   5. *O. divaricata.*

1. O. Clàytoni (Michx.) Clarke. Stem 3–10 dm. high; leaves twice or thrice ternate; leaflets 4–8 cm. long, acute or acuminate, coarsely toothed to divided; umbel-rays 4–6; bracts and bractlets linear to oblong-linear, acute; fruit 18–20 mm. long, with a slender strigose base. *O. brevistylis* DC. *W. Claytoni* Britt. [B]. Shaded hillsides: N.S.—Man.—S.D.—Kans.—N.C. *Canad. Allegh.* My–Jl.

2. O. villicaùlis (Fern.) Rydb. Stem stout, rather densely villous, 3–10 dm. high; leaves twice ternate, pubescent on both sides, ultimate leaflets 2–7 cm. long, short-acuminate; bracts 1–3, linear; bractlets 3–5, lanceolate; fruit 1.5–2 mm. long, attenuate at the base; stylopodium 0.6 mm. long; style 1.5–2 mm. long. *O. longistylis villicaulis* Fern. Woods: Pa.—D.C.—Okla.—N.D. *Allegh.* My–Je.

3. O. longístylis (Torr.) DC. *Fig. 409.* Stem stout, glabrous or nearly so, 3–10 dm. high; leaves twice to thrice ternate; ultimate leaflets 3–7 cm. long, acuminate, sparingly hirsute; involucre and involucels of few narrowly lanceolate, acuminate bracts; fruit 2–3 cm. long, with a slender attenuation at the base; stylopodium elongate-conical, 1 mm. long; styles 2 mm. long. *W. longistylis* Britton [B]. Damp woods: N.S.—Ga.—Colo.—Alta. *Canad.—Allegh.—Plain—Submont.* My–Jl.

4. O. obtùsa (Coult. & Rose) Fern. Stem 3–6 dm. high, glabrous or more or less pilose, with spreading or reflexed hairs; leaves twice or thrice ternate; ultimate leaflets ovate or lanceolate, mostly acuminate; umbels with spreading branches; fruit about 15 mm. long; stylopodium flattened, 0.5 mm.

f. 409.

long or less; styles very short, recurved. *W. obtusa* Coult. & Rose. Woods in the mountains: Newf.—Que.—Alta.—N.M.—Calif.—B.C. *Boreal.—Submont.— Subalp.* Je–Au.

5. O. divaricàta Nutt. Stem slender, 3–6 dm. high, sparingly pubescent; leaves twice or thrice ternate; ultimate leaflets 2–6 cm. long, ovate, acute or acuminate, more or less pubescent; branches of the umbels very long and divaricate in fruit; fruit 12–13 mm. long; stylopodium 0.25 mm. long; styles of about the same length. *W. divaricata* Britton. Woods: B.C.—Ore.—Utah— Man.—Minn.; also Gaspe, Que. *W. Boreal.—Submont.—Mont.* Je–Jl.

6. CRYPTOTAÈNIA DC. HONEYWORT.

Glabrous leafy-stemmed perennials. Leaves alternate, trifoliolate, the terminal leaflet petioluled. Bracts wanting. Bractlets minute or none. Calyx-teeth obsolete. Corolla white. Fruit flattened laterally, linear-oblong, glabrous. Carpels with equal obtuse ribs. Stylopodium conic. Oil tubes usually solitary in the intervals, as well as under the ribs, 2–4 on the commissural side. Seed-face plane or nearly so. *Deringa* Adans.

1. C. canadénsis (L.) DC. Stem erect, 3–10 dm. high; leaves trifoliolate; leaflets ovate, 5–13 cm. long, acute or acuminate, doubly serrate, often lobed; umbels irregular; fruit 4–6 mm. long, often somewhat falcate. *D. canadensis* Kuntze. Woods: N.B.—Man.—S.D.—Tex.—Ga. *Canad.—Allegh.* Je–Jl.

7. FALCÀRIA (Riv.) Bernh.

Biennial herbs. Umbel compound, involucres and involucels present, of several narrow bracts. Sepals minute. Petals white, emarginate. Disk broad, undulate on the margin, with a small stylopodium. Fruit oblong-cylindric, slightly compressed laterally; primary ribs thick, obtuse. Oil-tubes solitary in the intervals.

1. F. vulgàris Bernh. Biennial; stem 3–8 dm. high, glabrous, striate; lower leaves simple or ternate, the upper ternate with ternate or pinnate divisions; leaflets linear-lanceolate, often falcate, densely serrate; umbel compound; bracts and bractlets many, setaceous; petals white, emarginate; fruit cylindric, glabrous. *F. Rivini* Host. Near dwellings: Iowa; adv. from Eurasia. Jl–S.

8. ERIGENÌA Nutt. HARBINGER-OF-SPRING, TURKEY-PEA, PEPPER-AND-SALT.

Low subacaulescent perennials, with a deep-seated tuber-like root. Leaves ternately decompound, with oblong segments. Bracts wanting. Bractlets linear or spatulate. Calyx-teeth obsolete. Petals white, flat. Fruit strongly flattened laterally, nearly orbicular, notched at each end, the carpels flattened laterally, the ribs filiform. Oil-tubes 1–3 in each interval. Commissure with a corky projection connecting it with the gymnophore. Seed-face deeply sulcate.

1. E. bulbòsa (Michx.) Nutt. Stem low, 1–2 dm. high; basal leaves long-petioled; stem-leaves solitary, sessile, subtending the irregular umbel; fruit 2 mm. long, 3 mm. broad. Woods: Pa.—Minn.—Kans.—Ala. *Allegh.* F–My.

9. HYDROCÓTYLE L. MARSH PENNYWORT.

Low perennials, with slender creeping stems or rootstocks. Leaves petioled, peltate or reniform. Umbels simple or proliferous. Involucre none. Calyx-teeth minute or obsolete. Petals white. Fruit strongly flattened laterally, orbicular, the dorsal ribs high and thick, the intermediate filiform, the lateral ones low and broad, contiguous with those of the other carpel, or all ribs filiform. Oil-tubes minute or wanting, but a thick oil-bearing tissue often surrounding the seeds.

Leaves peltate ; fruit thin, except the corky dorsal and lateral ribs.　1. *H. umbellata.*
Leaves not peltate ; fruit rather thick ; ribs filiform.　　　　　　　　2. *H. americana.*

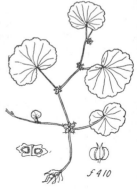

1. H. umbellàta L.　Branches of the rootstock bearing tubers; leaves peltate, the blades orbicular or rounded-reniform, 1–2.5 cm. wide, crenately lobed; umbels simple; fruit notched at each end, 2–3 mm. broad; dorsal ribs obtuse.　Wet places: Mass. — Minn. — Tex. — Fla.; Calif.; Mex.—S.Am. *Temp.—Trop.* Je–S.

2. H. americàna L.　*Fig. 410.*　Rootstock or stem filiform, creeping, often with small fusiform tubers; leaves not peltate, the blades reniform, crenately lobed and crenate, shining, 2–5 cm. wide; umbels few-flowered, axillary, almost sessile; fruit about 2 mm. broad, without oil-bearing layer; ribs small, filiform.　Wet places: N.S.—Minn.—N.C. *Canad.—Allegh.* Je–S.

ƒ 410

10. EÙLOPHUS Nutt.

Glabrous caulescent perennials, with fascicled tuberous roots.　Leaves pinnately or ternately compound, with narrow, linear to oblong-linear leaflets, of which the terminal one usually is elongate.　Bracts lanceolate or rarely wanting; bractlets lanceolate, several, subscarious.　Flowers white or pinkish, in long-peduncled umbels.　Calyx-teeth prominent.　Stylopodium conic; styles long, recurved.　Fruit laterally flattened, ellipsoid to linear-oblong, glabrous.　Ribs filiform; pericarp thin.　Oil-tubes 1–5 in the intervals, 4–8 on the commissural side.　Seed-face broadly concave, with a central longitudinal ridge.

1. E. americànus Nutt.　Stem 8–12 dm. high; branched above; lower leaves 1–4 dm. long, biternately compound, with linear or oblong-linear segments; upper leaves ternate, with elongate, narrowly linear segments; umbels 10–20-rayed; bracts reduced or none; bractlets linear-lanceolate, 2 mm. long; fruit oblong, 4–6 mm. long, 3–4 mm. broad; oil-tubes 3 in the intervals, 4 on the commissural side.　Low ground: Ohio—se Kans.—Ark.—Tenn. *Allegh.* Jl.

11. ATÈNIA H. & A.　Yamp, Squaw-root.

Smooth slender herbs, with tuberous or fusiform-fascicled roots.　Leaves pinnate, with few linear or linear-lanceolate divisions.　Flowers white.　Involucres of few or several bracts, rarely wanting; involucels present.　Calyx-teeth prominent.　Stylopodium conic. Fruit orbicular to oblong, flattened laterally.　Ribs filiform, inconspicuous.　Oil-tubes large and solitary in each interval, 2 on the commissural side.　Seeds somewhat flattened dorsally; face plane.

1. A. Gaìrdneri H. & A.　*Fig. 411.*　Stem 3–10 dm. high; leaves pinnate; leaflets narrowly linear to filiform, 5–15 cm. long; fruit nearly orbicular, usually less than 2 mm. long; calyx-teeth ovate, very small; stylopodium low-conic.　The roots constituted one of the Indian foods; they have a very agreeable

ƒ.411

taste.　*Carum Gairdneri* (H. & A.) A. Gray.　Meadows and valleys: Alta.— S.D.—N.M.—Calif.—B.C. *Plain—Submont.* Au–S.

12. BÉRULA Hoffm. WATER PARSNIP.

Smooth aquatic perennials, with rootstocks. Flowers white, umbellate. Leaves simply pinnate, with toothed or incised leaflets. Bract and bractlets conspicuous, but narrow. Calyx-teeth minute. Stylopodium conic. Fruit nearly suborbicular, flattened laterally, emarginate at the base, glabrous. Ribs slender, inconspicuous. Oil-tubes numerous and contiguous, closely surrounding the seed cavity. Seeds subterete.

1. **B. erécta** (Huds.) Coville. Stem 2–10 dm. high, glabrous; leaves pinnate, with 11–19 leaflets; leaflets ovate to linear, sharply serrate, incised or laciniately lobed, 1–7 cm. long; fruit scarcely 2 mm. long. *B. angustifolia* (L.) Mert. & Koch. Swamps and streams: Ont.—Ill.—N.M.—Calif.—B.C. *Temp.— Plain—Submont.* Jl–S.

13. CÀRUM L. CARAWAY.

Biennials or perennials with taproot, leafy-stemmed. Leaves twice or thrice pinnatifid, with filiform divisions. Involucres present. Flowers white, in many-rayed umbels. Calyx-teeth prominent. Stylopodium conic. Fruit oblong, somewhat flattened laterally. Ribs rather strong Oil-tubes solitary in the intervals, 2–6 on the commissural side. Seed-face plane.

1. **C. Cárui** L. Stem 3–6 dm. high; leaves 3–4 times pinnatifid, with lanceolate to filiform acute segments; involucres of 1–3 linear bracts; involucels usually none; fruit oblong, about 4 mm. long, with conspicuous ribs. Waste places: Newf.—Pa.—Colo.—Alta.; escaped from cultivation; nat. of Eur. My Jl.

14. AMMI L.

Perennial or annual herbs. Leaves finely dissected or divided. Flowers perfect, in compound umbels. Involucres and involucels present, the bracts of the former divided. Petals white, very unequal, cleft at the apex. Fruit short, truncate, laterally flattened; carpels 5-angled, with prominent ribs; oil-tubes solitary in the intervals. Seeds terete.

1. **A. màjus** L. Stem 3–8 dm. high, branched; leaves finely dissected, ultimate segments toothed; umbels open, the concreted bases of the rays 2 mm. broad; fruit 1.5–2 mm. long. Waste places: S.D.—Tex.; introd. from Eur.

15. CHAEROPHÝLLUM L. CHERVIL.

Leafy-stemmed annuals. Leaves ternately-pinnately decompound, the leaflets pinnatifid, with linear or oblong lobes. Involucre wanting. Involucels of many bractlets. Calyx-teeth obsolete. Petals white. Fruit flattened laterally, narrowly oblong, rounded at the base, acute or beaked at the apex. Carpels terete; ribs equal. Stylopodium conic; style short. Oil-tubes solitary in the intervals, 2 on the commissural side. Seed-face sulcate.

Fruiting pedicels much exceeding the elliptic bractlets; leaf-segments oblong; fruit beaked; the ribs narrower than the intervals. 1. *C. procumbens.*
Fruiting pedicels rarely exceeding the linear or oblong bracts.
  Fruit beaked.
    Leaf-segments oblong; ribs broader than the intervals; bractlets spreading. 2. *C. Tainturieri.*
    Leaf-segments linear; ribs scarcely broader than the intervals; bractlets reflexed. 3. *C. reflexum.*
  Fruit beakless; leaf-segments linear, bractlets spreading. 4. *C. texanum.*

1. **C. procúmbens** (L.) Crantz. Stem slender, spreading, 1.5–5 dm. high; leaves ternate, then pinnate; leaflets pinnatifid, with oblong, obtuse lobes; umbels few-rayed, each ray with 2–4 fruits on spreading pedicels; fruit narrowly oblong, 8 mm. long, glabrous, short-beaked; ribs filiform. Moist ground: N.Y.—Iowa—Kans.—Miss.—N.C. *Allegh.* Ap–Je.

2. C. Tainturièri Hook. Stem stouter, 3–6
dm. high, pubescent; leaves similar to those of the
preceding, but the lobes often acute; umbels more
compact, with 7 or 8 sessile or short-pedicelled fruits;
fruit narrowly lance-oblong, beaked, glabrous, 6–7
mm. long; ribs prominent. Dry soil: Tenn.—Mo.—
Kans.—Tex.—Fla. Mr–My.

3. C. refléxum Bush. *Fig. 412.* Stem slender,
2–4 dm. high, glabrous; lower petioles sometimes
ciliolate; leaves ternate, then pinnate, the ultimate
segments narrowly linear; fruit 6–8 mm. long,
beaked, acute at the base. Open woods and copses:
Mo.—Ark.—Kans. *Ozark.* Je.

4. C. texànum Coult. & Rose. Stem 2–6 dm.
high, glabrous above; ultimate segments linear but
short; fruiting umbels dense; fruits 8–15, oblong,
not beaked, glabrous; ribs broader than the intervals. Barrens and stony soil:
Mo.—Kans.—Tex. *Texan.* Ap–Je.

f 412.

16. CORIÁNDRUM (Tourn.) L. CORIANDER.

Glabrous leafy-stemmed annuals. Leaves pinnately divided or decompound.
Bracts wanting. Bractlets few, narrow. Sepals ovate, acute. Petals white.
Stylopodium conic. Fruit subglobose; ribs slender. Oil-tubes solitary in the
intervals, a few on the commissural side. Seed-face concave.

1. C. satìvum L. Stem about 6 dm. high; lower leaves pinnately divided; segments broad, ovate or obovate, toothed or cleft; upper leaves pinnately decompound, with narrowly linear segments; fruit 4 mm. thick. Waste
ground: Mass.—S.D.—N.C.; escaped; native of Eur. My–Jl.

17. SPERMÓLEPIS Raf.

Slender, branching, leafy-stemmed annuals. Leaves finely dissected, with
linear or filiform segments. Umbels unequally few-rayed. Bracts and bractlets
wanting. Calyx-teeth obsolete. Petals white. Fruit ovate, somewhat flattened
laterally, usually tuberculate or bristly. Carpels with thick pericarp, ribs not
strong. Stylopodium conic. Oil-tubes solitary in the intervals, and in one
species also under the ribs, 2 on the commissural side. Seed-face concave.
Leptocaulis Nutt. *Apiastrum* Nutt.

Fruit tuberculate.
  Oil-tubes under the ribs present.   1. *S. patens.*
  Oil-tubes under the ribs wanting.   2. *S. divaricata.*
Fruit with hooked bristles.   3. *S. echinata.*

1. S. pàtens (Nutt.) B. L. Robinson. *Fig. 413.*
Stem slender, 3–6 dm. high, branches above; leaves
biternately divided, with filiform divisions; umbels
long-peduncled, with unequal rays, ascending; fruit
1 mm. long, densely tubercled. *Leptocaulis patens*
Nutt. *Apiastrum patens* Coult. & Rose [B]. Sandy
soil: Ind.—Neb.—Tex.—Ark. *Ozark—Prairie.* Je.

2. S. divaricàta (Walt.) Britton. Stem 3–6
dm. high, with spreading branches; leaves twice or
thrice pinnately divided into narrowly linear or filiform segments; umbels with divaricate rays; fruit 1
mm. long, prominently ribbed and tubercled. *L. divaricatus* DC. Sandy soil: N.C.—Kans.—N.M.—
Tex.—Fla. *Austral.* Ap–My.

f 413

3. S. echinàta (Nutt.) Heller. Stem 1–3 dm.
high, erect; leaves as in the preceding; fruit with

hooked bristles and obsolete ribs, 1 mm. long; oil-tubes solitary in the intervals, 2 on the commisural side. *L. echinatus* Nutt. Prairies: Mo.—Kans.—
Ariz.—Miss.; Calif. *Austral. —Son.* Ap–My.

18. FOENÍCULUM Mill. FENNEL.

Biennial or perennial leafy-stemmed herbs, often aromatic. Leaves pinnately decompound. Bracts and bractlets wanting. Sepals obsolete. Petals
yellow or yellowish. Stylopodium stout, conic. Fruit elongate, nearly terete;
ribs all filiform. Oil-tubes solitary in the intervals, 2 on the commissural side.
Seed-face flat.

1. **F. vulgare** Gaertn. Stem 5–15 dm. high, branched; leaves finely dissected, glaucescent; segments filiform; fruit oblong, 5 mm. long. *F. Foeniculum* (L.) Karst. Waste places: Conn.—Neb—Tex. –Fla.; escaped from cultivation. Jl–S.

19. ÀPIUM (Tourn.) L. PARSLEY.

Annual or biennial leafy-stemmed herbs. Leaves once or twice pinnate.
Bracts and bractlets present, linear. Calyx-teeth obsolete. Petals yellowish.
Stylopodium short-conic. Fruit ovoid, somewhat flattened laterally; ribs all
filiform. Oil-tubes solitary in the intervals, 2 on the commissural side. Seed-
face plane.

1. **A. Petroselínum** L. Stem 3–10 dm. high, glabrous; leaves triangular
in outline, bipinnate; segments obovate, dentate to incised, or those of the
upper leaves linear and entire; bracts 2–4, linear; bractlets subulate; petals
greenish yellow; fruit 4 mm. long, glabrous. *Petroselinum hortense* Hoffm.
P. sativum Hoffm. *P. Petroselinum* (L.) Karst. [B]. Around dwellings: Ont.
—N.D.—Neb.—Md.; escaped from cultivation, nat. of Eur.

20. BUPLEÙRUM L.

Annuals or perennials, with entire, clasping or perfoliate stem-leaves.
Flowers yellow, in small umbels. Involucres present or wanting; involucels of
5 or more ovate bractlets. Calyx-teeth obsolete. Fruit oblong, flattened laterally. Ribs equal, slender or prominent. Stylopodium prominent and flat.
Oil-tubes wanting or continuous around the seed-cavity. Seed-face plane or
nearly so.

1. **B. rotundifòlium** L. Annual; stem 3–10 dm. high, branched above;
leaves oblong to ovate, perfoliate, palmately veined; bracts wanting; bractlets
oblong to ovate or orbicular, abruptly pointed; fruit 3–3.5 mm. long, 2.5 mm.
broad; ribs slender; oil-tubes wanting. Fields and roadsides: N.H.—S.D.—
Kans.—N.C.; Ariz.; nat. from Eur. Jl–Au.

21. ZÍZIA Koch. ALEXANDERS, MEADOW PARSNIP.

Smooth caulescent perennials. Leaves simple or ternately compound, with
broad serrate leaflets. Flowers yellow, umbellate. Bracts wanting; bractlets
small. Calyx-teeth prominent. Stylopodium wanting; styles long, erect. Fruit
flattened laterally, ellipsoid or oblong, glabrous. Ribs filiform. Oil-tubes
large, solitary in each broad interval, 2 on the commissural side. Seed pentagonal in cross-section; face plane.

Basal leaves, at least the earlier ones, simple, cordate.   1. *Z. cordata.*
Basal leaves ternate.   2. *Z. aurea.*

1. **Z. cordàta** (Walt.) Koch. *Fig. 414* Stem 3–7 dm. high, glabrous;
basal leaf-blades cordate or rounded-cordate, 2–10 cm. long, crenate; stem-
leaves ternate, with ovate to lanceolate, serrate or incised leaflets; fruit ovate,
3 mm. long. Wet meadows and open woods: Conn.—Ga.—Utah—Ore.—B.C.
Boreal.—Plain—Submont. My–Au.

2. Z. aùrea (L.) Koch. Stems 3–10 dm. high; leaves all except the uppermost twice or thrice ternate; leaflets ovate to lanceolate, sharply serrate, 2–10 cm. long; fruit oblong, about 4 mm. long. Fields and meadows: N.B.—Fla.—Tex.—Wyo.—Mont.—Sask. *E. Temp.—Plain—Submont.* My–Jl.

22. TAENÍDIA Drude.

Glabrous, glaucous perennials. Leaves bi- or triternately compound. Bracts and bractlets wanting. Calyx-teeth obsolete. Corolla yellow. Fruit flattened laterally, broadly oblong, glabrous. Carpels with equal, filiform ribs. Stylopodium wanting. Oil-tubes mostly 3 in each interval, 4 on the commissural side. Seeds rounded-pentagonal in cross-section, the face plane.

f. 414

1. T. integérrima (L.) Drùde. *Fig. 415.* Stem 3–10 dm. high; leaflets lanceolate to ovate, entire; umbels 10–20-rayed; fruit oblong, 4 mm. long. *Pimpinella integerrima* A. Gray. Rocky or sandy woods and copses: Que.—Minn.—Kans.—Miss.—N.C. *Canad.—Allegh.* My–Je.

23. MUSÍNEON Raf.

Low glabrous or scabrous perennials, with thick elongated roots and dichotomously branched stems. Leaves pinnately decompound, usually with more or less winged rachis. Flowers yellow, rarely white, in long-peduncled umbels. Bracts wanting; bractlets few and narrow. Calyx-teeth prominent. Stylopodium depressed. Fruit ovate or ovate-oblong, flattened laterally. Ribs equal and filiform. Oil-tubes usually 3 in each interval, unequal in size, 2–4 on the commissural side. Seed-face broadly concave. *Musenium* Nutt.

f 415

Fruit glabrous or slightly puberulent.
Fruit strongly scabrous-puberulent.
    Divisions of the leaves obtuse.
    Divisions of the leaves acute or acuminate.

1. *M. divaricatum.*

2. *M. trachyspermum.*
3. *M. angustifolium.*

1. M. divaricàtum (Pursh) Coult. & Rose. Stems ascending or decumbent, 1–2 dm. high; leaves bipinnatifid; segments obovate, 3–5-toothed; branches of the umbels 10–25, 1–2 cm. long; pedicels very short; fruit 4 mm. long. Dry ground: Man.—Neb.—Colo.—Alta. *Plain—Submont.* My–Jl.

2. M. trachyspérmum Nutt. *Fig. 416.* Perennial, with a thick fusiform deep-seated root; stems decumbent or ascending, 0.5–2 dm. long, somewhat puberulent; leaves bipinnatifid; divisions obovate, obtusish, 3–5-toothed; flowers yellow; fruit 2–3 mm. long, scabrous-tuberculate, with prominent ribs. *M. divaricatum Hookeri* T. & G. Dry plains: Sask.—Neb.—Colo.—Alta. *Plain—Submont.* My–Jl.

3. M. angustifòlium Nutt. Stem decumbent or ascending, 1.5 dm. long or less, puberulent; leaves

f 416

bipinnatifid; rachis with very narrow wings; segments lanceolate, acute, 3–5-toothed or cleft; fruit about 3 mm. long, scabrous-tuberculate, with rather prominent ribs. Sandy and gravelly plains: Sask.—Colo.—Alta. *Plain—Submont.* My–Jl.

24. DAUCOPHÝLLUM (Nutt.) Rydb.

Low cespitose perennials, acaulescent or nearly so, with a branched caudex. Leaves numerous, basal, or 1 or 2 cauline, pinnate or bipinnate, with filiform or narrowly linear divisions. Flowers cream-colored to yellow, in dense umbels. Bracts wanting; bractlets few, narrow, linear. Calyx-teeth prominent. Stylopodium wanting. Fruit ovoid or oblong, granular on the intervals. Ribs equal, rather strong, but not at all winged. Oil-tubes 2 or 3 in the intervals, 4–6 on the commissural side. Seeds terete or somewhat depressed; face plane.

1. **D. tenuifòlium** (Nutt.) Rydb. Leaves glaucous; peduncles 1–2.5 dm. high, glabrous; flowers cream-white or ochroleucous; fruit 3–4 mm. long, puborulent or nearly glabrous; oil-tubes 2 or 3 in the intervals. *Musenium tenuifolium* Nutt. Dry hills and plains: S.D.—Neb.—Wyo. *Plain—Submont.* My–Jl.

25. CONÌUM L. POISON HEMLOCK.

Tall glabrous leafy-stemmed perennials, with spotted stems. Leaves pinnately decompound. Bracts and bractlets ovate, acuminate. Calyx-teeth obsolete. Petals white. Fruit broadly ovate, glabrous, somewhat flattened laterally; ribs strong, wavy. Oil-tubes none. Seed-face deeply concave. Poisonous.

1. **C. maculàtum** L. Stem 6–15 dm. high; lower leaves petioled, with dilated sheaths, the upper sessile, pinnately dissected, the segments dentate to incised; fruit 3 mm. long, 2 mm. broad. Waste places: N.S.—Iowa—Del.; Calif.; W.Ind.; S.Am.; nat. from Eur. Je–Jl.

26. CYNOSCIÀDIUM DC.

Glabrous leafy-stemmed annuals. Leaves digitately or pinnately divided into narrow segments. Calyx-teeth prominent, persistent. Petals white. Fruit slightly flattened laterally, ovate or oblong; ribs prominent, the lateral ones stronger and corky. Stylopodium conic. Oil-tubes solitary in the intervals, 2 on the commissural side. Seed flattened dorsally, its face plane.

1. **C. pinnàtum** DC. *Fig. 417.* Stem slender, 3–6 dm. high; stem-leaves pinnately divided into few, narrow, entire, distant divisions, the terminal one the largest; fruit oblong, 3 mm. long. Wet places: Mo.—Kans.—La.—Tex. *Austral.* My–Au.

f. 417.

27. PTILÍMNIUM Raf.

Smooth branching annuals, with finely dissected leaves. Bracts foliaceous. Bractlets minute or narrow. Calyx-teeth usually small. Fruit flattened laterally, ovoid or elliptic, glabrous; dorsal and intermediate ribs filiform or broad, the lateral ones corky. Stylopodium conic. Oil-tubes solitary in the intervals, 2 on the commissural side. Seeds nearly terete. *Discopleura* DC.

Involucral bracts cleft or parted; fruit 2–3 mm. long.   1. *P. capillaceum.*
Involucral bracts entire; fruit 1–1.5 mm. long.   2. *P. Nuttallii.*

1. P. capillàceum (Michx.) Raf. Stem slender, spreading, 3–6 dm. high; leaves dissected into filiform divisions; calyx-teeth minute; fruit ovate, with filiform dorsal and intermediate ribs, the lateral ones forming a broad band around the fruit; style short. *D. capillacea* DC. Wet places: Mass.—S.D.—Tex.—Fla. *E. Temp.* Je–O.

2. P. Nuttállii (DC.) Britton. Stem stout, 6–10 dm. high; bractlets minute; calyx-teeth prominent; fruit ovate, with stout dorsal and intermediate ribs, the lateral ones distinct from those of the other carpel; style long, recurved. *D. Nuttallii* DC. Swamps: Ill.—Kans.—Tex.—Ala. *Ozark—Texan.* My–S.

28. AETHÙSA L. Fool's Parsley.

Glabrous leafy-stemmed annuals. Leaves pinnately dissected. Involucre wanting or of a single bract. Bractlets 1–5, reflexed. Calyx-teeth obsolete. Petals white. Fruit globose-ovoid, glabrous; ribs all prominent, corky, acute. Oil-tubes solitary in the intervals, 2 on the commissural side. Seed-face plane. Poisonous.

1. A. Cynàpium L. Stem 3–7 dm. high; leaves twice or thrice pinnate, the petioles with dilated bases; segments linear, acute; bractlets linear; fruit 3 mm. long. Waste places and as weed in gardens: N.S.—Minn.—Pa.; nat. from Eur. Je–S.

29. SÌUM (Tourn.) L. Water Parsnip.

Smooth caulescent perennials, with rootstocks, growing in water or wet places. Leaves pinnate, with serrate or pinnatifid leaflets. Flowers white in large umbels. Bract and bractlets numerous, narrow. Calyx-teeth minute. Stylopodium depressed; styles short, recurved. Fruit flattened laterally, oval in outline, glabrous. Ribs equal, prominent and corky. Oil-tubes 1–3 in each interval, 2–6 on the commissural side. Seed not compressed; face plane.

Stem stout; leaflets 7–17; fruit 3 mm. long.　　1. *S. cicutaefolium.*
Stem slender; leaflets usually 1–7; fruit 2 mm. long.　　2. *S. Carsonii.*

1. S. cicutafòlium Gmel. *Fig. 418.* Stem 6–10 dm. high; leaves pinnate, with 7–17 leaflets, or if growing in water the submerged leaves twice or thrice pinnatifid; leaflets in the emersed leaves linear or lanceolate, 3–10 cm. long, sharply serrate; fruit 3 mm. long, with prominent ribs. Water and wet places: Newf.—Va.—Calif.—B.C. *Temp.—Plain—Mont.* Je–Au.

2. S. Carsònii Durand. Stem 3–6 dm. high; leaflets 1–7, rarely more, linear to lanceolate, 2.5–5 cm. long, when submerged or floating thin, ovate or oblong, laciniate-toothed or dissected; umbels few-rayed; fruit with less prominent ribs than in the preceding; oil-tubes 2–4 on the commissural side. Water: Me.—R.I.—Pa.; Minn. *Allegh.* Jl–Au.

f.418.

30. CICÙTA L. Cowbane, Poison or Water Hemlock.

Smooth poisonous marsh plants, with short, often erect rootstocks and leafy stems. Leaves pinnate or bipinnate, with serrate leaflets. Flowers white, in compound umbels. Bracts few or none. Bractlets several and slender. Calyx-teeth rather prominent. Stylopodium low, but sometimes low-conic. Fruit oblong to orbicular, flattened laterally, glabrous. Ribs strong, corky, flattish, the lateral ones largest. Oil-tubes solitary in the intervals, 2 on the commissure. Seed-face plane or nearly so. Poisonous.

Axils of the leaves not bearing bulblets; leaflets lanceolate.
  Fruit 4 mm. long, not constricted on the commissure; lateral ribs the largest.
    1. *C. maculata.*
  Fruit 3 mm. long, constricted on the commissure; ribs all
    nearly equal. 2. *C. occidentalis.*
Axils of the leaves (especially the upper ones) bearing bulblets;
  leaflets narrowly linear. 3. *C. bulbifera.*

1. **C. maculàta** L. Stem stout, 1-2 m. high;
rootstock short; leaves twice or thrice pinnate; leaflets narrowly lanceolate or linear lanceolate, 5-20
cm. long, sharply serrate; lateral ribs of the fruit
wedge-shaped, closely in contact with those of the
other carpel; dorsal ribs slender; oil-tubes large.
(?) *C. arguta* Greene. Swamps and wet meadows:
N H,—Man.—S.D.— Tex.—Va. *E. Temp.* Je–Au.

2. **C. occidentàlis** Greene. *Fig. 419.* Stem
stout, 1-2 m. high; leaves twice pinnate; leaflets
lanceolate or linear-lanceolate, 5-8 cm. long, sharply
serrate; fruit ellipsoid, 3 mm. long, constricted at
the commissure; oil-tubes large. *C. dacotica* Greene,
(?) *C. subfalcata* Greene. Swamps and wet meadows: Alta.—S.D. –N.M.—Calif. –B.C. *W. Temp.–-
Plain—Submont.* Je–S.

f.419.

3. **C. bulbifera** L. Stem slender, 3-9 dm high; leaves twice or thrice
ternate; leaflets linear, sparsely toothed, 2-5 cm. long; fruit rather rare, orbicular, 2 mm long, constricted at the commissure; ribs broad and low; intervals
narrow. Swamps: Mo. –Md,—Ore.—B.C. *Boreal--Plain.* Jl–S.

31. AMMOSELÌNUM T. & G. Sand Parsley,

Low diffuse annuals. Leaves ternately and then pinnately divided into
linear or spatulate divisions. Bracts and bractlets present, entire or dissected.
Calyx-teeth obsolete. Petals white. Fruit flattened laterally, ovate to oblong;
ribs prominent, corky, scabrous, the lateral ones thick and contiguous with those
of the other carpel; pericarp hard. Stylopodium wanting; style short. Oil-
tubes solitary in the intervals, 2 on the commissural side. Seed-face plane.

1. **A. Pòpei** T. & G. Stem about 1 dm. high, scabrous on the angles;
leaf-segments linear; involucre usually of a dissected leaf; bractlets few; fruit
oblong, 4-5 mm. long. Sandy soil: Mo.—Kans.—N.M.—Tex. *Texan—Son.*
Ap–My.

32. PHELLÓPTERUS Nutt.

Perennial herbs, with deep-seated fleshy roots, a subterranean stem merely
reaching the surface of the ground and there bearing a cluster of leaves and
naked peduncles. Leaves from once to thrice pinnate, rather fleshy. Flowers
white or purplish. Calyx-teeth evident. Stylopodium wanting. Fruit oblong
to orbicular in outline, nearly orbicular in cross-section. Ribs or most of them
with thin broad wings, which sometimes are somewhat thickened at the insertion; the lateral distinct from those of the other carpel. Oil-tubes usually
more than one in each interval. Seeds more or less flattened dorsally; face
broadly and shallowly concave.

1. **P. montànus** Nutt. Leaves twice or thrice pinnate, with oblong
toothed divisions, somewhat glaucous; peduncles 1-5 cm. long; involucre inconspicuous, hyaline and lobed; involucels conspicuous, of obovate entire distinct bractlets; fruit broadly elliptic to nearly orbicular, 6-8 mm. long. *Cymop-
terus montanus* T. & G. Dry plains: Man.—S.D.—Kans.—Colo.—Wyo. *Plain
—Submont.* Ap–My.

33. THÁSPIUM Nutt. MEADOW PARSNIP.

Leafy-stemmed perennials. Leaves once to thrice ternate, or the basal ones simple. Bracts usually wanting. Bractlets small. Calyx-teeth conspicuous. Corolla yellow or purple. Fruit ovoid or oblong, slightly flattened dorsally if at all. Carpels with at least some of the ribs thickly winged. Stylopodium wanting. Oil-tubes solitary in the intervals, 2 on the commissural side. Seeds somewhat dorsally compressed, the face plane.

Basal leaves simple or once ternate.   1. *T. trifoliatum.*
Basal leaves twice or thrice ternate.   2. *T. barbinode.*

f.420.

1. T. trifoliàtum (L.) A. Gray. *Fig. 420.* Stem 6–15 dm. high, glabrous; basal leaves mostly cordate, serrate; stem-leaves mostly once ternate; leaflets ovate or lanceolate, serrate; flowers dark purple or yellow; fruit globose-ovoid, 4 mm. long; all ribs equally winged. *T. atropurpureum* Nutt. *T. aureum* Nutt. [G]. Woods: R.I.—Man.—Kans. S.C. *Canad.—Allegh.* Je–Jl.

2. T. barbinòde (Michx.) Nutt. Stem 6–15 dm. high, loosely branched, pubescent at the nodes; leaves once to thrice, mostly twice ternate; leaflets ovate to lanceolate, acute, coarsely toothed or incised, often cleft or parted; flowers light yellow; fruit broadly oblong, 6 mm. long; some of the ribs usually not winged or only slightly so. Along streams: Ont.—Minn.—Kans.—Fla. *E. Temp.* My–Je.

34. CYMÓPTERUS Raf.

Dwarf subacaulescent perennials, with deep-seated thick root. Leaves pinnate or bipinnate. Bracts wanting or rarely few, small and linear; bractlets conspicuous, foliaceous. Flowers white or yellow. Calyx-teeth obsolete or evident. Fruit flattened dorsally, oval, in ours glabrous. Dorsal and intermediate ribs filiform or some of them usually winged; the lateral ones with broad thickened corky wings. Stylopodium wanting. Oil-tubes 4–12 in the intervals, 8–14 on the commissure (in ours). Seed-face plane.

1. C. acaùlis (Pursh) Rydb. Stem above ground less than 1 dm. high; leaves clustered, bipinnatifid; segments entire or few-toothed; peduncles shorter than the leaves; bractlets linear, entire, more or less united, foliose; fruit 6–8 mm. long; wings of the fruit broad, moderately thick throughout. *C. glomeratus* (Nutt.) DC. Dry arid places: Man.—Okla.—Colo.—Alta. *Prairie—Plain —Submont.* Ap–Je.

35. COGSWÉLLIA Spreng. COUS, BISCUIT ROOT, WHISK-BROOM PARSLEY.

Acaulescent or short-stemmed perennials, with thickened, tuberous roots. Leaves ternately, rarely pinnately, dissected. Flowers yellow, white, or purple. Bracts wanting; bractlets usually present. Calyx-teeth usually obsolete. Stylopodium wanting. Fruit strongly flattened dorsally. Dorsal and intermediate ribs filiform, close together, the lateral ones winged, the wings coherent to those of the other carpel till maturity. Oil-tubes solitary or few in the intervals, 2–10 on the commissural side. Seed-face plane or rarely slightly concave. *Peucedanum* Am. auth. *Lomatium* Raf.

Bractlets obovate or spatulate; whole plant glabrous; corolla yellow; involucre wanting.   1. *C. montana.*
Bractlets lanceolate, oblong or linear.
   Corolla white.

Bractlets scarious-margined, as well as the whole plant
  puberulent or rarely glabrate.   2. *C. orientalis.*
Bractlets not scarious-margined, villous.   3. *C. macrocarpa.*
Corolla yellow or purple.
  Fruit villous or puberulent.   4. *C. villosa.*
  Fruit glabrous.
    Scape, petioles, and involucels decidedly villous;
      bractlets lanceolate, distinctly scarious-margined.   5. *C. daucifolia.*
    Scape, petioles, and involucels sparingly puberulent or
      short-pubescent; bractlets linear, scarcely scarious-
      margined.   6. *C. foeniculacea.*

1. **C. montàna** (Coult. & Rose) M. E. Jones. Corm napiform or fusi
form, or plant sometimes with thickened roots; scape 1–3 dm. high; leaves
usually ternate and then once or twice pinnate; leaflets short-oblong, crowded;
bractlets obovate or oblanceolate, with white or purplish scarious margins;
fruit elliptic, glabrous, 5–6 mm. long, 2–3 mm. wide; wings about half as broad
as the body. *Lomatium montanum* Coult. & Rose. *L. purpureum* A. Nels.
Hillsides: Ore.—Wyo.—N.D. *Submont.—Mont.* My–Je.

2. **C. orientàlis** (Coult. & Rose) M. E. Jones.
Fig. 421. Perennial, with a thick taproot; stem
often scapiform, 1–3 dm. high; leaves bipinnate;
segments oblong or lanceolate, pinnately cleft into
short oblong divisions, grayish-pilose with short
hairs; bractlets lanceolate, distinct, scarious-margined, fruit glabrous, oval, 5 mm. long, 4 mm. wide;
wings not as broad as the body. *P. nudicaule* Nutt.
[B], in part. *Lomatium orientale* Coult. & Rose [G].
Dry plains: Man.—Iowa—Kans.—N.M.—Ariz.—
Mont. *Prairie—Plain—Submont.* Ap–Je.

3. **C. macrocárpa** (Nutt.) M. E. Jones. Per
ennial, with an elongate fusiform root; stem 1–6
dm. high, more or less pubescent, branched at the
base; leaves three or four times pinnately (or the
first division ternately) dissected into short linear

f.421

or oblong divisions, more or less pubescent; bractlets somewhat foliaceous,
lanceolate or linear, often united and unilateral; calyx-teeth evident; fruit
elliptic, glabrous, 1–2 cm. long, 5–7 mm. wide; wings nearly as broad as the
body. *P. macrocarpum* Nutt. Hills and plains: Man.—Colo.—Calif.—B.C.
Plain—Submont. Ap–Je.

4. **C. villòsa** (Raf.) Schultes. Perennial, with a fusiform root; leaves
finely dissected, first ternate, then several times pinnate, villous; segments
numerous, linear, crowded; bractlets conspicuous, lanceolate, more or less
united, very tomentose; pedicels 2–6 mm. long; fruit oval, puberulent or pilose,
5–8 mm. long, 4–5 mm. wide. *Lomatium foeniculaceum* Coult. & Rose. *P. villosum* Nutt. [B]. *L. villosum* Raf. Dry plains and hills: Sask.— w Kans.—
Wyo.—Alta. *Plain—Submont.* My–Je.

5. **C. daucifòlia** (Nutt.) M. E. Jones. Perennial, with a fusiform root;
leaves finely dissected, first ternate, then pinnate several times; segments linear-
filiform, more or less villous; scape 1–3 dm. high; bractlets lanceolate, united
at the base, acuminate, densely white-villous; fruit glabrous, 7–8 mm. long,
4.5–5 mm. wide. *L. daucifolium* Coult. & Rose [G]. Prairies: Mo.—Neb.—
Tex. *Prairie.* Mr–My.

6. **C. foeniculàcea** (Nutt.) Coult. & Rose. Perennial with a fusiform
root; leaves finely dissected as in the preceding; segments linear, puberulent;
scape 1–2 dm. high; bractlets linear, acute or acuminate, nearly distinct; fruit
as in the preceding. *Ferula foeniculacea* Nutt. *P. foeniculaceum* Nutt. [B].
Prairies and plains: N.D.—S.D.—Man. *Plain.* My.

36. ANÈTHUM (Tourn.) L. Dill.

Leafy-stemmed, glabrous annuals. Leaves finely pinnately dissected. Bracts and bractlets wanting. Calyx-teeth obsolete. Petals yellow. Stylopodium wanting. Fruit oblong or elliptic, strongly flattened dorsally, the dorsal and intermediate ribs slender but sharp, the lateral ones winged. Seed-face plane.

1. **A. gravèolens** L. Stem 3–10 dm. high, striate, glabrous; leaves dissected into filiform divisions, the sheaths strongly nerved, scarious-margined; fruit 6 mm. long, 3 mm. wide. Waste places: Conn.—N.D.—Va.; escaped from cultivation; nat. of Eur. Jl–S.

37. PLEIOTAÈNIA Coult. & Rose. Prairie Parsley.

Glabrous leafy-stemmed perennials. Leaves twice pinnate. Bracts wanting. Bractlets narrow. Calyx-teeth conspicuous. Corolla yellow. Fruit obovate to oval, strongly flattened dorsally, glabrous, the dorsal and intermediate ribs small or obscure, the lateral ones forming a broad corky wing, contiguous with those of the other carpel. Stylopodium wanting. Oil-tubes many, irregularly arranged around the seed and scattered in the corky pericarp. Seed-face plane. *Polytaenia* DC. [G, B]; not Desv.

1. **P. Nuttállii** (DC.) Coult. & Rose. Stem 6–10 dm. high, glabrous; leaf-segments cuneate, incised; fruit 6–10 mm. long. Dry soil: Mich.—N.D.—Tex.—Ala. *Allegh.—Ozark.* Ap–My.

38. CYNOMARÁTHRUM (Nutt.) Coult. & Rose.

Acaulescent perennial, with multicipital caudices densely covered by old leaf-sheaths. Leaves narrow in outline, pinnately dissected, with very narrow divisions. Flowers yellow or perhaps sometimes white. Calyx-teeth evident. Stylopodium evident, flat, apparently somewhat spongy. Fruit strongly flattened dorsally, oblong. Dorsal and intermediate ribs sharp or more or less winged, the lateral ones broadly winged. Oil-tubes mostly several in each interval, or obscure. Seed-face plane. Poisonous.

1. **C. Nuttállii** (A. Gray) Coult. & Rose. *Fig. 422.* Scape 2–5 dm. high; leaves pinnate or bipinnate; leaflets cuspidate, 1–5 cm. long, ascending; bractlets lanceolate, often more or less united below; petals yellow; pedicels in fruit 2–6 mm. long; fruit elliptic, 8–10 mm. long, 4 mm. wide, narrowly winged. *Peucedanum graveolens* S. Wats. *P. Kingii* S. Wats. [B]. Dry hills: w Neb.—Wyo.—Utah. *Plain —Mont.* My–Jl.

f. 422.

39. ANGÉLICA L. Angelica.

Stout perennials, with a woody root. Leaves ternate-pinnately or pinnately compound, usually with broad segments. Flowers white, or rarely greenish yellow or purplish, in large umbels. Bracts scanty or none, in one species foliaceous, bractlets small or wanting. Calyx-teeth mostly wanting. Stylopodium conic. Fruit flattened dorsally, ovate or oblong, glabrous or pubescent, with a prominent crenulate disk. Dorsal and intermediate ribs strong, the lateral ones broadly winged; wings distinct from those of the other carpel. Oil-tubes one to several in each interval, or indefinite, 2–10 on the commissural side. Seed-face plane or slightly concave.

Oil-tubes several, distinct; seeds adherent to the pericarp.
Oil-tubes continuous about the seed, which is loose in the pericarp.

1. *A. villosa.*

2. *A. atropurpurea.*

1. A. villòsa (Walt.) B.S.P. Stem 6–15 dm.
high, tomentose above; leaves twice or thrice ternately or pinnately divided, the uppermost reduced
to the inflated sheaths; leaflets thickish, lanceolate
to oblong, 2–5 cm. long, serrate; bracts wanting;
bractlets linear; fruit nearly round, pubescent, 4
mm. broad, the dorsal and intermediate ribs prominent; lateral wings thin, as broad as the body; oil-
tubes 3–6 in the intervals, 6–10 on the commissure.
A. hirsuta Muhl. Dry soil: Conn.—Minn.—Miss.—
Fla. *E. Temp.* Jl-Au.

2. A. atropurpùrea L. *Fig. 423.* Stem stout,
dark purple, 1–2 m. high; leaves first ternate, then
pinnate; leaflets ovate to lanceolate, 2.5–7.5 cm.
long, sharply serrate; bracts wanting; bractlets few,
subulate; fruit oblong, glabrous, 6 mm. long; dorsal
and intermediate ribs prominent; lateral wings half as broad as the body; oil-
tubes 25–30. River banks: Lab.—Minn.—Iowa—Del. *Canad.* Je-Jl.

40. LEVÍSTICUM Koch. LOVAGE

Perennial herbs. Leaves decompound, with broad leaflets. Involucre and
involucels of numerous narrow bracts. Flower whitish or yellowish, in large
compound umbels. Fruit ovate-oblong, flattened dorsally; primary ribs winged;
oil-tubes solitary in the intervals.

1. L. officinàle (L.) Koch. Glabrous perennial; stem tall; leaves once to
thrice pinnately dissected into cuneate segments, 5–10 cm. long, acute, deeply
cleft or those of the upper leaves entire; umbels 7–10 cm. broad. *L. Levisticum*
(L.) Karst. Waste places: Vt.—Minn.—N.Y.; escaped from cultivation.

41. HERACLÈUM L. COW PARSNIP, COW CABBAGE.

Tall stout leafy-stemmed perennials. Leaves ternately compound, with
large, broad leaflets. Flowers white, in large umbels. Bracts deciduous; bractlets numerous. Calyx-teeth small or obsolete. Stylo-
podium thick, conic. Fruit broadly obovate, strongly
flattened dorsally, pubescent. Dorsal and intermediate ribs filiform; lateral ribs with broad wings,
contiguous to those of the other carpel, strongly
nerved towards the outer margin. Oil-tubes solitary
in the intervals, conspicuous, about half as long as
the carpels, 2–4 on the commissural side. Seeds very
strongly flattened; face plane.

1. H. lanàtum Michx. *Fig. 424.* Tall perennial; stem stout, 1–2.5 m. high, villous, especially
above; leaves ternate; sheaths much dilated; leaflets
stalked, round-cordate, 1–3 dm. broad, palmately
cleft and incised; bracts and bractlets subulate;
fruit obcordate, about 1 cm. long, somewhat pubescent. Wet ground: Vt. — N.C. — Calif. — Alaska.
Temp.—Plain—Mont. Je-Au.

42. PASTINÀCA L. PARSNIP.

Stout caulescent biennial, with thick taproot. Leaves pinnately compound,
with broad leaflets. Flowers yellow, in large umbels. Calyx-teeth obsolete.

Stylopodium flat. Dorsal and intermediate ribs slender, the lateral ones strongly winged. Fruit oval or elliptic, glabrous. Seed-face plane.

1. **P. satìva** L. Biennial, with a fusiform root; stem 6–15 dm. high; leaves pinnate; leaflets ovate or oval, sessile, 2–12 cm. long, lobed and incised or dentate; fruit broadly oval, 5–7 mm. long, 4–6 mm. broad, glabrous. Roadsides and waste places: Vt.—Fla.—Calif.—B.C.; escaped from cultivation; nat. of Eur.

43. CONIOSELÌNUM Fisch. Hemlock Parsley.

Glabrous perennials, with a thick root. Leaves ternate, and then pinnately decompound, with toothed leaflets. Flowers white, in large umbels. Bracts more or less conspicuous, or wanting; bractlets numerous, elongate, narrow. Calyx-teeth obsolete. Stylopodium slightly conic. Fruit oblong, dorsally flattened, glabrous. Dorsal and intermediate ribs prominent, sometimes narrowly winged, the lateral ones broadly winged and thick. Oil-tubes usually solitary, in the dorsal intervals, 1–several in the lateral ones, and 2–8 on the commissural side. Seed-face plane or slightly concave. Poisonous.

1. **C. chinénse** (L.) B.S.P. Stem 5–15 dm. high; leaflets twice pinnate; leaflets 2.5–5 cm. long, laciniately lobed, acute; bracts narrow or wanting; bractlets narrow, inconspicuous; fruit oblong or oval, 4–5 mm. long, prominently winged; oil-tubes 2 or 3 in the intervals, 4–8 on the commissural side. Wet places: Lab.—Minn.—Neb.—N.C. *Canad.—Allegh.* Au–S.

44. OXÝPOLIS Raf. Cowbane. Hemlock.

Smooth perennials, with fascicled roots. Leaves simply pinnate or ternate or reduced to the petioles. Bracts few or none. Bractlets numerous, small, or none. Calyx-teeth evident. Petals white. Fruit slightly flattened dorsally, ovate or obovate, the dorsal and intermediate ribs filiform, the lateral ones winged, closely contiguous to those of the other carpel. Stylopodium thick, short-conic. Oil-tubes solitary in the intervals, 2–6 on the commissural side. Seed-face plane. *Tiedemannia* DC. *Archemora* DC. Poisonous.

1. **O. rigídior** (L.) Raf. *Fig. 425.* Stems 6–15 dm. high; leaves pinnate; leaflets 3–9, ovate to linear-lanceolate, entire or remotely toothed; fruit oblong, 5–7 mm. long. *A. rigida* DC. Swamps: N.Y.—Minn.—La.—Fla. *E. Temp.* Au–S.

Family 125. CORNACEAE. Dogwood Family.

Shrubs or trees, rarely perennial herbs. Leaves alternate or opposite, without stipules, often firm, usually entire. Flowers perfect or unisexual, in cymes or heads, or the staminate ones in ament-like spikes. Sepals 4 or 5. Petals 4 or 5, or rarely numerous, imbricate or valvate, inserted at the base of the epigynous disk, or wanting. Stamens as many as the petals; anthers attached at the base or the back. Gynoecium of 1–4 united carpels; styles united; stigmas entire, lobed, or cleft. Ovules usually solitary in each cavity, anatropous, pendulous. Fruit a drupe; stone 1–4-celled. Endocarp fleshy.

Flowers in open cymes, not subtended by an involucre; shrubs.
  1. Svida.
Flowers in head-like umbels, subtended by an involucre of 4 or more white leaves.

Herbs with rootstocks; sepals tipped with a deciduous bristle; drupe globose.
2. CHAMAEPERICLIMENUM.
Trees; sepals without bristles; drupe ellipsoid. 3. CYNOXYLON.

1. SVÌDA Opiz. CORNEL, DOGWOOD, KINNIKINICK.

Shrubs or trees, with hard wood and mostly opposite branches. Leaves opposite or rarely alternate, entire. Flowers perfect in naked, open, dichotomous cymes. Sepals 4, usually small. Petals 4, valvate, white. Stamens 4; filaments filiform or subulate. Ovary 2-celled, or rarely 5-celled. Fruit drupaceous, with thin pulp; stone bony, usually 2-celled. Seeds flattened. *Cornus* L., in part.

Pubescence, at least on the young twigs, the inflorescence, and along the veins on the lower side of the leaves woolly and more or less spreading.
Fruit light blue.
Leaf-blades rounded oval, abruptly pointed, woolly beneath. 1. *S. rugosa.*
Leaf blades ovate or elliptic, gradually acuminate, silky-
downy beneath. 2. *S. Amomum.*
Fruit white.
Leaf-blades scabrous above; stone globose. 3. *S. asperifolia.*
Leaf-blades finely strigose above; stone globose.
Sepals minute; stone higher than broad. 4. *S. interior.*
Sepals prominent, erect; stone broader than high. 5. *S. Baileyi.*
Pubescence wholly appressed or none.
Leaves opposite; fruit white.
Branches reddish or purple; shrub stoloniferous; leaf-
blades broadly ovate or oval, abruptly short-acuminate. 6. *S. stolonifera.*
Branches gray; shrub not stoloniferous; leaf-blades
lanceolate to ovate, gradually acuminate.
Stone higher than broad, smooth; leaf-blades mostly
ovate. 7. *S. instolonea.*
Stone usually broader than high, grooved; leaves
mostly lanceolate. 8. *S. foemina.*
Leaves alternate; fruit bluish 9. *S. alternifolia.*

1. S. rugòsa (Lam.) Rydb. Shrub, 2–3 m. high; branches greenish, warty-dotted; leaf-blades slightly pubescent above, woolly beneath, 5–15 cm. long, 5–12 cm. broad; cymes flat; stone subglobose, about 5 mm. high, somewhat ridged. *Cornus rugosa* Lam. *C. circinata* L'Hér. [G, B]. Copses: N.S.—Man.—N.D.—Iowa—Va. Je–Jl.

2. S. Amòmum (Mill.) Small. Shrub, 1–4 m. high; branches purplish, slightly silky-downy; leaf-blades lanceolate to ovate, glabrous above, often somewhat rusty, pubescent beneath; cymes flat; sepals oblong or oblong-lanceolate; fruit 6–8 mm. high, subglobose; stone broader than high, pointed at each end, longitudinally ridged. *C. Amomum* Mill. [G, B]. *C. sericea* L. Along streams: Newf.—N.D.—La.—Fla. *E. Temp.* My–Jl.

3. S. asperifòlia (Michx.) Small. Shrub, 1–5 m. high; branches brown, rough-pubescent; leaf-blades firm, elliptic to oval or ovate, 4–15 cm. long, rough-pubescent above, downy beneath; sepals minute, deltoid; petals oblong-lanceolate; fruit subglobose, 5–6 mm. broad; stone furrowed along the edge. *C. asperifolia* Michx. [G, B]. Sandy soil: Ont.—Minn.—Tex.—Ala. *Allegh.* My–Je.

4. S. intèrior Rydb. *Fig. 426.* A shrub, 2–5 m. high; bark of the old stems grayish; leaves elliptic or oval, acute at both ends, 5–9 cm. long, finely short-strigose on both sides and more or less villous on the veins and in their angles beneath; sepals minute; fruit white, about 5 mm. in diameter; stone elliptic, slightly oblique, longer than broad, nearly smooth. River banks: N.D.—Kans.—Colo.—Yukon. *Plain—Submont.* Je–Jl.

f 426.

5. **S. Baìleyi** (Coult. & Evans) Rydb. Shrub, 1–3 m. high; branchlets brownish, villous; leaf-blades ovate or ovate-lanceolate, strigose above, woolly beneath; sepals lanceolate; petals ovate; fruit globose, 6 mm. broad; stone flattened, oblique, broader than high, channeled on the edge. *C. Baileyi* Coult. & Evans [G, B]. Sandy shores: Ont.—Man.—Minn.—Pa. *Allegh.* My–Je.

6. **S. stolonífera** (Michx.) Rydb. Shrub, 1–3 m. high; branches bright red-purple, glabrous or nearly so; leaf-blades ovate, rounded at the base, 3–12 cm. long, finely strigose on both sides, whitish beneath; cyme flat; sepals minute; fruit globose, 6–7 mm. broad; stone higher than broad. *C. stolonifera* Michx. [G, B]. Wet places: Newf.—Mack.—Iowa—D.C. *Canad.—Allegh.* Je–Au.

7. **S. instolònea** A. Nels. Shrubs, 2–5 m. high, not stoloniferous; bark of the old stems gray; young twigs brownish; leaves usually oval or elliptic, acute, thin, light green and less pale beneath than in the eastern *S. stolonifera,* lanceolate to oval, acute or short-acuminate, obtuse or acute at the base, strigose on both sides, 3–12 cm. long; fruit white; stone higher than broad, 5 mm. long, 3–3.5 mm. broad, smooth. *S. stolonifera riparia* Rydb. River banks: Man.— Kans.—N.M.—Calif.—Alaska. *W. Temp.—Plain—Mont.* My–Jl.

8. **S. foémina** (Mill.) Rydb. Shrub, 1–3 m. high; branches gray, glabrous; leaf-blades firm, elliptic or lanceolate, 5–15 cm. long, sparingly strigose or glabrous, pale beneath; cymes convex; sepals triangular; fruit depressed-globose, 4–5 mm. high, on bright-red pedicels; stone broader than high, 3–4 mm. broad, faintly ribbed and furrowed. *C. candidissima* Marsh. [B]; not Mill. *C. paniculata* L' Hér. [G]. *S. candidissima* Small. Thickets and river banks: Me.—Man.—Neb.—Ga. *Canad.—Allegh.* Je–Jl.

9. **S. alternifòlia** (L. f.) Small. Shrub or tree, 2–6 m. high; branches greenish, striped with white; leaf-blades elliptic, oval, or ovate, 5–15 cm. long, acuminate at each end or rounded at the base, sparingly strigose when young, glabrate in age, whitish beneath; sepals minute; fruit depressed-globose, 8–10 mm. broad; stone broader than high, obovoid, 5–6 mm. broad, longitudinally furrowed. *C. alternifolia* L. f. [G, B]. Hillsides: N.B.—Minn. —Iowa—Ala.—Ga. *Canad.—Allegh.* My–Je.

2. CHAMAEPERICLÍMENUM Aschers. & Graebn. BUNCHBERRY.

Low perennial herbs, with rootstocks. Leaves opposite, or the upper whorled, entire. Flowers perfect in close head-like clusters, subtended by 4 petal-like bracts. Sepals 4, tipped each with a short deciduous bristle. Petals 4, valvate. Stamens 4. Ovary 2-celled, sessile. Ovules solitary in each cell. Drupe globular, red; stone 2-celled. *Cornella* Rydb.

1. **C. canadénse** (L.) Aschers. & Graebn. *Fig. 427.* Stem simple, 5–20 cm. high; leaves subsessile, mostly in an apparent whorl of 4–6 at the summit, oval, ovate, or obovate, acute at each end, 3–7 cm. long, and a pair of smaller ones at about the middle of the stem; bracts 4, white or cream-colored. *Cornus canadensis* L. *Cornella canadensis* Rydb. Woods: Lab.—N.J.—Minn.—N.M.—Calif.—Alaska. *Boreal. —Submont.—Mont.* My–Au.

f. 427.

3. CYNÓXYLON Raf. FLOWERING DOGWOOD.

Trees, with spreading branches and rough bark. Leaves opposite, petioled, rather thick, entire. Flowers perfect, regular, in thick clusters subtended by 4-8 petal-like, conspicuous bracts. Sepals 4, erect, persistent. Petals 4, greenish, valvate. Stamens 4, exserted. Ovary 2-celled; styles united; stigmas depressed; ovules solitary in each cavity. Fruit a red, elongate drupe.

1. **C. flóridum** (L.) Raf. Tree up to 15 m. high; leaf-blades elliptic or oval, 5-18 cm. long, acute or short-acuminate, sparingly strigose above, more densely pubescent and pale beneath; bracts 4, showy, white or pink, 4-6 cm. long, notched at the apex. Fruits clustered, oblong, 1.5 cm. long, red; stone pointed at both ends, 2-grooved. *Cornus florida* L. [G, B]. Woods: Me.—Minn.—Tex.—Fla. *E. Temp.* My-Je.

Family 126. PYROLACEAE. WINTERGREEN FAMILY.

Perennial, mostly evergreen herbs, with elongate rootstocks. Leaves basal or crowded on the short stem or at the ends of the branches, thick and leathery, entire or toothed. Flowers perfect, often slightly irregular, in racemes or corymbs. Sepals 4-5, persistent. Petals 4-5, wax-like. Stamens twice as many as the petals; filaments usually subulate; anthers introrse, becoming inverted in anthesis, opening by pores or slits. Gynoecium of 4 or 5 united carpels; ovary superior, 4- or 5-celled; styles united; stigma 5-lobed. Capsule loculicidal, valvate. Seeds minute, numerous.

Plants leafy-stemmed ; flowers corymbose ; style very short and ending in the peltate
    stigma ; filaments dilated and hairy at the middle   1 CHIMAPHILA.
Plants scapose, with a basal rosette of leaves ; flowers racemose or solitary ; style
    evident ; filaments subulate, naked.
    Flowers solitary ; petals spreading ; valves of the capsule not cobwebby on the
        margins.   2. MONESES.
    Flowers racemose ; petals more or less converging, concave ; valves of the capsule
        cobwebby on the margins when opening.
        Flowers without hypogynous disk ; petals without tubercles.
            Style exserted, deflexed, curved ; stigma narrower than the style
                        3. PYROLA.
            Style short, straight ; stigma thicker than the style, peltate.
                        4. BRAXILIA.
    Flowers with a 10-lobed hypogynous disk ; petals with a pair of tubercles at
        the base.   5. ORTHILIA.

1. CHIMÁPHILA Pursh. PIPSISSEWA, PRINCE'S PINE.

Perennial herbs, with more or less cespitose, horizontal rootstock and short leafy stems. Leaves persistent, coriaceous, opposite or whorled, short-petioled and serrate. Flowers perfect, in terminal corymbs. Sepals 5, persistent. Corolla white or pink, wax-like; petals 5, broad, concave, sessile. Stamens 10; filaments very short; anthers incurved; sacs opening by pores at the ends of the basal ascending beaks. Ovary 5-celled; stigma orbicular, peltate, barely 5-crenate; style very short. Capsule 5-celled, the latter loculicidal from the top; valves not cobwebby on the margins; seeds many.

Leaves lanceolate, broadest below the middle, mottled along the veins ; dilated por-
    tion of the filaments hairy.   1. *C. maculata.*
Leaves narrowly oblanceolate, broadest below the middle, not
    mottled ; dilated portion of the filaments merely ciliate
    on the margins.
    Leaves faintly veined ; pedicels erect or strongly ascending.   2. *C. occidentalis.*
    Leaves prominently veined ; pedicels in anthesis recurved,
        in fruit ascending.   3. *C. corymbosa.*

1. C. maculàta (L.) Pursh. Stem 1 dm. high
or less; leaves in verticils of 2–4, short-petioled;
blades lanceolate, acute, sharply serrate with few
teeth, 1.5–7 cm. long, dark green above, mottled
along the veins, reddish brown beneath; flowers nodding; sepals elliptic, obtuse, ciliolate; petals white
or rose; capsule depressed-globose, 7 mm. broad.
Woods: Mass. — Minn. (?) — Ala. — Ga. *Canad.* —
Allegh. Je–Jl.

2. C. occidentàlis Rydb. *Fig. 428.* Stems 1–2
dm. high, terete; leaves whorled; blades 3–10 cm.
long, oblanceolate, acute, sharply serrate, dark green
and shining above, pale beneath; peduncles 5–10 cm.
long, corymbosely 4–7-flowered; bracts narrow;
sepals suborbicular or orbicular-ovate, 2 mm. long,
rounded and erose-ciliate; petals nearly orbicular,
ciliate on the margin; capsule depressed-globose, 5–6 mm. in diameter. *C. umbellata* Am. auth., in part; not Nutt. In dry woods: S.D.—Mont.—N.M.—
Calif.—Alaska. *Submont.—Mont.* Je–Au.

3. C. corymbòsa Pursh. Stem 5–15 cm. high, more or less angled; leaves
whorled; blades narrowly oblanceolate, 2.5–7 cm. long, dark green and shining
above, pale beneath, sharply serrate; pedicels in anthesis divergent, in fruit
erect; sepals broadly ovate, erose; petals elliptic, greenish white; capsule obo-
void-globose, 6 mm. broad. *C. umbellata* W. Barton [B, G]; not Nutt.
Woods: N.S.—Minn.—Ga. *Canad.—Allegh.* Jl–Au.

2. MONÈSES Salisb. One-flowered Wintergreen, Single Beauty.

Perennial herbs, scapose or short-stemmed, with slender rootstocks. Leaves
persistent, coriaceous, opposite or in 3's, short-petioled, crenate or serrate.
Flowers perfect, solitary, on a long peduncle, nodding. Sepals 5 (rarely 4),
persistent. Petals white or rose-colored, 5 (rarely 4), orbicular or broadly
ovate, spreading. Stamens 10 or 8; filaments short, subulate, somewhat dilated,
incurved. Ovary 4 or 5-celled; style straight; stigma peltate, with 4 or 5 narrow, at first erect, at length radiating lobes. Capsule depressed-globose, 4 or 5-
celled, loculicidally 4 or 5-valved from the summit; valves not woolly on the
edges.

1. M. uniflòra (L.) A. Gray. Stem 5–15 cm. high; leaves mostly crowded
near the base; blades 8–25 mm. long, acute or rounded at the base, thin;
sepals oval, obtuse, 3 mm. broad, ciliolate; petals ovate to orbicular, obtuse,
8–12 mm. long; capsule 6–8 mm. in diameter. In woods: Greenl.—Pa.—Colo.
—Ore.—Alaska; Eurasia. *Arct.—Boreal.—Subalp.* Je–Au.

3. PÝROLA (Tourn.) L. Wintergreen.

Scapose perennials, with slender, stoloniferous rootstocks. Leaves persistent, firm, subcoriaceous, mainly basal, or in one species usually none and
replaced by scales. Flowers racemose, nodding, perfect. Sepals 5, persistent,
spreading. Petals 5, concave, sessile. Stamens 10; filaments declined; anthers
contracted under the terminal pores. Ovary 5-celled; hypogynous disk none;
style long, declined and curved upwards towards the end, which is enlarged,
truncate and concave and forming a ring or collar; stigma much narrower,
5-lobed. Capsule 5-celled, 5-lobed, loculicidally 5-valved; valves opening from
the base and cobwebby on the margins.

Leaves not mottled.
  Petals pink or purplish.
    Leaf-blades round-reniform to orbicular.   1. *P. asarifolia.*
    Leaf-blades orbicular to rounded-ovate.   2. *P. uliginosa.*
  Petals white or greenish.

Sepals lanceolate, much longer than broad; leaf-blades
suborbicular.   3. *P. americana.*
Sepals broadly triangular, as broad as long.
Leaf-blades oval, longer than the petioles.   4. *P. elliptica.*
Leaf-blades orbicular, usually shorter than the petioles.   5. *P. chlorantha.*
Leaves mottled; leaf-blades ovate, acute.   6. *P. picta.*

1. **P. asarifòlia** Michx. Leaf-blades reniform
to orbicular, usually broader than long, shining,
finely crenulate, 2.5–6 cm. wide; scape 1–3 dm. high;
bracts lanceolate; sepals lanceolate, acute or acuminate, about 2 mm. long; petals oval, pink or purplish,
5–6 mm. long. Wet woods and swamps: N.B.—N.Y.
—Minn. -N.M.—B.C. *Submont.—Mont.* Je–Jl.

2. **P. uliginòsa** Torr. *Fig. 429.* Leaf-blades
broadly oval to orbicular, usually longer than broad,
shining, finely crenulate, 2–6 cm. broad, 2.5–8 cm.
long; scape 1 3 dm. high; bracts lanceolate, usually
purplish; sepals ovate or triangular-lanceolate, acute
or short-acuminate, about 2 mm. long; petals oval
or obovate, pink or purplish, 5–7 mm. long. *P. rotundifolia incarnata* A. Gray, scarcely DC. *P. asari-
folia incarnata* Fern. [G]. Wet woods and bogs:
N.S.—N.Y.— Minn.—Colo.- -Calif.- -D.C. *Submont.--Subalp.* Je–Jl.

f. 429.

3. **P. americàna** Sweet. Leaf-blades suborbicular, tapering at the base,
rounded or retuse at the apex, 2.5–8 cm. long, entire or crenulate, bracts lanceolate, 7–8 mm. long; sepals lanceolate, 3–3.5 mm. long; petals white, rounded
obovate, 7 mm. long. *P. rotundifolia* Michx. [B]; not L. Woods: N.S.—
Man.--Ky.—N.C. *Canad.—Allegh.* Je –Au.

4. **P. elliptica** Nutt. Leaf-blades oval, acute at the base, rounded, retuse
or short-acuminate at the apex, rather thin, crenulate and the teeth often cal-
lous-mucronate, 1.5–6 cm. wide, 2–8 cm. long; bracts narrowly lanceolate to
subulate; sepals triangular-ovate, acute or acuminate, often less than 2 mm.
long; petals greenish white, oval, 5–6 mm long. Rich woods: N.S.—D.C.—
Minn.—N.M.—B.C. *Boreal.—Submont.--Mont.* Je Au.

5. **P. chlorántha** Sw. Leaf-blades orbicular to broadly oval or obovate,
crenulate, thick, but rather dull, rounded or acute at the base, 1–4 cm. wide;
sepals triangular-ovate, about 1.5 mm. long, obtuse or acute; petals greenish-
white, oval, about 7 mm. long. In woods and swampy places: Lab.—D.C.—
Calif.—B.C.; n Eur. *Boreal.—Subalp.* Je–Jl.

6. **P. pícta** Smith. Leaf-blades ovate or oval, acute at both ends, cal-
lous-denticulate, firm and coriaceous, blotched white above and purple beneath,
1–5 cm. wide, 1–6 cm. long; sepals triangular-ovate, acute, about 1.5 mm. long;
petals oval, greenish or purplish, about 6 mm. long. Woods: B.C.—Calif.—
Ariz.—Colo.—S.D.--Mont. *Submont.—Mont.* Jl–Au.

4. BRAXÍLIA Raf.

Scapose perennials, with stoloniferous rootstocks. Leaves persistent, basal,
firm, petioled. Flowers racemose, perfect. Sepals 5, persistent, spreading.
Petals 5, concave. Stamens 10; filaments erect, connivent; anthers opening by
oblique pores at the end, not produced into tubes. Ovary 5-celled; style
straight, erect, in ours short, without a collar; stigma peltate, thicker than the
style, with 5 marginal papillae. Capsule as in *Pyrola*. *Erxlebenia* Opiz.

1. **B. mìnor** (L.) House. Leaf-blades oval or orbicular, 1–3 cm. wide,
1–4 cm. long, finely crenulate, light green; sepals ovate or triangular-ovate, 1.5
mm. long, acute or acutish; petals white or pinkish, orbicular, 4–5 mm. long;
capsule 5–6 mm. in diameter. *Pyrola minor* L. [G, B]. In woods: Greenl.—
Conn.—Colo.—Calif.—Alaska; Eurasia. *Arct.—Boreal.—Subalp.* Je–Au.

5. ORTHILÌA Raf.

Perennials, usually with a short leafy stem and slender, stoloniferous rootstocks. Leaves persistent, but thin. Flowers racemose, nodding and decidedly secund. Sepals 5, persistent, spreading. Petals oblong, erect, with a pair of tubercles at the base within. Stamens 10; filaments slender, not declined, all equally connivent; anthers not contracted beneath the pores. Ovary 5-celled; hypogynous disk present and 10-lobed; style long and straight; stigma peltate, 5-lobed. Capsule 5-celled, 5-lobed, loculicidally 5-valved; valves opening from below, cobwebby on the margins. *Ramischia* Opiz.

1. O. secúnda (L.) House. *Fig. 430.* Leaf-blades ovate, oval or lanceolate, thin, acute at both ends or rarely obtuse, 2–5 cm. long, 1–3 cm. wide, crenulate or serrulate; sepals less than 1 mm. long, triangular; petals oblong, greenish-white, about 4 mm. long. *Pyrola secunda* L. [G, B]. In woods: Lab.—D.C.—Calif.—Alaska; Eurasia. *Boreal.—Subalp.* Je–Au.

f. 430.

Family 127. MONOTROPACEAE. INDIAN PIPE FAMILY.

Saprophytic herbs or root-parasites, with densely matted roots. Leaves scale-like, destitute of chlorophyll. Flowers perfect, usually drooping. Sepals 2–6, distinct, imbricate, deciduous. Petals distinct or partially united, rarely wanting. Stamens 6–12, hypogynous; filaments distinct or united at the base; anthers 2-celled or with confluent sacs, opening by valves or pores. Gynoecium of 1–6 united carpels; ovary 1–6-celled, superior; styles united; stigma capitate, disk-like, or funnelform. Ovules numerous, anatropous. Fruit a 1-celled loculididal capsule. Seeds numerous.

Flowers in elongate racemes; corolla gamopetalous, globular or nearly so; anther 2-awned.   1. PTEROSPORA.
Flowers in short, few-flowered racemes, or solitary; corolla of 4–6 erect, distinct petals; anthers awnless.
Flowers solitary; stigma naked.   2. MONOTROPA.
Flowers racemose; terminal flower 5-merous, the lateral ones 3- or 4-merous; stigma glandular or hairy on the margins.   3. HYPOPITYS.

1. PTERÓSPORA Nutt. PINE-DROPS, GIANT BIRD'S-NEST.

Purplish or brown plants, without proper leaves, and with numerous roots forming rounded masses often 5 cm. or more in diameter. Flowers in long, many-flowered racemes. Sepals 5, oblong. Corolla globose, gamopetalous, with 5 reflexed lobes. Stamens 10, included; filaments subulate, glabrous; anthers introrse, the sacs longitudinally dehiscent. Ovary 5-lobed, 5-celled; style short, straight; stigma capitate, 5-lobed. Capsule depressed-globose, 5-valved. Seeds with a terminal reticulate wing.

1. P. Andromedèa Nutt. *Fig. 431.* Plant 2–15 dm. high, viscid-hairy; scales lanceolate or linear, numerous; sepals oblong, obtuse; corolla 6-8 mm. long, whitish; capsule 8–12 mm. in diameter. In rich woods: Que.—Pa.—S.D.—N.M.—Calif.—B.C.; n Mex. *Boreal.—Submont.* Je–Au.

f. 431.

2. MONÓTROPA L. INDIAN PIPE.

White or pink, leafless plants, with sessile scales. Flowers perfect, solitary, nodding, but in fruit erect. Sepals 2–4, similar to the upper bracts. Petals 5 or 6, somewhat dilated at the apex, erect, tardily deciduous. Stamens 10–12; filaments linear-subulate, more or less pubescent; anthers becoming equally and transversely 2-valved. Hypogynous disk 10–12-toothed. Ovary 5-celled; styles very short; stigma funnelform, crenate on the edge. Capsule 5-celled, loculicidally 5-valved, erect. Seeds numerous; testa produced at both ends.

1. **M. uniflòra** L. *Fig. 432.* Plant 1–3 dm. high, white or pink, turning black in drying; flowers 1.5 2 cm long; petals abruptly dilated above, strigillose within. Deep woods: Newf.—Fla.—Calif.—B.C.; Mex.; Japan to India. *Temp.—Submont.* Je–Au.

3 HYPÓPITYS (Dill.) Adans. PINESAP.

Yellowish or reddish, violet-scented, leafless plants, with sessile scales. Flowers perfect, few or several in a raceme, the terminal one 5-merous, the lateral ones 3–4 merous. Calyx regular; sepals not resembling the bracts. Petals distinct, as many as the sepals, saccate at the base. Stamens 6–10, anthers horizontal, opening by two transverse, unequal valves. Hypogynous disk, 6–10-toothed. Ovary 3–5-celled; style short; stigma funnelform or disklike. Capsule 3–5-celled, erect, loculicidally 3–5-valved. Seeds numerous.

1. **H. lanuginòsa** (Michx.) Nutt. Plant deep-pink or crimson, pubescent; scales of the stem ovate or lanceolate; sepals 6,5–7 mm. long, cuneato-spatulate, long-ciliate; petals oblong-cuneate, pubescent, ciliate; capsule ovoid, 4–6 mm. long; style copiously pubescent; stigma retrorsely bearded. *Monotropa Hypopitys* A. Gray [G]; not L. *Hypopitys Hypopitys* Small [B]. Woods: Newf.—Minn.—La.—Fla. *Temp.* Je–Au.

Family 128. ERICACEAE. HEATH FAMILY.

Perennial herbs, shrubs, or trees, with alternate, opposite, or whorled leaves, commonly leathery and persistent, without stipules. Flowers perfect. Sepals 4 or 5, or rarely 6–10, distinct or partially united. Corolla regular or slightly irregular; petals as many as the sepals, usually more or less united. Stamens as many or twice as many, hypogynous; anthers two-celled, sometimes prolonged into terminal tubes, opening by terminal pores or lengthwise, often with horn-like awns. Gynoecium of 2–5 united carpels; ovary with as many cells; styles united; stigma capitate or peltate. Fruit usually a capsule, sometimes a berry or drupe.

Fruit a capsule.
    Capsule septicidal, the valves separating from the central columella, which bears the placentae; anthers awnless.
        Petals distinct or nearly so, spreading.   1. LEDUM.
        Petals united, forming a gamopetalous corolla.
            Corolla without pouches.
                Corolla campanulate; leaves evergreen; stamens usually 5.
                    Stamens exserted; anthers opening by a chink at the top.
                         2. RHODODENDRON.
                    Stamens included; anthers opening lengthwise.
                         3. LOISELEURIA.
                Corolla urn-shaped, 4-toothed; leaves deciduous; stamens 5–10, mostly 8.   4. MENZIESIA.
        Corolla rotate or openly bell-shaped, with 10 pouches each enclosing an anther in bud; stamens 10.   5. KALMIA.
    Capsule loculicidal, the valves carrying with them the partitions; anthers awned, except in *Epigaea* and *Chamaedaphne*.

Calyx and hypanthium not accrescent, nor berry-like.
　Anther opening by terminal pores; corolla not salverform.
　　Calyx-lobes valvate in the bud; leaves neither imbricate nor scale-like;
　　　corolla globose-urceolate.　　　　　　　　　6. ANDROMEDA.
　　Calyx-lobes imbricate in the bud.
　　　Leaves neither scale-like nor imbricate; anthers awnless.
　　　　　　　　　　　　　　　　　　　　　7. CHAMAEDAPHNE.
　　　Leaves scale-like, densely imbricate; anthers awned.
　　　　　　　　　　　　　　　　　　　　　8. CASSIOPE.
　Anthers opening by longitudinal slits, awnless; corolla salverform.
　　　　　　　　　　　　　　　　　　　　　9. EPIGAEA.
Calyx and hypanthium accrescent, becoming fleshy and berry-like, enclosing
　the capsule; anthers 4-awned.　　　　　　　10. GAULTHERIA.
Fruit a berry or drupe.
　Leaves thin, deciduous; fruit fleshy.　　　　　12. ARCTOUS.
　Leaves evergreen, leathery; fruit mealy.　　　　11. ARCTOSTAPHYLOS.

1. LÈDUM L. LABRADOR TEA.

Resinous shrubs, with scaly buds and fragrant foliage. Leaves alternate, thick and leathery, with more or less revolute margins. Flowers in terminal corymbs. Bracts deciduous. Calyx persistent; sepals 5. Petals 5, obtuse, spreading, imbricate in the bud. Stamens 5–10, exserted; filaments filiform; anthers didymous, the sacs opening by terminal pores. Disk annular, 8–10-lobed. Ovary 5-celled, usually covered by scales; style filiform; stigma 5-lobed. Capsule oblong or ovate, 5-celled, septicidally 5-valved from the base.

Stamens 5–7; capsule oblong; leaves oblong to oval.　　　1. *L. groenlandicum.*
Stamens 10; capsule oval to obovoid; leaves linear.　　　2. *L. decumbens.*

1. L. groenlándicum Oeder. *Fig. 433.* Shrub 3–10 dm. high, with rusty-tomentose twigs; leaf-blades oblong or elliptic, obtuse at both ends, 1.5–5 cm. long, 0.5–2 cm. wide, dark green above, densely tomentose beneath, with rusty hairs; pedicels 2–2.5 cm. long, in fruit recurved; sepals minute, triangular; petals white, about 5 mm. long; capsule lance-ovoid, puberulent, 5–7 mm. long, 2–3 mm. thick. *L. latifolium* Ait. In bogs: Greenl.—N.J.—Wash.—Alaska. *Boreal.—Arctic.* Je–Jl.

2. L. decúmbens (Ait.) Lodd. Shrub depressed, 1–5 dm. high, with tomentulose twigs; leaves linear, strongly revolute, 1–1.5 cm. long, obtuse, glabrate above, tomentulose beneath; sepals ovate to half-or-bicular; petals oval, 6–7 mm. long; capsule 3–4 mm. long, slightly longer than thick. Swamps: Greenl.—Labr.—Man.—Alaska. *Arct.—Subarct.*

f. 433.

2. RHODODÉNDRON L. RHODODENDRON.

Shrubs or trees, with thick, alternate leaves. Flowers usually in clusters, perfect. Calyx 5-lobed, saucer-shaped. Corolla campanulate or rotate, deeply 5-lobed. Stamens 5–10, usually shorter than the corolla; filaments declined; anthers obovoid. Ovary 5-celled, lobed; style curved, enlarged at the apex. Capsule ovoid or oblong.

1. R. lappónicum (L.) Wahl. *Fig. 434.* A depressed shrub, less than 2 dm. high, much branched; leaves evergreen, oblong to oval, 1.5 cm. long, obtuse, resinous-dotted, slightly paler beneath; clusters 3–6-flowered; corolla purple, mottled, rotate-campanulate, 1 cm. wide; stamens 5–8, rarely 10; capsule 4–5 m. long. Rocky places: Greenl.—N.H.—N.Y.—Man.—Alaska; Eurasia. *Arctic.* Je–Jl.

f. 434.

3. LOISELEÙRIA Desv. ALPINE AZALEA.

Low prostrate shrubs. Leaves usually opposite, crowded, leathery, entire, revolute. Flowers in terminal, umbellate clusters. Calyx persistent; lobes 5, narrow. Corolla campanulate, white or pink, deciduous, the lobes broad, spreading. Stamens 5; filaments subulate, glabrous; anthers globular-didymous, opening by longitudinal slits. Ovary 2- or 3-celled, on a disk; style columnar; stigma truncate. Capsule ovoid or conic, 2 or 3-valved, the valves cleft. *Chamaecistus* Oeder.

1. L. procúmbens (L.) Desv. Shrub 1–3 dm. high; leaves oblong or oval, 3–8 mm. long, obtuse, revolute, 1-ribbed beneath, glabrous and glaucous beneath; corolla 4–5 mm. long, the lobes ovate; capsule 3–4 mm. long. *C. procumbens* Kuntze [B]. Arctic meadows: Greenl.—N.H. (White Mts.)—Que. (Gaspe)—Man.—Alaska. *Arct.—Alp.* Je–Jl.

4. MENZIÈSIA Smith.

Shrubs, with erect branching stems and alternate deciduous leaves. Flowers perfect, in terminal clusters from early buds. Calyx 4–5-lobed, often obsolete. Corolla urn-shaped, campanulate, or globose, usually 4-lobed. Stamens 5–10, mostly 8, included; filaments subulate, flattened; anthers linear-sagittate, awnless, opening by terminal oblique pores or chinks. Disk 8–10-lobed. Ovary 4- (rarely 5-) celled, style included; stigma truncate, with 4 or 5 lobes. Capsule septicidal, 4-valved.

1. M. glabélla A. Gray. Erect shrub, 2–3 m. high; leaf-blades obovate or elliptic, 3–6 cm. long, 1–3 cm. wide, green and sparingly short hairy above, glabrous or nearly so and pale beneath; corolla urceolate, 8–9 mm. long; capsule ovoid, about 6 mm. long, glabrous or minutely puberulent. Woods: B.C.—Ore.—Wyo.—Alta.—Minn. (?). *W. Boreal. Submont.* Je–Jl.

5. KÁLMIA L. AMERICAN LAUREL, SWAMP LAUREL.

Evergreen shrubs or trees, with alternate, opposite, or whorled leaves. Flowers perfect, in terminal or axillary corymbs or umbels, with early deciduous bracts. Sepals 5, persistent, leathery. Corolla rotate, pink or white, with 10 pouches at first enclosing the anthers, 10-keeled and rounded 5-lobed. Stamens 10; filaments shorter than the corolla, straightening elastically at maturity; anthers awnless, opening by terminal pores. Disk 10-lobed. Ovary 5-celled. Capsule globose, depressed at the apex, septicidal, 5-valved.

Flower-clusters lateral; leaf-blades oblong or lanceolate, not strongly revolute: corolla 1 cm. wide; pod 3–3.5 mm. broad, glandular.   **1. *K. angustifolia.***
Flower-clusters terminal; leaf-blades linear-oblong, revolute-marginal; corolla 1–1.5 cm. wide, pod 5–6 mm. basal, glabrous.   **2. *K. polifolia.***

1. K. angustifòlia L. Shrub, up to 1 m. high; leaf-blades 2–6 cm. long, bright green above, pale beneath, glabrate; calyx 5–6 mm. wide, the lobes ovate, abruptly pointed, somewhat glandular; corolla purple or crimson; capsule globose. Swamps: Newf.—Man.—Ga. *Canad.—Allegh.* Je–Jl.

2. K. polifòlia Wang. *Fig. 435.* Shrub, 3–7 dm. high, glaucous; leaf-blades 2–3 cm. long, obtuse, strongly white-glaucous beneath, subsessile; calyx 5–6 mm. wide, the lobes ovate to oblong; corolla rose-purple; pod 5–6 mm. broad. *K. glauca* Ait. [B]. Swamps: Lab.—Man.—Minn.—Pa. *Canad.* My–Jl.

6. ANDRÓMEDA L. BOG ROSEMARY.

Glabrous evergreen shrubs, with coriaceous, entire, revolute-margined leaves. Flowers perfect, in

f 435

terminal umbels. Sepals 5, persistent. Corolla globose-urceolate, 5-toothed, with recurved teeth. Stamens 10, included; filaments bearded, unappendaged; anthers short, ovate, obtuse; sacs opening by a terminal pore, each with an ascending awn. Disk 10-lobed. Ovary 5-celled; style columnar; ovules many. Capsule subglobose, 5-angled, 5-valved, loculicidal. Seeds with smooth testa, coriaceous, shining.

Leaf-blades glaucous beneath; pedicels in anthesis several times as long as the corolla or capsule, in fruit erect. 1. *A. polifolia.*
Leaf-blades canescent beneath; pedicels in anthesis once or twice as long as the corolla or capsule, in fruit recurved. 2. *A. glaucophylla.*

1. **A. polifòlia** L. Shrub, 1–3 m. high, with acid foliage; leaf-blades oblong to linear, 3–5 cm. long, 3–8 mm. wide, dark green above, white beneath, mucronulate, the margins usually strongly revolute; bracts ovate, glaucous; pedicels 1–2 cm. long; calyx-lobes triangular, acute, about 1 mm. long; corolla pink, about 6 mm. long. In swamps: Lab.—N.J.—Ida.—Wash.—Alaska; Eurasia. *Boreal.* Je–Jl.

2. **A. glaucophýlla** Link. A shrub, 1–3 m. high; leaf-blades linear-oblong, strongly revolute, 2–5 cm. long, 2–6 mm. wide, bright green above, finely canescent beneath; flowers nodding, in dense clusters on recurved branch-lets; pedicels less than 1 cm. long; calyx-lobes triangular, spreading; corolla pink, 5–6 mm. long; capsule globose, about 7 mm. broad. *A. canescens* Small. Swamps: Labr.—N.J.—Ind.—Minn.—Man. *Canad.* My–Jl.

7. CHAMAEDÁPHNE Moench. LEATHER LEAF.

Shrubs, with branching stems. Leaves alternate, leathery, slightly toothed. Flowers perfect, in 1-sided, leafy-bracted racemes. Calyx stellate, subtended by two bractlets; lobes 5, longer than the tube. Stamens 10, included; filaments subulate, glabrous; anther-cells awnless, produced into a tube, opening by terminal pores. Disk 10-lobed. Ovary 5-celled; style elongate; stigma entire. Capsule depressed, 5-valved, many-seeded. *Cassandra* D. Don.

1. **C. calyculàta** (L.) Moench. Shrub, 1–15 dm. high; leaves scurfy, the blades oblong to obovate, 1–5 cm. long, serrulate; racemes 2–12 cm. long; bracts conspicuous, resembling the leaves; calyx-lobes triangular, 1.5–2 mm. long, acute; corolla white, 6–7 mm. long, the lobes ovate, obtuse; capsule angular, spheroid, 4 mm. broad. *Cassandra calyculata* (L.) D. Don. Swamps: Newf.—Ga.—Minn.—B.C.—Alaska; Eur. *Boreal.* Ap–My.

8. CASSÌOPE D. Don. MOSS-PLANT, WHITE HEATHER.

Evergreen low branching shrubs, with thick, opposite, crowded, imbricate, 4-ranked leaves and axillary nodding flowers. Sepals usually 5, not bracted, imbricate, thickened at the base. Corolla campanulate, usually 5-lobed. Stamens 8–10, included; anthers attached to the filaments near the apex; sacs opening by large terminal pores and tipped with recurved awns. Disk 10-crenate. Ovary 4–5-celled; style slender, somewhat thickened below. Capsule globose or ovoid, loculicidal, 4–5-valved.

1. **C. tetragòna** (L.) D. Don. *Fig. 436.* Tufted shrub, with erect or ascending branches, 1–3 dm. high; leaves very thick, ovate, 2–5 mm. long, more or less pubescent; peduncles 1–2.5 cm. long, glabrous or nearly so; sepals ovate, 2–2.5 mm. long, acute; corolla white or rose-colored, 5–6 mm. long.

f.436.

Wet places: Greenl.—Lab.—Man.—B.C.—Alaska; Eurasia. *Arct.—Subalp.— Alp.* Jl–Au.

9. EPIGAÈA L. Trailing Arbutus, Ground Laurel.

Low evergreen shrubs, with creeping stems. Leaves alternate, leathery. Flowers perfect or dioecious, in axillary clusters. Calyx subtended by several bractlets; lobes 5, imbricate. Corolla salverform, the limb 5-lobed. Stamens 10, included; anthers awnless. Disk 10-lobed; style elongate; stigma 5-lobed. Capsule depressed-globose, 5-valved, many-seeded.

1. **E. rèpens** L. *Fig. 437.* Stem hirsute, 0.5–3 dm. long; leaves oblong to ovate or suborbicular, 2–10 cm. long, ciliate; bracts and calyx-lobes ovate or lanceolate, the latter 5–9 mm. long; corolla pink or white, the lobes spreading, half as long as the tube; capsule pubescent, 7–9 mm. broad. Woods: Newf.—Fla.—Ky.—Sask. *E. Temp.* Mr My.

10. GAULTHÈRIA (Kalm) L.
Creeping Wintergreen.

Shrubs or undershrubs with alternate evergreen leaves. Flowers perfect, solitary, axillary, or in axillary racemes. Hypanthium and calyx enlarging and becoming fleshy, enclosing the capsule and forming a berry-like fruit. Sepals 5. Corolla campanulate or urn-shaped. Stamens 10; filaments dilated below, included; anther-sacs opening by terminal pores, 2-awned or 2-pointed. Disk 10-toothed. Ovary and capsule 5-celled, 5-lobed.

f. 437.

1. **G. procúmbens** L. A low undershrub, minutely pubescent, with creeping rootstocks or stems, the branches ascending or erect, 3–15 mm. high; leaf-blades oval, ovate, or obovate, retuse, crenate or serrate, with bristle-tipped teeth, often variegated above; pedicels 4–6 mm. long, with two ovate scale-like bracts; calyx white, 3–4 mm. wide, with ovate lobes; corolla white, wax-like, ovoid-urceolate, 6–9 mm. long, the lobes ovate, recurved, villous within; fruit subglobose, 7–11 mm. broad, bright red or white. Checkerberry. Woods: Newf.—Ga.—Man. *Canad.—Allegh.* Jl–Au.

11. ARCTOSTÁPHYLOS Adans. Bearberry, Kinnikinnick, Manzanita.

Shrubs or small trees, with alternate, thick, evergreen leaves. Flowers perfect in terminal, bracteolate racemes, often pendulous. Sepals 5, persistent. Corolla urn-shaped, with 4–5 recurved lobes. Stamens 8–10; filaments dilated and hairy at the base; anthers with 2 reflexed awns on the backs, opening by terminal pores. Ovary 4–10-celled, with a single pendulous ovule in each cell. Fruit drupaceous, either with a 1–8-seeded stone or 1–8 one-seeded more or less coalescent stones. *Uva-ursi* Mill.

1. **A. Ùva-úrsi** (L.) Spreng. Depressed and trailing, diffusely branched shrub, forming patches 1–2 m. across; leaves short-petioled, coriaceous; blades spatulate, entire-margined, obtuse, glabrous or nearly so, 1–3 cm. long, finely reticulate; racemes short and crowded; sepals ovate, acute; corolla ovoid-urn-shaped, white, about 4 mm. long; fruit rather dry, insipid, mealy, red, 6–10 mm. in diameter. Woods: Lab.—N.J.—Colo.—Calif.—Alaska; Eurasia. *Boreal.—Mont.* My–Je.

12. ARCTOÙS (A. Gray) Niedzu. Alpine Bearberry.

Low cespitose shrubs, with shreddy bark and alternate deciduous leaves, clustered towards the ends of the branches. Flowers few, in fascicles from terminal scaly buds. Sepals 4–5, short. Corolla globose-urceolate, 4–5-toothed. Stamens 8–10, included; anther-sacs with 2 recurved dorsal awns. Fruit drupaceous, with 4–5 1-seeded nutlets. *Mairania* Neck.

Fruit black; corolla-lobes ciliate; leaves less than 3 cm. long.　　1. *A. alpina.*
Fruit red; corolla-lobes not ciliate; leaves usually more than 3
cm. long.　　　　　　　　　　　　　　　　　　　2. *A. erythrocarpa.*

1. A. alpìna (L.) Niedzu. *Fig. 438.* Depressed-
prostrate shrub, with branches 5–12 cm. high; leaf-
blades spatulate or obovate, tapering below into
short petioles, 2–3 cm. long, 1–1.5 cm. wide, crenate,
strongly veined, glabrous, except the ciliate margin;
corolla white or pink; fruit 5–7 mm. in diameter,
bluish-black when ripe. *Arctostaphylos alpina* (L.)
Spreng. [G]. *Mairania alpina* Desv. [B]. In alpine-
arctic localities: Greenl.—N.H.—B.C.—Alaska; Eurasia. *Arct.—Subalp.—Alp.* My–Je.

2. A. erythrocárpa Small. A depressed shrub,
1–2 cm. high; leaf-blades spatulate, 3–6 cm. long,
tapering below into short petioles, crenate, glabrous,
not ciliate; corolla white or pinkish, turning yellowish; fruit 7–10 mm. in diameter, bright red. Mountains: Man.—B.C.—Alaska. *Subarct.—Subalp.* My–
Je.

f. 438.

Family 129. **VACCINIACEAE.** Huckleberry Family.

Shrubs or small trees, or rarely delicate vines. Leaves alternate, simple, sometimes evergreen. Flowers perfect, clustered or solitary. Hypanthium well developed, more or less completely enclosing and adnate to
the ovary. Sepals 4 or 5. Corolla usually gamopetalous, 5- or 4-lobed,
rarely of free petals. Stamens twice as many as the lobes of the calyx or
corolla. Gynoecium of 4 or 5 united carpels. Ovary 4–10-celled, inferior,
crowned by an epigynous disk. Fruit a berry or drupe, pulpy.

Petals united into a gamopetalous corolla.
　Ovary half-inferior; berry white, acute; low creeping plants.　1. Chiogenes.
　Ovary wholly inferior; berry blue, black, or red, rounded; shrubs or undershrubs.
　　Ovary 10-celled; fruit berry-like; drupe with 10 nutlets.　2. Gaylussacia.
　　Ovary 4- or 5-celled, or incompletely 10-celled; fruit a many-seeded berry.
　　　Anthers exserted, 2-awned on the back; corolla open-campanulate.
　　　　　　　　　　　　　　　　　　　　　　　　　　3. Polycodium.
　　　Anthers included.
　　　　Corolla open-campanulate; hypanthium jointed to the pedicels.
　　　　　　　　　　　　　　　　　　　　　　　　　4. Batodendron.
　　　　Corolla globose, ovoid or urceolate.
　　　　　Filaments glabrous; anthers 2-awned on the back; leaves deciduous.
　　　　　　　　　　　　　　　　　　　　　　　　　5. Vaccinium.
　　　　　Filaments pubescent; anthers awnless.
　　　　　　Ovary and berry incompletely 10-celled, by a false partition
　　　　　　　intruding from the back of each cell; leaves thin, deciduous.
　　　　　　　　　　　　　　　　　　　　　　　　6. Cyanococcus.
　　　　　　Ovary and berry 4- or 5-celled; no false partitions; leaves
　　　　　　　leathery, persistent.　　　　　　　　7. Vitis-Idaea.
Petals distinct; delicate prostrate vines.　　　　　　　8. Oxycoccus.

1. CHIÓGENES Salisb. Creeping Snowberry, Moxie Plum,
Capillaire.

Evergreen shrubs, with prostrate stems. Leaves alternate, 2-ranked. Flowers solitary, subtended by two bracts. Sepals 4, free. Corolla campanulate,
with 4 rounded lobes. Stamens 8, included; filaments rough; anthers awnless;
sacs not prolonged into tubes at the apex. Ovary and berry 4-celled, the latter
white, mealy.

1. C. hispìdula (L.) T. & G. Stems slender, creeping, 1–3 dm. long, strigose; leaves short-petioled, dark green and glabrous above, pale and with scattered brown appressed hairs beneath, ovate, acute, 3–6 mm. long; pedicels very

short; corolla short-campanulate, 1.5–2 mm. long; berry hispid, 5–6 mm. in diameter. Cold woods: Newf.—N.C.—Minn.—Ida.—B.C. *Boreal.* My–Je.

2. GAYLUSSÀCIA H.B.K. Huckleberry.

Shrubs with alternate, deciduous or evergreen leaves; leaf-blades mostly entire, usually resinous-dotted. Flowers in axillary, drooping racemes. Sepals 5. Corolla campanulate to tubular, 5-lobed. Stamens 10, included; filaments distinct, winged; anther-cells prolonged into tubes. Ovary 10-celled. Ovules solitary in each cavity, pendulous. Fruit a drupe, with a 10-celled stone or 10 bony or horny nutlets.

1. G. haccàta (Wangenh.) C. Koch. Shrub, 1 m. high or less; twigs pubescent when young; leaf-blades oval to oblong, firm, entire-margined, ciliolate, 3–5 cm. long; racemes short, 1-sided; bracts and bractlets reddish; corolla obconic, red or reddish-green, 5–6 mm. long; fruit globose, black, 6–10 mm. broad, sweet. *G. resinosa* (Ait.) T. & G. [D]. Rocky woods and hillsides: Newf.—Sask.—Ill.—Ga. *Canad.—Allegh.* My–Je.

3. POLYCÒDIUM Raf. Buckberry, Deerberry, Squaw Huckleberry.

Shrubs, with alternate entire leaves. Flowers in simple or branched, bracted racemes. Sepals 5, persistent. Corolla open-campanulate, white, pink, or purplish-green. Stamens 10; filaments distinct; anthers exserted, 2-awned on the back, the cells prolonged into slender tubes. Ovary inferior, 5-celled. Berries subglobose, green or yellowish, mawkish. Seeds few.

Leaves and branchlets glabrous.   1. *P. neglectum.*
Leaves and branchlets pubescent.   2. *P. stamineum.*

1. P. neglèctum Small. Shrub branched, 1–1.5 m. high; leaf-blades thin, elliptic to oblong-lanceolate or oblanceolate, 3–10 cm. long, acute, narrowed at the base, bracts much smaller than the leaves; corolla white or pink; berry 5–8 mm. broad, green or yellow, hardly edible. *Vaccinium neglectum* Fern. [G]. Dry woods: Va.—Kans.—La.—Ala. *Austral.* Ap–My.

2. P. stamíneum (L.) Greene. *Fig. 439.* Shrub branching, 1–2 m. high; leaf-blades oblong or elliptic to oblong-lanceolate or oblanceolate, 3–7 cm. long, pale beneath; bracts leaf-like, but smaller; corolla greenish-white or purplish, berries globular, green, 1 cm. broad, tart. *V. stamineum* L. [G]. Dry woods and hillsides: Me.—Minn.—Kans.—Ala. —Ga. *Canad.—Allegh.* Ap–Je.

f. 439.

4. BATODÉNDRON Nutt. Farkleberry, Sparkleberry, Tree Huckleberry.

Shrubs or small trees. Leaves alternate, leathery. Flowers in leafy-bracted racemes or panicles. Calyx persistent; sepals 5. Filaments distinct; anthers 2-awned on the back, opening into slender tubes. Ovary 5-celled. Berry black, hardly edible.

1. B. arbòreum (Marsh.) Nutt. Shrub or small tree; leaves oval or obovate, 2.5–3 cm. long, entire or glandular-toothed, lustrous above, pale beneath; corolla open-campanulate, white or pink; berry 5–6 mm. broad. *Vaccinium arboreum* Marsh [G]. Sandy soil, often in woods: N.C.—Fla.—Tex.—Kans. *Austral.* May–Je.

5. VACCÍNIUM L. BILBERRY, WHORTLEBERRY, HUCKLEBERRY.

Low shrubs, with alternate, thin, deciduous leaves. Flowers on drooping pedicels, solitary or two to four together. Calyx-lobes 5 or 4, small. Corolla more or less urceolate, 5- or 4-toothed, rose-colored or white. Filaments glabrous; anthers 2-awned on the back, included. Stamens 8–10. Ovary and fruit 4–5-celled, without false partitions. Fruit sweet and edible, blue, black, or red, with or without a bloom.

Branches not angled.
  Leaves entire-margined.   1. *V. uliginosum.*
  Leaves serrate.   2. *V. caespitosum.*
Branches angled.
  Fruit purplish, black, or blue; leaves more than 1 cm. long.
    Leaves subentire, obtuse; fruit blue with a bloom.   3. *V. ovalifolium.*
    Leaves serrate, acute or acuminate; fruit black without a bloom.   4. *V. membranaceum.*
  Fruit bright red; leaves usually less than 1 cm. long, serrate, light green on both sides.   5. *V. scoparium.*

1. V. uliginòsum L. Shrub, 1–6 dm. high; leaves glabrous, green above, pale beneath, 1–2 cm. long; corolla pink, ovoid-urceolate, only slightly contracted at the throat, 4–5-toothed; berry blue, with a bloom, globose-ellipsoid, about 6 mm. in diameter. Bogs: Greenl.—N.Y.—Man.—B.C.—Alaska; Eurasia. *Arct.—Boreal.* My–Jl.

2. V. caespitòsum Michx. Shrub 0.5–3 dm. high; leaves rather thin, obovate, cuneate or oblanceolate, glabrous; corolla ovoid-urceolate, 4–5 mm. long; berry blue, with a bloom, 6–8 mm. in diameter. Alpine-arctic situations: Lab.—Minn.—N.H.—Colo. —B.C.—Alaska. *Arct.—Subalp.—Alp.*

3. V. ovalifòlium Smith. Shrub 2–3.5 dm. high; leaf-blades elliptic, oval, or oblong-oval, entire-margined, bright green above, pale beneath, 3–5 cm. long; corolla ovoid-urceolate, about 8 mm. long and 5–6 mm. in diameter; berry 8–10 mm. in diameter. Woods: Alaska—Ore.—Mich.—Que. *Boreal.* —Mont. My–Jl.

4. V. membranàceum Dougl. *Fig. 440.* Shrub, 3–15 dm. high; leaf-blades thin, oval, only slightly paler beneath, 3–5 cm. long; corolla ovoid-urceolate, about 5 mm. long, 4 mm. wide; berry 8–10 mm. in diameter. *V. myrtilloides* Hook. Woods: B.C.—Calif.—Wyo.—Mich. *Boreal.—Mont.* Je–Jl.

f. 440.

5. V. scopàrium Leiberg. Shrub, 1–2 dm. high; leaf-blades light green, thin, 1 cm. or less, ovate or lance-ovate, acute at both ends, serrulate; corolla ovoid-urceolate, 3 mm. or less long; berry red, about 5 mm. in diameter. *V. erythrococcum* Rydb. GROUSE-BERRY. Mountain sides: B.C.—Calif.—Colo.— S.D.—Alta. *Submont.—Subalp.* Je–Jl.

6. CYANOCÓCCUS (A. Gray) Rydb. BLUEBERRY.

Shrubs with alternate, thin, deciduous leaves. Flowers in fascicles or very short racemes, developed with the leaves, from separate scaly buds; bracts and bractlets scaly, mostly deciduous. Calyx-lobes 5, usually small. Corolla in ours campanulate, white or slightly rose-colored; lobes 5. Stamens 10; filaments hairy; anthers awnless. Ovary and fruit incompletely 10-celled, by false partitions or projections from the back of each carpel; berry blue or black, with a bloom, sweet and edible, many-seeded.

Corolla urn-shaped, 1–2 times as long as broad, shrub 2 m. high or less.
  Leaf-blades glaucous beneath, oval or elliptic.   1. *C. vacillans.*

Leaf-blades not glaucous beneath, lanceolate.
Leaves entire, as well as the branchlets pubescent.   2. *C. canadensis.*
Leaves serrate, glabrous, except the veins and mar-
gins; branchlets glabrous or pubescent in lines.
Leaf-blades lanceolate.   3. *C. pennsylvanicus.*
Leaf-blades linear-lanceolate.   4. *C. angustifolius.*
Corolla cylindro-urceolate, 2–3 times as long as broad; shrub
1–4 m. high.   5. *C. corymbosus.*

1. **C. vacíllans** (Kalm) Rydb. Shrub, 1–2 m. high; branchlets yellowish-
green, glabrous; leaf-blades 2–5 cm. long, apiculate, entire or slightly serrulate,
glabrous, glaucous beneath; racemes short and dense, few-flowered; corolla pink
or whitish, 5–7 mm. long; berries globose, blue, with a bloom, 4–7 mm. broad,
sweet. *Vaccinium vacillans* Kalm. [G, B]. Dry places: Me.—Iowa—Kans.—
Ga. *Canad.—Allegh.* Jl–Au.

2. **C. canadénsis** (Richards.) Rydb. Shrub, 3–6 dm. high; leaves broadly
lanceolate, acute at both ends, softly pubescent, especially beneath, 2–4 cm.
long, 8–18 mm. wide; corolla cylindro campanulate, about 4 mm. long and 3
mm. in diameter; berry bluish-black, with a bloom, 5–8 mm. in diameter. *Vaccinium canadense* Richards. [G, B]. Moist places: Lab.—Va.—Kans.—Sask.
Canad. My–Je.

3. **C. pennsylvánicus** (Lam.) Rydb. Shrub, 1–4 dm. high, with warty
branches; leaves oblong or oblong-lanceolate, slightly pubescent on the veins
beneath and the margins, acute at both ends, 2–4 cm. long; corolla as in the
preceding; berry 6–10 mm. in diameter, bluish-black, with a bloom, very
sweet. *V. pennsylvanicum* Lam. [G, B]. Hillsides and woods: Newf.—N.J.—
Ill.—Sask. *Canad.* My–Je.

4. **C. angustifólius** (Ait.) Rydb. Shrub, 1–3 dm. high, with warty, glabrous branches; leaves linear-lanceolate, 1–2 cm long, 2–6 mm. wide, glabrous
on both sides, or slightly ciliolate on the midrib beneath, acute at each end;
berry 5–6 mm. in diameter, bluish-black. *V. angustifolium* Ait. *V. pennsylvanicum angustifolium* A. Gray [B, G]. *V. fissum* Schrank. High mountains
or subarctic regions: Labr.—Mich.—Man. *Arct.—Huds.* Je–Jl.

5. **C. corymbòsus** (L.) Rydb. Shrub, 1–4 m. high; branchlets puberulent
or glabrous; leaf-blades oblong or elliptic, 2.5–8 cm. long, acute at both ends,
entire, ciliate, pubescent along the veins beneath; corolla white or pinkish,
6–10 mm. long; berries subglobose, blue-black, with a bloom, 7–10 mm. broad.
V. corymbosum L. [G, B]. Swamps: Newf.— Me.—Minn.—La.—Fla. My–Je.

7. **VÌTIS-IDAÈA** (Tourn.) Moench. Mountain Cranberry,
Swedish Cranberry.

Low shrubs, with coriaceous, persistent leaves. Flowers in short racemes
from separate buds. Calyx-teeth small, 4 or 5. Corolla urceolate or campanulate, 4- or 5-toothed. Stamens 8–10; filaments hairy; anthers awnless. Ovary
and fruit 4–5-celled, without false partitions. Fruit in ours red, sour.

1. **V. punctàta** Moench. Low shrub, 1–5 dm. high; leaves thick, persistent, dark green and shining above, pale and black-dotted beneath, glabrous,
obovate or oval, obtuse or rounded at the apex, 5–15 mm. long, with revolute
margins; corolla campanulate, about 4 mm. long, 4-lobed; stamens 8; berry
8–10 mm. in diameter, red. *Vaccinium Vitis-Idaea* L. [G, B]. *Vitis-Idaea
Vitis-Idaea* (L.) Britton [BB, R]. Rocky places and open woods: Greenl.—
Mass.—Minn.—Alta.—Alaska; Eurasia. Je–Jl.

8. **OXYCÓCCUS** (Tourn.) Hill. Cranberry.

Delicate trailing or creeping vines. Leaves alternate, persistent, entire,
leathery, with more or less revolute margins. Flowers solitary or few together,
on slender pedicels. Sepals 4. Petals 4, pink or red, distinct, narrow, and re-

curved.   Stamens 8; anther-sacs produced into slender tubes, opening with terminal pores.   Ovary and fruit 4-celled.   Fruit globose or ellipsoid, red, sour.

Leaves ovate or lanceolate, strongly revolute, acute, 4–8 mm. long; fruit globose, 6–9 mm. in diameter; flowers strictly terminal; bracts subulate or filiform. Stems glabrous; leaves 4–6 mm. long, obtuse; petals 5–7 mm. long.

                                                        1. *O. palustris.*

   Stems puberulent; leaves 2–4 mm. long, acutish; petals 3–4 mm. long.

                                                        2. *O. microcarpus.*

Leaves elliptic or oblong, obtuse, 6–14 mm. long; flowers mostly lateral; bractlets lanceolate or oblong, foliaceous; fruit rounded-ellipsoid.

                                                       3. *O. macrocarpus.*

f.441.

   1. O. palùstris Pers.   Stems slender, creeping and rooting, 1–4 dm. long; leaves thick, rounded or cordate at the base, dark green above, white beneath, 4–6 mm. long; flowers in small umbel-like corymbs from scaly terminal buds; petals oblong, 5–7 mm. long; berry globose, 6–9 mm. in diameter, usually spotted. *Vaccinium Oxycoccus* L. [G]. *O. Oxycoccus* MacMill.   Cold bogs: Greenl.—N.J.—Mich.— Wash.—Alaska; Eurasia. *Boreal—Subarctic.* My– Jl.

   2. O. microcárpus Turcz.   Stem creeping, puberulent, 1–3 dm. long; leaves as in the preceding but smaller and acutish; flowers and fruit also similar but smaller. *V. microcarpum* (Turcz.) Hook.   Bogs: Greenl.—Labr.—Man.—Alta.—Alaska. *Arctic.* Je–Jl.

   3. O. macrocárpus (Ait.) Pers. *Fig. 441.* Stems 3–10 dm. long; leaves oblong-elliptic, usually rounded at both ends, 8–15 mm. long, green above, pale beneath, slightly revolute; inflorescence racemose, from a terminal bud, nearly always produced into a leafy branch; berry nearly always longer than broad, 10–18 mm. in length, not spotted. Bogs: Newf.— N.C.—Minn.—Sask. *Boreal.* Je–Au.

Family 130. **PRIMULACEAE.** PRIMROSE FAMILY.

   Caulescent or scapose herbs.   Leaves opposite, alternate, or whorled, without stipules.   Flowers perfect, in ours regular.   Sepals 4–9, partially united.   Corolla hypogynous, usually gamopetalous, or rarely wanting; lobes 4–9.   Stamens as many as the calyx-lobes and alternate with them, partly adnate to the corolla-tube or the calyx; filaments sometimes united at the base.   Pistil single; ovary free or nearly so, 1-celled, with a central placenta.   Fruit a 1-celled capsule, opening by 2–8 valves.

Ovary wholly superior and free.
  Corolla present.
    Corolla-lobes erect or spreading; stamens distinct, except in *Lysimachia*.
      Corolla salverform or funnelform, its lobes imbricate; stamens included; plants scapose; leaves in rosettes, either on the basal crown or at the ends of short branches of the caudex.
        Corolla-tube equaling or exceeding the calyx; style filiform; scapose perennials, simple, bearing a single tuft of leaves on the short crown.   1. PRIMULA.
        Corolla-tube shorter than the calyx, its throat constricted; style very short; scapose annuals.   2. ANDROSACE.
    Corolla rotate; its lobes convolute or involute in bud; stamens exserted; plants leafy-stemmed.
      Capsule opening lengthwise; erect plants.
        Flowers mostly 7-merous; corolla white; proper leaves clustered near the top of the stem, the lower ones reduced.
                                    3. TRIENTALIS.
        Flowers mostly 5-merous; leaves scattered on the stem, opposite or verticillate.
          Staminodia wanting; filaments united at the base; flowers solitary or racemose.   4. LYSIMACHIA.

Staminodia present; filaments distinct.
  Flowers solitary; staminodia conspicuous; corolla-lobes
    broad. 5. STEIRONEMA.
  Flowers in short axillary spikes; staminodia tooth-like;
    corolla-lobes long and linear. 6. NAUMBURGIA.
Capsule circumscissile; low depressed herbs.
  Corolla longer than the calyx; filaments pubescent, adnate to the
    base of the corolla. 7. ANAGALLIS.
  Corolla shorter than the calyx; filaments glabrous, adnate to the
    corolla-tube. 8. CENTUNCULUS.
Corolla-lobes reflexed; stamens more or less monadelphous; plants scapose.
    10. DODECATHEON.
Corolla wanting; calyx with 5 petaloid lobes; flowers solitary, sessile, axillary.
    9. GLAUX.
Ovary partly inferior, the lower portion included in and adnate to the hypanthium;
  plant leafy-stemmed. 11. SAMOLUS.

1. PRÍMULA L. PRIMROSE, COWSLIP.

Perennial scapose herbs. Leaves all in a basal rosette. Flowers in ours
umbellate. Calyx tubular, campanulate or funnelform, persistent, usually
angled, 5-lobed, the lobes imbricate. Corolla salver-shaped or trumpet-shaped,
with a funnelform throat, the tube longer than or equaling the calyx. Stamens 5, distinct, inserted in the tube or the throat of the corolla. Style filiform; stigma capitate. Capsule 1-celled, 5-valved at the apex, many seeded.

Leaves and calyx slightly if at all farinose; scape very slender.
  Mature capsule less than twice as long as the calyx, 2-5 mm. thick; leaves
    usually dentate.
    Pedicels in anthesis little if at all surpassing the bracts; leaf-blades oblance-
      olate, gradually tapering at the base; corolla limb 5–8 mm. wide, the
      lobes shallowly notched. 1. P. stricta.
    Pedicels much exceeding the bracts, usually several times
      as long; leaf-blades obovate, abruptly contracted
      at the base; corolla-limb 8–20 mm. wide; the lobes
      deeply notched.
      Calyx about half as long as the corolla-tube, its lobes
        oblong, obtuse; leaves thin. 2. P. mistassinica.
      Calyx at least two thirds as long as the corolla-tube,
        lobes lanceolate, acutish; leaves somewhat thicker. 3. P. MacCalliana.
  Mature capsule 2–3 times as long as the calyx, about 1 mm.
    thick; leaves entire or merely undulate; corolla 5–9 mm.
    wide. 4. P. egaliksensis.
Leaves and calyx usually more or less farinose; scape more
  stout; leaf-blades oblanceolate. 5. P. laurentiana.

1. P. strícta Hornem. Leaf-blades oblanceolate or lance-obovate, 1–4 cm. long, 5–15 mm. wide;
scape 5–30 cm. high; bracts lance-subulate, gibbous
at the base, 3–8 mm. long; calyx-lobes oblong, obtuse, half as long as the tube; corolla lilac, the limb
5–8 mm. wide, the lobes 1–3 mm. wide, shallowly
notched. Bogs: Greenl. — Que. — Man. — Sask. —
Yukon. Arctic—Subarctic. Je–Jl.

2. P. mistassínica Michx. Fig. 442. Leaf-
blades spatulate or obovate, denticulate or repand,
1–3 cm. long, obtuse or rounded at the apex; scape
1–1.5 dm. high, slender; bracts subulate; calyx-lobes
nearly as long as the tube; corolla pink or pale purple; lobes of the limb usually 4–5 mm. long, obcordate. Wet banks: Greenl.—N.Y.—Iowa—Minn.
—Alta. Boreal. My–Jl.

f. 442.

3. P. MacCalliàna Wieg. Leaf-blades broadly spatulate or obovate-cuneate, rather thick, 1–2.5 cm. long, obtuse or rounded and crenate at the apex;
scape 5–12 cm. high; bracts subulate, 5 mm. long; corolla deep rich blue, with
a large orange eye; lobes 5–6 mm. long, broadly obcordate. Wet places in the
Rockies: Man.—Alta.—B.C. Mont. Je.

4. **P. egaliksénsis** Wormsk. Leaves thin, obovate or ovate, entire or obscurely undulate, slender-petioled, 1–5 cm. long, 5–15 mm. wide; scape 1–2.5 dm. high; bracts lanceolate, gibbous at the base, 3–7 mm. long; calyx 4–6 mm. long, the lobes less than half as long as the tube; corolla about 5–9 mm. wide, the lobes deeply lobed. Bogs: Greenl.—Newf.—Man.—Alaska. *Arctic.* Jl.

5. **P. laurentiàna** Fern. Leaf-blades spatulate to rhombic-ovate, 2.5–10 cm. long, long-petioled, usually whitened beneath; scape 1–4.5 dm. high; bracts lance-subulate, 6–11 mm. long; pedicels 1–5 cm. long; calyx 6–8 mm. long; corolla pale lilac, with a yellow eye, its tube barely exserted. *P. farinosa macropoda* Fernald [G]. Calcareous cliffs: Lab.—Mack.—Minn.—Me. Je.

2. ANDRÓSACE (Tourn.) L.

Small scapose annuals. Leaves rosulate at the base of the stem. Flowers umbellate, inconspicuous. Calyx campanulate or turbinate, persistent, 5-lobed or 5-parted. Corolla salver-shaped, almost funnelform; tube shorter than the calyx; throat constricted; lobes of the limb 5, imbricate in the bud. Stamens 5, included, with short filaments. Style very short; stigma capitate. Capsule short, 5-valved; seeds usually many.

Bracts of the involucre ovate or oblong.   1. *A. occidentalis.*
Bracts of the involucre lanceolate or subulate.
    Peduncles, pedicels, and calyx-lobes densely puberulent,
      the latter exceeding the fruit.   2. *A. puberulenta.*
    Peduncles and pedicels sparingly puberulent or glabrous;
      calyx-lobes glabrous or nearly so, not exceeding
      the fruit.
      Corolla longer than the calyx.
      Peduncles 1–2 dm. high, many times longer than the
       strongly ascending or suberect pedicels.   3. *A. septentrionalis.*
      Peduncles less than 3 cm. high, often equaled or
       exceeded in length by the spreading pedicels.   4. *A. subumbellata.*
      Corolla shorter than the calyx.   5. *A. diffusa.*

1. **A. occidentàlis** Pursh. Leaves oblanceolate or spatulate, 5–15 mm. long, entire or denticulate, puberulent; scapes 2–6 cm. high, puberulent; bracts spatulate, obtuse, about 5 mm. long; pedicels ascending or spreading; calyx-lobes lanceolate, ascending; corolla much shorter than the calyx. In dry or sandy soil: Man.—Ill.—Mo.—Tex.—B.C. *Prairie—Plain—Submont.* Ap–Je.

2. **A. puberulénta** Rydb. Leaves oblanceolate, 1–3 cm. long, acute, entire or sinuately denticulate; peduncles 3–10 cm. long; bracts narrowly lanceolate, 3–4 mm. long; pedicels spreading; calyx-lobes lanceolate, strongly keeled; corolla white, about equaling the calyx. Hills and mountains: Mack.—Man.—S.D.—N.M.—Ariz.—Yukon. *Plain—Submont.*

3. **A. septentrionàlis** L. Leaves oblanceolate or oblong, acute, dentate or entire, rather thick, 1–2 cm. long; scape 1–2 dm. high, slightly puberulent; bracts subulate, 3–4 mm. long; calyx-lobes triangular, scarcely 1 mm. long, glabrous or nearly so; corolla about 4 mm. long, exceeding the calyx. Sandy plains and banks: Man.—Alta.—Wyo.—Ida.—Alaska; Eurasia. *Boreal.* My–Je.

4. **A. subumbellàta** (A. Nels.) Small. Leaves oblanceolate, acute, thin, entire or denticulate, puberulent, 1–2 cm. long; bracts subulate, 2–3 mm. long; pedicels 1–5 cm. long; calyx-lobes lanceolate, scarcely 1 mm. long; corolla nearly 4 mm. long, slightly exceeding the calyx. Mountains, especially in gravelly soil: Que.—S.D.—Colo.—Ariz.—B.C. *Boreal.—Subalp.* Jl–S.

5. **A. diffùsa** Small. Leaves oblanceolate, acute, entire or denticulate, puberulent; scapes many, ascending, 3–15 cm. high; bracts subulate, 2–3 mm. long; pedicels glabrous, spreading, 2–7 cm. long; calyx-lobes lanceolate, nearly 1 mm. long; corolla scarcely 3 mm. long, shorter than the calyx. *A. elongata* Richardson; not L. Hills and plains: Man.—Mack.—N.M.—Ariz.—B.C. *Plain —Mont.* My–Jl.

3. TRIENTÀLIS L. Star Flower.

Low perennial herbs, with rootstocks and simple stems. Leaves mostly clustered at the end of the stem, the lower leaves being much reduced or even scale-like. Flowers in the upper axils, with filiform pedicels. Sepals mostly 7, nearly distinct, linear, imbricate. Corolla white or pinkish, rotate, deeply 5-7-parted to near the base, with convolute lobes. Stamens mostly 7; filaments long and slender, slightly united into a ring at the base. Style filiform. Capsule with about 5 revolute valves. Seeds few.

1. **T. americàna** Pursh. *Fig. 443.* Stem 1–2.5 dm. high, naked below or with a few scale-like lanceolate leaves, at the summit with a verticil of 5–10 short-petioled or sessile leaves; blades 3–10 cm. long, thin, entire-margined or minutely crenulate; corolla 8–12 mm. wide; lobes ovate or lanceolate, acuminate. Damp woods and thickets: Lab.—Va.—Ills.—Sask. *E. Boreal.* My–Je.

4. LYSIMÀCHIA (Tourn.) L. Loosestrife.

Perennial or rarely annual caulescent herbs. Leaves opposite or whorled, entire, glandular-punctate. Flowers axillary, solitary, or in terminal or axillary racemes. Sepals 5, practically distinct. Corolla rotate or short funnelform, yellow, its lobes entire, convolute. Stamens 5; staminodia wanting; filaments united at the base; anthers short. Styles filiform; ovary 1-celled. Capsule globose, few- several seeded.

Leaves mostly verticillate; flowers axillary.   1. *L. quadrifolia.*
Leaves opposite or rarely alternate.
  Flowers in terminal bracted racemes; leaf-blades lanceolate; stem erect.   2. *L. terrestris.*
  Flowers axillary; leaf-blades orbicular or broadly oval; stem prostrate.   3. *L. Nummularia.*

1. **L. quadrifòlia** L. Perennial, with a horizontal rootstock; stem simple, 3–10 dm. high; leaves in whorls of 3's–7's, lanceolate to oval, acute or acuminate; 3–10 cm. long; pedicels 2–4 cm. long; corolla 1–1.5 cm. wide, the lobes oblong, rounded at the apex, spotted or streaked. Woods and thickets: N.B.—Minn. (?)—Mo.—Ga. *Canad.—Allegh.* Je–Jl.

2. **L. terréstris** (L.) B.S.P. Glabrous perennial, with a rootstock, and often bulbiferous in the axils; leaves opposite or some of them alternate, lanceolate, acute at both ends, usually black-punctate, 2–8 cm. long; flowers in terminal bracted racemes; corolla rotate, deeply parted, yellow with purple streaks or dots, nearly as long as the sepals. *L. stricta* Ait. In swamps: Newf.—Ga. —Ark.—Sask.—B.C. *Boreal.* Jl–S.

3. **L. Nummulària** L. Stem creeping, up to 6 dm. long; leaves small, short-petioled, opposite, the blades sometimes subcordate at the base, 1–2.5 cm. long; corolla-lobes broadly ovate, about 1 cm. long, yellow, dark-dotted. Around dwellings: Newf.—Minn.—Iowa—Ind.—N.J.; escaped from cult.; nat. of Eur.

5. STEIRONÈMA Raf. Fringed Loosestrife.

Perennial, caulescent herbs. Leaves opposite, or whorled above, entire, with ciliate petioles, without glandular dots. Flowers axillary, solitary, on nodding pedicels. Sepals 5, essentially distinct. Corolla rotate, yellow, 5-lobed, the lobes erose-denticulate. Stamens and staminodia 5; filaments distinct; anthers linear. Style filiform; ovary 1-celled. Capsule round, 10-20-seeded.

Leaf-blades lanceolate to ovate, pinnately veined.
  Stem-leaves with broad, ovate or lanceolate blades; corolla
    2–3 cm. broad.
    Leaf-blades mostly ovate or ovate-lanceolate, rounded
      or subcordate at the base; sepals broadly lanceolate,
      abruptly acute, in fruit scarcely exceeding the cap-
      sule.   1. *S. pumilum.*
    Leaf-blades mostly lanceolate, cuneate at the base; sepals
      lanceolate, gradually acuminate, in fruit distinctly ex-
      ceeding the capsule.   2. *S. ciliatum.*
  Stem-leaves with narrow, oblong-lanceolate or linear-lance-
    olate blades; corolla 1.5–2 cm. broad.
    Blades of the stem-leaves oblong-lanceolate, those of the
      floral branches often broadly lanceolate or ovate, con-
      spicuously subverticillate; sepals ovate-lanceolate,
      short-acuminate.   3. *S. verticillatum.*
    Blades all lance-linear, those of the branches not con-
      spicuously subverticillate; sepals lanceolate, gradu-
      ally acuminate.   4. *S. hybridum.*
Leaf-blades linear or lance-linear, 1-ribbed, the lateral veins obsolete.   5. *S. quadriflorum.*

f.444.

1. S. pùmilum Greene. Stem 2–5 dm. high,
glabrous; petioles ciliate throughout; blades ovate or
ovate-lanceolate, rounded or subcordate at the base,
4–8 cm. long, acute or short-acuminate at the apex,
ciliolate on the margins; peduncles 2–5 cm. long,
often curved; calyx 6–7 mm. long; corolla bright
yellow, 2–2.5 cm. broad; lobes broadly ovate, caudate-acuminate; anthers as long as or longer than
the filaments. The type of the species represents a
stunted form, the more normal one is *S. pumilum
longepedicellatum* Lunell. *S. ciliatum occidentale*
Suksd. Damp meadows: Man.—Minn.—Colo.—Ore.
Wash. *W. Boreal.* Jl–Au.

2. S. ciliàtum (L.) Raf. *Fig. 444.* Stem 3–12
dm. high, glabrous; leaves with ciliate petioles;
blades ovate or lanceolate, acute or acuminate at the
apex, acute, rounded, truncate, or subcordate at the base, 5–15 cm. long, glabrous except the ciliolate margins; peduncles axillary, 1–5 cm. long; corolla
2–3.5 cm. broad, yellow; lobes erose-denticulate, broadly obovate. *S. mem-
branaceum* Greene, a low form with shorter filaments. Moist thickets and
swamps: N.S.—Ga.—Ariz.—B.C. *Temp.—Submont.* Je–Au.

3. S. verticillàtum Greene. Stem 3–6 dm. high, first simple, soon
branched below; petioles short, ciliate; blades of the stem-leaves lanceolate or
more often oblong-lanceolate, 3–5 cm. long, 7–15 mm. wide, ciliolate on the
margins, acute at each end; those of the branches shorter and comparatively
broader, short-petioled; corolla about 1.5 cm. broad, the lobes obovate, acute or
short-acuminate, erose above; anthers longer than the granuliferous filaments.
S. Lunellii Greene, a robust, but stunted form. Sloughs and ditches: N.D.—
Iowa—Minn.—Alta. *Prairie—Plain.* Je–Au.

4. S. hỳbridum (Michx.) Raf. Stem 1–4 dm. high, simple or sparingly
branched; petioles ciliate throughout; leaf-blades mostly narrowly lanceolate
to lance-linear, gradually tapering at each end, 3–8 cm. long, 8–15 mm. wide;
calyx-lobes 6–7 mm. long, acuminate; corolla bright yellow, 1.5–2 cm. broad;
capsule 4–5 mm. long. *S. lanceolatum hybridum* A. Gray [G]. Thickets and
meadows: Que.—Fla.—Ariz.—Kans.—S.D.—Alta. *Temp.* Je–Au.

5. S. quadriflòrum (Sims) Hitchc. Stem 2–9 dm. high, 4-angled; stem-
leaves sessile or nearly so, narrowly lance-linear to linear, 3–9 cm. long, 4–10
mm. wide, shining, sometimes slightly revolute; calyx-lobes lanceolate, acute,
5–7 mm. long; corolla 1.5–2 cm. broad. *S. longifolium* (Pursh) A. Gray.
Banks: N.Y.—Va.—Mo.—Man. *Canad.—Allegh.* Je–Jl.

6. NAUMBÚRGIA Moench. TUFTED LOOSESTRIFE.

Erect perennial, caulescent water-herbs. Leaves opposite, entire, narrow. Flowers in axillary spikes. Sepals 5–7, linear, imbricate. Corolla yellow, deeply 5–7-parted, with narrow segments. Stamens 5–7, alternating with as many tooth-like staminodia; filaments slender, glabrous. Style slender, stigma capitate. Capsule 5–7-valved, few-seeded.

1. **N. thyrsiflòra** (L.) Duby. Perennial, with a rootstock; stem 3–8 dm. high, simple, glabrous or slightly villous; leaves sessile, 5–10 cm. long, lanceolate to linear-lanceolate, acute; racemes 1–2.5 cm. long; flowers nearly sessile, crowded; corolla 4–6 mm. broad, yellow, with black spots; divisions linear. *Lysimachia thyrsiflora* L. [G]. In water and swamps: N.E.—Pa.—Colo.—Calif. —Alaska; Eurasia. *Plain—Submont.* My–Jl.

7. ANAGÁLLIS (Tourn.) L. POOR MAN'S OR SHEPHERD'S WEATHERGLASS, PIMPERNEL.

Annual or rarely perennial, decumbent, caulescent herbs. Leaves mostly opposite or whorled, entire. Flowers axillary, solitary, on slender pedicels. Calyx 5-cleft, persistent. Corolla blue, pink, or white, rotate; lobes 5, entire, convolute. Stamens 5; filaments slender, adnate to the base of the corolla, sometimes pubescent. Style filiform; stigma capitate; ovary 1-celled; ovules half-anatropous. Capsule circumscissile, subglobose; seeds numerous.

1. **A. arvénsis** L. *Fig. 445.* A glabrous annual; stems spreading, 5–30 cm. long, 4-angled; leaf-blades ovate or oval, 5–20 mm. long, entire, sessile or clasping; peduncles axillary, filiform, 1–2 cm. long; corolla scarlet, pink, or white, 5–7 mm. broad. Waste places: Newf.—Fla.—Calif.—B.C.; Mex.; nat. from Eurasia and Africa. Je–Au.

445.

8. CENTÚNCULUS L. FALSE PIMPERNEL, CHAFFWEED.

Small depressed caulescent annuals. Leaves alternate or the lower opposite. Flowers minute, axillary, solitary. Sepals 4 or 5, united below, persistent. Corolla not exceeding the calyx; tube subglobose; lobes entire, spreading. Stamens 4 or 5, inserted in the throat; filaments dilated; anthers cordate. Ovary 1-celled; style filiform; stigma capitate. Ovules numerous, half-anatropous. Capsule subglobose, circumscissile; seeds numerous.

1. **C. mínimus** L. Slender annual, 2–15 cm. high; leaves subsessile, obovate or oblong, 4–8 mm. long; flowers sessile or nearly so; sepals linear or linear-lanceolate, acuminate; corolla pink; lobes lanceolate. Moist soil: N.S.— Ills.—Fla.—L. Calif.—B.C.; Eur. and S. Am. *Temp.—Submont.* Ap–S.

9. GLAÚX (Tourn.) L. BLACK SALTWORT, SEA MILKWORT.

Small caulescent herbs, perennial. Leaves opposite, entire, sessile. Flowers small, axillary. Calyx campanulate, 5-lobed; the lobes petaloid, pink or white, imbricate in the bud. Corolla wanting. Stamens 5, inserted at the base of the calyx, alternate with its lobes; filaments filiform-subulate; anthers cordate. Capsule 5 valved at the top, few-seeded.

1. **G. marítima** L. Succulent leafy perennial; stems 0.5–3 dm. high, glabrous; leaves sessile, from oval to linear-oblong, 4–12 mm. long; calyx campanulate, about 3–4 mm. long, pinkish or white; lobes oval, rounded. Saline soil: Newf.—N.J.—Colo.—Ore.—Alaska; Eurasia. *Boreal.—Submont.* Je–Au.

10. **DODECÀTHEON** L. Shooting Star, American Cowslip, Bird-bills.

Scapose perennials, with short rootstocks. Leaves in basal rosettes. Flowers umbellate on solitary scapes, nodding. Calyx-lobes 4 or 5, longer than the tube, reflexed in flower. Corolla hypogynous; tube very short; lobes 5 or 4, imbricate in the bud, reflexed, many times as long as the tube. Stamens 5 or 4, exserted; filaments usually united; anthers attached by their bases. Ovary 1-celled, free; style filiform; stigma capitate. Ovules numerous, half-anatropous. Capsule partially 5-valved, the very apex in some species separating off as a lid. Seeds numerous.

Anthers at least twice as long as the short filaments.
  Bracts broadly ovate.   1. *D. thornense.*
  Bracts lanceolate.
    Leaves broadly ovate, abruptly contracted at the base, membranous, often denticulate; filaments free half their length; anthers auricled at the base.   2. *D. Frenchii.*
    Leaves oblanceolate or spatulate, gradually tapering into the winged petiole; filaments united to near the apex; anthers not conspicuously auricled.
      Calyx two thirds to three fourths as long as the mature capsule, its lobes twice as long as the tube; leaves mostly narrowly oblanceolate.   3. *D. radicatum.*
      Calyx less than two thirds as long as the capsule, its lobes only slightly longer than the tube; leaves broadly oblanceolate or spatulate to ovate.   4. *D. Meadia.*
Anthers less than twice as long as the filaments.
  Bracts lanceolate, acute; anthers 4–5 mm. long, half longer than the filament-tube.   5. *D. pauciflorum.*
  Bracts oblong, obtuse; anthers 3–4 mm. long, only slightly longer than the filament-tube.   6. *D. salinum.*

1. D. thornénse Lunell. Leaves oblanceolate, glabrous, entire, very obtuse, 10–15 cm. long, 1.5–3 cm. wide, tapering into a winged petiole; scape 2–6 dm. high, sometimes 10–15-flowered; calyx-lobes hyaline-margined, triangular-lanceolate; corolla-tube with a rose-colored limb and a dark rose ring, separated by a white and yellow field; lobes 12–15 mm. long, oblong; capsule subcylindric, 7–12 mm. long. Prairies: N.D. Je.

2. D. Frénchii (Vasey) Rydb. Leaves broadly ovate, 10–20 cm. long; scape 2–4 dm. high; bracts broadly lanceolate; calyx-tube 2 mm. long, the lobes 4 mm. long; corolla dark purple, with a very dark wavy ring in the bottom of the throat, bordered by a broader yellow one; filaments 2 mm. long; connective of the anthers dark purple; pod subcylindric, 15 mm. long, 4 mm. thick, valvate at the apex. *D. serratum* Raf. (?) *D. Meadia Frenchii* Vasey [G, B]. *D. Meadia membranacea* Knuth. Rich woods, at the bases of ledges: Ill.—Minn.—Ark.—Pa. (?). *Allegh.* My–Je.

3. D. radicàtum Greene. *Fig. 446.* Leaves 5–20 cm. long; blades oblanceolate, or rarely elliptic, entire-margined or (in the v. *sinuatum*) sinuate, obtuse or acutish at the apex, tapering below into short petioles; scape 1–4 dm. high, 5–15-flowered; calyx-tube 3–4 mm. long; lobes linear-lanceolate, about 6 mm. long; corolla purple, with a very dark wavy line in the throat; lobes linear or oblong, obtuse, 15–18 mm. long; anthers linear, acute, 5–6 mm. long; capsule oblong-ovoid, about 12 mm. long. Wet meadows: S.D.—Kans.—N.M.—Wyo. *Plain—Mont.* My–Jl.

f. 446.

4. D. Meàdia L. Leaves 5–20 cm. long, entire to coarsely crenate; scape 3–5 dm. high; corolla pink-purple to almost white; lobes 10–15 mm. long, acutish; tube of the filaments more than 1 mm.

long; anthers linear, 6–7 mm. long; capsule oblong-cylindric, 12–15 mm. long, 5-valved at the apex. Rocky bluffs: Pa.—Man.—S.D.—Tex.—Ga. *Allegh.—Austral.* My–Je.

5. D. pauciflòrum (Durand) Greene. Leaves glabrous, 3–10 cm. long; blades oblanceolate, entire-margined, obtuse at the apex, tapering into more or less elongate petioles; scape 1–4 dm. high, 1–10-flowered; calyx-tube 2 mm. long; lobes lanceolate, about 4 mm. long; corolla purple, with a very dark wavy line in the throat; anthers 4–5 mm. long, linear, usually obtuse; capsule elongate-oblong, 12–15 mm. long. Wet meadows: Mack.—Sask.—Neb.—Colo.—Wash.—B.C. *Plain—Mont.* Je–Au.

6. D. salìnum A. Nels. Leaves 2–4 cm. long, spreading, glabrous; blades elliptic or oblanceolate, rarely obovate, obtuse, thin, entire-margined, tapering below into petioles; scape slender, 1–2 dm. high, purplish above, glabrous, 3–12-flowered; calyx-tube about 2 mm. long; lobes lanceolate, 3 mm. long; corolla lilac-purple, with yellowish throat, often with a wavy purple line; lobes oblong, obtuse or acutish; anthers purple, often with whitish margins, acutish; capsule oblong, about 8 mm. long. Saline flats: Sask.—N.D.—Utah—Ida. *Plain—Mont.* My–Jl.

11. SÁMOLUS (Tourn.) L. WATER PIMPERNEL, BROOKWEED.

Annual or perennial, caulescent herbs. Leaves alternate, entire. Flowers in terminal racemes or panicles. Sepals united at the base, persistent. Hypanthium more or less developed. Corolla perigynous, short-salverform; tube very short; lobes rounded, imbricate. Stamens 5, adnate to the corolla-tube; staminodia sometimes present, narrow; anthers cordate. Ovary 1-celled; stigma capitate. Ovules numerous, half-anatropous. Capsule short, 5-valved at the apex. Seeds numerous.

1. S. floribúndus H.B.K. Stem 1–6 dm. high; leaves few, spatulate, or oval, obtuse, 3–15 cm. long, narrowed into winged petioles; upper stem-leaves subsessile and ovate; sepals ovate or triangular; corolla white, 3 mm. broad; staminodia 5, at the sinuses of the corolla. Wet places: Newf.—Fla.—Calif.—B.C.; Mex. *Temp.* Je–Au.

Family 131. **EBENACEAE.** EBONY FAMILY.

Shrubs or trees, with alternate, or rarely opposite or whorled, simple leaves. Flowers inconspicuous, dioecious, or rarely polygamous or perfect, solitary or cymose. Calyx 3–7-lobed, persistent, accrescent. Corolla gamopetalous; lobes 3–7, valvate or imbricate. Stamens 3 or 4 times as many as the corolla lobes, filaments adnate to the corolla-tube; anthers introrse. Gynoecium of 3 or more united carpels; ovary 3–several-celled; styles united at the base. Fruit a fleshy berry, or sometimes a capsule. Seeds flattened, endosperm copious. Embryo straight.

1. DIOSPÝROS L. PERSIMMON. DATE PLUM.

Shrubs or trees. Flowers dioecious, rarely polygamous, solitary or in axillary cymes. Calyx 3–7-lobed. Corolla urn-shaped to salverform, 3–7-lobed. Stamens mostly 16 in the staminate flowers, inserted in pairs at the base of the corolla, in the pistillate flowers represented by 8 staminodia; anthers opening by lateral chinks. Ovary 4–12-celled; styles 2–6. Fruit a berry, surrounded by the accrescent calyx.

1. D. virginiàna L. *Fig. 447.* Tree, up to 35 m. high; leaf-blades leathery, ovate, oval, or elliptic, 8–20 cm. long, entire, acute or acuminate at the apex, acute to cordate at the base, deep green above, pale beneath; calyx-lobes 4, triangular; corolla urn-shaped, 8–13 mm. long; lobes 4, more or less reniform, recurved; fruit depressed-globose, 3–4 cm. broad, astringent when green, sweet when ripe. Woods and fields: Conn.—Iowa—Kans.—Tex.—Fla. *Allegh. —Austral.* Je.

Family 132. **SAPOTACEAE.** Sapodilla Family.

Shrubs or trees, sometimes thorny. Leaves alternate, without stipules, entire. Flowers clustered, perfect or rarely polygamous. Sepals 4–12, imbricate, in 1 or 2 series. Corolla with 4 or more lobes, deciduous. Stamens as many as the corolla-lobes, adnate to the tube, opposite to the lobes, alternating with as many staminodia. Gynoecium of several united carpels; ovary 4–12-celled; styles united. Ovules solitary in each cavity. Fruit a berry. Seeds nut-like. Embryo straight.

1. **BUMÈLIA** Sw. Buckthorn.

Shrubs or small trees. Leaves conspicuously nerved, sometimes spine-bearing in their axils. Flowers perfect, in axillary clusters. Sepals unequal, imbricate. Corolla white, the lobes longer than the tube, appendaged on each side. Stamens 5; anthers versatile. Staminodia 5, petal-like. Ovary 5-celled. Berry drupe-like; seeds mostly solitary.

Leaves nearly glabrous.   1. *B. lycioides.*
Leaves rusty-woolly beneath.   2. *B. lanuginosa.*

1. B. lycioìdes (L.) Pers. A glabrous, spiny shrub or small tree, up to 8 m. high; leaves oblong, elliptic or oblanceolate, 4–12 cm. long, acute or acuminate, or rarely rounded at the apex, prominently reticulate, tapering at the base; clusters densely many-flowered; corolla 4 mm. long; staminodia boat-like, obtuse; berry ellipsoid, 1 cm. long. Low ground: Va.—Ill.—Kans.—Tex.—Fla. *Austral.* Je–Jl.

2. B. lanuginòsa (Michx.) Pers. *Fig. 448.* A spiny or spineless shrub or tree, 3–20 m. high; leaves oblong-oblanceolate to obovate or elliptic, from acute to retuse at the apex, tapering at the base, glabrous above; staminodia ovate, erose-dentate; clusters 6–12-flowered; fruit globular, 10–15 mm. long. Sandy soil: Ga.—Ill.—Kans.—Tex.—Fla. *Austral.* Jl.

f.448.

Family 133. **OLEACEAE.** Olive Family.

Shrubs or trees, rarely herbs. Leaves opposite, exceptionally alternate or whorled, without stipules, simple or pinnately compound. Flowers mostly paniculate, perfect, polygamous, or dioecious, regular, complete or incomplete. Sepals usually 4, partially united, rarely wanting. Petals 2–6, distinct or partially united, narrow, imbricate or valvate, or wanting. Stamens 2 or 4, adnate to the base of the corolla, if the latter is present. Gynoecium of 2 united carpels; ovary 2-celled, free; stigmas capitate or 2-lobed. Fruit a loculicidal, 2-valved capsule, a samara, or a berry. Seeds usually solitary, sometimes 2–4.

Flowers dioecious, monoecious, or polygamous, in ours apetalous; fruit indehiscent.
  Fruit a samara; leaves in ours pinnately compound.   1. FRAXINUS.
  Fruit a drupe; leaves simple.   2. ADELIA.
Flowers perfect; corolla present; leaves simple.
  Fruit a loculicidal capsule; seeds winged; stigma 2-cleft.   3. SYRINGA.
  Fruit a berry-like capsule; seeds not winged; stigma thickened. 4. LIGUSTRUM.

1. FRÁXINUS (Tourn.) L. ASH.

Trees or shrubs. Leaves opposite, usually odd-pinnate, rarely simple.
Flowers inconspicuous, polygamous or dioecious, usually in panicles. Calyx
small, 4-lobed, or wanting, Corolla in ours wanting, otherwise rarely present,
of 4 distinct or more or less united petals. Stamens 2, rarely 3 or 4; filaments
short or elongate, inserted at the base of the petals, or hypogynous; anthers
linear or oblong, introrse. Ovary 2-celled; styles united; stigma 2-cleft; ovules
2 in each cell, pendulous. Fruit a samara.

Body of the samara terete or nearly so; calyx present in the pistillate flowers, per
    sistent on the fruit.
  Wing of the samara wholly terminal, not decurrent on the body; branches and
    petioles glabrous.   1. F. americana.
  Wing of the samara decurrent halfway down on the body.
    Leaflets distinctly petioluled.
      Branchlets and petioles velvety pubescent.   2. F. pennsylvanica.
      Branchlets and petioles glabrous.   3. F. lanceolata.
    Leaflets subsessile.   4. F. campestris.
Body of the samara flattened; calyx wanting or a mere ring.
  Leaflets petioluled; branchlets 4-angled; wing of the sa-
    mara decurrent halfway down on the body.   5 F. quadrangulata.
  Leaflets sessile; branchlets terete; wing of the samara
    decurrent to the base of the body.   6. F. nigra.

1. F. americàna L. Tree, up to 40 m. high, with terete twigs; leaves
5–9-foliolate; leaflets lanceolate or elliptic, 5–15 cm. long, glabrous and lustrous above, pale, glabrous or with scattered hairs beneath, entire or sparingly
denticulate; samara 2.5–3.5 cm. long, with a linear-oblong or oblanceolate wing,
2–3 times as long as the cylindric body. Rich woods: N.S.—Minn.—S.D.—
Tex.—Fla. *Temp.* Ap–My.

2. F. pennsylvánica Marsh. Tree, up to 25 m. high, with pubescent,
terete twigs; leaves 5–7-foliolate; leaflets elliptic-lanceolate, 5–15 cm. long,
acuminate at each end, undulate or irregularly serrate, pubescent beneath, pale
green; samara 2.5–6 cm. long; wing linear-spatulate, longer than the cylindric
body. *F. pubescens* Lam. Along streams: N.B.—N.D.—Kans.--N.C. *Canad.*
—*Allegh.* Ap–My.

3. F. lanceolàta Borkh. Tree, up to 20 m.
high, with glabrous, terete branches; leaves 5–7-
foliolate; leaflets lanceolate to elliptic, 5–20 cm.
long, pale green, acuminate at each end, glabrous,
entire or serrulate; samara 3.5–6 cm. long; wing
linear-oblong, much longer than the body. *F. viridis*
Michx. *F. pennsylvanica lanceolata* (Borkh.) Sarg.
[G]. Along streams: Me.—Sask.--La.—Fla. *E.
Temp.* Ap–My.

4. F. campéstris Britton. *Fig. 449.* A tree,
8–12 m. high, with round twigs; leaves mostly 7-foli-
olate; leaflets ovate to lanceolate, usually long-
acuminate, toothed, more or less hairy beneath or
glabrous on both sides, 3–10 cm. long; calyx small,
4-toothed; samara spatulate, about 3 cm. long; wing
narrow, decurrent on the body to the middle or below. Valleys and hillsides: Man.—Kans.—Tex.—
Wyo.—Sask. *Prairie—Plain.*

f.449.

5. **F. quadrangulàta** Michx. Tree, up to 35 m. high, with glabrous 4-angled twigs; leaves 7–11-foliolate; leaflets lanceolate or ovate, 5–15 cm. long, acuminate, serrate, rounded at the base; samara 3–4 cm. long; wing oblong or oblong-cuneate, shorter than the body, notched at the end. Woods: Ont.—Minn.—Iowa—Okla.—Ark.—Ala. *Allegh.—Austral.* Mr–Ap.

6. **F. nìgra** Marsh. Tree, up to 30 m. high, with terete, glabrous twigs; leaves 7–11-foliolate; leaflets lanceolate, acuminate, 8–15 cm. long, serrate, more or less pubescent on the veins beneath, sessile; samara 3–4 cm. long, linear-oblong to elliptic, winged all around, the body half as long as the wing. *F. sambucifolia* Lam. Swamps: Newf.—Man.—Minn.—Ark.—Va. *Canad.—Allegh.* Ap–My.

2. ADÈLIA (P. Br.) Michx.

Shrubs or trees, often widely branching. Leaves opposite, simple. Flowers inconspicuous, polygamo-dioecious, appearing before the leaves. Calyx usually present, but small, 4–6-lobed. Corolla usually wanting, rarely with 1 or 2 deciduous petals. Stamens 2–4; anthers extrorse. Ovary 2-celled; stigma thick, sometimes 2-lobed; ovules 2 in each cavity. Fruit a drupe; seeds solitary, rarely 2. Endosperm fleshy. *Forestiera* Poir.

1. **A. acuminàta** Michx. A shrub or small tree, up to 10 m. high; leaves deciduous, membranous, elliptic or oval, acuminate at each end, slightly serrate about the middle; staminate flowers in fascicles, the pistillate ones in short panicles; drupe oblong, 12–15 mm. long, deep purple. *F. acuminata* Poir. [BB]. River banks and swamps: Ill.—Kans.—Tex.—Ga. *Austral.* Mr–Ap.

3. SYRÍNGA L. Lilac.

Shrubs, with opposite entire petioled leaves. Flowers in terminal panicles, perfect. Calyx-lobes unequal. Corolla lilac, purple or white; tube cylindric; lobes 4, more or less spreading. Stamens 2. Ovary 2-celled; styles united; stigma 2-cleft. Ovules 2 in each cavity. Fruit a loculicidal capsule, 2-valved. Seeds flat, winged.

1. **S. vulgàris** L. Shrub, 2–7 m. high; leaf-blades ovate, 5–10 cm. long, acuminate at the apex, acute to cordate at the base; pedicels club-shaped; calyx 2–2.5 mm. long, the lobes ovate or triangular; corolla lilac, rarely white, its lobes oval; capsule oblong, 1.5 cm. long. Around dwellings: N.Y.—N.D.—Kans.—Ga.; escaped from cultivation; nat. of Eur. Ap–My.

4. LIGÚSTRUM (Tourn.) L. Privet.

Shrubs or trees, with simple opposite leaves. Flowers in terminal panicles, perfect. Calyx short, tubular, with 4 short lobes. Corolla white or greenish, funnelform, the tube short, the lobes 4. Stamens 2, the filaments adnate to the corolla-tube, the anthers introrse. Ovary 2-celled, the ovules 2 in each cavity, pendulous. Fruit a 2-celled, 1–2-seeded berry.

1. **L. vulgàre** L. Shrub, 0.5–3 m. high, much branched; leaves glabrous, 1–6 cm. long, leathery, elliptic or lance-oblong, deep green, short-petioled, entire; corolla white, 4–6 mm. long, fruit globose, 6–8 mm. broad. Roadsides and thickets: Me.—Minn.—N.C.; nat. from Eur.; escaped from cultivation.

Family 134. GENTIANACEAE. Gentian Family.

Annual or perennial caulescent herbs, or in warmer climates rarely shrubs. Leaves normally opposite, sometimes connate at the base. Inflorescence cymose. Flowers regular, perfect. Calyx of 2, 4, or 5, more or less united sepals. Corolla of 4 or 5 more or less united petals; lobes convolute or imbricate, entire or fringed. Stamens as many as the corolla-lobes and alternate with them, partly adnate to the corolla. Gynoecium

of two united carpels; ovary 1-celled, superior, with 2 parietal placentae.
Fruit a capsule.

Corolla-lobes convolute in bud; normal leaves present.
  Corolla not spurred.
    Style filiform, mostly deciduous.
      Stigmas linear or nearly so, as long as the style; anthers coiled; corolla
        rotate.   1. SABBATIA.
      Stigma rounded, much shorter than the style; corolla funnelform to sal-
        ver-shaped.
        Corolla small, red, rose, or yellowish; tube surpassing the calyx; fila-
          ments spirally twisted.   2. CENTAURIUM.
        Corolla large, blue, purple, or white; tube much shorter than the
          calyx; stamens recurved.   3. EUSTOMA.
    Style stout, short, persistent, or none.
      Corolla campanulate, funnelform, or salver-shaped; calyx 4- or 5-lobed;
        stamens inserted in the corolla-tube.
        Corolla without plaits or lobes in the sinuses; calyx without an inter-
          calycine membrane; sepals imbricate.
          Flowers 4-merous, rather large, usually more than 3 cm. long;
            corolla-lobes more or less fringed or toothed; inner sepals
            broader, membranous-margined.   4. ANTHOPOGON.
          Flowers 5-merous (rarely 4-merous), small, less than 2 cm. long;
            outer sepals broader; corolla-lobes never fringed, rarely toothed.
                                  5. AMARELLA.
      Corolla plicate in the sinuses, the plaits more or less extended into mem-
        branous lobes or teeth; calyx with an inter-calycine membrane,
        its lobes valvate.   6. DASYSTEPHANA.
      Corolla rotate, calyx parted to near the base; stamens inserted on the
        base of the corolla.   7. PLEUROGYNE.
  Corolla spurred.   8. HALENIA.
Corolla-lobes imbricate in bud; leaves reduced to scales.   9. BARTONIA.

1. SABBATIA Adans.

Annual or biennial leafy-stemmed herbs. Leaves glabrous, opposite or
seldom whorled. Flowers perfect, solitary or in terminal cymes. Calyx-lobes
4, two of them narrower. Corolla from white to rose-colored, rotate, the lobes
4–12. Stamens 4–12, filaments adnate to the short corolla-tube; anthers coiled.
Ovary 1 celled, with intruding placentae. Capsule short, 2-valved. Seeds
numerous, reticulate.

Branches of the stem all opposite; leaves little if at all longer than broad.
                                             1. S. angularis.
Branches all alternate, or only the lowest opposite; leaves decidedly longer than broad.   2. S. campestris.

1. S. angularis (L.) Pursh. Stem 1–4 dm.
high, narrowly winged; leaves broadly ovate to orbicular; calyx-lobes linear, or linear-lanceolate, 10–14
mm. long, much shorter than the corolla; corolla
pink or white, with a lemon or greenish eye; lobes
oblong or cuneate, 15–20 mm. long, obtuse; capsule 5
mm. long. Rich soil: Ont.—Mich.—se Kans.—Okla.
—Fla. *Allegh.—Austral.* Jl–Au.

2. S. campestris Nutt. *Fig. 450.* Stem 1–4
dm. high; branched; leaves ovate or ovate-lanceolate, 1.5–4 cm. long, partly clasping; calyx deep
green, the tube strongly ribbed, the lobes linear, 2–3
cm. long; acute; corolla lilac, the lobes oval or
obovate. Capsule 8–10 mm. long, wholly or mostly
included in the calyx-tube. Prairies: Iowa—se
Kans.—Tex.—Ark. *Prairie.* Jl–Au.

f. 450.

2. CENTAURIUM Hill. CENTAURY.

Annual, biennial, or rarely perennial caulescent herbs. Leaves opposite,
entire, sessile or clasping. Flowers in terminal cymes, pedicelled. Calyx with

4 or 5 keeled lobes. Corolla salver-shaped, rose-colored, pink, white, or yellow; lobes 4 or 5, convolute. Stamens 4 or 5, partially adnate to the corolla-tube; filaments short; anthers at last twisted. Ovary 1-celled; placentae 2, sometimes intruding. Capsule oblong or fusiform, 2-valved. Seeds many, reticulate. *Erythraea* Neck.

Corolla-lobes oblong, obtuse.   1. *C. exaltatum.*
Corolla-lobes ovate or lanceolate, acute.   2. *C. texense.*

1. **C. exaltàtum** (Griseb.) W. F. Wight. Stem 1–3 dm. high, slender, glabrous, with erect or strongly ascending branches; leaves sessile, oblong to linear, 1.5–4 cm. long; corolla-tube yellow, 8–10 mm. long; lobes rose-colored, 4–5 mm. long. *Erythraea Douglasii* A. Gray. *E. exaltata* (Griseb.) Coville [B]. Wet meadows and river bars: Wash.—Calif.—Utah—Neb. *Plain—Submont.* My–Jl.

2. **C. texénse** (Griseb.) Fern. Stem 0.5–2 dm. high, branched above; leaves lanceolate or linear above, 0.5–1.5 cm. long, sessile; calyx 8–10 mm. long; lobes subulate; corolla rose-colored, its lobes oblong, acute, shorter than the tube. Rocky places: Tex.—Kans. *Texan.*

3. EÙSTOMA Salisb.

Annual caulescent herbs, or perennials by means of rosettes. Leaves opposite, sessile or clasping, glaucous. Flowers in terminal cymose panicles, or solitary. Calyx with 4 or 5 narrow keeled lobes. Corolla deeply campanulate, blue, purple, or white; lobes 5 or 6, convolute in bud, erose-denticulate. Stamens 5 or 6, adnate to the corolla-tube; filaments filiform; anthers versatile. Ovary 1-celled; stigmas 2. Seeds numerous, pitted.

1. **E. Russelliànum** (L.) Griseb. *Fig. 451.* Annual, or perennial with rosettes, with a taproot; stem 3–6 dm. high; leaves ovate to lanceolate or oblong, 3-ribbed, 2–4 cm. long; calyx 15–20 mm. long; lobes subulate; corolla deep purple, rarely white, 3–4 cm. long; lobes obovate. *E. Andrewsii* A. Nels. Meadows: S.D.—Tex.—N.M.—Colo. *Plain—Submont.* Je–Au.

J.451.

4. ANTHOPÒGON Necker. FRINGED GENTIAN.

Annual, biennial, or perennial caulescent herbs. Leaves opposite, entire, sessile. Flowers solitary at the end of the stem or its branches, perfect, 4-merous. Calyx large; lobes keeled, the inner distinctly wider than the outer, scarious- or hyaline-margined. Corolla showy, funnelform; lobes erose or fimbriate, without plaits in the sinuses, convolute. Stamens with glands at the base of the filaments. Ovary 1-celled; ovules numerous. Capsule stipitate.

Corolla enclosed in the ventricose wing-angled calyx.   1. *A. ventricosus.*
Corolla exserted from the merely angled calyx.
   Corolla-lobes fringed all around the summit; leaves lanceolate with rounded or
      subcordate bases.   2. *A. crinitus.*
   Corolla merely dentate at the summit; stem-leaves linear or linear-lanceolate.
      Filaments glabrous.
         Stem-leaves linear-lanceolate; corolla 3–5 cm. long, its lobes long-ciliate-
            erose along the sides.   3. *A. procerus.*
         Stem-leaves linear; corolla 2.5–3 cm. long, its lobes short-erose along the
            margins.   4. *A. tonsus.*
      Filaments hairy at the middle.   5. *A. Macounii.*

1. **A. ventricòsus** (Griseb.) Rydb. Stem simple or branched above, 2–3 dm. high, slightly angled; basal leaves spatulate, 1–1.5 cm. long; stem-leaves lanceolate to oblong-ovate, about 3 cm. long; calyx nearly 2 cm. long, ventricose, wing-angled; outer sepals lanceolate, long-acuminate, the inner broader, acute; corolla-lobes elliptic, serrate-fimbriate. *Gentiana ventricosa* Griseb. [B]. Wet places: Man.—Sask. *Boreal.*

2. **A. crinìtus** (Fröl.) Raf. Annual; stem 2–5 dm. high, branched; leaves 2–5 cm. long, partly clasping; calyx 2.5–3 cm. long; outer lobes lanceolate, acuminate, the inner ones much broader, scarious-margined; corolla deep sky-blue, 4–5 cm. long; lobes spreading. *Gentiana crinita* Fröl. [G, B]. Low ground: Que.—Man.—Iowa—Ga. *Canad.—Allegh.* S–O.

ſ452.

3. **A. procèrus** (Holm) Rydb *Fig. 452.* Stem branched above, angled, 3–5 dm. high; basal leaves spatulate or oblanceolate, obtuse; upper leaves linear-lanceolate, 2–6 cm. long; calyx-tube 12–15 mm. long; outer lobes 18–22 mm. long, narrowly lanceolate, the inner about three times as broad, 15–20 mm. long; corolla 3 5 cm. long, the lobes broadly obovate. *G. procera* Holm [B,G]. Meadows: Ont.—N.Y.—Minn.—Man. *Canad.*

4. **A. tónsus** (Lunell) Rydb. Stem branched above, 3–4 dm. high, slightly angled; basal leaves oblanceolate or spatulate, obtuse, 1–2 cm. long; stem-leaves 4–7 cm. long; calyx-tube about 1 cm long; outer lobes 10–14 mm. long, the inner 8–12 mm. long, 2–3 times as broad as the outer; corolla blue or purple, 3–4 cm. long; lobes oblong to obovate, fimbriate-toothed around the summit. *Gentiana detonsa tonsa* Lunell. Wet meadows: Sask.—Minn.—S.D.—Alta.—Mack *Plain.* Jl–S.

5. **A. Macoùnii** (Holm) Rydb Stem strict, angular, 0.5–3 dm. high, branched near the base; basal leaves spatulate or oblanceolate, 1–2 cm. long, obtuse; stem-leaves acute, 2–3 cm. long; calyx purplish green; tube about 1 cm long; lobes acuminate, the outer about 10 mm. long; corolla deep bluish, 1.5–3 cm. long; lobes broadly obovate, toothed at the summit and fringed on the sides, veiny. *G. Macounii* Holm. Prairies, gravelly soil, and edges of marshes: w Ont.—Alta.—Mont.—BC. *Boreal.—Submont.* Jl–S.

5. **AMARÉLLA** Gilib. GENTIAN.

Annual, biennial or perennial herbs. Leaves opposite, entire, sessile. Flowers perfect, solitary or cymose, 4-merous or 5-merous. Calyx-lobes imbricate, the outer two often somewhat larger, without a membrane within the lobes. Corolla funnelform or salverform; lobes entire or sparingly toothed, but without plaits in the sinuses. Ovary 1-celled; ovules numerous.

Corolla with a fringed crown in the throat, its lobes obtuse or acute; glands at the base of the corolla none or obsolete.
  Corolla 6–8 mm. long, the lobes obtuse; crown with few bristles.
                                                                1. *A. Gurliae.*
  Corolla 10–18 mm. long, the lobes acute, crown usually with many bristles (except in *A. acuta*).
    Flowers numerous, crowded, very short-pedicelled, the whole inflorescence dense and spike-like; corolla usually yellowish.
                                                        2. *A. strictiflora.*
    Flowers comparatively few, distinctly pedicelled; corolla blue or purplish.
      Upper stem-leaves elongate-lanceolate.
                                                    3. *A. scopulorum.*

Upper stem-leaves ovate or ovate-lanceolate; crown
usually with few bristles.   4. *A. acuta.*
Corolla without a fringed crown in the throat, but with glands
  at the base; lobes acuminate.
  Stem slender, less than 2.5 dm. high; stem-leaves lanceolate
    or oblong, 1-ribbed; corolla 8–15 mm. long.   5. *A. propinqua.*
  Stem stouter, 2.5–5 dm. high; stem-leaves ovate, 3–5-ribbed;
    corolla 20–25 mm. long.   6. *A. occidentalis.*

1. **A. Gúrliae** Lunell. Annual; stem simple or branched at the base, purplish, 1–3 dm. high; basal leaves spatulate; stem-leaves lanceolate or lance-
linear; flowers 1–5, on axillary branches; calyx-tubes 4 or 5, unequal, 3–4 mm.
long, linear, the longer nearly equaling the corolla-tube; corolla 6–8 mm. long,
4–5-merous, lilac; lobes ovate, obtuse. Springs: Butte, N.D. Au.

2. **A. strictiflòra** (Rydb.) Greene. Strict annual; stem 2–4 dm. high, simple or with erect
branches; basal leaves spatulate, 2–3 cm. long; upper stem-leaves lanceolate, 3–5-ribbed, 2–5 cm. long,
acute or acuminate; calyx-lobes moderately unequal,
linear or lanceolate, acute, 5–8 mm. long; corolla
8–12 (rarely 14) mm. long, varying from greenish
yellow to azure or white, with blue lobes; crown well
developed, but setae often rather few. *G. strictiflora*
(Rydb.) A. Nels. *A. conferta* Greene. *A. theiantha*
Lunell. Wet meadows and open woods: Alaska—
Calif.—Colo.—S.D.—Sask. *Submont.—Subalp.* Jl–S.

3. **A. scopulòrum** Greene. *Fig. 453.* Annual;
stem 2–4 dm. high, simple or with ascending
branches; basal leaves spatulate, 1–3 cm. long, obtuse; stem-leaves lanceolate, often acuminate at the
apex, 2–6 cm. long; flowers 1–4 in the axils of the leaves; usually on naked
pedicels; calyx-lobes linear, acute, the longer 10–12 mm. long; corolla 15–18
mm. long, blue or rarely greenish yellow; lobes ovate-lanceolate, acute; crown
well developed. *Gentianella Clementis* Rydb. Mountains: Mont.—Ariz.—N.M.
—N.D. *Submont.—Subalp.* Jl–Au.

4. **A. acùta** (Michx.) Raf. Annual; stem 2–5 dm. high; lower leaves
spatulate or obovate, 3–5 cm. long, the upper ovate-lanceolate or ovate, acuminate, rounded or subcordate at the base; flowers several on axillary branches;
calyx-lobes lanceolate; corolla blue, 10–15 mm. long; lobes acute; setae of the
crown comparatively few. *G. acuta* Michx. [B]. *G. Amarella acuta* Herder
[G]. Moist places: Lab.—Man.—Minn.—Me.

5. **A. propínqua** (Richardson) Greene. Annual, often branched near the
base; stem 0.5–2.5 dm. high, often purplish; basal leaves spatulate or oblanceolate, 1–2 cm. long; stem-leaves lanceolate, 1–2 cm. long, acute; flowers 1–3 in
the axils, on naked pedicels; calyx-lobes very unequal, the two outer ovate or
oblong, acute, 5–7 mm. long, the rest linear, 3–5 mm. long; corolla blue, 10–17
mm. long. *G. propinqua* Richardson [B]. Moist slopes and glades: Alaska—
B.C.—Man.—Lab.—Mack; e Asia. *Arct.—Subalp.—Alp.* Jl–Au.

6. **A. occidentàlis** (A. Gray) Greene. Annual; stem 3–5 dm. high,
branching, wing-margined; basal leaves oblanceolate or elliptic, rounded at the
apex; stem-leaves broadly ovate, 2–5 cm. long, 1.5–3 cm. wide, 3–5-ribbed, acute
at the apex, sessile or somewhat clasping at the base; flowers axillary in clusters of 3–5; calyx-tube 2–2.5 mm. long; corolla funnelform, bright blue; 15–20
mm. long, the lobes ovate-lanceolate, cuspidate; capsule slender-stipitate. *Gentiana quinqueflora occidentalis* A. Gray [G, B]. Moist places: Ohio—Tenn.—
Mo.—Minn. *Prairie.* Au–O.

6. DASYSTÉPHANA Adans. GENTIAN.

Mostly perennial, rarely annual, caulescent herbs. Leaves opposite, entire or merely erose-ciliolate. Flowers perfect, cymose, sessile or nearly so, rarely solitary and pedicelled. Calyx with a membrane inside the calyx-lobes, which vary from obsolete to foliaceous. Corolla salverform, funnelform, or clavate, without glands and crowns, but with cleft plaits between the convolute lobes. Stamens with converging or cohering anthers. Ovary 1-celled; ovules numerous. Capsule stipitate.

Corolla cylindric, oblong, or broadly clavate, its lobes connivent or erect; anthers
  cohering in a ring or short tube.
  Calyx-lobes and bracts ciliolate-scabrous ; leaves rough-margined.
    Calyx-lobes linear or spatulate, equaling or exceeding the tube, corolla-lobes
      evident, longer than the 2-cleft appendages of the plaits. 1. *D. Saponaria.*
    Calyx lobes lanceolate or ovate, shorter than the tube ;
      corolla lobes obsolete, shorter than the short-notched ap-
      pendages of the plaits.   2. *D. Andrewsii.*
  Calyx-lobes, bracts, and leaf-margins smooth or nearly so.
    Corolla-lobes ovate, much longer than the plaits ; upper
      leaves ovate to lanceolate.
      Calyx-lobes ovate ; leaves more or less clasping, all
        broad.   3. *D. flavida.*
      Calyx-lobes lanceolate, leaves not clasping, the lower
        often linear.   4 *D. Grayi.*
    Corolla-lobes rounded, not much exceeding the plaits ; leaves
      linear to linear-lanceolate.   5. *D. linearis.*
Corolla campanulate-funnelform, its lobes ascending-spreading ; an-
  thers distinct or soon separating.
  Calyx-lobes erect or none ; corolla-lobes ascending or nearly
    erect, not toothed ; plaits equally lobed.
    Calyx-lobes well developed ; calyx-tube truncate at the
      apex.   6. *D. affinis.*
    Calyx-lobes none or minute ; calyx-tube irregular, more or
      less lobed or cleft.   7. *D. Forwoodii.*
  Calyx-lobes widely spreading, linear-lanceolate ; corolla-lobes
    in anthesis strongly spreading, more or less denticulate ;
    plaits unequally 2-lobed.   8. *D. puberula.*

1. **D. Saponària** (L.) Small. Stem 3–8 dm. long, glabrous; leaves lanceolate to oblong, 2.5–7.5 cm. long; flowers near the end of the stem; calyx 1.5–2 cm. long; corolla blue or purplish blue, 3.5–4 cm. long; plaits many-toothed; capsule 1.5 cm. long. *G. Saponaria* L. [G, B]. Swamps and wet places: Ont.—Minn.—La.—Fla. *E. Temp.* Au–O.

2. **D. Andréwsii** (Griseb.) Small. Stem glabrous, simple, 3–6 dm. high, leafy; leaves lanceolate or ovate-lanceolate, 7–15 cm. long, 3–7-ribbed; flowers in terminal clusters and in the axils of the uppermost leaves; calyx-tube 8–12 mm. long; lobes lanceolate or ovate, ciliolate, spreading; corolla broadly clavate, bluish or white, closed; lobes obsolete, rounded; lobes of the plaits rounded, erose, yellowish. *G. Andrewsii* Griseb. [G, B]. Wet meadows and among bushes: Que.—Ga.—Mo.—Neb.—N.D.—Man. *Allegh.—Prairie—Plain.* Au–O.

3. **D. flávida** (A. Gray) Britton. Stem stout, 3–10 dm. high, glabrous; leaves ovate to lanceolate, more or less clasping, 5–12 cm. long; flowers in a terminal cluster, sometimes 1 or 2 in the upper axils; calyx-lobes ovate, acute; corolla 3–5 cm. long, open, greenish or yellowish white, the lobes twice as long as the erose-denticulate appendages. *G. flavida* A. Gray [B]. Moist soil: Ont.—Man.—Iowa—Mo.—Va. *Allegh.—Prairie.* Au–O.

4. **D. Gràyi** (Kusnez.) Britton. Stem 3–6 dm. high, glabrous; leaves distant, the lower linear or oblanceolate, obtuse, the upper lanceolate or ovate-lanceolate, 5–8 cm. long; flowers in a terminal cluster; calyx-lobes unequal; corolla greenish blue, 3–3.5 cm. long, broadly clavate, its lobes acute. *G. Grayi* Kusnez. *G. linearis latifolia* A. Gray [G, B]. *G. rubricaulis* Britton; not Schwein. Wet places: N.B.—Minn.—Neb.—N.Y. *Canad.—Allegh.* Au–S.

5. D. lineàris (Fröl.) Britt. Stem slender, 2.5–7 dm. high, glabrous; leaves linear or lanceolate, narrowed at the base; flowers in a terminal cluster; calyx-lobes linear, erect, subequal; corolla blue or white, 3–4 cm. long, open. *G. linearis* Fröl. [G, B]. *G. rubricaulis* Schw. Bogs and wet rocks: N.B.— Minn.—Md. *Canad.—Allegh.* Au–S.

6. D. affînis (Griseb.) Rydb. *Fig. 454.* Stems 1–3 dm. high, leafy; leaves 1.5–3 cm. long; calyx-tube 5–7 mm. long; lobes 2–10 mm. long, acute; corolla blue or purple, 2.5–3 cm. long; lobes ovate, acute; lobes of the plaits ovate, 2-cleft, acuminate, laciniate, two thirds as long as the corolla-lobe. *G. affinis* Griseb. [B]. Mountains and hills: B.C.— Calif.—Colo.—S.D.—Sask. *Mont.—Subalp.* Au–S.

f.454.

7. D. Forwoódii (A. Gray) Rydb. Closely related to *D. affinis;* stem 1–3 dm. high, leafy; leaves 1.5–3 cm. long, rather thick; flowers short-pedicelled in the upper axils, forming a dense spike-like inflorescence; calyx-tube 5–8 mm. long, usually purplish; corolla blue or purple, about 3 cm. long; lobes obovate, acute; lobes of the plaits 2-cleft, lanceolate, acuminate, a little shorter than the corolla-lobes. *G. Forwoodii* A. Gray. Hills and mountains: Alta.—Colo.—Ida. *Submont.—Mont.* Jl–Au.

8. D. pubérula (Michx.) Small. Stem strict, erect, 3–5 dm. high, more or less scabrous, leafy; leaves lanceolate, firm, scabrous-puberulent on the margins and midrib, paler beneath, 3–5 cm. long; flowers subsessile, forming a dense spike; calyx-tube not scarious, about 1 cm. long; lobes subulate, 8–12 mm. long, spreading; corolla blue, about 4 cm. long, open-funnelform; lobes obovate, obtuse or acute; lobes of the plaits 2-cleft, often laciniate, half as long as the corolla-lobes. *G. puberula* Michx. [B, G]. Prairies: Ohio—Ga.—Kans. —Sask. *Allegh.—Plain.*

7. PLEURÓGYNE Eschsch. MARSH FELWORT.

Low slender annuals. Leaves opposite, entire. Flowers perfect. Calyx deeply 4- or 5-cleft. Corolla rotate, 4- or 5-cleft, its divisions acute, with a pair of scale-like appendages on their bases. Stamens inserted on the base of the corolla; anthers introrse, versatile. Style none; stigmas decurrent on the sutures; ovary 1-celled. Capsule lanceolate or oblong, not stipitate. Seeds numerous.

1. P. rotâta Griseb. Stem 5–15 cm. high, simple or with erect branches; basal leaves spatulate; stem-leaves linear or lance-linear, 1–2 cm. long, 2–4 mm. wide; sepals linear or lance-linear, as long as the corolla; corolla-lobes about 1 cm. long; capsule oblong. Arctic regions: Greenl.—Que.—N.H.— Man. *Arct.* Jl–Au.

8. HALÈNIA Borkh. SPURRED GENTIAN.

Annual or perennial caulescent herbs. Leaves opposite, entire. Flowers perfect, in terminal or axillary cymes or panicles. Calyx deeply 4-cleft. Corolla white, yellow, blue, or purple, with a short tube, 4-lobed; each lobe produced into a spur or sack at the base. Stamens 4, inserted on the lower part of the corolla-tube; filaments filiform or dilated at the base; anthers versatile. Ovary 1-celled; placentae intruded; style short or wanting; stigma 2-lobed. Capsule ovoid or oblong. Seeds slightly flattened, smooth. *Tetragonanthus* J. F. Gmel.

f.455.

1. H. defléxa (Smith) Griseb. *Fig. 455.* An annual, usually simple, 1.5–4.5 dm. high; basal leaves obovate or spatulate, petioled; stem-leaves oblong or the upper ovate, acute, 3–5-ribbed, 2–4 cm. long; sepals lanceolate, acuminate; corolla yellowish or greenish white, or purplish, 6–8 mm. long; lobes triangular-ovate, acute; spurs slightly spreading, rather slender. *T. deflexus* Kuntze [B, R]. Damp woods: N.S.—N.Y.—S.D.—B.C. *Boreal.—Submont.* Jl–Au.

9. BARTÒNIA Muhl.

Small annuals or biennials, with scale-like leaves. Flowers small, perfect, racemose or paniculate. Calyx deeply 4-parted, the lobes keeled, acuminate. Corolla white or yellow, campanulate, deeply 4-cleft, the lobes imbricate in bud. Stamens 4, short, inserted in the sinuses of the corolla. Ovary 1-celled; stigma 2-lobed. Capsule 2-valved; seeds numerous.

1. B. virgínica (L.) B.S.P. Stem 5–20 cm. high, yellowish, straight or flexuose, sharply 4-angled; scale-like leaves opposite or subopposite, subulate, 2–4 mm. long; flowers mostly opposite, pedicelled; corolla 3–4 mm. long, yellow, the lobes oblong, denticulate, obtuse. Moist soil: Newf.–Minn.—La.—Fla. *Canad.—Allegh.* Jl–S.

Family 135. MENYANTHACEAE. Buckbean Family.

Perennial aquatic or bog plants, with horizontal rootstocks. Leaves alternate, but often all basal, simple or trifoliate. Flowers perfect, solitary or in terminal clusters or racemes. Calyx of 5, partially united sepals. Corolla from rotate to funnelform; lobes induplicate-valvate. Stamens 5; filaments adnate to the corolla tube. Gynoecium of 2 united carpels; ovary 1-celled with two parietal placentae; stigmas 2. Fruit a capsule, indehiscent, irregularly bursting, or valvate.

Corolla bearded within ; leaves trifoliolate ; erect herbs. 1. MENYANTHES.
Corolla not bearded ; leaves simple, the blades rounded ; plant floating.
　　　　　　　　　　　　　　　　　　　　　　　　　　2. NYMPHOIDES.

S.456.

1. MENYÁNTHES (Tourn.) L. Buckbean, Marsh Trefoil, Bog Bean.

Perennial bog plant. Leaves long-petioled, trifoliolate, basal. Flowers perfect, racemose or paniculate, on a long scape. Calyx 5-parted. Corolla short-funnelform, 5-cleft, its lobes fimbriate or bearded within, white, spreading. Stamens inserted on the tube of the corolla; filaments filiform; anthers sagittate. Capsule ovoid, indehiscent or at last ruptured. Seeds few, subglobose, shining.

1. M. trifoliàta L. *Fig. 456.* Perennial, with a stout rootstock; scape 1–3 dm. high; leaflets oval or elliptic, thick, glabrous, 5–10 cm. long; corolla white, tinged with rose, campanulate-funnelform, about 15 mm. long; lobes copiously fimbriate within. In water and bogs: Lab.—Newf.—Pa.—Colo.—Calif.—Alaska; Eurasia. *Boreal.—Mont.* My–Jl.

2. NYMPHOÌDES (Tourn.) Hill. Floating Heart.

Aquatic herbs, with floating or creeping stems. Leaves sometimes all basal and long-petioled, the stem-leaves with shorter petioles, the blades with a cordate base, or peltate, floating. Flowers solitary or in ours clustered at the nodes, perfect, regular. Calyx green, with 5, rather narrow lobes. Corolla white or yellow, nearly rotate, 5-lobed. Stamens 5, filaments very short, adnate to the base of the corolla. Ovary 1-celled. Capsule indehiscent or opening irregularly. *Limnanthemum* J. F. Gmel.

1. **N. lacunòsum** (Vent.) Fern. Leaf-blades ovate, oval, or rounded-ovate, deeply cordate at the base, 1.5–6 cm. long; flower-clusters with slender tubers; corolla white, its lobes broadly oval, 1 cm. long; seeds smooth. *L. lacunosum* (Vent.) Griseb. [B]. Shallow ponds and lakes: N.S.—Minn.—La.—Fla.

Family 136. APOCYNACEAE. Dogbane Family.

Perennial herbs (all ours), vines, shrubs, or trees, with milky juice. Leaves opposite or alternate, without stipules, simple. Flowers perfect, regular. Calyx of 5 persistent sepals, imbricate in the bud. Corolla of 5 partially united petals, with convolute lobes. Stamens 5, alternating with the corolla-lobes; anthers 2-celled. Gynoecium of two, distinct (in ours) or united carpels; if united, the ovary either 2-celled, or 1-celled with two parietal placentae; styles united. Fruit in ours of two follicles.

Leaves alternate; erect herbs.   1. Amsonia.
Leaves opposite.
    Erect or diffuse herbs; corolla campanulate to urceolate.   2. Apocynum.
    Plants creeping or trailing; corolla salverform.   3. Vinca.

1. AMSÒNIA (Clayton) Walt.

Perennial herbs. Leaves alternate, entire. Flowers in terminal corymbiform cymes. Calyx-lobes 5. Corolla salverform; tube slightly enlarging upwards, villous within; lobes narrow. Disk wanting. Stamens included; anthers unappendaged. Carpels 2, united by the styles. Stigmas appendaged by a reflexed membrane. Follicles 2, erect, several-seeded. Seeds unappendaged.

Calyx glabrous; follicles not torulose.
    Leaves ovate to oblong-lanceolate, not glaucous beneath; the bases of the lower
    ones obtuse or abruptly acute, the margins not callous.
        1. *A. Tabernaemontana.*
    Leaves narrowly lanceolate, glaucous beneath; the bases
      gradually acute, the margins more or less callous.   2. *A. salicifolia.*
Calyx pubescent; follicles subtorulose.   3. *A. illustris.*

1. **A. Tabernaemontàna** Walt. Stem 6–13 dm. high, more or less branched, glabrous or sparingly hairy above; leaf-blades oval to broadly oblong-oblanceolate, acute or acuminate, dull above, green and sometimes thinly pubescent beneath; calyx-lobes triangular, 1 mm. long; corolla purplish blue; tube 6–8 mm. long; lobes linear or lance-linear; follicles 9–12 cm. long, glabrous. *A. Amsonia* (L.) Britton [B]. River banks and woods: N.J.—Kans.—Tex.—Fla. *Austral.* My–Je.

2. **A. salicifòlia** Pursh. Stem 3–5 dm. high, glabrous, or slightly pubescent when young; leaves alternate, narrowly lanceolate, acute or acuminate, glabrous above, glaucous beneath; calyx about 1 mm. long, the teeth minute, deltoid; corolla blue-purple, the tube 6–10 mm. long, the lobes 5–7 mm. long, lanceolate; follicles slender, 5–10 cm. long, oblong, pitted. River banks: Va.—Ga.—Tex.—Kans. *Ozark.* My–Jl.

3. **A. illústris** Woodson. Stems 7–11 dm. high, glabrous, clustered; leaves alternate or the lower whorled, lanceolate to lance-linear, acuminate, short-petioled, the blades 5–7 cm. long, glabrous, glaucous beneath, shining above; calyx 1.5–3 mm. long, hirsute-strigillose, the teeth deltoid, corolla sky-blue, the tube 6–8 mm. long, the lobes spreading, 4–6 mm. long; follicles 8–12 cm. long, 2–3 mm. thick, acuminate, pitted. Sand-bars and banks: Mo.—Kans.—Tex. *Ozark.* My–Jl.

2. APÓCYNUM (Tourn.) L. DOGBANE, INDIAN HEMP.

Perennial herbs with branched stems. Leaves opposite, entire. Flowers in corymbose cymes. Sepals 5. Corolla campanulate; tube with five small appendages, alternating with the stamens; lobes 5. Stamens adnate to the base of the corolla-tube; anthers sagittate, at least coherent to the stigma. Follicles 2, slender. Seeds numerous, tipped with a long coma.

Corolla fully twice as long as the calyx, its lobes spreading or at least ascending in anthesis.
  Corolla open-campanulate, the lobes reflexed-spreading; cymes both axillary and terminal.
    Leaves more or less pubescent beneath.   1. *A. androsaemifolium.*
    Leaves perfectly glabrous on both sides.   2. *A. ambigens.*
  Corolla nearly cylindric, the lobes ascending; cymes ending leafy branches.   3. *A. medium.*
Corolla less than twice as long as the calyx, its lobes erect or nearly so.
  Leaves and cymes glabrous or nearly so.
    Leaves acute at the base, petioled.
      Leaf-blades oval or elliptic, equally green on both sides.   4. *A. cannabinum.*
      Leaf-blades narrowly lance-oblong, gradually tapering at both ends, pale beneath.   5. *A. album.*
    Leaves, at least those of the main stem, truncate or cordate at the base, subsessile or sessile.
      Leaf-blades elliptic; inflorescence ample.   6. *A. sibiricum.*
      Leaf-blades broadly oval; inflorescence short and dense, much surpassed by the branches.   7. *A. cordigerum.*
  Leaves and cymes and usually the whole plant densely pubescent.   8. *A. pubescens.*

1. **A. androsaemifòlium** L. Stem erect, 3–15 dm. high, glabrous; leaf-blades 4–10 cm. long, pale beneath; calyx-lobes ovate, acuminate, scarcely 1 mm. long; corolla pink; tube 4–5 mm. long; lobes ovate, obtuse. Thickets and fields: Anticosti—Ga.—Ariz.—Ida.—B.C. *Temp.—Plain—Mont.* Je–Π.

2. **A. ámbigens** Greene. Stem 3–6 dm. high, glabrous, often tinged with purple above, with ascending branches; leaf-blades thick, dark green above, paler beneath, rounded or sometimes subcordate at the base, 3–6 cm. long; calyx lobes lanceolate, acute, 2 mm. long, usually purple-tinged; corolla flesh-colored or rose, 4–4.5 mm. long; lobes 2 mm. long, oblong, obtuse, reflexed. *A. androsaemifolium* Woodson; not L. River bottoms and hillsides: Mont.— S.D.—Colo.—Utah—Wash. *Submont.—Mont.* Je–Au.

3. **A. mèdium** Greene. Stem stout, 7–15 dm. high, glabrous; leaves firm, ovate-oblong to elliptic, slightly pubescent beneath; cymes terminal at the ends of the stem and later at the ends of leafy branches; calyx-lobes half as long as the corolla-tube; corolla white, 6–7 mm. long; lobes acute, somewhat spreading; follicles 7–12 cm. long. *A. speciosum* G. S. Miller. Fields and roadsides: Que.—Minn.—Iowa—Md. *Allegh.* Je–Au.

4. **A. cannábinum** L. *Fig. 457.* Stem 5–15 dm. high, glabrous, with erect or ascending branches; leaf-blades oblong, lanceolate-oblong, or ovate-oblong, light green above, glaucous beneath, usually perfectly glabrous; calyx-lobes lanceolate, acute or acuminate, about 2.5 mm. long; corolla greenish-white; lobes ovate, about 1.5 mm. long. Thickets and fields: Anticosti—Fla.—L. Calif.—B.C. *Temp.—Plain—Submont.* Je–Au.

f.457.

5. **A. álbum** Greene. Stem slender, much branched, often purplish, glabrous; leaves short-petioled, glabrous, narrowly lance-oblong, sometimes

lance-linear, 3–8 cm. long, 5–20 mm. broad, dark green above, pale beneath, gradually tapering at each end, cuspidate; cymes terminating the stem and branches, rather open; bracts subulate; corolla white, 4–5 mm. long, the lobes acute; follicles 6–10 cm. long, 3–4 mm. thick. *A. cannabinum glaberrimum* A.DC. Grassy and sandy shores: N.S.—Minn.—Neb.—Tex.—Md. *Allegh.* Jl–Au.

6. **A. sibíricum** Jacq. Stem 3–6 dm. high, glabrous, with ascending branches; leaves oblong, oblong-lanceolate, or oval, 2–8 cm. long, cordate-clasping, rounded or truncate at the base, obtuse or acute and mucronate at the apex, pale green and glabrous; calyx-lobes lanceolate, acute, 2.5 mm. long; corolla greenish white, campanulate; lobes lance-oblong, erect; follicles 5–9 cm. long. *A. hypericifolium* Ait. [B]. *A. cervinum* Greene. *A. cannabinum hypericifolium* A. Gray [G]. River valleys and hillsides: Ont.—Ohio.—N.M.—B.C. *Temp.—Plain—Submont.* Je–Au.

7. **A. cordígerum** Greene. Stem stout, 4–7 dm. high, glabrous; lower leaves broadly oval, cordate and clasping at the base, glabrous, 5–8 cm. long, 2–5 cm. wide, firm, rounded and mucronate at the apex, the upper subsessile and rounded at the base; cyme at the end of the stem, overtopped by the branches, short and dense; calyx two-thirds as long as the corolla; corolla white, 3–4 mm. long, the lobes obtusish. Prairies and banks: Wis.—Minn.—N.D.—Ark. *Prairie.* Je–S.

8. **A. pubéscens** R. Br. Stem 3–6 dm. high, densely pubescent; leaves oval or elliptic, mucronate, obtuse at the base, white-pubescent beneath; petioles 2–4 mm. long; calyx-lobes as long as the corolla-tube, lanceolate, acute; corolla purplish, its lobes erect; follicles 6–7 cm. long. *A. cannabinum pubescens* DC. [G]. Roadsides: R.I.—Iowa—Mo.—Va. *Allegh.* Je–Au.

3. VÍNCA L. PERIWINKLE, MYRTLE.

Herbs, sometimes woody at the base. Leaves opposite, often evergreen. Flowers solitary in the axils. Calyx 5-parted, the lobes acuminate. Corolla funnelform or salverform, pubescent within, the limb more or less oblique. Stamens 5, short. Carpels 2; ovules several; styles filiform, the stigma annular.

1. **V. mìnor** L. Stem spreading, trailing, 1–6 dm. long; leaves ovate to oblong, entire, firm, shining, dark green; corolla blue, 1.5–3 cm. broad, the lobes obovate, truncate. Roadsides and around dwellings: Conn.—Minn.—Ga.; escaped from cult.; nat. of Eur. Ap–Je.

Family 137. ASCLEPIADACEAE. MILKWEED FAMILY.

Perennial herbs or vines, or in warmer climates shrubs, with milky juice. Leaves opposite, whorled, or alternate, without stipules. Flowers perfect, regular, mostly umbellate. Sepals distinct or nearly so, imbricate in bud. Corolla campanulate, urceolate, funnelform or rotate, 5-lobed, the lobes usually reflexed. A 5-lobed crown is usually present between the corolla and the stamens, adnate to either or both. Stamens 5, adnate to the base of the corolla; filaments monadelphous; anthers attached by the base, introrse. Gynoecium of 2 carpels. Styles 2, connected at the summit by the stigmas. Ovules numerous, anatropous, pendulous. Fruit of 2 carpels. Seeds many, compressed, appendaged by a coma.

Erect or decumbent herbs, not twining.
　Corolla-lobes reflexed during anthesis.
　　Hoods of the crown crestless or with an obscure crest within.
　　　　　　　　　　　　　　　　　　　　　　　1. ACERATES.

Hoods of the crown with a horn-like or tooth-like crest within.
2. ASCLEPIAS.
Corolla-lobes erect-spreading during anthesis. 3. ASCLEPIODORA.
Twining vines.
Crown of the corolla not appendaged ; corolla rotate. 5. VINCETOXICUM.
Crown of the corolla with appendages.
Plant herbaceous ; crown with 5 truncate lobes, each ending in a simple or 2-cleft awn-like appendage. 4. GONOLOBUS.
Plant woody ; crown with 10 broad and 5 narrow appendages.
6. PERIPLOCA.

1. ACERÀTES Ell. GREEN MILKWEED.

Perennial caulescent herbs. Leaves opposite or alternate. Flowers perfect, regular, umbellate. Calyx small, usually with 2 glands in the sinuses of the 5 acute lobes. Corolla rotate; lobes 5, reflexed in anthesis. Hoods mostly attached over the short or obsolete column, erect, involute-concave, saccate at the base, entire, emarginate, or 2- or 3 toothed at the apex, with broad auricles at the base, without horns or crests within. Follicles fusiform. Seeds with a coma.

Auricles of the hoods when present, concealed within ; leaves oval to linear.
Umbel solitary, terminal ; plant hirsute. 1. A. lanuginosa
Umbels several, lateral ; plant glabrate or tomentose when young.
Hood rounded, entire, much shorter than the anther ; crown on a column. 2. A, hirtella.
Hood auricled at the base, lanceolate-oblong, almost equaling the anthers ; crown sessile. 3. A. viridiflora.
Auricles of the hoods conspicuously spreading ; umbels lateral ; leaves narrowly linear.
Hoods emarginate or truncate at the summit, crestless within ; umbels distinctly peduncled. 4. A. auriculata.
Hoods trilobed at the summit, with an internal crest-like fildrib terminating in the middle lobe ; umbels subsessile or on very short peduncles 5. A. angustifolia.

1. **A. lanuginòsa** (Nutt.) DC. Stem erect, 1–4 dm. high; leaves oblong or elliptic, short-petioled, 3–10 cm. long; corolla greenish; hood oblong, about 5 mm. long; hood purplish, oblong, with an auricle or fold on the ventral margins. Prairies: Ill.—Minn.—Mont.—Wyo. *Prairie—Plain.* Je–Au.

2. **A. hirtélla** Pennell. Stem 5–10 dm. high; scabro-puberulent; leaves linear or elongate linear-lanceolate, 7–17 cm. long, 5–15 mm. wide; umbels several; pedicels with spreading hairs, 12–15 mm. long; corolla-lobes oblong, 5 mm. long, greenish yellow, dull purple without; hood purple, oval, obtuse, adnate at the base to the column; follicles erect, on reflexed pedicels, fusiform, 7–10 cm. long, minutely puberulent. *A. floridana* Hitchc., in part. Prairies and pinelands: Mich.—Wis.—Iowa—Okla.—Mo. *Prairie.* Je–S.

3. **A. viridiflòra** (Raf.) Eaton. Stem 2–6 dm. high, often decumbent; leaves oval or oblong and obtuse, retuse, or acute, or elongate lanceolate and acuminate (var. *Ivesii* Britton), or narrowly linear (var. *linearis* A. Gray), becoming leathery, with undulate margins; umbels sessile or nearly so, many-flowered; corolla-lobes oblong-lanceolate, about 5 mm. long; hood dull purple or at least tinged with purple, lance-oblong, entire except for the small infolded auricles at the base. Dry or sandy soil: Mass.—Fla.—N.M.—Mont. *E. Temp.—Submont.* Je–Jl.

4. **A. auriculàta** Engelm. Stem 2–8 dm. high, mostly solitary; leaves alternate, scattered, narrowly linear, 5–15 cm. long, with scabrous, often revolute margins; corolla-lobes greenish, about 4 mm. long; hood yellowish, with purplish keel, erect, emarginate or truncate, the involute sides with broad auricles at the base. Plains and prairies: Neb.—Utah—N.M.—Tex. *Plain—Submont.* Je–Au.

5. **A. angustifòlia** (Nutt.) Dec. Stem several, from a thick rootstock, 3–6 dm. high, puberulent or glabrate in age; leaves narrowly linear, 5–12 cm. long, revolute-margined; umbels subsessile; corolla-lobes 5 mm. long, oblong,

greenish; hood white, erect, as high as the anthers, laterally compressed, 3-dentate at the apex, the inner margins with an erose truncate lobe. *Asclepias stenophylla* A. Gray. Dry plains: S.D.—Colo.—Tex.—Mo. *Plain.* Je–Au.

2. **ASCLÈPIAS** (Tourn.) L. Milkweed, Silkweed, Butterfly-weed.

Perennial herbs, with deep taproots, sometimes woody at the base. Leaves opposite, alternate, or whorled. Flowers perfect, regular, in axillary or terminal umbels. Calyx small, usually with small glands at the base of the 5 lobes. Corolla rotate, deeply 5-cleft; lobes valvate-convolute in bud, reflexed in anthesis. Hoods of the crown involute, arising from the base of the corolla-lobes or on a short column, concave, hooded, bearing within a horn- or tooth-like projection, and sometimes additional processes between the anthers. Follicles fusiform, naked or with soft processes or warty. Seeds mostly with a coma.

Plant more or less hirsute; hoods orange; leaves mostly alternate; juice not milky.
    1. *A. tuberosa.*
Plant not hirsute; hoods greenish, purplish, yellowish, or
  white; leaves mostly opposite or verticillate; juice
  milky.
  Leaves orbicular to linear-lanceolate, opposite (except
    in *A. quadrifolia*).
    Follicles with soft spinulose processes, tomentose;
      leaves large and broad, transversely veined,
      oval or ovate; corolla purplish.
      Plant pubescent.
        Hoods oblong-ovate, obtuse, slightly exceeding
          the stamens.
          Hoods ascending; processes of the fruit
            numerous, almost filiform, 3–10 mm.
            long.
          Hoods erect; processes of the fruit few,
            conic-subulate.
        Hoods lanceolate, produced, about three times
          as long as the stamens.
      Plant glabrous.
    Follicles without processes.
      Umbels solitary on a naked terminal peduncle.
        Leaves oblong, with cordate clasping base,
          obtuse or retuse; corolla purple.
        Leaves ovate or oblong-ovate; corolla green-
          ish white.
      Umbels mostly more than one; peduncles not
        overtopping the leaves.
        Leaves broadly oval or rectangular-oval or
          nearly orbicular, obtuse or retuse at
          both ends; umbels sessile.
          Plant puberulent when young, glabrate in
            age; column very short.
          Plant tomentulose; column half as long
            as the anthers.
        Leaves ovate or lanceolate, or rarely oval,
          acute.
         Follicles erect or ascending, on reflexed
          pedicels.
          Leaves ovate or oval.
            Corolla white or yellowish.
              Plants glabrous or puberulent
                on the young parts; hood
                little exceeding the an-
                thers.
                Leaves oval; pedicels 2 cm.
                  long.
                Leaves elliptic; pedicels
                  drooping, 5 cm. long.
              Plant pubescent; hood twice as
                long as the anthers.
            Corolla dark purple.
           Leaves narrowly lanceolate; corolla
            greenish purple.
        Follicles erect, on erect fruiting pedicels.
          Leaves all opposite; corolla rose-purple.
            Plant glabrous or nearly so; leaves
              oblong-lanceolate.

2. *A. kansana.*

3. *A. syriaca.*

4. *A. speciosa.*
5. *A. Sullivantii.*

6. *A. amplexicaulis.*

7. *A. Meadii.*

8. *A. latifolia.*

9. *A. arenaria.*

10. *A. variegata.*

11. *A. exaltata.*

12. *A. ovalifolia.*
13. *A. purpurascens.*

14. *A. brachystephana.*

15. *A. incarnata.*

Plant hirsute-pubescent; leaves
    broadly lanceolate.   16. *A. pulchra.*
Leaves, at least some of them, in whorls
    of 4's or 6's.   17. *A. quadrifolia.*
Leaves narrowly linear, verticillate or scattered.
Plant tall, 4–6 dm. high, from a rootstock; leaves
    verticillate.
  Hoods entire, ovate, truncate.   18. *A. verticillata.*
  Hoods dorsally hastate-sagittate.   19. *A. galioides.*
Plant low, 1–2 dm. high, bushy, from a ligneous
    base; leaves scattered, crowded.   20. *A. pumila.*

1. A. tuberòsa L. Stem 3–6 dm. high, coarsely hirsute, very leafy; leaves usually alternate, short-petioled or subsessile, lance-linear or lance-oblong, subcordate to acute at the base, 5–10 cm. long, with revolute margins; corolla-lobes linear-oblong, obtuse, 6–8 mm. long, greenish or reddish orange; column about 1 mm. long; hoods linear-oblong or linear-lanceolate, bright orange or yellow, 5–6 mm. long; follicles 7–10 cm. long, minutely pubescent. Dry fields and meadows: Ont.—Fla.—Tex.—Ariz.—Minn. *Temp.—Plain.* Je–S.

2. A. kansàna Vail. Stem 2–6 dm. high, canescent or glabrate; lower leaf-blades broadly ovate, truncate or rounded at the base, obtuse or emarginate, 18–16 cm. long, the upper elliptic oblong, mucronate, tomentose, becoming glabrate above, canescent-tomentose beneath; corolla-lobes oblong, pink-purple, 7–8 mm. long, obtuse, the margins infolded, with a basal tooth on each side; follicles 8–10 cm. long, 3–3.5 cm. wide, densely white-tomentose. Fields and prairies: Kans.—Iowa—Mo. *Prairie.* Je.

3. A. syrìaca L. *Fig. 458.* Stem 5–15 dm. high, canescent or in age glabrate; leaf-blades elliptic to lance-oblong, 1–1.5 dm. long, 4–11 cm. wide, rounded, truncate, or acute at the base, obtuse or mucronate at the apex, green and glabrous or nearly so above, tomentulose beneath; corolla-lobes dull greenish purple, 8 mm. long, more or less pubescent; column less than 1 mm. long; hoods whitish or greenish purple; follicles 7–9 cm. long, lance-ovate, tomentose. *A. Cornuti* Dec. Fields and waste places: N.B.—Ga.—Kans.—Sask. *Canad.—Allegh.—Plain.* Je–S.

f.458.

4. A. speciòsa Torr. Stem tall, 1–2 m. high, finely canescent-tomentose; leaves oval, subcordate at the base, thick; corolla greenish purple; lobes oblong or ovate-oblong, 9–13 mm. long; column short or none; hoods lanceolate, 11–15 mm. long, with two blunt teeth below; follicles 7–10 cm. long, with numerous processes. Valleys: Sask.—Minn.—Kans.—Calif.—B.C. *Prairie—Plain—Submont.* My–Jl.

5. A. Sullivántii Engelm. Stem stout, 6–12 dm. high, glabrous; leaves ovate-oblong, cordate at the base, 10–15 cm. long, mucronate, subsessile, glabrous; corolla-lobes purplish, elliptic, 11–13 mm. long; hoods obovate, entire, obtusely 2-eared at the base; follicles nearly glabrous, 8–10 cm. long, with blunt processes. Rich ground: Ont.—Minn.—Kans.—Ohio. *Allegh.—Prairie.* Jl–S.

6. A. amplexicaùlis Smith. Stem glabrous, 3–10 dm. high; leaves opposite or rarely in 4's at the base of the stem, sessile, oblong or elliptic, 6–12 cm. long, clasping, retuse and apiculate at the apex, wavy-margined, glabrous; corolla lobes greenish or greenish purple, 6–10 mm. long, acutish; hoods erect, flesh-colored, sessile on a short column, gibbous or saccate at the base; follicles 10–16 cm. long, long-attenuate. Sandy soil: N.H.—Minn.—Neb.—Tex.—Fla. *E. Temp.* Je–Jl.

7. **A. Meàdii** Torr. Stem slender, 4–6 dm. high, glabrous; leaves ovate or oblong-ovate, 3–7 cm. long, pale green, scabrous on the margins; corolla-lobes greenish yellow, ovate, acute, 6–9 mm. long; hoods rounded, truncate at the apex, and with a sharp tooth on each margin; follicles narrow, puberulent, 10–12 cm. long. Dry ground: Wis.—Iowa—Ill. *Prairie.* Je–Au.

8. **A. latifòlia** (Torr.) Raf. Stem simple, stout, 3–8 dm. high, minutely puberulent when young; leaves sessile or short-petioled; blades rounded-oval, mucronate and often emarginate at the apex, cordate or subcordate at the base, 7–15 cm. long, minutely puberulent; corolla-lobes ovate, 7–10 mm. long, greenish; column thick, very short; hood truncate; follicles ovoid, 6 cm. long, glabrate. *A. Jamesii* Torr. Dry plains: Neb.—Tex.—Ariz.—Utah. *Plain.* Jl–Au.

9. **A. arenària** Torr. Stem decumbent or ascending, tomentose, 3–6 dm. long; leaf-blades rectangular-oval or obovate, truncate or retuse and often mucronate at the apex, subcordate or truncate at the base, 5–10 cm. long, tomentose; corolla-lobes oval-oblong, greenish white; hoods oblong, truncate, with a triangular tooth on each margin; follicles puberulent, 5–10 cm. long. In sandy soil: Neb.—Tex.—Colo.; n Mex. *Son.—Plain.* Je–S.

10. **A. variegàta** L. Stem glabrous or minutely puberulent, 3–10 dm. high, erect, from a woody rootstock; leaves in 3–7 pairs, the middle pairs rarely whorled, oval to obovate, 6–14 cm. long, acute at each end, or obtuse and mucronate, undulate, light green above, paler beneath, glaucescent and sometimes tomentulose; peduncles 1–4, longer than the pedicels; corolla-lobes 6–8 mm. long, white, acute; hoods erect, 4 mm. high, globose, with a claw-like base; follicles slender-fusiform, 1–1.3 dm. long, long-acuminate, tomentulose. Shaded grounds: Conn.—Kans.—Tex.—Fla. *E. Temp.* My–Je.

11. **A. exaltàta** (L.) Muhl. Stem green and glabrous, 1–1.5 m. high; leaves elliptic or elliptic-lanceolate, rarely obovate, acuminate at each end, glabrous above, sparingly and minutely pubescent beneath; corolla-lobes greenish, often tinged with purple without, 8 mm. long; hoods on a short column, white, flesh-colored, or bluish, erect, saccate at the base, truncate at the summit; follicles 12–16 cm. long, slender-fusiform, long-acuminate, minutely puberulent, glabrate. *A. phytolaccoides* Pursh [G]. Thickets and moist woods: Me.—Minn.—Mo.—Ga. *Canad.—Allegh.* Je–Au.

12. **A. ovalifòlia** Dec. Stem simple, erect, 3–6 dm. high, finely tomentose; leaf-blades oval, ovate, or ovate-lanceolate, 5–8 cm. long, acute at both ends, finely tomentose on the upper surface, glabrate in age; lobes of the corolla greenish white or purplish, ovate-oblong, obtuse, about 5 mm. long; hoods orange or yellow, oval-oblong, with a large tooth on each margin; follicles pubescent. Woods and prairies: Man.—Ill.—Neb.—Alta. *Prairie—Submont.* Je–Jl.

13. **A. purpuráscens** L. Stem glabrous or tomentulose, 3–10 dm. high; leaves ovate-oblong, 1–2 dm. long, truncate or acute at the base, acute or mucronate at the apex, glabrous above, tomentulose beneath; corolla-lobes dark purple, oblong, obtuse, 8–10 mm. long; hoods pale red or purple, oblong, round-lobed at the middle; follicles 10–12 cm. long, attenuate, tomentose. Dry ground: N.H.—Minn.—Kans.—N.C. *Canad.—Allegh.* Je–Jl.

14. **A. brachystéphana** Engelm. Stems several, usually branched, ascending or spreading, 1.5–2.5 dm. high, minutely puberulent; leaf-blades narrowly linear-lanceolate, long-attenuate, 5–8 cm. long, puberulent or in age glabrate; corolla greenish purple, its lobes about 4 mm. long, oblong, acute; column almost none; hood ovate, obtuse, with large triangular lobes on the margins; follicles lance-ovoid, acuminate, 5–7 cm. long, tomentose. Dry soil: Kans.—Tex.—Ariz.—Wyo.; n Mex. *St. Plains—Son.* Ap–Au.

15. **A. incarnàta** L. Stem glabrous or minutely puberulent, 6–20 dm.

high, very leafy; leaves opposite; leaf-blades 4–17 cm. long, oblong-lanceolate to linear-lanceolate, acute to subcordate at the base; corolla deep rose-purple to pink or white; lobes oblong, 4–6 mm. long; column 1–1.5 mm. long, slender; hoods oblong, obtuse at the apex, the rounded margins overlapping; follicles 5–7 cm. long, lance-ovoid, acuminate, glabrous or minutely puberulent. The southwestern plant is usually tall (over 1 m. high), and has long, narrow leaves (var. *longifolia* A. Gray). In swamps: N.B.—Fla.—N.M.—Wyo.—Man. *E. Temp.—Plain—Submont.* Je–Au.

16. **A. púlchra** Ehrh. Stem hirsute, 5–15 dm. high, branched above, leafy; leaves short-petioled or sessile, broadly lanceolate, 6–12 cm. long, acute or acuminate at the apex, mostly rounded at the base; umbels many, corymbose; corolla-lobes obtuse, often notched, rose-purple, rarely white; column slender, but distinct, 1 mm. long; hoods erect, 2 mm. high, oblong, obtuse, biauriculate-hastate at the base; follicles 4 cm. long, slender, acuminate, minutely hirsute. *A. incarnata pulchra* Pers. [G]. Fields and swamps: N.S.—Minn.—Ga. *Canad.—Allegh.* Jl–S.

17. **A. quadrifólia** Jacq. Stem glabrous or minutely puberulent, erect, 3–6 dm. high, leafless below; leaves distant, usually 3 or 4 pairs, the lower and upper pairs in whorls of 4, ovate or ovate-lanceolate, acute or acuminate, on short-margined petioles, glabrous; corolla-lobes pink to nearly white, oblong, 5 mm. long; hoods white, erect-spreading, 4–5 mm. long, ovate oblong, broadly obtuse at the apex, with a salient tooth on each side at the base; follicles very slender, 10–12 cm. long, glabrous. Dry soil, in woods and thickets: Me. –Minn. –Ark.—N.C. My–Jl.

18. **A. verticillata** L. Stem angled, glabrous or pubescent in lines, 4–10 dm. high; leaves narrowly linear to almost filiform, 5–10 cm. long, verticillate in whorls of 2's to 4's; corolla greenish, sometimes tinged with purple, lobes oblong, 4 mm. long; column about 1 mm. long; hood white, broadly ovate, truncate at the apex, lobed on the sides near the base; follicles linear lanceolate, 7–10 cm. long, glabrous or minutely puberulent. Dry and sterile soil: Me.—Fla.—Tex.—Mex.—Ariz.—N.D.—Man. *E. Temp.—Plain—Son.* Ap–Au.

19. **A. galioides** H.B.K. Stem erect, solitary, glabrous or minutely puberulent in lines, 4–10 dm. high; leaves in whorls of 3's to 6's, narrowly linear, 6–8 cm. long; corolla-lobes greenish white, oblong, obtuse, 4 mm. long; column nearly 1 mm. long; hoods erect-spreading, broadly rounded at the apex, entire-margined, hastate-sagittate on the back; follicles 5–7 cm. long, glabrous or minutely puberulent, lanceolate. Dry plains: Tex.—Kans.—Colo.—Calif.; Mex. *St. Plains—Son.* Je–Au.

20. **A. pùmila** (A. Gray) Vail. Stems tufted, glabrous or nearly so, 1–2.5 dm. high, simple; leaves very numerous, crowded, irregularly scattered on the stem, about 3.5 cm. long, linear-filiform, revolute, glabrous or minutely scabrous-puberulent; corolla greenish-white; lobes oblong, 4–5 mm. long; column about 0.5 mm. long; hoods oblong, erect, entire; follicles narrowly lanceolate, 3–5 cm. long, minutely puberulent. Dry plains: N.D.—Ark.—N.M.—Colo.—Mont. *Prairie—Plain—Submont.* Jl–S.

3. **ASCLEPIODÒRA** A. Gray. Spider Milkweed.

Perennial herbs. Leaves opposite or alternate. Flowers perfect, regular, purplish, greenish, or white, in terminal, solitary or corymbose umbels. Calyx-lobes narrow, acute. Corolla rotate, 5-lobed, the lobes acute, spreading or erect. Hoods inserted over the whole short column, arising from the base of the corolla-lobes, saccate at the base, obtuse at the apex, crested within, with 1 or 2 small appendages between the hoods, simulating an inner crown. Follicles fusiform, often soft-spinulose.

Umbels solitary ; hoods incurved.   1. *A. decumbens*.
Umbels 2–5 ; hoods spreading-ascending.   2. *A. viridis*.

1. **A. decúmbens** (Nutt.) A. Gray. Perennial, with a woody rootstock; stems usually several, 2–4 dm. high, angled; leaves 4–15 cm. long, lanceolate to linear, attenuate, thick, alternate or whorled in 3's; corolla globose in bud; lobes greenish; hood purplish, incurved, obtuse, 2-lobed on the ventral margins; follicles about 8 cm. long, pubescent, smooth, or with a few soft processes. Dry sandy soil: Kans.—Ark.—Tex.—Ariz.—Nev.; Mex. *Austral.—Son.—Submont.* Je–S.

2. **A. víridis** (Walt.) A. Gray. Stem nearly glabrous, decumbent or ascending, 2–6 dm. high, angled; leaves oblong or oblong-lanceolate, 6–10 cm. long, obtuse or acute, and mucronate at the apex, narrowed or rounded at the base; corolla globose-ovoid in bud; lobes greenish, oblong; hoods purplish or violet, entire on the ventral margins; follicles erect, 6–11 cm. long, with or without soft spinose processes. Dry soil: Ky.—Kans.—N.M.—Tex.—S.C. *Austral.* My–Je.

4. GONÓLOBUS Michx.

Perennial herbaceous vines. Leaves opposite, cordate, thin. Flowers in peduncled axillary cymes. Calyx-lobes 5, glandular within. Corolla white, campanulate; lobes 5, nearly erect, slightly contorted. Crown nearly sessile, of 5 membranous, erect, truncate lobes, each appendage ending in 2 awns. Stamens inserted at the base of the corolla; filaments united into a short tube; anthers ending in an erect or inflexed membrane. Pollinia oblong-elliptic, on broad caudicles, pendulous. Stigma slightly 2-lobed, conic. Follicles thick, sharply angled. Seeds comose.

1. **G. laèvis** Michx. *Fig. 459.* Stem high, slender, glabrous or pubescent in lines; leaves 5–15 cm. long, ovate, acuminate, deeply cordate at the base, palmately veined, glabrous above, the veins minutely pubescent beneath; corolla-lobes linear-lanceolate, 4 mm. long, with reflexed tips; crown-lobes ovate, merely truncate; follicles 10–15 cm. long, glabrous. *Enslenia albida* Nutt. Thickets: Pa.—Kans.—Tex.—Fla. *Allegh.—Austral.* Je–Au.

5. VINCETÓXICUM Walt. ANGLE-POD.

Herbaceous or shrubby vines, with opposite cordate leaves. Flowers perfect, corymbose-umbellate, in the axils of the leaves. Calyx 5-cleft, with 5 glands within. Corolla rotate, 5-parted, with a wing-like or cup-shaped crown, the lobes convolute in bud. Stamens inserted on the base of the corolla, the filaments connate; anthers without appendages, the sacs transversely dehiscent. Stigma flat-topped. Follicles thick, acuminate, angled or tubercled.

1. **V. carolinénse** (Jacq.) Britton. Stem hirsute; leaves broadly ovate-cordate, acuminate, 8–18 cm. long, pubescent at least beneath; peduncles 5–10 cm. long; corolla brownish purple, the lobes linear-oblong; crown cup-shaped, as high as the anthers, thin, lobed or toothed, obtusely 5-lobed, with a subulate bifid process in each sinus; follicles muricate. Thickets: Va.—Iowa—Mo.—La.—S.C. *Allegh.* My–Jl.

6. PERÍPLOCA L. SILK VINE.

Woody twining plants, with opposite leaves. Flowers in umbels, regular, perfect. Calyx 5-parted, glandular within. Corolla deeply 5-parted, the lobes obtuse; crown adnate to the corolla, with 10 broad and 5 slender elongate scales or appendages. Filaments short, free, connivent. Pollen granulose, in two masses in each cell. Follicles slender, glabrous, united at the apex. Seeds with a coma.

1. P. graèca L. Plant glabrous; leaf-blades ovate-oblong, acute, paler beneath, umbels cymose; corolla-lobes dull brownish or greenish, with a dark median line and a white spot at the base within; appendages of the crown slender, ligulate, 2-cleft. Around dwellings: Mass.—Kans.—Fla.; escaped from cultivation; native of Greece and the Orient.

Family 138. CONVOLVULACEAE. MORNING-GLORY FAMILY.

Annual or perennial, twining or trailing herbs or vines, or rarely trees or shrubs. Leaves alternate, without stipules. Flowers perfect, regular, either solitary, axillary, or in cymes. Sepals 5, rarely 4, more or less united. Corolla gamopetalous, usually funnelform, either plicate and the plaits convolute, or induplicate-valvate, rarely imbricate in the bud, 5-lobed, or rarely 4-lobed Stamens 5, alternate with the corolla-lobes, partially adnate to the corolla-tube; anthers 2-celled, erect or incumbent. Gynoecium of 2 or 3 united carpels; ovary 2- or 3-celled, or by false partitions 4 or 6-celled; styles usually united; ovules 1 or 2 in each cavity. Fruit a capsule, 2–6-celled, or by breaking down of the septum 1-celled. Seeds large; endosperm mucilaginous; embryo curved with large plaited or crumpled cotyledons.

Styles distinct or partly so.
  Styles 2, distinct, each 2-cleft; stigmas filiform.   1. EVOLVULUS.
  Style solitary, 2-cleft, the divisions simple; stigmas capitate.   2. STYLISMA.
Styles united up to the stigmas.
  Stigmas subglobose.
    Corolla salverform; stamens exserted.   3. QUAMOCLIT.
    Corolla funnelform; stamens included.   4. IPOMOEA.
  Stigmas linear to oblong-cylindric.   5. CONVOLVULUS.

1. EVÓLVULUS L.

Annual or perennial pubescent herbs. Leaves alternate, entire. Flowers perfect, regular, solitary in the axils, or in terminal racemes or panicles. Sepals 5, nearly equal. Corolla funnelform, campanulate, or rotate, white, pink, or blue, with 5-angled or 5-lobed limb. Stamens 5, with filiform filaments. Styles 2, distinct, each 2-cleft; stigmas slender. Capsule short, 2–4-valved; seeds 1–4, glabrous.

1. E. Nuttalliànus R. & S. *Fig. 460.* Perennial, with a more or less woody base; stems branched below, ascending or spreading, 1.5–2.5 dm. high, silky-hirsute; leaves subsessile, silky-hirsute on both sides, oblong or oblanceolate, 1–1.5 cm. long, acute; sepals silky, linear-lanceolate, acuminate; corolla about 1 cm. broad. *E. argenteus* Pursh [G]; not R. Br. *E. pilosus* Nutt. [B]; not Lam. Sandy soil: S.D.—Mo.—Tex.—Ariz.—Colo. *Plain—Submont.* My–Jl.

f 460

2. STYLÍSMA Raf.

Perennial herbs, with prostrate or twining stems. Leaves alternate, with entire blades. Flowers 1–5, on axillary peduncles. Calyx often pubescent; lobes 5. Corolla campanulate or funnelform, of various colors, its limb plaited, 5-angled or slightly 5-lobed. Stamens 5, included; filaments filiform or di-

648 CONVOLVULACEAE

lated at the base. Ovary 2-celled; styles partly united; stigmas capitate; capsule 2-celled, 2–4-valved, 1–4-seeded.

1. S. Pickeríngii (M. A. Curtis) A. Gray. Stem branched at the base, the branches prostrate or reclining; leaves linear-spatulate, densely pubescent, 2–6 cm. long, acute, entire; bracts linear, longer than the calyx; sepals lanuginose, oval, 4–5 mm. long; corolla white, 1–1.5 broad; capsule ovoid, 6–8 mm. long, pubescent. *Breweria Pickeringii* A. Gray [G, B]. Pinelands: N.J.—Iowa—Tex.—Fla. *Allegh.—Austral.* Je–Au.

3. QUÁMOCLIT Moench. CYPRESS VINE.
RED MORNING GLORY.

Annual or perennial vines. Leaves alternate, petioled, with entire or lobed blades, glabrous. Flowers in axillary cymes. Calyx-lobes 5, equal, membranous or herbaceous. Corolla scarlet, salverform, diurnal, the limb spreading, shorter than the tube. Stamens 5, exserted, the filaments filiform. Ovary 2- or 4-celled; styles wholly united; stigma capitate; ovules 4. Capsule globose or ovoid, usually 4-celled, 4-seeded.

Leaf-blades entire, cordate; sepals acuminate. 1. *Q. coccinea.*
Leaf-blades pectinate-pinnatifid; sepals obtuse. 2. *Q. vulgaris.*

1. Q. coccínea (L.) Moench. *Fig. 461.* Stem high-climbing; leaf-blades 3–10 cm. long, cordate, entire or 3-lobed; peduncles 2–several-flowered, stouter than the pedicels; corolla 2–4 cm. long, the limb 1.5–2 mm. broad; capsule subglobose, 7–8 mm. broad. *Ipomoea coccinea* L. [G]. Thickets and banks: Mass.—Pa.—Neb.—Ariz.—Fla.; trop. Am. *Temp.—Trop.* Jl–O.

2. Q. vulgàris Choisy. Stem high-climbing; leaf-blades 2–10 cm. long, pectinately pinnatifid; segments linear; peduncles 1–3-flowered, pedicels clavate; sepals 3–5 mm. long, appressed; corolla 2.5–3 cm. long, the lobes ovate or triangular; capsule ovoid, 8–10 mm. long. *I. Quamoclit* L. [G]. *Q. Quamoclit* Britt. [B]. Waste places: Va.—Kans.—Tex.—Fla.; nat. from trop. Am. *Austral.—Trop.* Jl–O.

f. 461.

4. IPOMOÈA L. MORNING GLORY.

Annual or perennial, usually climbing or trailing vines, or rarely erect. Leaves alternate, entire or lobed. Flowers perfect, regular, solitary and axillary, or cymose. Sepals 5, closely imbricate. Corolla funnelform, of various colors, with a spreading limb. Stamens 5, included. Ovary 2- or 4-celled; styles united. Capsule septicidal, 2- or 4-valved. Seeds often pubescent.

Plant bushy, not climbing; leaves linear. 1. *I. leptophylla.*
Plant climbing; leaf-blades cordate.
  Ovary 2-celled; capsule 4-seeded; stigma entire or 2-lobed.
    Sepals obtuse, glabrous; peduncles longer than the petioles; corolla 5–8 cm. long. 2. *I. pandurata.*
    Sepals acuminate, bristly-ciliate; peduncles short; corolla 1–2 cm. long. 3. *I. lacunosa.*
  Ovary 3–5-celled; capsule 6–10-seeded; stigma 3–5-lobed.
    Leaf-blades entire; sepals acute. 4. *I. purpurea.*
    Leaf-blades 3-lobed; sepals long-acuminate. 5. *I. hederacea.*

1. I. leptophýlla Torr. *Fig. 462.* Perennial, with a deep-seated enormous fleshy root; stems several, erect or ascending, 3–12 dm. high, glabrous;

leaves linear, 5–15 cm. long, acute, entire; sepals ovate or orbicular-ovate, 5–10 mm. long, the inner larger than the outer; corolla pink-purple, 5–7 cm. long. BUSH MORNING GLORY, BIG ROOT. Plains and prairies: S.D.—Tex.—N.M.—Mont. *Plain.* Je–Au.

2. **I. panduràta** L. Perennial, with a large root; stem 1–4 m. long; leaf-blades glabrate, ovate, often fiddle-shaped, 3–10 cm. long, acuminate to obtuse, cordate at the base; pedicels 5–10 cm. long, 1–5-flowered; bracts ovate; sepals ovate or oblong, 12–18 mm. long, colored, imbricate; corolla white or pinkish, 7–10 cm. broad, with 5 shallow lobes and purplish tube. Dry soil. Ont.—Conn.—Kans.—Tex. —Fla. *E. Temp.* Jo–S.

3. **I. lacunòsa** L. Annual, sparingly hirsute or glabrate; stem 5–25 dm. long; leaf-blades ovate, entire or 4-lobed, 2–7 cm. long, deeply cordate at the base; peduncles 3-flowered; sepals 10–12 mm. long; corolla 1.5–2 cm. long, white, often with purple-margined limb, 1–1.5 cm. broad; capsule globose, 1 cm. broad. Fields and low ground: Pa.—Kans.—Tex.—S.C. *Allegh.—Austral.* Jl–S.

4. **I. purpùrea** (L.) Roth. Annual; stem high, climbing, twining, retrorsely-hirsute; leaf-blades cordate, short-acuminate; flowers umbellate on long peduncles; calyx bristly-hairy; corolla purple or rarely pink or white, 4 5–7 cm. long. COMMON MORNING GLORY. *Pharbitis purpurea* Voigt. Cultivated ground, waste places, etc.: N.S.—Fla.—Tex.—Colo.—N.D.; escaped from cultivation; native of trop. Am.

5. **I. hederàcea** Jacq. Stem 5–15 dm. long, retrosely hairy; leaf-blades cordate, 3-lobed, 5–13 cm. long, the lobes ovate, acuminate or acute; peduncles 3-flowered; calyx densely hairy below, lobes lanceolate, with long recurved tips; corolla white, purple, or pale blue, 3–4.5 cm. long; capsule depressed-globose. Waste places: Me.—S.D.—Mex.—Fla.; introd. from trop. Am. *E. Temp.* Jl–O.

5. **CONVÓLVULUS** (Tourn.) L. BINDWEED, MORNING GLORY.

Annual or perennial, trailing or twining herbaceous vines, rarely shrubs. Leaves alternate, usually sagittate or cordate at the base. Flowers perfect, regular, solitary or clustered, on axillary peduncles. Sepals 5, rarely equal, or the outer larger. Corolla funnelform or campanulate, with plaited, entire, 5-angled or 5-lobed limb. Stamens 5, included; filaments filiform or dilated at the base. Ovary 1 2-celled; styles united; stigmas linear to oblong; ovules 4. Capsule subglobose, 2–4-valved. Seeds glabrous.

Calyx not closely subtended by the bracts.
  Plant not canescent; leaf-blades hastate, but otherwise entire.
    Whole plant glabrous or nearly so.   1. *C. arvensis.*
    Stem and lower surface of the leaves with scattered
      long hairs.   2. *C. ambigens.*
  Plant more or less canescent; leaves usually lobed or dissected.
    Main divisions of the leaf-blades ovate or oblong, obtuse, lobed; sepals auricled at the base.   3. *C. hermannioides.*
    Main divisions of the leaf-blades linear, entire, the
      basal lobes usually deeply cleft; sepals not auricled.   4. *C. incanus.*
Calyx closely subtended and enclosed by the large bracts.
  Plants climbing or trailing; bracts cordate.
    Leaves narrowly hastate; flowers usually double.   5. *C. japonicus.*
    Leaves triangular-hastate; flowers simple.
      Leaf-blades hastate; the basal lobes often sinuate-
      dentate, acute; stem and leaves glabrous or
      slightly hairy.

Leaves with more or less angular spreading basal lobes; corolla mostly pink. 6. *C. americanus.*
Leaves with rounded or scarcely angular basal lobes, which are not spreading; corolla mostly white. 7. *C. sepium.*
Leaf-blades more sagittate; basal lobes rounded, entire; stem and leaves densely pubescent. 8. *C. interior.*
Plants erect or decumbent; bracts oval or oblong.
    Plant glabrous. 9. *C. Macounii.*
    Plant finely pubescent. 10. *C. spithamaeus.*

1. **C. arvénsis** L. Stem branched at the base, prostrate or low-twining, 2–8 dm. long, glabrous; leaf-blades oblong or ovate-sagittate to linear-hastate, 1–5 cm. long, obtuse or acutish; basal lobes more or less spreading, acute; bracts subulate; corolla white or rose-tinged, 1.5–2 cm. long and broad. Waste places and fields: N.S.—N.J.—Nev.—Wash.—B.C.; nat. from Eur. *Temp.— Submont.* Je–Au.

2. **C. ámbigens** House. Stem prostrate or trailing, 2–10 dm. long, finely cinereous, or glabrate in age; leaf-blades ovate-oblong or triangular-oblong, 1.5–4 cm. long, abruptly acute, truncate or nearly so at the base; basal lobes somewhat spreading, often toothed below; flowers axillary, solitary; corolla 1.5–2 cm. long and about as broad, white or with pink stripes. River valleys and loose soil: S.D.—Kans.—N.M.—Calif.—Ore.; intr. east to Ill. and Iowa. *W. Temp.—Plain—Submont.* Je–Au.

3. **C. hermannioìdes** A. Gray. Stem branched at the base, spreading or procumbent, 3–15 dm. long, silky-tomentose; leaf-blades ovate, oblong, or oblong-lanceolate, 1.5–7 cm. long, sinuately-toothed, cordate or sagittate at the base; corolla white, 2–2.5 cm. long, 1.5–2 cm. wide, angulate. Dry plains: Tex.—s Colo.—Neb. *Plain—Son.* Ap–S.

4. **C. incánus** Vahl. Stem branched at the base, procumbent, 3–10 dm. long, cinereous; leaf-blades variable, linear-sagittate to lanceolate-hastate, with the basal lobes spreading, more or less lobed, or pedately cleft, 2.5–5 cm. long, often mucronate; sepals becoming 6–8 mm. long; corolla white or pink, about 12 mm. long, 15 mm. broad; lobes acuminate. Dry hills and plains: Neb.— Ark.—Tex.—Ariz. *Son.* Je–Au.

5. **C. japónicus** Thunb. Stem twining, up to 7 m. long, minutely pubescent; leaf-blades hastate, 5–8 cm. long, the main division lanceolate, or ovate, the basal divisions usually angularly lobed; peduncles often equaling or exceeding the leaves; bracts ovate, rounded at the apex; corolla pink, 3–5 cm. broad, usually double. Around dwellings: N.Y.—D.C.—Neb.—Mo.; escaped from cultivation; nat. of Japan and China.

6. **C. americánus** (Sims) Greene. *Fig. 463.* Stem twining, 1–3 m. high, glabrate; leaf-blades glabrous, broadly hastate or ovate-hastate, 5–12 cm. long, acuminate or apiculate; basal lobes more or less spreading, broadly triangular, with a smaller lobe below; bracts broadly ovate-cordate, obtuse or acutish; corolla 4–5 cm. long, 5–7 cm. broad, pink or rose-purple. Closely related to *C. sepium* L., of Europe. Thickets and fence-rows: N.S.—N.C.— N.M.—Wash.—B.C. *Temp.—Submont.* My–Au.

f 463.

7. **C. sèpium** L. Similar to the preceding in habit; leaves broader, more rounded at the apex, the basal lobes less angular, and directed downward; corolla usually white. Thickets and waste places: Me.—N.D.—Tex.—Fla.; nat. from Eur. Je–Au.

8. **C. intèrior** House. Stems prostrate or some-

what climbing, 2–8 dm. long, densely and softly pubescent; leaf-blades deltoid or hastate-ovate, 2.5–4 cm. long, glabrous above, pubescent beneath; base shallowly cordate or truncate; bracts broadly ovate, obtuse, closely investing the calyx; corolla white, 3.5–4 cm. long, the limb entire, 3–4 cm. broad. *C. repens* Auth. [B]; not L. Sandy soil: Minn.—Okla.—Ariz.—Colo.—Alta. *Prairie— Plain.* My–Jl.

9. **C. Macoûnii** Greene. Stem erect or decumbent, 7–15 cm. high, glabrous; leaves broadly sagittate, about 5 cm. long, glabrous; bracts broad, oval, obtuse, 2–2.5 cm. long, auricled at the base; corolla white, fully 5 cm. long. Sandy soil: Minn.—Sask.- B.C. *Plain.* Au.

10. **C. spithamaèus** L. Stem erect or decumbent, 1–7 dm. high, simple; leaf blades oblong-ovate or fiddle-shaped, 4–10 cm. long, rounded or cordate at the base, finely pubescent; bracts oblong or ovate-oblong; calyx delicate, the lobes acute or acuminate; corolla pink, or white, 3–4 cm. long, 4–6 cm. broad; capsule globose. Dry rocky soil: N.S.—Man.—Minn. Fla. *E. Temp.* My–Au.

Family 139. CUSCUTACEAE. DODDER FAMILY.

Herbaceous parasites, with twining stems, at first germinating in the ground, but soon attaching themselves to some other plant by means of aerial rootlets, and taking their nourishment from these, the lower part of the stem having decayed early. Leaves reduced to minute alternate scales. Flowers perfect, regular, in compact or open cymes. Calyx inferior; sepals 5, or rarely 4, distinct or more or less united. Corolla campanulate or subglobose, 5-lobed, or rarely 4-lobed, the lobes being imbricate, usually with crenulate or fringed scales in the throat, alternating with the lobes. Stamens 5, rarely 4, partly adnate to the corolla; anthers 2-celled. Styles 2, mostly distinct; stigmas capitate or elongate. Capsule subglobose or ovoid, 2-celled, circumscissile or bursting irregularly. Embryo curved or spiral, in a fleshy endosperm.

1. **CÚSCUTA** (Tourn.) L. DODDER, LOVE-VINE, CORAL-VINE.

Characters of the family.

Stigmas elongate, linear; capsule circumscissile.
  Calyx-lobes deltoid, not overlapping, longer than broad.   1. *C. Epithymum.*
  Calyx-lobes broadly triangular-ovate, broader than long,
    their edges overlapping at the base.   2. *C. planiflora.*
Stigmas capitate; capsule indehiscent or bursting irregularly.
  Flowers not bracteate; calyx-segments more or less united.
    Corolla-lobes not incurved.
      Capsule depressed-globose.
        Corolla remaining at the base of the mature
          capsule, its lobes acute; flowers mostly
          4-merous.   3. *C. polygonorum*
        Corolla surrounding the upper part of the ma-
          ture capsule or crowning it.
          Flowers mostly 4-merous; corolla-lobes ob-
            tuse.   4. *C. Cephalanthi.*
          Flowers mostly 5-merous; corolla-lobes acute.   5. *C. racemosa.*
      Capsule ovoid.
        Scales of the crown neither truncate nor bifid;
          style ⅓–½ as long as the ovary.   6. *C. Gronovii.*
        Scales of the crown truncate and bifid; style
          ¼–½ as long as the ovary.   7. *C. curta.*
    Corolla-lobes incurved at the tips.
      Flowers mostly 4-merous; scales of the crown
        poorly developed.   8. *C. Coryli.*
      Flowers mostly 5-merous; scales of the crown well
        developed.
        Capsule depressed-globose; flowers 2–3 mm.
          long; corolla not fleshy.   9. *C. pentagona.*

Capsule subglobose, rather pointed, not de-
pressed ; flowers 3–5 mm. long ; corolla some-
what fleshy.   10. *C. indecora.*
Flowers subtended by several bracts ; calyx-segments free.
 Flowers pedicelled, loosely paniculate.   11. *C. cuspidata.*
 Flowers sessile in dense clusters.
  Bracts obtuse, appressed.   12. *C. compacta.*
  Bracts acute, squarrose.   13. *C. paradoxa.*

1. C. Epíthymum Murr. Stem very slender; flowers in capitate clusters,
whitish or pinkish; calyx 4- or 5-lobed, with acute lobes; corolla-tube cylindric,
longer than the calyx, its lobes spreading, in age capping the circumscissile
capsule; stamens exserted. Parasitic on clover: Me.—S.D.—Neb.—N.J.; nat.
from Eurasia. Jl–Au.

2. C. planiflòra Tenore. Stem filiform, reddish; flowers in dense glomerules, 5–10 mm. in diameter; calyx obconic; lobes round, short, obtuse, carinate
beneath; corolla-tube very short; lobes 5, always reflexed, ovate, apiculate. *C.
Epilinum* and *C. Epithymum* of western reports. *C. calycina* Webb. & Berth. (?).
Parasitic on alfalfa and other herbs: S.D.—Colo.—Wyo.—Utah—Mont.; introduced in alfalfa from s Eur. Je–Au.

3. C. polygonòrum Engelm. Stem rather coarse, orange-yellow; flowers
sessile or nearly so, in small clusters; calyx broad, the lobes 4 or 5, ovate-
oblong, acute or acutish; corolla little exserted, the lobes triangular, longer
than the tube, acute; capsule subglobose, 2.5–3 mm. broad. *C. obtusiflora*
H.B.K. [G]. *C. chlorocarpa* Engelm. Parasitic on *Polygonum* and other
herbs: Pa.—Minn.—Kans.—Ark.—Del. *Allegh.* Jl–S.

4. C. Cephalánthi Engelm. Stem filiform; flowers cymose-paniculate,
about 2 mm. long, 4-merous, or even 3-merous; corolla deeply campanulate;
scales ovate or spatulate; capsule 2–2.5 mm. in diameter. *C. tenuiflora* Engelm.
On coarse herbs and shrubs. Me.—Pa.—Tex.—Calif.—Wash. *Temp.—Plain.*
Je–S.

5. C. racemòsa chiliàna Engelm. Stem slender; flowers in loose, racemose cymes; pedicels longer than the flowers; calyx-lobes short, deltoid; corolla campanulate, lobes short, spreading or reflexed, acutish; scales of the
crown short, fringed; style longer than the ovary; capsule depressed-globose.
On *Medicago sativa*, locally: Me.—S.D.—Tex.—Calif.; introduced from S. Am.

6. C. Gronòvii Willd. *Fig. 464.* Stem rather
coarse, high-climbing; flowers paniculate-cymose;
calyx-lobes elliptic; corolla deeply campanulate;
lobes elliptic, obtuse, spreading, flat; scales ovate,
fringed, spreading; capsule broadly ovoid, 4–5 mm.
in diameter. On coarse herbs or shrubs: N.S.—Fla.
—Tex.—Colo.—N.D.—Man. *E. Temp.—Plain.* Jl–O.

7. C. cúrta (Engelm.) Rydb. Stems rather
slender; flowers in small cymes; calyx short; lobes
rounded, obtuse; corolla deeply campanulate, 3 mm.
long; lobes short, broadly oval; scales short, bifid or
truncate, appressed to the tube; capsule ovoid, the
marcescent corolla surrounding or covering the upper
portion. On coarse herbs: Minn.—Kans.—Utah—
Calif. *W. Temp.* Au.

8. C. Córyli Engelm. Stem coarse, yellow or
in age orange; flowers short-pedicelled; calyx-lobes 4 or 5, triangular or deltoid,
equaling the corolla-tube; corolla campanulate, 2 mm. long, the lobes triangular, erose-crenulate, with inflexed tips; scales small, oblong, rounded at the
apex; capsule depressed, 2.5–3 mm. high. *C. inflexa* Engelm. Parasitic on
hazel and other shrubs: Va.—Mont.—N.Mex.—Neb.—Ga. *E. Temp.* Jl–Au.

9. **C. pentagòna** Engelm. Stems slender; flowers in head-like clusters, sessile, less than 2 mm. long; calyx-lobes rounded, obtuse; corolla-lobes longer than the tube, acute or acuminate, triangular; scales large and deeply laciniate. *C. arvensis* Beyr. [G]. On various herbs: Mass.—Fla.—Calif.—B.C.; Mex. and W.Ind. *Temp.—Trop.—Plain—Submont.* Ap–N.

10. **C. indecòra** Choisy. Stem filiform, high-twining; flowers loosely paniculato-cymose; calyx-lobes ovate or lanceolate, acute; corolla campanulate, 3–5 mm. long; lobes ovate-lanceolate, erect or more commonly spreading; scales large, broadly oval, fringed; capsule enveloped by the corolla, rounded-ovoid. On herbs and low shrubs, mostly *Carduaceae* and *Fabaceae:* Fla.— Ill.—Utah—Calif.; Mex., C.Am., W.Ind. and S.Am. *Temp.—Trop.* Jl–S, or in the south Ja–D.

11. **C. cuspidàta** Engelm. Stem filiform; flowers in loose panicles, with bracted pedicels; bracts and sepals ovate orbicular; corolla 3–5 mm. long, campanulate; lobes oblong, cuspidate or mucronate, rarely obtuse; scales large, deeply fringed; capsule capped by the marcescent corolla. On *Ambrosiaceae* and rarely *Fabaceae:* Mo.— Tex.—Utah.—N.D.—Minn. *Austral.—Son.* Au–S.

12. **C. compácta** Juss. Stem stout, yellowish white; flowers sessile, crowded in dense masses; bracts imbricate, ribbed, rhombic-orbicular, serrulate; sepals 5, oval, obtuse, crenulate; corolla salverform, the lobes oblong or ovate, obtuse, spreading; scales half as long as the corolla-tube, fringed near the top; capsule subglobose, 3–3.5 mm. long. Parasitic on shrubs: N.H.—S.D.—Tex.— Fla. *E. Temp.* Jl–S.

13. **C. paradóxa** Raf. Stem slender, yellowish white, the flowering branches often forming spiral masses around the host; flowers very numerous, sessile; bracts imbricate, strongly ribbed, serrulate; sepals 5, oblong, obtuse, broader than the bracts; corolla salverform, 3 mm. long; lobes oblong-lanceolate to ovate, obtuse, spreading or recurved; scales 2–3, as long as the corolla-tube, copiously fringed; capsule longer than thick, 2 mm. long. *C. glomerata* Choisy [G]. Parasitic on Compositae, *Persicaria,* and other coarse herbs: Ohio —S.D.—Tex.—Tenn. *Allegh.—Austral.* Jl–S.

Family 140. **POLEMONIACEAE.** Phlox Family.

Annual or perennial herbs, or rarely shrubby plants. Leaves alternate or opposite, entire, pinnatifid, or pinnately compound. Flowers perfect, regular or nearly so. Sepals 5, partly united. Corolla of 5 united petals, regular or nearly so, 5- or rarely 4-lobed, from rotate to salver- or trumpet-shaped; lobes convolute in aestivation. Stamens 5, rarely 4, often unequal; filaments adnate to the corolla-tube, often inserted at different heights. Gynoecium of 3 united carpels; ovary 3-celled; styles wholly united; stigmas 3. Fruit a 3-celled, loculicidal capsule. Seeds often with a mucilaginous coat.

Calyx wholly herbaceous; leaves pinnately compound.   1. POLEMONIUM.
Calyx more or less scarious between the lobes, or the lobes spinulose-tipped and pun-
    gent; leaves simple or pinnatifid, but not compound.
  Calyx distinctly enlarging in fruit, not ruptured by the capsule.
                                       2. COLLOMIA.
  Calyx soon ruptured by the fruit, or if not ruptured enlarged in fruit and calyx-
    teeth spinescent.
    Corolla strictly salver-shaped with a narrow throat.
      Perennials, with leaves opposite; seeds not mucilaginous when wetted.
                                       3. PHLOX.
      Annuals, with the floral leaves alternate; seeds mucilaginous when wetted.
                                     4. MICROSTERIS.

Corolla campanulate, funnelform, trumpet-shaped, or if somewhat salver-shaped, the throat open.
Corolla elongate, with a narrow tube and funnelform or spreading throat.
    Calyx-teeth either unequal in size or spinulose-laciniate, or both ; pungent-leaved annuals. 5. NAVARRETIA.
    Calyx-teeth neither unequal nor spinulose-laciniate ; perennials or non-pungent annuals.
      Calyx not at all scarious in the sinuses ; mostly undershrubs, with pungent leaves. 6. LEPTODACTYLON.
      Calyx decidedly scarious in the sinuses. 7. GILIA.
Corolla rotate. 8. GILIASTRUM.

1. POLEMÒNIUM (Tourn.) L. JACOB'S LADDER, SKUNK-WEED.

More or less caulescent perennial or annual herbs, with pinnate leaves. Flowers perfect, solitary or clustered. Calyx wholly foliaceous, campanulate, more or less glandular, 5-cleft, not ruptured by the fruit. Corolla campanulate or rotate-funnelform to almost tubular. Stamens 5; filaments adnate mostly to the lower part of the corolla-tube, rarely unequally adnate to the upper part. Style elongate. Ovules three to many (rarely 1 or 2) in each cell. Seeds remaining unchanged when wetted.

Stem erect ; anthers exserted. 1. P. occidentale.
Stem reclining ; anthers included. 2. P. reptans.

1. P. occidentàle Greene. Stems leafy, solitary and rather simple, glabrous below, glandular-puberulent above, 3–9 dm. high; leaves glabrous or sparingly pubescent; leaflets 15–27, ovate-oblong or lanceolate, rounded, oblique at the base, acute at the apex, 1–4 cm. long; calyx glandular-puberulent, 7–8 mm. long; lobes oval, obtusish; corolla blue or violet, 10–12 mm. long, campanulate; stamens about as long as or longer than the corolla. Valleys and open woods: Sask.—S.D.—Colo.—Calif.—Alaska. *W. Temp.—Submont.—Mont.* Je–Au.

f.465.

2. P. réptans L. *Fig. 465.* Stem 3–7 dm. long; leaf-blade shorter than the petiole; leaflets elliptic to oblong-lanceolate, acute or apiculate; pedicels 5–10 mm. long; calyx 5 mm. long, in fruit up to 10 mm. long; lobes triangular; corolla blue, 1.5–2 cm. broad; lobes spatulate. Thickets and woods: N.Y.—Minn.—Kans.—Ga. *Allegh.* Ap–My.

2. COLLÒMIA Nutt.

Simple or branched, caulescent annuals. Leaves alternate, entire or pinnatifid. Flowers perfect, in subcapitate clusters, at the end of the stem or axillary. Calyx obpyramidal, 5-cleft, scarious in the sinuses, not ruptured by the capsule. Corolla funnelform or salverform, 5-lobed. Stamens 5, unequally adnate to the tube of the corolla. Ovules solitary or rarely few in each cell. Seeds developing both mucilage and spiracles when wetted.

1. C. lineàris Nutt. Stems simple or in age sometimes branched, 1–4 dm. high, hirsutulous; leaves linear-lanceolate, 1.5–5 cm. long, 2–8 mm. wide, puberulent, the floral ones ovate or lance-ovate, often paler at the base and sometimes tinged with red; flowers in terminal head-like spikes; corolla trumpet-shaped, 10–15 mm. long; lobes oval; limb 4–5 mm. broad. *Gilia linearis* A. Gray [G]. *C. lanceolata* Greene. Dry and sandy soil: Man.—Iowa—Colo.—Calif.—B.C., adv. in Que., N.B., and N.Y. *W. Temp.—Mont.* My–Au.

3. PHLÓX L. PHLOX.

Perennial (all ours), or annual herbs, often shrubby at the base. Leaves opposite or sometimes alternate above, entire. Flowers in terminal corymbiform or paniculate cymes. Calyx of 5 sepals, united into a 5-ribbed tube; lobes often pungent. Corolla salverform, white, blue, purple, or red, with a slender tube; lobes spreading, obovate, orbicular, or obcordate. Stamens 5, included; filaments equally adnate to the tube of the corolla. Capsule included in the tube of the calyx, which is ruptured at maturity. Seeds not emitting spiral threads when wetted.

Leaves flat, usually broad, without thickened margins, not fascicled; stem herbaceous.
  Stem erect or ascending, 3 dm. high or more; corolla-lobes entire or notched.
    Calyx-lobes shorter than the tube.
      Calyx-lobes subulate-setaceous.   1. *P. paniculata.*
      Calyx-lobes deltoid-lanceolate.
        Flowers in an elongate panicle; calyx-lobes less
          than one third as long as the tube.   2. *P. maculata.*
        Flowers in corymbiform cymes; calyx-lobes more
          than one third as long as the tube.   3. *P. glaberrima.*
    Calyx-lobes longer than the tube.
      Plant without runners or leafy shoots; all leaves
        linear or lanceolate.   4. *P. pilosa.*
      Plant with prostrate or decumbent leafy shoots;
        lower leaves and those of the shoots oblong
        or ovate.   5. *P. divaricata.*
  Stem diffuse, 1–2 dm. high; corolla-lobes 2-cleft.   6. *P. bifida.*
Leaves subulate, rigid, or if flat with thickened margins,
  more or less fascicled; stem usually more or less
  woody at the base, low and cespitose.
  Flowers cymose; leaves subulate.   7. *P. subulata.*
  Flowers solitary or few together.
    Leaves beset with cobwebby hairs; plant densely pulvinate cespitose.
      Leaves ovate, distinctly 4-ranked, imbricate, arachnoid-lanate, mucronate-tipped   8. *P. bryoides.*
      Leaves subulate, loosely sparingly arachnoid.
        Corolla small, its limb 8–10 mm. wide.   9. *P. Hoodii.*
        Corolla larger, its limb 10–15 mm. wide.   10. *P. planitiarum.*
    Leaves not cobwebby, except sometimes at base.
      Leaves subulate, glabrous, the margin neither conspicuously thickened nor ciliate.
        Calyx more than 10 mm. long, calyx slightly
          arachnoid, its lobes longer than the tube.   10. *P. planitiarum.*
        Calyx less than 8 mm. long, glabrous, its lobes
          shorter than the tube.   11. *P. scleranthifolia.*
      Leaves oblong or linear, hispid-ciliate on the thickened margins.
        Leaves linear, 1–2 cm. long; corolla-tube twice
          as long as the calyx; young stems white-angled.   12. *P. Kelseyi.*
        Leaves oblong, 7–12 mm. long; corolla-tube not
          twice as long as the calyx; young stems not
          white-angled.   13. *P. alyssifolia.*

1. P. paniculàta L. A perennial herb; stem 6–12 dm. high; leaves elliptic or lance-elliptic, 8–20 cm. long, undulate, acuminate; panicle pyramidal; calyx pruinose or sparingly pubescent; corolla pink or white, the tube 2 cm. long, the limb 15–17 mm. broad, the lobe obovate. Open woods: Pa.—Kans.—La.—Fla. *Allegh.—Austral.* Jl–S.

2. P. maculàta L. *Fig. 466.* Stem erect, often clustered, 3–10 dm. high; leaves oblong-lanceolate to linear, 5–12 cm. long; panicle cylindric, 1–4 dm. long, often glandular pubescent; corolla pink-purple, the tube 1.5–2 cm. long, the limb 2–2.5 cm. broad, the lobes rounded-obovate or broadly cuneate. Low ground: N.J. — Minn. — Tenn. — Fla. *Allegh. — Austral.* Je–Au.

3. P. glabérrima L. Stem ascending, some-

f.466.

times tufted, 3–8 dm. high, rarely pubescent; leaves oblong, lanceolate, or linear, 2.5–10 cm. long, acute or acuminate; corolla purple or pink, the tube 2–3 cm. long, the limb 2–2.5 cm. broad, the lobes rounded or truncate, sometimes undulate. Open woods: Va.—Minn.—Ala.—Fla. *Allegh.—Austral.* My–Jl.

4. **P. pilòsa** L. Stem erect, often tufted, 2–5 cm. high, corymbose above; leaves linear or narrowly lanceolate, 3–8 cm. long, attenuate; corymb becoming lax; corolla pink-purple or white, the tube 1–1.5 cm. long, the limb 2 cm. broad, the lobes obovate-cuneate. Sandy soil: Ont.—Man.—Tex.—Fla. *E. Temp.* Ap–Je.

5. **P. divaricàta** L. Stem erect or ascending, 2–5 dm. high, usually tufted, with decumbent or creeping shoots; leaves oblong or lanceolate, or ovate-lanceolate, 1.5–5 cm. long, acute, the upper sometimes cordate, those of the shoots oblanceolate; corolla lavender or bluish, the tube 1–1.5 cm. long, the limb 2.5–3 cm. broad, the lobes apiculate or notched. Damp woods: Ont.—Minn.—Kans.—La.—Fla. *E. Temp.* Ap–Je.

6. **P. bìfida** Beck. Stem diffusely branched at the base, 1–3 dm. high; leaves linear, 2–5 cm. long, acute; calyx 6–8 mm. long, the lobes subulate, shorter than the tube; corolla violet-purple, the tube 1 cm. long, the limb 1.5 cm. broad, the divisions of the lobes oblong or linear. Prairies and barrens: Mich.—Iowa—Kans.—Tenn. *Allegh.* Ap–Je.

7. **P. subulàta** L. Stem decumbent, forming mats, the floral branches erect, 1–1.5 dm. high, pilose; leaves linear or linear-subulate, 1–1.5 cm. long, acute; calyx 8–9 mm. long, pilose, the lobes lanceolate-subulate, as long as the tube; corolla purple to white, the tube 1–1.5 cm. long, the limb 1.5–2 cm. broad, the lobes cuneate, emarginate. Rocky and barren places: N.Y.—Iowa—Ky.—Fla. *Allegh.—Austral.* Ap–Je.

8. **P. bryoìdes** Nutt. Plant less than 1 dm. high; leaves densely imbricate, ovate or triangular-lanceolate, 2–3 mm. long, densely cobwebby, with inflexed margins; flowers solitary at the ends of the branches, sessile; calyx-teeth subulate, shorter than the tube; corolla about 1 cm. long; lobes obovate, about 3 mm. long. Dry hills: Neb.—Colo.—Utah—Wyo. *Plain—Submont.* My–Je.

9. **P. Hoòdii** Richardson. Leaves subulate, sparingly lanate, 4–10 mm. long, apiculate; flowers sessile, solitary at the ends of the branches; calyx 5–7 mm. long, arachnoid; teeth similar to the leaves; corolla white, 8–10 mm. long; lobes obovate, entire or mucronate, 5 mm. long. MOSS PINK, MOSS PHLOX. Hills and plains: Mack.—Man.—Neb.—Wyo.—Ida.—Yukon. *Plain.* Ap–My.

10. **P. planitiàrum** A. Nels. Stem erect, branched at the ground, white, 5–10 cm. high, sparingly arachnoid at the nodes; leaves subulate, 1–2 cm. long, with strong midrib; flowers on short pedicels; calyx-teeth as long as the tube, subulate, pungent; corolla white, half longer than the calyx; lobes obovate, 5–7 mm. long. *P. andicola* (Britton) E. Nels. [R]; not Nutt. *P. Douglasii andicola* Britt. [G, B]. Plains and hills and sandy soil: N.D.—Colo.—Ida. *Plain.* Je–Jl.

11. **P. scleranthifòlia** Rydb. Branches depressed, glabrous or nearly so; leaves very narrow, about 1 cm. long, less than 0.5 mm. wide, spinulose-tipped, somewhat divergent, with prominent midrib; flowers sessile or nearly so; calyx about 5 mm. long; teeth lanceolate, bristle-pointed; corolla white; tube exceeding the calyx; lobes oblong. Dry hills: S.D. *Submont.* Je.

12. **P. Kélseyi** Britton. Branches 1–2 dm. long; leaves flat, lanceolate or linear, 1–1.5 cm. long, narrowly white-margined, usually ciliate, sometimes glandular, sometimes glabrate; flowers solitary, short-pedicelled; calyx nearly 1

cm. long, ciliate and often glandular; teeth subulate; corolla blue, lilac, or pink, rarely white, fully 2 cm. long; lobes broadly obovate, 7–8 mm. long. *P. abdita* A. Nels. Valleys: N.D.—Colo.—Mont. *Plain—Subalp.*

13. P. alyssifòlia Greene. Branches short, with white shreddy bark; leaves oblong or linear, or the earlier ones ovate, 5–15 mm. long, 2–4 mm. wide, usually ciliate on the margins and sometimes a little glandular, cuspidate; flowers sessile or nearly so; calyx ciliate, strongly ribbed, often considerably glandular; calyx 7–8 mm. long; teeth oblong, accrose-cuspidate, rather shorter than the tube; corolla bluish, rarely white, 15–18 mm. long; lobes obovate, 7–8 mm. long. *P. collina* Rydb. *P. variabilis* A. Brand. Mountains and dry benches: Sask.— S.D.—Colo.—Utah.—Mont. *Submont.—Mont.* My–Au.

4. MICRÓSTERIS Greene.

Small branched annuals. Leaves all except the floral ones opposite, narrow and entire. Flowers small, perfect, axillary. Calyx tubular, 5-cleft, ruptured by the fruit; lobes scarious-margined. Corolla salver-shaped, with a narrow tube; throat funnelform. Stamens unequally adnate to the corolla tube. Capsule 3-celled; seeds few, large, the coat, when wetted, mucilaginous but without spiricles.

1. M. micrántha (Kellogg) Greene. Stem glandular and puberulent; lower leaves 1–2 cm. long, puberulent; calyx tubular, 4–5 mm. long, puberulent; teeth subulate; corolla 5–7 mm. long; tube white; limb white, rose-colored, or violet. *Collomia micrantha* Kellogg. Dry or sandy soil: S.D.—Neb. Colo.—Ariz.—c Calif. *Plain—Son.—Mont.* Ap–Je.

5. NAVARRÈTIA R. & P.

Glabrous or viscid-puberulent, branched annuals. Leaves all alternate, setaceously or spinulosely pinnatifid, or the lowest subentire. Flowers perfect, in crowded, bracted, head-like clusters at the ends of the branches. Calyx-tube short, scarious in the sinuses; lobes unequal, costate, pungently tipped, some of them at least spinulosely toothed or cleft. Corolla salverform, with a funnelform throat. Stamens adnate up to or nearly to the throat of the corolla. Seeds one to many in each cell, commonly mucilaginous and producing spiricles, when wetted. Capsule 1–3-celled, partially dehiscent or indehiscent.

Corolla white; pericarp hyaline and indehiscent, closely adherent to the mucilaginous seeds.
  Calyx and the bases of the floral leaves long-hairy; flowers nearly 1 cm. long.
                                                   1. *N. intertexta.*
  Calyx and floral leaves glabrous, the former hairy only in the
    sinuses; flowers only about 7 mm. long.
                                       2. *N. minima.*
Corolla yellow; pericarp of the capsule firm, dehiscent.
                                       3. *N. Breweri.*

f.467.

1. N. intertéxta (Benth.) Hook. Stem erect, 1–2 dm. high, simple or widely branched, reflexed-strigose; leaves glabrate or sparingly hairy, pinnatifid or bipinnatifid, with filiform, spinescent divisions; bases of bracts and calyx densely arachnoid-villous with white hairs; some of the calyx-teeth dentate; corolla white, 9–10 mm. long; limb 2.5–3 mm. broad; stamens equally inserted, exserted. *Gilia intertexta* Steud. Plains and sandy places: Wash.—Mont.—Ida.—Calif.; introduced in Iowa. *Submont.* My–Au.

2. N. mínima Nutt. *Fig. 467.* Stem erect, simple, or branched near the base, 1–5 cm. high, glabrate or nearly so; leaves glabrous, pinnatifid with filiform-subulate, accrose divisions; bracts glabrous; calyx slightly hairy in the sinuses; corolla white, about 7 mm. long; limb 1–1.5 mm. broad; tube shorter than the calyx; stamens equally inserted in the throat. *G. minima* A. Gray. Bottom lands and sandy places: Wash.—Sask.—N.D.—Neb.—Ariz.—Calif. *W. Temp.—Plain —Submont.* My–Jl.

3. **N. Brèweri** (A. Gray) Greene. Stem simple or branched below, 1–10 cm. long, glandular-puberulent; leaves pinnatifid, glandular-puberulent, with subulate-acerose divisions; bracts and calyx not villous, merely puberulent; corolla 6–7 mm. long; limb 1.5 mm. broad; stamens equally inserted in the throat, exserted. Wash.—Wyo.—Kans. (?)—Colo.—Calif. *W. Temp.* My–Au.

6. LEPTODÁCTYLON Nutt.

Much branched undershrubs, or one species herbaceous except the base. Leaves alternate, or rarely opposite, with smaller ones fascicled in the axils, numerous, palmately 3–5-parted, usually acerose-subulate, pungent. Flowers showy, solitary or in small clusters, sessile. Calyx tubular or campanulate, with subulate teeth, ruptured by the fruit. Corolla trumpet-shaped, with a funnelform throat, yellow or reddish or lilac. Filaments equally adnate to the throat of the corolla; anthers included in the throat of the corolla. Capsule many- or few-seeded. Seeds neither mucilaginous nor with spiricles.

1. **L. caespitòsum** Nutt. Densely pulvinate-cespitose perennial, with woody caudex; leaves densely crowded, slightly ciliate below, about 5 mm. long, usually 3-parted, with subulate straight pungent divisions; calyx 6–8 mm. long; teeth subulate; corolla 4-merous, light yellow, trumpet-shaped, 12–15 mm. long; lobes 4 mm. long. *Gilia pungens caespitosa* A. Gray. *G. caespitosa* A. Nels. [B]; not A. Gray. Dry hills: w Neb.—Wyo.—Utah. *Plain—Submont.* My–Jl.

7. GÍLIA R. & P.

Annual, biennial, or perennial herbs. Leaves all or at least the upper ones alternate, entire to pinnatifid. Flowers perfect, mostly cymose. Calyx campanulate; tube more or less hyaline in the sinuses and bursted by the fruit; lobes subequal. Corolla trumpet-shaped or salverform, with an open salverform throat. Stamens 5, usually equally highly inserted. Capsule usually many-seeded; seeds changed when wetted, mucilaginous and usually emitting spiricles.

Flowers capitately or spicately glomerate.
  Plant strigose or villous; flowers leafy-bracted.
    Perennials; flowers in dense heads or spikes.
      Plant simple or branched only at the base; stem or
        branches strict, with single head-like or spike-
        like inflorescence.
        Inflorescence spike-like, more or less interrupted;
          corolla greenish-white, its lobes acute.   1. *G. spicata.*
        Inflorescence head-like; corolla white, its lobes
          obtuse.   2. *G. congesta.*
      Plant branched above, as well as at the base; inflor-
        escence of several heads.   3. *G. iberidifolia.*
    Annuals, repeatedly branched, with small flower-clusters
      in the axils of the leaves.   4. *G. pumila.*
  Plant glabrous below, viscid above; flowers in long-peduncled
    heads, not leafy-bracted.   5. *G. tricolor.*
Flowers paniculate or thyrsoid-paniculate.
  Corolla-tube more than 1 cm. long; calyx-lobes attenuate and
    spine-tipped.
    Inflorescence thyrsoid, narrow; corolla mostly red.
      Corolla 3–4 cm. long, its lobes lanceolate, attenuate.   6. *G. aggregata.*
      Corolla 2.5–3 cm. long, its lobes oblong or ovate-
        oblong.   7. *G. rubra.*
    Inflorescence open, inclined to be flat-topped; corolla
      white.
      Corolla-tube 3–4 cm. long; its lobes obtuse.   8. *G. longiflora.*
      Corolla-tube 1.5–2.5 cm. long; its lobes acutish.   9. *G. laxiflora.*
  Corolla-tube less than 1 cm. long.
    Stamens exserted; corolla salverform; biennials.   10. *G. calcarea.*
    Stamens not exserted; corolla with a funnelform throat;
      annuals.   11. *G. inconspicua.*

1. G. spicàta Nutt. Biennial or perennial, with a taproot; stem simple, erect, or branched at the base, arachnoid-villous, 1–3.5 dm. high; leaves linear, entire, or with few linear lobes, 2–5 cm. long, 1–2 mm. wide, sparingly arachnoid; calyx-lobes lanceolate, cuspidate; corolla 9–11 mm. long, greenish-white; tube one third longer than the calyx; lobes 3–4 mm. long, oval. Dry hills: S.D.—Neb.—Colo.–Utah—Wyo. *Plain—Mont.* My–Je.

2. G. congésta Hook. Perennial, with a branched caudex; stems simple, 1–1.5 dm. high, arachnoid-villous; basal leaves simple, linear, or 3-cleft, 2–4 cm. long; stem-leaves pinnately 3–5-cleft; calyx densely arachnoid; teeth lanceolate, cuspidate, shorter than the tube; corolla white, 7–8 mm. long. *G. cephaloidea* A. Brand; not Rydb. Hills and sandy plains: Sask.—S.D.—Wyo.—B.C. (?). *Plain—Submont.* Je–Jl.

3. G. iberidifòlia Benth. Suffruticose perennial; stems several from the woody caudex, erect or ascending, floccose, 1–2 dm. high, branched above; leaves pinnately divided into 3–7 linear lobes, 2–5 cm. long, or the lower entire; calyx teeth lance-subulate, cuspidate, shorter than the tube; corolla white; tube a little exserted; lobes oval, 2 mm. long or less, obtuse. *G. congesta* Britt. [B]; not Hook. Dry hills, plains, and badlands, N.D.—Neb.—Colo.—Utah. *Plain—Submont.* Je–Au.

4. G. pùmila Nutt. Stem erect, branched, 1–2 dm. high, crisp-hairy; leaves pinnatifid, with few linear divisions, cuspidate, sparingly crisp-hairy; flowers in small terminal heads; calyx-lobes subulate, shorter than the tube, cuspidate; corolla 9–10 mm. long; tube twice as long as the calyx; lobes oval, 2 mm. long. Dry hills: w Tex.—Neb.—Wyo.—Utah—Ariz. *Plain—Son.—Submont.* My–Au.

5. G. tricolor Benth. Annual; stem 1.5–5 dm. high, leaves 2–3-pinnately divided; flowers short-pedicelled, in simple cymes; calyx-lobes deltoid-lanceolate; corolla campanulate-funnelform, the tube yellowish, the throat brown-purple, the lobes lilac or violet, oval or orbicular. Waste places: Neb.; escaped from cultivation; native of Calif. Ap My.

6. G. aggregàta (Pursh) Spreng. Stem erect, 3–5 dm. high, more or less puberulent, simple up to the inflorescence; leaves pinnatifid into narrowly linear divisions, more or less crisp-hairy; calyx-lobes lanceolate, attenuate, spinulose-tipped; corolla 3–4 cm. long, crimson to white; lobes 7–9 mm. long, lanceolate. *Callisteris aggregata* and *C. flavida* Greene. SKYROCKET. Hills, glades, and meadows: Mont.—Kans.—N.M.—Calif.—B.C. *Plain—Submont.—Mont.* My–Au.

7. G. rùbra (L.) Heller. Biennial; stem 5–18 dm. high; leaves pinnately parted, the segments linear-filiform; calyx-lobes subulate, fully as long as the tube; corolla scarlet or yellow, dotted with red within, 2.5–3 cm. long. *G. coronopifolia* Pers. Sandy soil: Mass.—Kans.—Tex.—Fla. *E. Temp.* Je–Au.

8. G. longiflòra (Torr.) G. Don. *Fig. 468.* Stem erect, glabrous, branched above, 2–5 dm. high; leaves deeply pinnatifid, with few filiform divisions, or often entire and filiform; calyx glabrous; teeth subulate, much shorter than the tube; corolla white; lobes about 1 cm. long. Sandy plains: Neb.—Tex.—Ariz.—Utah. *Plain—Submont.* Je–Au.

f. 468.

9. G. laxiflòra (Coult.) Osterhout. Stem glabrous or glandular-puberulent above, about 3 dm. high; leaves pinnatifid, with filiform divisions, the upper few-lobed or entire; calyx-teeth subulate; corolla white or tinged with blue; lobes 4–5 mm. long. *Gilia Macombii laxiflora* Coult. Plains and tablelands: Colo.—Kans.—Tex.—N.M.—Utah. *Plains— Submont.* Je–S.

10. G. calcàrea M. E. Jones. Stem erect, usually branched, glandular-pilose, 0.5–5 dm. high; leaves pinnatifid, glandular; flowers paniculate; calyx densely glandular; teeth triangular-lanceolate, acuminate; corolla-tube yellowish, twice as long as the calyx; lobes violet, rounded-ovate. *G. pinnatifida* Nutt. [B]; not Moc. & Sess. *G. viscida* Woot. & Standl. Sandy soil: Neb.— Kans.—N.M.—Wyo. *Plain—Mont.* Je–S.

11. G. inconspícua (Smith) Dougl. Stem 2–4 dm. high, glandular-puberulent; leaves mostly near the base, pinnatifid, more or less crisp-pubescent or glabrate; lobes mostly oblong, cuspidate; calyx-tube about 2 mm. long, glabrous or glandular-puberulent; teeth subulate, about 1.5 mm. long; corolla 7–8 mm. long; limb purplish; lobes oval, 2 mm. long, obtuse. Hillsides: Kans.—N.M.— Wyo.—Wash.—Calif. *Submont.—Mont.* F–Au.

8. GILIÁSTRUM (A. Brand) Rydb.

Perennial or annual herbs. Leaves alternate, from toothed to pinnatifid. Flowers perfect, cymose. Calyx campanulate, 5-lobed or 5-cleft; tube scarious in the sinuses and burst by the fruit; corolla rotate-funnelform, deeply 5-cleft. Stamens inserted near the base of the corolla. Seeds several in each cell, mucilaginous when wetted. *Gilia § Giliastrum* A. Brand.

1. G. aceròsum (A. Gray) Rydb. Suffruticose perennial, 1–3 dm. high, very leafy, glandular-puberulent; leaves pinnatifid, with linear-subulate, acerose divisions, glandular-puberulent; calyx glandular-puberulent, 5–6 mm. long; teeth lanceolate, acuminate; corolla blue, nearly rotate, about 1 cm. long; lobes broadly ovate, much longer than the united portion. *Gilia rigidula acerosa* A. Gray. *G. acerosa* Britton [B]. Sandy plains and hills: Tex.—Kans.—Colo.— Ariz. *St. Plains—Son.* Ap–Jl.

Family 141. HYDROPHYLLACEAE. WATERLEAF FAMILY.

Perennial or annual, mostly caulescent herbs with watery sap. Leaves alternate, or rarely opposite or in basal rosettes, without stipules. Flowers perfect, regular or nearly so, in scorpioid racemes, or solitary and axillary. Calyx of 5 more or less united sepals, often appendaged in the sinuses. Corolla 5-lobed, mostly campanulate or funnelform, often appendaged inside, the lobes convolute or imbricate in aestivation. Stamens 5, alternate with the corolla-lobes; filaments adnate to the base of the corolla, distinct, often bearded; anthers introrse, versatile. Ovary of 2 united carpels, 1-celled or 2-celled, with parietal placentae; styles more or less united; stigmas often capitate; ovules numerous, or few. Fruit a capsule. Seeds amphitropous; endosperm cartilaginous. *Hydroleaceae.*

Corolla-lobes convolute in bud ; placentae dilated.
 Stamens exserted ; leaves alternate.
  Calyx without appendages.   1. HYDROPHYLLUM.
  Calyx with reflexed appendages in the sinuses. 2. DECEMIUM.
 Stamens included ; leaves opposite ; calyx without appendages. 3. ELLISIA.
Corolla-lobes imbricate in bud ; placentae not dilated.   4. PHACELIA.

1. HYDROPHÝLLUM (Tourn.) L. WATERLEAF.

Perennial (all ours) caulescent herbs. Leaves alternate, pubescent, lobed, or (in all ours) pinnately divided. Flowers in more or less scorpioid cymes. Calyx-tube short; lobes 5, narrow, with naked sinuses. Corolla white, blue or purple; tube short; lobes 5, convolute, each with a linear appendage within. Stamens 5, exserted; filaments partly pubescent. Ovary 1-celled, pubescent;

styles 2-cleft; ovules 4, enclosed by the fleshy placentae, attached at the bottom and top of the ovary. Capsule 2-valved; seeds 1–4, thick.

Stem glabrate or minutely pubescent; leaf-segments 5–7, acuminate; calyx-lobes
   linear, distinct to the base.
  Calyx-lobes spreading; corolla-lobes erect. — 1. *H. virginianum.*
  Calyx-lobes appressed; corolla-lobes with spreading tips. — 2. *H. patens.*
Stem rough-hairy; leaf-segments 9–13, obtusish; calyx divided
   two thirds its length into lanceolate lobes. — 3. *H. macrophyllum.*

1. **H. virginiànum** L. *Fig. 469.* Stem 2–7 dm. high, nearly glabrous; leaves 1–3 dm. long; pinnately divided, the segments coarsely toothed, acute or acuminate; lower petioles longer than the blades; cyme open; calyx bristly, the lobes linear, 5 mm. long, nearly distinct; corolla white or purplish, 1 cm. long. Moist woods: N.H.—Man.—N.D.—Kans. S.C. *Canad.—Allegh.* My–Au.

2. **H. pàtens** Britton. Similar to the preceding, lighter green, distinguished by the character given in the key. Deciduous woods: Minn.—N.Y. Je–Au.

3. **H. macrophýllum** Nutt. Stem 2–6 dm. high; leaves 2–3 dm. long, pinnatifid, the blades longer than the petioles; segments oblong, serrate or incised, the upper confluent; calyx bristly, the lobes lanceolate, acute; corolla white, 10–12 mm. long, the lobes notched. Woods: Ohio—Iowa—Ala.—Va. *Allegh.* Ap–Je.

f.469.

2. DUODMIUM Raf.

Biennial caulescent herbs. Leaves alternate, with orbicular-cordate toothed blades, or the lower ones somewhat pinnatifid below. Flowers in more or less scorpioid cymes. Calyx-tube short, the lobes 5, narrow, subulate-lanceolate, with reflexed appendages in the sinuses. Corolla violet or purple, 5-lobed, appendaged within, the lobes convolute. Stamens 5; filaments exserted, partly pubescent. Style 2-cleft. Fruit as in *Hydrophyllum.*

1. **D. appendiculàtum** (Michx.) A. Brand. Stem erect, 3–6 dm. high, usually branched; lower leaves with a large terminal segment and small and few lateral ones, the terminal segment or the blade of the upper leaves subpalmately 5–7-lobed and dentate; cymes open; calyx bristly; corolla 1–1.5 cm. broad. *Hydrophyllum appendiculatum* Michx. [G, B]. Woods: Ont.— Minn.—Kans.—N.C. *Canad.—Allegh.* My–Je.

3. ELLÍSIA L.

Annual caulescent herbs. Leaves alternate or opposite, pubescent, from once to thrice pinnatifid. Flowers perfect, solitary in the axils. Calyx enlarging in fruit, 5-lobed, appendaged in the sinuses. Corolla campanulate, usually with 5 appendages within; lobes convolute. Stamens 5, included. Ovary 1-celled; styles 2-cleft; ovules 2–4 on each fleshy placenta, which is attached to the top and bottom of the ovary. Seeds pitted. *Macrocalyx* Trew.

1. **E. Nyctèlea** L. *Fig. 470.* Stem erect, 0.6–3 dm. high, diffusely branched; leaf-segments spreading, entire, toothed or incised; calyx-lobes triangular-lanceolate, becoming 7–8 mm. long, acute, pubes-

f.470.

cent; corolla bluish white, 7–8 mm. broad.   *M. Nyctelea* (L.) Kuntze [B, R]. Alluvial soil and shady places: N.J.—N.C.—Colo.—Ida.—Alta.   *Plain—Mont.*

4. PHACÈLIA Juss. Scorpion Weed.

Perennial or annual caulescent herbs.   Leaves alternate or opposite below. Flowers perfect, in scorpioid racemes or cymes.   Calyx slightly enlarging in fruit, 5-lobed, without appendages.   Corolla white, purple, or blue; tube often appendaged within; lobes 5, imbricate.   Stamens 5; filaments adnate to the tube of the corolla.   Ovary 1-celled, with two narrow placentae attached to the walls of the ovary; styles 2-cleft; ovules 2 to many on each placenta.   Capsule 1-celled or falsely 2-celled, 2-valved.   Seeds reticulate or roughened.

Corolla-lobes entire or crenulate.
  Corolla-tube with 10 vertical appendages.
    Leaves all simple and entire, or some of the lower pinnately 3–5-divided, with
        entire divisions; capsule acute; ovules 4.
      Perennial; leaves grayish.   1. *P. leucophylla.*
      Annual; leaves green.   4. *P. linearis.*
    Leaves from sinuate-crenate to twice pinnatifid.
      Capsule globular-ovoid, obtuse; ovules 4; stamens
        long-exserted.   2. *P. integrifolia.*
      Capsule ovoid, acute; ovules 12–40; stamens slightly
        if at all exceeding the corolla.   3. *P. Franklinii.*
  Corolla-tube with no or obsolete appendages; ovules and
    seeds 2–8 on each placenta.
    Plant puberulent; corolla white or pink, 8–10 mm. broad.   5. *P. dubia.*
    Plant hirsute; corolla violet-purple, 12–15 mm. broad.   6. *P. hirsuta.*
Corolla-lobes fimbriate.   7. *P. Purshii.*

1. P. leucophýlla Torr.   Stem 1–3 dm. high, densely canescent and only slightly hirsute; basal leaves petioled, 5–15 cm. long, oblanceolate or elliptic; calyx-lobes linear or linear-oblong, obtuse, about two thirds as long as the corolla; filaments villous-bearded.   Dry hills: S.D.—Colo.—Ida.—B.C.   *Plain— Submont.*   Je–Au.

2. P. integrifòlia Torr.   Stem 2–6 dm. high, branched, glandular-puberulent, only slightly hirsute; leaves short-petioled or the upper sessile; blades 2–10 cm. long, round-lobed and crenate; inflorescence open, 5–10 cm. long; calyx-lobes elliptic or oblong; corolla bluish or whitish, 5–6 mm. long; filaments glabrous.   In gypsum soil: Kans.—w Tex.—Ariz.—Utah; n Mex.   *St. Plains— Son.*   Ap–Jl.

3. P. Franklínii R. Br.   Stem 2–4 dm. high, softly hirsute, erect, rather simple; leaves pinnatifid, densely pilose or hirsute; divisions many, oblanceolate or oblong, deeply cleft or pinnatifid into lanceolate divisions; inflorescence dense; calyx-lobes linear, about 5 mm. long; corolla pale blue or almost white, about 8 mm. long; stamens slightly exserted. Hillsides and banks: Ont.—Man.—Mich.—Wyo.— Ida.—Mack.   *Boreal.—Submont.*   My–Jl.

4. P. lineàris (Pursh) Holz.   *Fig. 471.*   Stem 1–4 dm. high, at least branched, grayish-puberulent; leaves 2–5 cm. long, simple, linear, or cleft into 3–7 linear divisions; inflorescence thyrsoid-paniculate; sepals narrowly linear, hispid-ciliate; corolla bright blue, rarely white, about 1 cm. long; filaments glabrous, scarcely exserted.   *P. Menziesii* (R. Br.) S.

f. 471.

Wats.   Valleys and hillsides, in sandy soil: Alta.—S.D.—Wyo.—Calif.—B.C. *Plain—Submont.—W. Temp.*   Ap–Jl.

5. P. dùbia (L.) Small. Annual; stem branched at the base, 0.5–4 dm. high, minutely glandular as well as puberulent; basal leaves petioled, 1–3 cm. long, the segments oval to orbicular; stem-leaves sessile, up to 1.5 cm. long, with 3–5 segments, the terminal one obovate, acute; calyx-lobes oblong, 3–4 mm. long, obtuse; corolla 7–11 mm. long, greenish near the center, the lobes broadly ovate, undulate, blotched near the base; filaments pubescent. *P. parviflora* Pursh. Banks: Pa.—Kans.—Tex.—Ga. *Allegh.—Austral.* Ap–Je.

6. P. hirsùta Nutt. Annual or biennial; stem simple or branched at the base, 5–30 cm. long; leaves 2–7 cm. long, the earlier ones spatulate, the lower leaves pinnatifid, slender-petioled, the upper sessile, the segments falcate, acute; calyx bristly, the lobes oblong or linear; corolla-lobes rounded, undulate; filaments barely exserted. Dry soil and open woods: Va.—Kans.—Tex.—Ga. *Allegh.—Austral.* Ap–Je.

7. P. Púrshii Buckley. Annual; stem 1–3 dm. high; cauline leaves 1–8 cm. long, pinnatifid, short-petioled or sessile; segments oblong or lanceolate, acute or acuminate, entire or toothed; calyx bristly, the lobes linear or oblong-spatulate, 4–5 mm. long, ciliate; corolla light blue or rose, 1.5 cm. broad; lobes rounded, fimbriate; appendages present; filaments villose. Thickets and woods: Pa.—Minn.—Okla.—Ala.—N.C. *Allegh.* Ap–Je.

Family 142. HELIOTROPIACEAE. HELIOTROPE FAMILY.

Herbs, shrubs, or vines. Leaves alternate, without stipules, usually entire-margined. Flowers perfect, regular, mainly in scorpioid racemes or spikes, rarely axillary. Calyx of 5 slightly united sepals. Corolla of 5 united petals, funnelform, salver-shaped, or campanulate. Stamens 5, wholly adnate or nearly so to the corolla-tube. Gynoecium of 2–4 united carpels; ovary 2–4-celled, often 4-lobed; styles united; stigmas annular, surpassed by a 2-lobed appendage; ovules pendulous. Fruit dry, separating into 2 or 4 one- or two-seeded nutlets, drupaceous.

Flowers in scorpioid bractless spikes.
 Fruit 4-lobed, separating into 4 1-seeded nutlets.   1. HELIOTROPIUM.
 Fruit 2-lobed, separating into 2 2-seeded nutlets.   2. TIARIDIUM.
Flowers axillary, subtended by foliar leaves; fruit didymous, each half at least
  separating into 2 hemispheric nutlets.
 Stigma-tips subulate, elongate, 3–4 times as long as the style. 3. LITHOCOCCA.
 Stigma-tips truncate, penicillate-bristly; style long and filiform.
     4. EUPLOCA.

1. HELIOTRÒPIUM L. HELIOTROPE.

Annual or perennial herbs, or in warmer climate shrubs. Leaves alternate, or rarely opposite, pubescent. Flowers perfect, regular, in terminal scorpioid racemes or spikes. Calyx-lobes 5, narrow. Corolla salver-shaped; tube cylindric, usually naked in the throat; lobes 5, induplicate or plicate. Stamens 5, included; filaments short; anthers blunt, distinct. Ovary 4-celled or 2-celled, with 2 more or less introding placentae; ovules pendulous. Fruit 4-lobed, separating into 4 nutlets.

Leaves pubescent; anthers obtuse or mucronate.   1. *H. europaeum.*
Leaves glaucous; anthers acuminate.   2. *H. spathulatum.*

1. H. europaèum L. Annual, cinereous-pubescent; stem 1–5 dm. high; leaves oblong, oval, ovate, or obovate, 2–6 cm. long, undulate; racemes solitary or a few together; corolla white or bluish, the tube 2 mm. long, the limb 3–4 mm. broad. Waste places: Mass.—Neb.—Fla.; nat. from Eur. Je–O.

2. H. spathulàtum Rydb. *Fig. 472.* A glabrous, more or less glaucous, fleshy perennial, 3–5 dm. high; leaves spatulate, fleshy, indistinctly nerved,

2–5 cm. long, obtuse or rounded at the apex; inflorescence branched into 2–5 racemes, these often
starting from one point at the end of the common
peduncle; corolla white or slightly tinged with blue,
6–8 mm. long; nutlets 2.5–3 mm. long, scarcely
rugose. *H. curassavicum* Hook. [B]; not L. *H.
curassavicum obovatum* DC.; not *H. obovatum* D.
Don. River valleys: Man.—N.M.—Calif.—Wash.
Plain.

2. TIARÍDIUM Lehm.

Annual herbs. Leaves alternate. Flowers perfect, regular, in terminal scorpioid spikes or racemes.
Calyx 5-lobed. Corolla salver-shaped; tube cylindric, naked in the throat; lobes 5, induplicate.
Stamens 5, included; filaments short; anthers blunt,
distinct. Ovary 2-lobed, 4-celled. Fruit 2-lobed,
separating into 2 2-seeded nutlets.

1. **T. índicum** (L.) Lehm. Annual, more or less hirsute; stem 1–7 dm.
high; leaves ovate, oval or oblong, 2–10 cm. long, repand or undulate; racemes
solitary, 1–3 dm. long; corolla blue, the limb 4–6 mm. broad. Waste places:
Va.—Ind.—Kans.—Tex.—Fla.; adv. from India. *Heliotropium indicum* L.
[G, B]. My–N.

3. LITHOCÓCCA Small.

Annual herbs, with narrow alternate leaves. Flowers perfect, scattered,
terminal or axillary on slender peduncles, bractless. Calyx 5-lobed, the lobes
unequal. Corolla salver-shaped; tube elongate, canescent; limb short; stamens
5; anthers blunt, almost naked, scarcely coherent. Stigmas elongate-subulate.
Fruit didymous, each half ultimately separating into 2 nutlets, without pits
on the inner faces.

1. **L. tenélla** (Nutt.) Small. Annual, strigose; stems 1–3 dm. high, forking above; leaves linear or nearly so, 1.5–5 cm. long, more or less revolute;
calyx-lobes linear, unequal; corolla white, pubescent in the throat, the tube
about as long as the calyx, the lobes oblong; fruit 4-lobed, pubescent. *Heliotropium tenellum* Torr. [G]. Dry soil: Ky.—Kans.—Tex.—Ala. *Austral.*
Ap–Au.

4. EÚPLOCA Nutt.

Annual caulescent herbs. Leaves alternate,
broad, entire, petioled, pubescent. Flowers perfect,
regular, solitary in the axils of leaf-like bracts.
Calyx-lobes 5, narrow. Corolla salver-shaped; tube
cylindric, naked in the throat; limb 5-angled,
strongly plicate in aestivation. Stamens 5, included;
anthers subsessile, with a beard at their tip; ovary
4-celled; style elongate; cone of stigma penicillate-
setose; ovules pendulous. Fruit didymous, the two
halves separating into 2 hemispheric, 1-seeded nutlets.

1. **E. convolvulàcea** Nutt. *Fig. 473.* Diffusely
branched annual; stem 1–4 dm. high, hispid-strigose;
leaf-blades lanceolate or ovate, 2–4 cm. long, hirsute-
strigose; flowers usually supra-axillary; calyx-lobes
linear-subulate; corolla white, salver-shaped to funnelform, 10–15 mm. long; nutlets almost hemispheric, 3 mm. broad, more or
less strigose. *Heliotropium convolvulaceum* A. Gray [B]. Sandy plains: Tex.
—Neb.—Wyo.—Calif.; Mex. *Plain—Son.* My–Au.

Family 143. BORAGINACEAE. BORAGE FAMILY.

Herbs (all ours), or shrubby plants. Leaves alternate, without stipules, simple, entire, bristly. Flowers perfect, regular, or sometimes irregular, in scorpioid racemes or spikes. Calyx of 5, rarely 4, more or less united sepals, persistent. Corolla salver-shaped or campanulate, rarely nearly rotate, deciduous; throat often with a crown of 5 small scales (fornices); lobes 5, convolute, plicate or induplicate, usually equal. Stamens 5, adnate to the corolla-tube; filaments often appendaged; anthers introrse. Gynoecium of 2 united carpels; styles united, rising between the four lobes of the ovary; stigma simple or 2-lobed; ovules anatropous, solitary in each cell of the ovary. Fruit usually of 4 nutlets.

Nutlets with hooked prickles, at least on the margins.
 Nutlets spreading or divergent on the low receptacle, short, prickly all over, attached horizontally or obliquely on the receptacle.   1, CYNOGLOSSUM.
 Nutlets erect on the elevated receptacle, prickly on the margin, rarely along the back.   2. LAPPULA.
Nutlets unarmed, or if prickly the prickles not hooked.
 Receptacle conic or elongate; nutlets attached laterally.
  Calyx in fruit much enlarged, veiny-reticulate and folded.   3. ASPERUGO.
  Calyx in fruit neither much enlarged nor conspicuously veiny.
   Corolla blue or white (yellowish only in a few species of Oreocarya), with fornices in the throat; cotyledons entire.
   Pedicels and calyx persistent in fruit.
    Calyx-lobes spreading in fruit; leaves alternate. perennials, with bracted racemose or thyrsoid inflorescence.   4. OREOCARYA.
    Calyx nearly closed in fruit; leaves proper opposite, with connate bases; annuals.   5. ALLOCARYA.
   Pedicels in fruit falling off with the calyx; the latter closed; branched but not dichotomous annuals.   6. CRYPTANTHA.
   Corolla yellow, often with naked throat, cotyledons 2-cleft.
     7. AMSINCKIA.
 Receptacle flat or merely convex.
  Scar of the nutlets small and marginless.
   Nutlets obliquely attached; flowers mostly bractless; corolla blue or white, with funnelform throat.   8. MERTENSIA.
   Nutlets attached by the very base.
    Corolla salverform or funnelform, its lobes rounded and spreading.
     Racemes not bracted; corolla in our blue, its lobes convolute in the bud.   9. MYOSOTIS.
     Racemes bracted; corolla yellow or yellowish, its lobes imbricated in the bud.   10. LITHOSPERMUM.
    Corolla tubular, its lobes erect, acute, otherwise as in Lithospermum.   11. ONOSMODIUM.
  Scar of the nutlets large and excavated, bordered by a prominent margin.
   Corolla regular.
    Corolla rotate; anthers connivent into a cone.   12. BORAGO.
    Corolla campanulate; or salverform.
     Fornices in the throat of the corolla short, blunt.   13. ANCHUSA.
     Fornices long, linear or lanceolate.   14. SYMPHYTUM.
   Corolla irregular, the throat and limb oblique, the lobes of the latter unequal.
    Throat of the corolla closed by the fornices; stamens included.   15. LYCOPSIS.
    Throat of the corolla open and naked, without fornicles; stamens exserted.   16. ECHIUM.

1. CYNOGLÓSSUM (Tourn.) L. HOUND'S TONGUE.

Annual, biennial, or perennial caulescent herbs. Leaves alternate, entire, mostly decurrent, pubescent. Flowers perfect, regular, in scorpioid racemes. Calyx with 5 spreading or reflexed lobes, slightly accrescent in age. Corolla white, blue, or purple, salverform or funnelform; tube short; throat closed by 5 scales; lobes 5, imbricate. Stamens 5, included. Nutlets 4, flat, distinct, equally divergent, armed with short barbed prickles, attached by an oval or rounded scar at the upper end to a flat or low-pyramidal receptacle.

Soft-pubescent; lower leaf-blades oblong or spatulate; stem-leaves lanceolate; inflorescence many-flowered and leafy.   **1.** *C. officinale.*
Hispidulous; lower leaf-blades oval; stem-leaves oblong to ovate,
clasping; inflorescence few-flowered and naked.
 Corolla 10–12 mm. broad, the lobes orbicular, the sinuses closed;
 nutlets 7–9 mm. long.   **2.** *C. virginianum*
 Corolla about 7 mm. broad, the lobes oblong-ovate, the sinuses
 open; nutlets 4–5 mm. long.   **3.** *C. boreale.*

1. C. officinàle L. Biennial; stem 6–8 dm. long, softly pilose, somewhat
canescent; basal leaves petioled; blades 1–3 dm. long, velvety-pilose; racemes
panicled; calyx-lobes oval or oblong; corolla dull red, rarely white, 7–10 mm.
broad; nutlets depressed on the back, margined, about 7 mm. long; prickles in
4–5 rows on the margin and scattered on the back. Pastures and waste places:
Que.—N.C.—Kans.—Utah—Mont.; nat. from Eur. My–Jl.

2. C. virginiànum L. Perennial; stem 2–9 dm. high, naked and branched
above; leaves few, oval to oblong, the lower narrowed below into a margined
petiole, the upper sessile; racemes elongating in fruit; fruiting pedicels recurved; calyx soft-pubescent, the lobes ovate to oblong, obtuse; corolla white
or purplish, 8–10 mm. broad; nutlets echinate all over. Woods: N.B.—Minn.
—Kans.—La.—Fla. *E. Temp.* Ap–My.

3. C. boreàle Fern. Perennial; stem 4–8 dm. high, hirsute and in the
inflorescence strigose; basal leaves long-petioled; blades 1–3 dm. long, hispid;
racemes geminate; fruiting pedicels spreading, recurved; calyx strigose, 2–2.5
mm. long; lobes ovate; corolla blue, 6–8 mm. wide; nutlets ovoid-pyriform, 4–5
mm. long, glochidiate-prickly all over. Woods: N.B.—Mass.—Minn.—B.C.
Boreal. Je–Jl.

2. LÁPPULA (Riv.) Moench. Stickseed, Burseed, Stick-tight,
Beggar-ticks.

Annual, biennial, or perennial caulescent herbs. Leaves alternate, entire,
narrow, pubescent. Flowers perfect, regular, in terminal scorpioid racemes.
Calyx with 5 narrow lobes, slightly if at all accrescent. Corolla salverform,
blue or white; tube very short; throat closed by 5 scales; lobes 5, imbricate,
obtuse. Stamens 5, included; filaments very short. Fruit of 4, finally distinct
nutlets, these with barbed prickles along the edge, and sometimes with smaller
ones on the dorsal faces, muricate on the sides, obliquely attached by the inner
angle to a conical receptacle. *Echinospermum* Sw.

Fruit globose; nutlets equally prickly all over the back; basal leaves with ovate or
cordate blades.   **1.** *L. virginiana.*
Fruit pyramidal; nutlets with strong prickles on the margins,
not prickly on the back or the prickles there much
shorter.
 Inflorescence leafy-bracted only at the base; bracts minute
 above; gymnobase short-pyramidal; scar of the nutlets ovate or triangular; perennials or biennials.
  Marginal prickles free to the base or nearly so.
   Corolla 1.5–3 mm. wide; leaves thin; stem-leaves
   lanceolate; nutlets usually with a few prickles
   on the back.   **2.** *L. americana.*
   Corolla 4 mm. wide; leaves firmer; nutlets without
   prickles on the back.   **3.** *L. floribunda.*
  Marginal prickles united for one third to one half their
  length into a distinct wing.
   Stem-leaves linear-lanceolate, densely strigose. **4.** *L. angustata.*
   Stem-leaves oblong-lanceolate, hispidulous, the hairs
   with papillose bases.   **5.** *L. scaberrima.*
 Inflorescence leafy, the floral leaves, although smaller, re-
 sembling those of the stem; annuals.
  The annular margin connecting the bases of the prickles
  inconspicuous in all four nutlets.
   Marginal prickles in two rows.
    Corolla almost 3 mm. wide; prickles shorter than
    the length of the fruit.

Nutlets ovate, 2.5–3 mm. long, their backs
muricate throughout. — 6. *L. echinata.*
Nutlets lance-ovate, 3–3.5 mm. long, their
backs with a row of papillae or short
spines along the middle. — 7. *L. erecta.*
Corolla less than 2 mm. wide; prickles nearly
as long as the fruit.
Fruit 4–5 mm. long; sepals ovate-lanceolate,
obtuse; floral leaves ample, ovate or
oblong. — 8. *L. cenchrusoides.*
Fruit 3–4 mm. long; sepals lanceolate or
linear-lanceolate, acute; floral leaves much
reduced, lanceolate. — 9. *L. Fremontii.*
Marginal prickles in one row. — 10. *L. occidentalis.*
The annular margin connecting the bases of the prickles
at least in three of the nutlets broadened and
forming a cup.
Cup deep, more or less saccate. — 11. *L. heterosperma.*
Cup shallow, with revolute margins. — 12. *L. foliosa.*

1. **L. virginiàna** (L.) Greene. Annual or biennial; stem 5–9 dm. high,
somewhat hirsute; leaves thin, oblong, elliptic or lanceolate, acute or short-
acuminate, the lower 1–2 dm. long, petioled, the upper sessile, shorter; corolla
white or pale blue, 2–3 mm. broad; fruit about 4 mm. broad; nutlets papillose
and prickly. Woods and thickets: N.B.—O.D.—Kans.—La.—Ga. *Canad.—
Allegh.* Je–S.

2. **L. americàna** (A. Gray) Rydb. Biennial; stem 3–10 dm. high, retrorsely hirsute; lower leaves petioled; blades oblanceolate, acute, 5–10 cm.
long, sparingly hirsute, thin; upper leaves sessile, lanceolate, acute or acuminate; corolla pale blue, 1.5–2 mm. wide; back of the fruit convex, but not
keeled, muricate, *L. deflexa americana* [G]. Copses: Man.—Neb.—Ida.—B.C.
Plain—Submont. Je–Jl.

3. **L. floribùnda** (Lehm.) Greene. Biennial or perhaps perennial; stem
3–15 dm. high, strigose or pilose; lower leaf-blades narrowly oblanceolate, 5–15
cm. long, strigose; upper leaves linear or linear lanceolate, sessile; nutlets 5–6
mm broad; back slightly ridged, glabrous or scabrous, muricate. Hillsides and
among bushes: Man.—N.M.—Calif. B.C. *W. Temp.—Plain—Mont.* Je–Au.

4. **L. angustàta** Rydb. Short-lived perennial; stem strict, hirsute, 6–8
dm. high; lower leaves petioled, 9–15 cm. long, strigose or hirsute on the petioles; blades narrow, linear-oblanceolate, mostly acute; corolla blue, about 4
mm. wide; fruit 4–5 mm. wide; back slightly keeled, minutely hispidulous or
glabrous. Mountain valleys: S.D.—Colo. *Submont.* Je–Jl.

5. **L. scabérrima** Piper. Perennial; stem softly hirsute below; basal
leaf-blades spatulate, obtuse, harshly hispid-strigose; stem-leaves oblong-lanceolate; corolla blue, 5 mm. broad; fornices broader than long, papillose; fruit
6 mm. wide; back smooth, with prominent keel. Hills and mountains: w Neb.
—Colo. *Plain—Submont.* Jl.

6. **L. echinàta** Gilib. Stem 2–6 dm. high, branching above, leafy; leaves
linear or linear-spatulate, 2–5 cm. long, acute; bracts rather large; pedicels
1–4 mm. long; fruit 3 mm. broad; nutlets granulose or tubercled on the back.
E. Lappula Lehm. *L. Lappula* (L.) Karst. [B]. Waste places: N.S.—Alta.—
N.D.—Kans.—Tex.—N.J.; nat. from Eur. My–S.

7. **L. erécta** A. Nels. Stem 2–4 dm. high, strigose; basal leaves numerous, rosulate, oblanceolate, 1–3 cm. long; stem-leaves linear or linear-oblong,
sessile, 1–4 cm. long, strigose; corolla blue; fruit about 5 mm. broad. Foothills and sandy places: Alta.—S.D.—Wyo.—Yukon. *Submont.—Mont.* Je–S.

8. **L. cenchrusoìdes** A. Nels. Stem intricately branched, 2–4 dm. high,
strigose; leaves and bracts numerous, small, oblong or ovate, obtuse, 1–2 cm.
long, strigose above, hispid below; back of the nutlets papillose-tuberculate.
Hills: Wyo.—S.D. *Submont.* S.

9. **L. Fremóntii** (Torr.) Greene. Stem erect, strigose, with erect or ascending branches, 2–4 dm. high; leaves 1–3 cm. long, oblong or oblanceolate, strigose above, hispid beneath; bracts smaller, lanceolate, acute; back of the fruit strongly muricate. *E. Fremontii* Torr. Cañons and river banks: S.D.—Wyo.—Calif. *Submont.* Ap–S.

10. **L. occidentàlis** (S. Wats.) Greene. *Fig. 474.* Stem 2–6 dm. high, usually branched above, with ascending branches, more or less hirsute, with spreading hairs; basal leaves spatulate, 2–4 cm. long; stem-leaves linear or oblong, more or less hirsute; corolla blue, 1.5–2 mm. wide; fruit about 3 mm. wide. *E. Redowskii occidentalis* S. Wats. Sandy places and dry plains: Sask.—Man.—Mo.—N.M.—Alaska. *W. Temp.* My–Au.

11. **L. heterospérma** Greene. Stem much branched, 1–3 dm. high, hirsute, green; basal leaves 1–2 cm. long, oblanceolate; stem-leaves lanceolate or oblong, hirsute on both sides; bracts lanceolate, longer than the fruit; corolla blue, 1.5–2 mm. wide; fruit about 5 mm. wide. *L. texana* Britton [B]. *L. heterosperma homosperma* A. Nels., a form with all four nutlets cupulate. River valleys: N.D.—Kans.—Tex.—Ariz.—Mont. *Submont.* My–Au.

f.474.

12. **L. foliòsa** A. Nels. Stem diffusely and profusely branched from the base, 1–2 dm. high; leaves broadly linear, 1–3 cm. long; bracts oblong-lanceolate, longer than the fruit; corolla blue; nutlets nearly alike, papillose, scabrous and distinctly ridged on the face. *L. desertorum foliosa* A. Nels. *Plain —Submont.* Gravelly slopes: N.D.—Neb.—Wyo.—Colo. Je–Jl.

3. ASPERÙGO (Tourn.) L. Catchweed, German Madwort.

Low annuals, with hispid foliage and few axillary flowers on recurved pedicels. Calyx foliaceous, strongly reticulate-veiny, enlarged in fruit; lobes 5, flat, with 5 smaller ones intervening. Corolla with a short tube, enlarged throat, and spreading 5-lobed limb. Stamens 5, included. Nutlets ovoid, granular-tubercled, keeled, laterally attached above the middle to an elongate-conic receptacle.

1. **A. procúmbens** L. Stem slender, 2–4.5 dm. long; leaves oblong or the lower spatulate, 1–4 cm. long; corolla about 2 mm. broad; fruiting calyx 8–12 mm. broad. Waste places: N.Y.—D.C.—Colo.—Mont.; adv. from Eur. My–Au.

4. OREOCÀRYA Greene.

Perennials or at least biennials, with taproots or short rootstocks. Leaves alternate, the earlier in basal rosettes, more or less hispid. Calyx wholly persistent, 5-cleft; the lobes ascending or spreading in fruit. Corolla salverform, white or yellowish; tube often elongate, annulate within near the base; throat closed by conspicuous fornices; limb 5-lobed. Inflorescence thyrsoid or paniculate, leafy-bracted. Nutlets triangular, sometimes somewhat wing-margined, attached for the larger part of their length by a slender scar on the inner angle to a subulate or columnar receptacle.

Fruit depressed; nutlets smooth, at the margins separated by an open space; caulescent cespitose perennials.   1. *O. suffruticosa.*
Fruit conical or ovoid; nutlets touching each other.
  Floral leaves long, many times longer than the flower-clusters; inflorescence spike-like; biennials or short-lived perennials.   2. *O. virgata.*

Floral leaves comparatively short, little if at all surpassing
the flower-clusters.
  Inflorescence a broad, open, round-topped thyrsus;
    branches usually again branched.   3. *O. thyrsiflora.*
  Inflorescence, at least in flowers, narrow, almost spike-
    like, if more open, the branches simple.
    Plants green, very hispid, only the basal leaves
      canescent; bristles with conspicuous pustulate
      bases, much longer than the short strigose or
      tomentose pubescence.
      Leaves linear or narrowly linear-lanceolate; plant
        1–2 dm. high, very slender.   4. *O. Macounii.*
      Leaves, at least the basal ones, broadly oblance-
        olate or spatulate.
        Corolla 7–10 mm. broad; nutlets ovate, more
          than half as broad as long.
          Basal leaves broadly spatulate, usually can-
            escent; stem 1–3 dm. high.   5. *O. glomerata.*
          Basal leaves oblanceolate, greener; stem
            usually 3–4 dm. high.   6. *O. affinis.*
        Corolla 5–7 mm. broad; nutlets lanceolate, less
          than half as broad as long; basal leaves
          oblanceolate, somewhat canescent; plant 1–3
          dm. high.   7. *O. perennis.*
  Plant canescent; basal leaves narrowly linear-oblance-
    olate, glossy, closely appressed-canescent; nutlets
    mamillate-papillose.   8. *O. cana.*

   1. O. suffruticòsa (Torr.) Greene. Stems 1–3 dm. high, decumbent or
ascending at the base, cinereous and hirsute; leaves linear-oblanceolate or linear, 3–10 cm. long; calyx-lobes ovate-lanceolate; corolla 5–6 mm. long; nutlets
more or less tubercled and white-spotted. *Eritrichium Jamesii* Torr. *Krynitzkia
Jamesii* A. Gray. Plains and foot-hills: S.D.—Tex. -N.M.—Utah—Wyo. *Plain
—Submont.* My–Jl.

   2. O. virgàta (Porter) Greene. Stem 3–6 dm. high, hispid with yellowish
hairs; basal leaves oblanceolate, hirsute and hispid, 3–8 cm. long; stem-leaves
linear; corolla 6–7 mm. long and fully as broad; nutlets in age brown or gray;
margins sharp, almost winged. *K. virgata* A. Gray. Dry hills and mountains:
Colo.—Wyo.—w Neb.—S.D. (?). *Submont.—Mont.* My–Jl.

   3. O. thyrsiflòra Greene. Stem branched above, very hispid, 2–4 dm.
high; basal leaves numerous, spatulate or oblanceolate, very hispid, 4–10 cm.
long; inflorescence 1–2 dm. long, 5–8 cm. wide, often yellowish-hispid; corolla
about 5 mm. long, 3 mm. wide; nutlets 3 mm. long. *O. hispidissima* (Torr.)
Rydb. Hillsides and river valleys: Neb.—N.M.—Utah.—Wyo. *Plain—Mont.*
Je–Au.

   4. O. Macoùnii Eastw. Biennial or perennial, with a slender taproot;
stem sparingly hirsute; leaves sparingly hirsute; corolla white, 5 mm. long,
4 mm. wide; nutlets ovate, obtuse, 2 mm. long,
acutely margined, rounded on the back and coarsely
muricate. Plains: Sask.—Man. *Plain.* Jl.

   5. O. glomeràta (Pursh) Greene. *Fig. 475.*
Perennial or biennial; stem erect, 1–3 dm. high; basal
leaf-blades 2–6 cm. long; stem-leaves oblanceolate;
corolla white, 7–8 mm. long, 7–10 mm. wide; nutlets
ovate, 3 mm. long, acute-margined, obtuse, transversely rugose and muricate on the back. *K. glomerata* A. Gray. Dry hills: Sask.—Man.—Neb.—
(? Colo.)—Utah—B.C. *Plain—Submont.* My–Jl.

   6. O. affìnis Greene. Biennial; stems 2–4 dm.
high, hispid; leaves oblanceolate, somewhat grayish-
strigose and hirsute, 3–10 cm. long; stem-leaves
linear or linear-oblanceolate, greener and more hirsute; corolla 7–9 mm. long, 7–10 mm. wide; nutlets
ovate, rounded on the back, coarsely tuberculate on

f.475

the back (rarely somewhat rugose), finely so towards the margin. *Krynitzkia pustulata* Blankinship. Dry hills: S.D.—Colo.—Ida.—Mont. *Plain—Submont.* Je–Jl.

7. **O. perénnis** (A. Nels.) Rydb. Perennial, with a taproot; stem hispid; basal leaves grayish-strigose and hirsute; corolla 5–7 mm. long; nutlets about 3 mm. long, rounded and muricate on the back, not at all cross-ridged. *O. affinis perennis* A. Nels. Dry hills: N.D.—Colo.—Ida. *Submont.* Je–Jl.

8. **O. càna** A. Nels. Stem 1 dm. high or less, silvery-strigose; leaves 1–3 cm. long; corolla about 4 mm. long and 3 mm. wide; nutlets broadly ovate, 2 mm. long. *Krynitzkia sericea* Nutt.; A. Gray, in part. *O. sericea* Greene, in part [B]. Dry hills: w Neb.—S.D. *Submont.* Je–Au.

5. ALLOCÀRYA Greene.

Diffusely branched annuals. Lower leaves opposite, the upper alternate, narrow, more or less hirsute. Flowers in scorpioid racemes, some leafy-bracted, others not, perfect, regular. Calyx wholly persistent, 5-parted; lobes connivent and enclosing the fruit. Corolla salver-shaped, white; tube usually short; throat furnished with more or less conspicuous scales (fornices); limb 5-cleft. Nutlets ovoid or oblong, or somewhat trigonous, obliquely attached by an oblong or rounded scar to the low-conical receptacle. Seeds amphitropous, ascending.

Bracts oblong, scarcely more than twice as long as the flowers; nutlets uncinate-bristly on the back.   1. *A. Nelsonii.*
Bracts (often lacking under some of the flowers) linear and resembling the leaves; nutlets merely rugose on the back.   2. *A. californica.*

1. **A. Nelsònii** Greene. Stem diffuse, branched, strigose, 1–2 dm. high; leaves linear or linear-oblanceolate, obtuse, strigose, 1–3 cm. long; nutlets 1.5 mm. long, transversely rugose, somewhat muriculate. Saline flats and river banks: S.D.—Wyo.—Ida. Jl–Au.

2. **A. califórnica** (F. & M.) Greene. Stem ascending or spreading, branched, strigose, 5–20 cm. high; leaves linear, strigose, 3–5 cm. long; nutlets ovate-lanceolate in outline, muricate and transversely rugose. *A. scopulorum* Greene [B]. Sandy soil: Sask.—S.D.—Neb.—Colo.—Utah—Nev.—Wash. *Son.* —*Mont.* Je–Au.

6. CRYPTÁNTHA Lehm.

Annual caulescent branched herbs. Leaves alternate, narrow, entire. Flowers perfect, regular, in scorpioid spikes or racemes. Calyx usually deciduous with the fruit, 5-cleft; the lobes in fruit connivent around the nutlets. Corolla salver-shaped or funnelform, white; tube short; throat closed by 5 scales (fornices); lobes 5, imbricate. Stamens 5, included; filaments short. Nutlets ovoid, attached for at least the lower half by a narrow, simple or bifurcate groove to the pyramidal receptacle. Seeds more or less amphitropous, pendulous. *Krynitzkia* Fisch. & Mey.

Nutlets, one smooth and three tuberculate or muricate.
  Sepals in fruit strongly thickened on the back.   1. *C. crassisepala.*
  Sepals in fruit not strongly thickened on the back.   2. *C. Kelseyana.*
Nutlets all smooth.
  Groove of the nutlets simple, continuous to the base.   3. *C. confusa.*
  Groove two-forked at the base.
    No open areola between the forks of the grooves; nutlets acute.   4. *C. calycosa.*
    A distinct open areola between the forks of the groove; nutlets lanceolate, acuminate.
      Leaves spatulate or oblanceolate, or the upper sometimes linear.   5. *C. Pattersoni.*
      Leaves narrowly linear.   6. *C. Fendleri.*

1. **C. crassisépala** (T. & G.) Greene. Stem diffusely branched, 1–2 dm. high, hispid; leaves oblanceolate or linear-spatulate, 1–3 cm. long; calyx-lobes

linear, 5 mm. long, in fruit 8–10 mm. long; corolla 5 mm. long; limb 1 mm. wide; groove of the nutlets open, forming an open lance-ovate areola below. *Krynitzkia crassisepala* A. Gray. Loose soil, ''prairie-dog towns,'' and waste places: Sask.—Tex.—Utah—Alta.; n Mex. *Plain—Submont.* My–Jl.

2. **C. Kelseyàna** Greene. Stem branched, 1–2 dm. high, hispid; leaves linear-oblanceolate, obtuse, 3–5 cm. long, hispid; calyx-lobes subulate, 3–4 mm. long, in fruit 6–8 mm.; corolla 4 mm. long; limb scarcely 1 mm. wide; nutlets lanceolate; groove narrow, opening only at the base into a small areola. Dry sandy soil: Sask.—Colo.—Mont. *Plain— Submont.* Je–Jl.

3. **C. confùsa** Rydb. Stem rather stout, hispid, branched; leaves 2–5 cm. long, 3–8 mm. wide, coarsely hirsute; calyx-lobes 2 mm. long, lance-subulate, hispid, in fruit 3–4 mm. long; corolla about 3 mm. long; limb scarcely 1 mm. wide; nutlets ovate, 2 mm. long, light-colored, thin-walled, attached by the lower half or two thirds. River valleys and open woods: B.C.—Mont.—S.D. Wyo.—Utah. *Plain—Submont* Je–Au.

4. **C. calycòsa** (Torr.) Rydb. Stem branched, 2–4 dm. high, hispid; leaves oblong or oblanceolate, 2–4 cm. long, 4–8 mm. wide, hispid; racemes short, but elongating in fruit; calyx-lobes about 3 mm. long, in fruit 6–7 mm., subulate, hispid; corolla 3–4 mm. long, about 1 mm. broad; nutlets 2.5 mm. long. *C. flexuosa* A. Nels. *K. Torreyana calycosa* A. Gray. Stony and sandy places: Mont.—S.D.—Wyo.—Nev.—Wash. *Submont.* Jl–Au.

5. **C. Pattersòni** (A. Gray) Greene. Stem branched at the base, 1–2 dm. high, hirsute; leaves 2–5 cm. long, 3–5 mm. wide, hispid; calyx-lobes 2.5 mm. long, in age 4 mm. long; corolla 4 mm. long, 1–1.15 mm. wide; nutlets lanceolate, more or less mottled, attached by the lower half. *K. Pattersoni* A. Gray. Hillsides and dry places: S.D.—Colo.—Wyo.—Ida. *Submont.—Mont.* My–Jl.

6. **C. Féndleri** (A. Gray) Greene. Stem erect, hirsute with ascending hairs; calyx-lobes linear, 2–3 mm. long, very hispid; corolla 3 mm. long; limb scarcely 1 mm. wide; nutlets ovate, acuminate, mottled, attached by their whole length. *K. Fendleri* A. Gray. River valleys and sandy places: Sask.—Neb.—N.M.—Ariz.—Mont. *Plain—Submont.* Je–Au.

7. **AMSÍNCKIA** Lehm. FIDDLE NECK, BUCKTHORN WEED.

Rough-hispid biennials. Leaves alternate, linear or oblong to ovate, entire. Flowers perfect, regular, in scorpioid spikes, leafy-bracted. Calyx persistent, 5-lobed, or some of the lobes united to near the apex. Corolla salver-shaped, yellow; tube mostly rather elongate; throat in some flowers with and in others without fornices, in the latter the stamens inserted high up, in the former low down in the tube; limb often plicate in the sinuses. Nutlets bony, more or less triangular, attached below the middle to the pyramidal receptacle.

1. **A. idahoénsis** M. E. Jones. Stem 3–7 dm. high, white-hirsute; leaves lanceolate, 4–7 cm. long, the lower tapering at the base, the upper sessile; racemes 1–2 dm. long, rather lax; sepals lance-linear, in fruit 7–9 mm. long, densely hispid; corolla pale yellow, about 3 mm. long; nutlets broadly lanceolate, slightly keeled on the back, densely and strongly papillose, somewhat closely cross-ribbed. Along railroads: N.D.—Minn.—Iowa; adventive from the West. Jl.

8. **MERTÉNSIA** Roth. LUNGWORT, BLUEBELL, LANGUID LADY.

Perennial caulescent herbs. Leaves alternate, entire. Flowers perfect, regular, in racemes or panicles. Calyx deeply 5-cleft, persistent, essentially unchanged in age. Corolla tubular-funnelform or trumpet-shaped, blue or purple, rarely white; throat rarely appendaged; lobes 5, imbricate. Stamens included or barely exserted; filaments often flattened. Nutlets obliquely attached by a small scar near the base to the convex receptacle, wrinkled at maturity. Style filiform; stigma entire.

Corolla trumpet-shaped, with spreading, nearly entire limb and naked throat.
    1. *M. virginica.*
Corolla with cylindric tube and funnelform throat, which is crested
  within.
    Tube of the corolla at least twice as long as throat and limb. 2. *M. foliosa.*
    Tube of the corolla only slightly if at all longer than the
      throat and limb.
      Plant tall, usually more than 4 dm. high ; leaves distinctly
        veined, in most feather-veined, but in a few triple-
        veined with anastomosing veins.
        Stem, calyx-tube, and lower surface of the leaves long-
          hairy.
          Corolla about 1.5 cm. long ; stem densely pubescent. 3. *M. pilosa.*
          Corolla about 1 cm. long ; stem sparingly pubescent. 4. *M. paniculata.*
        Stem glabrous ; calyx strigose, or ciliate on the margins. 5. *M. coronata.*
      Plant low, usually less than 4 dm. high ; leaves with a
        strong midrib, without distinct lateral veins (except
        sometimes the basal leaves, which then however are
        thick and firm).
        Upper surface of the leaves short-pubescent. 6. *M. linearis.*
        Upper surface of the leaves not hairy. 7. *M. lanceolata.*

1. **M. virgínica** (L.) Pers. Stem clustered, 2–6 dm. high; leaves 5–20 cm. long, oblong-elliptic or oval, glaucous, the lower with winged petioles, the upper sessile; panicle short; calyx 2–2.5 mm. long, the lobes oblong or lanceolate, longer than the tube; corolla blue, rarely white, 2.5–3 cm. long, the limb 1–1.5 cm. broad. Banks: Ont.—N.Y.—Minn.—Kans.—Ga. Mr–My.

2. **M. foliòsa** A. Nels. Stem erect, glabrous, 2–4 dm. high; leaves thick, glabrous or slightly pustulate above, the lower ones oblong or spatulate, 4–7 cm. long, petioled; stem-leaves lanceolate or the upper ones ovate; calyx-lobes lanceolate, ciliolate; corolla 15–18 mm. long. Hills and cañons: N.D.—S.D.—Wyo. —Utah. *Submont.* My–Je.

3. **M. pilòsa** (Cham.) DC. Stem simple, 3–4 dm. high, villous; leaves lanceolate, 5–8 cm. long, acuminate at each end, villous-hirsute beneath, strigose above; pedicels and calyx canescent-strigose; lobes linear-lanceolate, 5 mm. long, strigose; corolla-tube equaling the limb. *Pulmonaria pilosa* Cham. *Lithospermum corymbosum* Hook. Wet places: Man.—Alta.—Alaska. *Boreal.*

4. **M. paniculàta** (Ait.) Don. Stem more or less hirsute, 3–6 dm. high; leaves lanceolate, acuminate at each end, 5–10 cm. long, hirsute on both sides; pedicels and calyx strigose; lobes of the latter 4 mm. long; corolla-tube 4–5 mm. long, about equaling the throat and limb. *Pulmonaria paniculata* Ait. Along streams and shady banks: Que.—Minn.—Iowa—Ida.—Wash.—Alaska. *Boreal.—Mont.* My–Jl.

5. **M. coronàta** A. Nels. Stem glabrous, shining, assurgent, 2–4 dm. high; leaves scabrous-hispidulous, or nearly pustulate above, the basal ones petioled; blades oblong, obtuse, 5–10 cm. long; upper stem-leaves ovate-lanceolate; calyx-lobes lanceolate, ciliate-margined; corolla-tube 5 mm. long, a little longer than the limb. Rocky hills: Ida.—Wyo. —S.D. *Submont.* Je.

6. **M. lineàris** Greene. Stems 2–4 dm. high, leafy, glabrous; leaves narrowly linear, 2–7 cm. long, roughly strigillose above, glabrous beneath; pedicels glabrous or pustulate; calyx-lobes lanceolate, ciliolate on the margins, otherwise glabrous; corolla about 1 cm. long; tubes about equaling the throat and limb. Dry hillsides: Sask.—Neb.—Colo.—Alta. *Plain—Mont.* My–Jl.

f. 476.

7. **M. lanceolàta** (Pursh) DC. *Fig. 476*. Stem 2–4 dm. high, glabrous; leaves glabrous on both sides or merely pustulate above, the lower petioled, 3–10 cm. long; stem-leaves oblong or lanceolate; pedicels pustulate-muricate; calyx-lobes lanceolate, ciliolate on the margins, otherwise glabrous; corolla about 3 cm. long; tube equaling the limb. *M. coriacea* A. Nels. *M. papillosa* Greene. Hills: Sask.—N.D.—Colo.—Ida.—B.C. *Submont.—Subalp.* Ap–Au.

9. MYOSÒTIS (Rupp.) L. Forget-me-not.

Annual, biennial, or perennial caulescent herbs. Leaves alternate, entire, narrow. Flowers perfect, regular, in 1-sided racemes. Calyx persistent, 5-lobed; lobes erect or spreading. Corolla salverform, blue, pink, or white, often with an eye; tube short; throat with transverse crests; lobes rounded, convolute. Stamens included. Nutlets small, ovoid, smooth, attached by the very base to the flat receptacle.

Perennials, tufted; corolla 4–6 mm. wide.   1. *M. alpestris.*
Annuals; corolla less than 2 mm. wide.
  Fruiting pedicels longer than the calyces.   2. *M. arvensis.*
  Fruiting pedicels not longer than the calyces.
    Stem tall, 3–5 dm. high; calyx nodding in fruit, 5–7
      mm. long; nutlets about 2 mm. long.   3. *M. macrosperma.*
    Stem lower, 1.5–3 dm. high; calyx not nodding in fruit-
      ing, less than 5 mm. long; nutlets about 1.5 mm. long.   4. *M. virginica.*

1. **M. alpéstris** F. W. Schmidt. Stem 1–3 dm. high, hirsute; basal leaves spatulate or oblanceolate, 3–7 cm. long, hirsute; stem-leaves linear-lanceolate; calyx-lobes lanceolate, erect in fruit; corolla bright blue, 4–6 mm. wide. Wet places in the mountains: Alta.—S.D.—Colo.—Alaska; Eurasia. *Mont.—Subalp.* My–Au.

2. **M. arvénsis** (L.) Willd. Stem hirsute, erect, 1–3 dm. high; leaves oblong or oblong-lanceolate, 1–5 cm. long, entire, sessile or nearly so; racemes loosely flowered, elongate; pedicels spreading; calyx bristly, the lobes lanceolate, 2–3 mm. long; corolla usually white, 2–3 mm. broad. Waste places: N.B.—Minn.—W.Va.—N.C.; nat. from Eur. Je–Au.

3. **M. macrospérma** Engelm. Stem 3–5 dm. high, hirsute with ascending hairs, leafy; basal leaves oblanceolate, 2–8 cm. long, sparingly hirsute; stem-leaves lanceolate or oblong; racemes 1–2 dm. long; calyx-lobes 2.5 mm., in fruit 6–7 mm. long; corolla light blue or white, 3 mm. long. River banks and open woods: D.C.—Fla.—Tex.—Calif.—B.C.—Ida. *E. Temp.—Plain— Submont.* Ap–Je.

4. **M. virgínica** (L.) B.S.P. Stem branched from the base, 0.5–3 dm high, hirsute; basal leaves spatulate, 1–3 cm. long; stem-leaves linear-oblong; corolla white, very small. Dry banks and rocky woods: Me.—Ga.—Tex.—Ore. —B.C. *Temp.* Ap–Je.

10. LITHOSPÉRMUM (Tourn.) L. Gromwell, Puccoon, Indian Paint.

Annual or perennial caulescent herbs. Leaves alternate, entire, narrow, hirsute. Flowers perfect, regular, in bracted spikes or racemes. Calyx persistent, 5-lobed. Corolla salverform or funnelform, white or yellow; tube often elongate; throat either appendaged or pubescent; lobes 5, entire, erose, or toothed. Stamens included. Stigma capitate or 2-lobed. Nutlets white, shining and smooth, or brown and wrinkled, attached by their very bases to the flat receptacle.

Annuals; nutlets dull, brownish, coarsely wrinkled and pitted.   1. *L. arvense.*
Perennials, with thick taproots; nutlets white and shining.
  Corolla-lobes neither fimbriate nor toothed; flowers all well
    developed.

Corolla greenish or pale yellow, 1 cm. or less long; plant
  very leafy.
Fornices in the throat of the corolla obsolete; flowers
  close together on short branches.   2. *L. ruderale.*
Fornices in the throat of the corolla prominent; flow-
  ers scattered.
Leaves lanceolate, acute; nutlets ovoid; corolla
  longer than the calyx.   3. *L. officinale.*
Leaves ovate, acuminate; nutlets globose-ovoid;
  corolla shorter than the calyx.   4. *L. latifolium.*
Corolla bright yellow or orange, more than 1 cm. long;
  crest of the tube prominent.
Hispid-pubescent; corolla-tube bearded at the base
  within.   5. *L. Gmelini.*
Canescent; corolla-tube not bearded at the base
  within.   6. *L. canescens.*
Corolla-lobes of the early flowers fimbriate or dentate; later
  flowers cleistogamous.
Corolla of the early flowers 2.5 cm. (rarely only 2 cm.)
  long; lobes distinctly fimbriate.   7. *L. linearifolium.*
Corolla of the early flowers 2 cm. or less long; lobes den-
  tate.
Corolla 15–20 mm. long; limb 8–10 mm. wide; stem
  low, ascending or decumbent.   8. *L. mandanense.*
Corolla 10 mm. or less long; limb 6–8 mm. wide; stem
  strict, 2–4 dm. high, mostly simple, or with erect
  branches.   9. *L. breviflorum.*

1. **L. arvénse** L. Stem 2–5 dm. high, strigose, simple or branched at the
base; leaves lanceolate, linear, or oblong, sessile or the lower short-petioled,
canescent-strigose; corolla white, about 6 mm. long; calyx equaling or exceeding the corolla-tube; nutlets 2 mm. long. Waste places and fields: Que.—Ga.—
Kans.—Minn.; Utah; Mont.; nat. from Eur. My–Au.

2. **L. ruderàle** Lehm. Stem 3–5 dm. high; leaves 3–8 cm. long, canescent, pilose, or strigose; corolla 10–12 mm. long; nutlets slightly wrinkled, 5
mm. long. *L. pilosum* Nutt. [B]. Hills: Alta.—S.D.—Colo.—Nev.—B.C. *Plain*
—Submont. Ap–Jl.

3. **L. officinàle** L. Stem 5–12 dm. high, finely puberulent; leaves numerous, sessile, 3–10 cm. long, rough above; corolla yellowish white, 4 mm. long,
funnelform, crested in the throat; nutlets 3 mm. long, ovoid, obtuse, seldom
all ripening. Waste places: Que.—Minn.—Kans.—N.J.; nat. from Eur. My–
Au.

4. **L. latifòlium** Michx. Stem 6–9 dm. high, rough-pubescent; leaves
5–13 cm. long, undulate, 4–8-ribbed, short-petioled; calyx strigillose, the lobes
linear; corolla yellowish white, 5 mm. long, funnelform, the limb 6–7 mm.
broad; nutlets ovoid, 3–3.5 mm. long, white, shining, sparingly pitted. Dry
soil: Ont.—Minn.—Kans.—Va. *Canad.—Allegh.* My.

5. **L. Gmelìni** (Michx.) Hitchc. Stem hispid, 3–
6 dm. high; leaves lanceolate, or oblong, hispid, 4–8
cm. long; corolla 15–20 mm. long; limb 15–20 mm.
wide; nutlets 4 mm. long. *L. hirtum* Lehm. Sandy
places, dry plains, and open woods: N.Y.—Fla.—
N.M.—Mont. *E. Temp.—Plain—Submont.* Ap–Jl.

6. **L. canéscens** (Michx.) Lehm. *Fig. 477.*
Stem 2–4 dm. high, hirsute or pilose; leaves oblong
to linear, 1–4 cm. long, canescent-strigose; corolla
10–12 mm. long; limb about 1 cm. wide; nutlets 3
mm. long. Plains and prairies: Ont.—Ala.—Tex.—
Colo.—Sask. *E. Temp.—Plain.* Ap–Jl.

7. **L. linearifòlium** Goldie. Stems strict, 1–5
dm. high, canescent-strigose; leaves all linear, 1–5
cm. long, strigose; corolla of the earlier flowers
2.5–3 cm. long; limb 10–18 mm. wide, erose-fimbri-

f.477.

ate; cleistogamous flowers 1 cm. long or less; nutlets more or less pitted. *L. angustifolium* Michx. [G]. Dry soil: Man.—Ill.—Tex.—Ariz.—B.C. *Prairie—Plain—Submont.* Ap–Jl.

8. **L. mandanénse** Spreng. Stems many, 1–2 dm. high; leaves linear, 3–4 cm. long, mostly strigose; corolla of the early flowers 1.5–2 cm. long; nutlets sparingly pitted. Dry plains: N.D.—Tex.—Ariz.—Alta. *Plain—Submont.* Ap–Jc.

9. **L. breviflòrum** Engelm. & Gray. Stems several, erect, 2–4 dm. high, finely strigose; leaves narrowly linear, 3–5 cm. long, ascending; corollas of the early flowers 1 cm. long or less, deep yellow; nutlets scarcely pitted except near the margins. *L. albicans* Greene. Dry plains: Kans.—Ark.—Tex.—N.M.—Colo. *Austral.—Son.* My–Je.

11. ONOSMÒDIUM Michx. FALSE GROMWELL.

Coarse perennial caulescent herbs. Leaves alternate, entire, usually broad, strongly nerved, pubescent. Flowers perfect, regular, in terminal scorpioid racemes or spikes. Calyx persistent, with 5 narrow lobes. Corolla tubular, yellow, white, or greenish; tube elongate; throat without appendages; lobes erect. Stamens included; anthers often sessile. Nutlets white, shiny, smooth or sparingly pitted, attached by their very base to the flat receptacle, often only one or two maturing.

Nutlets rounded, not at all constricted at the base; pubescence of the leaves mostly appressed.   1. *O. occidentale.*
Nutlets with a short constriction or neck; pubescence mostly spreading.   2. *O. hispidissimum.*

1. **O. occidentàle** Mackenzie. *Fig. 478.* Stem 2–6 dm. (rarely 6–12 dm.) high, branched above, white-hispid; leaves about 5 cm. long, prominently 5–7-ribbed, coarsely hirsute-strigose; corolla 12–20 mm. long, canescent, its lobes broadly triangular, acute, 3–4 mm. long; nutlets 3.5–4 mm. long, ovoid, not constricted at the base, dull, little if at all pitted. *O. molle* Britt., in part [B, G]. Plains and prairies: Man.—Ill.—Mo.—Tex.—N.M.—Utah—Alta. *Prairie—Plain—Submont.*

2. **O. hispidíssimum** Mackenzie. Stem stout, 10–12 dm. high, green, coarsely hirsute throughout; stem-leaves ovate, 8–10 cm. long; corolla about 1 cm. long; nutlets about 3 mm. long, brownish. *O. carolinianum* Britton, in part [B]. River banks: N.Y.—Minn.—Neb.—Tex.—Ga. *Canad.—Alleyh.* My Jl.

f. 478.

12. BORÀGO L. BORAGE.

Pubescent annuals or biennials. Leaves alternate, entire. Flowers in leafy racemes. Calyx deeply 5-cleft. Corolla rotate, the tubes very short, the throat closed by fornices; the limb 5-lobed. Stamens 5; filaments short-lobed, the anthers connivent. Nutlets 4, ovoid, attached by the bases, the scar concave.

1. **B. officinàlis** L. Stem erect, 3–7 dm. high; leaves ovate or oblong, 5–12 cm. long, the lower narrowed below into winged petioles; flowers 15–20 mm. broad; calyx-lobes lanceolate; corolla bright blue, the lobes ovate-lanceolate. Waste places: N.S.—Pa.—Iowa; adv. from Eur. Je–S.

13. ANCHÙSA L. ALKANET.

Annual or perennial herbs, with hispid or villous foliage. Flowers bracteate, in scorpioid cymes. Calyx 5-cleft. Corolla blue or purple; tube straight; throat closed by fornices or hairs. Nutlets rugose or granular, inserted by a broad concave base on the flat receptacle.

1. A. officinàlis L. Perennial, with a taproot; stem hispid, 3–8 dm. high; basal leaves oblanceolate, 1–2 dm. long, very hispid; stem-leaves lanceolate; corolla dark blue, fully 1 cm. long. Waste places and roadsides: N.Y.—N.D.—S.D.—Utah; sparingly introduced from Eur. Jl–S.

14. SÝMPHYTUM (Tourn.) L. COMFREY.

Large hairy perennials. Leaves alternate, entire, or the uppermost opposite; the lower long-petioled. Flowers in terminal scorpioid racemes. Calyx deeply 5-cleft. Corolla blue, purple, or yellow, campanulate, with short lobes; fornices in the throat lanceolate or linear, papillose-margined. Nutlets obliquely ovoid; scar large, concave, bordered by a prominent denticulate ring-margin.

1. S. officinàle L. Stem 6–10 dm. high; leaves lanceolate to ovate, the lower petioled, the upper decurrent on the stem; calyx-lobes lance-linear; corolla purplish or yellowish, 1–2 cm. long. Waste places: Newf.—Md.—Mont.; adv. or escaped from cultivation; native of Eur. Je–Au.

15. LYCÓPSIS L. BUGLOSS.

Annual caulescent herbs. Leaves alternate, bristly-hispid, narrow, entire. Flowers in terminal scorpioid racemes. Sepals 5, nearly distinct. Corolla blue, salverform, irregular, the tube curved, the throat clothed with hispid hairs, the limb unequally 5-lobed. Stamens 5, filaments adnate to the middle of the corolla-tube. Nutlets 4, erect on the flat receptacle, wrinkled, the scar of attachment concave.

1. L. arvénsis L. Stem often diffusely branched, 3–7 dm. high, often procumbent; lower leaves oblanceolate, the upper oblong or lanceolate, 2.5–10 cm. long, undulate; calyx becoming 7–8 mm. long; sepals linear, acute; corolla 5–7 mm. long, blue; nutlets 3 mm. long, coarsely wrinkled. Waste places: N.S.—Minn.—Neb.—Va.; nat. from Eur. Je–Jl.

16. ÈCHIUM L. BLUEWEED, VIPER'S BUGLOSS.

Biennial or perennial caulescent herbs. Leaves alternate, entire, pubescent. Flowers in scorpioid racemes or spikes. Calyx-lobes 5, narrow. Corolla white, blue, or violet, tubular-funnelform, irregular; throat open, unappendaged, the limb with 5 unequal, rounded lobes. Stamens 5, exserted, the filaments unequally long, dilated at the base. Nutlets wrinkled, attached to the flat receptacle by a flat scar.

1. E. vulgàre L. Stem 3–7 dm. high, simple; leaves linear or linear-lanceolate, 2–15 cm. long, bristly on both sides; corolla pale blue or purplish, rarely white, 1.5–1.8 cm. long, obliquely campanulate, the lobes ovate-triangular, ciliolate. Waste places: N.B.—Minn.—Kans.—Ga.; nat. from Eur. Je–Jl.

Family 144. VERBENACEAE. VERVAIN FAMILY.

Annual or perennial herbs, or in warmer climates sometimes shrubs or trees. Leaves usually alternate, rarely opposite or whorled, usually simple. Flowers perfect, more or less irregular, zygomorphic, in ours spicate. Calyx of 4 or 5 more or less united sepals, usually bracteolate. Corolla of 4 or 5 partially united petals, usually 2-lipped. Stamens 2, or 4, and then didynamous; filaments more or less adnate to the corolla; anthers erect or incumbent. Gynoecium of 2 or 4 carpels; styles united; stigma entire or 2- or 4-lobed; ovules 1 or 2 in each cavity, anatropous. Fruit drupaceous or berry-like, or separating into 2 or 4 nutlets. Seeds commonly solitary in each cavity; endosperm scant or wanting; embryo straight.

Corolla-limb 5-lobed ; nutlets 4 ; flowers in terminal spikes.
Corolla 4-lobed ; nutlets 2 ; flowers in short dense axillary spikes.

1. VERBENA.
2. PHYLA.

1. VERBÈNA (Tourn.) L. VERVAIN, VERBENA.

Annual or perennial herbs. Leaves opposite, entire, toothed, or dissected.
Flowers in terminal spikes. Calyx-tube funnelform. Corolla more or less bilabiate; tube narrow, straight or curved. Limb 5-lobed. Stamens 4, didynamous, rarely only 2; connective of the anthers often bearing a gland. Ovary 4-celled; stigma 2-lobed, only one lobe stigmatose; ovules solitary in each cell. Fruit of 4 nutlets, enclosed in the calyx.

Anthers not appendaged ; flowers in elongate spikes, less than 8 mm. long.
  Spike very slender and interrupted at maturity.
    Leaves once or twice pinnately cleft or incised.   1. *V. officinalis*.
    Leaves merely toothed.   2. *V. urticifolia*.
  Spike stout, continuous.
    Leaves not pinnatifid, sometimes merely lobed at the
      base in *V. hastata* ; plant tall, erect, strict.
    Spike peduncled ; corolla-limb 3–6 mm. broad ; pubescence sparse, coarse.
      Leaves linear, sessile.   3. *V. angustifolia*.
      Leaves lanceolate, ovate, or hastate, petioled.   4. *V. hastata*.
    Spike sessile ; corolla-limb 8–9 mm. broad ; pubescence dense, soft.
         5. *V. stricta*.
    Leaves more or less pinnatifid, at least incised ; bracts
      much longer than the calyx.   6. *V. bracteosa*.
Anthers of the longer stamens appendaged by a gland on the
  connective ; flowers 8 mm. long or more.
  Corolla-tube about twice as long as the calyx, the limb
    12–15 mm. wide.
    Leaf-blades incised or incised-lobed.   7. *V. canadensis*.
    Leaf-blades pinnatifid or bipinnatifid.   8. *V. Drummondii*.
  Corolla-tube about half longer than the calyx.
    Bracts narrowly lanceolate, shorter than the calyx ;
      leaf-segments oblanceolate.   9. *V. ambrosifolia*.
    Bracts setaceous, longer than the calyx ; leaf-segments
      linear.   10. *V. bipinnatifida*.

1. **V. officinàlis** L. Annual; stem 3–6 dm. high, branching, nearly glabrous; leaves 2–6 cm. long, ovate to oblong or lanceolate in outline, once to twice pinnatifid, more or less incised; calyx minutely puberulent, becoming 2 mm. long; corolla blue or purple, the tube 3 mm. long, the limb 4 mm. wide. Waste places and dry soil: Me.—Minn.—Tex.—Fla.; nat. from Eur. Je–S.

2. **V. urticifòlia** L. Annual or perennial; stem 4–15 dm. high, sparingly hirsute or glabrous; leaf-blades broadly lanceolate or ovate, 8–20 cm. long, coarsely and doubly crenate-serrate; calyx 2 mm. long; corolla white, 4 mm. long, the tube curved, pubescent in the throat, the limb 2 mm. wide. Roadsides and waste places: N.B.—S.D.—Tex.—Fla. *E. Temp.* Je–S.

V. hastata × urticifolia. Resembling in habit and the small flowers mostly *V. urticifolia*, but the spikes are denser, the corolla purplish, and the toothing of the leaves is coarser and often double. Mich.—Mo.—Kans.

V. stricta × urticifolia. In habit resembling *V. urticifolia*, but the spike denser, flowers larger, leaves broader and coarsely long-pubescent. Ill.—Iowa—Ark.

3. **V. angustifòlia** Michx. Perennial, puberulent or sparingly pubescent; stem 1–5 dm. high, with erect branches; leaves linear to linear-spatulate, 2–10 cm. long, distantly serrate; calyx nearly glabrous, becoming 4 mm. long; corolla blue, pubescent, about 6 mm. long, the limb 5–6 mm. wide. Dry soil: Mass.—Minn.—Okla.—Miss.—Fla. *Allegh.—Austral.* Je–Au.

V. angustifolia × hastata. In the general habit and the slender spike resembling *V. angustifolia*, but the leaves are generally broader, more acute, and with coarser pubescence and larger teeth. Vt.—Minn.

4. **V. hastàta** L. *Fig. 479.* Perennial; stem 4–15 dm. high; hispidulous, branched above, 4-angled; leaf-blades scabrous-hispidulous, lanceolate or ovate-

lanceolate, 4–12 cm. long, sharply and doubly serrate, often hastate at the base; spikes erect, narrow, 5–15 cm. long; calyx 2.5 mm. long, pubescent;
corolla purplish blue; nutlets smooth. River valleys,
thickets, and waste places: N.S.—Fla.—N.M.—Calif.
—B.C. *Temp.—Submont.* My–Au.

V. hastata × stricta. In habit resembling *V. hastata,*
but the leaves more veiny-reticulate and softer pubescent and the corolla larger. Ind.—Mo.—Colo.

5. **V. strícta** Vent. Perennial; stem stout,
3–12 dm. high, simple or branched above, densely
pilose, almost velutinous, terete or nearly so; leaf-
blades oblong, oval, 3–10 cm. long, doubly serrate or
incised, rugose, nearly sessile, densely soft-pubescent;
spike stout, strict; calyx hirsute, 5 mm. long; corolla dark blue; nutlets strongly reticulate. Dry soils,
pastures, and river valleys: Minn.—Ky.—Tex.—N.M.—Wash.; introduced eastward to Conn. *Prairie—W. Temp.* Je–Au.

V. angustifolia × stricta. In habit more like *V. stricta,* but the spike more slender, the flowers smaller, and the leaves narrower. Ind.—Mo.—Kans.

6. **V. bracteòsa** Michx. Annual or perennial; stem branched at the base,
prostrate or procumbent, 1–5 dm. long, diffusely branched, angled, more or less
hirsute; leaf-blades spatulate in outline, pinnately lobed and incised or double-
toothed, 1–6 cm. long, prominently veined beneath, hirsute; bracts lanceolate
or linear-lanceolate; calyx bristly, 3 mm. long; corolla purple or bluish; limb
2 mm. broad; nutlets strongly reticulate on the back. *V. rudis* Greene. Waste
places, prairies and plains: Ont.—Ill.—Fla.—Tex.—Calif.—B.C. *Prairie—Plain
W. Temp.—Submont.*

V. bracteosa × urticifolia. Like *V. bracteosa* in habit, but the spikes are long
and slender, and the bracts smaller. Ill.—Mo.—Kans.—Minn.

V. bracteosa × hastata. Resembling most *V. bracteosa* in habit, but stouter, more
erect, with broad, laciniate rather than dissected leaves and shorter bracts. Ill.—
Mo.—Neb.—Colo.

V. bracteosa × stricta. Resembling *V. bracteosa* in habit, but stouter, with broad,
laciniate, strongly reticulate leaves and larger flowers. Ill.—Mo.—Kans.—Neb.

7. **V. canadénsis** (L.) Britton. Perennial; stem 3–6 dm. long, minutely
pubescent; leaves 3–8 cm. long, ovate in outline, rounded at the apex, truncate
at the base, incised; calyx becoming 12–14 mm. long, the lobes setaceous-
tipped; corolla 2–2.5 cm. long, the limb 12–15 mm. wide. *V. Aubletia* Jacq.
Dry soil: Ill.—Kans.—N.M.—Fla.; Mex. *Austral.—Trop.* My–Au.

8. **V. Drummóndii** (Lindl.) Baxter. Perennial; stem 2–4 dm. long, often
branched at the base, hirsute; leaf-blades 2–5 cm. long, pinnatifid or bipinnatifid, with narrow segments; calyx becoming 9–10 mm. long, the lobes seta-
ceous-tipped; corolla fully 2 cm. long, the limb 11–12 mm. wide. Prairies and
bottom-lands: Va.—Kans.—Tex.—Fla. *Austral.* My–Au.

9. **V. ambrosifòlia** Rydb. Perennial; stem branched and decumbent at
the base, sparingly hirsute, 2–4 dm. long; leaf-blades obovate in outline, bipinnatifid or biternately divided, with oblanceolate divisions, sparingly hirsute;
calyx becoming 8–9 mm. long; corolla-limb 6–8 mm. wide; nutlets coarsely
wrinkled all over the back. Plains and stony soil: S.D.—Tex.—Ariz. *Plain—
Son.—Submont.* Je–Au.

10. **V. bipinnatífida** Nutt. Perennial; stems often branched at the base,
1–4 dm. high, decumbent at the base; leaf-blades 2–5 cm. long, bipinnatifid;
segments linear; calyx becoming 9–10 mm. long; corolla-limb 7–9 mm. wide;

nutlets 3 mm. long, prominently wrinkled above. Plains and dry ground: S.D.—La.—Tex. *Son.—Plain.* My–Jl.

2. PHȲLA Lour.

Perennial, caulescent, prostrate or creeping herbs. Leaves opposite, peti-oled, toothed or lobed. Flowers perfect, in dense, sometimes head-like, pe-duncled axillary spikes, conspicuously bracted. Calyx-tube flattened and 2-keeled. Corolla 2-lipped; tube curved; limb 4-lobed, oblique. Stamens 4, didynamous; connective of the anthers not appendaged. Ovary 2-celled; stigma oblique; ovules solitary in each cell. Fruit tardily separating into two nutlets.

Leaves from linear-oblanceolate to cuneate, 2–8-toothed above; peduncles slightly if
  at all exceeding the leaves.   **1.** *P. cuneifolia.*
Leaves lanceolate, oblong, or oval, sharply serrate; peduncles much
  exceeding the leaves.   **2.** *P. lanceolata.*

1. **P. cuneifòlia** (Torr.) Greene. *Fig. 480.*
Stems branched at the base, procumbent or creeping, 2–10 dm. long; leaves opposite and more or less fascicled in the axils; blades 1–3 cm. long; midvein prominent, the lateral veins obsolete; corolla purplish or whitish, 4 mm. long. *Lippia cuneifolia* (Torr.) Steud. [G, B]. Plains and prairies: S.D.—Tex.—Ariz.—Wyo. *Prairie—Plain—Submont.* Je–Au.

2. **P. lanceolàta** (Michx.) Greene. Stem branched from the base, procumbent, 1–4 dm. long; leaves opposite, rarely fasciculate; blades ellliptic-lanceolate, 1.5–5 cm. long, acute, with about 4 pairs of lateral veins; corolla pale blue, 2–2.5 mm. long. *L. lanceolata* Michx. [G, B]. Valleys: Ont.—Minn.—Fla.—Tex.—Neb. *E. Temp.—Plain—Son.* Je–O.

Fig. 480

Family 145. LAMIACEAE. MINT FAMILY.

Aromatic herbs or shrubs, with 4-angled stems. Leaves opposite or whorled, simple, usually toothed or lobed, glandular-punctate. Flowers perfect, irregular, zygomorphic, or nearly regular, pseudo-verticillate, *i. e.*, in congested axillary reduced cymes. Calyx of 5 united sepals, sometimes regular, sometimes irregular, and then usually 2-lipped. Corolla bilabiate, or nearly regular. Stamens 4, didynamous, or one of the pairs abortive; anthers 2-celled. Gynoecium of united carpels; ovary 4-lobed and 4-celled; styles united, arising between the lobes; stigma terminal; ovules solitary in each cell. Fruit of 4 nutlets, included in the persistent calyx. Endo-sperm fleshy or wanting. *Labiatae.*

Ovary of 4 united nutlets; style not basal; nutlets laterally attached.
  Corolla strongly bilateral, the upper lip very small, the lower large.
    Flowers in small congested cymes, axillary to small bracts, and forming a
      raceme-like panicle; calyx-lobes shorter than the tube; leaves toothed.
                               1. TEUCRIUM.
    Flowers solitary in the axils of bracts similar to the leaves; calyx-lobes longer
      than the tube; leaves laciniate.   2. MELOSMON.
  Corolla almost regularly 5-lobed; stamens short-exserted.   3. ISANTHUS.
Ovary of 4 distinct or nearly distinct nutlets; styles basal; nutlets basally attached.
  Corolla bilabiate.
    Calyx 2-lipped; both lips entire; stamens 4.   4. SCUTELLARIA.
    Calyx either 2-lipped and at least one of the lips toothed, or regularly 4–5-
      toothed.
      Stamens included in the corolla-tube.   5. MARRUBIUM.
      Stamens exserted from the tube.

Upper lip of the corolla concave.
  Anther-bearing stamens 4.
    Upper stamens longer than the lower.
      Calyx 5-toothed.
        Anther-sacs parallel or nearly so ; stamens divergent.
                                    6. AGASTACHE.
        Anther-sacs divaricate ; anthers approximate in pairs.
          Flowers in terminal spikes ; floral leaves reduced ;
            plant erect.   7. NEPETA.
          Flowers in axillary verticils ; floral leaves like the
            rest ; plant spreading.   8. GLECOMA.
      Calyx distinctly 2-lipped.   9. MOLDAVICA.
    Upper stamens shorter than the lower.
      Calyx distinctly 2-lipped, closed in fruit.
                                 10. PRUNELLA.
      Calyx 5-toothed, not 2-lipped, open in fruit.
        Calyx membranous, inflated in fruit, faintly nerved.
                         11. DRACOCEPHALUM.
        Calyx not membranous, not inflated in fruit, strongly
          5–10-nerved.
          Anther-sacs not transversely 2-valved.
            Nutlets 3-sided, truncate above.
              Calyx-teeth awn-pointed, spreading.
                          12. LEONURUS.
              Calyx-teeth not awn-pointed.
                          13. LAMIUM.
            Nutlets ovoid, nearly terete, rounded above.
                        14. STACHYS.
        Anther-sacs transversely 2-valved.
                    15. GALEOPSIS.
  Anther-bearing stamens 2.
    Connective of the anthers very long, articulated to the filaments,
      bearing a perfect anther at the ascending end and a reduced
      one or none at the other ; calyx 2-lipped.
                    16. SALVIA.
    Connective of the anthers short ; anther-cells confluent.
      Calyx 2-lipped, the tube 13-ribbed, the teeth equal.
                  17. BLEPHILIA.
      Calyx equally 5-toothed, the tube 15-ribbed.
                  18. MONARDA.
Upper lip of the corolla flat.
  Stamens curved, often converging.
    Anther-bearing stamens 2.   19. HEDEOMA.
    Anther-bearing stamens 4.
      Corolla-tube strongly bent.   20. MELISSA.
      Corolla-tube straight.
        Calyx naked in the throat ; flowers sessile in dense heads,
          subtended by subulate-setaceous bracts.
                  21. CLINOPODIUM.
        Calyx hairy in the throat ; flowers few in the axils, pedi-
          celled ; bracts minute.   22. CALAMINTHA.
    Stamens straight, distant and diverging ; calyx 5-toothed ; anther-
      bearing stamens 4.   23. PYCNANTHEMUM.
Corolla nearly regular, 4–5-toothed.
  Anther-bearing stamens 2.   24. LYCOPUS.
  Anther-bearing stamens 4.   25. MENTHA.

1. TEÙCRIUM (Tourn.) L. GERMANDER, WOOD SAGE.

Perennial caulescent herbs, rarely shrubs. Leaves toothed. Flowers perfect, in raceme-like inflorescence. Calyx campanulate, 10-ribbed; lobes 5, unequal. Corolla purplish, pink, or white; upper lip very short; lower lip 3-lobed, the middle lobe elongate. Stamens 4, inserted between the lobes of the upper lip, the anterior pair the longer; anther-sacs divergent, confluent at the base.

Upper calyx-teeth obtuse ; stem appressed-pubescent.   1. *T. canadense.*
Upper calyx-teeth acute or acuminate ; stem villous.
  Calyx and bracts glandular-pubescent as well as villous ; stem
    loosely villous.   2. *T. occidentale.*
  Calyx and bracts scarcely glandular ; stem more closely pu-
    bescent.   3. *T. borealc.*

1. **T. canadénse** L. Stem 2–12 dm. high; leaves thin, ovate to oblong-ovate, 6–14 cm. long, acute or acuminate, closely dentate-serrate, rounded or subcordate at the base, glabrate or hispidulous above, pale and pubescent beneath; bracts shorter than the calyx; calyx canescent, often purplish; corolla pink, 15–20 mm. long, granular-puberulent. River banks and low ground; Me.—Minn.—Kans.— Tex.—Ga. *E. Temp.* Je–S.

f.481.

2. **T. occidentàle** A. Gray. *Fig. 481.* Perennial, with a rootstock; stem 3–8 dm. high, villous-hirsute; leaf-blades ovate-oblong, oblong, or lanceolate, 4–9 cm. long, sharply serrate, strigose above, paler and softly villous beneath; calyx 5–6 mm. long, viscid-villous, often purplish; corolla purplish, 8–10 mm. long. Thickets and wet grounds: Ont.— Pa.—N.M.—Calif.—B.C. *Temp.—Plain—Submont.* Je–Au.

3. **T. boreàle** Bickn. Perennial; stem 3–8 dm. high, pubescent with recurved hairs; leaves thin, ovate-oblong or oblong-lanceolate, 5–10 cm. long, acute, serrate, hispidulous above, tomentulose beneath; bracts shorter than the calyx, ciliate; corolla purplish, 12–15 mm. long. *T. occidentale boreale* Fern. [G]. Wet ground: N.H.—Minn.—Mo.—N.Y. *Canad.—Allegh.* Au.

2. MELÓSMON Raf.

Perennial or annual herbs. Leaves opposite, incised or parted. Flowers solitary in the axils of the leaves. Calyx-tube very short, 10-ribbed; lobes 5, equal or nearly so. Corolla blue, lilac, or white; upper lip small; lower lip 3-lobed, with the middle lobe the longer. Stamens 4, as in *Teucrium*.

1. **M. laciniàtum** (Torr.) Small. Perennial, with a woody root and short cespitose caudex; stems 1–2 dm. high, glabrous or nearly so; leaf-blades pinnately 3–7-parted into linear divisions which are sometimes again lobed; calyx 10–12 mm. long; lobes linear-subulate, acuminate, corolla pale blue or lilac, 1.5–2 cm. long. *T. laciniatum* Torr. [B]. Plains: Kans.—Tex.—Ariz.—Colo. *St. Plains—Son.* My–Au.

3. ISÁNTHUS Michx.

Annual herbs with branching stems. Leaves opposite, narrow, viscid-pubescent. Flowers in axillary cymes. Calyx campanulate, the tube 10-ribbed, the teeth narrow, nearly equal. Corolla blue, nearly regular, with spreading lobes. Stamens 4, longer than the corolla-tube; anther-cells divergent. Nutlets reticulate.

1. **I. brachiàtus** (L.) B.S.P. Stem 1–4 dm. high; leaves elliptic, 1–4 cm. long, acute, mostly entire, 3-ribbed; peduncles 1–3-flowered; corolla blue, 5–8 mm. long. Dry soil: Ont.—Minn.—Tex.—Ga. *Allegh.—Austral.* Jl–Au.

4. SCUTELLÀRIA L. SKULLCAP.

Annual or perennial herbs, sometimes shrubby. Leaves opposite, entire or toothed. Flowers perfect, in terminal racemes or panicles, often leafy-bracted. Calyx campanulate, 2-lipped; lips entire, the upper with a crest. Corolla blue, violet, or white; throat dilated; limb 2-lipped, the upper lip arched, the lower 3-lobed; lateral lobes small, the middle one large and spreading. Stamens 4, didynamous; anthers ciliate, those of the upper pair 2-celled, those of the lower 1-celled. Nutlets papillose or tubercled.

Flowers in axillary and terminal racemes.
  Flowers less than 1 cm. long.   1. *S. lateriflora.*
  Flowers more than 1 cm. long.

Leaves cordate, slender-petioled, crenate; lateral lobes of
the corolla nearly equaling the upper lip.   2. *S. ovata.*
Stem-leaves not cordate; lateral lobes of the corolla
shorter than the upper lip.   3. *S. canescens.*
Flowers solitary in the axils of the leaves.
Annuals.   4. *S. Drummondii.*
Perennials.
Plants with horizontal rootstocks.
Leaf-blades lanceolate, ovate, or cordate, more or less
distinctly toothed, at least the lower ones.
Corolla more than 15 mm. long.   5. *S. epilobifolia.*
Corolla less than 15 mm. long.
Foliage glabrous or puberulent.   6. *S. ambigua.*
Foliage viscid-pubescent.   7. *S. parvula.*
Leaf-blades entire or nearly so, ovate, oblong or linear.   8. *S. Brittonii.*
Plants with a woody caudex.   9. *S. resinosa.*

1. **S. lateriflòra** L. Perennial, with a creeping rootstock; stem 2–10 dm.
high, glabrous or sparingly puberulent above, branched; leaves slender-petioled;
leaf-blades thin, ovate or ovate-lanceolate, 2–6 cm. long, coarsely serrate,
rounded or cordate at the base; calyx 2–4 mm. long; corolla blue, 6–7 mm.
long. Along streams and swamps: Newf.—Fla.—Ida.—B.C. *Temp.—Submont.*
Je–S.

2. **S. ovàta** Hill. Stem erect or ascending, 1–6 dm. high, softly pubescent, rounded-cordate or oblong-cordate, 3–10 cm. long, crenate; calyx 4–5
mm. long, glandular-pilose; corolla bright blue, nearly 2 cm. long, the lower
lip conspicuously spotted. *S. cordifolia* Muhl. [B]. *S. versicolor* Nutt. [G].
Moist banks: Pa.—Minn.—Tex.—Fla. *Allegh.—Austral.* Je–Jl.

3. **S. canéscens** Nutt. Stem erect, finely pubescent, 5–12 dm. high; lower
leaf-blades truncate or cordate at the base, 4–12 cm. long, serrate, finely pubescent; calyx 3–5 mm. long, glandular; corolla blue, 15–18 mm. long, the lip
undulate, often notched. *S. incana* Muhl. [B]. Woods and copses: Ont.—Ill.
—Kans.—Ala.—Ga. *Allegh.—Austral.* Jl–Au.

4. **S. Drummóndii** Benth. Stem erect, diffusely branched at the base,
hirsute, 0.5–2 dm. high; leaves ovate or oval, 1–1.5 cm. long, obtuse, entire or
crenately undulate; calyx 2–4 mm. long; corolla blue or purple, pilose, 10–12
mm. long. Prairies: Kans.—Tex.; Mex. *Texan—Son.* Ap–Je.

5. **S. epilobifòlia** A. Hamilt. *Fig. 482.* Stem
strigose, at least on the angles, 2–9 dm. high; leaf-
blades oblong or oblong-lanceolate, acute, crenate,
truncate or cordate at the base, more or less densely
strigose and often glandular-granuliferous beneath;
calyx 3.5–5 mm. long, minutely pubescent; corolla
blue, or rarely pink or white, 15–18 mm. long, pubescent. *S. galericulata* Am. auth.; not L. [G,B,R].
River banks and swamps: Newf.—N.C.—Ariz.—
Alaska. *Temp.—Submont.* Je–Au.

6. **S. ambígua** Nutt. Stem 0.5–3 dm. high,
simple or diffusely branched; leaves broadly obovate
to lanceolate, 1–2 cm. long, prominently nerved beneath, truncate or subcordate at the base, sessile;
calyx 2–4 mm. long, puberulent; corolla blue, 4–8
mm. long, minutely pilose. *S. parvula* Britt. [B];
not Michx. *S. parvula ambigua* Fern. [G]. Dry sandy soil: Que.—N.D.—Tex.
—Fla. *E. Temp.* Ap–Jl.

7. **S. párvula** Michx. Stem often diffusely branched, 0.5–2 dm. high,
pubescent; leaves ovate to reniform, obtuse, entire or shallowly toothed,
strongly nerved beneath; calyx 2–4 mm. long, minutely glandular; corolla blue,

6–7 mm. long. *S. campestris* Britton [B]. Sandy soil: Que.—Ill.—Minn.—
Kans.—Okla.—N.C. *Allegh.* Ap–Jl.

8. **S. Brittônii** Porter. Stem puberulent, 1–3 dm. high; leaves entire.
nervose, finely puberulent, oblong to oval, 1–3 cm. long; calyx puberulent, 5–6
mm. long; corolla about 2 cm. long, gradually dilated upwards. *S. resinosa* A.
Gray; not Torr. *S. virgulata* Nels. Hillsides and valleys: Neb. (?)—Wyo.—
Colo. *Submont.—Mont.* My–Jl.

9. **S. resinôsa** Torr. Stems many, diffusely branched, 1–3 dm. high,
grayish puberulent; leaf-blades ovate, oval, elliptic, or oblong-spatulate, 1–1.5
cm. long, grayish-puberulent, strongly veined beneath; calyx 3–5 mm. long,
merely resinous; corolla blue or purplish, minutely pilose, 1 cm. long. *S.
Wrightii* A. Gray. Plains and hills: Neb.—Tex.—Ariz.—Colo. *Plain—Son.*

5 MARRÙBIUM (Tourn.) L. HOREHOUND.

Perennial caulescent herbs. Leaves toothed, woolly. Flowers perfect, in
axillary clusters. Calyx tubular, 5–10-ribbed; lobes 10, equal, or alternately
longer and shorter, recurved at maturity. Corolla white or purplish, 2-lipped;
upper lip erect, entire or notched; lower lip 3-lobed, the lateral lobes small, the
middle one large and broad. Stamens 4, included, didynamous, the anterior pair
the longer; anthers 2-celled, with divergent sacs. Nutlets smooth or granular.

1. **M. vulgàre** L. Stem white-woolly, 2–10 dm. high, usually strict; leaf-
blades suborbicular, oval or ovate, 1–4 cm. long, crenate, strongly rugose; bracts
subulate, with hooked ends; calyx 4–5 mm. long, 10 ribbed; lobes ridged,
hooked; corolla white, 5–6 mm. long, puberulent. Waste places: Me.—N.C.—
Calif.—B.C.; nat. from Eurasia.

6. AGASTACHE Clayton. GIANT HYSSOP.

Tall perennial herbs, with rootstocks. Leaves petioled, broad, toothed.
Flowers perfect, in dense terminal spike-like panicles. Calyx narrowly campanulate or tubular, slightly 2-lipped; lobes nearly equal or the upper somewhat longer. Corolla purplish, rose, or yellowish, 2-lipped; upper lip erect,
notched; lower lip 3-lobed, the middle lobe toothed or undulate. Stamens 4,
exserted, didynamous; anthers 2-celled; sacs almost parallel. Nutlets smooth.
Lophanthus Benth. *Vleckia* Raf.

Corolla greenish yellow; calyx-teeth ovate, obtuse or acutish; plant glabrous.
                                                      1. *A. nepetoides.*

Corolla blue or purple; calyx-teeth lanceolate, acute.
  Plant glabrous or nearly so; leaves pale beneath; co-
    rolla blue.   2. *A. anethiodora.*
  Plant finely hirsute; leaves green or both sides; co-
    rolla purple.   3. *A. scrophulariaefolia.*

1. **A. nepetoides** (L.) Kuntze. Stem glabrous
or nearly so, 1–2 m. high; leaf-blades ovate to lanceolate, 6–12 cm. long, acute or acuminate, serrate;
panicle 0.5–4 dm. long, 1.5–2 cm. thick, erect; corolla 7–9 mm. long. *L. nepetoides* Benth. Hillsides:
Ont.—S.D.—Kans.—Ga. *Canad.—Allegh.* Jl–S.

2. **A. anethiodòra** (Nutt.) Britton. *Fig. 483.*
Stem glabrous or nearly so, 6–15 dm. high; leaf-
blades ovate or deltoid-ovate, acute or acuminate,
serrate, green above, pale beneath, 5–8 cm. long,
anise-scented; spike dense, 4–10 cm. long, scarcely
2 cm. thick; corolla blue, 8–10 mm. long. *L. anisatus*
Benth. *A. Foeniculum* (Pursh) Kuntze [G]. Among
bushes: Ont.—Ill.—Colo.—Alta.—Mack. *Canad.—
Allegh.—Plain—Submont.* Jl–S.

3. **A. scrophulariaefòlia** (Willd.) Kuntze. Stem

f.483.

finely hirsute, 1–2 m. high, sharply angled; leaf-blades ovate, ovate-lanceolate, or cordate, acute or short-acuminate, coarsely serrate; panicles spike-like, 5–30 cm. long, 1–1.5 cm. thick; corolla purplish, 6–8 mm. long. *L. scrophulariae-folia* Benth. Hillsides and thickets: Conn.—Minn.—Kans.—N.C. *Canad.—Allegh.* Jl–O.

7. NÉPETA L. CATNIP, CAT MINT.

Perennial or annual caulescent herbs. Leaves broad, toothed or incised, pubescent. Flowers perfect, in axillary cymes, clustered. Calyx inflated in age, 15-ribbed, incurved; lobes 5, the upper slightly longer than the lower. Corolla white or blue; throat enlarged; upper lip erect, entire or notched; lower lip 3-lobed, with large middle lobe. Stamens 4, didymous, exserted; anthers with 2 diverging sacs. Nutlets flattened, smooth.

1. **N. Catària** L. *Fig. 484.* Perennial; stem densely tomentulose, 6–10 dm. high; leaf-blades ovate or cordate, coarsely crenate, 2–8 cm. long, green above, densely white-tomentose beneath; flower-clusters spicate at the ends of the branches; calyx densely puberulent, its lobes subulate; corolla white or pale purple, dark-dotted, 10–12 mm. long.

f.484.

Waste places and around dwellings: N.B.—Va.—Utah.—Ore.; nat. from Eur. Jl–N.

8. GLECÒMA L. GROUND IVY, GILL-OVER-THE-GROUND.

Low creeping perennial herbs. Leaves opposite, long-petioled. Flowers perfect, in axillary verticils. Calyx-tube short-tubular, 15-nerved, oblique, unequally 5-toothed. Corolla 2-lipped; tube exserted, enlarged above; upper lip erect, 2-lobed or emarginate; lower lip spreading, 3-lobed, the middle lobe broad, emarginate. Stamens 4; anther-sacs divergent. Nutlets ovoid, smooth.

1. **G. hederàcea** L. Stem pubescent, 3–5 dm. long; leaf-blades orbicular or reniform, crenate, 1–4 cm. broad; calyx-teeth lanceolate, acuminate; corolla 1.5–2 cm. long, light blue, three times as long as the calyx. *Nepeta hederacea* (L.) Trevisan [G]. Waste places, roadsides, and thickets: Newf.—Ga.—Colo.—Mont.; nat. from Eurasia. Mr–My.

9. MOLDÀVICA (Tourn.) Adans. DRAGON HEAD.

Herbs, with blue or purple flowers, in axillary and terminal bracted clusters. Leaves mostly toothed or dissected. Calyx tubular, 15-nerved, 5-toothed or 2-lipped, with the upper 3 teeth more or less united. Corolla enlarged above, 2-lipped; upper lip erect, emarginate; lower lip spreading, 3-lobed. Stamens 4, the upper the longer, all fertile; anthers more or less approximate in pairs; sacs divergent. Nutlets ovoid, smooth. *Dracocephalum* L., in part.

Corolla 2–3 times as long as the calyx.	1. *M. punctata.*
Corolla little if at all longer than the calyx.	
Flowers mostly at the ends of the branches; stem-leaves ovate or broadly lanceolate, sharply serrate; calyx hirsute.	2. *M. parviflora.*
Flowers verticillate in the axils: stem-leaves narrowly lanceolate or oblong, denticulate or nearly entire; calyx short-pubescent.	3. *M. thymiflora.*

1. **M. punctàta** Moench. Annual; stem 3–8 dm. high; leaves oblong or linear-oblong, dentate or incised, obtuse, 2–5 cm. long; verticils of flowers mostly axillary; bracts narrowly oblong, shorter than the calyx, deeply pectinately aristate-toothed; calyx curved, the lower teeth shorter; corolla nearly white. *D. Moldavica* L. [B]. *M. Moldavica* (L.) Britton [BB]. Sandy soil: Neb.—Mex.; adv. from Eur. Je–Au.

2. M. parviflòra (Nutt.) Britton. *Fig. 485.*
Annual or biennial; stem finely puberulent, 2–6 dm.
high; leaf-blades lanceolate, ovate, or oblong,
coarsely serrate, the upper with spinulose-tipped
teeth; bracts pectinate, with awn-pointed teeth; upper tooth of the calyx longer than the rest; corolla
light blue, scarcely longer than the calyx. *D. parviflorum* Nutt. [G, B]. Hillsides and valleys: N.Y.—
Neb.— N.M.— Ariz.— Alaska. *Submont.— Mont.*
Je–Au.

f.485.

3. M. thymiflòra (L.) Rydb. Perennial; stem
3–5 dm. high, angled, puberulent; lower leaf-blades
ovate, crenate, usually 5-ribbed, crenate, the upper
oblong-lanceolate to linear-lanceolate, entire or denticulate; calyx-tube 7 mm. long, hispidulous, purple-
tinged, the uppermost tooth oval, abruptly cuspidate,
the lateral ones lanceolate, and the lowest two subulate, gradually acuminate;
corolla slightly exceeding the calyx. *D. thymiflorum* L. Waste places: N.D.;
adv. from Eur. Je–Au.

10. PRUNÉLLA L. Self-heal, Heal-all, Carpenter Weed.

Perennial caulescent herbs, with rootstocks. Leaves opposite, petioled,
toothed, pubescent. Flowers perfect, irregular, in axillary clusters, toward the
end of the stem forming a leafy spike-like inflorescence. Calyx 4-lipped, closed
at maturity; tube 10-ribbed; upper lip truncate, slightly 2-lobed, the lower with
3 narrow lobes. Corolla purple or white, 2 lipped; upper lip arched; lower lip
3-lobed, with broad middle lobe. Stamens 4, but 2 sterile; fertile stamens with
forked filaments, one prong bearing the 2-celled anthers. Nutlets smooth.

1. P. vulgàris L. Stem procumbent or ascending or erect, 0.5–5 dm.
high; leaf-blades ovate, oblong, or lanceolate, entire or crenate, 2–10 cm. long;
spike terminal, very dense, 1–3 cm. long in flower, or 5–10 cm. in fruit; bracts
broadly ovate-orbicular, cuspidate; corolla violet or purple, 8–12 mm. long.
Woods and wet places among bushes: Newf.—Fla.—Calif.—B.C.; nat. from
Eurasia. *Temp.—Mont.* My–O.

11. DRACOCÉPHALUM (Tourn.) L. False Dragon-head.

Perennial herbs, with mostly simple stems. Leaves opposite, glabrous,
rather narrow, usually toothed. Calyx campanulate, 5-ribbed, more or less
inflated; lobes 5, nearly equal. Corolla purple, pink, or white, 2-lipped; tube
dilated upwards; upper lip entire; lower lip 3-lobed, with the middle lobe
emarginate. Stamens 4, all fertile; anthers 2-celled; sacs parallel. Nutlets
smooth, 3-angled. *Physostegia* Benth.

Stem with ample leaves up to the inflorescence.
  Corolla 18–22 mm. long, strongly ampliate; calyx-teeth abruptly short-acuminate.
    Leaves oblanceolate or lanceolate, decidedly narrowed at the base.
                                                   1. *D. speciosum.*
    Leaves linear or linear-oblanceolate, slightly narrowed
      at the base.   2. *D. virginianum.*
  Corolla 10–18 mm. long, less strongly ampliate; calyx-teeth
    acute or obtuse.
    Leaves tapering at the base, strongly serrate; corolla
      15–18 mm. long.   3. *D. formosius.*
    Leaves broad at the base, dentate or denticulate; corolla
      10–15 mm. long.   4. *D. Nuttallii.*
Stem with the upper leaves much reduced, the inflorescence
  therefore seemingly long-peduncled; leaf-blades narrowly
  oblanceolate.
  Corolla 2–3 cm. long; leaf-blades crenate-dentate.   5. *D. denticulatum.*
  Corolla 1–1.5 cm. long; lower leaf-blades rather sharply
  repand-denticulate.   6. *D. intermedium.*

1. D. speciòsum Sweet. *Fig. 486.* Stem 3–10 dm. high, simple or with a few branches; leaves oblanceolate or lanceolate, acute, serrate or incised, at least above the middle, 1–2 dm. long; calyx 6–8 mm. long, the tube cylindric, the teeth lanceolate; corolla bright pink, 2.5–3 cm. long. *P. virginiana speciosa* (Sweet) A. Gray. Low ground: Ill.—Man. —Kans.—Tex. *Prairie.* Jl–S.

f.486.

2. D. virginiànum L. Stem erect, 3–12 dm. high; leaves linear or lance-linear, sharply serrate, 5–15 cm. long, 4–15 mm. wide; calyx about 8 mm. long, the tube narrowly funnelform; corolla light purple or rose, 2–3 cm. long. *P. virginiana* Benth. [G]. Moist soil: Que.—Fla.—Tex.—Kans. *E. Temp.* Jl–S.

3. D. formòsius (Lunell) Rydb. Stem about 1 m. high, glabrous, shining, often branched above; leaves sessile, oblong-lanceolate or oblanceolate, narrowed at the base, acuminate, sharply serrate, the larger 1–2 dm. long; calyx puberulent, 7 mm. long, the lobes triangular, acute, 2 mm. long; corolla rose, 15–18 mm. long. *P. formosior* Lunell. Bottom-lands: Ind.—Wis.—N.D. *Prairie.* Ap–S.

4. D. Nuttàllii Britton. Stem simple, 3–10 .dm. high, glabrous; leaves lanceolate, oblong, or linear, dentate, the upper sessile, 7–10 cm. long; spike 2–10 cm. long; calyx campanulate, with ovate, obtuse teeth; corolla purple, about 1 cm. long. *Physostegia parviflora* Nutt. [G, B]. Banks of streams and among bushes: Sask.—Minn.—Iowa—Colo.—Ore.—B.C. *W. Temp.—Plain— Prairie—Submont.* Jl–Au.

5. D. denticulàtum Ait. Stem slender, 3–5 dm. high, usually simple; lower leaves petioled, the blades oblong-oblanceolate, usually obtuse, crenate-dentate, 3–7 cm. long, the upper sessile, mostly linear; spike elongate, rather lax, usually simple; corolla rather strongly ampliate. *P. denticulata* Britton [B, G]. Prairies and banks: Pa.—Ill.—Kans.—Fla. Je–Au.

6. D. intermèdium Nutt. Stem slender, 3–10 dm. high; leaves remote, the lower linear-oblanceolate or linear, usually repand-denticulate, 5–7 cm. long, the upper much reduced and linear; spike very slender, mostly simple; corolla ampliate. *P. intermedia* A. Gray [B, G]. Prairies: w Ky.—Ill.—Kans.—Tex. *Ozark.* My–Jl.

12. LEONÙRUS L. MOTHERWORT.

Annual or perennial caulescent herbs. Leaves opposite, toothed, or palmately cleft or parted. Flowers perfect, irregular, in dense axillary clusters. Calyx tubular-campanulate, 5-ribbed; lobes 5, nearly equal, subulate or awn-tipped. Corolla white or pink, 2-lipped; upper lip erect, nearly flat; lower lip spreading, 3-lobed; middle lobe broad, truncate, or notched. Stamens 4, all fertile; anthers 2-celled; sacs mostly parallel. Nutlets smooth, 3-angled.

Lobes of the leaves coarsely toothed or entire; corolla white or pale purple, with a ring of hairs in the tube.   1. *L. Cardiaca.*
Lobes of the leaves again cleft or incised; corolla red or purplish, the tube glabrous within.   2. *L. sibiricus.*

1. L. Cardìaca L. Perennial; stem erect, 3–10 dm. high, strigose-puberulent; leaf-blades 2–10 cm. long, palmately 3–5-cleft, with lanceolate, more or less acuminate, entire or toothed lobes; calyx 5-angled; lobes spinulose; corolla pale purple or white, villous without, 9–10 mm. long. Waste places: N.S.— N.C.—Tex.—Utah—Mont.; nat. from Eur. *Plain—Submont.* Je–Au.

2. L. sibìricus L. Biennial; stem 6–15 dm. high, puberulent or glabrate; leaves long-petioled; blades deeply 3-parted, 4–7 cm. long, the divisions acuminate, deeply cleft into linear or lanceolate lobes; calyx-lobes subulate, with spreading bristle-tips; corolla purple or red, 9–12 mm. long. Waste places: Pa.—Del.—Iowa; Bermudas; adv. from Asia. Je–Au.

13. LÀMIUM L. DEAD NETTLE, HENBIT.

Annual or perennial caulescent herbs, with branched stems. Leaves oppo-site, petioled, usually broad, toothed or incised. Flowers perfect, irregular, in remote axillary clusters. Calyx campanulate, 5-nerved; lobes 5, equal or the upper somewhat longer. Corolla purple, blue or white, 2 lipped; upper lip erect, concave, entire; lower lip 3-lobed; lateral lobes narrow, the middle one dilated and notched. Stamens 4, all fertile; anthers 2-celled; sacs divergent. Nutlets smooth or tuberculate.

1. L. amplexicaule L. Annual or biennal; stem slender, weak, branched from the base and lower axils, usually decumbent, 1–5 dm. long, glabrous; leaf-blades orbicular, cordate, or reniform, 1–4 cm. broad, crenate, the upper sessile and more or less clasping; flowers axillary; corolla purplish or red, 12–10 mm. long; middle lobe of the lower lip spotted. Waste places and cul-tivated ground: N.B.—Fla.—Utah—Calif.—B.C.; adv. from Eur.

14. STÀCHYS (Tourn.) L. HEDGE NETTLE.

Annual or perennial herbs, ours with rootstocks. Leaves opposite, in ours petioled and toothed. Flowers perfect, in axillary clusters, sometimes approxi-mate, forming leafy spikes. Calyx mostly campanulate, 5–10-ribbed, 5-lobed. Corolla purple, pink, or white, 2-lipped; upper lip erect, entire or notched; lower lip spreading, 3-lobed; middle lobe largest, often 2-lobed. Stamens 4; anthers all fertile, 2-celled; anther-sacs divergent or parallel; filaments naked. Nutlets obtuse, not truncate at the apex, smooth.

Annuals.
  Corolla yellow, twice as long as the calyx. .... 1. *S. annua.*
  Corolla purplish, scarcely exceeding the calyx .... 2. *S. arvensis.*
Perennials, with rootstocks.
  Stem glabrous, except on the angles.
    Petioles less than 1 cm. long; stem bristly on the angles.
      Leaf-blades linear or lance-linear, narrowed at the base. .... 3. *S. Grayana.*
      Leaf-blades lanceolate, truncate or subcordate at the base. .... 4. *S. aspera.*
    Petioles 1–2 cm. long; stem usually wholly glabrous; leaf-blades subcordate or truncate at the base. .... 5. *S. tenuifolia.*
  Stem more or less pubescent on the sides as well as on the angles.
    Bristles at least on the angles of the stem very strong and reflexed.
      Calyx-lobes much shorter than the tube. .... 6. *S. brevidens.*
      Calyx lobes about as long as the tube.
        Sides of the stem scabrous-hispidulous. .... 7. *S. pustulosa.*
        Sides of the stem hispid, hirsute, or pilose.
          Leaf-blades linear-lanceolate or linear. .... 8. *S. arenicola.*
          Leaf-blades lanceolate or ovate.
            Leaf-blades lanceolate, acute; corolla about 10 mm. long. .... 9. *S. homotricha.*
            Leaf-blades ovate or the upper ovate-lance-olate, short-acuminate; corolla about 12 mm. long. .... 10. *S. Schweinitzii.*
    Bristles on the stem slender, soft, and mostly spreading.
      Teeth of the leaves large and comparatively few (10–15 on each side); lateral lobes of the co-rolla large. .... 11. *S. ampla.*
      Teeth of the leaves fine and numerous (20 or more on each side); lateral lobes of the corolla small.
        Corolla 10–12 mm. long; stem slender, 3–4 dm. high.
          Leaf-blades linear or lance-linear, acute; bracts mostly exceeding the flowers. .... 12. *S. borealis.*

Leaf-blades lanceolate, elliptic or oblong-lanceolate, the lower obtuse or rounded at the
apex; bracts small, the upper reflexed.   13. *S. scopulorum.*
Corolla 12–15 mm. long; stem tall and stout, 4–10
dm. high; bracts much exceeding the flowers, spreading; lower leaves acute.
Stem densely and finely pilose; leaf-blades
ovate to lanceolate, rarely subcordate at
the base; hairs without pustulate bases.   14. *S. puberula.*
Stem sparingly long-hairy; leaf-blades oblong-
lanceolate, the larger cordate at the base;
hairs on the upper surface of the leaves
with pustulate bases.   15. *S. arguta.*

1. **S. ánnua** L. Stem erect, 1–3 dm. high, glabrous; leaves oblong to
oblanceolate, the lower crenate, the upper subentire; calyx pilose and glandular-puberulent, 6–8 mm. long, the teeth subequal and equaling the tube; corolla 15 mm. long. Waste places and ballast: Atlantic ports; Kans.; adv. from
Eur.

2. **S. arvénsis** L. Stem slender, diffusely branched, 1–6 dm. high, hirsute; leaf-blades ovate or subcordate, hirsute, about 2.5 cm. long, crenate, the
lower long-petioled; calyx 4.5–6 mm. long, hirsute, the teeth lanceolate, acuminate. Waste places and ballast: Me.—Va.; reported from Neb.; adv. from
Eur. Jl–O.

3. **S. Grayàna** House. Stem erect, 3–10 dm. high, branched; leaf-blades
oblong to linear-lanceolate, 3–6 cm. long, acute, serrate, short-peticled; calyx
6–8 mm. long, the teeth lanceolate, subulate-tipped, ciliate; corolla light purple,
1 cm. long; upper lip pubescent on the back. *S. ambigua* (A. Gray) Britton;
not Smith [G, B]. Moist sandy soil: N.Y.—Minn.—Ga. *Allegh.* Je–Au.

4. **S. áspera** Michx. Stem erect, hirsute, 5–12 dm. high; leaf-blades
oblong-lanceolate to ovate-oblong, 3–10 cm. long, acute, crenate-serrate, truncate or cordate at the base; calyx 5–6 mm. long, the teeth triangular, acuminate; corolla 10–12 mm. long, purple, pubescent along the back, the upper lip
glandular-pilose on the back. *S. tenuifolia aspera* Fern. [G]. Moist ground:
Ont.—Minn.—La.—Fla. *E. Temp.* Jl–Au.

5. **S. tenuifòlia** Willd. Stem erect, 3–9 dm. high; leaf-blades oblong or
oblong-lanceolate, acute, serrate, cordate or truncate at the base; petioles 1–2
cm. long; calyx 5–6 mm. long, the lobes lanceolate, acuminate; corolla about
1 cm. long; upper lip glandular-pubescent on the back. Copses and moist
ground: Ont.—Minn.—Kans.—Fla. *E. Temp.* Jl–Au.

6. **S. brévidens** Rydb. Perennial; stem 4–6 dm. high, the angles hispid
with retrorse hairs, the sides finely pubescent; leaves short-petioled, the blades
oblong, obtuse or rounded at each end, crenate, appressed-pilose above, finely
and densely soft-pubescent beneath, 5–10 cm. long, 1.5–2.5 cm. wide; inflorescence branched; bracts lanceolate, spreading, the upper shorter than the
flowers; calyx soft-pubescent, the tube 3 mm. long, the teeth 2 mm. long;
deltoid, acuminate; corolla purplish, about 12 mm. long. Meadows: Tower,
Minn.

7. **S. pustulòsa** Rydb. Perennial; stem 6–10 dm. high, the angles hispid
with stout and rather short hairs, pustulate at the base, the sides scabrous-hispidulous; leaves short-petioled or the upper subsessile; leaf-blades lanceolate,
5–12 cm. long, 2–4 cm. wide, serrate with numerous teeth, short-hispidulous on
both sides, and hispid on the veins beneath, the hairs often with pustulate
bases; inflorescence 1–2 dm. long, the bracts lanceolate, the lower ones longer
than the flowers and spreading, the upper shorter and reflexed; calyx hispid
or hispidulous, the tube 3 mm. long, the teeth lance-subulate, attenuate, 3 mm.
long; corolla purplish, 10–12 mm. long, the upper lip puberulent without.
Swamps and beeches: Miss.—Kans.—Minn. *Allegh.* Jl–S.

8. S. arenícola Britton. Perennial; stem 4–10 dm. high, the angles coarsely retrorsely hispid, the sides with softer hairs; leaf-blades linear-lanceolate, narrowed at the base, crenate-dentate, 5–10 cm. long, 1–1.5 cm. wide, densely and softly pubescent on both sides, more coarsely so on the veins beneath; inflorescence simple or in age branching; calyx long-pilose, the tube 3 mm. long, the teeth lance-subulate, attenuate, 3 mm. long; corolla purple, about 1 cm. long, the upper lip pubescent. Sandy soil: N.Y.—Mo.—Iowa (?) —Upper Mich. *Allegh.* Jl.–Au.

9. S. homótricha (Fern.) Rydb. Stem 4–10 dm. high, densely retrorse-hispid on both the angles and sides; leaves lanceolate, mostly subsessile, 5–10 cm. long, 1.5–3 cm. wide, acute, crenate-serrate, densely and softly long-hairy on both sides, with stronger hairs on the veins beneath; inflorescence simple or compound, 1–2 dm. long; bracts, except the lowermost, short, ovate, scarcely exceeding the flowers; calyx long-hirsute, the tube 3 mm. long, the teeth lance-subulate, 3 mm. long; corolla about 10 mm. long. *S. palustris homotricha* Fern. [G]. Swamps: N.B.—N.Y.—Mo.—Iowa—Minn. *Canad.—Allegh.*

10. S. Schweinitzii Rydb. Stem 5–10 dm. high, densely retrorse-hispid on the angles and sides; leaves short-petioled, the blades ovate, acuminate at the apex, rounded or subcordate at the base, 5–10 cm. long, 3–4 cm. wide, densely long-hairy on both sides, the hairs on the upper surface and on the veins beneath with pustulate bases; bracts usually long, attenuate, spreading; calyx long-hirsute, the tube 4 mm. long, the teeth subulate, 4 mm. long; corolla purplish, 12–15 mm. long. *S. velutina* Schw.; not Willd. Swamps· Minn. —Iowa—S.D.

11. S. ámpla Rydb. Stem 5–7 dm. high, softly hirsute and more or less viscid, leaves ovate to lanceolate, 5–10 cm. long, sessile or nearly so, crenate-serrate, softly pubescent on both sides; corolla rose-colored or pink, about 15 mm. long, somewhat puberulent without. Meadows: S.D. *Submont.* Au.

12. S. boreàlis Rydb. Stem rather slender, 3–6 dm. high, often reddish, both the sides and the angles softly pilose with long hairs; leaves linear-lanceolate, subsessile, crenate-serrate, acute at the apex, mostly cuneate at the base, 5–10 cm. long, 1–2 cm. wide, softly pubescent with appressed hairs; inflorescence mostly simple, 1–2 dm. long; bracts linear-lanceolate, mostly exceeding the flowers; calyx softly long-pilose, decidedly viscid, the tube 3 mm. long, the teeth lance-subulate; corolla purple, about 1 cm. long. *S. palustris* Hook., in part [G, B]. Closely resembling *S. palustris* L. of Europe, but pubescence longer and softer, and calyx-teeth with shorter and weakly aristate tips. Swamps: Mack.—Minn. *Boreal.*

13. S. scopulòrum Greene. *Fig. 487.* Stem 3–6 dm. high, softly hirsute with spreading hairs and more or less glandular; leaf-blades serrate, 5–10 cm. long, softly pubescent on both sides; corolla rose, lilac, or purplish, blotched. *S. palustris* Coult.; not L. Wet meadows: Minn.—N.M.—Utah—Wash.— Mack. *Plain—Submont.* Jl–S.

5.487.

14. S. pubérula (Jennings) Rydb. Stem stout, 5–10 dm. high, densely soft-pubescent with spreading hairs, somewhat viscid, and with stronger hairs on the angles; leaves short-petioled or subsessile, the blades thin, 7–15 cm. long, 2–4 cm. wide, acute at the apex, obtuse to subcordate at the base, crenate-serrate, softly appressed-pubescent and somewhat viscid on both sides; inflorescence 1–2 dm. long; bracts lanceolate or ovate, longer than the flowers; calyx densely soft-pilose; the tube 4 mm. long, the teeth lance-subulate, 4 mm.

long; corolla 12–15 mm. long, purplish. *S. palustris puberula* Jennings. Swamps: Minn.—Iowa—w Ont. *Prairie.* Jl–Au.

15. S. argùta Rydb. Stem stout, perhaps 1 m. high, hirsute both on the sides and the angles; leaves oblong-lanceolate, sharply serrate, acute or acuminate at the apex, often truncate at the base, 8–10 cm. long, appressed-hirsute on both sides; inflorescence branched, 1–2 dm. long; lower bracts leaf-like, the rest lanceolate and longer than the flowers; calyx hirsute, the tube and teeth each 4 mm. long; corolla purplish, about 12 mm. long. Minn.

15. GALEÓPSIS L. HEMP NETTLE.

Annual herbs. Leaves opposite, petioled. Flowers perfect, in several axillary verticils. Calyx 5-ribbed, with 5 subequal lobes. Corolla dilated at the throat; upper lip ovate, arched, entire; lower lip 3-cleft, with an obcordate middle lobe; palate with 2 teeth at the sinuses. Stamens 4; anthers all 2-celled. Nutlets ovoid, slightly flattened, smooth.

1. G. Tetràhit L. Stem 3–10 dm. high, swollen below the nodes; leaf-blades ovate, coarsely serrate, 5–15 cm. long; corolla purple or pink, variegated with white, 15–25 mm. long, twice as long as the calyx. Waste places: Newf. —N.C.—S.D.—Alaska; nat. or adv. from Eur. Je–S.

16. SÁLVIA (Tourn.) L. SAGE.

Annual or perennial herbs (ours), or rarely shrubby plants. Leaves opposite, sometimes mostly basal, entire, toothed, or lobed. Flowers perfect, in axillary clusters, disposed in spikes, racemes, or panicles. Calyx 2-lipped; upper lip 3-toothed; lower lip with 2 longer and narrower lobes. Corolla strongly 2-lipped, upper lip arched, lower lip longer, spreading, with 3 lobes. Stamens 4, but only 2 anther-bearing, or sometimes only 2; lower branch of the connective deflexed, destitute of an anther-cell. Nutlets smooth.

Leaf-blades closely crenate, the lower petioled, oblong, the upper lanceolate or ovate,
   sessile ; upper lip of the calyx 3-toothed.   1. *S. sylvestris.*
Leaf-blades entire or sinuately toothed, oblanceolate to linear ; upper lip of the calyx entire.
     Corolla 15–30 mm. long ; tube exserted.   2. *S. Pitcheri.*
     Corolla 8–12 mm. long ; tube included in the calyx.   3. *S. lanceolata.*

1. S. sylvéstris L. Stem 3–8 dm. high, densely puberulent, branched; lower leaves petioled, the blades oblong, 5–10 cm. long, subcordate at the base, the upper sessile, ovate or lanceolate, finely puberulent; upper lip of the calyx 3-toothed, the lower 2-cleft; corolla violet-blue, 10–14 mm. long. Pastures: Council Bluffs, Iowa; adv. from Eur. Jl.

2. S. Pítcheri Torr. *Fig. 488.* Stem 4–12 dm. high, finely retrosely strigose; leaves short-petioled, linear-lanceolate or linear, 3–12 cm. long, toothed or entire, strigillose; calyx densely canescent, 6–8 mm. long; upper lip barely pointed; corolla blue or white; upper lip densely bearded on the back. *S. azurea grandiflora* Benth. [G]. Dry plains and prairies: Neb.—Tex.—Colo. *Prairie—Plain.* Jl–S.

F. 488.

3. S. lanceolàta Willd. Perennial; stem puberulent, erect, branched, 1–4 dm. high; leaf-blades oblong-lanceolate to oblong or linear, 2–8 cm. long, remotely serrate or undulate or entire, puberulent or glabrate; calyx 7–8 mm. long, puberulent; upper lip abruptly pointed; corolla purplish; upper lip minutely puberulent. *S. lanceaefolia* Auth. [G]; not Poir. Prairies, plains, and hillsides: N.D.—Tex.—N.M.—Mont.; Mex. *Plain—Submont.* My–S.

17. BLEPHÍLIA Raf.

Perennial caulescent herbs. Leaves opposite, pubescent, usually shallowly lobed. Flowers in axillary clusters. Calyx 2-lipped, glabrous in the throat, the tube 13-ribbed, the lobes of the lower lip much longer than those of the upper. Corolla bluish or purplish, longer than the calyx, the tube glabrous within; upper lip erect, entire; lower lip 3-lobed, spreading. Two of the filaments anther-bearing, the other two reduced to staminodia or wanting. Nutlets smooth.

Leaves shallowly toothed, softly short-pubescent.   1. *B. ciliata.*
Leaves sharply serrate, hirsute or villous-hirsute.   2. *B. hirsuta.*

1. **B. ciliàta** (L.) Raf. Stem softly pubescent, 3–6 dm. high; leaf-blades lanceolate, 5–10 cm. long, whitish-downy beneath, acute at the apex, rounded or subcordate at the base, short-petioled or sessile; verticils mostly contiguous, 2–2.5 cm. thick; outer bracts ovate, colored; corolla pink or purplish, 1 cm. long, pubescent. Woods and thickets: Mass.—Minn.—Kans.—Ga. *Canad.—Allegh.* Je–Au.

2. **B. hirsùta** (Pursh) Torr. Stem hirsute, or villous-hirsute, 3–10 dm. high, producing runners with ovate leaves in the spring; leaf-blades ovate or lanceolate, 5–12 cm. long, rounded or subcordate at the base; verticils about 2 cm. thick; bracts linear-subulate; corolla pale, with purple spots, 1 cm. long. Shaded places. Vt. Minn.—Kans.—Tex.—Ga. *Canad.—Allegh.* Je–S.

18. MONÁRDA L. HORSE MINT, WILD BERGAMOT, LEMON MINT.

Perennial or annual, caulescent herbs. Leaves opposite, petioled, with broad usually toothed blades. Flowers perfect, in dense, remote or contiguous, axillary clusters. Calyx tubular or nearly so, mostly pubescent in the throat, 15-ribbed, nearly regularly 5-lobed. Corolla 2-lipped; throat dilated; upper lip narrow, erect or arched, entire or notched; lower lip spreading, 3-lobed, the middle lobe much the larger. Anther-bearing stamens 2, and with 2 or no rudimentary filaments; anthers narrow, 2-celled, versatile; sacs divergent. Nutlets smooth.

Heads solitary at the ends of the stem and branches; stamens conspicuously exceed-
   ing the acute upper lip of the corolla.
  Corolla scarlet.   1. *M. didyma.*
  Corolla purple, pink, or white.
    Leaves distinctly petioled.
      Pubescence of spreading hairs or none.
        Plant with few scattered hairs or glabrate; peti-
          oles 1–3 cm. long.   2. *M. fistulosa.*
        Plant decidedly hairy, at least on the stem be-
          low the nodes; petioles less than 1 cm. long.   3. *M. comata.*
      Pubescence of short appressed hairs.
        Petioles seldom more than 5 mm. long.   4. *M. menthaefolia.*
        Petioles 10–30 mm. long.   5. *M. mollis.*
    Leaves sessile or nearly so.   6. *M. Bradburiana.*
Heads or verticillate glomerules several in the upper axils of
   the leaves; stamens scarcely exceeding the emarginate
   or cleft upper lip of the corolla.
  Calyx-lobes triangular to lanceolate; corolla yellowish;
   perennials.   7. *M. punctata.*
  Calyx-lobes subulate-setaceous; corolla pink or white.
    Calyx puberulent, the lobes spreading; bracts merely
     ciliate or puberulent.
      Bracts oblong, abruptly narrowed into the bristle-
       tip; corolla about 2.5 cm. long.   8. *M. dispersa.*
      Bracts lanceolate, gradually tapering into the bris-
       tle-tip; corolla 15–18 mm. long.   9. *M. pectinata.*
    Calyx-tube pubescent, the lobes erect; bracts pubescent
     on the back and ciliate.   10. *M. clinopodioides.*

1. **M. dídyma** L. Perennial; stem somewhat hairy, 6–10 dm. high; leaf-blades ovate-lanceolate, acuminate, pubescent on both sides, sharply serrate,

7–15 cm. long, 2.5–7 cm. wide, the floral ones and the bracts usually tinged with red; calyx glabrous without, slightly hirsute in the throat within, the teeth subulate; corolla 3–5 cm. long, puberulent. BEE BALM, OSWEGO TEA. Along streams: Que.—Ga.—Iowa—Mich. *Canad.—Allegh.* Jl–S.

2. **M. fistulòsa** L. Perennial; stem sparingly villous, 5–12 dm. high, purple or purple-spotted; leaf-blades ovate or lanceolate, 2–12 cm. long, remotely and sharply serrate; bracts pale or purple; corolla lilac, 3 cm. long. WILD BERGAMOT. Dry hills and thickets: Me.—Man.—Kans.—Tex.—Fla. *E. Temp.* Je–S.

3. **M. comàta** Rydb. Stem 4–6 dm. high, simple; lower leaves with petioles, which are about 1 cm. long, the upper subsessile; blades ovate or lance-ovate, more or less cordate at the base, serrate, 4–6 cm. long, sparingly silky-pilose on both sides; corolla red-purple, almost wine-color, densely villose-puberulent. Hillsides: Colo.—S.D. *Submont.* Jl–Au.

4. **M. menthaefòlia** Benth. Stem 3–10 dm. high, finely puberulent; leaf-blades short-petioled or the upper subsessile, ovate or ovate-lanceolate, at least the lower cordate at the base, more or less cinereous, with short pubescence; corolla rose or lilac, 2–2.5 cm. long. Hillsides and valleys: Sask.—Man.—Ill.—Tex.—Utah—Ida.—Alta. *Prairie—Plain—Mont.* Je–Au.

5. **M. móllis** L. Stem 3–10 dm. high, finely pubescent; leaf-blades elongate deltoid-lanceolate, sharply serrate, 2–5 cm. long, finely puberulent, often paler beneath; corolla lilac, about 2.5 cm. long. Prairies and among bushes: Me.—Ga.—Tex.—B.C. *Temp.—Submont.* Je–Au.

6. **M. Bradburiàna** Beck. Perennial; stem villous-hirsute or glabrate, 3–6 dm. high; leaf-blades ovate or lanceolate, 3–10 cm. long, ciliate, serrate or nearly entire; bracts more or less deeply colored, ciliate; corolla 2–2.5 cm. long, pale purple or white. Thickets and hills: Ind.—Kans.—Ala. *Ozark.* My–Jl.

7. **M. punctàta** L. Perennial; stem 3–10 dm. high; finely pubescent; leaf-blades linear-oblong to lanceolate, 2–8 cm. long, serrate; outer bracts foliaceous, more or less colored; corolla about 2 cm. long. HORSE MINT. Sandy soil: N.Y.—Minn.—Tex.—Fla. *E. Temp.* Jl–S.

8. **M. dispérsa** Small. Annual; stem 2–8 dm. high, often branched, puberulent; leaf-blades oblong-spatulate to oblong-lanceolate, the upper linear, 2–10 cm. long, shallowly serrate; bracts often colored within; corolla about 2.5 cm. long. Plains and prairies: Tenn. — Mo. — Kans. — Tex. — Fla.; Mex. *Austral.* Je–S.

f.489.

9. **M. pectinàta** Nutt. *Fig. 489.* Annual or perennial (?), with a taproot; stems simple or branched at the base, 2–4 dm. high, canescent-strigose; leaf-blades lanceolate or oblanceolate, distantly serrulate, minute strigillose or glabrate, ciliate on the margins, densely punctate, 2–4 cm. long; bracts 12–16 mm. long, green or purplish, lanceolate or oblong, spinulose-cuspidate; corolla yellowish white, 15–18 mm. long, puberulent. *M. citriodora* Coult. [G, B]; not Cerv. *M. Nuttallii* A. Nels. LEMON MINT. Plains, especially in sandy soil: Neb.—Tex.—Ariz.—Utah. *Plain—Submont.*

10. **M. clinopodioìdes** A. Gray. Annual; stem 3–6 dm. high, closely pubescent; leaf-blades oblong to linear, 2–5.5 cm. long, sharply serrate; bracts elliptic or lanceolate, gradually tapering into the bristle-tip, pubescent on the back, strongly ciliate; corolla 2–2.5 cm. long. Plains and prairies: Kans.—Tex. *St. Plains—Texan.* Je–S.

19. HEDEÒMA Pers. MOCK PENNYROYAL.

Annual or perennial herbs, usually aromatic. Leaves opposite, small, entire or sparingly toothed. Flowers perfect, in remote or contiguous axillary clusters. Calyx 13-ribbed, commonly gibbous at the base, pubescent in the more or less constricted throat, somewhat 2-lipped, the 2 lower calyx-lobes being different and usually longer than the 3 upper ones. Corolla blue or purple, 2-lipped; throat but little enlarged; upper lip erect, entire or 2-lobed; lower lip 3-lobed and spreading. Stamens 4, or usually only 2 fertile, the other with much reduced filaments or wanting. Nutlets smooth.

Calyx strongly 2-lipped, with the teeth of the two lips very dissimilar; filaments of the posterior stamens manifest.   1. *H. pulegioides.*
Calyx-teeth similar or nearly so, the lower often somewhat longer; filaments of the posterior stamens rudimentary.
  Calyx-teeth about equal in length; bracts spreading or reflexed, hispid-ciliate.   2. *H. hispida.*
  Calyx teeth of the lower lip much longer than those of the upper; bracts mostly erect, cinereous-hispidulous.   3. *H. camporum.*

1. H. pulegioìdes (L.) Pers. Annual; stem finely pubescent, erect, 1–4 dm. high, branched; leaf-blades oval to oblong, 1–2 cm. long, obtuse, sparingly serrate; upper three calyx-lobes deltoid or deltoid-ovate, the lower two subulate, ciliate; corolla 4–5 mm. long. AMERICAN PENNYROYAL. Dry soil: N.S.—S.D.—Kans.—Fla. *E. Temp.* Jl–S.

2. H. hispida Pursh. Annual; stem 1–2 dm. high, hispidulous; leaves linear, entire, thickish, hispid-ciliate; upper calyx-teeth subulate, the lower more aristiform, equaling the bluish corolla, 6 mm. long. Sandy soil: Sask.—Wis.—Ky.—Colo.—Alta.—Ont. *Prairie—Plain.* My–Jl.

3. H. campòrum Rydb. *Fig. 490.* Perennial, with a slender taproot; stems branched at the base, decumbent or ascending, finely puberulent; leaves spreading, linear-lanceolate, lanceolate, or oblong, 1–2 cm long, grayish-puberulent, subsessile; calyx 7–8 mm. long, somewhat saccate below; teeth subulate, the lower 2 mm., the upper 1 mm. long; corolla puberulent, with an ample limb, 10–12 mm. long, more than half longer than the calyx. *H. longiflora* Rydb.; not Briq. *H. Drummondii* Coulter [B]; not Benth. Cañons: N.D.—Kans. *Plain.*

f.490

20. MELÌSSA L. BEE BALM, LEMON BALM.

Perennial herbs, with branching stems. Leaves opposite, usually toothed. Flowers in axillary one-sided clusters. Calyx at last reflexed, 2-lipped, the tube 13-ribbed, the upper lip with 3 short, the lower with 2 longer lobes. Corolla white or yellowish, the tube curved, glabrous within; upper lip erect, entire or notched, lower lip 3-lobed, spreading. Stamens 4, the anther-cells spreading. Nutlets smooth.

1. M. officinàlis L. Perennial, lemon-scented; stem 3–6 dm. high; leaf-blades ovate, 2–8 cm. long, crenate-serrate, ciliate; calyx 8 mm. long, hairy without and within; corolla 1 cm. long, the upper lip inflated; lower lip bearded near the base. Waste places: Me.—Kans.—Fla.; also Ore.—B.C.; adv. from Eur. Je–Au.

21. CLINOPÒDIUM L. BASIL.

Annual or perennial herbs or rarely shrubs. Leaves opposite, entire or toothed, narrow or broad. Flowers in ours in dense axillary clusters subtended by linear-filiform bracts. Calyx 13-ribbed, often swollen at the base; upper lip with 3 short lobes; lower lip with 2 longer lobes. Corolla 2-lipped, abruptly dilated at the throat; upper lip erect; lower lip spreading, 3-lobed. Stamens 4, all anther-bearing; anther 2-celled, awnless. Nutlets smooth.

1. **C. vulgàre** L. Perennial, with a rootstock; stem 1–5 dm. high, simple or nearly so, sparingly hairy; leaf-blades oval or ovate, 1–4 cm. long, obtuse, undulate or crenate; calyx 8–9 mm. long, villous-hirsute, ribbed; lobes subulate, aristate; corolla white or purple, 10–12 mm. long. *Satureja vulgaris* Fritsch. [G]. Thickets and waste places: N.S.—N.C.—Colo.—Man.; nat. or adv. from Eur. *Plain—Mont.* Je–Au.

22. CALAMÍNTHA Moench. CALAMINT.

Annual or perennial herbs or shrubs. Leaves opposite, entire or toothed. Flowers few in the axils, with minute bracts. Calyx hairy in the throat, 2-lipped, the tube 13-ribbed, swollen at the base; upper lip with 3 short, the lower with 2 longer lobes. Corolla longer than the calyx, the throat abruptly dilated; upper lip entire or notched, the lower spreading, 3-lobed. Stamens 4, anther-bearing. Nutlets smooth.

1. **C. Nuttállii** Benth. Perennial, stoloniferous; stem 1–3 dm. high, tufted; leaves of the stolons obovate, the lower stem-leaves spatulate, the upper numerous, linear, entire or sparingly toothed; flower-clusters 2–6-flowered; pedicels longer than the calyx; calyx 4–5 mm. long, the lobes lance-subulate; corolla 8–9 mm. long, puberulent, purple. *Clinopodium glabrum* (Nutt.) Kuntze [B]. *Satureja glabra* Fernald [G]. Rocky banks: Ont.—Minn.—Tex. —N.Y. *Allegh.—Ozark.* My–Au.

23. PYCNÁNTHEMUM Michx. MOUNTAIN MINT, HORSE MINT.

Perennial caulescent herbs. Leaves opposite, but often close together on the short branches, entire or toothed. Flowers in dense axillary or terminal clusters. Calyx ovoid or cylindric, 10–13-ribbed, not conspicuously pubescent in the throat, with 5 lobes. Corolla white or purplish, longer than the calyx, 2-lipped; upper lip erect, entire or notched; lower lip 3-lobed. Stamens 4, nearly equal, the filaments usually exserted; anthers 2-celled, the cells parallel. Nutlets various, smooth, roughened, or pubescent. *Koellia* Moench.

Leaf-blades linear or lanceolate, at least three times as long as broad.
  Leaves entire.
    Calyx-lobes deltoid.   1. *P. virginianum.*
    Calyx-lobes subulate to lanceolate.   2. *P. flexuosum.*
  Leaves, at least the larger ones, toothed.
    Flower-clusters very woolly at maturity; stem loosely
      pubescent; stamens exserted.   3. *P. pilosum.*
    Flower-clusters not woolly at maturity; stem closely
      pubescent; stamens included.   4. *P. verticillatum.*
Leaf-blades ovate, mostly about twice as long as broad.   5. *P. muticum.*

1. **P. virginiànum** (L.) Durand & Jackson. *Fig. 491.* Stem puberulent, 4–9 dm. high, branched above; leaves lanceolate or linear-lanceolate, 2–5 cm. long, acute, mostly entire, subsessile; flower-clusters 8–10 mm. broad; bracts linear or lance-linear; calyx 3.5–4 mm. long, the lobes one-third as long as the tube; corolla 6–7 mm. long, the upper lip ovate, notched, the three lobes of the lower lip oblong or ovate. *Pycnanthemum lanceolatum* Pursh. *K. virginiana* Britton [B]. Thickets: Me.—S.D.—Kans.— Ala.—Ga. Jl–S.

2. **P. flexuòsum** (Walt.) B.S.P. Stem glabrous, 4–7 dm. high; leaves firm, linear or lance-linear, glabrous or nearly so, sessile, 3–5 cm. long, 1–3 mm. wide; flower-clusters dense, 6–10 mm. broad; bracts appressed, acuminate; calyx-lobes one

f. 491.

third as long as the tube. *P. linifolium* Pursh. *K. flexuosa* MacMill. [B]. Fields and thickets: Me.—Fla.—Tex.—Kans.—Minn. *E. Temp.* Jl–S.

3. P. pilòsum Nutt. Stem pubescent, 3–7 dm. high; leaves lanceolate, short-petioled, entire or sparingly denticulate, 3–5 cm. long, 6–12 mm. wide, pubescent; flower-clusters dense, about 8 mm. broad; bracts lanceolate, villous-pubescent, at least equaling the clusters; calyx-teeth ciliate, one fourth as long as the tube. Prairies and open woods: Ont.—Ga.—Ark.—Kans.—Iowa. *Allegh.* —*Austral.* Jl–S.

4. P. verticillàtum (Michx.) Pers. Stem closely pubescent, 3–8 dm. high; leaves lanceolate or lance-linear, subsessile, serrulate or entire, 2–7 cm. long, 6–20 mm. wide; flower-clusters dense, canescent, 10–12 mm. broad; bracts appressed, at least equaling the clusters; calyx-teeth one fourth as long as the tube; corolla pubescent. *K. verticillata* Kuntze [B]. Moist places: Vt.—N C. —Mo.—Minn. *Allegh.* Jl–S.

5. P. mùticum (Michx.) Pers. Stem puberulent, 3–8 dm. high; leaves short-petioled or subsessile, ovate or lance-ovate, rounded at the base, sharply serrate, 3–7 cm. long, 1–4 cm. wide, the lower glabrate, the upper white-canescent; flower-clusters dense, 8–12 mm. broad; bracts appressed, lanceolate; calyx teeth one fifth as long as the tube, pubescent. *K. mutica* Britton [B]. Sandy soil: Me.—Fla.—Mo.—Iowa. *E. Temp.* Jl–S.

24. LÝCOPUS (Tourn.) L. WATER HOREHOUND, BUGLE WEED.

Perennial caulescent herbs, mostly odorless, often stoloniferous. Leaves opposite, entire, toothed, or pinnatifid. Flowers perfect, in remote axillary verticils. Calyx regular or nearly so, 4- or 5-toothed. Corolla funnelform or campanulate; lobes 4, nearly equal, or one of them broader. Anther-bearing stamens 2, with or without 2 additional rudimentary filaments. Anther-sacs 2, parallel. Nutlets 3-angled, truncate, smooth.

Calyx-teeth ovate, obtusish, shorter than the nutlets.
  Corolla 1.5–2 mm. long; calyx-teeth acute; style included.   1. *L. virginicus.*
  Corolla 2–3 mm. long; calyx-teeth obtuse; style exserted.   2. *L. uniflorus.*
Calyx-teeth lanceolate or subulate, acute or acuminate, longer than the nutlets.
  Stem and lower surface of the leaves densely and finely pubescent, the former often velvety.   3. *L. velutinus.*
  Stem sparingly and coarsely pubescent or glabrous; leaves glabrous or nearly so.
    Leaves merely coarsely serrate.
      Bracts minute; corolla twice as long as the calyx.   4. *L. rubellus.*
      Bracts, at least the outer ones, conspicuous; corolla scarcely exceeding the calyx.
        Leaves tapering at the base, rather thin.   5. *L. lucidus.*
        Leaves rounded at the base, thicker.   6. *L. asper.*
    Leaves sinuately pinnatifid.   7. *L. americanus.*

1. L. virgínicus L. Stem 1–8 dm. high, purplish, puberulent, obtusely angled, not tuberous at the base, but the stolons tuber-bearing; leaves ovate to elliptic, 2–14 cm. long, acuminate at both ends, coarsely toothed, petioled; calyx cylindric, the lobes erect, oblong-lanceolate; corolla less than twice as long as the calyx; lobes almost equal, erect; stamens included. Moist soil: N.Y.—Neb.—Kans.—Ala.—Ga. *Canad.—Allegh.* Jl–S.

2. L. uniflòrus Michx. Rootstock tuberous-thickened at the base of the stem; stem 1–8 dm. high, acutely angled; leaves 2.5–8 cm. long, lanceolate or oblong-lanceolate, sharply serrate; calyx campanulate; corolla 2–3 mm. long; lobes spreading. *L. communis* Bickn. [B]. Wet places: Newf.—Va.—Neb.— Ore.—Alaska. *Temp.—Submont.* Au–S.

3. **L. velùtinus** Rydb. Stem 3–5 dm. high; leaf-blades lance-elliptic, coarsely toothed, 4–10 cm. long, short-petioled; calyx campanulate, copiously pubescent, 2.5 mm. long; lobes lanceolate, acuminate, the upper two recurved; corolla 3.5 mm. long, glandular-punctate. Low ground and swamps: Colo.— Tex.—Miss.—Ark. *Austral.—Son.* Au.

4. **L. rubéllus** Moench. Stem glabrous or finely puberulent, green or purplish, 5–10 dm. high, sharply angled, with scaly or leafy stolons; leaf-blades oblong, lanceolate or elliptic, 3–15 cm. long, acuminate, sharply serrate; bracts minute; calyx 2.5–3 mm. long, the lobes deltoid or deltoid-subulate; corolla fully twice as long as the calyx, sparingly glandular-punctate. Low ground: N.Y.—Minn.—La.—Fla. *Allegh.—Austral.* Jl–O.

5. **L. lùcidus** Turcz. Stem stout, strict, 3–9 dm. high, sparingly pubescent on the angles; leaves oblong-lanceolate, sessile or short-petioled, 5–15 cm. long, sharply serrate; calyx-teeth subulate-lanceolate, nearly as long as the tube; corolla little longer than the calyx. Wet soil, especially in thickets: Neb.— Kans.—Ariz.—Calif.—B.C.; Asia. *W. Temp.—Plain —Submont.* Jl–S.

6. **L. ásper** Greene. Stem 3–6 dm. high, obtusely angled; leaves firm, narrowly lanceolate, acute, sessile, strongly serrate, 4–6 cm. long, very veiny, rough on both sides; calyx-teeth ovate-lanceolate, acuminate. Wet places: w Ont.—Neb.—B.C. *Boreal. —Plain.* Jl–Au.

7. **L. americànus** Muhl. *Fig. 492.* Stem sharply angled, puberulent or glabrous, 1–10 dm. high; leaves lanceolate in outline, petioled, acuminate, pinnatifid or incised, 3–10 cm. long; calyx-teeth triangular-subulate, cuspidate, rigid; corolla little exceeding the calyx. *L. sinuatus* Ell. WATER HOARHOUND. Swamps and wet meadows: Newf. —Fla.—Calif.—B.C. *Temp.—Submont.* Je–O.

f. 492.

25. **MÉNTHA** (Tourn.) L. MINT, SPEARMINT, PEPPERMINT.

Aromatic caulescent perennial herbs, with rootstocks. Leaves opposite, punctate, toothed, usually petioled. Flowers perfect, in dense, axillary clusters. Calyx campanulate, 10-ribbed, regular or nearly so, 5-lobed. Corolla funnelform or campanulate, nearly regular; lobes 4, the upper larger than the rest. Stamens 4, anther-bearing, erect; anther-sacs 2, parallel. Nutlets smooth.

Whorls of flowers forming terminal spikes.
  Spikes less than 1 cm. thick; leaves sessile.   1. *M. spicata.*
  Spikes more than 1 cm. thick; leaves more or less petioled.
    Leaves lanceolate, narrowed at the base; calyx-teeth
      hirsute.   2. *M. piperita.*
    Leaves ovate, rounded at the base; calyx glabrous.   3. *M. citrata.*
Whorls of flowers all axillary.
  Calyx-tube glabrous, the teeth ciliate.   4. *M. gentilis.*
  Calyx more or less pubescent throughout.
    Leaf-blades ovate, rounded at the base.   5. *M. arvensis.*
    Leaf-blades lanceolate or lance-ovate, more or less cuneate
      at the base.
      Stem and petioles densely pubescent, with villous
        hairs; leaves decidedly pubescent.   6. *M. canadensis.*
      Stem and petioles retrorsely strigose or crisp-hairy
        with short hairs, or glabrous below; leaves
        sparingly puberulent or glabrous.
        Leaf-blades thin, dark green and not strongly
          veined, tapering gradually into slender peti-
          oles, which equal or exceed the flower-clusters.   7. *M. glabrior.*

Leaf-blades thick, strongly veined, abruptly contracted into short petioles, which are much
shorter than the flower-clusters.   8. *M. Penardi.*

1. M. spicàta L. Stem erect, 3–5 dm. high, glabrous; leaves lanceolate,
sessile or nearly so, sharply serrate, 3–7 cm. long; spikes 5–10 cm. long, bracts
subulate-lanceolate, ciliate; calyx-teeth subulate; corolla glabrous. SPEARMINT.
Fields and waste places: N.S.—Fla.—Kans.—Minn.; Calif.—Utah—B.C.; nat.
from Eur. Jl–S.

2. M. piperìta L. Stem glabrous, 2–7 dm. high; leaf-blades 1.5–5 cm.
long, acute, sharply serrate, short-petioled; spikes 2–15 cm. long, often interrupted; calyx glabrous, 2.5 mm. long, cylindro-campanulate, the lobes subulate from a broad base, shorter than the tube; corolla 4 mm. long. PEPPERMINT.
Wet places: N.S.—Minn.—Ala.—Fla.; nat. from Eur. Jl–S

3. M. citràta Ehrh. Stem glabrous, 3–11 dm. high; leaf-blades ovate
to orbicular-ovate, 1.5–4 cm. long, acute, sharply serrate, rounded or subcordate at the base; calyx 4 mm. long, cylindro-turbinate, the lobes subulate from
a broad base; corolla pink or white, 4.5–5 mm. long. BERGAMOT MINT. Wet
places: N.Y.—Iowa—Ga.; nat. from Eur. Jl–S.

4. M. gentìlis L. Stem sparingly pubescent, 2–7 dm. high, much
branched; leaf-blades ovate to oval or elliptic, 1.5–5 cm. long, serrate, cuneate
at the entire base; calyx 1.5 mm. long, campanulate, the lobes deltoid, acute;
corolla deep-pink or lilac, 2.5 mm. long. Waste places: Me. Iowa—Ga.; nat.
from Eur. Au–O.

5. M. arvénsis L. Stem 1.5–6 dm. high, closely pubescent; leaf-blades
2–6 cm. long, shallowly serrate; calyx 2–3 mm. long, finely pubescent, campanulate, the lobes deltoid, as broad as long, one-third as long as the tube;
corolla bright pink, 4–5 mm. long. FIELD MINT. Waste places and river banks:
N.B.—N.D.—Neb.—Fla.; nat. from Eur. Jl–S.

f. 493.

6. M. canadénsis L. Stem 2–6 dm. high, pubescent with recurved hairs; leaf-blades elliptic to
lanceolate, 2–7 cm. long, acute, serrate, cuneate at
the entire base; calyx 2 mm. long, pubescent, campanulate, the lobes lanceolate-subulate, half as long
as the tube; corolla pink or white, 3 mm. long. *M.
arvensis canadensis* Briq. [G]. WILD MINT. Wet
meadows and among bushes: N.B.—Man.—Neb.—
N.C. *Canad.—Allegh.* Jl–O.

7. M. glàbrior (Hook.) Rydb. *Fig. 493.* Stem
3–6 dm. high, glabrate or minutely puberulent on the
angles; leaf-blades ovate, elliptic or lanceolate,
acute, serrate, glabrous or nearly so; calyx 3 mm.
long, finely pubescent; lobes triangular-lanceolate,
acuminate, slightly longer than broad; corolla narrowly campanulate, 4 mm. long. *M. canadensis gla-
brata* Benth. *M. arvensis glabrata* Fern. [G]. Wet places: Me.—Pa.—Colo.—
Utah—Mont. *Boreal.—Submont.* Jl–Au.

8. M. Penárdi (Briq.) Rydb. Stem 2–4 dm. high, short-pubescent, especially on the angles; leaf-blades ovate-elliptic or ovate-lanceolate, 2–5 cm.
long, serrate, acute; calyx about 3 mm. long, softly pilose; teeth lanceolate,
acute; corolla usually pink, 4 mm. long. Wet places, among bushes: S.D.—
Neb.—N.M.—Utah—B.C.—Mack. *Submont.* Je–Au.

Family 146. **SOLANACEAE.** Potato Family.

Herbaceous plants or rarely shrubs, with narcotic or stimulant properties. Leaves alternate, without stipules. Flowers perfect, usually regular. Calyx of 5 (rarely 4 or 6) more or less united sepals, mostly persistent. Corolla hypogynous, rotate, campanulate, funnelform, or salverform; limb mostly lobed. Stamens 5, rarely 4 or 6, fertile; filaments
adnate to the tube of the corolla and alternate with its lobes; anthers
introrse, opening lengthwise or by pores. Gynoecium usually of 2, rarely
3–5, united carpels; ovary 2- (rarely 3–5-) celled; style terminal; stigma
entire. Ovules and seeds numerous, amphitropous. Fruit a berry or
capsule. Seeds crustaceous, often tuberculate, flattened; endosperm
copious, fleshy.

Fruit a berry.
  Corolla plicate ; lobes usually induplicate ; all our species herbs or vines.
    Calyx inflated and bladder-like in fruit.
      Sepals nearly distinct, auricled at the base ; ovary 3–5-celled.
                                          1. Nicandra.
      Sepals united to near the tip ; not auricled ; ovary 2-celled.
        Corolla open-campanulate, yellow or whitish, often with a darker cen-
          ter ; seeds finely pitted ; flowers nodding in anthesis.
                                          2. Physalis.
        Corolla rotate, violet or purple ; seeds rugose-tuberculate ; flowers erect
          in anthesis.   3. Quincula.
    Calyx not bladder-like in fruit.
      Calyx closely investing the berry.
        Stamens alike, not declined ; unarmed herbs.
          Lobes of the fruiting calyx much exceeding the berry ; annuals.
                                    4. Leucophysalis.
          Lobes of the fruiting calyx not exceeding the berry ; perennials.
                                    5. Chamaesaracha.
        Stamens dissimilar, declined ; prickly annuals.   6. Androcera.
      Calyx not inclosing the berry.
        Anthers short, opening by a terminal pore or short slit.
                                    7. Solanum.
        Anthers long, tapering to the summit, opening longitudinally.
                                  8. Lycopersicon.
  Corolla little if at all plicate, its lobes imbricate ; shrubs. 9. Lycium.
Fruit a capsule.
  Capsule circumscissile near the top, which separates as a lid ; corolla irregular.
                                      10. Hyoscyamus.
  Capsule opening by valves ; corolla regular.
    Capsule prickly ; seeds flat.   12. Datura.
    Capsule not prickly ; seeds scarcely flattened.   11. Nicotiana.

1. NICÁNDRA Adans. Apple of Peru.

Annual caulescent herbs. Leaves alternate, with broad, sinuately toothed
or lobed blades. Flowers solitary in the axils, nodding. Calyx angled, accrescent, inflated in fruit; sepals nearly distinct, cordate, auricled at the base.
Corolla blue or violet, campanulate, slightly 5-lobed, plicate in bud. Stamens 5;
filaments adnate to the corolla, dilated at the base; anthers opening lengthwise. Ovary 3–5-celled; stigma 3–5-lobed. Berry rather dry, enclosed in the
calyx. *Physalodes* Boehm.

1. **N. Physalòdes** (L.) Pers. Plant glabrous or nearly so; stem 3–9 dm.
high, branched; leaf-blades ovate, oval, or oblong, 5–15 cm. long, sinuately
lobed; calyx becoming 3.5 cm. long in fruit; corolla 2.5 cm. broad. *P. Physalodes* Britton [B]. Waste places: N.S.—N.D.—Kans.—Fla.; nat. from Peru.
E. Temp.—Trop. Jl–S.

2. PHÝSALIS L. Ground Cherry, Strawberry Tomato, Tomatillo.

Annual or perennial herbs. Leaves alternate, entire or sinuately toothed.
Pedicels usually solitary from the axils of the leaves. Calyx campanulate,
5-lobed, accrescent and becoming bladder-like in fruit, 5-angled or prominently

10-ribbed. Corolla yellow or whitish, campanulate or funnelform, often with a darker brownish or purplish center, plicate in the bud. Stamens adnate to the base of the corolla; anther-sacs opening by longitudinal slits. Stigma minute. Seeds numerous, kidney-shaped, flattened, finely pitted.

Annuals, with branching roots.
  Plant more or less viscid-pubescent.
    Fruiting calyx sharply 5-angled, deeply sunken at the base; calyx-lobes at the
      flowering time lanceolate, subulate-tipped.
      Leaf-blades ovate, subentire at the base; stem slender, diffuse, sharply
        angled.   1. *P. pubescens.*
      Leaf-blades cordate, strongly sinuate-toothed to the
        base; stem obtusely angled.   2. *P. pruinosa.*
    Fruiting calyx obtusely or indistinctly 5–10-angled;
      calyx-lobes at flowering time triangular, acute.   3. *P. missouriensis.*
  Plant glabrous or only the upper parts short-hairy, not
    viscid.
    Peduncles shorter than the flowers; calyx-lobes triangu-
      lar; corolla with a brown center.   4. *P. ixocarpa.*
    Peduncles longer than the flowers; calyx-lobes lance-
      olate; corolla wholly yellow.
      Peduncles much longer than the fruiting calyx;
        corolla 5–8 mm. broad.   5. *P. pendula.*
      Peduncles scarcely exceeding the fruiting calyx;
        corolla 8–10 mm. broad.   6. *P. angulata.*
Perennials, with horizontal rootstocks, or rarely with woody
  caudices.
  Pubescence, if any, not stellate, although in *P. pumila*
    with some branched hairs.
    Leaves and stem glabrous, or the veins of the former
      and the upper part of the latter with scattered
      appressed hairs.
      Fruiting calyx ovoid, nearly filled with the berry,
        scarcely sunken at the base.
        Leaves ovate, ovate-lanceolate, or oval.   7. *P. subglabrata.*
        Leaves lanceolate, oblanceolate, or linear.   8. *P. longifolia.*
      Fruiting calyx pyramidal, very much inflated and
        deeply sunken at the base.   9. *P. macrophysa.*
    Leaves and stem more or less pubescent with spreading
      hairs.
      Pubescence sparse, consisting of flat, sometimes
        jointed hairs, scarcely viscid.
        Fruiting calyx ovoid, scarcely angled and
          scarcely sunken at the base; leaves thick,
          oblanceolate or spatulate to rhombic, sub-
          entire.
          Leaves oblanceolate or spatulate; hairs all
            simple.   10. *P. lanceolata.*
          Leaves broader, often rhombic; hairs on the
            lower surface branched.   11. *P. pumila.*
        Fruiting calyx pyramidal-ovoid, obtusely 5-angled
          and deeply sunken at the base; leaves ovate
          to lanceolate, generally more or less toothed.   12. *P. virginiana.*
      Pubescence dense, viscid, partly of fine and short,
        partly of long, flat, jointed hairs.
        Leaves large; blades over 5 cm. long, more or
          less cordate.
          Stem densely and leaves sparingly pubescent
            with long (2 mm. or more) white hairs.
            Stem erect; anthers purple; leaf-blades
              rounded or subcordate at the base.   13. *P. ambigua.*
            Stem spreading; anthers yellow; leaf-
              blades cuneate or acute at the base.   14. *P. nyctaginea.*
          Stem and leaves densely pubescent with
            shorter viscid hairs, the stem and veins
            beneath also with fewer long hairs; an-
            thers yellow.   15. *P. heterophylla.*
        Leaves smaller; blades less than 5 cm. long;
          long flat hairs few, mostly confined to
          the calyx.
          Plant erect or ascending; leaf-blades rounded-
            ovate or rhombic.   16. *P. comata.*
          Plant prostrate, diffuse; leaf-blades nearly
            orbicular.   17. *P. rotundata.*
  Pubescence more or less stellate.   18. *P. mollis.*

1. **P. pubéscens** L. Annual, more or less villous and viscid; stem spreading, branched, swollen at the nodes; leaf-blades thin, 2–6 cm. long, ovate, acute or acuminate, oblique at the entire base, pubescent; peduncles short, in fruit about 1 cm. long; corolla 5–10 mm. broad, yellow, with a dark center; fruiting calyx 2–3 cm. long. Sandy soil: Pa.—Kans.—Calif.—Fla.; Mex.— S.Am. *Austral.—Trop.* Jl–S.

2. **P. pruinòsa** L. Stem stout, finely villous and somewhat viscid, erect or spreading, 3–6 dm. high; leaf-blades 3–10 cm. long, finely pubescent, sinuately toothed; calyx viscid, in fruit 2–3 cm. long, sunken at the base; corolla 3–8 mm. in diameter; berry yellow or green. Cultivated ground and waste places: Mass.—Fla.—Tenn.—Iowa; adv. in Colo. *Plain—E. Temp.* Jl–S.

3. **P. missouriénsis** Bush. Annual, finely pubescent and viscid; stem spreading, branched, often zigzag, villous with short hairs; leaf-blades 1–8 cm. long, ovate, oblique at the base, acute, sinuately dentate, hairy at least on the veins; peduncles 5–10 mm. long in fruit; calyx-lobes triangular, acute; corolla 3–8 mm. broad, yellow, usually with a darker center; anthers yellow; fruiting calyx 1.5–2 cm. long, not sunken at the base. *P. Lagascae* Rydb. [B]; not R. & S. Rich soil: Mo.—Kans.—Okla.—Ark. *Ozark.* Jl–S.

4. **P. ixocárpa** Brot. Stem first erect, later widely spreading, angled, glabrous or the younger parts sparingly pubescent; leaves from cordate to ovate, with a cuneate base, sinuately dentate or entire, 2.5–7 cm. long; corolla 1–1.5 cm. broad; fruiting calyx rounded-ovoid, obscurely 10-angled; berry purple. TOMATILLO. Cultivated ground and waste places: introd. in Mass.— Iowa—Va.; naturalized in the south; nat. of Mex.—S.Am.; cult.

5. **P. péndula** Rydb. Annual, glabrous; stem angled, about 5 dm. high; leaf-blades lanceolate or ovate-lanceolate, sinuate-dentate, rarely entire; calyx-teeth broadly triangular; peduncles in fruit 4–5 cm. long, drooping; corolla 5–8 mm. broad, wholly yellow; fruiting calyx 2 cm. long, indistinctly 10-angled, purple-veined, nearly filled with the berry. *P. lancifolia* Rydb., in part [B]; not Nees. Rich soil: Ill.—Kans.—Tex. *Ozark.*

6. **P. angulàta** L. Annual; stem slender, 5–10 dm. high, glabrous; leaf-blades ovate, sharply sinuate-dentate, thin, 4–7 cm. long; peduncles 2–4 cm. long in fruit; calyx-lobes lanceolate; corolla 8–10 mm. broad, yellow; anthers usually purple-tinged; fruiting calyx about 3 cm. long, obtusely 5–10-angled, nearly filled with the berry. Rich soil: Va.—Iowa—Tex.—Fla.; Mex.—Brazil; W.Ind. *Austral.—Trop.* Jl–S.

7. **P. subglabràta** Mack. & Bush. Stem 3–9 dm. high; leaf-blades 3–7 cm. long, entire or undulate, glabrous or very sparingly strigose; calyx strigose, in fruit ovoid, reticulate, 2.5–3 cm. long; corolla yellow, with darker center, 1–1.5 cm. broad. River valleys, sandy or cultivated ground: Ont.—Pa.—Ark.—Colo. —Ida. *E. Temp.—Plain—Submont.* Je–S.

8. **P. longifòlia** Nutt. Stem tall, 5–10 dm. high; leaf-blades entire-margined or somewhat repand; calyx glabrous or nearly so, in fruit ovoid, about 3 cm. long; corolla 1–2 cm. broad, yellow, with brownish center; berry yellow, the lower portion and the stipe glutinous. River valleys and rich soil: Iowa—Ark.— Ariz.—Mont.; Mex. *Prairie—Plain—Submont.* My–Au.

9. **P. macrophỳsa** Rydb. Perennial, with a rootstock; stem 5–10 dm. high, glabrous; leaf-blades thin, 4–8 cm. long, the lower obtuse, the upper acute or acuminate, slender-petioled; peduncles reflexed at maturity; calyx-teeth ovate-triangular; corolla yellow, with a dark center, 2 cm. broad; fruiting calyx 3–4 cm. long, indistinctly 10-angled; berry small, in the center of the inflated calyx. Rich soil: Ark.—Kans.—Tex. *Ozark.* My–Jl.

10. **P. lanceolàta** Michx. Stem 3–6 dm. high, erect or spreading, sparingly hirsute; leaf-blades acute or obtuse, entire or rarely wavy, thickish, spar-

ingly short-hirsute; calyx hirsute or strigose; corolla dull yellow, with brownish center, about 1.5 cm. broad; berry yellow or greenish. Prairies and plains: Ill. —S.C.—Ariz.—S.D. *Prairie—Plain—Submont.* Jl–S.

11. P. pùmila Nutt. Stem 3–10 dm. high, obscurely angled, hirsute; leaf-blades thick, acute at both ends, 5–10 cm. long, entire or sinuate; calyx densely hirsute, 4–5 cm. long; corolla yellow, with brown center, 1.5–2 cm. broad. Prairies and river valleys: Ill.—Ark.—Tex.—Colo. *Prairie—Plain—Submont.* My–Jl.

12. P. virginiàna Mill. Stem 3–10 dm. high, erect, angular, more or less hirsute; leaf-blades 3–6 cm. long, hirsutulous; calyx hirsute or puberulent; corolla 1.5–2.5 cm. broad, sulphur-yellow, with purplish center; berry reddish. Prairies, river valleys, and cultivated ground: N.Y.—Fla.—Tex.—Mont. *E. Temp.—Submont.* Ap–O.

13. P. ambígua (A. Gray) Rydb. Perennial, with a rootstock, long-villous, but slightly viscid; stem 3–8 dm. high; leaf-blades more than 5 cm. long, thin, light green, round-ovate or subcordate, sinuately toothed or entire, acute; calyx-lobes triangular; corolla 1.5–2 cm. broad, yellow, with a darker center; fruiting calyx ovoid, slightly sunken at the base. Rich soil: Vt.—Iowa—Tenn.—Va. *Canad. Allegh.*

14. P. nyctagínea Dunal. Perennial, with a rootstock; stem long-villous, somewhat zigzag, spreading, branched; leaf-blades dark green, ovate, oblique, rather thick, entire or somewhat sinuate-dentate; peduncles long-villous, 1–2 cm. long and reflexed in fruit; corolla 1–1.5 cm. broad, yellow, with a darker center; fruiting calyx ovoid, 5-angled, sunken at the base. Dry soil: R.I.—Iowa—La. *Allegh.* Je–S.

15. P. heterophýlla Nees. *Fig. 494.* Stem 3–10 dm. high, erect or decumbent and spreading, puberulent and villous, viscid; leaf-blades 5–10 cm. long, usually broadly cordate, acute or acuminate, thick, more or less sinuately toothed, densely pubescent and very viscid; calyx viscid-villous, in fruit ovoid, somewhat angled and sunken at the base;  corolla 1.5–2 cm. wide, yellow, with purple center; berry yellow. Cultivated ground, sandy or loose soil, etc.: N.B.—Fla.—Tex.—Utah—Sask. *E. Temp.—Submont.* My–S.

16. P. comàta Rydb. Stem about 5 dm. high, finely short-pubescent and viscid, with a few scattered long hairs on the upper part; leaf-blades 3–5 cm. long, somewhat repand or entire, finely pubescent; calyx finely pubescent and villous, in fruit 3–4 cm. long, rounded-ovoid, scarcely sunken at the base; corolla greenish yellow, with brown center. Hillsides, cliffs, and dry plains: Neb.—Colo.—Tex. *Plain—Submont.* Jl–Au.

17. P. rotundàta Rydb. Stem diffuse or spreading, dichotomously zigzag-branched, densely and finely viscid-pubescent; leaf-blades with truncate or subcordate base, 2–4 cm. broad, sinuate-dentate; fruiting calyx ovoid, scarcely angled, not sunken at the base; corolla 1.5 cm. broad, greenish yellow, with brownish center. Plains: S.D.—Tex.—N.M.—Colo. *Plain.* Jl–S.

18. P. móllis Nutt. Perennial, with a horizontal rootstock; stem 3–6 cm. high, stellate-tomentose; leaf-blades rounded-cordate or the upper ovate, coarsely toothed; peduncles 2–4 cm. long, or in fruit 4–6 cm. long; calyx densely stellate, the lobes triangular; corolla 1.5–2 cm. broad, bright yellow, with a purplish center; fruiting calyx ovoid, acuminate, 3–5 cm. long, slightly

5-angled and somewhat sunken at the base. Thickets and banks: Ark.—Kans. —Calif.—Tex.—Mex. *Ozark.—Son.*

3. QUÍNCULA Raf.

f.495.

Herbaceous, scurfy-granuliferous perennials, with diffuse stems. Leaves alternate, sinuately lobed or pinnatifid, somewhat fleshy. Peduncles commonly in pairs from the axils of the leaves. Flowers perfect, regular, erect during anthesis. Calyx campanulate, accrescent and becoming bladder-like in fruit, 5-angled, reticulate, with 5 converging lobes. Corolla flat, rotate, pentagonal in outline, violet or purple. Anthers opening by longitudinal slits. Seeds comparatively few, flattened, kidney-shaped, rugose-tuberculate.

1. **Q. lobàta** (Torr.) Raf. *Fig. 495.* Stem low, spreading or prostrate; leaf-blades oblanceolate or spatulate to oblong, sinuately lobed or more or less pinnatifid, with rounded lobes; calyx in fruit sharply 5-angled, ovoid, sunken at the base; corolla purplish, 2–3 cm. broad. *Physalis lobata* Torr. High plains and bluffs: Kans. —Tex.—Ariz.; Mex. *Plain—Son.—Submont.* Ap–Au.

4. LEUCOPHÝSALIS Rydb.

Erect, villous and viscid annuals. Leaf-blades entire, decurrent on the petioles. Flowers usually 3 or 4, fascicled in the axils. Calyx campanulate, 5-lobed, soon filled with berry, thin, neither angled nor ribbed, faintly veined. Corolla rotate, white, with a cream-colored or yellow center, plicate. Stamens adnate to the base of the corolla; anthers opening by longitudinal slits. Ovary 2-celled. Seeds kidney-shaped, flat, punctate.

1. **L. grandiflòra** (Hook.) Rydb. Stem erect, 3–9 dm. high, more or less villous; leaf-blades 1–2 dm. long, ovate or lanceolate, acute, villous and viscid; calyx-lobes lanceolate, fully equaling the tube; corolla 3–4 cm. broad, *Physalis grandiflora* Hook. [G]. Rich soil: Vt.—Sask.—Minn. *Canad.* My–Jl.

5. CHAMAESARÀCHA A. Gray.

Scurfy perennial herbs. Leaves alternate, entire or pinnatifid, with winged petioles. Peduncles solitary or 2 or 3 in the axils of the leaves. Calyx campanulate, 5-lobed, only slightly accrescent, not becoming bladder-like in fruit, close-fitting to the berry. Corolla rotate, white or ochroleucous, often tinged with purple, plicate. Stamens 5, filaments adnate to the base of the corolla; anthers opening by longitudinal slits. Seeds reniform, flattened, rugose-favose.

Pubescence dense, hirsute as well as puberulent.   1. *C. conioides.*
Pubescence sparse, puberulent or stellate, hirsute if at all only on
  the calyx.   2. *C. Coronopus.*

1. **C. conioìdes** (Moric.) Britton. Stems erect, decumbent or prostrate, viscid; leaf-blades oblanceolate to obovate-rhombic, acute and tapering into short petioles, usually deeply lobed, varying from subentire to pinnatifid; calyx-lobes triangular, acutish; corolla about 1 cm. broad, white or ochroleucous. *C. sordida* (Dunal) A. Gray. Dry clayey soil: Kans.—Colo.—Ariz.—Tex.; Mex. *St. Plains—Son.* My–Jl.

2. **C. Corónopus** (Dunal) A. Gray. Stem obtusely angled; leaf-blades linear or lanceolate, sinuately lobed to pinnatifid, pruinose; calyx-lobes triangular, acute; corolla white or ochroleucous. Dry clayey soil: Kans.—Colo.—Ariz. —Tex. *St. Plains—Son.—Submont.* My–D.

6. ANDRÓCERA Nutt. BUFFALO BUR.

Annual prickly herbs. Leaves alternate, pinnatifid. Flowers perfect, cymose. Calyx campanulate, prickly, somewhat accrescent and enclosing the berry in fruit, 5-lobed. Corolla yellow, violet, or purple, 5-lobed, plaited. Stamens 5, dissimilar, declined, adnate to near the throat of the corolla; anthers narrow, opening by terminal pores. Ovary 2-celled. Fruit a berry, enclosed in the prickly calyx. Seeds flattened.

Corolla yellow ; plant stellate-pubescent.	1. *A. rostrata.*
Corolla violet ; plant glandular-pubescent.	2. *A. citrullifolia.*

1. A. rostràta (Dunal) Rydb. Stem yellowish-hirsute and prickly, 2–7 dm. high, widely branching; leaf-blades 5–12 cm. long, once or twice pinnatifid; segments oval to oblong, obtuse, entire or undulate, yellowish hirsute; calyx bristly; corolla yellow, about 2.5 cm. broad. *Solanum rostratum* Dunal [G, B]. Waste places, prairies, and river valleys: N.D.—Tex.—N.M.—Wyo.; Mex.; adv. eastward to N.H. and Fla. *Prairie—Plain,—Submont.* Mr–S.

2. A. citrullifòlia (A. Br.) Rydb. Annual, glandular-pubescent; stem densely armed with yellow prickles, 3–10 dm. high; leaves irregularly bipinnatifid, 5–15 cm. long; corolla 2.5–3.5 cm. broad, violet, the lobes ovate, acuminate; lowest anther the largest, purple, the other four yellow. *Solanum citrullifolium* A. Br. [G]. It has been mistaken for *S. heterodoxum* Dunal of Mex. [B]. *S. citrullifolium* Dunal [G]. Sandy soil: Iowa—Kans.—Tex.—N.M.; Mex. *Tex.—Son.*

7. SOLÀNUM (Tourn.) L. NIGHTSHADE, BITTERSWEET, POTATO, HORSE NETTLE.

Annual or perennial, often prickly, or most of ours unarmed herbs, rarely vines. Leaves alternate, entire, toothed or pinnatifid. Flowers perfect, regular, in cymes, racemes, umbels, or panicles. Calyx from campanulate to rotate; lobes 5. Corolla variously colored, rotate, 5-angled or 5-lobed, plicate. Stamens 5, adnate to near the throat of the corolla. Anthers narrow, converging or united into a cone; sacs opening by terminal pores, rarely lengthwise. Ovary mostly 2-celled. Fruit a berry, seated in the calyx. Seeds flattened.

Annuals.	
Leaves pinnatifid.	1. *S. triflorum.*
Leaves sinuately dentate or entire.	
Leaves glabrous or nearly so ; sepals very obtuse.	2. *S. nigrum.*
Leaves decidedly strigose beneath ; sepals abruptly	
acutish.	3. *S. interius.*
Perennials.	
Plant green, glabrous or pubescent, but not stellate,	
never prickly.	
Low plants with tuber-bearing rootstocks ; leaves pin-	
nately divided.	
Corolla 5-angled, its lobes as broad as or broader	
than long.	4. *S. tuberosum.*
Corolla distinctly star-shaped, its lobes longer	
than broad.	5. *S. Jamesii.*
Tall plants, more or less woody below, not tuberifer-	
ous ; leaves simple or pinnately 3-lobed.	
Berry ellipsoid ; plant climbing.	6. *S. Dulcamara.*
Berry globose ; plant reclining.	7. *S. triquetrum.*
Plants with branched hairs ; stem often prickly.	
Leaves silvery-white or canescent, oblong to linear.	8. *S. elaeagnifolium.*
Leaves green, ovate.	
Leaves sinuately lobed or toothed.	
Plant long-pubescent, green ; calyx somewhat	
prickly.	9. *S. carolinense.*
Plant with short, stellate pubescence, canes-	
cent ; calyx not prickly.	10. *S. Torreyi.*
Leaves bipinnately lobed ; prickles very stout,	
orange-brown.	11. *S. sisymbriifolium*

1. **S. triflòrum** Nutt. Stem branched at the base, spreading, 2–9 dm. long, sparingly hirsute; leaf-blades oblong or ovate in outline, pinnatifid, 3–9 cm. long; lobes acute, entire or toothed; calyx-lobes oblong or oblong-lanceolate; corolla white, 8–10 mm. broad; berry globose, 10–15 mm. thick. Prairies, cultivated ground, waste places, ''prairie-dog towns'': w Ont.—Kans.—N.M.—Ariz.—B.C. *Prairie—Plain—Mont.* Je–S.

2. **S. nìgrum** L. Stem erect, branched, 1–12 dm. high; leaf-blades ovate or oblong-ovate, 2–8 cm. long, undulate to sinuately lobed; inflorescence um- belliform; calyx-lobes ovate or oblong-ovate; corolla white or bluish, about 1 cm. broad; lobes oblong or linear; berries subglobose, 5–8 mm. thick, black. Sandy soil, waste places: N.S.—Fla.—Tex.—Wash.; nat. or adv. from Eur. Mr–D.

3. **S. intèrius** Rydb. *Fig. 496.* Stem 3–6 dm. high, usually with narrow denticulate margins or wings; leaf-blades deltoid or rhombic, 3–7 cm. long, acuminate, usually sinuately lobed or dentate, with acute or acuminate lobes or teeth; inflorescence corymbiform; calyx-lobes ovate, 2 mm. long; corolla yellowish white; lobes ovate, acute, 3–4 mm. long; berry greenish black, nearly 1 cm. in diameter. River valleys among bushes: Neb.—Kans.—N.M.—Ariz.—Utah. *Plain—Submont.* Je–Au.

4. **S. tuberòsum** L. A perennial, with tubers; stem weak, branched; leaves unequally and interruptedly pinnatifid, with 5–9 oblong-ovate divisions; corolla lilac or white, the lobes lanceolate; berry green. Occasionally escaped from cultivation, but not persistent; native of the Andes of South America. POTATO.

5. **S. Jamèsii** Torr. Stem 1–3 dm. high, erect, sparingly hairy or gla- brous; leaves pinnately divided; divisions 5–7, lanceolate, subentire, 2–4 mm. long, glabrate, or with scattered hairs; inflorescence cymose; corolla white; lobes lanceolate, 7–8 mm. long. WILD POTATO. Mountains: Tex.—Colo.—Utah —Ariz.; introduced in Iowa; Mex. *Son.—Submont.* Je–S.

6. **S. Dulcamàra** L. Stem climbing or twining, 4–15 dm. long; leaf- blades ovate, 3–10 cm. long, acute or acuminate, undulate, entire or with a lobe or two on one or both sides at the base; calyx glabrous, its lobes triangu- lar; corolla blue or white, about 1.5 cm. wide; lobes oblong-lanceolate, pubes- cent near the tips. BITTERSWEET. Thickets: N.B.—Fla.—Kans.—Ida.; adv. or nat. from Eur. Je–Au.

7. **S. triquétrum** Cav. Perennial; stem reclining or spreading, somewhat shrubby below, 3–10 dm. high; leaf-blades ovate or lanceolate, 2–6 cm. long, entire or hastately 3-lobed, truncate or cordate at the base; calyx glabrous, the lobes triangular-ovate; corolla purplish, 1.5 cm. broad; berry globose, 8–10 mm. in diameter. Thickets: Kans.—Tex.—Mex. *Texan—Trop.* Je–S.

8. **S. elaeagnifòlium** Cav. Perennial, with a cespitose caudex or root- stock; stem 3–11 dm. high, silvery stellate-canescent, often more or less prickly; leaf-blades oblong to linear, 5–15 cm. long, usually obtuse, undulate to deeply sinuate, silvery-stellate; calyx 5-ribbed; corolla violet or white, 2–2.5 cm. broad; berries globose, 1–1.5 cm. thick, yellow or black. Dry soil: Mo.—Tex.— Calif.; Mex. *Son.* My–O.

9. **S. carolinénse** L. Stem erect, 3–6 dm. high, hirsute and prickly; leaf- blades mostly ovate, sinuately toothed or lobed, 1–2 dm. long, green, but rather densely stellate-pubescent, prickly on the petioles, midribs, and the stronger

veins beneath; corolla blue, about 2 cm. in diameter; berry 1–1.5 cm. in diameter, globose. HORSE NETTLE. Sandy soil and waste places: Mass.—Fla.—
Tex.—Neb.; adv. in Utah and Idaho. *E. Temp.* Je–S.

10. **S. Tórreyi** A. Gray. Perennial, cinereous, with stellate scurfy pubescence, and sparingly prickly; stem 3–8 dm. high; leaf-blades oblong to oval
in outline, 6–15 cm. long, sinuately lobed; calyx densely pubescent, the lobes
ovate, subulate-tipped; corolla 2.5–3 cm. broad, blue or white, the lobes triangular; berry globose, 1–1.5 cm. thick, yellow. Sandy soil: Ark.—Kans.—
Tex. *Texan.*

11. **S. sisymbriifòlium** Lam. Annual; stem erect, 3–7 dm. high, branching, very prickly; leaf-blades oblong or oval in outline, 8–20 cm. long, deeply
pinnatifid, the divisions again sinuately lobed; pedicels bristly and glandular;
calyx bristly, its lobes lanceolate; corolla white or light blue, 3–4 cm. broad;
berry globose, 1.5–2 cm. broad. Waste places: Ga.—Iowa—Tex.; introd. from
the tropics.

8. **LYCOPÉRSICON** Mill. TOMATO, LOVE APPLE.

Annual herbs, with once or twice pinnately divided leaves. Flowers perfect,
in small racemes, opposite the leaves. Calyx mostly 5-parted. Corolla rotate,
5-lobed, plicate. Stamens usually 5; filaments short; anthers elongate, connate
or connivent, introrsely longitudinally dehiscent. Ovary 2–3-celled, many-
seeded. Fruit a berry.

1. **L. esculéntum** Mill. Viscid-pubescent, much branched; stem 3–10
dm. high; leaves 2-pinnatifid, lobed and dentate; corolla yellow, 10–15 mm.
broad; fruit through cultivation very variable, subglobose, ellipsoid, or pear-
shaped, red or yellow. *L. Lycopersicon* Karst. [B, R]. Waste places and
around dwellings: N.Y.—Fla.—Tex.—Colo.; Calif.; escaped from cultivation.

9. **LÝCIUM** L. MATRIMONY VINE.

Shrubs or woody vines, often spiny. Leaves alternate, thick, entire, often
with smaller ones clustered in their axils. Flowers perfect, regular, solitary or
clustered in the axils. Calyx enlarged and persistent under the fruit, deeply
5 cleft. Corolla whitish, yellowish, or purplish, funnelform, salverform, or campanulate; lobes 5 or rarely 4, imbricate, obtuse. Stamens 5 or 4; filaments adnate up to the mouth of the corolla-tube; anthers opening lengthwise. Ovary
2-celled. Berry rather dry.

1. **L. halimifòlium** Mill. Tall shrub, often climbing or trailing; stems
2–8 m. long, sometimes spiny; leaves lanceolate, oblong, or spatulate, 1–4 cm.
long; corolla purplish, changing to greenish, 8–12 mm. wide. *L. vulgare* Dunal
[B]. Thickets and waste places: Ont.—Conn.—Utah—Alta.; escaped from cultivation, native of Eurasia and n Africa.

10. **HYOSCÝAMUS** (Tourn.) L. HENBANE.

Viscid-pubescent herbs. Leaves alternate, lobed or pinnatifid. Flowers
perfect, regular, solitary in the upper axils and in terminal racemes. Calyx urn-
shaped, 5-cleft, striate. Corolla funnelform, with slightly oblique, 5-lobed limb.
Stamens declined, mostly exserted; anthers opening longitudinally. Ovary
2-celled; stigma capitate. Capsule 2-celled, circumscissile above the middle.

1. **H. nìger** L. Biennial, with a fusiform root; stem viscid-villous, 3–10
dm. high; leaves oblong to ovate, sinuately toothed or lobed, the upper clasping,
viscid-villous; calyx campanulate, strongly veined, in fruit 2–2.5 cm. long; corolla campanulate, lurid-yellowish, purple-veined, 2 cm. long. Waste places and
around dwellings: N.S.—N.Y.—Man.—Colo.—Utah—Mont.; nat. from Eur.
My–S.

11. **NICOTIÀNA** (Tourn.) L. TOBACCO.

Annual or perennial, viscid-pubescent herbs. Leaves alternate, entire or
repand. Flowers perfect, regular, in terminal racemes or panicles. Calyx

campanulate or tubular-ovoid, 5-lobed. Corolla funnelform, salver-shaped, or nearly tubular; lobes 5, spreading. Stamens 5, included; filaments adnate to the base of the corolla-tube or free; anthers opening lengthwise. Ovary 2-celled, rarely 4-celled; stigma capitate. Fruit a 2- or rarely 4-valved capsule. Seeds numerous.

Corolla cylindric; capsule 2-celled.   1. *N. rustica.*
Corolla funnelform; capsule 4-celled.   2. *N. quadrivalvis.*

1. N. rústica L. Annual; stem 5–12 dm. high, finely puberulent; leaf-blades ovate, 5–20 cm. long, entire; panicle open; calyx 5–7 mm. long, becoming 1 cm. long; lobes broadly triangular, shorter than the tube; corolla lurid-yellow or greenish, 1.5–2 cm. long, the tube constricted at the throat, the limb 10–15 mm. broad, the lobes rounded, veiny. Waste places: N.Y.—Minn.—La.—Fla.; cult. by the Indians. *Allegh.—Austral.*

2. N. quadriválvis Pursh. *Fig. 498.* Annual; stem 3–6 dm. high, viscid-pubescent, branched; lower leaves petioled, the upper sessile; blades ovate-lanceolate, lanceolate, or oblong, acute at both ends, viscid-pubescent; flowers few; corolla about 2 cm. long, funnelform. Valleys and cultivated ground: Wash.—Mont.—Ore.; cultivated by the Indians, and escaped in N.D.

f. 498.

12. DATÙRA L. THORN APPLE, JIMSON WEED, JAMESTOWN WEED, STRAMONIUM.

Annual or perennial herbs, or in the tropics shrubs or trees, narcotic. Leaves alternate, mostly lobed. Flowers perfect, regular, solitary in the axils. Calyx prismatic-funnelform, 5-lobed. Corolla funnelform, plaited, 5-lobed; lobes acuminate. Stamens 5, included, adnate to near the middle of the corolla-tube. Ovary 2-celled or falsely 4-celled; stigma 2-lobed. Capsule more or less prickly, 4-valved or splitting irregularly.

Corolla 1.2–2 dm. long; leaves subvelutinous beneath.   1. *D. Metel.*
Corolla about 1 dm. long; leaves glabrous.
   Corolla white; lower prickles of the capsule shorter.   2. *D. Stramonium.*
   Corolla violet; prickles all alike.   3. *D. Tatula.*

1. D. Mètel L. Annual; stem finely puberulent, 8–15 dm. high; leaves ovate, 1–2 dm. long, sinuately toothed, subcordate, rounded or cuneate at the base, finely puberulent; calyx 6–12 cm. long, the lobes lanceolate, acuminate; corolla 1.2–2 dm. long, white, 1 dm. broad, 10-lobed; capsule subglobose, 3 cm. long, densely armed with sharp prickles. Waste places: R.I.—Minn.—Kans.—Fla.; escaped from cultivation and nat. from Trop. Am.

2. D. Stramònium L. *Fig. 497.* Annual; stems glabrous, 5–15 dm. high, widely branching; leaf-blades ovate to oblong, glabrous, 1–2 dm. long, acute, sinuately or laciniately toothed; calyx 3–5 cm. long; lobes triangular or triangular-lanceolate, 5–7 mm. long; corolla white, 6–10 cm. long; lobes with slender tips; capsule 4–6 cm. long, densely prickly. Waste

f. 497.

places and cultivated ground: N.S.—Fla.—Tex.—Colo.—Minn.; W.Ind. and S.Am.; nat. from Asia. My–D.

3. D. Tátula L. Annual; stem 5–15 dm. high, glabrous, purplish, branched; leaf-blades ovate to oblong, sinuately or angularly toothed, glabrous; calyx 3–6 cm. long; lobes triangular-lanceolate, 4–7 mm. long; corolla violet or lavender, 8–11 cm. long; lobes with slender tips; capsule oval or ovoid, 4–6 cm. long, densely prickly. Waste places and fields: Conn.—Fla.—Tex.—Colo.—Calif.—Wash.; W.Ind. and Mex.; nat. from S.Am. Je–O.

Family 147. **SCROPHULARIACEAE.** FIGWORT FAMILY.

Annual or perennial herbs, rarely shrubs or trees, with mostly round stems. Leaves usually opposite, sometimes whorled, or alternate, without stipules. Flowers perfect, more or less irregular. Calyx of 5 or 4, more or less imbricate and united sepals. Corolla usually 2-lipped, sometimes almost regular. Stamens usually 4, rarely 5, or only 2 fertile, one pair longer than the rest, the fifth stamen sometimes represented by a sterile filament, all partly adnate to the tube of the corolla. Gynoecium of 2 united carpels; ovary 2-celled; styles usually united, rarely distinct. Fruit a 2-celled, 2-valved capsule, rarely baccate. Seeds numerous; endosperm present, surrounding the embryo, fleshy or cartilaginous. [RHINANTH-ACEAE.]

Anther-bearing stamens 5.
  Corolla rotate.
  Corolla funnelform, 2-lipped. (Occasional forms of)
                                  1. VERBASCUM.
                                    8. PENTSTEMON.
Anther-bearing stamens 4 or 2.
  Corolla spurred, saccate or gibbous at the base on the lower side.
    Corolla distinctly spurred at the base.
      Capsule regular; flowers in terminal racemes.   2. LINARIA.
      Capsule inequilateral, one cell larger than the other; flowers axillary.
                                    3. CHAENORRHINUM.
    Corolla merely saccate or gibbous at the base.   4. ANTIRRHINUM.
  Corolla neither spurred, nor saccate, nor gibbous on the lower side.
    Stamens 5, 4 anther-bearing; the fifth sterile and often rudimentary.
      Sterile stamen rudimentary, represented by a scale or gland on the upper
        inside of the corolla-tube or throat; corolla short.
        Corolla gibbous at the base on the upper side; ovules and seeds few or
          solitary; annuals.   5. COLLINSIA.
        Corolla not gibbous at the base, but more or less ventricose especially
          on the lower side; ovules and seeds numerous; 4 upper lobes of the
          corolla erect, the lower spreading; perennials. 6. SCROPHULARIA.
      Sterile stamen elongate, filiform to spatulate; corolla-tube elongate, tubu-
        lar or funnelform.
        Filaments longer than the sterile stamen; seeds winged; inflorescence
          spicate.   7. CHELONE.
        Filaments shorter than the sterile stamen; seeds not winged; inflor-
          escence cymose-paniculate.   8. PENTSTEMON.
    Stamens 4 or 2.
      Upper lip or lobes external in the bud.
        Anther-bearing stamens 4.
          Corolla more or less bilabiate; plants leafy-stemmed.
            Sepals united into an angled tube.   9. MIMULUS.
            Sepals distinct or nearly so.   10. LEUCOSPORA.
          Corolla nearly regular.
            Calyx unequally 5-parted: flowers solitary in the axils of the
             leaves; plant caulescent.   11. MACUILLAMIA.
            Calyx regularly 5-lobed; flowers solitary on scape-like peduncles
             from the basal rosette of leaves; plant acaulescent.
                                12. LIMOSELLA.
        Anther-bearing stamens 2; calyx of 5 almost distinct sepals.
          Sterile filaments short, stout or almost wanting.
                                13. GRATIOLA.
          Sterile filaments elongate, 2-lobed at the apex.
                                14. ILYSANTHES.
      Upper lip or lobes internal in the bud.
        Stamens 2.
          Corolla almost regularly 4-lobed.
            Corolla rotate; stamens not exserted.   15. VERONICA.
            Corolla short-tubular; stamens exserted.   16. VERONICASTRUM.

Corolla none, or 2-lipped, cleft to near the base; upper lip entire;
lower irregularly cleft or toothed; basal leaves ample, stem-
leaves bract-like, reduced and alternate.　　17. BESSEYA.
Stamens 4.
Corolla slightly 2-lipped; stamens not ascending under the upper
lip.
Corolla with an open mouth; stamens all with 2-celled anthers; stamens and style nearly equaling the corolla;
pedicels not bracted.
Anther-cells glabrous or with a few bristles at the apex;
style short, persistent, and reflexed on the capsule;
stigma 2-lobed; corolla yellow.　　21. DASYSTOMA.
Anther-cells lanate on the valvular surface; stigma linear,
consisting of a line on each side of the style-apex.
Corolla purple, pink, often with red spots inside, or
white; capsule rounded with a mucro; leaves
linear to filiform, sessile.
Anthers uniform; sepals linear to subulate; leaves
not auricled, in ours filiform; seeds closely reticulate.　　18. AGALINIS.
Anther-cells of the upper stamens shorter; leaves
auricled at the base, linear; seeds with raised
ridges.　　19. OTOPHYLLA.
Corolla yellow; capsule acute or acuminate; leaves lanceolate to ovate, petioled.　　20. AUREOLARIA.
Corolla salver-shaped, the tube narrow, densely villous within;
stamens with only one anther-cell developed; filaments and
style less than half as long as the corolla-tube; pedicels bi-
bracteolate.　　22. BUCHNERA.
Corolla distinctly 2-lipped; stamens ascending under the upper lip.
Anther-sacs dissimilar, the inner one pendulous by its apex;
leaves mostly alternate.
Calyx deeply cleft in front and behind, less deeply so on the
sides; upper lip of the corolla much longer than the 3-
lobed lower one.　　23. CASTILLEJA.
Calyx almost equally 4-cleft; upper lip of the corolla slightly
if at all longer than the 1–3-saccate lower one, which is
minutely or obsoletely toothed.　　24. ORTHOCARPUS.
Anther-sacs alike, parallel; leaves mostly opposite.
Margins of the 2-lobed upper lip of the corolla recurved;
calyx 4-cleft.　　25. EUPHRASIA.
Margins of the upper lip of the corolla not recurved.
Ovules several or numerous; capsule several- or many-
seeded.
Calyx split below, or below and above, not inflated;
capsule ovoid or oblong, oblique.
Galea prolonged into a filiform recurved beak;
throat with a tooth on each side.
26. ELEPHANTELLA.
Galea if prolonged into a beak, the latter not
filiform, straight or incurved; throat without
teeth.　　27. PEDICULARIS.
Calyx 4-toothed, inflated and veiny in fruit.
28. RHINANTHUS.
Ovules 2 in each cell; fruit 1–4-seeded.
29. MELAMPYRUM.

1. VERBÁSCUM (Tourn.) L. MULLEIN.

Annual, biennial, or rarely perennial, caulescent herbs, more or less glandular or densely pubescent with branched hairs. Leaves alternate, entire, toothed,
or pinnatifid, sometimes decurrent. Flowers perfect, in terminal spikes, racemes,
or panicles. Sepals 5, partly united. Corolla rotate, 5-lobed, slightly irregular,
the upper lobe exterior in the bud. Stamens 5, all fertile, exserted, the upper 3,
or all of the filaments pubescent. Ovary 2-celled; styles dilated or flattened at
the apex. Capsule oblong to subglobose, 2-valved, septicidal. Seeds rugose,
wingless.

Plant densely woolly; flowers in a dense spike.
　Leaves strongly decurrent; corolla 1.5–2 cm. wide.　　　　　　1. V. Thapsus.
　Leaves not decurrent; corolla more than 2 cm. wide.　　　　　2. V. phlomoides.
Plant glabrous or sparingly glandular; flowers racemose.　　　3. V. Blattaria.

1. V. Thápsus L. Biennial; stem stout, 3–20 dm. high, woolly with densely
matted branched hairs, winged; lower leaves spatulate or elliptic, 1–4 cm. long;
upper stem-leaves oblanceolate, densely woolly; raceme dense and spike-like, continuous, 1–10 dm. long; calyx-lobes triangular-lanceolate; corolla deep yellow,

rarely white, 1–2.5 cm. wide, pubescent without. Waste places, fields, and road-sides: N.S.—Fla.—Calif.—B.C.; nat. from Eur. *Temp.—Submont.* Je–Au.

2. V. phlomoìdes L. Biennial; stem simple, 3–12 dm. high, woolly; leaves oblong to ovate-lance-olate, crenate or entire, woolly-tomentose, sessile and usually clasping, or the lower petioled and with truncate or cordate base; corolla yellow or cream-colored. Waste places: Mass.—Iowa—Tenn.; adventive from Eur. Je–Au.

3. V. Blattària L. *Fig. 499.* Biennial; stem rather slender, 4–12 dm. high, glabrous below, gland-ular above; basal leaves obovate or spatulate; stem-leaves oblong or ovate, the upper clasping, 2–12 cm. long, dentate, incised or lobed, glabrous; raceme slender, interrupted, 1–5 dm. long; calyx-lobes linear or oblong, recurved; corolla white or yellow, 3–4 cm. broad; filaments ma-genta. Fields and waste places: Que.—Fla.—Calif.—B.C.; nat. from Eur. *Temp.—Submont.* My–S.

f. 499.

2. LINÀRIA (Tourn.) L. Butter-and-eggs, Toadflax.

Annual or perennial herbs, with erect stems. Leaves alternate or opposite, or whorled on the shoots. Flowers in terminal racemes, spikes, or panicles; sepals 5, partially united. Corolla irregular, 2-lipped; tube spurred at the base; throat partly closed by a convex fold. Stamens 4, didynamous, included. Cap-sule short, opening by 3-toothed pores below the apex. Seeds angled or wrin-kled, sometimes winged.

Corolla yellow, with an orange throat, 2–3 cm. long.	1. *L. vulgaris.*
Corolla blue or white, 12 mm. or less long.	
Corolla of the earlier flowers, including the spur, 2–2.5 cm. long; sepals lanceolate, acute or obtuse; seeds round-angled, densely tuberculate.	2. *L. texana.*
Corolla of the earlier flowers, including the spur, about 1 cm. long; sepals linear-lanceolate, acuminate; seeds sharply angled, smooth or nearly so.	3. *L. canadensis.*

1. L. vulgàris Mill. *Fig. 500.* Perennial; stem glabrous or puberulent above, 1–10 dm. high; leaves linear or nearly so, 2–7 cm. long; racemes dense; calyx-lobes ovate, acute or acuminate. *L. Linaria* (L.) Karst. Waste places and fields: Newf.—Ga.—N.M.—Man.; nat. from Eur. *E. Temp.—Submont.* My–O.

2. L. texàna Scheele. Annual or biennial; stem 2–8 dm. high, glabrous; leaves linear-spatulate, 1–2.5 cm. long; racemes slender, 5–30 cm. long; calyx-lobes linear-lanceolate; corolla blue or white, 10–12 mm. long. *L. canadensis* Coult. & Nels.; not Dumort. Sandy soil: Kans—Fla.—Tex.—Calif.—Ore.—Mex. *Son.* Mr–Je.

f. 500.

3. L. canadénsis (L.) Dumort. Annual or bi-ennial; glabrous or nearly so; stem 1–6 dm. high, sometimes branched at the base; leaves narrowly linear, acute, 1–2.5 cm. long; racemes 4–15 cm. long; corolla light blue. Sandy soil: N.S.—Neb.—Fla. *E. Temp.* My–S.

3. CHAENORRHÌNUM (D.C.) Lange. SMALL SNAPDRAGON.

Annual herbs. Leaves alternate, usually entire. Flowers axillary, violet, blue, or white. Calyx 5-parted, the lobes narrow, corolla irregular, closed at the throat, spurred at the base, 2-lipped, the upper lip 2-lobed, the lower 3-lobed. Stamens 4; filaments slender. Capsule inequilateral, one carpel larger than the other. *Linaria* (L.) Desf.

1. **C. mìnus** (L.) Lange. Annual; stem glandular-pubescent, 1–3 dm. high, branched; leaves linear-spatulate or linear, 1–3 cm. long, glandular; calyx-lobes linear; corolla blue or bluish, 6–8 mm. long, spur short. Waste places: N.B.—Pa.—Iowa; adv. from Eur.

4. ANTIRRHÌNUM (Tourn.) L. SNAPDRAGON.

Annual or perennial caulescent herbs. Leaves alternate, or the lower opposite, entire. Flowers solitary in the upper axils, or in terminal racemes or panicles. Sepals 5, partially united. Corolla irregular, 2-lipped; tube more or less saccate below; throat closed by a convex fold. Stamens 4, included; filaments sometimes dilated above. Styles united. Ovules numerous. Capsule opening by 2 or 3 pores below the apex. Seeds wingless, smooth or wrinkled.

1. **A. màjus** L. Perennial, glandular-pubescent and viscid; stem 3–10 dm. high; leaves oblong, lanceolate, or linear, entire, petioled, glabrous, 2–8 cm. long; calyx-lobes short, oblong or ovate; corolla crimson, white or variegated, 2–3 cm. long. Waste places: N.S.—Kans.—Va.; escaped from cultivation or adventive from Eur. Je–S.

5. COLLÍNSIA Nutt. BLUE-EYED MARY, BLUE-LIPS.

Annual caulescent herbs. Leaves opposite or whorled, narrow, entire or toothed. Flowers solitary or clustered in the upper axils, perfect, irregular. Calyx-lobes 5. Corolla 2-lipped, violet, pink, or white, open; tube short; lower lip 3-lobed, the middle lobe involute. Stamens 4, didynamous, declined, enclosed in the middle lobe of the lower lip of the corolla; staminodium present, but gland-like, near the base of the corolla. Styles united. Capsule ovoid or globose, 2-valved, the valves 2-cleft.

Corolla 4–6 mm. long, the throat longer than the limb.   1. *C. parviflora.*
Corolla 10–15 mm. long, the throat shorter than the limb.
  Pedicels as long as the corolla; corolla-lobes deeply notched
    or cleft.   2. *C. violacea.*
  Pedicels longer than the corolla; corolla-lobes shallowly retuse
    or emarginate.   3. *C. verna.*

1. **C. parviflòra** Dougl. Annual; stem 1–3 dm. high, at last diffuse and spreading, minutely puberulent; leaves oblong, lanceolate, oblanceolate, or linear, entire or nearly so, the floral ones often in whorls of 3–5; corolla blue, 4–6 mm. long. *C. tenella* Piper, in part. Shaded hillsides: Ont.—S.D.—Ariz.—Calif.—B.C. *W. Temp.—Submont.—Mont.* Ap–Jl.

2. **C. violàcea** Nutt. Stem 0.5–2 dm. high, branched; leaves suborbicular to oblong, lanceolate or linear, 0.5–4 cm. long, undulate or sparingly serrate, the upper sessile; calyx finely pubescent; corolla violet, 1 cm. long; upper lip smaller than the lower. Prairies: Mo.—Kans.—Okla.—Ark. *Ozark.* Ap–Je.

3. **C. vérna** Nutt. Stem 1–5 dm. high, sometimes branched at the base; lower leaves suborbicular or ovate, the upper oblong or lanceolate, 1–5 cm. long, undulate or serrate, the upper half-clasping; calyx glabrous; corolla 1 cm. long, the lower lip bright blue, the upper white or purplish. Woods and copses: N.Y.—Minn.—Okla.—Ky. *Allegh.* Ap–Je.

6. SCROPHULÀRIA (Tourn.) L. FIGWORT.

Perennial, caulescent herbs. Leaves opposite, usually petioled, with broad,

FIGWORT FAMILY 711

toothed or incised blades. Flowers perfect, in terminal, paniculate cymes.
Calyx with 5 short broad lobes. Corolla 2-lipped, purple, yellowish, or green;
tube usually short; upper lip erect; lower lip with a spreading or reflexed middle lobe. Stamens 4, short; staminodium scale-like, in ours spatulate. Styles
united. Capsule ovoid, 2-valved. Seeds numerous, marginless, rugose.

Corolla dull; sterile filament deep purple.
  Leaves glabrous beneath or nearly so; corolla 5–8 mm. long; capsule mostly
    less than 6 mm. long.   1. *S. marilandica.*
  Leaves densely pubescent beneath; corolla 8–9 mm. long;
    capsule more than 6 mm. long.   2. *S. neglecta.*
Corolla lustrous without; sterile filament greenish yellow.
  Upper lip of the corolla shorter than the tube; inflorescence
    densely glandular; leaf-blades mostly dentate-hastate at
    the base.   3. *S. occidentalis.*
  Upper lip of the corolla as long as the tube; inflorescence
    sparingly glandular; leaf blades not dentate-hastate. 4. *S. lanceolata.*

f. 501.

1. S. marilándica L. Stem 1–2 m. high, glabrous or nearly so; leaf-blades oblong-ovate to lanceolate, 5–20 cm. long, serrate or crenate-serrate; calyx-lobes oblong or rounded, obtuse; corolla green or yellowish. Woodlands: Me.—Man.—S.D.—Kans.
—Ga. *Canad.—Allegh.* Jl–S.

2. S. neglécta Rydb. Stem minutely pubescent; leaf-blades oblong-ovate to oblong-lanceolate, serrate, 1–1.5 dm. long; calyx-lobes ovate, obtuse; corolla yellowish, broader than in the preceding. Woods: Ky.—Kans.—Okla.—Ark. *Ozark.* Jl–S.

3. S. occidentàlis (Rydb.) Bickn. *Fig. 501.* Stem 1–2 m. high, more or less puberulent, or glabrate below, glandular in the inflorescence; leaves ovate; corolla lurid-greenish; tube gibbous below, 5 mm. long; upper lip about 3 mm. long; sterile stamen greenish yellow, very broad, kidney-shaped, on a distinct claw; capsule ovoid, 8–10 mm. long. Low ground in thickets and woods: N.D.—Okla.—N.M.
—Calif.—Wash. *W. Temp.—Submont.—Mont.*

4. S. lanceolàta Pursh. Stem 5–15 dm. high, nearly glabrous; leaf-blades ovate to lanceolate, 5–20 cm. long, incised-serrate; calyx-lobes ovate; corolla 8–10 mm. long; capsule 7–10 mm. long, attenuate at the apex. *S. leporella* Bickn. Open woods: Vt.—Minn.—Kans.—Va. *Canad.—Allegh.* My–Au.

7. CHELÒNE L. Turtlehead.

Perennial, leafy-stemmed herbs. Leaves opposite, toothed. Flowers in terminal, spike-like racemes. Calyx 5-lobed, subtended by sepal-like bracts. Corolla white to rose-purple or red, with a dilated throat, the upper lip concave, the lower lip pubescent within, its lateral lobes larger than the middle one. Fertile stamens 4, filaments pubescent; anthers woolly; sterile filament small. Seeds flattened, winged.

Calyx-lobes and bracts glabrous, not ciliate; corolla white or cream-colored.
  Corolla purplish within; leaf-blades lanceolate or elliptic, 2–4 cm.
    wide.   1. *C. glabra.*
  Corolla white within, distantly greenish yellow; leaf-blades linear
    to narrowly lance-linear, 1–2 cm. wide.   2. *C. linifolia.*
Calyx-lobes and bracts puberulent and ciliate; corolla pink. 3. *C. obliqua.*

1. C. glàbra L. *Fig. 502.* Stem 3–18 dm. high; leaves oblong or lanceolate, 6–15 cm. long, acuminate, petioled, serrate; corolla 2.5–3 cm. long. Swamps and wet thickets: Newf.—Man.—Kans.—Fla. *E. Temp.* Jl–S.

2. **C. linifòlia** (Coleman) Pennell. Stem 5–10 dm. high, glabrous; leaves linear or lance-linear, glabrous or short-puberulent beneath (*C. velutina*), 10–15 cm. long, 1–2 cm. wide, serrate; corolla about 3 cm. long. *C. glabra linifolia* Coleman. Swampy places: Mich.—Ind.—Iowa—Minn.

3. **C. oblìqua** L. Stem 5–8 dm. high, with spreading branches; leaves broadly lanceolate to oblong, 5–20 cm. long, short-petioled, mostly laciniate-serrate, dull and veiny; corolla fully 2 cm. long, lower lip less bearded than in the preceding. Damp thickets and along streams: Va. — Iowa — Fla. *Allegh.—Austral.* Jl–S.

8. **PENTSTÈMON** (Mitchell) Schmidel. BEARD-TONGUE, MAYFLOWER, PRIDE-OF-THE-MOUNTAIN.

f. 502.

Perennial, caulescent herbs. Leaves opposite, from petioled to sessile and clasping, entire or toothed, or rarely pinnatifid. Flowers irregular, in terminal racemes or panicles. Calyx deeply 5-cleft. Corolla 2-lipped, elongate, open; the tube often somewhat ventricose above, but not gibbous at the base; upper lip most erect, with 2 spreading lobes. Stamens 4; filaments nearly equal; staminodium well developed, spatulate, usually bearded. Styles united; stigma capitate. Capsule usually ovoid, 2-valved. Seeds numerous, wingless. Anther-cells either united or confluent.

Leaves not linear-filiform.
  Sterile filament glabrous or puberulent; plant glabrous.
    Calyx-lobes ovate or lanceolate, acuminate, with narrow or no scarious margin.   1. *P. alpinus.*
    Calyx-lobes orbicular or broadly ovate, abruptly short-acuminate, with broad scarious erose margins.   2. *P. glaber.*
  Sterile filament long-bearded.
    Corolla 3–5 cm. long.
      Leaves strongly serrate; plant puberulent above.   3. *P. Cobaea.*
      Leaves subentire; plant glabrous.   4. *P. grandiflorus.*
    Corolla 2 cm. long or less.
      Plants perfectly glabrous.
        Corolla 15–20 mm. long.
          Inflorescence interrupted; bracts, except the lower, shorter than the flowers; basal leaves spatulate or oblanceolate.
          Bracts ovate or cordate.
            Bracts sessile, not strongly reticulate; stem 2–4 dm. high; corolla blue.   5. *P. nitidus.*
            Bracts cordate-clasping, strongly reticulate; stem 3–6 dm. high; corolla pale lavender.   6. *P. Buckleyi.*
          Bracts lanceolate to lance-linear.
            Calyx-lobes lanceolate, acute or gradually acuminate; plant 2.5–4 dm. high.   7. *P. secundiflorus.*
            Calyx-lobes ovate, decidedly scarious-margined; plant less than 2.5 dm. high.   8. *P. Fendleri.*
         Inflorescence dense; bracts large, long-acuminate, most of them exceeding the flowers; basal leaves linear or nearly so, narrower than the stem-leaves.
          Bracts strongly veined, the upper ovate.   9. *P. Haydeni.*
          Bracts not strongly veined, linear-lanceolate.   10. *P. angustifolius.* 19. *P. procerus.*
        Corolla less than 10 mm. long.
      Plants glandular, at least in the inflorescence.
        Corolla-tube decidedly gibbous-ventricose; corolla purple; sterile filament densely long-bearded.

Leaves and stem pubescent or puberulent.
Stem viscid-villous as well as puberulent.
  especially the upper part; corolla
  2.5–3 cm. long.   11. *P. eriantherus.*
Stem merely puberulent; corolla 2–2.5
  cm. long.   12. *P. Cleburni.*
Leaves and stem glabrous; plant pubescent
  only in the inflorescence.   13. *P. Digitalis.*
Corolla-throat decidedly gibbous, the lobes
  erect or ascending.
Corolla-throat funnelform, not gibbous,
  the lobes widely spreading.   14. *P. tubiflorus.*
Corolla-tube nearly cylindric, only slightly wid-
  ening upwards.
Sterile filament short-bearded; corolla white
  or nearly so, with spreading lobes.
Inflorescence dense; stem glabrous below.   15. *P. albidus.*
Inflorescence open; stem puberulent
  throughout.   16. *P. pallidus.*
Sterile filament long-bearded on the upper
  part; corolla-lobes ascending
Stem glabrous or nearly so.   17. *P. gracilis.*
Stem densely puberulent.   18. *P. radicosus.*
Leaves linear-filiform; corolla-tube narrow, the limb widely
spreading.   20. *P. ambiguus.*

1. P. alpìnus Torr. Stem 2–4 dm. high; basal leaves oblanceolate, obtuse, short-petioled; stem-leaves lanceolate and sessile, 5–10 cm. long, acute or acuminate; inflorescence dense and usually one-sided; calyx 8–10 mm. long; corolla funnelform, more or less ventricose, bluish purple; sterile stamen with yellow beard at the end. *P. oreophilus* Rydb. *P. riparius* A. Nels. Mountains: Colo.—s Wyo. *Submont.—Mont.* Je–Au.

2. P. glàber Pursh. Stems stout, glabrous and glaucescent, 3–6 dm. high; lower leaves oblanceolate, 5–15 cm. long, glabrous, the upper oblong or lanceolate; calyx glabrous, about 5 mm. long; corolla bluish purple, about 3 cm. long, somewhat ventricose; sterile stamen sparingly hirsute. *P. eriantherus* Nutt. *P. Gordoni* Hook. Plains and hills: N.D.—Neb.—Ida. *Plain—Submont.* Je–Jl.

3. P. Cobaèa Nutt. Stem 3–6 dm. high, short-pubescent; basal leaves oblanceolate, petioled, 6–12 cm. long; lower stem-leaves lanceolate, sharply serrate, finely pubescent, 5–10 cm. long; upper leaves ovate, cordate-clasping; calyx finely and densely pubescent, about 1 cm. long; corolla 3.5–5 cm. long, puberulent, white or purple, strongly ventricose. Open places: Mo.—Kans.—Tex.—Ark. *Ozark—Texan.* My–Je.

4. P. grandiflòrus Nutt. Stem 6–12 dm. high, glabrous; basal leaves glaucous, glabrous, obovate, entire; upper stem-leaves broadly ovate, oval, or rounded; calyx about 5 mm. long, glabrous; lobes ovate, acute; corolla about 4 cm. long, pink, strongly ventricose, with a narrow tube; sterile stamen hooked and minutely pubescent at the apex. *P. Bradburyi* Pursh. Prairies and plains: Wis.—Mo.—Okla.—Colo.—Wyo. *Plain.* My–Jl.

5. P. nítidus Dougl. Stem 2–3 dm. high, glabrous, glaucous; lower leaves oblanceolate, 3–5 cm. long, glabrous, entire; upper leaves and bracts lanceolate or ovate, acute or abruptly acuminate; calyx glabrous, 4–6 mm. long; corolla blue, 15–18 mm. long, obliquely funnelform. Plains: Man.—S.D.—Wyo.—Wash.—Alta. *Plain.* My–Jl.

6. P. Búckleyi Pennell. Stem 3–6 dm. high, glabrous; basal leaves oblanceolate, 5–10 cm. long, glabrous, glaucous, entire; stem-leaves lanceolate, 6–10 cm. long, sessile; bracts broadly ovate, strongly reticulate, cordate-clasping; calyx glabrous, the lobes lanceolate; corolla pale lavender, 15–20 mm. long. *P. amplexicaulis* Buckley; not Moench. Sandhills: w Kans.—Tex. *St. Plain.* My–Je.

7. P. secundiflòrus Benth. *Fig. 503.* Stem glabrous and glaucous; basal leaves oblanceolate, 3–7 cm. long, glaucous and glabrous; stem-leaves lanceolate to ovate-lanceolate, more or less acuminate; inflorescence secund; calyx 5–6 mm. long, glabrous, often somewhat bluish; corolla about 2 cm. long, funnelform, bluish purple; sterile stamen club-shaped, curved, densely bearded. Dry plains or hills: S.D.(?)—Wyo.—N.M.—Utah. *Submont.—Mont.*

ſ 503.

8. P. Féndleri A. Gray. Stem glabrous and glaucous, 1–2 dm. high; basal leaves petioled, 3–6 cm. long; blades oval or ovate; stem-leaves ovate to lanceolate; bracts small, lanceolate; calyx 4 mm. long; corolla 15–18 mm. long, almost salverform; sterile stamen club-shaped, densely yellow-bearded. Dry plains: Colo.—Kans.—w Tex.—N.M.—(? Calif.). *St. Plains—Son.* My–Jl.

9. P. Hàydeni S. Wats. Stem 2–5 dm. high, stout, leafy, glabrous; lower leaves, especially those of the shoots, narrowly linear, the upper stem-leaves linear-lanceolate; calyx glabrous, about 8 mm. long; lobes lanceolate, acuminate; corolla blue, funnelform; sterile stamen bearded at the apex. Sand hills: Neb.—Wyo. *Plain.* Je–Jl.

10. P. angustifòlius Pursh. Stem 1–3 dm. high, glabrous; leaves linear or the upper linear-lanceolate, 5–10 cm. long, glabrous; calyx glabrous, about 5 mm. long; lobes lanceolate, acuminate, often with narrow scarious margins; corolla about 15 mm. long, blue, lilac, or white, funnelform, scarcely ventricose; sterile stamen bearded above. *P. caeruleus* Nutt. Plains: S.D.—Colo.—Mont. *Plain.* My–Jl.

11. P. eriántherus Pursh. Stem stout, 1–4 dm. high, leafy, canescent and more or less villous, especially above; basal leaves spatulate or oblanceolate, petioled, 5–10 cm. long; stem-leaves oblong or linear, entire or nearly so, gray-ish-puberulent and sometimes villous on the veins; calyx fully 1 cm. long; lobes lance-subulate; sterile stamen with very long yellow hairs two thirds its length; throat long-bearded within. *P. cristatus* Nutt. Plains: N.D.—Neb.—Nev.—Wash. *Colum.—Plain.* My–Jl.

12. P. Clèburni A. Nels. Stem low, densely puberulent, 1–2 dm. high; basal leaves entire or nearly so, petioled, ovate or elliptic, 2–4 cm. long; stem-leaves narrowly oblong; calyx minutely glandular, about 8 mm. long; lobes lanceolate; corolla-throat short-villous within; sterile stamen long-villous at the apex. *P. Jamesii* A. Nels.; not Benth. *P. auricomus* A. Nels. Dry plains and hills: Wyo.—S.D.—Mont. *Plain.* My–Je.

13. P. Digitàlis (Sweet) Nutt. Stem 5–15 dm. high, glabrous; basal leaves oblanceolate, petioled, glabrous, entire or repand; stem-leaves ovate or lanceolate, 5–15 cm. long, the upper cordate-clasping; inflorescence open; calyx glandular-pubescent, 6–10 mm. long; corolla white, 2.5–3 cm. long, the throat ventricose. *P. laevigatus Digitalis* A. Gray [G]. Fields and thickets: Me.—Va.—Ark.—Kans.—Iowa. *Allegh.* My–Je.

14. P. tubiflòrus Nutt. Stem 5–10 dm. high, glabrous; leaves glabrous, the basal ones oblanceolate, entire; stem-leaves lanceolate or oblong-lanceolate, entire or denticulate, the upper ovate and clasping; inflorescence open; calyx glandular, the lobes lanceolate; corolla 1.5–2 cm. long, white or nearly so. Moist soil: Mo.—Ark.—Kans. *Ozark.* My–Jl.

15. P. álbidus Nutt. Stem 1.5–4 dm. high, puberulent, viscid-villous above; lower leaves petioled, lanceolate or oblanceolate, 3–8 cm. long, densely puberulent; upper leaves linear or linear-lanceolate, entire or dentate; calyx viscid-villous, 5–7 mm. long; lobes lanceolate; corolla about 2 cm. long, viscid-puberulent. Prairies and plains: Man.—Kans.—Colo.—Ida.—Alta. *Plain.* My–Jl.

16. **P. pállidus** Small. Stem 3–5 dm. high, puberulent; basal leaves oblanceolate or spatulate, 4–10 cm. long, entire, puberulent; stem-leaves lanceolate, long-acuminate, denticulate or entire; panicle with ascending branches; calyx puberulent, the lobes ovate; corolla white or purplish, 18–20 mm. long, the tube gradually dilated. Sandy soil: Conn.—N.Y.—Ark.—Iowa. *Allegh.— Ozark.* Je–Jl.

17. **P. grácilis** Nutt. Stem 2–4 dm. high, viscid in the inflorescence; basal leaves short-petioled; blades oblanceolate or elliptic, entire; stem-leaves mostly linear-lanceolate, usually denticulate; corolla almost cylindric, lilac or whitish, 18–20 mm. long, slightly bearded within; sterile stamen bearded three fourths its length. Plains and prairies: Man.—Tex.—Colo.—Alta. *Plain— Submont.* Je–Jl.

18. **P. radicòsus** A. Nels. Stem strict, erect, 2–3 dm. high puberulent; basal leaves small or none; stem-leaves 2–4 cm. long; calyx viscid, 5 mm. long; corolla nearly tubular, dark blue, 14–18 mm. long, sparsely bearded within; sterile stamen bearded half its length. Dry plains: Mont.—S.D.—Colo.—Utah —Ida. *Plain—Submont.* My–Jl.

19. **P. procèrus** Dougl. Stem decumbent at the base, 1–3 dm. high, glabrous or puberulent; basal leaves petioled, oblanceolate or linear-oblanceolate, glabrous, 4–8 cm. long; inflorescence usually compact, but in large specimens interrupted below; calyx glabrous, 5 mm. long; lobes obovate or cuneate, abruptly cuspidate, scarious-margined and erose; corolla dark purplish-blue; lower lip bearded within. *P. micranthus* Nutt. *P. confertus caeruleo purpureus* A. Gray. Hills and mountains: Man.—Colo. Calif.—D.C.—Yukon. *Plain— Subalp.* Je–Au.

20. **P. ambíguus** Torr. Stem glabrous, 3–6 dm. high, diffusely branched; bracts subulate; peduncles opposite, 1–3-flowered, calyx glabrous, 8 mm. long; lobes ovate, short-acuminate; corolla 2 cm. long, rose-colored or white; fifth stamen often anther-bearing. Plains: Tex.—Okla.—Colo.—Utah—Ariz.; Mex. *Son. St. Plains.* My–S.

9. MÍMULUS L. Monkey-flower.

Annual or perennial herbs. Leaves opposite, mostly toothed. Flowers perfect, axillary, solitary, sometimes disposed in leafy racemes or panicles. Calyx angled; tube longer than the unequal 5 lobes. Corolla 2-lipped, yellow, blue, or red; tube with two ridges within on the lower side; upper lip spreading or reflexed; lower lip erect. Stamens 4; filaments partly adnate to the corolla-tube. Styles glabrous, united up to the 2-lobed stigma. Capsule loculicidal; placentae separating from the valves, remaining united in the middle, seldom separating.

Calyx neither very oblique nor strongly inflated; its lobes nearly equal.
  Corolla not yellow; stem erect.
    Corolla rose-colored, 3.5–5 cm. long, open; leaves several-ribbed from the
      base; plant viscid-pilose.   1. *M. Lewisii.*
    Corolla blue or violet, rarely white, 2–3 cm. long, closed
      at the mouth; leaves pinnately ribbed; plant
      glabrous.
      Leaves petioled; calyx longer than the pedicel.   2. *M. alatus.*
      Leaves sessile; corolla shorter than the pedicel.   3. *M. ringens.*
  Corolla yellow; stem decumbent; plant viscid-villous.
    Perennials; flowers 1.5–4 cm. long; sepals linear-lance-
      olate.   4. *M. moschatus.*
    Annuals; flowers 0.5–2 cm. long; sepals ovate, triangu-
      lar or broadly lanceolate.   5. *M. floribundus.*
Calyx very oblique, decidedly inflated in fruit; upper lobe much
  longer than the rest, obtuse; corolla yellow.

Calyx-teeth obtuse; stem decumbent or floating, rooting at
the nodes.   6. *M. Geyeri.*
Calyx-teeth acute; stem neither floating nor rooting at the
nodes.   7. *M. guttatus.*

1. **M. Lewísii** Pursh. Stem 3–10 dm. high, more or less viscid-pilose;
leaves lanceolate, oblong or ovate, dentate, 4–8 cm. long, viscid-pubescent; calyx
glandular-pubescent, about 2 cm. long; teeth broadly-triangular, subulate-
acuminate; corolla crimson, rose-red, or paler, 3.5–5 cm. long, bearded within.
Along streams: Minn.—Colo.—Ariz.—Calif.—B.C. *W. Temp.—Prairie—Plain
—Mont.* Je–Au.

2. **M. alàtus** Solander. Rootstock often tuberous-thickened; stem 2–10
dm. high, 4-winged; leaves elliptic to lanceolate, 5–15 cm. long, serrate, thickened; calyx 12–15 mm. long, each plait ending in a mucro 1 mm. long; corolla
2 cm. long. Swamps and meadows: Conn.—Minn.—Kans.—Tex.—Fla. *E.
Temp.* Je–S.

3. **M. ríngens** L. Rootstock elongate; stem 3–12 dm. long, angled,
grooved; leaves lanceolate or narrowly elliptic, 5–20 cm. long, remotely serrate; calyx 12–20 mm. long, each plait ending in a folded mucro, 3–6 mm.
long; corolla 2.5–3 cm. long. Swamps and wet banks: N.S.—Man.—N.D.—
Tex.—Ga. *E. Temp.* Je–S.

4. **M. moschàtus** Dougl. Stem 1–4 dm. long, decumbent or creeping; leaf-
blades petioled, ovate or oblong-ovate, villous and viscid; calyx prismatic, 7–10
mm. long; teeth broadly lanceolate, acuminate, somewhat unequal; corolla 15–40
mm. long, yellow, funnelform. MUSK FLOWER. Wet places: Newf.—N.Y.—
Colo.—Calif.—B.C. *Temp.—Submont.—Mont.* My–Au.

5. **M. floribúndus** Dougl. Stem 1–3 dm. high, erect or diffuse, villous
and viscid; leaf-blades ovate or subcordate, viscid-villous; calyx short-campanulate, 4–5 mm. long; teeth triangular; corolla yellow, 6–10 mm. long, tubular-
funnelform. Wet places, sandy soil: Mont.—S.D.—Colo.—Ariz.—Calif.—B.C.
W. Temp.—Submont. My–Au.

6. **M. Géyeri** Torr. Perennial; stem 1–4 dm.
long, glabrous; leaves rounded or reniform, denticulate to entire. 8–25 mm. long, all except the uppermost with margined petioles; calyx glabrous or
slightly puberulent, 5–8 mm. long; upper lobe
broadly ovate, twice as long as the very short
rounded other lobes; corolla yellow, 8–12 mm. long.
M. Jamesii T. & G. *M. glabratus Jamesii* A. Gray
[G]. In water: Mich.—Ill.—Colo.—Wyo.—N.D.—
Mont. *Prairie—Plain—Submont.* Je–Au.

7. **M. guttàtus** DC. *Fig. 504.* Stem erect or
ascending, 1–10 dm. high, glabrous; lower leaves
petioled, the upper sessile or clasping; blades ovate,
rounded, or subreniform, sinuately dentate or denticulate, 2–7 cm. long, glabrous; calyx 10–15 mm.
long, puberulent, the teeth broadly triangular-ovate,   *f. 504.*
acute; corolla 2–3.5 cm. long, yellow, rarely spotted, bearded within. *M. luteus*
A. Gray; not L. *M. Langsdorfii* Donn. Swamps and along streams: Sask.—
S.D.—N.M.—Calif.—Alaska; Mex. *W. Temp.—Submont.—Subalp.* Ap–Au.

10. LEUCÓSPORA Nutt.

Annual or perennial, often glandular-pubescent herbs. Leaves opposite,
toothed, incised, or pinnatifid. Flowers solitary or 2 together in the axils.
Calyx 5-lobed, the lobes unequal, longer than the tube. Corolla white or blue,
2-lipped. Stamens 4, included; anther-cells contiguous. Style incurved. Capsule septicidal, the valve often 2-cleft. Seeds striate.

1. **L. multífida** (Michx.) Nutt. Annual; stem 1–2 dm. high, viscid-pubescent; leaves pinnately parted into narrow entire or toothed segments; calyx sessile, shorter than the two bractlets, the lobes linear; corolla 3 mm. long. *Conobea multifida* Benth. Prairies and river banks: Pa.—Iowa—Kans.—Tex.—Tenn. *Allegh.—Austral.* Jc S.

11. MACUILLÀMIA Raf. WATER HYSSOP.

Perennial succulent herbs, with creeping or floating stems. Leaves opposite, entire or somewhat toothed, broadest above the middle, palmately veined, sessile. Flowers perfect, axillary, solitary, peduncled. Calyx subtended by 2 small bractlets; sepals nearly distinct, unequal, the uppermost the broadest. Corolla blue or white, nearly regular, 5-lobed; lobes spreading. Stamens 4, slightly didynamous, included; filaments adnate to the tube of the corolla. Capsule ovoid or oval, septicidal, the valves 2-cleft. Seeds numerous. *Monniera* Auth.; not P. B.

1. **M. rotundifòlia** Michx. Stem more or less densely piloso, creeping or floating, rooting at the nodes; leaves sessile, rounded-obovate or suborbicular, 1–2 cm. long; calyx glabrous, 4 mm. long, lobes elliptic, obtuse, the upper larger; corolla white or yellowish, 5–6 mm. long, campanulate; lobes rounded-oval. *Herpestes rotundifolia* Pursh. *Bacopa rotundifolia* Wettst. [G]. *Monniera rotundifolia* Michx. [B, R]. Water: Ill.—Va.—Tex.—Mont. *Prairie—Plain.* Jl–S.

12. LIMOSÉLLA L. MUDWORT.

Low stemless, somewhat succulent annuals, or perennials by means of stolons. Leaves rosulate at the base, narrow, entire. Flowers perfect, solitary on scape-like peduncles. Calyx campanulate, 5-lobed. Corolla nearly regular, open-campanulate, 5-cleft. Stamens 4, with the short filaments adnate to the corolla-tube; anthers confluently 1-celled. Style short; stigma capitate. Capsule 2-celled at the base, 1-celled above, many-seeded. Seeds ovoid, rugulose.

1. **L. aquática** L. Small annual, with runner; leaf-blades 0.5–3 cm. long; glabrous; pedicels 0.5–3 cm. long; calyx 1–1.5 mm. long; teeth triangular, acute; corolla 2–2.5 mm. long, exserted, white or purplish. Water and mud: Lab.—Colo.—Calif.—B.C.; Eurasia. *Temp.—Plain—Mont.* Je–O.

13. GRATÌOLA L. HEDGE HYSSOP.

Annual or perennial, somewhat succulent herbs. Leaves opposite, entire or toothed. Flowers solitary on axillary peduncles. Calyx usually subtended by 2 bractlets; sepals 5, nearly distinct. Corolla white, yellow, or purplish, 2-lipped. Stamens 2, included; staminodia 2, scale-like or filiform, or wanting; anther-sacs transverse, separated by a dilated membranous connective. Capsule globose or ovoid. Seeds striate.

Peduncles as long as the bracts or longer; plant glandular-puberulent; capsule ovoid. 1. *G. neglecta.*
Peduncles shorter than the bracts; plant glabrous; capsule globose. 2. *G. virginiana.*

1. **G. negléota** Torr. *Fig. 505.* Stem 1–2 dm. high, viscid-puberulent or glabrate below; leaves usually glabrous, oblong-lanceolate to linear, 1–5 cm. long, entire or denticulate, sessile; bractlets 2, linear, 5–6 mm. long; calyx about 5 mm. long; sepals lanceolate; corolla light yellow or nearly white, 8–10 mm. long. *G. virginiana* Benth.; not L. [B, R]. Mud and shallow water: Me.—Fla.—Calif.—B.C. *Temp.—Plain—Submont.* My–O.

2. **G. virginiàna** L.   Stem glabrous, 1–4 dm. high, diffusely branched; elliptic to lanceolate, often cuneately narrowed at the base, 1.5–5 cm. long, repand to dentate-serrate; sepals linear-lanceolate; corolla white, 8–12 mm. long.   *G. sphaerocarpa* Ell. Low ground: N.J.—Iowa—Kans.—Tex.—Fla.; Mex. *Austral—Trop.* Je–S.

14. ILYSÁNTHES Raf.   FALSE PIMPERNEL.

Annual or biennial, leafy-stemmed herbs; leaves opposite, glabrous, mostly sessile, entire or shallowly toothed.   Flowers solitary in the axils, slender-petioled.   Sepals 5, united at the base.   Corolla white or purplish, 2-lipped, the lower lip spreading.   Stamens 2; anther-cells divergent.   Sterile stamens 2, 2-lobed, one lobe glandular, capitate, the other smaller, glabrous.   Seeds wrinkled.

f 505

Pedicels shorter than the subtending leaves; sepals as long as the capsule or longer.
1. *I. dubia.*
Pedicels longer than the subtending leaves; sepals shorter than the capsule.
2. *I. inequalis.*

1. **I. dùbia** (L.) Barnhart.   Stem stout, 1–5 dm. high; leaves oblong to ovate, 1–3 cm. long, obtuse, distantly serrate, 5-nerved; corolla 8–10 mm. long. *I. gratioloides* (L.) Benth.   *I. attenuata* (Muhl.) Small [B].   Wet places: Me.—N.D.—Tex.—Fla.   *E. Temp.*   My–O.

2. **I. inequàlis** (Walt.) Pennell.   Stem wiry, 1–2 dm. high, the lower branches often decumbent; leaves obovate to oblong or the upper ovate, 0.5–2 cm. long, entire or nearly so, sessile or partly clasping; corolla 5–7 mm. long. *I. riparia* Raf.   *I. gratioloides* Small, in part.   *I. dubia* Britton [B].   *I. anagallidea* (Michx.) B. L. Robinson [G].   Wet places: Mass.—N.D.—Tex.—Fla. Jl–S.

15. VERÓNICA (Tourn.) L.   SPEEDWELL, BROOKLIME.

Annual or perennial, caulescent herbs.   Leaves opposite or rarely alternate or whorled, entire or toothed.   Flowers perfect, axillary, racemose, spicate, or solitary.   Calyx of 4, rarely 5, slightly united sepals.   Corolla rotate, only slightly irregular, 4-lobed, the lower lobe usually narrower than the rest.   Stamens 2, one on each side of the upper corolla-lobe.   Styles wholly united.   Stigmas capitate.   Capsule flat, usually notched or 2-lobed at the apex, loculicidal. Seeds flattened or concave on the sides.

Flowers in axillary racemes.
  Leaves and stem pubescent; dry-ground plants.   1. *V. officinalis.*
  Leaves and stem glabrous; water- and bog-plants.
    Leaves all short-petioled; leaf-blades ovate, oblong, or oval.   2. *V. americana.*
    Leaves of the flowering shoots at least sessile, lanceolate to linear.
      Pedicels less than twice as long as the fruit, the latter not strongly flattened; leaves broadly lanceolate.   3. *V. catenata.*
      Pedicels several times as long as the fruit, the latter strongly flattened; leaves linear-lanceolate or linear.   4. *V. scutellata.*
Flowers in terminal spikes or racemes, or solitary in the axils of the leaves.
  Perennials; flowers in terminal spikes or racemes; bracts reduced and unlike the leaves.
    All leaves sessile, ovate or oval to oblong; capsules obovate or oval, merely emarginate.   5. *V. Wormskjoldii.*
    Lower leaves petioled; blades rounded-oval or the upper oblong; capsule obcordate.
      Racemes dense, spike-like; leaves often in whorls of 3's, the blades linear-lanceolate or elongate-lanceolate, long-attenuate.   6. *V. maritima.*

Racemes lax, not spike-like; leaves opposite, the
  blades oval, ovate or oblong.
Peduncles and pedicels appressed-pubescent;
  corolla 3-4 mm. broad, pale blue or whitish.   7. *V. serpyllifolia.*
Peduncles and pedicels with spreading viscid
  gland-tipped hairs; corolla 5-10 mm. broad,
  deep blue.   8. *V. humifusa.*
Annuals; flowers solitary in the axils of the leaves, *i. e.*,
  bracts resembling the other leaves and only slightly
  reduced.
Peduncles shorter than the leaves.
Leaf-blades spatulate, oblong, or linear; capsule
  emarginate.
Plant glandular-puberulent.   9. *V. xalapensis.*
Plant glabrous or nearly so.   10. *V. peregrina.*
Leaf-blades oval or ovate, or the upper lanceolate;
  capsule obcordate.   11. *V. arvensis.*
Peduncles as long as the leaves or usually longer.   12. *V. persica.*

1. **V. officinàlis** L. Perennial; stem creeping,
5-30 cm. long, branched; leaf-blades oblong to oval,
or obovate, 1-4 cm. long, serrate, short-petioled; racemes spike-like; corolla blue, 3-3.5 mm. broad;
capsule broadly cuneate or obovate-cuneate, 3-3.5
mm. broad, truncate or retuse, minutely glandular.
Woods and stony places: Newf.—S.D.—Tenn.—Ga.;
Eurasia. *Canad.—Allegh.* My–Au.

2. **V. americàna** Schwein. *Fig. 506.* Stem
erect or decumbent, 1-6 dm. high, sometimes
branched; leaves petioled; blades ovate to oblong-
lanceolate, crenate or serrate or sometimes loosely-
rounded or truncate at the base; racemes loosely-
flowered; corolla blue or nearly white, 4-5 mm.
broad; capsule thick, slightly notched, broader than
long. *V. oxylobula* Greene. *V. crenatifolia* Greene,
a reduced form. In water: Newf.—Va.—N.M. Calif.—Alaska. *Boreal.—
Plain—Subalp.* My–Au.

f.506.

3. **V. catenàta** Pennell. Stem glabrous or glandular-pubescent above,
2-9 dm. high, branched; leaves sessile, lanceolate or oblong, 2-12 cm. long,
acute, finely serrate or entire, often clasping; racemes spreading, many-flowered; calyx-lobes obtusish; corolla blue, 4-5 mm. broad; capsule minutely
notched, suborbicular, wider than long. *V. Anagallis-aquatica* Am. auth., in
part. Wet places and shallow water: N.Y.—Kans.—N.M.—Colo.—B.C. *E.
Temp.—Plain—Submont.* My–S.

4. **V. scutellàta** L. Stem ascending or decumbent at the base, 1-5 dm.
high; leaves sessile, linear or lance-linear, acute, entire or remotely denticulate;
racemes slender, divergent, zigzag; corolla 8-10 mm. wide; capsule flat, deeply
emarginate. Swamps and wet places: Newf.—N.Y.—Colo.—Calif.—B.C.—
Yukon; Eurasia. *Plain—Submont.* Je–Au.

5. **V. Wormskjòldii** R. & S. Stem 1-3 dm. high, glandular-villous above,
glabrate below, strict; leaves ovate, oval, or elliptic, 1-3 cm. long, sessile, entire
or crenulate, drying black; corolla dark blue, campanulate, 4-5 mm. broad;
capsule elliptic-obovate, emarginate. *V. alpina unalaschcensis* C. & S. [G].
Wet meadows: Greenl.—N.H.—S.D.—N.M.—Ariz.—Alaska. *Mont.—Subalp.*
Je S.

6. **V. marítima** L. Stem 3-10 dm. high; more or less pubescent;
branched towards the top; leaves or some of them often in whorls of 3's, short-
petioled, long-attenuate, serrate with the teeth directed forward, 5-12 cm. long,
usually pubescent; racemes spike-like, often branched, 1-2 dm. long; corolla
blue, 4 mm. long; capsule suborbicular, slightly retuse. *V. longifolia* L. Roadsides and waste places: N.S.—N.Y.—N.D.; adv. from Eur. Jl–Au.

7. **V. serpyllifòlia** L. Stems decumbent or creeping, 0.5–3 dm. high; leaves short-petioled or sessile; blades oblong, oval or suborbicular, 5–15 mm. long, entire or crenulate, obtuse; corolla white or pale blue, 3–4 mm. broad; capsule retuse at the apex, obreniform, ciliate and puberulent. Open woods, fields, and thickets: Lab.—Ga.—Mo.—Minn.; B.C.; Eurasia. *Canad.—Submont. —Subalp.* My–Au.

8. **V. humifùsa** Dickson. Similar to the preceding; stem stouter, 2–4 dm. high; leaf-blades 1–2.5 cm. long; capsule 4–6 mm. broad. *V. serpyllifolia borealis* Laest. Springy places: Lab.—N.Y.—Mont.—Colo.; Calif.—Alaska; Eur. *Boreal.—Mont.* My–Au.

9. **V. xalapénsis** H.B.K. Stem 1–3 dm. high, branched, with ascending branches, glandular-pubescent throughout; leaves thick, the lower ones petioled, spatulate, crenate, the upper linear or oblong; racemes spike-like; corolla whitish, 2–3 mm. wide; capsule orbicular, slightly notched, glandular. *V. peregrina* Coult.; not L. Sandy soils: Ont.—Tex.—Calif.—B.C.; Mex. *Plain—Submont.* Je–S.

10. **V. peregrìna** L. Stem simple or branched below, 0.5–4 dm. high; leaves leathery, fleshy, oblanceolate to oblong or linear, 1–3 cm. long, crenate or serrate, or entire; corolla 2–3 mm. broad, whitish; capsule broader than high, 4 mm., deeply notched, glabrous. Low ground and waste places: N.S.—Man.—Neb.—La.—Fla. *E. Temp.* My–O.

11. **V. arvénsis** L. Stem pubescent, 0.4–5 dm. high, at first simple, later branched; lower leaves ovate or oval, crenate, 5–15 mm. long, petioled and opposite, the upper sessile, alternate and often entire; corolla blue or nearly white, 2 mm. broad; capsule broadly obcordate, 2 mm. high. Fields and waste places: Newf.—N.S.—Fla.—Tex.—Calif.—B.C.; nat. from Eur. Mr–S.

12. **V. pérsica** Poir. Stem finely pubescent, 1–4 dm. high, usually branched at the base; leaves opposite or alternate, short-petioled; blades ovate, oval, or suborbicular, 8–15 mm. long, coarsely serrate; sepals pubescent, elliptic or lanceolate; corolla blue, 9–11 mm. broad; capsule obreniform, 7–8 mm. broad, pubescent. *V. Tournefortii* C. C. Gmel. [G]. *V. byzantina* B.S.P. [B]. Fields and waste places: N.Y.—Ga.—Colo.—Utah.—S.D.—Man.; nat. from Eur. Ap–N.

16. **VERONICÁSTRUM** Benth. CULVER'S-ROOT.

Perennial leafy-stemmed herbs. Leaves whorled or opposite, toothed. Flowers in terminal spike-like racemes. Sepals 4, slightly united. Corolla white, short-tubular, slightly 2-lipped, 4-lobed, the upper lip entire. Stamens 2, exserted. Stigma minute. Capsule narrow, subterete, 4-valved at the apex. Seeds minutely reticulate. *Leptandra* Nutt.

1. **V. virgínicum** (L.) Farwell. Stem 3–18 dm. high; leaf-blades oblong or lanceolate, acute or acuminate, finely serrate; corolla 5–6 mm. long; filaments villous above the middle; capsule oblong, 3–4 mm. long. *Veronica virginica* L. *Leptandra virginica* Nutt. [B]. Meadows and thickets: N.S.—Man.—Tex.—Ala. *E. Temp.* Je–S.

17. **BESSÈYA** Rydb. KITTEN-TAILS.

Perennial herbs with rootstocks. Leaves mostly basal with oblong or ovate, crenate blades and short petioles; stem-leaves reduced, alternate, sessile. Flowers perfect, in terminal spikes. Sepals 4, or sometimes 1–3, almost distinct. Corolla irregular, 2-lipped, cleft nearly to the base, or wanting; upper lip broad, entire, concave; lower lip irregularly 3-cleft. Stamens 2; filaments exserted, adnate to the tube of the corolla, or, if the latter is lacking, inserted on the outside of a hypogynous disk; anther-sacs parallel or nearly so. Capsule rather turgid, short, emarginate. Seeds several, flat.

Corolla present, 2-lipped.   1. *B. Bullii.*
Corolla wanting.   2. *B. wyomingensis.*

1. B. Búllii (Eat.) Rydb. Stem 3–8 dm. high, pubescent; basal leaf-blades ovate or orbicular, rounded at the apex, truncate or cordate at the base, crenulate, 5–7-ribbed; corolla greenish yellow, 4–6 mm. long, the upper lip broad, entire, the lower entire or 3-lobed. *Synthyris Houghtoniana* Benth. *S. Bullii* Heller [B, G]. Prairies and oak barrens: Mich.—Minn.—Iowa—Ohio. My–Je.

2. B. wyomingénsis (A. Nels.) Rydb. Stem 2–4 dm. high; leaf-blades ovate to oblong, crenate, rounded or subcordate at the base, 3–5 cm. long, villous to glabrous; filaments exserted. *Synthyris rubra* Britton [B]; not Benth. *B. gymnocarpa* (A. Nels.) Rydb. *Wulfenia wyomingensis* and *W. gymnocarpa* A. Nels. Hills and mountains: S.D.—Neb.—Colo.—Utah—Ida.—Alta. *Mont.* My–Au.

18. AGALÌNIS Raf. GERARDIA.

Annual or perennial, caulescent herbs, with slender stems. Leaves mainly opposite, narrow or scale-like, entire. Flowers perfect, solitary in the axils of the leaves. Calyx 5-toothed to the middle. Corolla in ours rose-purple, 2-lipped; lower lip exterior in the bud, 5-lobed. Stamens 4, didynamous, included; filaments pubescent; anthers all alike. Styles wholly united, filiform. Capsule ovoid, rounded at the apex, loculicidal. Seeds angled. *Gerardia* L., in part.

Capsule oblong or ovoid-oblong; calyx-lobes about half as long as the tube; corolla 1.5–2.5 cm. long.
  Corolla 2–2.5 cm. long, deep purple; stem much branched with long lower branches.   1. *A. aspera.*
  Corolla 1.5–1.8 cm. long, rose-purple; stem simple or with few short branches.
Capsule subglobose; calyx-lobes less than half as long as the tube.   2. *A. Greenei.*
  Upper corolla-lobes, as well as the lower, spreading.
    Plants blackening in drying; calyx-tube not evidently reticulate-venose; seeds black; pedicels shorter than the calyx and capsule.
      Calyx-lobes triangular-lanceolate to lanceolate, ½–⅝ as long as the tube; corolla 12–20 mm. long.   3. *A. paupercula.*
      Calyx-lobes triangular to triangular-subulate, less than half as long as the tube; corolla 20–40 mm. long.   4. *A. purpurea.*
    Plants not blackening in drying; calyx-tube reticulate-venose; seeds yellowish brown; pedicels longer than the calyx and capsule.
      Stem conspicuously striate-angled; flowers racemose.   5. *A. Skinneriana.*
      Stem terete; flowers scattered, apparently terminal.   6. *A. Gattingeri.*
  Upper corolla-lobes ascending over the stamens.
    Calyx-lobes less than 1 mm. long; capsule 3–4 (rarely 5) mm. long; leaves 1–3.5 mm. wide, smooth.   7. *A. tenuifolia.*
    Calyx-lobes 1–2 mm. long; capsule 5–7 (rarely 4) mm. long; leaves 1–6 mm. long; scabrous.   8. *A. Besseyana.*

1. A. áspera (Dougl.) Britton. Annual; stem 1–6 dm. high, more or less scabrous; leaves linear, 1.5 mm. wide or less, 1–3 cm. long; calyx-tube 5–7 mm. long; lobes deltoid; corolla rose-purple; capsule elliptic or ovoid, 8–10 mm. long. *G. aspera* Dougl. [G, B]. Plains and prairies: Man.—Ill.—Ark.—Okla.—Sask. *Prairie—Plain.* Au–O.

2. A. Greènei Lunell. Stem 1–3 dm. high, mostly simple, the branches, if any, short; leaves linear, hispidulous-scabrous along the margins; pedicels 5–10 mm. long; calyx-lobes deltoid; corolla 1.5–1.8 cm. long, rose-purple; capsule about 7 mm. long, obovoid. Ditches: Leeds, N.D. Au.

3. A. paupércula (A. Gray) Britton. Annual; stem stout, 1–5 dm. high, smooth; leaves thickish, 1–3 cm. long, scabrous, often fascicled; calyx-tube campanulate, 3–4 mm. long; corolla rose-purple; capsule globose, 5–7 mm. long. *G. purpurea paupercula* A. Gray. *G. paupercula* Britt. [G, B]. Bogs and wet meadows: Que.—Ga.—Okla.—Sask. *Canad.—Allegh.—Plain.* Jl–S.

4. A. purpùrea (L.) Britton. *Fig. 506a.* Annual; stem 2–7 dm. high, smooth or slightly scabrous; leaves less than 3 mm. wide, 1–5 cm. long, rarely fasciculate; calyx-tube 3–4 mm. long; corolla rose-purple; capsule 4–5 mm. broad. *Gerardia purpurea* L. [G, B]. Swamps and low ground: Me.— Minn.—Tex.—Fla. *E. Temp.* Au–O.

5. A. Skinneriàna (Wood) Britton. Stem 2–7 dm. high, slightly scabrous, angled; leaves few, linear-spatulate, more or less revolute, 0.5–1 cm. long; calyx-tube 2–3 mm. long, the lobes minute, triangular; corolla light rose, about 1 cm. long; capsule 3–4 mm. long. *G. Skinneriana* Wood. [G, B]. Sandy soil: Mass.—Minn.—Kans.—La.—Fla. *E. Temp.* Au–O.

6. A. Gattíngeri Small. Stem 2–5 dm. high, wiry, smooth, the branches slender; leaves many, filiform, 1.5–3 cm. long; calyx-tube 2.5–3.5 mm. long, the lobes minute; corolla rose-purple, 1 cm. long; capsule 3.5–4 mm. broad. *G. Gattingeri* Small. Dry soil or woods: Mich.—Minn. —Tex.—Ala. *Allegh.—Ozark.*

7. A. tenuifòlia (Vahl) Raf. Stem 1–6 dm. high, smooth, much branched, the branches 4-angled; leaves not fasciculated, linear, 1–4 cm. long; calyx-tube 2–3 mm. long, the lobes ⅓–¼ as long as the tube; corolla rose-purple, 1 cm. long; capsule 4 mm. broad. *G. tenuifolia* Vahl [G, B]. Dry woods and thickets; Que.—Man.—Kans.—La.—Ga. *Canad.—Allegh.* Au–O.

8. A. Besseyàna Britton. Annual; stem 2–6 dm. high, somewhat scabrous, branching above; leaves linear or narrowly lance-linear, 1–5 cm. long, acute, scabrous; calyx-tube campanulate, 4–5 mm. long; corolla rose-purple, fully 1 cm. long; capsule globose, 5–7 mm. long. *G. Besseyana* Britton. River bottoms and prairies: Ont.—Minn.—Wyo.—Colo.—La. *Prairie—Plain.* Jl–S.

19. OTOPHÝLLA Benth. GERARDIA.

Annual leafy-stemmed herbs. Leaves opposite, sessile, auricled at the base, entire to pinnately divided. Flowers in terminal spikes. Calyx deeply 5-cleft. Corolla purple or rarely white, broadly dilated at the throat, the lobes spreading. Stamens 4, didynamous, included; filaments glabrous; anthers of the shorter stamens much smaller. Style slender; stigma entire. Capsule oval or globose. Seeds angled.

Leaves entire, except the basal auricles; corolla less than 2 cm. long.
1. *O. auriculata.*
Leaves parted into 3–7 linear segments; corolla more than 2 cm. long.
2. *O. densiflora.*

1. O. auriculàta (Michx.) Small. Stem 3–6 dm. high; leaves lanceolate, hirsute, 1–4 cm. long, at least the upper ones with lanceolate basal auricles; corolla 1.5–2 cm. long; capsule oval, 11–13 mm. long. *Gerardia auriculata* Michx. [G, B]. Low ground: N.J.—Minn.—Kans.—N.C. *Allegh.* Jl–S.

2. O. densiflòra (Benth.) Small. Stem 2–5 dm. high, much branched; leaves 1.5–3 cm. long, hispid, the segments linear, ciliate; spike continuous; corolla more than 2 cm. long; capsule oval, about 8 mm. long. *G. densiflora* Benth. [G, B]. Prairies: Kans.—Tex. *Ozark.* Au–O.

20. AUREOLÀRIA Raf. FALSE FOXGLOVE.

Annual or perennial leafy-stemmed herbs. Leaves mostly opposite, from entire to bipinnatifid, pubescent or glandular. Flowers in leafy-bracted racemes or panicles. Sepals 5, partly united, the tube campanulate, the lobes entire or toothed. Corolla yellow, slightly irregular; tube funnelform; lobes 5, spreading. Stamens 4, didynamous, included; filaments pubescent; anthers uniform; cells awned at the base. Capsule acute, beaked. *Dasystoma* Benth.; not Raf.

Annuals, more or less glandular; leaves pectinately bipinnatifid; corolla more or less
  marked with red-purple; capsule ellipsoid; seeds not winged.   1. *A. pedicularia.*
Perennials, not glandular; leaves entire to pinnatifid and lobed;
  corolla not tinged with purple; capsule broadly ovoid or
  subglobose; seeds winged.
Pedicels 1.5–3 mm. long; capsule rusty-pubescent.   2. *A. virginica.*
Pedicels 3–25 mm. long; capsule glabrous.
  Plant densely cinereous-puberulent.
    Calyx-lobes linear to lanceolate, entire or minutely
      dentate; petioles 10–15 mm. long.   3. *A. serrata.*
    Calyx-lobes broadly lanceolate, evidently dentate; peti-
      oles 5–6 mm. long.   4. *A. grandiflora.*
  Plant glabrous or minutely puberulent.   5. *A. flava.*

1. **A. pediculària** (L.) Raf. Annual or bien-nial, glandular-pubescent above; stem 3–12 dm. high, widely branched; leaves lanceolate to ovate in out-line, bipinnatifid and toothed; calyx-lobes coarsely toothed; corolla 2 3 cm. long. *Gerardia pedicularia* L. [G]. *Dasystoma pedicularia* Benth. [B]. Dry copses and wooded hillsides: Me.—Minn.—Fla. E. Temp. Au–S.

2. **A. virgínica** (L.) Pennell. *Fig. 507.* Per-ennial, grayish-pubescent; stem 3–12 dm. high; leaves lanceolate to oblong, 4–15 cm. long, toothed or rarely pinnatifid, or the upper ones entire; calyx-lobes as long as the tube; corolla 3–4 cm. long; cap-sule ovoid-conic, 1.5 mm. long, long beaked. *G. vir-ginica* (L.) B.S.P. [G]. *G. flava* Am. auth.; not L. *D. flava* Wood. Dry woods: N.H.—Mich.—Fla. E. Temp. Je–Jl.

J 507

3. **A. serràta** (Torr.) Rydb. Perennial, grayish puberulent; stem 4–12 dm. high; lower leaves pinnatifid, with entire segments, the upper smaller, serrate or entire, calyx-lobes linear-lanceolate, entire; corolla 2.5–4 cm. long; capsule ovoid, 1–1.5 cm. long, short-beaked. *D. serrata* (Torr.) Small [B]. Dry soil: Mo.—Kans.—Tex.—La. *Ozark.* Jl–S.

4. **A. grandiflòra** (Benth.) Pennell. Perennial, grayish-pubescent; stem 5–10 dm. high; leaves lanceolate or ovate in outline, pinnatifid throughout, the segments of the lower coarsely toothed; calyx-lobes lanceolate or oblong; cor-olla 3–5 cm. long; capsule ovoid, 1 cm. long, beaked. *G. grandiflora* Benth. [G]. *D. grandiflora* Wood. [B]. Open woods: Wis.—Minn.—Kans.—Tex.— Tenn. *Allegh.—Prairie.* Jl–Au.

5. **A. flàva** (L.) Farwell. Perennial, glabrous; stem 8–15 dm. high, glaucous; leaves pinnatifid or bipinnatifid throughout, 1–2 dm. long, the upper smaller and narrower; calyx-lobes shorter than the tube; corolla 4–5 cm. long; capsule oval-ovoid, 10–15 mm. long. *G. virginica* B. S. P., in part. *D. vir-ginica* Britton [B]. *G. quercifolia* Pursh. *G. flava* L. [G]. Dry woods and thickets: Me.—Minn. (?)—Ala.—Ga. *Canad.—Allegh.* Jl–Au.

21. DASÝSTOMA Raf. MULLEIN FOXGLOVE.

Perennial leafy-stemmed herbs. Leaves opposite, large, pinnately parted or pinnatifid, with toothed segments. Flowers solitary and sessile in the axils. Calyx 5-lobed, the tube campanulate or hemispheric. Corolla yellow, the tube short, funnelform, the 5 lobes spreading. Stamens 4, nearly equal, included. Stigma dilated, 2-lobed. Capsule compressed at the summit. *Brachygyne* Small.

1. D. macrophylla (Nutt.) Raf. Stem finely pubescent, 10–15 dm. high; leaves 1–4 dm. long, the lower pinnatifid, with pinnatifid or incised divisions, the upper lanceolate, entire or toothed; calyx-lobes oval to lanceolate, obtuse; corolla about 1 cm. long, its lobes reniform. *Seymeria macrophylla* Nutt. [G]. *Afzelia macrophylla* Kuntze [B]. *Brachygyne macrophylla* Small. River banks: Ohio—Neb.—Tex.—Mo. *Allegh.—Ozark.*

22. BÚCHNERA L. BLUE HEARTS.

Biennial or perennial leafy-stemmed herbs, blackening in drying. Leaves opposite or mainly so, usually toothed. Flowers in elongate terminal spikes. Calyx 5–10-ribbed, 5-lobed. Corolla white, blue or purple, salverform, the tube curved, the lobes 5, slightly unequal, spreading. Stamens 4, didynamous; anther-cells confluent. Stigma entire or slightly notched. Capsule short, loculicidal. Seeds reticulate.

1. B. americàna L. Stem 3–10 dm. high, hispid; basal leaves oblong, elliptic, or ovate, short-petioled, the upper oblong to lanceolate, 2–9 cm. long, sharply toothed; calyx hispid; corolla deep purple, 10–12 mm. long, the tube villous. Meadows and pine-lands: N.J.—Minn.—Kans.—La.—Va. *Allegh.* Je–S.

23. CASTILLÈJA Mutis. PAINTED CUP, INDIAN PAINT-BRUSH, PAINTER'S BRUSH, SQUAW FEATHER.

Annual or perennial caulescent herbs, often partly root-parasites. Leaves alternate, sessile, entire, toothed, or pectinately lobed, those subtending the flowers partly or wholly brightly colored. Flowers perfect, in terminal leafy spikes. Calyx laterally flattened, 4-lobed, more deeply cleft above and below than on the sides. Corolla highly colored, laterally flattened, strongly 2-lipped; upper lip arched; lower lip much shorter, 3-lobed, only slightly saccate; lobes most commonly lanceolate. Stamens 4, didynamous, surrounded by the upper lip of the corolla; anther-sacs unequal; the outer attached by the middle, the inner by its apex, pendulous. Styles wholly united, slender; stigmas 2-lobed or entire. Capsule loculicidal. Seeds reticulate.

Annuals or biennials.
  Leaves and bracts entire or nearly so.
    Bracts linear-lanceolate.   1. *C. exilis.*
    Bracts spatulate.   2. *C. indivisa.*
  Leaves pinnatifid ; bracts lobed.   3. *C. coccinea.*
Perennials.
  Galea several times longer than the very short lip, at least two-thirds as long as the corolla-tube ; bracts mostly tinged with scarlet or crimson.   4. *C. rhexifolia.*
  Galea less than three times as long as the lip, rarely half as long as the corolla-tube ; bracts pale, whitish, yellow, or pink.
    Corolla 1–2.5 cm. long ; leaves entire ; calyx equally cleft in front and behind.
      Upper part of the stem arachnoid-pilose ; bracts usually greenish white or pinkish.   5. *C. acuminata.*
      Upper part of stem puberulent ; bracts usually pale yellow.   6. *C. sulphurea.*
    Corolla 4–5 cm. long ; leaves pinnatifid ; calyx deeper cleft in front than behind.
      Corolla 4–5 cm. long ; calyx-lobes attenuate to acute ; stem tomentose.   7. *C. sessiliflora.*
      Corolla about 2 cm. long ; calyx-lobes obtuse ; stem appressed-lanate.   8. *C. citrina.*

1. C. éxilis A. Nels. Stem strict, 3–8 dm. high, more or less glandular-hirsute or villous; leaves linear-lanceolate or lanceolate, long-attenuate, 3–10 cm. long, 3-nerved; inflorescence long, spike-like; bracts linear-lanceolate, attenuate, only the uppermost tipped with scarlet or crimson; calyx more deeply cleft above than below, about equaling the corolla; corolla about 2 cm. long, yellowish or tinged with pink on the margin. *C. stricta* Rydb.; not DC. It has been confused with *C. minor* A. Gray [B]. In swampy places, especially in saline soil: Wash.—Nev.—Colo.—Nebr. (acc. to Gray)—Mont. *Plain—Mont.* Jl–Au.

2. **C. indivìsa** Engelm. Stem 2–4.5 dm. high; leaves linear or linear-lanceolate, 3–10 cm. long, rarely with a few lateral lobes; bracts spatulate, rounded at the apex, bright red; calyx as in the next species; corolla about 2.5 cm. long, greenish yellow, not longer than the bracts. Sandy soil: Kans.—Tex. *Texan.* My–Je.

3. **C. coccìnea** (L.) Spreng. Stem usually simple, 3–6 dm. high, more or less short-villous; leaves of the basal rosette oblong or obovate, entire, 2–5 cm. long; stem-leaves 3–8 cm. long, with 3–7 linear, obtuse divisions; bracts 3–5-cleft, tipped with crimson; lateral lobes usually lanceolate, acute or obtuse, the middle one broadly oblong, rounded at the apex; calyx 1.5–2 cm. long, almost equally deeply cleft above and below into two oblong, truncate or retuse divisions; corolla greenish yellow, about 2 cm. long. Meadows and thickets: Me.—N.C.—Tex.—Sask. *Canad.—Allegh.—Plain.* My–Jl.

4. **C. rhexifòlia** Rydb. Stem 3–6 dm. high, usually glabrous below, more or less villous above; leaves oblong-lanceolate to ovate, 3–5-nerved, about 5 cm. long, from glabrous to densely puberulent, bracts crimson, scarlet, or rose; calyx about 2.5 cm. long, green below, colored above; corolla about 3 cm. long, puberulent, green and tinged and bordered with crimson or scarlet; galea about 1 cm. long, lip about 4 mm. long. Open woods and mountain sides: Alaska—Utah—Colo.—S.D.—Sask. *Submont. Subalp.* Je–Au.

5. **C. acuminàta** (Pursh) Spreng. Stem 1.5–6 dm. high, slender; leaves linear to lanceolate, mostly entire, 5–10 cm. long, 3–5-ribbed, the upper broader than the lower; bracts oblong to obovate, corolla greenish yellow, 12–16 mm. long, the galea about 3 times as long as the lip. *C. pallida septentrionalis* A. Gray [G]. Moist thickets: Lab.—Newf.—Me.—Minn.—Man. *Canad.—Subarct.* Je–Au.

6. **C. sulphùrea** Rydb. Stems several, 3–5 dm. high, glabrous or finely puberulent; leaves lanceolate or the upper ovate, 3-ribbed, 4–5 cm. long, finely puberulent; bracts broadly ovate, obtuse, entire or with small teeth, puberulent, light yellow; calyx about 1.5 cm. long; corolla greenish, 22–25 mm. long; galea about 1 cm. long; lip 3–4 mm. long. Mountains: (?) B.C. — Mont. — Utah—Colo.—S.D. *Submont.—Mont.* Je–Au.

f.508.

7. **C. sessiliflòra** Pursh. *Fig. 508.* Stems 1–3 dm. high, more or less densely villous; leaves entire, linear or dissected into linear lobes, 3–5 cm. long, puberulent; bracts broader, 3-cleft, with lanceolate or linear-lanceolate divisions, finely canescent; calyx 3–4 cm. long; corolla 4–5 cm. long, yellow, puberulent; galea 10–15 mm. long; lip about 5 mm. long. Dry plains and hills: Man.—Ill.—Mo.—Tex.—Ariz.—Sask. *Plain.* Je–Au.

8. **C. citrìna** Pennell. Stem 1–3 dm. high; leaves 3–6 cm. long, narrow, 3-ribbed, trifid, with linear falcate lobes, cinereous-pubescent; bracts broader, trifid, with lemon-yellow tips; sepals united on the sides less than one third their length; corolla 18–23 mm. long; upper lip deep green with yellowish margins; lower lip green with yellow tips. Bluffs and plains: Kans.—Tex. *Texan.* My–Jl.

24. **ORTHOCÁRPUS** Nutt. Owl's Clover.

Annual caulescent herbs. Leaves alternate, sessile, pectinately cleft or entire, those subtending the flowers usually highly colored. Flowers perfect in terminal leafy spikes. Calyx tubular or deeply campanulate, nearly equally 4-cleft. Corolla very irregular, 2-lipped; upper lip erect, slightly arched, equaling or slightly shorter than the more or less 3-lobed saccate lower lip. Stamens 4, didynamous, ascending under the upper lip; anther-sacs dissimilar, the outer one attached by its middle, the inner pendulous by its apex. Styles wholly united, slender; stigmas entire. Capsule oblong, loculicidal. Seeds many, reticulate.

1. **O. lùteus** Nutt. Stem 1–4 dm. high, more or less hirsute, strict; leaves narrowly linear-lanceolate to almost subulate, entire or rarely 3-cleft, attenuate, 1.5–5 cm. long; bracts 3–5-cleft, with lanceolate acute divisions, puberulent; calyx 5–6 mm. long, hirsute; its lobes lanceolate, about 1 mm. long; corolla yellow, 10–15 mm. long; galea and lip nearly 5 mm. long, lip slightly gibbous. Sandy soil: B.C.—Wash.—Ariz.—N.M.—Neb.—Man. *Submont.—Mont.* Je–Au.

25. EUPHRÀSIA (Tourn.) L. EYEBRIGHT.

Low herbs, parasitic on other plants. Leaves opposite, dentate or incised. Flowers perfect, small, in terminal leafy spikes. Calyx 4-cleft, or rarely 5-cleft and with one lobe smaller than the other 4. Corolla very irregular, 2-lipped; upper lip 2-lobed, the margins recurved; lower lip much larger, 3-lobed, spreading. Stamens 4, didynamous, ascending under the upper lip; anther-sacs equal, parallel. Capsule oblong, loculicidal. Seeds many, oblong, longitudinally ribbed.

Bracts with acute or obtuse teeth.
  Inflorescence open ; bracts spreading, the lower in age distant ; corolla 4–5.5 mm. long.   1. *E. disjuncta.*
  Inflorescence denser ; bracts ascending, the lower even in fruit approximate ; corolla 5–7 mm. long.   2. *E. arctica.*
Bracts with subulate- or bristle-tipped teeth.   3. *E. hudsoniana.*

1. **E. disjúncta** Fern. & Wieg. Stem slender, 6–30 cm. high, branched below, puberulent; leaves ovate or orbicular, 8–18 mm. long, crenate-dentate; spike interrupted, at last 5–25 cm. long; bracts large, coarsely dentate; corolla 4–5.5 mm. long, with a yellow eye; upper lip purple, the lower white with purple lines. Open places: Lab.—Me.—N.D.—Alta.—Alaska. *Subarctic.* Je–S.

2. **E. árctica** Lange. *Fig. 509.* Stem 5–20 cm. long, pubescent; branches few, ascending; leaves 5–15 cm. long, pubescent; spike dense, not interrupted, 3–15 cm. long; bracts small; corolla white, with pale lavender lines, the upper lip often darker. Dry calcareous soil: Greenl.—Newf.—Que.—Minn.—Man.; Iceland, Scandinavia. *Arctic—Subarct.*

f 509

3. **E. hudsoniàna** Fern. & Wieg. Stem 5–20 cm. high, pubescent, simple or branched below; leaves oblong, 10–15 mm. long, sparingly pubescent, with few acute teeth; spike elongate, at last 5–15 cm. long; bracts oblong, 7–15 mm. long; corolla 5.5–6 mm. long, whitish, with pale violet lines. Grassy places: Hudson Bay—Alta.—Mack. *Subarctic.* Au.

26. ELEPHANTÉLLA Rydb. LITTLE RED ELEPHANT, ELEPHANT'S HEAD, ELEPHANT FLOWER.

Perennial herbs, with rootstocks, blackening in drying. Leaves alternate, pinnately divided, with toothed segments. Flowers perfect, in terminal spikes. Calyx obliquely campanulate, 5-toothed. Corolla 2-lipped; upper lip strongly arcuate, produced into a long filiform upcurved beak, enclosing the long styles; lower lip 3-lobed, the lateral lobes reflexed. Stamens 4, didynamous, ascending under the upper lip; anthers approximate in pairs; sacs transverse, parallel. Capsule compressed, beaked, loculicidal.

1. E. groenlàndica (Retz.) Rydb. Stem 2–6 dm. high, glabrous; leaves petioled, or the upper sessile, lanceolate in outline, pinnatifid, 5–20 cm. long; lobes linear or lanceolate, doubly dentate; spike dense, 1–3 dm. long; calyx glabrous, 5–7 mm. long; teeth triangular-ovate; corolla proper reddish purple or claret-colored, about 1 cm. long, the long curved beak 12–15 mm. long. Swamps and wet meadows: Greenl.—Lab.—Man.—N.M.—Calif.—Alaska. *Arctic.— Mont.—Subalp.* Je–Au.

27. PEDICULÀRIS (Tourn.) L. LOUSEWORT, INDIAN WARRIOR, DUCK-BILL.

Perennial (all ours) or annual herbs. Leaves alternate, opposite or whorled, pinnately veined, pinnatifid, lobed, or merely crenate. Flowers perfect, in terminal spikes or racemes. Calyx cleft on the lower side, 2–5-lobed. Corolla strongly 2-lipped; upper lip laterally compressed, sometimes short-beaked, toothed, or blunt; lower lip erect or ascending, 3-lobed, the lobes mostly spreading. Stamens 4, didynamous, ascending under the upper lip; anthers approximate in pairs; anther-sacs transverse, parallel. Capsule compressed, curved, beaked, loculicidal. Seeds many, pitted, striate, or ribbed.

Beak of the galea conical, without teeth.	1. *P. lapponica.*
Beak of the galea none or very short and thick, or the summit of the galea often bidentate.	
Annuals; stem branching, the flowers axillary as well as in terminal spikes.	
Galea truncate at the apex.	2. *P. euphrasioides.*
Galea rounded at the apex.	3. *P. parviflora.*
Perennials; stem simple; flowers in dense terminal head-like spikes.	
Corolla purple; plant low, seldom more than 1 dm. high.	4. *P. sudetica.*
Corolla yellow; plant taller, 1.5–20 dm. high	
Leaves divided to the midrib or nearly so into narrow, acute, dentate or serrate or incised divisions; galea with two lateral teeth; lip almost reaching the tip of the galea.	5. *P. Grayi.*
Leaves pinnately lobed two thirds to the midrib or less, with broadly oblong or rounded, obtuse and crenate lobes.	
Stem 1–4.5 dm. high; lower lip of the corolla much shorter than the upper.	6. *P. canadensis.*
Stems 3–10 dm. high; lower lip of the corolla reaching almost to the tip of the galea.	7. *P. lanceolata.*

1. P. lappónica L. Stems tufted, merely puberulent, 1–2 dm. high, simple; leaves lanceolate in outline, 2–4 cm. long, pinnately lobed, the lobes ovate to oblong, sharply serrate; spike short and dense; calyx cleft in front, 2-toothed behind, corolla about 1.5 cm. long. Wet places: Greenl.—Lab.— Man.—Mack; n Eur. *Arctic-Subarct.* Jl–Au.

2. P. euphrasioìdes Stephan. Annual; stem 1–3 dm. high, usually much branched, puberulent; leaves lanceolate, 2–5 cm. long, pinnately lobed, the lobes oblong to ovate, crenate-serrate; calyx cleft in front, 2- or 3-dentate behind; corolla about 1.5 cm. long, purple or the lower part yellowish. Barren lands and open woods: Greenl.—Lab.—Man.—B.C.—Alaska. *Arctic—Subarct.* Jl–Au.

3. P. parviflòra Smith. Annual; stem branched, about 3 dm. high, glabrous; leaves lance-oblong, 3–4 cm. long, pinnately parted, the lobes oblong, crenate; calyx 2-lobed, the lobes incised-lacerate; corolla about 12 mm. long. *P. palustris Wlassoviana* Bunge. Wet places: Man.—Alaska; Siberia. *Subarct. —Arctic.* Jl.

4. P. sudética Willd. Perennial; stems tufted, 1–2 dm. high, glabrous; leaves long-petioled, the blades ovate or lanceolate in outline, deeply pinnately divided, the lobes oblong or linear, serrate; spike dense and short; calyx more or less villous, the teeth lanceolate; corolla nearly 2 cm. long; teeth of the galea prominent. Tundras: Franklin—Man.—Alaska. *Arctic.* Jl–Au.

5. P. Gràyi A. Nels. Stem 5–20 dm. high, glabrous or pubescent above; leaves 2–6 dm. long, somewhat pubescent, especially when young; divisions lanceolate, often 5–7 cm. long, with triangular dentate lobes; spike 2–4 dm. long; bracts lance-linear, about equaling the flowers; calyx about 1 cm. long, villous-puberulent, its lobes lance-linear; corolla nearly 3 cm. long; galea 12–15 mm. long, cucullate; lip 1 cm. long with broad rounded lobes. *P. procera* A. Gray; not Adams. Mountain woods: N.M.—S.D.—Wyo—e Utah. *Mont.—Subalp.* Jl–Au.

6. P. canadénsis L. *Fig. 510.* Stems 1.5–4.5 dm. high, more or less pubescent, or glabrate below; leaves 7–13 cm. long, more or less pubescent; lobes oblong, obtuse, incised and dentate; spike 1–2 dm. high; the lower bracts foliaceous; calyx oblique, deeply cleft on the lower side, lobes minute, triangular; corolla yellow or reddish, 2–2.5 cm. long; galea nearly 1.5 cm. long, arcuate, beakless, but with 2 teeth near the apex; lip about 8 mm. long, 3-lobed. Woods and thickets: N.S. —Fla.—N.M.—Colo.—S.D.—Man.; Mex. *E. Temp.—Submont.—Mont.* My–Jl.

f 510.

7. P. lanceolàta Michx. Stem glabrous or nearly so, 3–10 dm. high; leaves often opposite, lanceolate, 5–13 cm. long; lobes oblong or ovate, obtuse, crenate-dentate; spikes 5–10 cm. long; calyx about 1 cm. long, cleft above and below, the two lobes somewhat foliaceous and crenate; corolla 2–2.5 cm. long, yellow, sometimes tinged with rose; galea 12–14 mm. long, arcuate, truncate at the apex; lower lip 10–12 mm. long; lobes rounded. In swamps: Ont.—Conn. —Va.—Neb.—Sask. *Canad.—Plain.* Jl–O.

28. RHINÁNTHUS L. Rattle-box, Yellow Rattle.

Annual, erect, caulescent herbs, blackening in drying. Leaves opposite, toothed or lobed. Flowers perfect, in terminal, leafy, one-sided spikes or solitary in the upper leaf-axils. Calyx compressed, 4-toothed, much inflated in age and reticulate. Corolla irregular, 2-lipped; upper lip compressed, arched, minutely 2-toothed below the apex; lower lip shorter, with 3 spreading lobes. Stamens 4, didynamous, ascending under the upper lip; anthers pilose; sacs transverse. Capsule orbicular, loculicidal. Seeds several, orbicular, winged.

Branches of the stem, when present, short and scarcely developed at the flowering time, later bearing reduced flowers. 1. *R. oblongifolius.*
Branches of the stem usually well developed at the flowering time, bearing well developed flowers. 2. *R. Kyrollae.*

1. R. oblongifòlius Fernald. Stem 1–4 dm. high, usually simple, green, leaves oblong or linear, obtuse, crenate-dentate, scabrous above, pilose beneath; calyx glabrate, the margin ciliate, greenish yellow; corolla yellow, 10–12 mm. long. Meadows: Lab.—N.H.—Man. *Canad.* Jl–Au.

2. R. Kyróllae Chab. *Fig. 511.* Stem 3–7 cm. high, usually without black lines, glabrous or pilose at the nodes and in decurrent lines, often branched at flowering time, yellowish; leaves elongate-lanceolate, oblong, or linear, dentate, with acute teeth, scabrous; calyx short-hairy, ciliate on the margins; corolla 8–9 mm. long; tube yellow. *R. Crista-Galli* Britton, in part [B]. Thickets and meadows: N.S. —Md.—Colo.—Wash.—Alta. *Boreal—Submont.* Jl–Au.

f. 511.

29. MELAMPỲRUM (Tourn.) L. COW-WHEAT.

Annual herbs, often blackening in drying. Leaves opposite, entire or few-toothed. Flowers perfect, solitary in the upper leaf-axils, or in leafy-bracted spikes. Calyx 4-toothed, the upper 2 teeth somewhat longer. Corolla irregular, 2-lipped; tube gradually enlarged upwards; upper lip compressed, obtuse or retuse; lower lip 3-lobed, 2-grooved beneath. Stamens 4, didynamous, ascending under the upper lip. Anther-sacs parallel. Capsule flat, oblique, loculicidal. Seeds 2–4, smooth, strophiolate.

1. **M. lineàre** Lam. Stem 1–5 dm. high, 4-angled, with puberulent lines; leaves lanceolate or linear-lanceolate, acuminate, short-petioled, 2–7 cm. long, entire, bracts broader, often ovate-hastate, or with salient bristle-pointed teeth; calyx-teeth subulate; corolla 8–12 mm. long, white or whitish, turning purplish. Dry woods: N.S.—N.C.—Ida.—B.C. *Boreal.—Allegh.—Mont.* My-Au.

Family 148. LENTIBULARIACEAE. BLADDERWORT FAMILY.

Small scapose, caulescent herbs, growing in water or wet places. Leaves when submerged dissected into linear or filiform segments, often bladder bearing, or when aerial basal and entire, or rarely wanting. Scapes naked or with minute scales, 1–many-flowered. Flowers perfect, irregular. Calyx of 2 or 5 herbaceous sepals. Corolla more or less 2-lipped; tube spurred or saccate. Stamens 2, the filaments adnate to the base of the corolla-tube on the upper side, flattened, twisted. Anthers 1-celled. Pistil solitary, with free central placenta; style thick and short; stigma 2-lobed, with unequal lobes. Fruit a 2 valved or irregularly dehiscent capsule. Seeds numerous, anatropous, rugose; endosperm wanting.

Calyx of 2 sepals; corolla-tube closed by a palate; plants submerged.
  Bracts at the base of the pedicels not accompanied by bractlets; calyx not closing in fruit.
    Branches verticillate and verticillately decompound; lateral lobes of the lower corolla-lip saccate.   1. VESICULINA.
    Branches alternate or none; lateral lobes of the lower corolla-lip not saccate.   2. UTRICULARIA.
  Bracts at the base of the pedicels accompanied by a pair of bractlets; calyx closing in fruit.   3. STOMOISIA.
Calyx of 5 sepals; corolla with an open throat; plants aerial, with basal entire leaves.   4. PINGUICULA.

1. VESICULÌNA Raf.

Aquatic herbs, with horizontal submerged stems and verticillate and verticillately decompound branches. Leaves none. Bladders terminating the ultimate branches, naked at the mouth. Inflorescence racemose, the scape without scales below; bractlets none. Calyx with two concave herbaceous lips. Corolla strongly 2-lipped, the upper lip entire, the lower 3-lobed, the lateral lobes of the latter saccate, forming a 2-lobed palate. Capsule many-seeded, the seeds tuberculate.

1. **V. purpùrea** (Walt.) Raf. Stem 3–10 dm. long, the internodes 2–5 cm. long, the branches in whorls of 5–7; bladders 2–3 mm. long, the valve glandular-hairy without; scape 5–15 cm. high, 2–4-flowered; pedicels in fruit 4–6 mm. long, erect; corolla red-purple; upper lip rhomboid, about 8 mm. long, 12 mm. broad; lower lip 8–12 mm. long, with a yellow spot at the base; spur conic, appressed to the lower lip. *Utricularia purpurea* Walt. [G, B]. *U. saccata* LeConte. Ponds: Me.—Fla.—La.—Minn. *E. Temp.* Jl S.

2. UTRICULÀRIA L. BLADDERWORT.

Aquatic plants. Stems horizontal, mostly submerged, finely dissected and in ours bladder-bearing; bladders urn-shaped, the mouth closed by a lid. Flowers perfect, racemose. Calyx-lobes 2, slightly united. Corolla 2-lipped, in ours yellow, with the throat usually closed by a palate; upper lip 2-lobed or entire; lower lip 3-lobed or entire, with a spur at the base. Capsule irregularly dehiscent, many-seeded.

Stem creeping on the bottom, in shallow water; corolla 4–12 mm. long.
　　Spur and palate of the corolla conspicuous; pedicels ascending in fruit.
　　　　Segments of the leaves capillary; upper lip of the corolla equaling the lower,
　　　　　　which is about 6 mm. long.　　　　　　　　　　　　　　　　　1. *U. gibba.*
　　　　Segments of some of the leaves linear, flat, serrulate; up-
　　　　　　per lip of the corolla about half as long as the lower,
　　　　　　which is 10–15 mm. long.　　　　　　　　　　　　　　　　2. *U. intermedia.*
　　Spur a mere sack; palate obsolete; pedicels recurved in
　　　　fruit; corolla 4–8 mm. long.　　　　　　　　　　　　　　　　3. *U. minor.*
Stem submerged or free-floating; corolla 14–20 mm. long.　　　4. *U. macrorrhiza.*

1. U. gíbba L. Stem creeping on the bottom in shallow water, radiating from the base of the scape; leaves alternate, 1–2-dichotomous, the segments bladder-bearing; scape 2–10 cm. high, 1–3-flowered; corolla yellow, the upper lip subtriangular, 4–6 mm. long, 6–8 mm. broad, the lower 6 mm. long, the palate prominent; spur conic, obtuse, shorter than the lip. Shallow water: Me.—Mich.—Minn. (?)—Tex.—Fla. *E. Temp.* Je–S.

2. U. intermèdia Hayne. Stem creeping on the bottom in shallow water; leaves alternate, some of them trichotomous at the base, then tri-dichotomously branched, with linear, serrulate branches and without bladders, the rest with shorter capillary branches and a few large bladders about 5 mm. long; scape 5–20 cm. high, 1–4-flowered; corolla yellow, the upper lip broadly triangular, 5–6 mm. long, 7–8 mm. broad, the lower slightly 3-lobed, 10–15 mm. long; palate prominent; spur conic at the base, cylindric above, acute, three fourths as long as the lip. Shallow water: Newf.—B.C.—Calif.—Ind.—N.J.; Eur. *Temp.* My–Au.

3. U. mìnor L. Stem leafy, submerged, short; leaves with few divisions, 1–3 mm. long; bladders few, 2 mm. long or less; scape 0.5–1.5 dm. high; corolla pale yellow, 4–6 mm. broad; spur much shorter than the lower lip. Shallow water: Greenl.—N.J.—Iowa—Calif.—B.C. *Arctic —Temp.—Mont.* Jl–Au.

4. U. macrorrhìza LeConte. *Fig. 512.* Stem submerged, leafy, 3–12 dm. long; leaves 2–4 mm. long, dichotomous; bladders 3–5 mm. long; scape 1–3 dm. long; racemes 5–10-flowered; corolla yellow, 15–20 mm. broad; upper lip entire, erect; lower lip 3-lobed, spreading; spur not appressed, horn-like, slightly curved, shorter than the lower lip. *U. vulgaris americana* A. Gray [G]. *U. vulgaris* Am. auth. [B]. Shallow water; Newf.—Fla.—Calif.—B.C. *E. Temp.—Plain—Subalp.* Je–Au.

3. STOMOÌSIA Raf.

Scapose mud-plants, with a tuft of root-like bladder-bearing branches, arising at the base of the scape, and a few leafy branches from the same place. Leaves delicate, some linear, grass-like, others root-like and bladder-bearing. Bladders minute, beaked, without bristles. Inflorescence racemose or spicate, with several scales on the scape. Calyx 2-lobed, the lobes thin, veiny, appressed to the capsule. Corolla 2-lipped, the upper lip erect, with a distinct claw, the lower with a prominent palate, the spur ciliate at the mouth.

1. S. cornùta (Michx.) Raf. Scape 2–30 cm. high, with several scales, 1–5-flowered; bracts and bractlets 1–2 mm. long; calyx yellowish, the upper lobe acuminate, the lower acute, shorter; corolla yellow; upper lip 12–16 mm. long; spur subulate, pendent, 7–12 mm. long. *Utricularia cornuta* Michx. [G, B]. Borders of ponds or in bogs: Newf.—Minn.—Tex.—Fla.; W.Ind. Jl–Au.

4. PINGUÍCULA (Tourn.) L. BUTTERWORT.

Terrestrial scapose herbs of wet places. Leaves in basal rosettes, flat, entire, producing above a mucilaginous secretion. Scapes 1-flowered. Sepals 5, often more or less united. Corolla yellow, violet or in ours purple, more or less 2-lipped, with an open throat; lobes entire or cleft; tube produced below into a nectariferous spur. Stamens 2. Capsule 2-valved. Seeds many.

f. 513.

1. P. vulgàris L. *Fig. 513.* Leaves elliptic or oval, 2–5 cm. long, 1 0 cm. wide; scape 3–12 cm. high; calyx about 3 mm. long; upper lip not cleft to the middle, with triangular lobes; corolla paler, less than 1 cm. wide; spur 6–8 mm. long. Bogs: Greenl.—Newf.—Vt.—Minn.—Mont.—Alta.—Yukon; Eurasia. *Subalp.—Subarctic.* My–Jl.

Family 149. OROBANCHACEAE. BROOM-RAPE FAMILY.

Root-parasites, destitute of green foliage. Leaves scale-like. Flowers perfect or rarely dioecious. Calyx of 4 or 5 more or less united sepals. Corolla more or less bilabiate, persistent, withering. Stamens 4, didynamous, mostly included; filaments terete, adnate to the tube of the corolla; anthers 2-celled or rarely 1-celled. Styles wholly united; stigma capitate or 2-lobed. Fruit a capsule, enclosed in the withering corolla. Seeds very many, anatropous, furrowed or tuberculate; endosperm transparent.

Flowers all perfect and complete.
  Flowers subtended by bractlets.   1. MYZORRHIZA.
  Flowers without bractlets.   2. ANOPLANTHUS.
Flowers diversified, the lower cleistogamous and fertile, the
  upper complete and well developed, but sterile.   3. EPIPHEGUS.

1. MYZORRHÌZA Philippi. BROOM-RAPE.

Whitish or pinkish herbs, parasitic on the roots of other plants. Leaves scale-like, usually pubescent. Flowers spicate, racemose or corymbose, subtended by bractlets. Calyx of 5, nearly equal, narrow lobes. Corolla 2-lipped; tube more or less curved; throat open; lips erect. Capsule 1-celled, 2-valved. *Aphyllon* A. Gray, in part.

1. M. ludoviciàna (Nutt.) Rydb. Stem 0.5–2 cm. high, glandular-puberulent, simple or branched; scales ovate, scarcely 1 cm. long; calyx-tube 3–4 mm. long; lobes lanceolate, attenuate, 7–8 mm. long; corolla usually purple, rarely yellowish; upper lip often only slightly cleft; lobes oblong, acute. *Orobanche ludoviciana* Nutt. [G, B]. Parasitic on *Artemisia* and other composites, etc., in sandy soil: Sask.—Minn.—Ill.—Tex.—Calif.—Wash.—B.C. *Plain—Submont.* Je–Au.

2. ANOPLÁNTHUS Endl. CANCER-ROOT.

Pale or pink root-parasites. Stem simple or branched. Leaves scale-like. Calyx 5-cleft, with nearly equal lobes. Corolla slightly bilabiate; tube more or less curved; upper lip 2-lobed; lower lip 3-lobed. Stamens included. Stigma peltate. Capsule 1-celled, 2-valved, with 4 placentae. *Thalesia* Raf. *Aphyllon* Mitchell.

Calyx-lobes subulate, much longer than the tube; stem very short, 1–3-flowered.
　　　　　　　　　　　　　　　　　　　　　　　　　　　1. *A. uniflorus.*
Calyx-lobes lanceolate or triangular, about equaling the tube;
　　stem evident, several-flowered.
　　Corolla purplish, its lobes rounded or merely mucronate;
　　　sepals more or less acuminate.　　　　　　　　　　　2. *A. fasciculatus.*
　　Corolla sulphur-yellow, its lobes acute; sepals acute.　　3. *A. luteus.*

1. A. uniflòrus (L.) Endl. Stem very short,
often subterranean; peduncles solitary or a few together, 5–15 cm. high, simple; flowers violet-scented;
calyx campanulate, the lobes lanceolate, acuminate;
corolla cream-colored, tinged with purple, 12–20
mm. long; lobes oblong to obovate, ciliate. *Orobanche uniflora* L. [G]. *Aphyllon uniflorum* A. Gray
T. uniflora Britt. Woods, parasitic on tree-roots:
Newf.—N.D.—Kans.—Tex.—Ga. *E. Temp.* Ap–Jl.

2. A. fasciculàtus (Nutt.) Walp. *Fig. 514.*
Stem 2–10 cm. long; pedicels 3–12, 2–8 cm. long;
scales brownish, ovate, obtuse; calyx-tube 3–4 mm.
long, campanulate; lobes triangular-lanceolate, acuminate, 3–4 mm. long; corolla 2–3 cm. long. *O.
fasciculata* Nutt. [G]. *Aphyllon fasciculatum* A.
Gray. *Thalesia fasciculata* Britt. [B]. Parasitic on
Artemisia, other composites, *Eriogonum, Phacelia,* etc.: Sask.—Ind.—N.M.—
Calif.—Yukon; Mex. *Temp.—Plain—Subalp.* My–Au.

3. A. lùteus (Parry) Rydb. Stem 5–10 cm. long, rather slender; scales
ovate, acute, glandular-puberulent; pedicels 2–5 cm. long; calyx-tube broadly
campanulate, 3 mm. long; lobes triangular or lance-triangular; corolla about
2 cm. long; lobes elliptic. *Aphyllon fasciculatum luteum* A. Gray. *T. lutea*
Rydb. [R]. Parasitic on grasses: Neb.—Wyo.—Ariz. *Plain.* Ap–Jl.

3. **EPIPHÈGUS** Nutt. BEECH-DROPS.

Parasitic caulescent herbs, with scale-like leaves. Flowers in elongate
racemes, of two kinds: cleistogamous and fertile on the lower part, and well
developed, perfect and complete, but usually sterile, above. Calyx 5-lobed,
nearly regular. Corolla gamopetalous; tube curved, slightly enlarged upwards;
limb 4-lobed, the uppermost lobe somewhat larger, concave or arching, the other
three nearly equal. Stamens slightly exserted. Ovary 1-celled, with 4 broad
placentae; stigma subcapitate, slightly 2-lobed. Capsule short, 2-valved at the
apex. *Leptamnium* Raf.

1. E. virginiàna (L.) W. Barton. Plant purplish or yellowish brown, sometimes puberulent; stems solitary or tufted, branched, 1–4 dm. high, sparingly
scaly; raceme spike-like; pedicels 1–2 mm. long; calyx variegated; lobes triangular, about equaling the tube, acute; corolla of the complete flowers 10–13
mm. long; upper lip notched, its lobes obtuse; lower three corolla-lobes (lower
lip) acute. *L. virginianum* Raf. Beech-woods: N.B.—Wis.—Minn. (?)—La.—
Fla. Au–O.

Family 150. **MARTYNIACEAE.** UNICORN-PLANT FAMILY.

Annual or perennial herbs, with branching stems. Leaves opposite,
rarely alternate, with broad blades. Flowers perfect, irregular. Calyx
4- or 5-lobed. Corolla gamopetalous, bilabiate; tube more or less curved;
upper lip 2-lobed, exterior in the bud; lower lip 3-lobed. Stamens 4,
didynamous, all anther-bearing, or the posterior two without anthers; filaments elongate. Gynoecium of 2 united carpels; ovary 1-celled, with 2

parietal placentae; style slender; stigmas 2. Fruit a beaked capsule. Seeds flattened; endosperm wanting.

1. PROBOSCÍDEA (Schmid.) Moench.
UNICORN-PLANT.

Characters of the family.

1. **P. louisiàna** (Mill.) Woot. & Standl. *Fig. 515*. A densely viscid annual; stem with spreading branches, 2–9 dm. long; leaves petioled; blades suborbicular or rounded-ovate, 5–30 cm. broad, sinuate, cordate at the base, calyx cleft on the lower side; corolla white or pink, spotted with yellow and purple within; pod 8–15 cm. long, strongly curved, separating into two elastically spreading valves; the beak longer than the body. *Martynia louisiana* Mill [R]. *M. proboscidea* Glox. Waste places and banks: Me.—Ga.—Tex.—Colo.—Neb.—Minn. *E. Temp.*

f.515

Family 151. BIGNONIACEAE. TRUMPET-CREEPER FAMILY.

Trees or woody climbers, rarely herbs. Leaves mostly opposite, simple or pinnately compound, without stipules. Flowers showy, perfect, irregular. Calyx 5-lobed or 2-lipped. Corolla more or less 2-lipped, the tube funnelform or tubular. Stamens 5, but 1 or 3 usually sterile, the fertile ones 2 or 4 and then didynamous; filaments filiform and adnate to the corolla-tube, the anthers with 2 divergent cells. Pistil 2-carpellary, 1-celled, with 2 parietal placentae, or by the union of the placentae 2-celled. Fruit a leathery or woody capsule. Seeds winged, numerous.

Woody vines, with pinnately compound leaves; fertile stamens 4.   1. TECOMA.
Trees, with simple leaves; fertile stamens usually 2.   2. CATALPA.

1. TÉCOMA Juss. TRUMPET CREEPER.

Woody vines, climbing with aerial rootlets. Leaves pinnately compound, with toothed leaflets. Flowers clustered, often paniculate, perfect. Calyx tubular-campanulate, 5-lobed. Corolla tubular-funnelform, slightly irregular, 5-lobed. Stamens 4, included. Capsule flattened contrary to the partition, firm-leathery. Seeds numerous, winged. *Campsis* Lour.

1. **T. radìcans** (L.) Juss. *Fig. 516*. Liana several meters long; leaves pinnate; leaflets 7–11, oval, ovate, or elliptic, 2–6 cm. long, acute or acuminate, serrate; flowers corymbose; calyx 14–18 mm. long, the lobes deltoid; corolla red or orange, often yellow within, 5–9 cm. long, the limb 3–5 cm. broad; capsule 1–1.8 dm. long. Woods and thickets: N.J.—Iowa—Tex.—Fla. *Allegh.—Austral.* Jl–S.

f.516.

2. CATÁLPA Scop. CATALPA, INDIAN BEAN, CIGAR TREE.

Trees or rarely shrubs, with opposite or whorled simple leaves, in ours with cordate blades. Flowers perfect, large, corymbose or paniculate. Calyx deeply

2-lipped. Corolla oblique, the upper lip 2-lobed, the lower 3-lobed. Fertile stamens usually 2, adnate to the base of the corolla, filaments flattened; anthers introrse. Capsule elongate, subterete, 2-celled, 2-valved, the partition at right angles to the valves. Seeds numerous, 2–4-ranked, winged on each side, the wings cut into a fringe.

1. C. speciòsa Warder. *Fig. 517.* A tree, 20–40 m. high, with thick bark. Leaf-blades ovate or cordate, 1–3 cm. long, acuminate, tomentulose beneath; panicle few-flowered; corolla white, inconspicuously spotted, 3.5–5 cm. long, 6–8 cm. broad. Damp soil: Ill.—Kans.—Ark.; cultivated and escaped in Iowa—Minn.—Neb. *Ozark.* My–Je.

Family 152. **ACANTHACEAE.** Acanthus Family.

Herbs or rarely shrubby plants, the stems usually with swollen nodes. Leaves simple, opposite or alternate, without stipules. Flowers perfect, irregular, often subtended by conspicuous bracts. Sepals 5, often nearly distinct, or variously united. Corolla usually 2-lipped, rarely nearly regularly 5-lobed. Stamens 4 and didynamous, or only 2 fertile, the others abortive or obsolete, the anthers 2-celled. Pistil 2-carpellary, styles united. Fruit a membranous or leathery capsule, longitudinally dehiscent, often flattened contrary to the valves and partition. Seeds usually flattened.

Corolla distinctly 2-lipped; fertile stamens 2.
  Placentae separating from the valves of the capsule; flowers in conspicuously bracted spikes.   3. DICLIPTERA.
  Placentae permanently attached to the valves of the capsule; flowers in axillary peduncled spikes or heads.   1. DIANTHERA.
Corolla scarcely 2-lipped; fertile stamens 4.   2. RUELLIA.

1. **DIANTHÈRA** L. Water Willow.

Perennial caulescent herbs. Leaves opposite, entire or rarely toothed. Flowers perfect, irregular, several on axillary peduncles. Calyx-lobes slender, 4 or 5. Corolla variously colored, strongly 2-lipped, the tube straight or curved, the upper lip erect, entire or 2-toothed, the lower one spreading, 3-lobed. Stamens 2, the filaments adnate to near the mouth of the corolla-tube; anthers 2-celled, the cells separated by a broad connective. Capsule with a stipe-like base, 2-celled, 4-seeded, the seeds flat, ovate to suborbicular.

1. D. americàna L. Perennial, with a creeping rootstock; stem 3–10 dm. high, 4-angled; leaves linear or lanceolate, 5–15 cm. long, acute at the apex, attenuate at the base; spikes 1–3 cm. long; calyx-lobes linear, 4–5 mm. long; corolla white, pink, or purple; lips as long as the tube, the upper notched, the lower 3-lobed; capsule 1.5–2 cm. long. Marshes and streams: Que.—Wis.—Kans.—Tex.—Ga. Jl–S.

2. **RUÉLLIA** (Plumier) L. Ruellia.

Perennial herbs or shrubs. Leaves opposite, simple, toothed or entire. Flowers perfect, solitary or clustered in the axils, or in terminal cymes or panicles. Calyx narrow, the lobes narrow, linear-filiform or lanceolate. Corolla funnelform or salver-shaped, the tube narrow, dilated into an ample throat, the lobes 5, slightly unequal, convolute. Stamens 4, the cells parallel and nearly equal. Style recurved at the apex. Capsule oblong or clavate, constricted below into a stipe-like base.

Calyx-lobes filiform-subulate, decidedly longer than the capsule.   1. *R. ciliosa.*
Calyx-lobes lance-linear, not longer than the capsule.   2. *R. strepens.*

1. **R. ciliòsa** Pursh. *Fig. 518.* Stem erect, 2–6 dm. high, hirsute; leaves subsessile, 2–5 cm. long, ovate to oblong, ciliate; calyx bristly, the lobes subulate-filiform, 1.5–2 cm. long; corolla blue, 5–7 cm. long, the tube as long as the throat, the limb 3–4 cm. broad. Dry soil: N.J.—Neb.—Tex.—Fla. *Allegh.* —*Austral.* Je–S.

2. **R. strèpens** L. Stem erect, 3–10 dm. high, glabrous or nearly so; leaves short-petioled, the blades oblong to oval, 7–15 cm. long; calyx-lobes 15–25 mm. long; corolla blue, 4–5 cm. long, the tube about as long as the throat and limb. Dry rich soil: Pa.—Fla.—Tex.—Iowa. *Allegh.—Austral.* My–Jl.

5. DICLÍPTURA Juss.

Annual or perennial caulescent herbs. Leaves opposite, entire. Flowers perfect, irregular, in conspicuously bracted spikes. Calyx hyaline, the lobes bristle-like. Corolla blue, violet, or red, 2-lipped, the tube slightly dilated above, the upper lip erect, concave, entire or notched, the lower spreading, entire or 3 toothed. Stamens 2, the filaments adnate up to the mouth of the corolla-tube, the anthers 2-celled; sterile stamens wanting. Ovary 2-celled; styles united, filiform. Capsule with a stipe-like base and suborbicular or ovoid body. Seeds 1–4. *Diapedium* Konig.

1. **D. brachiàta** (Pursh) Spreng. Stem 3–7 dm. high, with divergent branches, sparingly pilose; leaves ovate to oblong, 2–10 cm. long, acute or acuminate, with slender petioles; calyx 4–5 mm. long, the lobes linear-subulate; corolla pink or purple, 1.5–2 cm. long, the upper lip 3-toothed, the lower entire. Capsule about 5 mm. long. Dry soil: N.C.—Kans.—Tex.—Fla. *Austral.* Jl–O.

Family 153. **PHRYMACEAE.** LOPSEED FAMILY.

Perennial leafy-stemmed herbs. Leaves opposite, petioled, membranous, simple. Flowers spicate, perfect, irregular, subtended by small bracts. Calyx 2-lipped, reflexed in fruit; tube cylindric; upper lip 3-lobed, the lobes awl-shaped; lower lip 2-lobed, shorter. Corolla 2-lipped; tube cylindric; upper lip erect, notched, concave; lower lip larger, spreading, convex, with 3 obtuse lobes. Stamens 4, included, didynamous. Ovary oblique, 1-celled; style slender; stigmas 2. Ovules solitary. Fruit a narrow achene, surrounded by the accrescent, ribbed, closed, reflexed calyx.

1. PHRỲMA L. LOPSEED.

Characters of the family.

1. **P. leptostàchya** L. *Fig. 519.* Stem 3–10 dm. high, branched above, reflexed-hairy, enlarged and purple above the nodes; leaf-blades ovate or elliptic-ovate, 3–15 cm. long, acuminate to obtusish, serrate; spike 2–20 cm. long; calyx 3–5 mm. long, accrescent; upper 3 lobes hooked at the apex; corolla 8 mm. long, white and usually tinged with magenta; achene 4–5 mm. long. Woods and copses: N.B.—Fla.—Kans.—S.D.—Man. Je–Au.

Family 154. **PLANTAGINACEAE.** PLANTAIN FAMILY.

Annual or perennial herbs. Leaves basal or alternate. Flowers perfect, monoecious, or dioecious, bracteate. Sepals 4, persistent, often scarious-margined, slightly united. Corolla campanulate or tubular, with 4 erect or spreading lobes, scarious, nerveless, persistent. Pistil solitary; ovary superior, 1–4-celled; stigma single, filiform. Stamens 2 or 4; filaments adnate to the tube of the corolla; anthers 2-celled, versatile, opening lengthwise. Fruit a circumscissile capsule; seeds one to several in each cell, amphitropous; endosperm fleshy.

Scape 1–2-flowered ; ovary 1-celled, 1-seeded ; aquatic plants.   1. LITTORELLA.
Scape several- or many-flowered ; ovary 2-celled, 2–many-seeded ; land-plants.
                                                          2. PLANTAGO.

1. **LITTORÉLLA** Berg. SHORE-WEED.

Low, acaulescent, stoloniferous, aquatic plants, monoecious. Staminate flowers solitary, on naked scapes; calyx 4-parted; corolla cylindric, 4-lobed; stamens 4, exserted, with capillary filaments. Pistillate flowers usually 2, sessile at the base of the scape; calyx of 3 or 4 unequal sepals; corolla urn-shaped, 3- or 4-toothed; style long, laterally stigmatic. Fruit an achene.

1. **L. uniflòra** (L.) Aschers. Leaves 2–8 cm. long, 1–2 mm. wide, usually longer than the scape, linear-subulate; pistillate flowers very small; achene 2 mm. long. *L. lacustris* L. Shallow water along the shores: Newf.—Me.—Minn.; Eur. *Canad.* Jl–Au.

2. **PLANTÀGO** (Tourn.) L. PLANTAIN, RIBGRASS.

Annual or perennial herbs, in ours acaulescent and with basal leaves only. Blades often broad, prominently ribbed. Flowers arranged in an elongate spike, sometimes conspicuously bracted. Calyx-lobes equal, or 2 longer. Corolla mostly salver-shaped, constricted at the throat. Capsule more or less membranous. Seeds with flattened or concave faces.

Flowers all perfect ; corolla not closed over the fruit ; stamens 4.
  Leaves lanceolate to ovate ; neither leaves nor spike silky-pubescent or lanate ;
    stamens in all the flowers long-exserted.
    Spike cylindrical ; seeds not concave on the faces.
      Leaves ovate, abruptly contracted at the base.
        Lateral ribs of the leaf-blades confluent with the midrib below ; seeds
        1–2 in each cell.   1. *P. cordata.*
        Lateral ribs of the leaf-blades free from the mid-
        rib to the base ; seeds more than 2 in each
        cell.
        Pyxis dehiscent at the middle, round-ovoid,
        obtusish ; leaves usually thick and the
        dense spike obtuse.   2. *P. major.*
        Pyxis dehiscent far below the middle, elon-
        gate-ovoid, very acute ; leaves thin and
        the lax spike acute.
          Sepals not keeled ; pyxis ovoid, 3–4 mm.
          long.   3. *P. asiatica.*
          Sepals strongly keeled ; pyxis 4–6 mm.
          long.   4. *P. Rugelii.*
    Leaves lanceolate, gradually tapering into the peti-
    oles.
    Spike short, oblong, 1–8 cm. long ; seeds concave on the
    faces ; leaves narrowly lanceolate.   5. *P. eriopoda.*
  Leaves linear.
    Leaves and peduncles not long-villous ; stamens exserted ;
    seeds 1–2 in each cell, plump.   6. *P. lanceolata.*
    Leaves or peduncles or both more or less pubescent
    with long silky hairs ; anthers in the more fertile
    flowers included ; seeds solitary in each cell, con-
    cave on the face.
    Bracts aristate or herbaceous, at least the lower
    ones 2–many times as long as the flowers.   7. *P. decipiens.*

Plant dark green; spike compact.   8. *P. aristata.*
Plant light green; spike more slender.   9. *P. spinulosa.*
Bracts linear-subulate, neither aristate nor folia-
ceous, slightly if at all exceeding the flowers;
spike long-cylindric.   10. *P. Purshii.*
Flowers subdioecious or polygamo-dioecious; corolla in the
fertile plant remaining closed or early closing over the
capsule; leaves spatulate to filiform.
Stamens 4; leaf-blades oblong to spatulate or obovate.
Bracts and sepals obtuse; seeds golden yellow.   11. *P. virginica.*
Bracts and sepals acuminate; seeds dark red.   12. *P. rhodosperma.*
Stamens 2; leaves linear or filiform.
Capsule 10–30-seeded; leaves often distantly toothed. 13. *P. heterophylla.*
Capsule 4-seeded; leaves entire.
Bracts saccate-keeled; seeds 2 mm. long, four times
as long as broad; capsule nearly twice as long
as the sepals.   14. *P. elongata.*
Bracts not keeled; seeds 1 mm. long, two and a half
times as long as broad; capsule one third longer
than the sepals.   15. *P. pusilla.*

1. P. cordàta Lam. Perennial, deep green, glabrous; leaf-blades ovate to suborbicular, 5–30 cm. long, 7–9-ribbed, rounded or cordate at the base, acute or obtuse at the apex; scape 1.5–5 dm. long; spike cylindric, loosely flowered; bracts orbicular or reniform, 1.5–3 mm. long; sepals ovate or orbicular; corolla-lobes ovate, reflexed in age; pyxis globose-ovoid, circumcissile near the middle, 4–5 mm. long. Wet places: N.Y.—Ala.—La.—Minn. *Allegh.* Mr–Jl.

2. P. màjor L. Perennial; leaves petioled; blades oval or ovate, sparingly pubescent or glabrous, 5–35 cm. long, 5–7-ribbed, entire or coarsely toothed; scape 8–15 cm. long; spike dense, 4–20 cm. long; sepals ovate, acute, not keeled; corolla-lobes spreading or reflexed; pyxis ovoid, 3 mm. long; seeds 6–18. Waste places: Newf.—B.C.—Calif.—Fla., nat. from Eur. My–S.

3. P. asiática L. *Fig. 520.* Perennial; leaves petioled; blades oval or ovate, sparingly pubescent or glabrate, thin, 1–2 dm. long, 5–7-ribbed, dentate or subentire; scape 1–2.5 dm. high; spike laxer than in the preceding; bracts ovate, scarious-margined, round-keeled; sepals suborbicular, scarious, with green midrib, not keeled; corolla-lobes spreading; pyxis ovoid, 3–4 mm. long; seeds 14–20. *P. nitrophila* A. Nels. Alkaline ground and waste places: Man.—Neb.—Colo.—B.C. *Plain—Submont.* My–S.

f.520.

4. P. Rugélii Decne. Perennial, bright green, glabrous or nearly so; leaf-blades ovate or oblong, 5–7-ribbed, 1–2 dm. long, entire or shallowly toothed, mostly cuneate at the base; scape 1–5 dm. long; spike cylindric, 2–30 cm. long, acute; bracts fully half as long as the calyx, acute; sepals oblong, 2–2.5 mm. long, acute, often scarious-margined; corolla-lobes spreading; pyxis conic, 4–6 mm. long, circumscissile below the middle. Woods and waste places: N.B.—Fla.—Tex.—N.D.—Man. *E. Temp.* Je–S.

5. P. eriópoda Torr. Perennial; leaves petioled; blades oblanceolate, lance-oblong or elliptic, entire, 3–9-ribbed, pubescent, 0.5–2 dm. long; scape 1.5–4 dm. high; spike 2–10 cm. long, dense above; sepals oblong-obovate, scarious-margined; pyxis ovoid-oblong, obtuse, circumscissile below the middle. *P. retrorsa* Greene. Alkaline or saline soil: N.S.—Quo.—Minn.—N.M.—Nev.—Alta. *Plain—Submont.* Je–S.

6. P. lanceolàta L. A biennial or perennial; leaves 4–30 cm. long, linear-elliptic to elliptic, acute, 3–7-ribbed, entire or denticulate, hairy at the base; scapes 1–7 dm. long; spike dense, 1–8 cm. long; calyx-lobes 2–3 mm. long,

broadly oblong, obtuse; capsule oblong, circumscissile at the middle, 2-seeded.
Lawns and waste places: N.B.—Alaska—Colo.—Fla.; nat. from Eur. Ap–N.

7. **P. decípiens** Barneoud. Leaves linear, 1–10 mm. broad, 5–20 cm.
long, fleshy; scape 5–30 cm. long; spike cylindric, slender, 5–12 cm. long;
bracts and sepals purplish brown or drab; corolla-tube pubescent. *P. maritima*
Auth. [G, B]; not L. Beaches: Greenl.—N.J.; Que.—Man.; Calif.—Alaska.
Arct.—Boreal. Je–S.

8. **P. aristàta** Michx. Annual; leaves linear, acuminate, pubescent,
3-ribbed, 1–5 mm. wide; scape stout, erect, 7–30 cm. long; spike 3–12 cm. long,
pubescent; bracts 1–3 cm. long; sepals spatulate-oblong, obtuse; corolla-lobes
ovate, spreading; pyxis 2.5–3 mm. long, circumscissile at the middle; seeds 2.
Dry soil: Me.—B.C.—N.M.—Fla. *Temp.—Plain.* My–O.

9. **P. spinulòsa** Decne. Annual; leaves linear, acuminate, 3-ribbed, 3–6
mm. wide, villous; scape 5–10 cm. long; spike interrupted, 3–7 cm. long, villous;
bracts 7–15 mm. long, spreading; sepals spatulate-oblong, obtuse; corolla-lobe
ovate, spreading; capsule 2.5 mm. long, obtuse, circumscissile at the middle,
2-seeded. Minn.—Alta.—Tex. *Prairie—Plain.* My–O.

10. **P. Púrshii** R. & S. Annual; leaves ascending, linear or nearly so,
acute, short-petioled, 3-ribbed, 3–8 cm. long, entire, woolly; scape 5–35 cm.
long; spike dense, cylindric, 2–12 cm. long, very villous; sepals oblong, obtuse,
scarious-margined; corolla-lobes ovate, spreading; pod oblong, circumscissile
about the middle, 2-seeded. *P. gnaphaloides* Nutt. Plains and river valleys, in
sandy soil: Ont.—Sask.—B.C.—Ariz.—Tex.—Mo. *W. Temp.—Prairie—Plain—*
Mont. My–Au.

11. **P. virgínica** L. Annual or biennial, glabrous or pubescent; leaf-
blades spatulate to obovate or elliptic, obtuse or acute, entire or repand-dentate,
3–5-ribbed, 1–12 cm. long; scape 0.5–20 dm. high; spike dense; flowers dioecious; bracts lance-linear; sepals oblong or ovate, 2–2.5 mm. long, obtuse,
scarious-margined, longer than the bracts; corolla of the staminate flowers with
spreading lobes, that of the pistillate flowers with unequal lobes, which are
erect after fertilization; pyxis ovoid, 1.5–2 mm. long, circumscissile at the
middle. Dry soil: Conn.—Mich.—Kans.—Ariz.—Fla. *Allegh.—Austral.*

12. **P. rhodospérma** Decne. Annual or biennial, densely and coarsely pubescent; leaf-blades oblong or spatulate, 5–18 cm. long, spreading, obtuse,
3–5-ribbed, entire or repand; scape erect or decumbent, 2.5–20 cm. high; spike
elongate, dense; flowers dioecious; bracts narrowly ovate, shorter than the
calyx; sepals oblong to ovate, 2–2.5 mm. long, acuminate, scarious-margined;
corolla glabrous or papillose, its lobes ovate or deltoid, in the pistillate flowers
connivent after fertilization; capsule oblong, 3 mm. long, obtuse, circumscissile
below the middle; seeds dark red. Sandy soil: Mo.—Kans.—Okla.—Ariz.—
Tex.—La.

13. **P. heterophýlla** Nutt. Annual, glabrous or sparingly puberulent;
leaves 2–18 cm. long, linear to filiform, entire or with remote teeth or spreading narrow lobes; scapes several, ascending or spreading, 2–2.5 cm. long; spike
slender, 2–15 cm. long, lax and interrupted; bracts ovate, 1.5–2 mm. long;
flowers dioecious or polygamous; sepals oval to obovate, 1.5 mm. long, obtuse,
scarious-margined, mostly shorter than the bracts; corolla-lobes ovate, acute,
erect; capsule oblong, 3–3.5 mm. long, circumscissile below the middle. Moist
soil: N.J.—Kans.—S.C.—Calif.—Fla. *Allegh.—Austral.—Son.*

14. **P. elongàta** Pursh. Annual; leaves cinereous-pubescent, linear, entire or nearly so, 1-ribbed, 3–10 cm. long, 0.5–2 mm. wide; scape 3–8 cm. high;
spike 1–10 cm. long; bracts triangular-ovate, scarious-margined, 2 mm. long;
corolla-lobes triangular, 0.5 mm. long; pyxis oblong-ovate, rounded at the apex;
circumscissile just below the middle. *P. myosuroides* Rydb. Wet places: Man.
—Okla.—Utah—Alta. *Prairie—Plain.*

15. P. pusílla Nutt. Annual; leaves linear to filiform, 1–5 cm. long, with a callous apex; scape erect to spreading-ascending, 3–10 cm. high, striate above; spike cylindric, in age lax; bracts triangular-ovate, carnose, at last spreading, scarious-margined, 1.25–1.5 mm. long; pyxis ovate-ovoid, truncate-rounded at the apex, circumscissile one-third from the base. Wet places: Mass.—Kans.—Tex.—Ga.; also Wash.—Calif. *Temp.*

Family 155. RUBIACEAE. MADDER FAMILY.

Herbaceous or woody plants, with opposite or apparently verticillate leaves and usually perfect, but often dimorphous or trimorphous, regular and symmetrical flowers. Ovary inferior, sunken into and adnate to the hypanthium. Corolla in ours funnelform or rotate, gamopetalous and 3–5-lobed. Stamens as many as the lobes of the corolla and alternate with them; filaments adnate to the corolla; anthers mostly linear-oblong. Ovary in ours 2-celled, ripening into a didymous, indehiscent, dry or fleshy fruit. Ovules solitary in each cell. Endosperm in our genera fleshy or horny.

Ovary several-ovuled; fruit 2-lobed and loculicidal above.   1. HOUSTONIA.
Ovary with a single ovule in each cell.
  Flowers in dense heads; fruit of 2–4 nutlets; shrubs.   2. CEPHALANTHUS.
  Flowers solitary or in open cymes; herbs.
    Leaves opposite; stipules inconspicuous.
      Flowers paired, with united ovaries; evergreen herbs. 4. MITCHELLA.
      Flowers not paired; ovaries distinct; herbs not evergreen.
        Fruit separating into 2 or 3 indehiscent carpels.   3. DIODIA.
        Fruit of 2 distinct carpels, of which at least one is dehiscent ventrally.   5. SPERMACOCE.
    Leaves apparently verticillate, the stipules conspicuous, resembling the leaves.   6. GALIUM.

1. HOUSTÒNIA L. BLUETS.

Annual or perennial herbs. Leaves opposite, entire. Flowers perfect, often dimorphous, solitary or in corymbiform cymes. Hypanthium globose or obovoid. Sepals 4, distinct. Corolla funnelform or salver-shaped; lobes 4, valvate. Stamens 4; filaments adnate to the throat of the corolla. Ovary sessile, 2-celled; styles united; stigmas narrow. Ovules several in each cell. Capsule partly inferior, more or less 2-lobed and loculicidal above the hypanthium. Seeds rounded, rough.

Annual; flowers solitary.   1. *H. minima.*
Perennial; flowers cymose.
  Sepals at least equaling the hypanthium in length; fruit half superior.
    Fruit broader than high, much shorter than the sepals; leaf-blades ovate or lanceolate.   2. *H. purpurea.*
    Fruit as high as broad, slightly shorter than the sepals.
      Leaf-blades oblong to spatulate, at least the basal ones ciliate.   3. *H. canadensis.*
      Leaf-blades linear or linear-oblong, glabrous, not ciliate.   4. *H. longifolia.*
  Sepals shorter than the hypanthium; fruit free only at the apex; leaf-blades linear.   5. *H. angustifolia.*

1. H. mínima Beck. Plant glabrous or rough-pubescent; stem branched at the base, diffuse, 1–10 cm. long; leaves mainly basal, spatulate, obovate or ovate, 2–8 mm. long, acute, ciliate; pedicels 1–1.5 cm. long; hypanthium 1–1.5 mm. high; sepals oblong-lanceolate, longer than the hypanthium; corolla lilac or bluish, the tube equaling or exceeding the sepals; lobes oblong, as long as the tube. Dry hills: Ill.—Iowa—Kans.—Tex.—Ark. *Ozark.—Allegh.* Mr–Ap.

2. H. purpùrea L. Plant glabrous or somewhat pubescent; stem 5–10 cm. high, simple or sparingly branched; leaf-blades ovate or lanceolate, or the lower oval or spatulate, 1–6 cm. long, ciliate; pedicels slender, 1–6 cm. long; sepals subulate, 2–3 mm. long; corolla lilac, light purple or white, 6–7 mm.

long; the lobes ovate to oblong-lanceolate. Open places: Md.—Minn. (?)—Ark.—Ala.—Ga. *Canad.—Allegh.* My–S.

3. **H. canadénsis** Willd. Stem tufted, 5–20 cm. high, branched; leaves mainly basal, 1–3 cm. long, oblong to spatulate, ciliate, short-petioled; pedicels 1–5 cm. long; sepals oblong to lanceolate, 2–3 mm. long; corolla blue, 7–9 mm. long, the lobes oblong-lanceolate, twice as long as the tube. *H. ciliolata* Torr. Rocky places and open woods: Me.—N.D.—Ark.—W.Va. *Canad.—Allegh.* My–Au.

4. **H. longifòlia** Gaertn. *Fig. 521.* Plant puberulent or hispidulous below; stem tufted, 20–40 cm. high, 4-angled; leaves linear or linear-oblong, 1–3 cm. long, slightly revolute, sessile; pedicels 3–10 mm. long; sepals subulate, 2 mm. long; corolla pinkish, 8–9 mm. long, the lobes ovate, acutish. Sandy soil: Me.—Sask.—Mo.—Ga. *Canad.—Allegh.* My–S.

f. 521.

5. **H. angustifòlia** Michx. Plant glabrous; stem much branched at the base, diffuse, 10–60 cm. long; leaves numerous, linear or the lower linear-spatulate, 1–4 cm. long; pedicels 1–4 mm. long; sepals lanceolate; corolla 4–6 mm. long, white or purplish, the lobes oblong, pubescent within. Prairies: Ill.—Neb.—Tex.—Fla. *Prairie—Austral.* My–Jl.

2. **CEPHALÁNTHUS** L. Button-bush.

Shrubs or small trees, with 4-angled branches. Leaves opposite or rarely whorled, entire. Flowers perfect, in axillary or terminal, solitary or panicled heads. Hypanthium enlarged above. Sepals 4, subtended by bristle-like or chaffy bractlets. Corolla white or yellowish, tubular-funnelform; lobes 4, imbricate. Stamens 4, adnate up to the throat of the corolla; anthers with 2 cusps. Ovary 2-celled, with a single ovule in each cell; styles united, filiform; stigma capitate. Fruit dry, broadened upward, of 2–4 nutlets. Seeds elongate with a white aril.

1. **C. occidentàlis** L. *Fig. 522.* A shrub, 1–4 m. high; leaves opposite or in 3's, oblong or lanceolate, 5–20 cm. long, acute, undulate, acute to subcordate at the base, soft-pubescent to glabrous; heads 2.5–5 cm. in diameter; corollas white. Swamps and meadows: N.B.—se Minn.—Kans.—Ariz.—Fla.; Cuba. *E. Temp.—Trop.* Je–S.

f.522.

3. **DIÒDIA** L. Button-weed.

Annual or perennial herbs. Leaves opposite, entire, sessile, longitudinally striate, with sheathing stipules. Flowers perfect, axillary. Hypanthium obovoid or obconic. Sepals 2–4. Corolla white or lilac, funnelform or salver-shaped; lobes 4, short. Stamens 4; filaments adnate up to the throat of the corolla. Ovary 2-celled; styles united; stigma 2-lobed. Ovules solitary in each cell, amphitropous. Fruit oblong or globose, of 2 or rarely 3 or 4 indehiscent carpels, separating in fruit.

1. **D. tères** Walt. Finely pubescent or hispidulous annual; stem 1–4 dm. high, with spreading branches; leaves linear-lanceolate, 1–4 cm. long, acute; sepals 4, triangular, 1–2 mm. long; corolla about 6 mm. long, funnelform; the lobes ovate; capsule obovoid. Sandy soil: Conn.—Ill.—Kans.—N.M.—Fla. Jl–S.

4. MITCHÉLLA L. Partridge-berry. Twin-berry.

Perennial evergreen herbs, with creeping stems. Leaves opposite, entire. Flowers perfect, dimorphous, 2 together on terminal or axillary peduncles. Sepals mostly 4. Corolla white or pink, funnelform; the lobes mostly 4, spreading, pubescent within. Stamens mostly 4, the filaments short and united. Ovary 4-celled, styles long and exserted or short and included; stigmas 4. Ovules solitary in each cell. Fruit a double drupe formed from the united ovaries of the two flowers, each drupe with 4 nutlets.

1. M. repens L. *Fig. 523.* Stem prostrate and rooting, 1–4 dm. long; leaf-blades suborbicular, ovate or oval, 8–30 mm. long, dark green, shining, truncate or subcordate at the base; corolla white or pinkish, 10–13 mm. long; fruit 7–10 mm. in diameter, red or white. N.S.—Minn.—Tex.—Fla. Ap–Je.

5. SPERMACÓCE L. Button-weed.

Annual or perennial herbs or low shrubs, usually with angled stems. Leaves opposite, with sheathing stipules. Hypanthium obovoid or turbinate. Sepals 2–4, or rarely 5. Corolla white, pink, or blue, funnelform or salvershaped, the lobes usually 4, spreading, valvate. Stamens 4, adnate to the corolla tube. Ovary 2-celled; styles wholly or partly united. Ovules solitary in each cell, amphitropous. Fruit a capsule of 2 carpels, one opening along the ventral suture, the other usually indehiscent.

1. S. glábra Michx. A perennial herb, glabrous or sparingly pubescent; stem erect, the branches spreading or decumbent, 1–6 dm. long; leaves elliptic to lanceolate, 2–7 cm. long, acute or acuminate at each end; flowers in subglobose clusters, 5–12 mm. in diameter; corollas white, as long as or longer than the calyx, villous in the throat; fruit 3 mm. long. River banks: Ohio—se Kans.—Tex.—Fla. *Allegh.—Austral.* Je–S.

6. GÁLIUM L. Bedstraw, Cleavers.

Annual or perennial herbs, some somewhat woody at the base, with 4-angled stems and branches, apparently verticillate leaves* and small, mostly white flowers in cymes or panicles. Flowers perfect or in some species dioecious. Hypanthium globose or ovoid. Calyx obsolete. Corolla wheel-shaped, 4-parted, or rarely 3-parted. Stamens 4, rarely 3; filament short. Styles 2; stigmas capitellate; ovary 2-celled, with one ovule in each cell. Fruit didymous, of two indehiscent carpels, often bristle-hairy, dry.

Annuals.
  Stem coarse, reclining; leaves apparently in whorls of 6–8.
    Leaves 2–7 cm. long; nutlets, when ripe, 3–5 mm. broad; flowers white.
                                                                   1. *G. Aparine.*
    Leaves 1–2 cm. long; nutlets, when ripe, 2–3 mm. broad; flowers ochroleucous.
                                                                    2. *G. Vaillantii.*
  Stem slender, erect or ascending; leaves 2–4 in the whorls, linear-oblong or linear, often 10–20 mm. long.
                                                                    3. *G. bifolium.*
Perennials; leaves mostly in whorls of 4.
  Leaves 3-nerved, not cuspidate; fruit hispid.
    Leaves lanceolate, ovate or oval; flowers in open axillary cymes.
      Plant glabrous or nearly so; leaves lanceolate, acute or acuminate.
                                                                  4. *G. lanceolatum.*
      Plant pubescent; leaves oval or oval-oblong, obtuse.
                                                                    5. *G. circaezans.*
    Leaves linear to lanceolate; flowers in dense leafy terminal panicles.
  Leaves 1-nerved.
                                                                       6. *G. boreale.*
    Leaves not cuspidate.

* The leaves are really opposite; but the interposed stipules are large and leaf-like.

Leaves oval, more or less hirsute; fruit hispid. — 7. *G. pilosum.*
Leaves obovate or spatulate to linear, glabrous or
    nearly so; fruit glabrous or granular.
  Corolla white or whitish.
    Corolla greenish white, usually with 3 obtuse
      lobes.
      Pedicels straight, smooth; flowers usually
        2 or 3 together. — 8. *G. Claytoni.*
      Pedicels arcuate, scabrous; flowers usu-
        ally solitary on the peduncles.
        Leaves obovate or oblong, somewhat
          fleshy. — 9. *G. subbiflorum.*
        Leaves linear-oblong or linear-oblance-
          olate, thin. — 10. *G. trifidum.*
    Corolla white, 2–3.5 mm. broad, with 4 acute
      lobes.
      Leaves mostly ascending; fruit 2.5–3.5
        mm. broad. — 11. *G. tinctorium.*
      Leaves mostly reflexed; fruit 1–1.5 mm.
        broad. — 12. *G. labradoricum.*
  Corolla yellow. — 16. *G. verum.*
Leaves cuspidate-pointed.
  Fruit uncinately long-hispid; stem glabrous or
    nearly so. — 13. *G. triflorum.*
  Fruit smooth or granular; stem hispid or scabrous.
    Leaves oblong-spatulate or elliptic, retrorsely
      scabrous. — 14. *G. asprellum.*
    Leaves linear, slightly upward-scabrous on the
      margins. — 15. *G. concinnum.*

1. **G. Aparìne** L. Stem weak, prostrate or scrambling over bushes, 3–15 dm. long, retrosely hispid on the angles; leaves 6–8 in the whorls, oblanceolate, sometimes almost linear, 2–7 cm. long, 4–10 mm. wide, cuspidate, the margins and midribs retrorsely hispid; flowers white in 1–3-flowered cymes in the upper axils. Copses: N.B.—Fla.—Tex.—Calif.—B.C.; nat. from Eur. *Temp.—Plain —Submont.* My–S.

2. **G. Vaillántii** DC. Stem prostrate, retrorsely hispid as the preceding, but the internodes shorter, and the whole plant smaller; leaves 6–8 in the whorls, usually oblong-linear, sometimes slightly oblanceolate, cuspidate-pointed; flowers in 2–9-flowered cymes. Low ground among bushes: Ont.—Fla.—Tex.— Calif.—B.C.; Eurasia. *Temp.—Plain—Submont.* My–Au.

3. **G. bifòlium** S. Wats. Stem slender, erect, glabrous, 1–1.5 dm. high; leaves 2–4 in the whorls, often very unequal, thin, acutish, 1-nerved; flowers solitary on axillary peduncles, horizontal and curved under the fruit; carpels about 2 mm. in diameter. In mountain meadows and around springs: B.C.— S.D.—Colo.—Calif. *Submont.* My–Jl.

4. **G. lanceolàtum** Torr. Stem glabrous or nearly so, 3–6 dm. high; leaves lanceolate, 3–7.5 cm. long, 3-ribbed, ciliate on the margins and nerves; cymes few-flowered, loose, divaricate; flowers nearly sessile; corolla glabrous, yellowish, turning purplish, its lobes acuminate; fruit hispid with long hairs. Dry woods: Me.—Minn.—Ky.—Va. *Canad.* Je–Au.

5. **G. circaèzans** Michx. Stem more or less puberulent, 3–6 dm. high; leaves oval or ovate-oblong, obtuse, 3-ribbed, ciliate, 1.5–4.5 cm. long; peduncles usually only once forked, spreading or in fruit reflexed; corolla greenish, its lobes hairy without, acute or acuminate; fruit 3 mm. broad. Woods: Me.— Minn.—Tex.—Fla. *E. Temp.* My–Jl.

6. **G. boreàle** L. Stem erect, glabrous, 2–7 dm. high; leaves 4 in the whorls, from linear to broadly lanceolate, obtuse or acutish, 3–5 cm. long, the margins sometimes ciliate and slightly revolute, often with fascicles of smaller leaves in the axils; flowers white or ochroleucous (var. *linearifolium*) in terminal, densely many-flowered compound cymes; carpels about 2 mm. in diameter. Rocky places, hillsides, and along streams: Que.—N.J.—Neb.—N.M.— Calif.—Alaska; Eurasia. *Plain—Mont.* My–Au.

7. **G. pilòsum** Ait. Stem hirsute, 3–8 dm. high, 4-angled above, swollen at the nodes; leaves oval, 1–2 cm. long, obtuse, sparingly hairy, punctate; peduncles 2–3-forked, the flowers all pedicelled; corolla greenish yellow or purplish, the lobes 4, acuminate; fruit 4 mm. broad, uncinnate-hispid. Open woods: Mass.—Kans.—Tex. Fla. *E. Temp.* Je–Au.

8. **G. Clàytoni** Michx. Stem diffuse, 1.5–6 dm. high, retrorsely scabrous, leaves usually in whorls of 5's or 6's, linear-spatulate or linear-oblong, 8–15 mm. long, obtuse, glabrous and dull; flowers in terminal clusters of 2 or 3; corolla minute, greenish white, its lobes 3, oval, obtuse; fruit small, glabrous. Marshes and ditches: Mass.—Minn.—Neb.—Tex.—N.C. *Canad.—Allegh.* Jl–S.

9. **G. subbiflòrum** (Wieg.) Rydb. Stem slender, prostrate, 1–3 dm. long, somewhat scabrous on the angles; leaves mostly obtuse, slightly fleshy and with a faint midrib, 5–15 mm. long; peduncles axillary, usually solitary, or more rarely geminate, 1-flowered, or less commonly 2–3-flowered; corolla white, 3-lobed; fruit glabrous, its carpels 1 1.5 mm. in diameter. *G. trifidum subbiflorum* Wieg. Wet places: Minn.—N.M.—s Calif.—Wash. *W. Temp.—Plain—Mont.* Je–Au.

10. **G. trifidum** L. Stem slender, ascending or depressed, 2–4 dm. long, branched, scabrous on angles; leaves 5–15 mm. long, obtuse, 1-nerved, thin, dark green and dull on both sides; midrib and margins scabrous; flowers on slender 1–3-flowered axillary peduncles; corolla white, 3-lobed, its lobes about 0.5 mm. long; fruit glabrous, its carpels about 1.5 mm. in diameter. Wet meadows and swamps: Lab.—Newf. N.Y.—Colo.—Alaska. *Subarct.—Canad. —Plain—Submont.* Je–Au.

11. **G. tinctòrium** L. Stem 1.5–2.5 dm. high, glabrous or nearly so; leaves mostly in 4's, linear-lanceolate, 1.5–2.5 cm. long, slightly scabrous on the margins and the midrib; flowers few, in terminal clusters of 2's or 3's; pedicels straight, slender; corolla white, 2–3.5 mm. broad; fruit 2.5–3.5 mm. broad. Damp places: Que.—Minn.—Tex.—N.C. My–Jl.

12. **G. labradóricum** Wieg. Stem slender, 0.5–3 dm. high, glabrous; leaves small, 0.5–1.5 cm. long, soon reflexed, scabrous beneath on the margins and midrib; corolla 3.5 mm. broad. Tamarack swamps: Lab.—Newf.—Minn.—N.Y.—Conn. *Subarct.—Canad.* My–Jl.

13. **G. triflòrum** Michx. *Fig. 524.* Stem diffuse, 3–10 dm. long, glabrous or very sparingly hirsute, shining; leaves mostly 6 in the whorls, narrowly oval or slightly oblanceolate, 1-ribbed, narrowed at both ends, 3–7 cm. long; peduncles axillary or terminal, often exceeding the leaves, mostly 3-flowered; corolla greenish white, 4-lobed; fruit long-hispid with uncinate hairs; carpels 3–4 mm. in diameter. Damp open woods: Newf.—Ala. Colo.—Calif.—Alaska. *Temp.—Plain—Mont.* Je–Au.

14. **G. aspréllum** L. Stem 3–16 dm. high, reclining, retrorsely scabrous; leaves of the main stem mostly in 6's, 1–2 cm. long, oblong-spatulate or elliptic, cuspidate, scabrous on the margins and the midrib; flowers numerous, in terminal, leafy-bracted cymes; corolla 2 mm. broad, white, its lobes 4, acute; fruit 2.5 mm. broad, smooth. Swamps and thickets: Newf.—Minn.—Nebr.—N.C. *Canad.—Allegh.* Je–Au.

f 5 24.

15. **G. concínnum** T. & G. Stem low and slender, 1.5–3 dm. high, the angles minutely roughened; leaves linear, with minute cusps, veinless, the margins upwards roughened; flowers diffusely panicled, 2 to 3 times forked, the

pedicels short; corollas white, minute; fruit 2 mm. broad, muriculate, glabrous. Dry hills: N.J.—Minn.—Kans.—Ark.—Va. *Allegh.* Je–Au.

16. G. vèrum L. Stem smooth, 1.5–7 dm. high, erect; leaves in 8's or 6's, linear, rough on the margins, soon deflexed; flowers very numerous in leafy paniculate cymes; fruit glabrous. Yellow Bedstraw, Our Lady's Bedstraw. Fields: Me.—Pa.—N.D.; nat. from Eurasia. My–S.

Family 156. CAPRIFOLIACEAE. Honeysuckle Family.

Shrubs, trees, vines, or perennial herbs, with opposite leaves and perfect, regular or irregular flowers, mostly in cymes. Stipules generally none. Calyx 3–5-lobed or 3–5-toothed. Corolla gamopetalous, from rotate to tubular, often gibbous at the base; limb 5-lobed and sometimes 2-lipped. Stamens 5, in *Linnaea* only 4, adnate to the corolla and alternate with its lobes; anthers versatile. Ovary inferior, enclosed in the hypanthium, 1–6 celled. Fruit a 1–6-celled berry, drupe, or capsule. Ovules anatropous. Seed with a fleshy albumen; embryo small.

Style deeply 3–5-cleft; shrubs or trees with compound cymose inflorescence and drupaceous fruit.
  Leaves pinnate; ovary 3–5-celled, each cell with one ovule.   1. Sambucus.
  Leaves simple; ovary 1-celled and 1-ovuled.   2. Viburnum.
Style slender, undivided; stigma capitate.
  Stamens 4, didynamous; trailing evergreen herb; flowers long-peduncled, geminate.   3. Linnaea.
  Stamens 5.
    Flowers solitary, axillary; herbs.   4. Triosteum.
    Flowers in axillary or terminal cymes or fascicles; shrubs.
      Fruit a few-seeded berry.
        Corolla rarely gibbous at the base, regular or nearly so; fruit 2-seeded.   5. Symphoricarpos.
        Corolla gibbous at the base, mostly irregular and bilabiate; fruit several-seeded.
          Flowers in heads at the ends of the branches or in verticils in the upper axils; upper leaves connate; vines.   6. Lonicera.
          Flowers in pairs on axillary peduncles; leaves not connate; erect shrubs.
            Bracts and bractlets minute, not foliaceous; berries of the two flowers more or less united.   7. Xylosteon.
            Bracts and bractlets foliaceous; berries of the two flowers distinct.   8. Distegia.
      Fruit a 2-celled capsule; corolla slightly gibbous and somewhat irregular.   9. Diervilla.

1. SAMBÙCUS (Tourn.) L. Elder.

Shrubs or trees with opposite, odd-pinnate leaves, large pith in the young branches, small whitish flowers in compound cymes. Hypanthium ovoid or turbinate. Calyx-lobes minute, generally 5. Corolla rotate or saucer-shaped, regular, 5-lobed. Stamens 5, inserted at the base of the corolla; anthers oblong. Style short, 3–5-cleft; ovary and berry-like drupe 3–5-celled, each cell containing 1 ovule or seed.

Cyme not flat-topped, thyrsoid-paniculate, the axis continuous.
  Young branches, inflorescence, and the lower surfaces of the leaves pubescent.   1. *S. pubens.*
  Whole plant glabrous.   2. *S. microbotrys.*
Cyme flat-topped, umbelliform, 4–5-rayed, the rays again variously compound; fruit blackish.   3. *S. canadensis.*

1. S. pùbens Michx. Shrub, 6–35 dm. high; leaflets 5–7, dark green, ovate-lanceolate or oval, long-acuminate, generally narrowed and slightly oblique at the base, 5–12 cm. long, sharply serrate; cyme about 5–6 cm. high and 4–5 cm. broad; corolla white, turning brownish in drying; drupe scarlet or red or occasionally amber yellow, 4–6 mm. in diameter. *S. racemosa* Hook.; not L. [G]. Damp rocky places: N.S.—Ga.—Colo.—B.C.—Alaska. *Boreal.—Submont.* Ap–My.

2. S. microbòtrys Rydb. Shrub, low, 5–20 dm. high, with pale green foliage; leaflets ovate or rarely ovate-lanceolate, acute or short-acuminate, 3–9 cm. long, mostly rounded and oblique at the base, coarsely serrate; cyme small, about as long as broad, about 3 cm. in diameter; flowers whitish; fruit bright red, 4–5 mm. in diameter. Damp places in mountains: S.D.—Colo.—Ariz.—Utah. *Submont.— Subalp.* My–Je.

3. S. canadénsis L. *Fig. 525.* A shrub, 1–3 meters high, glabrous or nearly so, with dark green foliage; leaflets 5–11, mostly 7, ovate or oval, sharply serrate, mostly short-acuminate at the apex, acute or obtuse at the base, sometimes pubescent on the veins beneath, 5–12 cm. long, often with linear stipels; cyme flat-topped, 5–15 cm. high and 1–2 dm. broad; flowers white; fruit purplish-black, 4–5 mm. in diameter. In wet soil: N.S.— Fla.—Tex.—Colo.—Mont.—Sask. *E. Temp.—Plain—Submont.* Je–Jl.

2. VIBÚRNUM (Tourn.) L. Arrow-wood, Cranberry Tree, Snow-balls.

Shrubs or small trees with simple, often stipulate leaves. Flowers in compound cymes. Hypanthium ovoid, hemispherical, or turbinate. Calyx-teeth 5. Corolla rotate to short-campanulate, regular, 5-lobed. Stamens 5; anthers oblong. Style short, 3-cleft; ovary 1–3-celled; each cell with a single ovule. Drupe 1-seeded; seed compressed.

Leaves palmately veined, usually 3-lobed.
  Outer flowers of the cymes large, neutral and radiant; fruit red.
                                                 1. *V. trilobum.*
  None of the flowers neutral and radiant.
    Fruit red; cymes 1–3 cm. broad.   2. *V. eradiatum.*
    Fruit black; cymes 3–6 cm. broad.   3. *V. acerifolium.*
Leaves pinnately veined, not lobed; fruit blue or black.
  Cymes peduncled.
    Leaves coarsely dentate.
      Leaves short petioled.
        Leaves densely pubescent beneath; petioles
          2–7 mm. long.   4. *V. Rafinesquianum.*
        Leaves glabrous beneath, except on the veins
          and in their axils; petioles 7–12 mm. long.   5. *V. affine.*
      Leaves long-petioled.
        Lower surface of the leaves glabrous or merely
         with hair-tufts in the axils.   6. *V. dentatum.*
        Lower surface of the leaves stellate-pubescent. 7. *V. molle.*
    Leaves entire or crenulate.   8. *V. cassinoides.*
  Cymes sessile or nearly so.
    Upper leaves prominently acuminate.   9. *V. Lentago.*
    Upper leaves obtuse or acutish.
      Winter-buds, leaf-blades beneath, and petioles
        green and glabrous; petioles scarcely winged. 10. *V. prunifolium.*
      Winter-buds, leaf-blades beneath, and petioles
        red-tomentose; petioles winged.   11. *V. rufidulum.*

1. V. trílobum Marsh. Shrub, sometimes 3.5 m. high; leaves glabrous or nearly so above, more or less pubescent on the veins beneath, deeply 3-lobed, rounded or truncate at the base, the lobes divergent, long-acuminate, coarsely toothed; cyme peduncled, flat-topped, the exterior neutral flowers 1–2 cm. broad; fruit globose, red, very acid, 8–10 mm. in diameter; stone flat, orbicular, not grooved. *V. Opulus americanum* Ait. Low places: N.B.—N.J.—Iowa— Ore.—B.C. *Boreal.—Submont.* Je–Jl.

2. V. eradiàtum (Oakes) House. Shrub 1–2 m. high; leaves 3–5-ribbed, truncate or cordate at the base, more or less pubescent, 4–10 cm. broad, generally with 3 shallow lobes, these acute or short-acuminate, coarsely dentate; cyme rather few-flowered, short-rayed, 1–3 cm. broad; drupe round or ovoid,

8–10 mm. long; stone flat, orbicular, scarcely grooved. *V. pauciflorum* Pylaie; not Raf. Woods: Newf.—Pa.—Iowa—Colo.—Alaska. *Boreal.—Submont.* Je–Jl.

3. **V. acerifòlium** L. Shrub, low, branched, 1–2 m. high, its branches soft-pubescent, becoming glabrate; leaf-blades suborbicular to ovate, 4–10 cm. long, usually with 3 acute or acuminate lobes and coarsely toothed, rounded or cordate at the base, sparingly pubescent or glabrous above, more or less softly tomentose beneath; petioles 1–4 cm. long; cymes 7–10 cm. broad, long-peduncled; drupe black, 9–10 mm. long; stone flattened, slightly 2-grooved on one side, 2-ridged on the other. Rocky woods: N.B.—Minn.—Ga. *Canad.—Allegh.* My–Je.

4. **V. Rafinesqueànum** Schultes. Shrub, 5–15 dm. high, with gray branches; leaf-blades ovate, acute or acuminate, rounded or subcordate at the base, 3–8 cm. long, velvety pubescent beneath; cymes 3–7 cm. broad; drupe nearly black, 6–7 mm. long; stone 2-grooved on the faces. *V. pubescens* Torr.; not (Ait.) Pursh. *V. villosum* Raf., not Sw. *V. affine hypomalacum* Blake. Rocky woods: N.B.—Ga.—Ia.—S.D.—Man. *Canad.—Allegh.* Je–Jl.

5. **V. affine** Bush. Shrub, 5–10 dm. high, with yellowish brown branches; leaf-blades acuminate, 4–7 cm. long, coarsely dentate, shining above, light green beneath, rounded at the base; cymes 4–5 cm. wide; drupe black, 6–7 mm. long. Closely related to the preceding. Rocky woods and thickets: Ohio—Mo.—Minn. *Allegh.* Je–Jl.

6. **V. dentàtum** L. Shrub, erect, 2–5 m. high, with glabrous twigs; leaf-blades suborbicular, oval or ovate, 3–8 cm. long, acuminate, sharply dentate, rounded or cordate at the base, glabrous above, pubescent in the axils of the veins beneath; petioles 1–2 cm. long; cymes 5–8 cm. broad; fruit dark blue, subglobose, 5–6 mm. long; stone rounded on one side, grooved on the other. Around meadows and swamps: N.B.—Minn.—Ga. *Canad.—Allegh.* My–Je.

7. **V. mólle** Michx. Shrub, branched, 2–4 m. high; leaves suborbicular or ovate, 3–14 cm. long, short-acuminate to obtuse, crenate-dentate, rounded, truncate, or cordate at the base, glabrous or sparingly pubescent above, densely stellate-pubescent beneath; petioles 1–2 cm. long; cymes 4–10 cm. broad; drupe deep blue, globose-ovoid, 8–9 mm. long; stone grooved on one side, rounded on the other. Low ground: Pa.—Iowa—Tex.—Fla. *Allegh.—Austral.* My.

8. **V. cassinoìdes** L. Shrub, slender, 2–5 m. high, the branches, petioles and inflorescence scurfy; leaf-blades thick, ovate-elliptic, crenate, rounded to cuneate at the base, glabrous above, scurfy on the veins beneath; petioles 1–2 cm. long; cyme 3–8 cm. broad; fruit blue or pink, ovoid, 6–9 mm. long; stone flattened. Swamps: Newf.—Man.—Ala.—Ga. *Canad.—Allegh.* Je–Jl.

9. **V. Lentàgo** L. *Fig. 526.* Bush or small tree, sometimes 10 m. high; leaf-blades ovate, 4–10 cm. long, glabrous or slightly pubescent beneath, sharply serrulate, mostly rounded at the base and acuminate at the apex; cyme broad, sessile, 5–12 cm. in diameter; drupe oval, sweet, bluish black, with a bloom, 10–14 mm. long; stone flat, circular or oval. NANNYBERRY. In rich soil, in woods and along streams: Me.—Ga.—Colo.—S.D.—Man. *Canad.—Allegh.—Submont.* My.

f.526

10. **V. prunifòlium** L. Shrub 2–8 m. high; leaf-blades ovate or obovate, very finely serrate, obtuse or truncate at the base; petioles 1–2 cm. long; cyme 3–10 cm. broad; fruit blue or black, with a bloom, oblong; stone flat on

one side, convex on the other. Woods and thickets: Conn.—Mich.—Kans.—
Ark.—Ga. *Allegh.* Ap–Je.

11. V. rufídulum Raf. Shrub or small tree, up to 6 m. high; leaf-blades
leathery, oblong or elliptic, mostly obtuse, finely serrulate, the veins red-tomen-
tose beneath; cymes 8–12 cm. broad; drupe deep blue, with a bloom, ellipsoid,
10–14 mm. long; stone as in the preceding, but larger. *V. rufotomentosum*
Small [B]. Woods and thickets: Va.—Kans.—Tex.—Fla. *Austral.* Ap–My.

3. LINNAÈA (Gronov.) L. TWIN-FLOWER.

Trailing or creeping evergreen, with slender
branches, and opposite petioled leaves. Flowers pink
or purplish, borne in pairs at the end of long terminal peduncles. Hypanthium ovoid. Calyx-teeth 5,
subulate lanceolate, deciduous in fruit. Corolla funnelform, almost regular, not gibbous at the base;
limb 5-lobed. Stamens 4, didynamous, included.
Ovary 3-celled; two of the cells with several abortive
ovules, the third with a single perfect ovule; style
slender, exserted; stigma capitate. Fruit coriaceous,
3-celled, but 1-seeded.

f.527.

1. L. americàna Forbes. *Fig. 527.* Stem trailing, with pubescent branches; leaves somewhat coriaceous; blade round, oval, or orbicular, 8–15 mm. in
diameter, generally crenate above the middle; peduncles slender, often 1 dm. long; bracts and bract-
lets linear subulate; calyx-lobes lance-linear, not longer than the ovary; corolla
rose colored, fragrant, about 1 cm. long, funnelform. *L. borealis* Am. auth.;
not L. In cold woods: Greenl.—N.J.—Mich.—Iowa—Colo.—Alaska. *Boreal.*
—*Submont.—Subalp.* Je–Au.

4. TRIÓSTEUM L. HORSE GENTIAN.

Perennial herbs, with simple stems. Leaves opposite, sessile or connate,
entire. Flowers solitary in the axils of the leaves, subtended by 2 bractlets.
Hypanthium ovoid. Sepals 5, often foliaceous, persistent. Corolla yellowish,
purplish, or red, tubular-campanulate; limb oblique, with 5 imbricate lobes.
Stamens 5, adnate to the corolla-tube; filaments short; anthers linear. Ovary
3–5-celled; ovules solitary in each cell. Fruit a fleshy drupe, usually 2-celled.
Seed elongate, ribbed or angled.

Leaves narrowed to the sessile or slightly connate bases.
   Stem densely pubescent with short glandular hairs and a few glandless longer
      ones.   1. *T. aurantiacum.*
   Stem rather sparingly pubescent with long reflexed hairs not
      glandular.   2. *T. illinoense.*
Leaves, or at least some of them, with broadly dilated connate-
   perfoliate bases.   3. *T. perfoliatum.*

1. T. aurantìacum Bicknell. Stem 5–12 dm. high, glandular-puberulent
and hirsute; leaves ovate-oblong to oblong-lanceolate, 1.5–2.5 dm. long, acuminate, minutely soft-pubescent beneath, sparingly appressed-pubescent or glabrate above; sepals 12–20 mm. long, obtuse; corolla dull red, 14–20 mm. long;
drupe oblong-ovoid, 12–14 mm. long, orange-red. Thickets: Que.—Mass.—
Minn.—Kans.—N.C. *Allegh.* My–Je.

2. T. illinoénse (Wieg.) Rydb. Stem 5–10 dm. high, hirsute with reflexed hairs; leaves obovate, 8–12 cm. long, short-acuminate, narrowed at the
base, rarely perfoliate, densely and softly short-pubescent beneath, sparingly
pilose above; sepals 12–18 mm. long, linear, densely pubescent, acutish; corolla

15–17 mm. long, orange, viscid-pubescent; drupe orange, finely pubescent. *T. perfoliatum illinoense* Wieg. Thickets: Ohio—Iowa—Mo. *Allegh.* My–Je.

3. **T. perfoliàtum** L. Stem 5–12 dm. high, finely crisped-pubescent; leaves ovate or oval, 1–2.5 dm. long, acute or acuminate at the apex, the middle ones with broad connate bases, softly pubescent beneath; sepals linear, 11–15 mm. long, acute; corolla purplish, 12–15 mm. long, viscid-pubescent; drupe obovoid or globose, 8–12 mm. long, orange-yellow, finely pubescent. Thickets: N.J.—Minn.—Kans.—Ala. *Allegh.* My–Jl.

5. **SYMPHORICÁRPOS** (Dill.) Ludw. Coral-berry, Snow-berry, Wolf-berry, Stag-berry.

Shrubs, with opposite short-petioled leaves and white or pink flowers in small terminal or axillary clusters. Hypanthium globose or ellipsoid. Calyx-teeth 4–5. Corolla from open-campanulate to salverform, regular, rarely slightly gibbous at the base, 4–5-lobed. Stamens 4–5, adnate to the corolla. Ovary 4-celled; two of the cells with several abortive ovules, the other two each with a single pendulous ovule; style filiform; stigma capitate. Fruit a globose or ovoids 4-celled, but only 2-seeded berry.

Fruit red; style bearded.	1. *S. orbiculatus.*
Fruit white; style glabrous.	
Style and stamens somewhat exserted; leaves thick.	2. *S. occidentalis.*
Style and stamens not exserted; leaves rather thin.	
Erect shrub; leaves 2–5 cm. long; clusters several-flowered.	3. *S. albus.*
Diffuse shrub; leaves 1–2.5 cm. long; clusters 1–3-flowered.	4. *S. pauciflorus.*

1. **S. orbiculàtus** Moench. Shrub, 5–15 dm. high, with pubescent purplish branches; leaf-blades oval or ovate, entire or wavy, soft-pubescent beneath, 2–4 cm. long; flower-clusters shorter than the leaves; corolla campanulate, pinkish, 4 mm. long; berry ovoid-globose, 3–4 mm. long. *S. vulgaris* Michx. *S. Symphoricarpos* (L.) MacMill. Coral-berry, Buckbrush. River banks: N.Y.—Ga.—Tex.—Colo.—S.D.—Mont. *E. Temp.—Plain.* Jl.

2. **S. occidentàlis** Hook. Shrub, 1–1.5 m. high, with slender, light-colored branches; leaves rounded-oval, broadly ovate or almost orbicular, 3–7 cm. long, obtuse or rounded at the apex, more or less pubescent beneath, rather thick and firm, the margins entire or on vigorous shoots deeply sinuately toothed; flower-clusters spicate, 10–25 mm. long; corolla 6 mm. long; fruit globular, white, 8–10 mm. in diameter. Hillsides, open woods, and river banks: Mich.—Mo.—Colo.—B.C. *Plain—Submont.* Je–Jl.

3. **S. álbus** (L.) Blake. Shrub erect, 5–15 dm. high, with slender glabrous branches; leaves rather thin, oval, obtuse at both ends, sometimes slightly pubescent beneath, subentire or on young shoots sinuate-dentate; terminal flower-clusters often interruptedly spicate, the axillary few-flowered; corolla thin, slightly gibbous, 6 mm. long, bearded within; fruit white, globose, 6–10 mm. in diameter. *S. racemosus* Michx. [G, B]. Rocky places and on river banks: N.S.—Va.—Wyo.—Colo.—Calif.—B.C.—*Boreal.—Submont.* Je–Au.

4. **S. pauciflòrus** (Robbins) Britton. Shrub low, diffusely branched, 15–25 dm. high; leaves broadly oval or orbicular, softly pubescent beneath, 12–25 mm. long, entire; flowers 2–3 in terminal spikes and solitary in the upper axils; corolla open-campanulate, 4–6 mm. long, bearded within; fruit oval to globose, 4–6 mm. in diameter. *S. racemosus pauciflorus* Robbins. Rocky places: Vt.—Pa.—Neb.—N.M.—B.C. *Boreal.—Submont.* Je–Jl.

6. **LONICÈRA** L. Honeysuckle.

Mostly climbing shrubs, with opposite simple leaves, the uppermost in ours connate. Flowers usually irregular, in interrupted spikes or heads in the axils of the leaves. Hypanthium ovoid or globose. Calyx-lobes 5, small. Corolla in ours funnelform or trumpet-shaped, more or less gibbous at the base and more or less irregular; limb 5-lobed and more or less 2-lipped. Stamens adnate to the tube of the corolla; filaments slender; anthers linear or oblong. Ovary 2–3-celled; ovules numerous in each cell, pendulous; style slender; stigma capitate. Berry fleshy, 2–3-celled, few-seeded.

Corolla tubular-trumpet shaped, almost regular; stamens slightly exserted.
                                                       1. *L. sempervirens.*
Corolla funnelform, 2-lipped, the upper lip 4-lobed, the lower
   entire; stamens long-exserted.
   Leaves pubescent on both sides; branches glandular-villous.   2. *L. hirsuta.*
   Leaves glabrous above; branches glabrous.
      Corolla-tube more than 1 cm. long.
         Corolla pubescent without; leaves pubescent beneath.   3. *L. glaucescens.*
         Corolla glabrous without; leaves glabrous on both
           sides.   4. *L. prolifera.*
      Corolla-tube less than 1 cm. long; corolla glabrous without; leaves glabrous.   5. *L. dioica.*

1. **L. sempérvirens** Ait. An evergreen twining vine, 1–6 m. long; leaf-blades leathery, oval or oblong, or obovate, 3–10 cm. long, obtuse or apiculate, glaucous or sometimes softly pubescent beneath, the upper pairs more or less connate; flower-clusters forming terminal spike-like panicles; corolla scarlet or yellow, 3–4 cm. long, glabrous; berries scarlet, clustered. River banks: Conn. –Neb.– Tex.–Fla. *Allogh.—Austral.* Ap–S.

2. **L. hirsùta** Eat. A twining, high-climbing vine; leaf-blades oval or ovate, the lower short-petioled, the upper sessile and the uppermost connate, 5–9 cm. long, obtuse, ciliate, appressed-setulose above, downy beneath, dull; flowers in approximate whorls; corolla 2–2.5 cm. long, orange-yellow, clammy-pubescent, the tube slender, somewhat gibbous below. Copses and rocks: Vt.—Man.—Minn.— Ohio—Pa. *Canad.* Je–Jl.

3. **L. glaucéscens** Rydb. *Fig. 528.* Shrub, more or less twining, with light-colored sheddy bark; leaves obovate or oval, mostly subsessile, glabrous above, glaucous and more or less hairy beneath, the margins chartaceous, not ciliate; flowers in a short terminal interrupted spike; corolla yellow, changing into reddish, 2–2.5 cm. long, slightly hairy or pu-

f 5·28

berulent without, pubescent within, funnelform, strongly gibbous at the base; style and stamens more or less hairy, and exserted; fruit red. Hillsides and among bushes: Ont.—Pa.—Okla. –S.D.—Mack. *Canad.—Allegh.—Plain— Submont.*

4. **L. prolífera** (Kirchner) Rehder. A twining vine, 2–5 m. high; leaf-blades oblong, oval, or obovate, 2.5–8 cm. long, obtuse, entire, glaucous, the uppermost connate; flowers in terminal clusters; corolla pale yellow, the tube 1–1.5 cm. long, slightly gibbous at the base; filaments glabrous, berries subglobose, yellow. *L. Sullivantii* A. Gray [G, B]. Rocky woods and banks: Ont. —Man.—Iowa—Tenn. My–Je.

5. **L. dioìca** L. A twining vine or shrub, 1–3 m. high; leaf-blades thin, oblong to obovate or ovate, obtuse, very glaucous beneath, deep green above, sessile or the uppermost connate, 5–10 cm. long; flowers in terminal clusters; corolla yellowish green, tinged with purple, barely 1 cm. long, hirsute within, gibbous at the base; filaments hairy; berries subglobose, red. *L. parviflora* Lam. Hillsides and rocky places: Me.—Man.—Neb.—Ga. *Canad.—Allegh.* My–Je.

7. XYLÓSTEON Adans. BUSH OR FLY HONEYSUCKLE, TWIN-BERRY.

Erect branching shrubs, with opposite simple leaves. Flowers axillary in pairs, the common peduncle bibracteate at the summit. Bracts and bractlets if present small, lanceolate or subulate. Hypanthium ovoid or globose. Calyx-teeth 5, minute. Corolla funnelform, campanulate, more or less gibbous at the base and more or less irregular; limb 5-lobed, more or less 2-lipped. Stamens adnate to the corolla-tube; anthers linear or oblong. Ovaries 2–3-celled, those of the two flowers often more or less united; ovules numerous in each cell, pendulous; style slender; stigma capitate. Berry fleshy, 2–3-celled, few-seeded.

Corolla-lobes subequal.
    Peduncles 3–7 mm. long; fruit blue or black, with a bloom; leaves pale and glaucous, oblong.   1. *X. caeruleum.*
    Peduncles 10–30 mm. long; fruit orange or red.
        Corolla white or rose-colored, its lobes nearly as long as the tube; berries united at the base; leaf-blades cordate-oval, glabrous.   2. *X. tataricum.*
        Corolla greenish yellow, its lobes much shorter than the tube; berries distinct; leaf-blades ovate-oblong, downy beneath.   3. *X. canadense.*
Corolla strongly bilabiate, yellowish white; upper lip shallowly 4-lobed, the lower entire; fruit red or purplish; leaf-blades oblong.   4. *X. oblongifolium.*

1. X. caerùleum (L.) Dum.-Cours. *Fig. 529.*
Shrub erect, 5–10 dm. high; leaves 2–4 cm. long, rounded at the apex, pale and more or less pubescent beneath, glabrate above; flowers short-peduncled; bracts linear-subulate, longer than the ovaries; corolla yellow, 12–15 mm. long, often pubescent; ovaries of the two flowers wholly united, forming in fruit an oblong, bluish-black berry. *Lonicera caerulea* L. [G, B]. Low grounds: Lab.—Pa.—Minn.—Wyo.—Calif.—Alaska; Eurasia. *Boreal— Submont.—Mont.* Je.

f 529

2. X. tatáricum (L.) Medic. Shrub glabrous, 1.5–3 m. high; leaf-blades thin, entire, cordate-ovate-lanceolate, 2.5–7 cm. long, on short petioles; corolla white or rose-colored, its lobes widely spreading, 14–16 mm. long; berries united at the base, red or orange. *L. tatarica* L. [G, B]. Rocky shores and banks: Me.—N.D.—Ky.—N.J.; escaped from cultivation. My–Je.

3. X. canadénse (Marsh.) Duham. Shrub, 1–2 m. high, with glabrous branches; leaves thin, bright green, ovate, rounded or cordate at the base, pubescent beneath when young, more glabrate in age, strongly ciliate on the margins; bracts about half as long as the divergent, almost distinct ovaries; bractlets minute, rounded; corolla light yellow, almost regular, about 16 mm. long; fruit ovoid, red, about 6 mm. in diameter. *L. ciliata* Muhl. [B]. *X. ciliatum* (Muhl.) Pursh. Moist woods: N.S.—Pa.—Mich.—Sask. *Boreal.*

4. X. oblongifòlium Goldie. Shrub, 5–15 dm. high, with upright branches; leaf-blades 2–7 cm. long, downy when young, 2–5 cm. long; corolla 1–1.5 cm. long, yellowish, purplish within, gibbous at the base; berries red or purplish, more or less united. *L. oblongifolia* Hook. [G, B]. Forest swamps: N.B.—Man.—Minn.—Pa.—N.Y. *Canad.* My–Jl.

8. DISTÈGIA Raf. SWAMP HONEYSUCKLE, BEARBERRY.

Erect shrubs, with opposite leaves and 2–3-flowered axillary peduncled clusters. Bracts and bractlets close under the flowers, broadly oval or ovate, often cordate at the base, much enlarged in fruit. Corolla cylindric-campanulate, strongly gibbous at the base; limb almost regular, with 5 short rounded almost erect lobes. Fruit of the flowers distinct and surrounded by the bractlets and the bracts. Otherwise as in *Xylosteon*.

1. D. involucràta (Richards.) Cockerell. Shrub, 1–3 m. high; leaves short-petioled, ovate, oval, or obovate, 5–15 cm. long, acute or acuminate at the apex, rounded or acute at the base, glandular-dotted and more or less pubescent; bracts and bractlets foliaceous, and enlarged and reddish in fruit; bracts ovate or cordate, obtuse or acute; corolla yellow, pubescent; stamens and style slightly exserted; fruit globose or oval, dark purple or black, about 8 mm. in diameter. *Lonicera involucrata* (Richards.) Banks [G, B]. On wooded river banks: Que.—Mich.—Colo.—Calif.—Alaska. *Boreal.—Submont.—Subalp.*

9. DIERVÍLLA (Tourn.) Mill. BUSH HONEYSUCKLE.

Shrubs, with simple opposite leaves and yellow flowers in small terminal or axillary cymes, or solitary. Hypanthium slender, Calyx-lobes 5, linear. Corolla narrowly funnelform, slightly gibbous at the base; limb 5 lobed, nearly regular. Stamens 5, adnate to the tube of the corolla; anthers linear, Ovary 2-celled, ovules numerous; style filiform; stigma capitate. Fruit a linear-oblong septicidal, 2-valved, many-seeded capsule.

1. D. Lonicèra Mill. Shrub, 1–1.5 m. high, generally glabrous; leaves ovate or oval, acuminate, generally rounded at the base, serrulate, 5–12 cm. long; flower-clusters 1–5-flowered, terminal or in the upper axils; corolla yellow, pubescent without and within, about 18 mm. long, three of its lobes somewhat united; capsule slender, beaked, and crowned with the persistent sepals. *Diervilla trifida* Moench. *D. Diervilla* (L.) MacMill. Dry or rocky woods: Newf.—N.C.—Ky.—Sask, *Canad.—Allegh.*

Family 157. ADUÆACHAE, MOSCHATEL FAMILY.

Herbs with scaly or tuberiferous rootstocks. Leaves basal, opposite, cauline, ternately compound. Flowers small, greenish, in terminal head-like clusters. Hypanthium hemispheric, adnate to the ovary. Calyx-lobes 2 or 3. Corolla rotate, regular, 4–6-lobed. Stamens twice as many, inserted in pairs in the tube; anthers 1-celled. Ovary 3–5-celled; styles 3–5-parted; ovules 1 in each cell. Fruit a small drupe.

1. ADÓXA L. MOSCHATEL, MUSK-ROOT.

Characters of the family.

1. A. Moschatéllina L. *Fig. 530.* Stem erect, 5–15 cm. high; basal leaves 1–4, long petioled, ternately compound; divisions broadly ovate or orbicular, thin, glabrous, 3-cleft; stem-leaves a single pair, ternate; flowers 3–6; corolla of the terminal flower 4- or 5-lobed, that of the others 5- or 6-lobed; drupe green. Mossy woods: w Ont.—Wis.—Ia.—Colo.—Utah—Mack; Eur. *Boreal.— Submont.—Subalp.* Je–Jl.

f. 530

Family 158. SANTALACEAE. SANDALWOOD FAMILY.

Herbs, shrubs, or trees, mostly root-parasites or saprophytes. Leaves without stipules, simple. Flowers perfect, monoecious, or dioecious, mostly greenish. Hypanthium well developed and enclosing the ovary, adnate at least to the base. Calyx 3–6-lobed; lobes valvate. Corolla wanting. Stamens as many as the calyx-lobes, opposite and adnate at the base. Ovary 1-celled, 2–4-ovuled. Fruit a drupe or nut.

1. COMÁNDRA Nutt. BASTARD TOADFLAX.

Smooth sometimes hemi-parasitic perennials. Leaves alternate, mostly sessile. Flowers perfect, greenish, cymose. Hypanthium campanulate or urn-shaped, enclosing and adnate to the ovary. Calyx 5-lobed. Corolla wanting. Stamens 5, inserted at the base of the calyx-lobes; anthers 2-celled, connected to the lobes by a tuft of hairs. Fruit drupe-like or nut-like.

Flowers in corymbiform cymes at the ends of the branches; leaves sessile; style slender.
  Hypanthium in fruit globose, constricted above into a distinct neck.
    Leaves thin, pale beneath, with a prominent pale midrib; inflorescence ellipsoid.
        1. *C. umbellata.*
    Leaves firmer, not paler beneath, obscurely veined; inflorescence flat-topped.
        2. *C. Richardsoniana.*
  Hypanthium in fruit ellipsoid, the neck less distinct; leaves glaucous.
        3. *C. pallida.*
Flowers in 1–3-flowered lateral cymes; leaves short-petioled; style short.
        4. *C. livida.*

1. C. umbellàta (L.) Nutt. Rootstock mostly underground; stem 1.5–4 dm. high; leaves oblong, 1–3.5 cm. long; flowers on divergent branches; calyx green or purplish; sepals oblong; fruit globose, 5–6 mm. thick. Dry ground: Me.—Sask.—(?) Man.—Ga. *Canad.—Allegh.* My–Jl.

2. C. Richardsoniàna Fern. Rootstock mostly superficial, freely branching; stem 1–2.5 dm. high; leaves strongly ascending, green, lanceolate to ovate; inflorescence 1–3 cm. broad, the flowers on ascending branches; fruit globose, 5 mm. broad. Gravelly soil: Que.—Sask.—Kans.—Mo. *Canad.* My–Au.

3. C. pállida A. DC. *Fig. 531.* Stems 2–4 dm. high, branched above, from a stout rootstock; leaves ascending, pale and glaucous, those of the stem 1–5 cm. long, 2–8 mm. wide, those of the branches much narrower; hypanthium in flower 4 cm. long; calyx-lobes lanceolate, 3 mm. long; fruit about 8 mm. long, 5–6 mm. thick, green. Sandy soil: Man.—Tex. —Ariz.—Wash.—B.C. *W. Temp.—Plain—Submont.* Je–Au.

f. 531.

4. C. lívida Richards. Stem slender, 1–3 dm. high; leaf-blades oval, thin, bright green, 1–2.5 cm. long, 6–12 mm. wide; hypanthium in flower 1.5–2 mm. long; calyx-lobes deltoid, 1 mm. long; fruit subglobose, 6 mm. long, red. Bogs and wet places: Lab.—Vt.—Mich.—B.C.—Yukon. *Boreal.* Je–Jl.

Family 159. LORANTHACEAE. MISTLETOE FAMILY.

Mostly shrubby hemi-parasites or parasites, growing on woody plants. Leaves mostly opposite, sometimes reduced to green scales. Flowers dioecious or monoecious, regular. Calyx entire-margined or lobed. Corolla of 2–6 more or less united petals. often green and calyx-like, or wanting. Stamens 2–6; anthers 2-celled or confluently 1-celled. Gynoecium of a compound pistil, but 1-celled. Fruit a berry; seed solitary.

1. ARCEUTHÒBIUM Bieb.

Small shrubby fleshy parasites, mostly dioecious. Leaves opposite, reduced to scales. Flowers solitary in the axils of the upper scales. Calyx of the staminate flowers 2–5-lobed, usually 3-lobed. Stamens as many as and adnate to the lobes; anthers sessile, 1-celled, opening by a transverse circular slit. Hypanthium of the pistillate flowers enclosing the ovary. Calyx-lobes 2, persistent on the fruit. Berry ovoid, more or less flattened. *Razoumofskya* Hoffm.

Stems 5–10 cm. high, branched ; staminate flowers at ends of branches, pedicelled.
                                                     1. *A. americanum*.
Stem 0.5–2 cm. high, simple ; all flowers axillary.   2. *A. pusillum*.

1. A. americânum Nutt. *Fig. 532.* Stems dichotomously and verticillately branched, greenish yellow; staminate plant 5–10 cm. high, 1–2 mm. thick at the base; pistillate plant 2–5 cm. high; staminate flowers 2 mm. wide; lobes round-ovate; pistillate flowers somewhat smaller; fruit 2 mm. long, bluish. *A. americana* (Nutt.) Kuntze. Parasitic on *Pinus Murrayana* and *divaricata:* Sask.—Colo.—Ore.—B.C. *Mont.* Au–O.

2. A. pusillum Peck. Stems 0.5–2 cm. high, simple or sparingly branched, olive-green; scales suborbicular, appressed, obtuse, 1 mm. wide; flowers solitary in the axils, fruit 2 mm. long, nodding. *R. pusilla* Kuntze. Parasitic on *Picea* and *Larix:* Newf.—Pa.—Minn.—Man. *Canad.* Je.

f. 532.

Family 160. ARISTOLOCHIACEAE. BIRTHWORT FAMILY.

Perennial herbs or vines. Leaves alternate, or basal, petioled. Flowers perfect, regular or irregular. Hypanthium usually well developed, often wholly enclosing the ovary. Calyx regular or irregular, mostly of 3 sepals. Corolla wanting. Stamens 6 to many, filaments either free or adnate to the style column. Styles united; ovary 4–6-celled; ovules many in each cavity. Fruit a capsule. Seeds numerous; endosperm fleshy.

Caulescent herbs, or lianas ; calyx irregular, deciduous ; capsule dry.
                                                 1. ARISTOLOCHIA.
Acaulescent herbs, calyx regular ; capsule fleshy.   2. ASARUM.

1. ARISTOLÔCHIA L. SNAKEROOT, DUTCHMAN'S PIPE.

Caulescent herbs or lianas. Leaves alternate, the blades often cordate at the base, palmately 3–many-ribbed. Flowers irregular. Hypanthium well developed, often curved. Calyx corolla-like, oblique. Stamens 6, rarely 4 or 5, or more than 6, adnate to the stylar column. Ovary inferior, 4–6-celled, with as many parietal placentae. Styles united into a 4–6-angled column. Ovules numerous. Seeds horizontal, flattened.

Leaves leathery, densely tomentose ; hypanthium tomentose.
                                             1. *A. tomentosa.*
Leaves membranous, puberulent or glabrous ; hypanthium glabrous.
                                             2. *A. macrophylla.*

1. A. tomentôsa Sims. A twining liana, up to 10 m. long; leaf-blades ovate or suborbicular, 10–18 cm. long, cordate at the base; hypanthium yellowish without; calyx abruptly bent above the ovary, contracted into a narrow neck, the limb expanded, 3-lobed, greenish purple, with a dark brown center. Woods: N.C.—Ill.—Kans.—Okla.—Fla. *Austral.* My–Je.

2. A. macrophýlla Lam. A twining liana, up to 10 m. long; leaf-blades reniform to broadly ovate, 5–25 cm. broad, glabrous above, sparingly pubescent beneath; hypanthium yellowish green; calyx abruptly bent and contracted above, the limb, 3-lobed, 2–3 cm. broad, purplish brown. *A. Sipho* L'Hér. Woods: Pa.—Minn.—Kans.—Ga. *Allegh.* My–Je.

2. ÁSARUM (Tourn.) L. WILD GINGER.

Perennial acaulescent herbs, with elongated rootstocks. Leaves 2, basal, petioled; blades cordate or reniform. Flowers perfect, regular, solitary, on a scape arising between the leaves. Hypanthium well developed, campanulate, enclosing the ovary, angled. Sepals 3, inflexed in bud. Stamens 12, free. Ovary inferior, 6-celled; styles 6, united.

Calyx-lobes lanceolate, acuminate, longer than the hypanthium.
    Calyx much longer than the hypanthium, the tubular portion 10–20 mm. long.
                                              1. *A. acuminatum.*
    Calyx slightly longer than the hypanthium, the tubular portion 4–8 mm. long.
Calyx-lobes triangular, merely acute, scarcely as long as the   2. *A. canadense.*
    hypanthium.
                                              3. *A. reflexum.*

1. A. acuminàtum (Ashe) Bicknell. Leaf-blades reniform-cordate, densely pubescent, abruptly short-pointed or blunt, 7–14 cm. broad, tomentulose beneath; hypanthium pubescent; sepals gradually caudate-acuminate, the tips recurved-spreading, 1–2 cm. long. *A. canadense acuminatum* Ashe [G]. Rich woods: Ohio—Man.—Iowa—Tenn. *Allegh.* My–Je.

2. A. canadénse L. *Fig. 533.* Leaf-blades thin, reniform, 6–15 cm. broad, rounded or abruptly acute at the apex; hypanthium thinly pubescent; sepals abruptly acuminate, 1–1.5 cm. long, with revolute tips, 4–8 mm. long, curving upwards. Rich woods: N.B.—Man.—Kans.—S.C. *Canad.—Allegh.* Ap–My.

f 533

3. A. refléxum Bicknell. Leaf-blades reniform, 6–14 cm. broad, undulate with an open sinus, glabrate above, loosely pubescent beneath; sepals 8–10 mm. long, early reflexed, the obtuse tips 2–4 mm. long. *A. canadense reflexum* Robinson. Rich woods: Conn.—Iowa—Kans.—N.C. *Allegh.* Ap–My.

Family 161. **CUCURBITACEAE.** GOURD FAMILY.

Annual or perennial succulent herbs, trailing or climbing by means of tendrils. Leaves alternate, palmately veined or lobed, usually rough-hairy. Flowers usually axillary, monoecious or dioecious. Calyx of 4–6 more or less united sepals, imbricate. Corolla of as many petals which are more or less united. Stamens 1–5, often 3, two with 2-celled and one with 1-celled anthers; filaments distinct or united; anthers extrorse, often twisted. Gynoecium of a compound pistil; ovary 1–3-celled; styles terminal, united; stigma thick, dilated, or ringed. Fruit a pepo (large fleshy or dry berry with thick rind). Seeds usually numerous, flat and horizontal; endosperm wanting; embryo straight, with thick cotyledons.

Flowers large; corolla campanulate, yellow; fruit 1-celled, with 3–5 placentae, many-seeded; plant trailing.   1. PEPO.
Flowers small; corolla rotate, white or greenish; fruit 1–3-celled, few-seeded; plant climbing.
    Fruit 2- or 3-celled; ovules few, erect or ascending.
        Stamens 2 or 3, monadelphous; anthers horizontal; leaves lobed.
                                      2. MICRAMPELIS.
        Stamens syngenecious, wholly united to the column; leaves in ours divided.
                                          3. CYCLANTHERA.
    Fruit 1-celled; ovules solitary, pendulous; leaves lobed.   4. SICYOS.

1. **PÈPO** L. GOURD, PUMPKIN, SQUASH.

Annual or perennial, rough-pubescent trailing vines. Leaves petioled, lobed or entire, often cordate. Tendrils more or less branched. Flowers large, solitary in the axils of the leaves, monoecious. Hypanthium in the staminate flowers campanulate, in the pistillate ones urn-shaped. Sepals 5. Corolla

yellow, campanulate, 5-lobed, the lobes recurved
at the ends. Stamens 3, in the staminate flowers
with distinct filaments and linear, coherent, contorted anthers, in the pistillate flowers reduced to
staminodia. Pistil in the former wanting, in the
latter with a single ovary and 3–5 stigmas, which are
2-branched or at least 2-lobed. Pepo 1-celled, with
3–5 placentae, many-seeded, usually large, fleshy,
with a tough rind. Seeds flattened. *Cucurbita*
(Tourn.) L.

1. P. foetidíssimus (H.B.K.) Britt. *Fig. 534.*
Perennial, with a thick, long taproot; stem trailing, 1–8 m. long; leaves with stout petioles; blades
ovate-triangular or ovate, thick, canescent-scabrous,
truncate or cordate at the base. 1–4 dm. long, often
sinuately lobed, denticulate; hypanthium of the
staminate flowers hemispheric, that of the pistillate flowers ellipsoid; corolla 6–15 cm. long, campanulate, greenish orange; fruit globose, 5–10 cm. in
diameter, smooth, bitter. *C. foetidissima* H.B.K. [G,B]. *C. perennis* A.
Gray. Dry plains and valleys: Tex.—Mo.—S.D.—Colo.—Calif.—Mex. *Plain*
—*Son.* My–S.

2. MICRÁMPELIS Raf. BALSAM APPLE, MOCK APPLE.

Annual or perennial herbaceous climbing vines. Leaves petioled, membranous, 5–7-lobed. Tendrils simple or compound. Flowers monoecious, small,
the staminate ones racemose or paniculate, the pistillate ones solitary or clustered in the axils of the leaves. Hypanthium of the
staminate ones broadly campanulate, that of the pistillate ones urn-shaped. Sepals 5 or 6. Corolla
white or greenish, rotate, 5- or 6-lobed. Stamens in
the former 2 or 3 with united filaments and horizontal anthers, in the latter reduced to more or less
prominent staminodia. Pistil in the former wanting, in the latter with an inferior ovary, 2- or
3-parted, or -lobed. Fruit covered with soft spines,
becoming papery, spongy and fibrous within, 2- or
3-celled. Seeds few, 1–4 in each cavity, smooth, flattened, erect. *Echinocystis* T. & G.

1. M. lobàta (Michx.) Greene. *Fig. 535.* Annual; stem angular, nearly glabrous, climbing, 4–8
m. high; leaf-blades thin, scabrous on both sides,
3–7-lobed, with triangular-lanceolate lobes, cordate
at the base; staminate flowers 4–5 mm. long, numerous in compound racemes,
light greenish yellow; pistillate flowers solitary or two together; fruit globose-
ovoid, 4–5 cm. long, 3–4 cm. thick. *Echinocystis lobata* T. & G. [B]. Alluvial
soil among bushes: Me.—Va.—Tex.—Colo.—Ida.—Sask. *E. Temp.* Jl–S.

3. CYCLANTHÈRA Schrad.

Annual or perennial climbing vines, with tendrils. Leaf-blades usually
compound, often pedately 3–13-foliolate. Flowers monoecious, small, the staminate ones racemose or paniculate. Hypanthium flat. Sepals 5 or 6, subulate
or filiform, or wanting. Corolla greenish yellow, rotate, 5- or 6-lobed; stamens
united into a column. Pistillate flowers solitary. Ovary 2- or 3-celled; ovules
few in each cavity. Fruit oblique, spiny, elastically dehiscent.

1. **C. dissécta** (T. & G.) Arn. Annual; stem 1–1.5 m. high; leaf-blades 3–7-divided, 4–7 cm. long, the lobes acute or acuminate, toothed or incised; corolla 5–6 mm. broad; fruit 2.5–3 cm. long, armed with slender spines. Woods and thickets: La.—Kans.—Tex.—Mex. *Texan—Trop.* Jl–S.

4. SÍCYOS L. STAR-CUCUMBER, BUR-CUCUMBER.

Vines, with compound tendrils. Leaf-blades angularly 3–5-lobed. Flowers monoecious. Staminate flowers in racemes or corymbs, the pistillate ones in clusters at the ends of peduncles arising from the same nodes as the staminate clusters. Hypanthium broadly campanulate. Sepals 5. Corolla rotate, 5-lobed, white or greenish. Stamens monadelphous. Ovary 1-celled, 1-ovuled, often bristly.

1. **S. angulàtus** L. Plant viscid-pubescent; stem 2–6 m. long, angled, often branched; leaf-blades thin, 5–15 cm. broad, cordate, 5-angled or 5-lobed, the lobes acute or acuminate; corolla of the staminate flowers 10–12 mm. broad, white, striped with green. Thickets: Que.—S.D.—Kans.—Tex.—Fla. *E. Temp.* Je–S.

Family 162. CAMPANULACEAE. BELLFLOWER FAMILY.

Annual or perennial caulescent herbs (all ours) or rarely shrubs. Leaves without stipules, alternate, simple. Flowers perfect, racemose or spicate or solitary, terminal or axillary, regular. Hypanthium well developed, enclosing the ovary. Sepals 5, partially united, persistent. Corolla of 5 united petals, campanulate, tubular, or rotate. Stamens 5, inserted with the corolla; filaments distinct; anthers introrse. Gynoecium of 2–5 united carpels; ovary inferior or nearly inferior, 2–5-celled; styles united; stigmas with 2–5 lobes or rarely capitate. Fruit a 2–5-celled capsule, opening loculicidally or by pores. Seeds numerous, angled or flattened; endosperm fleshy; embryo straight.

Corolla campanulate or funnelform; inflorescence racemose or paniculate.
                     1. CAMPANULA.
Corolla rotate; inflorescence spicate.
Style declined; flowers all well developed; capsule clavate. 2. CAMPANULASTRUM.
Style straight; earlier flowers cleistogamous; capsule linear or oblong.
                     3. SPECULARIA.

1. CAMPÁNULA (Tourn.) L. BELLFLOWER, HAREBELL, BLUEBELL.

Perennial (all ours with rootstocks) or sometimes annual herbs. Leaves alternate, usually more or less toothed. Flowers perfect, all alike, in racemes or panicles, rarely solitary. Sepals 5, narrow, united only at the base. Corolla campanulate or funnelform, usually blue, with 5 lobes. Stamens 5; filaments dilated at the base; anthers oblong. Ovary, 3–5-celled; styles united; stigmas 3–5. Capsule turbinate, obconic, or obpyramidal, opening by pores, usually formed by the uplifting of small lids, either near the apex or near the base of the capsule. Seeds smooth.

Flowers long-pedicelled in open cymes.
  Flowers more than 15 mm. long; basal leaves with ovate or cordate blades.
    Early basal leaf-blades very thin, cordate with a deep sinus; plant perfectly
      glabrous.   1. *C. intercedens.*
    Early basal leaf-blades rather firm. ovate or cordate.
      but not with a deep sinus; basal petioles and lower
      part of the stem usually pubescent when young.   2. *C. petiolata.*
    Flowers 5–12 mm. long; stem retrorsely hispid; basal
      leaves all linear or narrowly lanceolate.
      Corolla mostly 10–12 mm. long, blue; branches ascend-
        ing; leaves serrulate.   3. *C. uliginosa.*
      Corolla mostly 5–8 mm. long, white, or merely tinged
        with blue; branches divaricate; leaves crenulate. 4. *C. aparinoides.*
Flowers short-pedicelled in dense racemes.   5. *C. rapunculoides.*

1. C. intercèdens Witasek. Stem 3–5 dm. high, slender; basal leaves long-petioled, the blades cordate with a deep sinus, 1–4 cm. long, usually fully as broad, coarsely dentate; stem-leaves linear, 5–10 cm. long, 2–6 mm. wide; sepals subulate, 5–6 mm. long, ascending; corolla 12–20 mm. long, the lobes broadly ovate, acute. *C. rotundifolia* Am. auth.; not L. Rocky places: N.S.— Pa.—Iowa—S.D. *Canad.* Je–S.

2. C. petiolàta A. DC. *Fig. 536.* Stem 1–4 dm. high, glabrous or retrorsely short hirsute all around; basal leaves long-petioled; blades ovate or ovate-cordate, 1–3 cm. long, usually longer than broad, rather thick, sinuate-dentate, with rounded or acute, but not sharp teeth; stem-leaves linear or linear-lanceolate or the upper almost filiform; hypanthium glabrous, turbinate; sepals subulate, 6–7 mm. long, ascending; corolla campanulate, 15–20 mm. long; lobes broadly ovate, acutish. *C. rotundifolia* Am. authors. Hills and mountains: Mack.—S.D.—N.M. —Ariz.—w Calif.—B.C. *Plain –Subalp.* Je–S.

f 536

3. C. uliginòsa Rydb. Stem 3–6 dm. high, retrosely hispidulous; leaves linear, 2.5–6 cm. long, 1–4 mm. wide; hypanthium subglobose; corolla similar to that of the next, but larger, blue, with darker veins, its lobes lanceolate. Wet meadows: N.B.—Sask.– Neb.—N.Y. *Canad.* Je–Au.

4. C. aparinoides Pursh. Stem filiform, 2–6 dm. high, retrorsely scabrous on the angles; leaves lanceolate or linear-lanceolate, 2–5 cm. long; hypanthium hemispheric; sepals lanceolate, 1.5–2 mm. long; corolla open-campanulate, white or tinged with blue, cleft beyond the middle, 5–8 mm. long. Wet meadows: N.B.—Ga.—Colo.—Sask. *Canad.—Allegh.—Plain.* Je–Au.

5. C. rapunculoìdes L. Stem 6–10 dm. high, finely pubescent above; lower leaves long-petioled, ovate-cordate, irregularly serrate, hispidulous beneath, upper stem-leaves ovate-lanceolate, short-petioled; flowers nodding; calyx-lobes linear, spreading; corolla campanulate, 2–3 cm. long, blue or violet. Around dwellings and roadsides: N.B.—Pa.—N.D., escaped from cultivation; nat. of Eur. Jl–S.

2. CAMPANULÁSTRUM Small.

Annual erect herbs. Leaves alternate, dentate, tapering at the base. Flowers perfect, spicate. Hypanthium turbinate, in age elongate. Corolla rotate, deeply 5-cleft. Sepals 5. Stamens 5, exserted; filaments declined. Ovary inferior, 3-celled; styles united, declined; stigmas 3. Capsule clavate, with valvular openings near the top.

1. C. americànum (L.) Small. Stem 2–20 dm. high, angled; lower leaf-blades orbicular or ovate, petioled, the upper lanceolate, tapering at the base, serrate, acuminate; sepals 6–10 mm. long, subulate; corolla blue or white; lobes lanceolate, 1–1.3 cm. long. *Campanula americana* L. [G, B]. Hillsides: N.B.—S.D.—Kans.– Ark.—Ga. *Canad.—Allegh.* Je–Au.

3. SPECULÀRIA (Heist.) Fabricius. VENUS'S LOOKING-GLASS.

Annual or biennial herbs. Leaves alternate, sessile or clasping. Flowers of two kinds, the earlier ones cleistogamous, close-fertilized in the bud, with a calyx of 3–4 sepals and an undeveloped corolla, the later ones with 5 sepals and a rotate corolla. Stamens 5; filaments distinct, linear. Ovary 3-celled; stigmas 3. Capsule prismatic or cylindric, opening by pores near the top or the middle.

Capsule oblong, less than 1 cm. long; leaf-blades cordate or ovate to oblong; sepals
  lanceolate.
    Leaves clasping, fully as broad as long.   1. *S. perfoliata.*
    Leaves not clasping, longer than broad.   2. *S. biflora.*
Capsule linear, more than 1 cm. long; sepals subulate; leaves linear to oblong.   3. *S. leptocarpa.*

1. **S. perfoliàta** (L.) A. DC. Stem 2–5 dm. high, retrorsely hispid on the
angles, leafy; leaves round-cordate, crenate, 7–20 mm. long, clasping or the
lower merely sessile; sepals of the earlier flowers 3 or 4, 2–3 mm. long, longer
than the rudimentary corolla, those of the later flowers 5, lanceolate-subulate,
4–5 mm. long; corolla 7–10 mm. long, almost rotate; capsule opening at or
below the middle. Hillsides and dry woods: Me.—Fla.—Ariz.—Ore.—B.C.;
Mex. *Temp.—Plain—Mont.* My–S.

2. **S. biflòra** (R. & P.) A. Gray. Stem 1–3 dm. high, simple or branched,
retrorsely hispid; leaves oblong to ovate, crenate, sparingly pubescent on the
veins; flowers 1 or 2 in the axils; sepals of the apetalous flowers 3 or 4, ovate,
1–2 mm. long, those of the petaliferous flowers 4 or 5, longer, lanceolate; corolla blue, 1.5–2 cm. broad; capsule more or less ribbed. Open places: Va.—
se Kans.—Tex.—Fla.; Ore.—Calif.—S.Am. *Austral.—Trop.* Ap–Jl.

3. **S. leptocárpa** (Nutt.) A. Gray. Stem erect, 1–3 dm. high, retrorsely
hispidulous; leaves lanceolate or linear-lanceolate, 1–3 cm. long; sepals of the
earlier flowers 3, subulate, 5 mm. long; sepals of the later flowers 5, subulate,
8–10 mm. long; corolla broadly funnelform, 6–8 mm. long; capsules open at
the apex. Dry soil: Iowa—Ark.—Tex.—Colo.—Mont. *Austral.* My–Jl.

Family 163. LOBELIACEAE. Lobelia Family.

Annual or perennial herbs, or in warmer climates trees. Leaves alternate, without stipules, in ours simple. Flowers perfect, rarely dioecious,
irregular. Hypanthium enclosing the ovary. Calyx of 5 almost distinct
sepals. Corolla 2-lipped; upper lip 2-cleft, often to or nearly to the base
of the corolla; lower lip longer, less deeply 3-lobed. Stamens 5, epigynous; filaments often coherent into a tube; anthers usually united or
merely connate. Gynoecium of 2–5 united carpels. Ovary single, 2–5
celled; styles terminal, wholly united; stigmas fringed. Fruit a 1–several-
celled capsule or rarely a berry. Seeds numerous, smooth or furrowed;
endosperm fleshy; embryo straight.

1. LOBÈLIA (Plum.) L. Lobelia.

Terrestrial herbs with leafy stems, or aquatic ones with long scapes. Leaves
ternate, or in the aquatic species mostly basal. Flowers perfect, in spikes,
racemes, or panicles. Hypanthium short, oblong to turbinate. Sepals 5, practically distinct. Corolla split to or nearly to the base on the upper side; upper
two lobes narrow, spreading or reflexed, the lower 3 lobes united to a broad
spreading lip. Stamens 5; filaments monadelphous, free from the corolla;
anthers united, two or all bearded at the apex. Ovary 2-celled; stigmas 2-lobed.
Capsule 2-valved. Seeds many.

Corolla red.   1. *L. cardinalis.*
Corolla blue or whitish.
  Corolla 1.5–2.5 cm. long; sepals with reflexed auricles.
    Anthers glabrous at the tip; corolla 2–2.5 cm. long.   2. *L. syphilitica.*
    Anthers bearded at the tip; corolla about 1.5 cm. long.   3. *L. puberula.*
  Corolla less than 1 cm. long (except *L. Dortmanna*).
    Stem leafy.
      Stem-leaves oblanceolate, oblong, lanceolate or
        ovate; racemes many-flowered.
        Capsule partly inferior, not inflated.
          Sepals not auricled, or scarcely so.
            Sepals and bracts puberulent or glabrous.   4. *L. spicata.*

Sepals and bracts hirsute-ciliate, the for-
mer subulate.
Sepals distinctly auricled at the base.
Capsule wholly inferior, inflated in fruit.
Stem-leaves linear; raceme few-flowered.
Branches and pedicels erect or nearly so; cap-
sule acutish.
Branches and pedicels more or less spreading;
capsule rounded at the apex.
Stem scapiform; leaves all in a basal rosette, usually
submerged.

5. *L. hirtella.*
6. *L. leptostachys.*
7. *L. inflata.*

8. *L. strictiflora.*

9. *L. Kalmii.*

10. *L. Dortmanna.*

f.537.

1. **L. cardinàlis** L. Perennial by means of
slender offsets; stem 5–12 dm. high, glabrous or
nearly so; leaves irregularly serrate or serrulate;
hypanthium hemispheric, 3 mm. long; sepals subulate, about 1 cm. long; corolla red or crimson, rarely
rose-colored, 3–4 cm. long. CARDINAL FLOWER. Wet
ground: N.B.—Fla.—Tex.—Colo.—Sask. *Plain.* Jl–O.

2. **L. syphilitica** L. *Fig. 537.* Perennial with
offsets; stem glabrous or with scattered hairs, 5–10
dm. high; leaves thin, lanceolate or oblong, sinuato-
dentate, 5–15 cm. long, more or less puberulent; raceme spike-like; sepals subulate, about 1 cm. long,
ciliate, with auricles in the sinuses; corolla 2–2.5 cm.
long, with thick tubes; hypanthium hemispheric. Wet
places: Me.—Ga.—La.—Colo.—S.D. *Plain.* Au–O.

3. **L. pubérula** Michx. Perennial; stem erect,
3–10 dm. high, simple; leaves 2–10 cm. long, finely toothed, the lower ones
oblanceolate or obovate, short-petioled, the upper lanceolate or ovate, sessile;
racemes 1–5 dm. long, the lower bracts sometimes leaf-like; sepals lanceolate,
revolute, with short rounded auricles; corolla deep blue, rarely white; capsule
conic, 7–9 mm. long. Meadows and wet woods: N.J.—Iowa—Kans.—Tex.—
Fla. *Allegh.—Austral.* Au–O.

4. **L. spicàta** Lam. Perennial or biennial; stem 3–10 dm. high, strict,
minutely pubescent below; leaves 3–7 cm. long, the lower obovate or spatulate,
repand-denticulate, the upper narrow, much reduced; hypanthium obconic,
becoming subglobose; sepals subulate, spreading; corolla 7–8 mm. long. Dry
sandy soil: N.S.—Man.—La.—N.C. *Canad.—Allegh.* Jl–Au.

5. **L. hirtélla** (A. Gray) Greene. Perennial, with offsets; stem simple,
2–4 dm. high, finely short-pilose; lower leaves spatulate or obovate, often sinu-
ate-denticulate, 2–4 cm. long, puberulent; bracts, pedicels, hypanthium, and
calyx hirsutulous; hypanthium turbinate, 2 mm. long; sepals 4 mm. long, subulate, minutely auricled at the sinuses; corolla light blue, 7–8 mm. long. *L.
spicata hirtella* A. Gray [B]. Hills and prairies: Minn.—Ia.—Kans.—Sask.
Prairie. Jl–Au.

6. **L. leptóstachys** A. DC. Annual or biennial; stem erect, 3–12 dm.
high, somewhat angled; leaves firm, 3–10 cm. long, the lower ones oblanceolate
or obovate, short-petioled, entire or toothed, the upper lanceolate or oblong,
sessile; pedicels erect, scabrous; hypanthium turbinate, becoming subglobose;
sepals subulate, 5–6 mm. long; auricles linear or subulate, deflexed; corolla
blue, 6–8 mm. long. Dry soil: Va.—Kans.—Ga. *Allegh.* Je–Au.

7. **L. inflàta** L. Annual; stem 1–10 dm. high, narrowly winged; leaves
2–9 cm. long, crenate, with the nerves prominent beneath, the lower obovate, the
rest ovate or oval; hypanthium campanulate, in age much inflated, purple-
ribbed; corolla 7–8 mm. long, pale lilac. Hillsides and dry soil: Lab.—Sask.
—se Kans.—Ark.—Ga. *Canad.—Allegh.* Jl–Nov.

8. L. strictiflòra (Rydb.) Lunell. Biennial; stem slender, glabrous or puberulent at the base, 1–3 dm. high, simple or with a few erect branches; basal leaves petioled; blades spatulate or oblanceolate, puberulent or glabrate; pedicels erect, 5–8 mm. long, usually glabrous, sometimes glandular-pilose; hypanthium turbinate, 2–3 mm. long; sepals subulate, 3 mm. long; corolla 7–8 mm. long. *L. Kalmii strictiflora* Rydb. Meadows and bogs: Man.—N.D.—Mont.—Wash.—B.C. *W. Boreal.—Submont.* Jl–Au.

9. L. Kálmii R. & S. Biennial, stem slender, 1–3 dm. high, branching, with more or less spreading branches, glabrous or nearly so, basal leaves spatulate or oblanceolate, 1–2.5 cm. long, the upper stem-leaves narrower and sessile; flowers light blue, 8–10 mm. long; pedicels 8–25 mm. long, spreading and often curved, with 2 glands above the middle; capsule subglobose, 2 mm. long. Wet meadows: N.S.—N.J.—Iowa—Man. *Canad.* Jl–S.

10. L. Dortmánna L. Acaulescent perennial; leaves all basal, tufted, linear, fleshy, terete, fistulose, with two longitudinal tubes, 1–5 cm. long; scape 1–4 dm. high, simple; hypanthium turbinate, in fruit clavate; sepals lanceolate, 2–3 mm. long; corolla blue, 12–15 mm. long. In water: Lab.—N.J.—Wash.—B.C.; Eur. *Boreal.—Subarctic.* Jl–Au.

Family 164. VALERIANACEAE. Valerian Family.

Annual or perennial herbs, usually heavy-scented, usually with dichotomously branched stems. Leaves opposite, entire or pinnately divided. Flowers perfect, monoecious or dioecious, in cymes. Hypanthium well developed, enclosing the ovary. Calyx of 3–5 sepals, sometimes elongating in fruit and pappus-like. Corolla tubular, funnelform, or salver-shaped; lobes 3–5, imbricate. Stamens distinct; anthers introrse. Gynoecium of 3 united carpels, 1–3-celled, but only one cell perfect; styles and sometimes also stigmas united. Fruit achene-like. Seed solitary, pendulous, endosperm usually wanting. Embryo straight.

Sepals minute or wanting; fruit 3-celled, but only one cell seed-bearing.
1. VALERIANELLA.
Calyx-lobes inrolled in flower, in fruit expanding and pappus-like; fruit strictly 1-celled.
2. VALERIANA.

1. VALERIANÉLLA (Tourn.) Hill. Corn Salad, Lamb's Lettuce.

Annual herbs, with dichotomously branched stems. Leaves opposite, entire or toothed, or rarely pinnatifid. Flowers perfect in head-like or corymbiform cymes. Calyx-teeth small or obsolete. Corolla funnelform; tube usually gibbous below; limb 5-toothed. Stamens 3. Ovary 3-celled, but only one ovule-bearing, the other two empty and more or less enlarged in wingless fruit.

Fruit triangular in cross-section; fertile cell broader than the empty ones.
1. V. chenopodifolia.
Fruit quadrate in cross-section; fertile cell about as broad as the empty ones.
Fruit ovate-tetragonal, downy-pubescent, with a broad shallow groove between the empty cells.
2. V. radiata.
Fruit oblong-tetragonal, usually glabrous, with a narrow groove between the empty cells.
3. V. stenocarpa.

1. V. chenopodifòlia (Pursh) DC. Stem 3–6 dm. high, glabrous; lower leaves entire or repand, spatulate, the upper oblong or lanceolate, 2.5–8 cm. long; corolla white, 2 mm. long; fruit triangular-pyramidal, 4 mm. long, glabrous or puberulent. Moist soil: N.Y.—Minn.—Ky.—Va. *Allegh.* My–Jl.

2. V. radiàta (L.) Dufr. Stem 2–7 dm. high, glabrous; leaves 2–10 cm. long, the lower obovate or fiddle-shaped, the upper oblong or oblanceolate, obtuse, entire or toothed at the base; corolla white, 2 mm. long. Meadows and low ground: N.Y.—Minn.—Tex.—Fla. *Allegh.—Austral.* My–Jl.

3. **V. stenocárpa** (Engelm.) Krok. Stem 2–5 dm. high, glabrous; plant very similar to the preceding; lower leaves oblanceolate or spatulate, the upper oblong, dentate below. Low ground: Mo.—se Kans.—Tex. Mr–Je.

2. VALERIÀNA (Tourn.) L. VALERIAN, TOBACCO-ROOT.

Perennial herbs, heavy-scented. Leaves opposite, entire to pinnately divided. Flowers perfect or polygamous or polygamo-dioecious. Calyx-limb at first inrolled, at last spreading, developing into 5–15 plumose bristles. Corolla funnelform or salver-shaped, 5-lobed; tube often gibbous or saccate at the base. Stamens usually 3. Ovary 1-celled; style slightly 2–3 cleft at the apex. Fruit achene-like, flattened, 1-nerved on one side and 3-nerved on the other.

Leaves thick, the lower entire, narrow, or with linear entire divisions; veining almost parallel; plants with thick fleshy taproots.
  Fruit scurfy, muricate or rugose, broadly ovate; corolla of the staminate plant 2.5–3 mm. wide.   1, *V. trachycarpa.*
  Fruit smooth, narrowly ovate; corolla of the staminate plant 4 mm. wide.   2. *V. ciliata.*
Leaves thin, the cauline ones pinnate; the veining distinctly pinnate; plants with rootstocks.
  Lower leaves spatulate, entire; stem-leaves with 1–7 divisions.
    Corolla of the pistillate plants 2–3 mm. long and slightly wider; inflorescence in fruit elongate and narrow.   3. *V. septentrionalis.*
    Corolla of the pistillate plants 3 mm. or more long; inflorescence even in fruit short, corymbiform or subcapitate.   4. *V. acutiloba.*
  Lower leaves as well as stem-leaves pinnately divided with 7–25 dentate divisions.   5. *V. officinalis.*

1. **V. trachycárpa** Rydb. Stem glabrous or nearly so, 5–7 dm. high; basal leaves 7–18 cm. long, finely puberulent; stem-leaves 1–2 pairs, pinnatifid with linear lobes; corolla of the essentially pistillate plant 1–1.5 mm. wide; fruit not hairy, 5 mm. long, 2.5 mm. wide. Mountains: S.D.—N.M.—Utah. *Submont.—Mont.* Jl S.

2. **V. ciliàta** T. & G. *Fig. 538.* Stem 3–12 dm. high, puberulent; leaves thick, puberulent and ciliate, especially on the petioles, the basal ones spatulate or oblanceolate; stem-leaves pinnately parted, with 3–7 narrowly lanceolate divisions; staminate corollas white, 4 mm. long, the pistillate ones smaller; fruit obovate, 4 mm. long, glabrous. *V. edulis* A. Gray [G, B]; not Nutt. Wet prairies: Ont.—Minn.—Iowa—Ohio. My–Je.

3. **V. septentrionàlis** Rydb. Stem weak and slender, 2–5 dm. high, glabrous or nearly so; basal leaves usually entire; blades spatulate or elliptic, 2–5 cm. long; stem-leaves usually 2 pairs, with 3–7 leaflets; terminal leaflet usually elliptic, twice as large as the lanceolate lateral ones; fruit 4 mm. long, 1.5 mm. wide. *V. wyomingensis* E. Nels. Hillsides and meadows: Lab.—Que.—Nev.—B.C. *Temp.—Submont.—Mont.* My–Jl.

f.538.

4. **V. acutíloba** Rydb. Stem of fertile plant 4–5 dm. high; basal leaves entire, 5–7 cm. long; blade spatulate or obovate, acute; stem-leaves usually 3 pairs, pinnately divided; lateral divisions lanceolate to linear, long-acuminate, the terminal one large, oblanceolate or of the upper leaves linear-lanceolate, entire or saliently toothed; cyme dense, contracted; corolla about 3–5 mm. long; fruit broadly ovate, about 4 mm. long; staminate plant lower, 3–4 dm. high, with more sterile shoots; stem-leaves usually only 2 pairs, less divided, with

only 1–2 pairs of lateral divisions; cymes denser; corolla 5–6 mm. long, more oblique, about 5 mm. wide. Mountains: S.D.—N.M.—Utah. *Mont.—Alp.* Je–Au.

5. V. officinàlis L. Stem 5–15 dm. high, pubescent, at least at the nodes; leaf-segments lanceolate, acute or acuminate, sharply dentate; flowers pink or white; inflorescence compound; fruit glabrous, 3 mm. long. Roadsides and waste places: N.Y.—N.J.—Ohio—Man.; escaped from cultivation; nat. of Eurasia. Je–Au.

Family 165. **DIPSACEAE.** Teasel Family.

Herbs with opposite or rarely whorled leaves. Flowers perfect, in involucrate heads, each usually with an involucel. Calyx cup-like or of several bristles. Corolla gamopetalous, tubular-funnelform with 2–5-lobed limb. Stamens 2–4, adnate to the upper part of the corolla-tube; filaments distinct; anthers versatile, opening lengthwise. Gynoecium of a single carpel; ovary inferior, 1-celled; style elongate, entire; stigma globose to elongate. Fruit an achene; embryo straight; endosperm fleshy.

Receptacle elongate, chaffy, the palea prickle-pointed.   1. Dipsacus.
Receptacle merely convex, pubescent, not chaffy.   2. Scabiosa.

1. **DÍPSACUS** L. Teasel.

Coarse herbs. Leaves opposite, coarsely toothed or pinnatifid. Involucral bracts and paleae of the receptacle rigid. Involucel calyx-like, enclosing the ovary. Calyx cup-shaped or 4-lobed. Corolla 4-lobed, slightly irregular, blue or white. Stamens 4. Achenes 8-ribbed.

1. D. sylvéstris Huds. *Fig. 539.* Coarse biennial herbs, armed with stout prickles; stem 1–2.5 m. high; basal leaves oblanceolate, crenate-serrate, 2–4 dm. long; stem-leaves lanceolate, entire, sparingly spiny on the margins, clasping; heads ovoid, 5–6 cm. long; bracts linear-lanceolate; paleae with a long straight spine-point; corolla 10–13 mm. long, white with lilac lobes. Waste places and fields: Me.—Va. —Mich.; Colo.—Utah; adv. or nat. from Eur. Jl–S.

f.539.

2. **SCABIÒSA** L.

Herbs with opposite leaves. Heads long-peduncled. Involucre of many herbaceous bracts, united at the base. Each flower subtended by a compressed involucre. Calyx cup-shaped, 5–10- (mostly 8-) awned. Corolla funnelform, 4–5-cleft, oblique. Stamens 4 or 2. Achenes crowned with the persistent calyx.

1. S. arvénsis L. Stem 4–9 dm. high, pubescent; basal leaves long-petioled, pubescent, oblanceolate in outline, entire to pinnatifid; upper stem-leaves reduced, sessile, pinnatifid with narrow lobes; heads depressed-hemispheric; corollas lilac. Waste places: Que.—Pa.—N.D.; adv. from Eur. Je–S.

Family 166. **AMBROSIACEAE.** Ragweed Family.

Annual or perennial herbs, rarely shrubs, monoecious or dioecious. Leaves alternate, or opposite on the lower part of the stem. Flowers in small heads, the staminate and pistillate ones in the same head or in different heads. Involucral bracts few, distinct or more or less united. Pistillate flowers when in distinct heads usually enclosed in a bur-like or nut-like involucre. Staminate flowers with tubular, obconic, or funnelform corollas, 4–5-lobed; pistillate flowers corolla-less or with a small border or crown. Calyx none or rudimentary. Ovary inferior. Stamens 5, distinct. Stigmas 2, hairy at the top. Ovules and seeds solitary.

Staminate and pistillate flowers in the same heads, the latter few (rarely solitary or none), marginal.
    Heads paniculate; pistillate flowers subtended and partly embraced by broad hyaline paleae, the corolla rudimentary or none.   2. CYCLACHAENA.
    Heads spicately or racemosely disposed; pistillate flowers not enclosed by hyaline paleae, the corolla evident.   1. IVA.
Staminate and pistillate flowers in different heads, the latter 1–7, without corolla and enclosed in a nut-like or bur-like involucre.
    Involucres of the staminate heads with united bracts; receptacles low; rudimentary styles penicillate or fimbriate at the apex.
      Spines or tubercles of the 1-flowered pistillate heads in a single row.
        3. AMBROSIA.
      Spines of the 1–4-flowered pistillate heads in more than one row.
        4. FRANSERIA.
    Involucres of the staminate heads with distinct bracts; receptacle cylindraceous, spines of the 2-flowered pistillate heads in several rows, uncinate.
        5. XANTHIUM.

1. ÍVA L. MARSH ELDER, BOZZLEWEED, SALT SAGE, POVERTY WEED.

Annual or perennial herbs, or shrubs. Leaves opposite or alternate above, glabrous or coarsely pubescent. Heads axillary or in terminal bracteate spikes. Involucres campanulate or hemispheric, or turbinate; bracts 3–6, distinct or partially united. Pistillate flowers 1–8, marginal; corolla a short truncate tube. Disk flowers perfect but sterile, with funnelform, 5-lobed corollas. Anthers entire at the base, mucronate above. Achenes broadest above the middle, without pappus.

Heads in terminal spikes, the subtending leaves reduced; leaves dentate.
        1. I. ciliata.
Heads short-peduncled in the axils of normal leaves, which are subentire.
        2. I. axillaris.

1. I. ciliàta Willd. Annual; stem erect, 5–15 dm. high, commonly mottled, hispid; leaves opposite, short-petioled, ovate or oval, 4–10 cm. long, serrate, acuminate; spike 3–20 cm. long; floral leaves ovate-lanceolate to rhombic, acuminate, ciliate; heads drooping; bracts of the involucre 3–5, obovate to suborbicular, 2 mm. long; pistillate flowers 3–5, the staminate ones 10–15; achenes 3 mm. long. Alluvial soil: Ill.—Neb.—N.M.—La. Prairie—Austral. Au–O.

f. 540.

2. I. axillàris Pursh. Fig. 540. Perennial suffrutescent at the base; stems much branched, 2–6 dm. high, sparingly hirsute or glabrate; leaves sessile, entire, obovate, oblong, or linear-oblong, 1–3 cm. long, fleshy, pubescent or glabrate, the upper alternate; heads axillary, 4–5 mm. broad, hemispheric; bracts of the involucres 4–5, connate at least at the base; pistillate flowers 4–5; staminate flowers 12–15. Alkaline or saline meadows: Man.—Okla.—N.M.—Calif.—B.C. W. Temp.—Plain—Submont. My–S.

2. CYCLACHAÈNA Fresen. HORSEWEED, CARELESS WEED.

Annual herbs, with mostly opposite petioled leaves. Heads paniculate, not leafy-bracteate. Involucres of 5 obovate bracts. Receptacle chaffy, the paleae subtending the pistillate flowers broad, nearly as large as the bracts, partly embracing the achenes. Fertile flowers 5, marginal, their corollas none or rudimentary. Staminate flowers 10–15, with funnelform corollas. Filaments monadelphous. Achenes pyriform, without pappus.

1. **C. xanthifòlia** (Nutt.) Fresen. Stem 1–2 m. high, puberulent above; leaf-blades ovate or ovate-lanceolate, cuneate to subcordate at the base, 5–10 cm. long, canescent beneath, scabrous above; heads 4–5 mm. broad, hemispheric; bracts ovate. *Iva xanthifolia* Nutt. [G]. Waste places and along streams: Sask.—Mich.—Okla.—N.M.—Wash.—Alta.; introduced eastward to Mass.—Del. *Prairie—Plain—Submont. Jl–S.*

3. **AMBRÒSIA** (Tourn.) L. RAGWEED, ROMAN WORMWOOD, BITTER-WEED.

Annual or perennial, often more or less woody, caulescent herbs, mostly monoecious, rarely dioecious. Leaves opposite or alternate, usually much divided, or rarely merely toothed. Staminate heads in terminal spikes or racemes. Involucres saucer-shaped or hemispheric, with 5–12 partly united bracts; receptacle naked or with filiform paleae. Corollas funnelform, 5-lobed. Anthers mucronate at the apex; style rudimentary, brush-like at the apex. Pistillate involucres enclosing the single flower, turbinate or subglobose, with several tubercles or spines in a single series. Corolla wanting. Stigmas filiform. Achenes ovoid or obovoid; pappus wanting.

Staminate heads sessile, the upper lobe of the involucre prolonged, hispid, lanceolate.
                                                                    1. *A. bidentata.*
Staminate heads peduncled, the upper lobe of the involucre
    not produced.
  Involucres of the staminate heads 3-ribbed; leaves pal-
      mately 3–5-cleft or entire.
    Leaves and their lobes long-acuminate; tubercles of the
        fruit about 1 mm. long, very acute; plant usu-
        ally stout.
      Stem and bracts subtending the pistillate heads not
          black-striate; staminate involucres usually less
          than 4 mm. broad.   2. *A. trifida.*
      Stem black-striate; bracts subtending the pistillate
          heads, and usually also the fruit, with black
          streaks; staminate involucres 4 mm. broad.   3. *A. striata.*
    Leaves and their lobes acute or short-acuminate; tuber-
        cles of the fruit 0.5 mm. long; plant usually slender.   4. *A. variabilis.*
  Involucres of the staminate heads not ribbed; leaves once
      to thrice pinnatifid.
    Leaves distinctly petioled; tubercles of the fruit sharp;
        annuals.
      Lower leaves mostly bipinnatifid, with narrow seg-
          ments; staminate heads about 3 mm. broad.   5. *A. elatior.*
      Lower leaves mostly pinnatifid, with broader seg-
          ments; staminate heads 4–5 mm. broad.
        Staminate involucres and lower surface of the
            leaves hirsute.
          Lobes of the staminate involucres acute; beak
              of the fruit nearly as long as the body;
              upper leaves entire.   6. *A. diversifolia.*
          Lobes of the staminate involucres rounded;
              beak of the fruit shorter than the body;
              upper leaves pinnatifid or cleft.   7. *A. media.*
        Staminate involucres, stem, and leaves hispidu-
            lous, the hairs with pustulate bases.   8. *A. longistylis.*
    Leaves sessile, pinnatifid; fruit unarmed or with small
        blunt tubercles; plant perennial with a running
        rootstock.   9. *A. coronopifolia.*

1. **A. bidentàta** L. Annual; stem erect, 3–9 dm. high, rough-hirsute; leaves alternate, lanceolate, 2–8 cm. long, acute, usually hastately toothed at the base; staminate heads numerous, involucre 3–3.5 mm. broad; pistillate heads solitary or clustered; fruit oblong, 6–8 mm. long, 4-angled, with 4 teeth at the top. Prairies: Ky.—Minn.—Kans.—Tex.—La. *Prairie—Austral. Jl–S.*

2. **A. trífida** L. Annual; stem 1–4 m. high, more or less hispid and sulcate; petioles usually more or less winged, and dilated at the base; blades

scabrous on both sides, broadly ovate or obovate in outline, 3–5-lobed, with rather narrow sinuses, or entire, the lobes lanceolate and serrate, rarely lobed; staminate racemes many; heads nodding; involucre as in the next species, but smaller; fruit 6–7 mm. long, rounded, 5–7-angled and with as many sharp tubercles, slightly sulcate and sometimes slightly pitted, but not black-streaked between the ribs. Moist places: Que.—Fla.—Mo.—Kans.—Man. *Canad.—Allegh.* Jl–O.

3. **A. striàta** Rydb. Annual; stem 1–3 m. high, sulcate and with a black streak in each groove, scabrous-hispidulous; lower leaves 5-cleft, the rest 3-cleft to near the base, with rounded, rather open sinuses, the middle lobe of the 5-cleft leaves often longer and 3-lobed, all serrate and long-acuminate, scabrous-hirsutulous on both sides; staminate racemes numerous; the involucre 4 mm. broad, round lobed, 3-ribbed on the outer third and hispidulous between the ribs, the other two thirds glabrous; fruit about 7 mm. long, bluntly 5- or 6-angled, more or less sulcate and black-streaked between the ridges, with 5 or 6 sharp tubercles. Bottom-lands: Mo. –Iowa—Colo. *Prairie—Plain.* Au–O.

4. **A. variàbilis** Rydb. Annual; stem 3–10 dm. high, slender, hispidulous, sulcate, and usually black-striate, leaves entire or 3-lobed to the middle, with narrow sinuses, hispidulous on both sides, the lobes ovate, acute or rarely short acuminate, serrate, light green; staminate racemes several; involucres 4 mm. broad, similar to that of the preceding species; fruit 6 mm. long, short-beaked, rounded, 5-angled, sulcate and black-streaked between the angles, with 5 short, broadly conic tubercles. Draws and waste places: Sask.—Neb.—Wyo. *Plain.* Jl–S.

5. **A. elàtior** L. Annual; stem 2–10 dm. high, hirsute and ciliate; leaves with ciliate petioles, thin, hispidulous or glabrate above, more or less grayish-strigose beneath, bipinnatifid, with lanceolate, acute divisions; staminate heads nodding, oblique, about 3 mm. broad, puberulent or in the western form more ciliate-pilose; fruit obovoid; body slightly pubescent or glabrate, 2.5–3 mm. long, with 5–7 sharp spines about 0.5 mm. long; beak 1–1.5 mm long. *A. artemisiaefolia* A. Gray [G]; not L. Waste places: N.S.—Va.—Colo.—Wash.—B.C. *Temp.—Plain –Submont.* Au–O.

6. **A. diversifòlia** (Piper) Rydb. Annual; stem 5 dm. high, strigose-hirsute; lower leaves pinnatifid, with oblong or lanceolate divisions, hirsutulous on both sides; petioles 1–2 cm. long; upper leaves lanceolate or ovate, entire, subsessile, acute; involucre of staminate heads 5–6-lobed, hispid-strigose; body of the fruit obovoid, 2.5 mm. long, hirsutulous, the beak 2 mm. long; tubercles 4 or 5, sharp. Valleys: Wash.—Wyo. *Plain.* Au–S.

7. **A. mèdia** Rydb. Annual; stem 4–6 dm. high, hispid and strigose; leaves pinnatifid, or the upper merely cleft, 5–10 cm. long, scabrous above, hispid-strigose beneath; divisions lanceolate, the lower lobed or toothed; staminate heads numerous; involucre 5-lobed and crenulate, 4–5 mm. broad, hispid-strigose; fruit obovoid; body 3 mm. long, puberulent; beaks 1 mm. long, pubescent; spines 5–7, sharp, subulate, 0.5 mm. long. Dry places: Minn.—Iowa — Kans. — N.M. — Nev. — Wash. *W. Temp.—Prairie—Plain—Submont.* Jl–O.

8. **A. longìstylis** Nutt. Annual; stem 2–3 dm. high, scabrous-hispidulous; leaves pinnatifid, 3–7 cm. long, scabrous-hispidulous, dark green, strongly ribbed; petioles 2–3 cm. long, wing-margined; staminate involucres 5-lobed; style 5 mm. long; body

f. 547.

of the fruit obovoid, 3 mm. long, hispidulous, variegated; beak 2 mm. long; tubercles 5–8, erect, acute. Plains: Wyo.—Neb. *Plain.* S–O.

9. A. coronopifòlia T. & G. *Fig. 541.* Stem 3–8 dm. high, canescent-strigose; leaves sessile or the lower with short, broadly-winged petioles, grayish-strigose on both sides, thick and strongly veined, pinnatifid; divisions entire, toothed, or cleft, lanceolate, acute; staminate heads nodding, oblique, 3 mm. wide; fruit round-elliptic; body pubescent, 3–3.5 mm. long, usually unarmed, sometimes with obtuse tubercles. *A. psilostachya* Gray [G, B]; not DC. Prairies and plains: Ill.—Sask.—La.—Calif.—Ida. *Plain—Submont.* Jl–O.

4. FRANSÈRIA Cav.

Annual or perennial herbs, rarely shrubby, resembling *Ambrosia*. Leaves mostly alternate, lobed or pinnatifid. Heads monoecious, the staminate ones in terminal racemes or spikes. Involucres hemispheric, open, with 5–12 more or less united bracts; receptacle chaffy. Corollas short, 5-lobed. Anthers scarcely coherent. Styles rudimentary, simple. Pistillate heads solitary or in small clusters below the staminate or rarely mixed with them. Involucres closed, globose or ovoid, beaked, enclosing 1–4 pistils, with several processes in more than one series, becoming bur-like at maturity. Stigmas 2. Achenes obovoid. *Gaertneria* Medic.

Spines of the fruit flat, never hooked; pistillate heads 1-flowered and 1-beaked; annuals.	1. *F. acanthocarpa.*
Spines of the fruit terete, usually hooked.	
Annuals; fruit 1-beaked, 1–2-flowered; leaves not tomentose beneath.	2. *F. tenuifolia.*
Perennials, with rootstocks; fruit 2-beaked, 2-flowered; leaves tomentose beneath.	
Leaves interruptedly pinnatifid, with lobed or coarsely toothed divisions.	3. *F. discolor.*
Leaves pinnately 3–5-parted or entire; divisions or blades merely serrulate or entire.	4. *F. tomentosa.*

1. F. acanthocárpa (Hook.) Coville. Annual; stem erect, diffusely branched, 3–6 dm. high, hirsute or hispid, with white hairs; leaves petioled or the upper sessile, bi- or tri-pinnatifid; divisions or their lobes oblong, elliptic, or linear, hispid-strigose on both sides or glabrate above; staminate heads nodding, about 3 mm. broad; involucres dark brown with 3 blackish ribs, cleft beyond the middle into 6–7 oval lobes; fertile heads 1-flowered; fruit 12–20 mm. long, with 7–12 flattened lance-subulate divergent spines. *F. Hookeriana* Nutt. *Gaertneria acanthocarpa* Britton [B]. Plains and sandy valleys: Man. —Mo.—Tex.—Calif.—B.C. *Son.—Plain—Submont.* Jl–O.

2. F. tenuifòlia Harv. & Gray. Perennial; stem 3–15 dm. high, pubescent or nearly glabrous; leaves bi- or tri-pinnatifid into oblong or linear divisions, strigose; staminate heads about 4 mm. broad; involucre pilose, lobed to near the middle; bur 1–2-flowered, but with a single beak, glandular-puberulent and with a depression above each hooked spine. Valleys: Tex.—Kans.—Colo.— s Calif.; Mex. *Son.* My–N.

3. F. díscolor Nutt. Perennial, with a rootstock; stem 2–4 dm. high, sparingly pubescent; leaves petioled, interruptedly pinnatifid, strigose above, white-tomentose beneath; larger division oblanceolate, cleft, and with triangular teeth; staminate racemes usually solitary; heads about 5 mm. wide; involucre finely pubescent, with 5–8 short lobes; pistillate heads 2-flowered; fruit ovoid, 4–5 mm. long; spines short with conic bases and usually slightly curved tips; beaks 2, short, hooked at the apex. *Gaertneria tomentosa* (Nutt.) Heller [B]. Dry soil: Iowa—S.D.—Kans.—N.M.—Ariz.—Wyo. *Plain—Submont.* Jl–Au.

4. F. tomentòsa A. Gray. Perennial; stem 3–10 dm. high, strict, purplish, somewhat tomentose or glabrate below; leaves pinnately 3–5-parted or entire, white-tomentose on both sides or grayish above; blades or divisions lanceolate-serrate; staminate racemes paniculate; heads about 4–5 mm. broad; involucre tomentose, round-lobed; fruit 6–8 mm. long, glandular-puberulent,

2-flowered, 2-beaked, with terete hooked spines. *Gaertneria Grayi* A. Nels. *G. tomentosa* (A. Gray) Kuntze [B]. River-bottoms: Neb.—Tex.—Colo. *Plain.* Au–S.

5. XÁNTHIUM L. COCKLEBUR, CLOTBUR.

Annual (all ours) or perennial herbs, monoecious. Leaves alternate, with toothed or lobed blades. Staminate heads in terminal spikes or racemes; involucres of several distinct bracts, in 1–3 series; receptacle chaffy; corolla tubular, 5-lobed; filaments monadelphous; anthers free, mucronate; style rudimentary, simple. Pistillate heads axillary; involucres closed, 1–3-beaked, usually 2-beaked, oblong or ovoid, with usually many spines; flowers 1–3, mostly 2; corolla wanting; stigmas 2; pappus none.

Plant unarmed; leaf-blades cordate at the base.
  Body of the fruit and bases of its prickles glabrous or nearly so.
    Fruit-body cylindro-ellipsoid, 1.5–2 cm. long.   1. *X. chinense.*
    Fruit-body broadly ovoid or subglobose, about 1 cm.
      long.   2. *X. globosum.*
  Body of the fruit and its prickles glandular-hispid or
    glandular-puberulent.
    Fruit glandular-puberulent, only slightly hispid.   3. *X. pennsylvanicum.*
    Fruit and especially its prickles very hispid.
      Body of the fruit oval or ovate, not twice as long
        as broad.
        Prickles very numerous and dense, brownish-
          pubescent.   4. *X. echinatum.*
        Prickles less numerous and more scattered,
          yellowish-pubescent.   5. *X. glanduliferum.*
      Body of the fruit oblong, twice as long as thick,
        densely prickly.
        Prickles not longer than the width of the
          body.   6. *X. commune.*
        Prickles much longer than the width of the
          body.
          Prickles densely and softly long-pubescent
            to near the tip.   7. *X. acerosum.*
          Prickles coarsely hispid on the lower half,
            the upper half glabrous.   8. *X. speciosum.*
Plant with triple spines at the axils; leaf-blades acute at
  each end.   9. *X. spinosum.*

1. **X. chinénse** Mill. Stem erect, 2–13 dm. high, glabrous; leaf-blades ovate or deltoid, 5–30 cm. long, lobed and coarsely toothed, truncate or cordate at the base; fruit oblong, 1.5–2 cm. long; prickles numerous, hooked; beaks nearly straight. *X. glabratum* (DC.) Britton [B]. *X. americanum* Walt. Waste places: Mass.—Man.—Neb.—Mex.—Fla. *Allegh.—Austral.* Au–O.

2. **X. globósum** Shull. Stem 3–10 dm. high, often purple-streaked; leaf-blades subdeltoid, 3–5-lobed, cordate at the base, serrate, scabrous-hispidulous on both sides; body of the fruit broadly ovoid or subglobose, densely glandular-puberulent, but not hispid, about 1 cm. long; prickles numerous, glabrous, straight, hooked at the apex, equaling the nearly straight beaks, 3–4 mm. long. Waste places: Mo.—Kans. Au–S.

3. **X. pennsylvánicum** Wallr. Stem 3–10 dm. high, glabrous below, rough above; leaf-blades thick, roughish, ovate, more or less distinctly 3–5-lobed, with broad blunt teeth; fruit puberulent and glandular, oblong, 15–20 mm. long, 5–8 mm. thick; spines 3–4 mm. long, sparingly hispid below; beaks stout, incurved and hooked, 6–7 mm. long. *X. canadense* Mill., in part [G]. Waste places: Que.—Mass.—Va.—Colo.—Utah—Tex. *Allegh.—Plain.* Au–O.

4. **X. echinátum** Murr. Stem 3–6 dm. high, rough, purple-blotched; leaves sinuately-lobed, firm, scabrous; fruit usually clustered, ovoid or oval, 15–20 mm. long, 8–12 mm. thick, glandular; prickles about 5 mm. long, about equaling the thick, conical, hispid beaks. Valleys and river banks: N.S.—N.C. —Colo.—Sask. *Canad.—Allegh.—Plain—Submont.* Au–O.

5. X. glanduliferum Greene. Stem 3–6 dm. high; rough and angled, sometimes purplish; leaves very thick and very scabrous, sinuately lobed and dentate, with triangular teeth; fruit broadly ellipsoid, 10–15 mm. long, 6–9 mm. thick, yellow; prickles 3–4 mm. long; beak very stout, conic, 4–5 mm. long. Valleys, especially in alkaline soil: Man. —Mo.—Neb.—Calif.—B.C. *W. Temp.—Plain.* Au–S.

j 542

6. X. commùne Britton. *Fig. 542.* Stem rather slender, 3–8 dm. high, roughish; leaf-blades broadly ovate, more or less sinuately lobed and dentate, scabrous, especially above; fruit oblong, 15–25 mm. long, 7–12 mm. thick; prickles 4–5 mm. long; beaks slightly incurved and hooked, about 5 mm. long. *X. italicum* Millsp. & Sherff.; not Mow. Waste places, valleys and marshes: Que.—Va.—Tex.—Ariz.—Alta. *Plain—Submont.* Au–O.

7. X. aceròsum Greene. Stem 4–9 dm. high, often spotted, glabrous below, scabrous above; leaf-blades cordate-ovate, obtuse, crenate-dentate, scabrous-hispidulous on both sides; fruit cylindric-ellipsoid, moderately glandular-pubescent, about 2 cm. long, 6–7 mm. thick; prickles not very numerous, softly pilose nearly to the curved apex, 6–9 mm. long; beaks slender, softly pubescent, irregularly incurved, 7–8 mm. long. Waste places: Wis.—N.D.—Neb.; N.Y. *Prairie—Plain.* Au–S.

8. X. speciòsum Kearney. Stem stout, 8–15 dm. high, angled above, scabrous; leaf-blades usually distinctly 3–5-lobed and dentate, very scabrous on both sides; fruit large, 25 mm. long or more; prickles 8–10 mm. long; beaks 10–12 mm. long, somewhat incurved, strongly hooked. Waste places and valleys: N.Y. (introduced ?)—Wis.—Mo.—Tex.—Ida.—Wash. Au–O.

9. X. spinòsum L. Stem 3–10 dm. high; leaves lanceolate, lobed or the upper entire, white-canescent beneath, and with pale veins above; spines yellow, 2–2.5 cm. long; fruit about 1 cm. long, oblong-cylindric; prickles numerous, weak; beak usually single, small or none. SPINY CLOTBUR. Waste places: Me. —Fla.—Tex.—Calif.; Mex., Cent.Am.; S.Am.; introd. from trop. Am. Au–N.

Family 167. **CARDUACEAE.** THISTLE FAMILY.

Herbs or shrubs, in the tropics sometimes trees. Leaves various, without stipules. Flowers aggregate in heads, inserted on a common receptacle and surrounded by an involucre of distinct or partly united bracts, in one or several series, the disk-flowers in the center (*i.e.,* all except those of the marginal series) or all flowers hermaphrodite or by the abortion of the pistil staminate, very rarely pistillate, the ray-flowers or marginal flowers pistillate or neutral. Plants sometimes dioecious or monoecious. Calyx reduced to a pappus consisting of bristles, awns, scales, or a cup-like crown, or wanting. Corolla of the disk flowers usually tubular or trumpet-shaped, 5- (rarely 4-) lobed or -toothed; that of the ray-flowers usually with a limb cleft on one side and drawn out into a strap-shaped or oblong, usually 3–5-toothed ligule. When the ray-flowers are present, the head is said to be radiate, or when they are wanting, discoid. Androecium of 5 stamens; filaments more or less adnate to the corolla-tube; anthers more or less united into a ring (syngenecious) except in *Kuhnia.* Gynoecium 2-carpellary, but with a single ovule and seed; style single; branches in the fertile flowers 2, often appendaged. Fruit an achene. Endosperm wanting. [COMPOSITAE.]

Anther-sacs not tailed at the base.
  Stigmatic lines at the base of the style-branches or below the middle; heads always discoid, never yellow or brown.
    Style-branches filiform or subulate, hispidulous.   Tribe 1. VERNONIEAE.
    Style-branches more or less clavate, papillose-puberulent.
                                  Tribe 2. EUPATORIEAE.
  Stigmatic lines extending to the tips of the style branches or to the appendage thereof, if present; heads most commonly radiate and with yellow or brown disk-flowers.
    Style-branches of the perfect flowers with more or less distinct appendages, these usually strongly hairy outside, glabrous inside, but never with a ring of longer hairs.   Tribe 3. ASTEREAE.
    Style-branches of the perfect flowers without appendages, or if with appendages, these hairy on both sides and with a ring of longer hairs.
      Pappus never capillary; style-branches rarely appendaged.
        Bracts of the involucres herbaceous or foliaceous.
          Receptacle with chaffy paleae subtending the flowers.
                            Tribe 7. HELIANTHEAE.
          Receptacle naked, or in *Gaillardia* with bristles, but not chaffy bracted.
            Plant-tissues without oil glands.   Tribe 8. HELENIEAE.
            Plant-tissues, especially the leaves and the bracts, with oil-tubes; plants therefore heavy-scented.
                            Tribe 9. TAGETEAE.
        Bracts of the involucres dry and scarious.   Tribe 10. ANTHEMIDEAE.
      Pappus capillary; style-branches often appendaged.
                            Tribe 11. SENECIONEAE.
Anther-sacs caudate at the base; corollas yellow only in a few species of *Ursinia*.
  Anthers not appendaged at the top; heads heterogamous or dioecious.
    Heads not radiate.
      Pistillate flowers with filiform corollas.   Tribe 4. GNAPHALIEAE.
      Pistillate flowers with tubular-ampliate corollas.
                            Tribe 6. ADENOCAULEAE.
    Heads radiate.   Tribe 5. INULEAE.
  Anthers with elongated, cartilaginous, mostly caudate appendages at the top; flowers all hermaphrodite or the marginal neutral; corolla not filiform.
                            Tribe 12. CYNAREAE.

TRIBE 1. VERNONIEAE.

Heads several- or many-flowered, separate; involucre of many imbricate bracts; pappus double, the outer of minute chaffy bristles, the inner of capillary bristles.
                              1. VERNONIA.
Heads 1–5-flowered, several crowded together; involucre of 8 bracts; pappus of chaffy bristles.   2. ELEPHANTOPUS.

TRIBE 2. EUPATORIEAE.

Achenes 5-angled without intervening ribs; pappus of wholly capillary bristles, mostly uniserial.   3. EUPATORIUM.
Achenes 10–20-ribbed or striate.
  Bracts of the involucres not herbaceous, striate-nerved.
    Pappus-bristles scabrous or barbellate; anthers united; bracts usually many.
                            4. BRICKELLIA.
    Pappus-bristles plumose; anthers distinct; bracts few.
                            5. KUHNIA.
  Bracts of the involucres herbaceous or partly colored, not striate.
                            6. LIATRIS.

TRIBE 3. ASTEREAE.

A. Plants not dioecious.
  a. Ray-flowers yellow or none.
    1. Pappus of scales or awns or lacking, never of numerous capillary bristles.
      Heads small, not over 4 mm. high, few-flowered; pappus more or less paleaceous.
        Disk-flowers usually fertile; pappus of the ray-flowers paleaceous.
                          7. GUTIERREZIA.
        Disk-flowers sterile; pappus of the ray-flowers coroniform.
                          8. AMPHIACHYRIS.
      Heads large, many-flowered; involucre viscid, hemispherical; pappus of a few deciduous awns.   9. GRINDELIA.
    2. Pappus at least in part of numerous capillary bristles.
      a. Pappus double, the inner of capillary bristles, the outer of paleae or short bristles; involucres many-flowered, hemispherical with narrow imbricated bracts.
        Achenes all alike, with pappus present.   10. CHRYSOPSIS.
        Achenes of the rays much thicker and without pappus.
                          11. HETEROTHECA.
      b. Pappus wholly of capillary bristles.

Heads discoid.
Involucres narrowly turbinate, their bracts more or less chartaceous, keeled, arranged in definite (usually 5) vertical ranks;
achenes elongate-linear; tips of style-branches subulate-filiform. 12. CHRYSOTHAMNUS.
Involucres broadly turbinate to hemispheric, their bracts more or
less imbricate, but not in definite vertical ranks; achenes
scarcely elongate-linear.
Style-tips obtuse; involucral bracts narrow and poorly imbricate. (Rayless species of) 34. ERIGERON.
Style-tips acute; bracts broad and well imbricate.
13. OONOPSIS.
Heads radiate.
Heads solitary, terminal or peduncled in the upper axils, or in
loose, open, cyme-like clusters.
Receptacle flat; bracts slightly if at all unequal and slightly
imbricated. 34. ERIGERON.
Receptacle alveolate; bracts usually well imbricated.
Leaves pinnate or toothed; lobes or teeth spinulose-
tipped.
Pappus of the fertile achenes deciduous in a ring;
bracts more or less foliaceous.
14. PRIONOPSIS.
Pappus persistent; bracts chartaceous, not foliaceous,
merely with a green tip. 15. SIDERANTHUS.
Leaves entire or toothed, but teeth not spinulose-tipped.
Bracts abruptly acuminate; stems leafy up to the
sessile heads. 13. OONOPSIS.
Bracts not abruptly acuminate.
Plants low-cespitose, with a woody caudex, more
or less evergreen leaves, and solitary peduncled
heads. 16. STENOTUS.
Plants with wholly herbaceous stem, woody, if at
all, only at the caudex; leaves not evergreen.
Perennials. 17. PYRROCOMA.
Annuals or biennials; rays with soft curled
ligules. 18. ISOPAPPUS.
Heads small, numerous in crowded racemes, thyrses, or axillary
clusters.
Bracts not longitudinally striate.
Rays not more numerous than the disk-flowers; receptacles alveolate; leaves not punctate.
19. SOLIDAGO.
Rays more numerous than the disk-flowers; receptacle
fimbriolate; heads corymbose; leaves punctate.
20. EUTHAMIA.
Bracts of the involucres longitudinally striate; heads in congested corymbs. 21. OLIGONEURON.
b. Ray-flowers blue, pink, purple, or white.
1. Pappus a mere crown or of few scales or awn-like bristles, or wanting.
Pappus wanting or coroniform.
Bracts of the involucre of nearly equal length; achenes obovate and
compressed; style-tips triangular. 22. BELLIS.
Bracts imbricate; achenes prismatic or terete, truncate; style-tips
blunt. 23. APHANOSTEPHUS.
Pappus of a series of awns.
Pappus of several setulose squamellae and usually 2 (rarely 3 or
4) subulate awns, one on each margin; tall, leafy-stemmed
perennials. 24. BOLTONIA.
Pappus of a single series of long squamellae or awns; low peren-
nials or biennials. 25. TOWNSENDIA.
2. Pappus of numerous capillary bristles.
a. Rays only slightly if at all exceeding the pappus; annuals.
Outer bracts foliaceous; stigma-tips acute. 26. BRACHYACTIS.
Bracts narrow, not foliaceous; stigma-tips obtuse.
35. LEPTILON.
b. Rays conspicuous, longer than the pappus, usually equaling or exceeding the width of the disk.
Style-tips lanceolate or oblong to filiform.
Pappus simple, of similar bristles.
Perennials, with a rootstock or caudex.
Bracts acuminate, as well as the leaves tipped with callous points or spines; plants with solitary heads at
the ends of the stems or branches, and cespitose caudices. 27. XYLORRHIZA.

Bracts not acuminate, or if long-attenuate with soft tips.
Pappus dilated at the apex; bracts narrow, more or
less keeled.   28. UNAMIA.
Pappus not dilated at the apex.
                                    29. ASTER.
Annuals or biennials, or if short-lived perennials, with a taproot; bracts in many series, with herbaceous spreading
or reflexed tips; stigma-tips linear to filiform.
                                    30. MACHAERANTHERA.
Pappus double, the inner of capillary bristles, the outer of short
bristles or scales.
Inner pappus-bristles more or less thickened above; bracts
thin-coriaceous, without herbaceous tips, shorter than the
disk.   31. DOELLINGERIA.
Inner pappus-bristles filiform, not thickened above; bracts
equaling the disk.   32. IONACTIS.
Style-tips triangular or ovate, obtuse or rarely acutish; bracts not
foliaceous.
Involucres turbinate; bracts well imbricate, in several series,
scarious-margined; disk-flowers white.
                                    33. LEUCELENE.
Involucres hemispherical or broader; bracts not scarious-margined, in 1–3 series; disk-corollas yellow, rarely ochroleucous.
                                    34. ERIGERON.
B. Heads unisexual, dioecious, discoid; pappus of the staminate flowers with clavate
tips.   36. BACCHARIS.

TRIBE 4. GNAPHALIEAE.

Bracts coriaceous; receptacle naked; pistillate flowers numerous; corolla reduced to
a short slender tube; hermaphrodite flowers few and sterile.
                                    37. PLUCHEA.
Bracts more or less scarious.
Receptacle chaffy; style or style-branches of the hermaphrodite sterile flowers not
truncate; paleae in fruit open and merely subtending the achenes.
                                    38. DIAPERIA.
Receptacle not chaffy, style or style-branches of the hermaphrodite flowers mostly
truncate.
Plants dioecious, or the pistillate heads with a few hermaphrodite flowers in
the center.
Pappus-bristles of the pistillate flowers falling off in a ring; central her-
maphrodite flowers none.   39. ANTENNARIA.
Pappus-bristles of the pistillate flowers falling off separately; central hermaphrodite flowers present in the pistillate heads.
                                    40. ANAPHALIS.
Plants not dioecious; flowers fertile throughout the heads, hermaphrodite in
the middle, surrounded by pistillate ones.   41. GNAPHALIUM.

TRIBE 5. INULEAE.

One genus.   42. INULA.

TRIBE 6. ADENOCAULEAE.

One genus.   43. ADENOCAULON.

TRIBE 7. HELIANTHEAE.

A. Bracts in 2 or more series, or if in a single series, not enclosing the achenes of
the rays; plants not glandular-viscid.
1. Disk-flowers hermaphrodite but sterile.
Marginal pistillate flowers with conspicuous rays.
Achenes not flattened; double involucres of very dissimilar sets of bracts.
Marginal achenes wholly enclosed in the hooded inner bracts and de-
ciduous with them; receptacle convex or conic.
                                    44. MELAMPODIUM.
Marginal achenes only partly enclosed by the inner bracts; receptacles
flat.   45. POLYMNIA.
Achenes flattened; ligules in ours yellow.
Ray-flowers in 2 or 3 series; achenes falling away free.
                                    46. SILPHIUM.
Ray-flowers in 1 series; achenes adnate to 2 or 3 paleae and falling
away with them.
Innermost series of bracts thin and reticulate; pappus obsolete or
of 2 deciduous awns.   47. BERLANDIERA.
Innermost series of bracts firm-coriaceous and enclosing the
achenes; pappus a conspicuous-toothed crown.
                                    48. ENGELMANNIA.
Marginal pistillate flowers with few, very broad, inconspicuous, persistent
ligules.   49. PARTHENIUM.
II. Disk-flowers fertile.
a. Ray-flowers fertile, the ligules with very short tube, persistent on the

  achenes and becoming papery in texture.
  Achenes of the disk compressed ; leaves entire. 50. ZINNIA.
  Achenes obtusely 4-angled ; leaves toothed. 51. HELIOPSIS.
 b. Ray-corollas deciduous from the achenes or wanting.
  Chaffs of the flat receptacle bristle-like ; ligules short, white.
              52. ECLIPTA.
Chaffs of the receptacle scale-like.
 1. Pappus a crown or none, or of a few squamellae on the angles of
  the achenes, with rarely minute ones between.
  a. Achenes of the disk-flowers not obcompressed (except in *Lepachys*
   and *Ximenesia*) ; paleae usually more or less concave and
   clasping.
  Receptacle conic, subulate, or columnar.
   Achenes 4-angled.
    Ray-flowers purplish or rarely whitish ; disk-corollas al-
    most without a tube ; pappus coroniform.
             54. ECHINACEA.
    Ray-flowers yellow or none ; disk-corollas with a short but
    manifest tube.
    Achenes 4-angled ; leaves not clasping.
             53. RUDBECKIA.
    Achenes terete, striate ; leaves cordate-clasping.
             55. DRACOPSIS.
   Achenes flattened, broad-margined or winged.
             56. LEPACHYS.
   Achenes 5-angled.
    Ligules present, white ; leaves opposite ; disk-flowers
    yellow.   57. GALINSOGA.
    Ligules absent ; leaves alternate ; disk-flowers white or
    purplish.   58. MARSHALLIA.
  Receptacle from flat to convex.
   Achenes of the disk neither sharp-angled, margined, nor
   winged.
    Rays fertile, their achenes commonly 3-angled or obcom-
    pressed ; plants with thick balsamiferous taproots.
           59. BALSAMORRHIZA.
    Rays sterile or wanting ; plants not with fleshy taproot.
           60. HELIANTHUS.
   Achenes of the disk thin-edged, margined or winged.
    Ray-flowers, if present, mostly neutral ; pappus of 1 or
    more awns or scales with accompanying squamellae.
    Leaves ovate-lanceolate ; ray-flowers 4–14.
           62. ACTINOMERIS.
    Leaves linear-lanceolate ; ray-flowers 10–24.
     Heads solitary, showy, terminating the stems or
     branches.   61. HELIANTHELLA.
     Heads small, axillary. 63. VERBESINA.
    Ray-flowers fertile ; pappus-awns without intermediate
    squamellae.   64. XIMENESIA.
 b. Achenes obcompressed ; paleae flat or hardly concave ; involucres
  distinctly double.
  Bracts of the involucres distinct or nearly so.
   Pappus in ours of small teeth, a mere border, or wanting.
           65. COREOPSIS.
   Pappus of 2–6 barbed or hispid awns.
    Achenes truncate above, tapering towards the base, flat-
    tened or angled ; marsh-plants with one kind of leaves.
           66. BIDENS.
    Achenes truncate at both ends with 3–6 long, subulate
    awns ; partly submerged aquatics with the submerged
    leaves filiformly dissected.
           68. MEGALODONTA.
  Bracts of the inner involucre united at least to near the middle.
           67. THELESPERMA.
 2. Pappus of 5–many squamellae, with thickened axis and hyaline margins ; paleae of the receptacle bristle-form ; pappus-squamellae not
  fimbriate.   79. GAILLARDIA.
B. Bracts of the involucres uniserial, partly or wholly enclosing the achenes of the
 fertile ray-flowers ; plants glandular-viscid. 69. MADIA.

       TRIBE 8. HELENIEAE.

A. Ligules persistent and becoming papery on the striate achenes ; plants more or
 less woolly.   70. PSILOSTROPHE.
B. Ligules deciduous or none.
 1. Receptacle naked.
  a. Bracts of the involucres pale or colored, at least the margins and tips
   scarious.
   Corollas of the disk-flowers with reflexed or spreading lobes ; bracts of the

broadly campanulate involucres obovate or broadly oblong.
71. HYMENOPAPPUS.
Corollas of the disk-flowers with linear, erect lobes ; bracts of the turbinate
involucres spatulate to linear-oblanceolate in two series ; ligules if
present deeply cleft, purple. 72. OTHAKE.
b. Bracts of the involucres neither colored nor scarious.
Achenes elongate, either linear-prismatic or clavate-obpyramidal.
Heads radiate ; flowers yellow ; foliage impressed, punctate.
73. PICRADENIOPSIS.
Heads discoid ; flowers in ours flesh colored ; foliage not punctate.
72. CHAENACTIS.
Achenes obpyramidal, less than 4 times as long as broad.
Bracts of the involucres erect, not spreading nor reflexed.
Involucres many-flowered ; pappus present ; achenes tapering below.
Bracts of the involucres nearly equal and similar, all distinct.
75. TETRANEURIS.
Outer bracts united below ; leaves usually pinnatifid with nar-
row divisions. 76. HYMENOXYS.
Involucres few-flowered ; pappus wanting ; achenes linear, 8–10-
striate. 77. FLAVERIA.
Bracts of the involucres spreading or reflexed ; leaves decurrent on the
stem ; tubes of the disk-corollas very short or reduced to a ring.
78. HELENIUM.
2. Receptacle with bristle-like chaffs. 79. GAILLARDIA.

TRIBE 9. TAGETEAE.

Bracts of the involucres more or less united ; style-branches of the disk-flowers elongate.
Bracts of the involucres united only at the base ; style-branches with conical tips ;
squamellae of the pappus many-aristate. 80. BOEBERA.
Bracts of the involucres united into a cup ; style-branches obtuse.
81. THYMOPHYLLA.
Bracts of the involucres distinct ; style-branches very short, obtuse, without an appendage. 82. PECTIS.

TRIBE 10. ANTHEMIDEAE.

Receptacle chaffy.
Achenes flattened.
Heads small ; involucres campanulate to hemispheric ; ligules few, short and
broad, in ours white or pinkish ; pappus wanting. 83. ACHILLEA.
Heads large ; involucres saucer-shaped ; ligules 10–30, linear, in ours yellow ;
pappus a membranous crown. 84. COTA.
Achenes terete, at least not flattened ; involucres hemispherical, large ; ligules
elongate.
Ray-flowers fertile ; paleae of the receptacle membranous, subtending all the
flowers. 85. ANTHEMIS.
Ray-flowers neutral ; paleae of the receptacle subulate, stiff, subtending only
the inner disk-flowers. 86. MARUTA.
Receptacle naked or merely pubescent.
Heads radiate, with well-developed ligules.
Pappus represented by a disk or margin, achenes with 3 ribs on the inner
half, nerveless on the back. 87. CHAMOMILLA.
Pappus wanting, at least in the disk-flowers ; achenes 10-ribbed or 10-angled.
Ligules white, flat, spreading, longer than the disk-corollas.
88. LEUCANTHEMUM.
Ligules, if present, yellow, concave, erect, not longer than the disk-flowers.
Bracts in 2 or 3 series. 89. TANACETUM.
Bracts in 4 or 5 series. 90. BALSAMITA.
Heads in ours discoid ; achenes 2–5-ribbed.
Anthers obtuse ; heads solitary, peduncled ; flowers all hermaphrodite.
87. CHAMOMILLA.
Anthers with pointed tips ; heads racemose, rarely solitary ; marginal pistil-
late flowers, if present, without ligules. 91. ARTEMISIA.

TRIBE 11. SENECIONEAE.

Plants scapose, with large basal leaves appearing after flowering, hermaphrodite
flowers sterile.
Heads solitary, all alike. 92. TUSSILAGO.
Heads racemose or corymbose, of two kinds on different plants, on some plants
the flowers are all pistillate and fertile without ligules or with imperfect ones,
on other plants the flowers are mostly hermaphrodite and sterile, with a few
marginal, pistillate, ligulate flowers. 93. PETASITES.
Plants not scapiform ; hermaphrodite flowers fertile.
Leaves mostly opposite.
Bracts 4 or 5, the edges overlapping ; shrubs, with filiform leaves.
94. HAPLOESTES.
Bracts 6–20, narrow, not overlapping ; herbs with flat leaves.
95. ARNICA.

Leaves all alternate.
 Heads discoid or practically so.
  Marginal pistillate filiform flowers wanting.
   Corolla-lobes longer than the throat; corollas whitish; receptacle small.
    usually with conic point in the middle. 96. MESADENIA.
   Corolla-lobes shorter than the throat; receptacle without a central
    point.
    Corollas ochroleucous or dirty-white; the throat short-cylindric.
                  97. SYNOSMA.
    Corollas yellow, the throat campanulate (rayless species of).
                  99. SENECIO.
  Marginal pistillate flowers present, their corollas filiform.
                  98. ERECHTITES.
                  99. SENECIO.
 Heads radiate, with well-developed ligules. 99. SENECIO.

TRIBE 12. CYNAREAE.

Achenes attached by the base; flowers all alike, hermaphrodite.
 Leaves not prickly; filaments glabrous; style-branches free; bracts transformed
  into hooked spines.   100. ARCTIUM.
 Leaves more or less prickly; filaments bearded; style-branches more or less
  united.
  Heads proper one-flowered, aggregated into a spherical cluster.
                101. ECHINOPS.
  Heads many-flowered, not aggregated.
   Receptacle densely bristly.
    Pappus plumose.   102. CIRSIUM.
    Pappus merely scabrous.   103. CARDUUS.
   Receptacle deeply honeycombed. 104. ONOPORDUM.
Achenes obliquely attached by one side.
 Outer bracts foliaceous; leaves and bracts spinose; marginal flowers not enlarged,
  fertile.   105. CARTHAMUS.
 Outer bracts not foliaceous; leaves not spiny; marginal corollas usually neutral
  and usually enlarged and oblique. 106. CENTAUREA.

Tribe 1. **VERNONIEAE.** Heads homogenious, *i.e.*, flowers all alike, discoid. Flowers hermaphrodite, in ours with regular tubular corollas, red, purple, pink, or white, never yellow. Involucral bracts imbricate. Anthers not caudate at the base. Style-branches slender, filiform, attenuate-subulate, acute, hispidulous, with stigmatic lines only at the base. Leaves usually alternate.

1. **VERNÒNIA** Schreber. IRONWEED.

Coarse erect perennial herbs, with alternate leaves. Inflorescence corymbose-paniculate. Involucres hemispheric to oblong-cylindric or turbinate, their bracts imbricate in several series. Receptacle flat. Corollas mostly purple or rose-colored, 5-cleft, with narrow lobes. Achenes 8–10-ribbed, truncate at the apex, callous at the base. Pappus in our species in two series, the outer of small squamellae or stout bristles, the inner of capillary scabrous bristles.

Involucres usually 15–20 mm. broad; bracts with subulate or filiform tips.
                1. *V. crinita.*
Involucres not more than 12 mm. broad; bracts not subulate-
  tipped.
 Leaves glabrous or sparingly pubescent, pitted beneath.
  Leaves linear, 1-ribbed, with indistinct lateral veins, en-
   tire or denticulate.   2. *V. marginata.*
  Leaves lanceolate to ovate, feather-veined, if linear-lance-
   olate, then sharply dentate.
   Leaves broadly lanceolate or ovate-lanceolate, acute,
    4-7 cm. long.   3. *V. corymbosa.*
   Leaves narrowly lanceolate or linear-lanceolate, acumi-
    nate, 8–15 cm. long.   4. *V. fasciculata.*
 Leaves velvety, not pitted beneath.
  Involucral bracts acuminate.
   Principal bracts erect or slightly spreading, glabrous within.
               5. *V. interior.*
   Principal bracts squarrose or recurved, pubescent within. 6. *V. Baldwini.*
  Involucral bracts acute to rounded at the apex, appressed.
   Involucres 6 mm. high; bracts few; pappus purple. 7. *V. illinoensis.*
   Involucres 8 mm. high; bracts more numerous; pappus tawny.
               8. *V. missourica.*

1. **V. crinìta** Raf. Stem glabrous, 1–3 m. high; leaves narrowly lanceolate or linear, acuminate, 8–15 cm. long, entire or denticulate, glabrous or nearly so; heads 50–90-flowered; involucre hemispheric; bracts with a lanceolate base; achenes glabrous or nearly so. *V. arkansana* DC. River valleys and banks: Mo.—Kans.—Okla.—Ark. *Ozark.* Jl–O.

2. **V. marginàta** (Torr.) Raf. Stem glabrous or nearly so, 4–8 dm. high; leaves linear or linear-lanceolate, 5–15 cm. long, punctate on both sides, puberulent above; heads in a flat-topped corymb, about 1 cm, high; bracts purple, ovate or lance-ovate, short-acuminate; achenes glabrous. *V. Jamesii* T. & G. Plains: Neb.—Okla.—Tex.—Colo. *Plain - St. Plains.* Au.

3. **V. corymbòsa** Schwein. *Fig. 543.* Stem glabrous, usually red, 4–8 dm. high; leaves sessile, ovate-lanceolate, 4–7 cm. long, acute, regularly serrate, scabrous above, glabrous and pitted beneath; heads corymbose, 8–12 mm. high, about 20-flowered; involucres campanulate or hemispheric; bracts purple, appressed; achenes glabrous or nearly so. *V. fasciculata* Coult.; not Michx. Valleys: Man.—Kans.—N.D. *Prairie.* Au

4. **V. fasciculàta** Michx. Stem glabrous, purplish, 6–10 dm. high; leaves sessile, sharply dentate, with faint lateral veins, 8–15 cm. long; heads crowded, about 20 flowered; involucres campanulate, 6 mm. high; bracts acute, glabrous or sparingly ciliate; achenes puberulent. Prairies and valleys: Ohio—Sask.—Okla. *Prairie.* Jl–S.

5. **V. intèrior** Small. Stem 1–2 dm. high, finely and closely pubescent; leaves elliptic or elliptic-lanceolate, 6–20 cm. long, acuminate, sharply serrate, glabrous or scabrellous above, thinly tomentose and resinous-dotted beneath; heads 18–30-flowered; involucre campanulate, 6–7 mm. high; bracts purple, or green with purple margins, acuminate, resinous on the back. Meadows: Neb.—Iowa—Ark.—Tex. *Prairie.* Jl–S.

6. **V. Baldwíni** Torr. Stem 6–20 dm. high, velvety; leaves oblong- or ovate-lanceolate, acute or acuminate, sharply serrate, scabrous-hirsutulous above, velvety beneath, 1–2 dm. long; heads 15–30-flowered; involucre hemispheric, 6–8 mm. broad; bracts somewhat canescent and ciliate. Woods: Ill.—Neb.—Tex. *Ozark.* Jl–S.

7. **V. illinoénsis** Gleason. Stem 10–15 dm. high, finely velutinous; leaves lanceolate or ovate-lanceolate, 6–15 cm. long, long-acuminate, sharply serrate, scabrous above, paler and sub-velutinous beneath; heads 35–55-flowered; involucres broadly campanulate; bracts appressed, ovate-oblong, obtuse or sub-acute; achenes glabrous. Prairies: Ont.—Ohio—Iowa. *Prairie.*

8. **V. missoùrica** Raf. Stem 10–15 dm. high, grayish-velutinous; leaves thick, lanceolate or ovate-lanceolate, 10–15 cm. long, long-acuminate, sharply serrate, scabrous-hirsutulous above, velutinous beneath; heads 35–55-flowered; involucres rounded-campanulate; bracts closely imbricate, ovate, obtuse or acute, arachnoid-ciliate; achenes puberulent. Low ground and prairies: Ill.—Kans.—Tex.—Ark. *Prairie.* Jl–S.

2. **ELEPHÁNTOPUS** L. ELEPHANT'S-FOOT.

Perennial herbs, with alternate or basal leaves. Heads in corymbed small clusters, 1–5-flowered. Involucres oblong, flattened; bracts 8, in 4 series, the outer two series much shorter. Receptacle flat, naked. Ray-flowers wanting.

Disk-corollas regular, 5-cleft, with a slender tube. Anthers sagittate at the base. Achenes 10-ribbed, truncate. Pappus of about 5 bristles, dilated at the base.

1. **E. caroliniànus** Willd. Stem leafy, erect, 3–10 dm. high, hirsute or glabrate above; leaves ovate to obovate, narrowed into a winged petiole, crenate-dentate, 7–20 cm. long; pappus-bristles with lanceolate bases. Woods: N.J.—Ill.—Kans.—Tex.—Fla. *Austral.* Au–O.

Tribe 2. **EUPATORIEAE.** Heads homogenious, discoid. Flowers hermaphrodite and fertile, never yellow. Involucres various. Anthers not caudate at the base. Style-branches thickened upwards or club-shaped, obtuse, uniformly puberulent; stigmatic lines indistinct.

3. **EUPATÒRIUM** L. Joe-Pye Weed, Thoroughwort, Boneset, White Snake-root, Mist-flower.

Erect perennial herbs, with opposite or verticillate leaves and cymose-paniculate heads. Involucres oblong to hemispheric, their bracts in 2–several series. Receptacle from flat to conic, naked. Ray-flowers wanting. Disk-corollas regular, white to purple, with slender tube and 5-lobed limb. Anthers obtuse at the base, appendiculate at apex. Style-branches elongate, flattened or thickened above, stigmatic at the base. Achenes 5-angled, truncate at the apex. Pappus simple, of numerous, scabrous, capillary bristles.

Receptacle flat.
  Involucral bracts imbricate in several series.
    Leaves in verticils of 3's–6's ; corolla flesh-colored.
      Teeth of the leaves deltoid, not strongly directed forward ; leaves glabrous above, short-villous to glabrate beneath ; stem not mottled.
        Leaf-blades lanceolate, broadest below the middle, abruptly contracted at the base, short-villous beneath.   1. *E. Holzingeri.*
        Leaf-blades lance-elliptic, broadest at the middle, cuneate at the base, long-villous on the veins beneath, puberulent or glabrous between them.   2. *E. falcatum.*
      Teeth of the leaves lanceolate, strongly directed forward, often incurved ; leaves usually hispidulous on both sides ; stem often mottled.
        Leaves neither strongly reticulate nor densely pubescent beneath ; lower part of the stem usually glabrate.   3. *E. maculatum.*
        Leaves strongly reticulate and densely grayish-pubescent beneath, ovate-lanceolate, acuminate ; stem pubescent throughout.   4. *E. Bruneri.*
    Leaves opposite or the upper ones alternate ; corollas white.
      Leaves clasping and united at the base around the stem, linear-lanceolate.   5. *E. perfoliatum.*
      Leaves not clasping-perfoliate.
        Leaves linear-lanceolate, sessile or nearly so, with a tapering base.   6. *E. altissimum.*
        Leaves ovate, long-petioled, with a broadly cuneate base.   7. *E. serotinum.*
  Involucral bracts subequal, not imbricate ; corollas white ; leaves ovate, petioled.   8. *E. urticaefolium.*
Receptacle conical ; bracts subequal ; corollas bluish.   9. *E. coelestinum.*

1. **E. Holzíngeri** Rydb. Perennial; stem 5–20 dm. high, solid, glabrous or nearly so up to the inflorescence; leaves 3–5 at each node, the blades 1–2 dm. long, lanceolate, dark green and glabrate above, grayish-villosulous beneath, acuminate at the apex, coarsely serrate; heads numerous, 5–7-flowered; involucre 7 mm. high, the inner bracts linear, acute. Meadows: Iowa—Mo.—Okla.—Minn. *Prairie.* Jl–S.

2. **E. falcàtum** Michx. Perennial; stem 5–20 dm. high, usually unspotted, purplish at the nodes, glabrous, solid; leaves mostly in 4's, lance-elliptic or lance-oval, cuneate at the base, coarsely serrate with salient teeth; lower

branches of the inflorescence usually spreading; heads numerous, 3–6-flowered; involucre oblong; bracts narrowly oblong, the inner acutish.   Open woods: N.H.—Wis.—Okla.—N.C.   *Allegh.—Ozark.*   Au–S.

3. **E. maculàtum** L.   Perennial; stem scabrous or puberulent, at least above, 5–20 dm. high, striate, often spotted with purple; leaves ovate, or the upper ovate-lanceolate, acute, coarsely serrate, puberulent beneath; bracts purplish, the outer ovate, rounded at the apex, the inner linear-oblong, acutish; heads about 1 cm. high.   *E. purpureum maculatum* A. Gray [G].   Moist ground: N.Y.—Ky.—Sask. (?)   *Canad. —Allegh.*   Au–S.

f 544.

4. **E. Brùneri** A. Gray.   *Fig. 544.*   Perennial; stem more or less canescent, 5–15 dm. high; leaves lanceolate or ovate-lanceolate, serrate, velutinous beneath, 10–15 cm. long; bracts more or less purplish, the outer ovate, pubescent, the inner linear-oblong, glabrous, all rounded or obtuse at the apex; heads nearly 1 cm. high.   *E. Rydbergii* Britton [B].   *E. atromontanum* A. Nels.   Moist ground: Sask.—Minn.—Mo.—N.M.—Utah—B.C.   *Plain—Submont.*   Jl–S.

5. **E. perfoliàtum** L.   Perennial; stem 6–15 dm. high, hirsute-villous; leaves elongate-lanceolate, long-attenuate, crenato-serrate, rugose-reticulate and densely pubescent beneath, 1–2 dm. long; heads 4–6 mm. high, cymoso-paniculate, bracts linear-lanceolate, pubescent.   BONESET.   Wet places: N.B.—Man.—Neb.—Texas—Fla.   *Canad. Allegh.*   Jl–S.

6. **E. altíssimum** L.   Perennial; stem 1–2 m. high, densely and finely pubescent; leaves narrowly lanceolate, tapering at each end, subsessile, triple-ribbed, entire or serrate above the middle, 5–12 cm. long, hirsutulous, the upper alternate; heads cymose-paniculate, about 5 flowered, 6–8 mm. high; bracts oblong, densely pubescent.   THOROUGHWORT.   Open places: Pa.—S.D.—Tex.—N.C.   *Allegh.—Prairie—Ozark.*   S–O.

7. **E. serótinum** Michx.   Perennial; stem 10–25 dm. high, densely pubescent, much branched; leaves petioled; blades lanceolate or ovate-lanceolate, acuminate, sharply serrate, 5–15 cm. long, 3–5-ribbed, puberulent; heads numerous, cymose-paniculate, 7–15-flowered, 4–6 mm. high; bracts linear-oblong, obtuse or truncate, pubescent.   Moist ground: Md.—Minn.—Tex.—Fla.   *Allegh.—Gulf.*   S–N.

8. **E. urticaefòlium** Reichard.   Perennial; stem 3–12 dm. high, glabrous or nearly so, much branched; leaves petioled; blades broadly ovate, acuminate, coarsely dentate, glabrous or sparingly hairy on the veins, triple-ribbed; heads cymose-paniculate, 4–5 mm. high, 10–30-flowered; bracts linear, acuminate. WHITE SNAKE-ROOT.   *E. ageratoides* L. f. [B].   Woods: N.B.—Sask.—Okla.—La.—Ga.   *E. Temp.*   Jl–N.

9. **E. coelestìnum** L.   Perennial, stoloniferous; stem branched, 3–10 dm. high, puberulent; leaves petioled; blades ovate, acute or obtuse at the apex, truncate at the base, crenate-dentate, 4–7 cm. long, more or less puberulent; heads in small compound cymes, 4–6 mm. high; bracts linear-lanceolate.   MIST-FLOWER.   Moist ground: N.J.—Kans.—Tex.—Fla.;   Cuba.   *Allegh.—Trop.*   Au–O.

4. BRICKÉLLIA Ell.   THOROUGHWORT.

Herbs or shrubs, with opposite or alternate leaves.   Heads discoid, with white, ochroleucous, or pink flowers, paniculate-cymose, or rarely solitary.   Involucre campanulate or oblong, its bracts striate, imbricate in several series. Receptacle flat or convex, naked. Disk-corollas regular, slender, 5-lobed. Anthers obtuse at the base. Style-branches elongate, obtuse, thickened upwards. Achenes 10-ribbed. Pappus simple, of numerous, long, scabrous or barbellate bristles. *Coleosanthus* Cass.

1. **B. umbellàta** (Greene) Rydb. Perennial herbs, with a short caudex; stem 3–6 dm. high, minutely puberulent; leaf-blades deltoid, hastate, or cordate, crenate-dentate, 3–6 cm. long, scabrous-puberulent, or nearly glabrous; heads nodding, in congested umbel-like clusters, 10–12 mm. high; outer bracts ovate, pubescent and ciliate, acuminate, the innermost linear. *Brickellia grandiflora minor* A. Gray; *Coleosanthus umbellatus* Greene [R]; *C. congestus* A. Nels.; *C. minor* Daniels. Hillsides and cañons: Kans.—Neb.—Wyo.—Ariz.—N.M. *Plain—Submont.* Jl–S.

5. KÙHNIA L. False Boneset.

Perennial herbs, with alternate resinous-dotted leaves. Heads discoid, cymose-corymbose, with white or purplish flowers. Involucre narrow, turbinate-campanulate; its bracts striate, imbricate in several equal series. Disk-flowers regular; tube slender; limb 5-cleft. Anthers obtuse and entire at the base, almost distinct. Receptacle flat, naked. Style-branches thickened upwards, obtuse, stigmatic only at the base. Achenes 10–20-ribbed. Pappus simple, of numerous plumose bristles.

Inner involucral bracts abruptly acute or cuspidate.	
Leaves densely puberulent beneath.	1. *K. suaveolens.*
Leaves glabrous or nearly so beneath.	2. *K. reticulata.*
Inner involucral bracts gradually attenuate at the apex.	
Leaves all lance-linear or linear.	3. *K. Hitchcockii.*
Leaves of the stem, at least the lower, broadly ovate.	4. *K. Jacobaea.*

1. **K. suavèolens** Fresen. Perennial; stem 3–6 dm high, puberulent and somewhat viscid; leaves sessile, lanceolate or ovate-lanceolate, puberulent on both sides, usually sharply dentate, veiny, 3–7 cm. long; heads numerous, 12–16 mm. high; outer bracts lanceolate, attenuate; pappus tawny. *K. eupatorioides corymbosa* T. & G. [G]; *K. Maximiliani* Sinning; *K. glutinosa* W. auth.; not Ell. [R]. Dry prairies and plains: Ill.—Ala.—Tex.—Colo.—Mont. *Prairie—Plain—Submont.* Je–O.

2. **K. reticulàta** A. Nels. Stem 3–4 dm. high, finely puberulent; leaves puberulent on the upper surface and on the veins beneath, punctate, strongly reticulate beneath, oblong-lanceolate, 2–4 cm. long, irregularly toothed or entire; bracts in 4 or 5 series, shorter than the disk, the outer short, lanceolate, the inner broadly linear, acute; pappus white. Cañons: Wyo.—N.D. *Submont.* Au.

3. **K. Hitchcóckii** A. Nels. Perennial, with a tufted woody caudex; stems decumbent, divaricately branched, minutely puberulent; leaves lance-linear or linear, puberulent, entire, or with a few sharp teeth, 3–4 cm. long; outer bracts linear-lanceolate, long-attenuate, the inner narrowly linear; pappus tawny. Dry plains: S.D.—Kans.—Colo. *Plain.* Au–S.

4. **K. Jacobaèa** Lunell. Stem stout, 4–6 dm. high, branched from the base, minutely puberulent; stem-leaves broadly ovate, 3-ribbed, coarsely toothed, 4–5 cm. long, 2–2.5 cm. wide, minutely scabrous-puberulent, reticulate; involucres cylindric, 10 mm. high; bracts linear, all gradually attenuate at the apex; achenes 5–5.5 mm. long, 15-striate; pappus white. River banks: Jamestown, N.D. *Prairie.* Au.

6. LIÀTRIS Schreb. Blazing Star, Button Snake-root.

Perennial herbs, with a globular corm or thickened rootstock. Leaves narrow, entire, alternate, more or less punctate. Heads discoid, racemose or spicate, with rose-purple, or rarely white flowers. Involucre oblong to hemispheric; its bracts more or less herbaceous, not striate, imbricate in several series. Receptacle flat or nearly so, naked. Anthers obtuse at the base. Style-branches elongate, obtuse, flattened at the end, stigmatose at the base. Achenes 10-ribbed, slender, tapering to the base. Pappus of 1 or 2 series of equal, firm, plumose or barbellate bristles. *Laciniaria* Hill.

Pappus long-plumose.
  Heads 15–60-flowered.
    Heads uniform, cylindric or cylindro-campanulate; corolla-lobes hairy within.
      Bracts rather firm, all except the innermost series more or less acuminate or cuspidate, with a stiff tip.
        Bracts and leaves glabrous, the former slightly squarrose.
          Bracts lanceolate, the outer narrowly so and more or less leaf-like, gradually acuminate.   1. *L. compacta.*
          Bracts broader and shorter, the outermost lanceolate, the middle ones ovate, and all except the innermost abruptly short-acuminate.   2. *L. glabrata.*
        Bracts distinctly hirsute-ciliate, strongly squarrose or recurved, the middle ones ovate and abruptly short-acuminate; leaves usually also hirsute.   3. *L. hirsuta.*
      Bracts rather thin, the outer softly cuspidate, the inner obtuse or mucronate, none squarrose.   4. *L. cylindracea.*
    Heads broadly campanulate, not uniform, the uppermost 1–3 much larger, often more than twice as large as the rest; corolla-lobes glabrous within.   5. *L. fallacior.*
  Heads 4–6-flowered; corolla lobes glabrous within.
    Leaves linear, 4–8 mm. wide; spike leafy-bracted throughout or nearly so; bracts strongly ciliate.   6. *L. punctata.*
    Leaves linear-filiform, 2–4 mm. wide; spike leafy-bracted only near the base; bracts ciliate or nearly so.   7. *L. acidota.*
Pappus barbellate, short-plumose, or scabrous.
  Heads oblong, 3–15-flowered, numerous in long spikes.
    Bracts scarcely ciliate, at least most of them obtuse or rounded at the apex, all appressed, not crisp; rachis of the spike glabrous.
      Bracts dark rose-purple, rather thick, with numerous and large gland-dots, all rounded at the apex; leaves strongly dotted.   8. *L. kansana.*
      Bracts pale or rose-colored, thin, sparingly small dotted, the lowermost ovate, acute; leaves not strongly dotted.   9. *L. spicata.*
    Bracts ciliate, all acute, with spreading crisp tips; rachis of the spike crisp-hirsute.
      Outer bracts ovate, merely acute.   10. *L. Bebbiana.*
      Outer bracts lanceolate, long-acuminate.   11. *L. pychnostachya.*
  Heads hemispheric or campanulate, 15–45-flowered.
    Leaves glabrous or nearly so; inflorescence elongate, spike-like, with many heads, but lax; bracts slightly tinged with purple.   12. *L. sphaeroidea.*
    Leaves more or less pubescent or hispidulous, at least on the margins.
      Outer bracts elliptic or oblong, about twice as long as broad, scarcely squarrose; inflorescence elongate.
        Heads usually 2 cm. broad or more; bracts with greenish white or slightly pinkish borders.   13. *L. Haywardii.*
        Heads 1–1.5 cm. broad; bracts with dark crimson borders.   14. *L. Herrickii.*
      Outer bracts broadly oval or suborbicular, nearly as broad as long.
        Inflorescence elongate, dense, with numerous, usually subsessile heads; bracts decidedly squarrose, the outer slightly if at all lacerate on the margins.   15. *L. aspera.*
        Inflorescence mostly short and with few, usually peduncled heads, rarely elongate; bracts scarcely squarrose, all lacerate on the margins.
          Involucre hemispheric, rounded at the base;

leaves linear-oblanceolate, light green,
sparingly pubescent.   16. *L. ligulistylis.*
Involucre campanulate, often acutish at
the base ; leaves very numerous, densely
pubescent, the lower oblanceolate.   17. *L. Rosendahlii.*

1. **L. compácta** (T. & G.) Rydb. Corm 1.5–2.5 cm. in diameter; stem
sulcate, glabrous, 3–7 dm. high, very leafy; leaves narrowly linear, glabrous,
1-ribbed or indistinctly 3-ribbed, 5–15 cm. long; bracts glabrous, strongly
ascending to slightly spreading, sometimes tinged with purple, the outer narrowly lanceolate, 1.5–2 cm. long, long-attenuate, the innermost oblong, short-
cuspidate; corolla reddish-purple, 1 cm. long; achenes 5 mm. long, hirsute;
pappus plumose, brownish, 8 mm. long. *Liatris squarrosa compacta* T. & G.
Prairies and open woods: Ark.—Neb.—n Tex. *Prairie.* Je–S.

2. **L. glabràta** Rydb. *Fig. 545.* Corm 1.5–2
cm. thick; stem sulcate, glabrous, 3–6 dm. high;
leaves glabrous, 5–15 cm. long, 3–7 mm. wide,
1-ribbed or indistinctly 3-ribbed, with a callous margin; spike lax, 1–2 dm. long, leafy; heads sessile,
about 1.5 cm. high; bracts callous-margined, glabrous, the outer ovate, short-acuminate, the tips
slightly spreading, the rest erect, and the innermost
oblong and cuspidate; corollas 12 mm. long, red-
purple; achenes 5–6 mm. long, hirsute; pappus
plumose, slightly purplish. *Liatris squarrosa intermedia* A. Gray, in part [G]; not *L. intermedia*
Lindl. *Laciniaria squarrosa* Britt., in part [B, R].
Sand-hills and bluffs: S.D.—Tex.—Colo. *Plain.*
Jl–S.

f.545.

3. **L. hirsùta** Rydb. Corm 1.3 cm. thick; stem
3–6 dm. tall, sulcate, more or less hirsute, at least above; leaves numerous, more or less hirsute, or in age glabrate, 5–10 cm. long, 3–6 mm. wide;
spike lax, 3–10 cm. long, with 3–10 heads, which are subsessile, 1.5–2 cm. high,
and 9–10 mm. broad; bracts very squarrose or even reflexed, strongly hirsute-
ciliate, the outer ovate, short-acuminate, the innermost elliptic, obtuse or
mucronate; corolla 1 cm. long, red-purple; achenes 6 mm. long, hirsutulous;
pappus plumose, brownish. Prairies and open woods: Iowa—Neb.—Okla.—
Ark. *Prairie.* Jl–S.

4. **L. cylindràcea** Michx. Stem single, 2–5 dm. high, glabrous, striate;
leaves linear, rigid, scarcely punctate, 5–7 cm. long, 2–4 mm. wide; heads rather
few, racemose; individual peduncles 2–4 cm. long; involucres cylindric, 1.5–2.5
cm. high; bracts broadly ovate, in 5–6 series. *Laciniaria cylindracea* (Michx.)
Kuntze [B]. Prairies: Ont.—Minn.—Mo. *Prairie.* Jl–S.

5. **L. fallàcior** (Lunell) Rydb. Corm horizontal, 6 cm. long, 2 cm.
wide; stem 4–6 dm. high; leaves numerous, rough, linear, the lower 4–6 mm.
wide; heads 15–30-flowered, 15–20 mm. long, crowded in a dense spike, the
upper much larger than the lower ones; bracts broadly ovate, acuminate, ciliate
on the margins, in about 5 series; achenes hirsute; pappus plumose. *Laciniaria
fallacior* Lunell. Prairies: N.D. S.

6. **L. punctàta** Hook. Stem glabrous or nearly so, 2–7.5 dm. high;
leaves strongly punctate, 1-nerved, ciliate on the margins, otherwise glabrous;
spike dense, elongate, 5–15 cm. long; heads 12–16 mm. high, sessile; involucres
narrowly campanulate; bracts cuspidate or acuminate, ciliate on the margins;
flowers purple or rose. *Laciniaria punctata* (Hook.) Kuntze [B, R]. Dry
plains and hills: Man.—Iowa—Tex.—Ariz.—Mont. *Plain—Mont.* Au–O.

7. L. acidòta Engelm. & Gray. Stem single or 2 or 3 from the same corm, glabrous, 3–5 dm. high; leaves narrowly linear, conspicuously punctate, glabrous; spike dense, 5–15 cm. long; heads 12–18 mm. high, sessile; involucres campanulate; bracts lanceolate, acuminate, sparingly ciliate; flowers purple. *Laciniaria acidota* (Engelm. & Gray) Kuntze [B]. Prairies: Kans.—Mo.—Tex. *Texan.* Au–O.

8. L. kansàna (Britton) Rydb. Corm woody, 1–1.5 cm. thick, crowned with fibrous remnants of old leaves; stem 5–10 dm. high, glabrous, sulcate; lower leaves linear or linear-oblanceolate, 1–2 dm. long, 7–25 mm. wide, faintly 5-ribbed, glabrous, conspicuously glandular-punctate, the upper gradually decreasing in size, the uppermost subulate, 1–2 cm. long; spike 1–3.5 dm. long, very dense; involucre nearly 1 cm. high; bracts dark purple, rounded at the apex, the outer oval, the inner oblong; achenes 4 mm. long, hirsutulous; pappus tawny, 6 mm. long, subplumose. *Laciniaria kansana* Britton. Bottom-lands: c Neb.—w Kans.—e Colo. *Plain.* Au.

9. L. spicàta (L.) Willd. Stem 3–18 dm. high, glabrous or nearly so; leaves numerous, linear, obtuse, 5–30 cm. long, obscurely punctate, glabrous; heads numerous, 5–13-flowered; involucre campanulate, about 1 cm. high; spike 2–4 dm. long; bracts oblong, scarious-margined at the apex; flowers purple, rarely white; achenes 5 mm. long, hispid; pappus tawny, barbellate. *Laciniaria spicata* Kuntze [B, R]. Moist soil: Mass.—Minn. La.—Fla. *Allegh.—Ozark.* Au–S.

10. L. Bebbiàna Rydb. Corm about 1.5 cm. thick, crowned with fibrous remains of leaves; stem 5–10 dm. high, glabrous below, crisp-hairy above; lower leaves narrowly linear-oblanceolate, 1–1.5 dm. long, 4–5 mm. wide, glabrous, inconspicuously punctate; spike 1.5–2 dm. high, very dense; heads about 1 cm. high, 7–10-flowered; bracts tinged with purple, puberulent and ciliate, acute, the outer ovate and scarcely squarrose, the inner lance-oblong, with spreading crisp tips; corolla rose-purple; achenes 4 mm. long, hirsutulous, pappus barbellate, tawny, 6–7 mm. long. Prairies: Ill.—Neb. *Prairie.* Au.

11. L. pychnostàchya Michx. Stem more or less hirsute above, 6–15 dm. high, very leafy; leaves linear or lance-linear, attenuate at the apex, often 3 dm. long; heads 3–6-flowered; involucre 8–12 mm. high; bracts oblong, pubescent and ciliate, purple-tinged; flowers purple; achenes hirsute. *Laciniaria pychnostachya* (Michx.) Kuntze. [B, R]. Prairies: Wis.—Ky. La.—Tex.—S.D. *Prairie.* Au–S.

12. L. sphaeroìdea Michx. Corm 1–1.5 cm. thick; stem 5–10 dm. high, glabrous up to the inflorescence; leaves firm, the lower linear-oblanceolate, petioled, 1–2 dm. long, 1–2 cm. wide, the upper linear, much reduced; inflorescence 1.5–3 dm. long; heads many, usually short-peduncled; involucre round-campanulate, 12–15 mm. high and broad; bracts with pale or rose-tinged margins; inner bracts erose on the margins; achenes 5 mm. long, hirsute; pappus tawny, tinged with rose. Dry soil: Mich.—Minn.—Okla.—Ga.—N.C. *Allegh.* Au–O.

13. L. Haywárdii Rydb. Corm rather small, about 1 cm. thick; stem 4–8 dm. high, glabrous up to the inflorescence, pale; leaves hirsutulous, the lower linear-oblanceolate, petioled, 1–2 dm. long, 1–2 cm. wide, the upper linear, reduced; inflorescence lax, 1–3 dm. long; heads short-peduncled; involucre turbinate-campanulate, 1.5–2 cm. broad, about 1.5 cm. high; outer bracts elliptic, the inner oblong, the lower erose-margined; achenes 7 mm. long, hirsute; pappus purple-tinged, 1 cm. long. Slopes: Black Hills, S.D.— Minn. (?). *Submont.* Jl–Au.

14. L. Herríckii (Steele) Rydb. Corm about 1 cm. thick; stem slender, glabrous below, 4–5 dm. high; leaves sparingly hirsutulous, or in age glabrate, the lower narrowly oblanceolate, 1–1.5 dm. long, 1–2 cm wide, the upper linear; inflorescence lax, 1–1.5 dm. long; heads 8–10, short-pedun-

782 CARDUACEAE

cled; involucre turbinate, 1 cm. high and broad, outer bracts elliptic, the inner oblong, slightly erose; pappus tipped with purple. *Laciniaria Herrickii* Steele. Prairies: Minn. Au.

15. L. áspera (Michx.) Greene. Corm 1–2 cm. thick; stem 3–10 dm. high, crisp-hairy; leaves numerous, scabrous-hispidulous, especially on the margins, the lower oblanceolate, 1–2 dm. long, 1–2.5 cm. wide, the upper linear, reduced; inflorescence spike-like, dense, 2–4 dm. long; heads numerous, mostly sessile; involucre rounded-campanulate, about 1.5 cm. broad and high; bracts with crimson or rose margins, the outer suborbicular, the inner oblong, erose; achenes 6 mm. long, hirsute; pappus tawny, tinged with rose. *L. scariosa* Auth., not L. *Laciniaria aspera* Greene [R]. Dry prairies: Ill.—Minn.—Neb.—Kans.—Ark. *Prairie.* Au–S.

16. L. ligulístylis (A. Nels.) Rydb. Corm 1.5–2 cm. thick; stem glabrate below, crisp-hairy above, 3–5 dm. high; leaves bright green, ciliate on the margins and hirsutulous, the lower linear-oblanceolate; inflorescence short and often dense; heads 1–10, seldom more, mostly short-peduncled; involucre hemispheric, 15–20 mm. high, 2–3 cm. broad; bracts with dark purple, erose margins, the outer rounded; pappus corded, 1 cm. long. *Laciniaria ligulistylis* A. Nels. [R]. *L. formosa* Greene. *Liatris borealis* Paxt.; not Nutt. Hills: Man.—S.D.—Colo.—Alta. *Plain—Submont.* Jl–Au.

17. L. Rosendáhlii Rydb. Corm 1.5–2 cm. thick; stem low, 3–4 dm. high, glabrous below, thick, very leafy; leaves densely hirsutulous, the lower lanceolate or oblanceolate, 1–2 dm. long, 1.5–3 cm. wide, the upper linear, reduced; inflorescence short, elliptic in outline, 5–15 cm. long; the heads 7–20, mostly peduncled; involucre campanulate, about 15 mm. broad and high; bracts with broad, scarious, dark purple, and erose borders, the outer rounded, the inner oblong; achenes 6 mm. long, hirsute; pappus tinged with purple. Dry soil: Itasca State Park, Minn. *Canad.* Au.

Tribe 3. **ASTEREAE.** Heads radiate or discoid; flowers all or the central ones with tubular corollas, hermaphrodite, and usually fertile, the marginal ones usually ligulate and pistillate. Receptacle naked. Anthers not caudate at the base. Style-branches in the hermaphrodite flowers flat, the lower part glabrous and with stigmatic lines, the upper part lanceolate or deltoid, short-pubescent. Leaves mostly alternate.

7. GUTIERRÈZIA Lag.

Annual or perennial herbs, or shrubs, with glutinous foliage. Leaves alternate, narrow, entire. Heads radiate, usually many, clustered in terminal corymbs. Involucre campanulate, oblong, or clavate, few- or many-flowered; bracts leathery, imbricate in few series. Ray-flowers few, pistillate, fertile; ligules yellow. Disk-flowers hermaphrodite, perfect, or the central ones sometimes sterile; corollas yellow. Anthers obtuse at the base. Style-branches flattened, with narrow appendages. Achenes obovate or oblong, terete or 5-angled. Pappus of many short scales, or that of the ray-flowers sometimes wanting.

Plant ligneous only at the short persistent caudex; stem 1–3, rarely 4–6 dm. high.
  Ligules at least three-fourths as long as the involucres.
    Ray- and disk-flowers each 3 or 4; involucres turbinate; leaves less than 1 mm. wide. . . . . 1. *G. filifolia.*
    Ray-flowers 6–8; disk-flowers usually 5; involucres campanulate; leaves 1–2 mm. wide. . . . . 2. *G. fulva.*
  Ligules about half as long as the involucre.
    Leaves linear, 1.2–3 mm. wide. . . . . 3. *G. diversifolia.*
    Leaves linear-filiform, less than 1 mm. wide.
      Plant low, 1–2 dm. high, branches green, leaves 1–2 cm. long. . . . . 4. *G. juncea.*

Plant 2–5 dm. high ; branches straw-colored ; leaves 3–4 cm. long.   **5. G. Sarothrae.**

Plant shrubby, 3–6 dm. high ; leaves 1–2 mm. wide ; ray- and disk-flowers each 3–4.   **6. G. linearis.**

1. G. filifòlia Greene. Stems 3–6 dm. high, puberulent; leaves 1–4 cm. long, scabrous-puberulent; involucres obovoid-turbinate, 3–4 mm. long; outer bracts ovate, acute, green-tipped, the inner obtuse, apiculate, scarcely at all green; ligules about 3 mm. long; ray- and disk-flowers each 3 or 4. Dry plains: Ida.—N.M.—Ariz.—Nev. *Plain—Submont.* Au–O.

2. G. fúlva Lunell. Stems 0.5–2 dm. high, numerous; leaves linear or lance-linear, 1–3 cm. long, 1–2 mm. wide; heads 4–5 mm. high in glomerules of 2–4 together; bracts in 3 or 4 series, ovate with an obtuse green tip; ligules 6–8, 3 mm. long. Dry hillsides: N.D. Au.

3. G. diversifòlia Greene. Stems 1–2 dm. high, angled, scabrous; lower leaves often oblanceolate, the rest linear, 2–4 cm. long, about 2 mm. wide, scabrous; inflorescence corymbiform; heads sub-sessile, clustered; bracts oblong, with thick, obtuse or acutish tips; ray-flowers 5 or 6; disk-flowers 8–10. (?) *G. viridiflora* Greene. Plains: Sask.—N.D.—Wyo.—Kans.—N.M. *Plain—Submont.* Jl.–S.

4. G. júncea Greene. Stems 1–2 dm. high; leaves 0.5 mm. wide, puberulent, spreading, early deciduous; inflorescence corymbiform; involucre turbinate, 3 mm. long; bracts ovate, acute, green tipped except the innermost; ray- and disk-flowers each 4 or 5. Dry hills and plains: Okla.—Colo. Ariz—Tex. *Plain—Mont.* Jl–Au.

5. G. Saróthrae (Pursh) Britton & Rusby. Perennial, with a cespitose woody caudex; leaves 3–4 cm. long, puberulent; heads numerous, sessile, 2 or 3 together; involucres turbinate, 4 mm. high; bracts oblong, acute; rays 3–6; ligules 1.5–2 mm. long; disk-flowers 2–5. *G. Euthamiae* T. & G.; *G. myriocéphala* A. Nels. Dry plains: Man.—Kans.—Utah—Mont. *Plain—Submont.* Jl–S.

6. G. lineàris Rydb. A shrub, 3–6 dm. high; leaves linear, about 3 cm. long, 1–2 mm. wide, puberulent, punctate; involucre campanulate, 5–6 mm. high; bracts scarious-margined, obtuse; ray- and disk-flowers each 3 or 4; heads subsessile, 2 or 3 together. Plains: Neb.—Kans.—N.M.—Colo. *Plain—Submont.* Jl–S.

8. AMPHIÁCHYRIS (DC.) Nutt.

Erect, glabrous, much-branched, annual herbs. Leaves small, alternate, entire. Heads radiate. Involucres turbinate-hemispheric; bracts coriaceous, imbricate. Receptacle naked. Ray-flowers pistillate, fertile; ligules yellow. Disk-flowers staminate or hermaphrodite but sterile; corollas yellow. Achenes terete, ribbed, strigose. Pappus of the rays minute, coroniform, that of the disk-flowers of 5–20 subulate scales or bristles dilated and united at the base.

1. A. dracunculoìdes (DC.) Nutt. Stem 1.5–4.5 dm. high, corymbosely branched above; leaves linear or linear-oblanceolate, 1–3.5 cm. long, 2–4 mm. wide; heads solitary at the ends of the branches; involucre about 3 mm. high; bracts oval, obtuse; ligules 5–10, about 2 mm. long. Prairies and hills: Mo.—Kans.—N.M.—Tex. Au–O.

9. GRINDÈLIA Willd. GUM-PLANT.

Coarse biennial or perennial herbs, or some western species even shrubby, usually with glabrous and glutinous foliage. Leaves alternate, usually spinulose-toothed. Heads relatively large, radiate or rarely discoid, often corymbose. Involucres hemispheric or depressed; bracts imbricate, mostly gummy, usually with subulate, spreading or recurved tips. Receptacle pitted. Ray-flowers pistillate and fertile, with yellow ligules, or wanting. Disk-flowers hermaphrodite and fertile; corollas yellow. Style-branches narrow, flattened, with linear or lanceolate appendages. Achenes 4- or 5-ribbed, sometimes flattened. Pappus of 2–8 awns or bristles, early deciduous.

Tips of the outer bracts spreading, none reflexed.
Bracts subulate, attenuate, the outer as long as the inner; achenes 2-toothed at the apex.   1. *G. lanceolata.*
Bracts linear, acute, the outer much shorter than the inner; achenes truncate.   2. *G. decumbens.*
Tips of the bracts squarrose, those of the outer bracts strongly reflexed.
Stem-leaves oblanceolate or rarely oblong.
Leaves finely and closely serrate.   3. *G. serrulata.*
Leaves rather remotely dentate or subentire.   4. *G. perennis.*
Stem-leaves ovate or oval, clasping, coarsely dentate.
Heads radiate.   5. *G. squarrosa.*
Heads discoid.   6. *G. nuda.*

1. G. lanceolàta Nutt. A biennial; stem 4–7 dm. high, glabrous; leaves linear or lanceolate, sessile and somewhat clasping, sharply serrate, acute, 2–5 cm. long; involucres 12–15 mm. high, 15–20 mm. broad; ligules 10–15 mm. long; pappus-awns 5 mm. long, smooth, somewhat curved. Prairies and barrens: Tenn.—Kans.—Tex.—La. *Austral.* Jl–S.

2. G. decúmbens Greene. A perennial, with a cespitose caudex; stems 2–5 dm. high, more or less decumbent at the base; basal leaves oblanceolate, obtuse, serrate towards the summit; stem-leaves oblong, acute, entire or denticulate; involucre 8 mm. high, 10–12 mm. broad; ligules 8 mm. long; pappus-awns straight, 4 mm. long, barbellate above. Plains: Kans.—Colo. *Plain—Submont.* Jl–Au.

3. G. serrulàta Rydb. Biennial or perhaps short-lived perennial; stem glabrous, about 6 dm. high; leaves sessile, oblanceolate or oblong, acutish, very viscid, 3–5 cm. long; involucre very viscid, about 15 mm. broad, ligules 7–8 mm. long, 1–1.5 mm. wide. Plains and hills: N.M.—S.D.—Wyo.—Utah. *Plain —Submont.* Jl–S.

4. G. perénnis A. Nels. A perennial, sometimes cespitose at the base; stems paniculately branched, 3–5 dm. high, glabrous; leaves 4–7 cm. long, the upper sessile and clasping; involucres nearly 1 cm. high, 12–20 mm. broad; bracts with rather long terete tips; ligules 8–10 mm. long. Plains and hills: Man.—Neb.—Colo.—Ida.—Alta. *Plain—Submont.* Jl–S.

5. G. squarròsa (Pursh) Dunal. *Fig. 547.* A biennial or perennial; stem 3–6 dm. high, corymbosely branched; basal leaves oblanceolate; stem-leaves oval, oblong-ovate or ovate, broad at the base; involucres 8–10 mm. high, 15–20 mm. broad; ligules 8–10 mm. long; pappus-awns smooth, straight. Prairies and plains: Ind.— ne Mich.—N.D.—Kans. —Ariz.—Ida. *Plain—Mont.* Au–S.

f. 547.

6. G. nùda Wood. A biennial; stem 3–6 dm. high, much branched; leaves 2–3 cm. long, crenate-dentate, acute; heads subsessile; involucre about 1 cm. high, 1–1.5 cm. broad; ligules wanting; pappus-awns 5 mm. long, smooth, straight or nearly so. *G. squarrosa nuda* A. Gray [B, G]. Dry plains: Mo.—Kans.—Tex. *Texan.* Au–S.

G. texana Scheele has been reported from Kansas, but the plants on which the record was based belong to *Prionopsis ciliata* Nutt.

10. **CHRYSÓPSIS** Nutt. Golden Aster.

Annual, biennial, or perennial caulescent herbs, usually conspicuously pubescent. Leaves alternate. Heads showy, radiate, or rarely discoid, mostly cymose-corymbose, many-flowered. Involucral bracts narrow, in several series, more or less imbricate; receptacle pitted. Ray-flowers many pistillate; ligules golden yellow. Disk-flowers hermaphrodite and fertile. Style-branches with linear or subulate appendages. Achenes flattened. Pappus double; the outer series of small scales or bristles, the inner of numerous scabrous capillary bristles.

Annual; leaves oblong-oblanceolate to broadly linear, with a broad, somewhat clasping base, pilose.   1. *C. pilosa.*
Perennials.
  Involucres strigose-pubescent.
    Leaves narrowly linear-oblanceolate.
      Leaves canescent-hirsute; disk fully 1 cm. high and 12 mm. broad.   2, *C. angustifolia.*
      Leaves silvery-white; disk 6–8 mm. high and 8–10 mm. broad.   3. *C. Berlandieri.*
    Leaves oblanceolate to obovate or lanceolate, strigose, at least when young.
      Heads sessile or subsessile, subtended by foliage-leaves.
        Leaves oblong or lanceolate, decidedly acute; those subtending the heads linear.   4. *C. foliosa.*
        Leaves oblanceolate, hirsutulous, mostly obtuse.   5. *C. imbricata.*
      Heads peduncled, naked or subtended by a small leaf.
        Most of the leaves sessile.
          Leaves oblanceolate or obovate, obtuse or merely apiculate; disk 15–20 mm. broad.   6. *C. villosa.*
          Leaves narrowly oblanceolate, decidedly acute; disk 10–12 mm. broad.   7. *C. hirsutissima.*
        Most of the leaves distinctly petioled, oblanceolate; disk 10–12 mm. broad.   8. *C. Bakeri.*
  Involucres hirsute, or both hirsute and resinous-granuliferous.
    Involucres densely hirsute, scarcely at all resinous-granuliferous.
      Plant 1–2 dm. high, heads usually solitary and sessile.   9. *C. pumila.*
      Plant 3–4 dm. high; heads several, usually peduncled.   10. *C. Ballardi.*
    Involucres sparingly hirsute, copiously resinous-granuliferous; plant 2–4 dm. high.
      Upper leaves obovate, oval, or rarely oblong, sessile.
        Leaves densely hispid, not copiously granuliferous.   11. *C. horrida.*
        Leaves conspicuously granuliferous, sparingly hispidulous.   12. *C. viscida.*
      Upper leaves as well as the lower oblanceolate, most of them petioled.
        Leaves broadly oblanceolate; heads peduncled.
          Leaves decidedly hairy, sparingly granuliferous.   13. *C. asprella.*
          Leaves sparingly hairy, conspicuously granuliferous.   14. *C. hispida.*
        Leaves narrowly linear-oblanceolate; heads subsessile.   15. *C. stenophylla.*

1. C. pilòsa Nutt. Stem 3–6 dm. high, branched above, pilose; leaves 2.5–8 cm. long, 4–10 mm. wide, entire or with a few coarse teeth; heads short-peduncled; involucre 8–10 mm. high; bracts linear, acuminate, hirsute and viscid; ligules 7–10 mm. long; achenes obovoid, ribbed, hirsute. *C. Nuttallii* Britt. [B]. Open sandy places: Mo.—Kans.—Tex.—La. *Ozark.* Jl–S.

2. C. angustifòlia Rydb. Stem erect, 2–4 dm. high, canescent and more or less hirsute; leaves 2–5 cm. long, 2–5 mm. wide, acute, grayish-strigose, hispid-ciliate; involucre turbinate-campanulate, 9–10 mm. high; bracts narrowly linear, acute; ligules about 1 cm. long. *C. stenophylla* Britt. & Brown [B, G]; not Greene. Sandy places: Alta.—Minn.—Mo.—Kans. *Prairie—Plain.* Au–S.

3. **C. Berlandièri** Greene. Stem 1–3 dm. high, canescent-strigose and hirsute; leaves 1.5–4 cm. long, 2–5 mm. wide, acute, linear-oblanceolate, silver-white; involucre turbinate, 6–10 mm. high; bracts linear, acute, white-strigose; ligules 4–6 mm. long; achenes 2 mm. long, white-strigose. Sandy and rocky soil: Kans.—Tex.—N.M. *St. Plains—Texan.* Je–S.

4. **C. foliòsa** Nutt. Stem 3–6 dm. high, very leafy, canescent, strigose and hirsute; lower leaves oblanceolate, acute, 3–5 cm. long, white, silky-strigose, subsessile; heads subtended by linear bract-like leaves, about 1 cm. high; bracts linear, acute, strigose. Sandy soil: Wis.—Minn.—Kans.—Utah—Wash. *Plain—Submont.—Basin.* Jl–Au.

5. **C. imbricàta** A. Nels. Stems 3–4 dm. high, subcanescent and with some long hairs; leaves numerous, the lower oblanceolate or spatulate, petioled, the upper broadly oblong or oblong-oblanceolate, sessile and mostly obtuse or apiculate, grayish-hirsutulous; involucres 8–10 mm. high; bracts lance-linear, hirsute-strigose. Mountains: w Kans.—Colo.—Utah—Mont. *Plain—Submont.* Jl.

6. **C. villòsa** (Pursh) Nutt. *Fig. 546.* Stem 2–5 dm. high, strigose and more or less hirsute; leaves 2–4 cm. long, obovate, elliptic or oblong, appressed-silky, hirsute-ciliate, usually sessile; heads peduncled, sometimes subtended by a small leaf; involucre about 1 cm. high. Dry hills: Man.—Tex. —N.M.—Ida. *Plain—Mont.—Son.* Jl–S.

f 546.

7. **C. hirsutíssima** Greene. Stem 1–3 dm. high, densely long-hirsute, as well as hirsutulous; leaves oblanceolate, somewhat spreading, the lower short-petioled, the upper sessile, densely grayish-strigose; heads fastigiate-corymbose; involucre 8 mm. high; bracts linear. Sandy soil: Sask.—Minn.—Colo.— Ariz. *Plain—Son.* Je–Au.

8. **C. Bàkeri** Greene. Stems 2–4 dm. high, often purplish, strigose and only sparingly hirsute; leaves mostly acute, strigose, only slightly resinous-granuliferous; heads few, peduncled; involucres about 8 mm. high; bracts often purplish-margined. Hills: Minn.—Kans.—N.M.—Utah—Ida. *Prairie—Submont.—Mont.* Jl–Au.

9. **C. pùmila** Greene. Stems decumbent at the base; leaves oblanceolate, 1–3 cm. long, mostly petioled, hirsute but not viscid; involucres about 8 mm. high and 1 cm. broad; bracts linear, hirsute but not viscid; ligules about 6 mm. long. Cañons and mountain-sides: Neb.—Colo.—Ida. *Plain—Subalp.* Je–S.

10. **C. Ballárdi** Rydb. Perennial, in habit resembling *C. camporum,* but more hispid; stem 3–4 dm. high, branched above, hirsute and hispid; leaves oblanceolate, 2–4 cm. long, obtuse or acute, hirsute with ascending hairs often pustulate at the base, the lower petioled, the upper sessile; involucre hemispheric, 15 mm. broad, the bracts densely hirsute, but scarcely resinous-granuliferous; ligules 10–12 mm. long; achenes 3 mm. long, silky-hirsute. Dry ground: Minn.—Kans. *Prairie.* Je–S.

11. **C. hórrida** Rydb. Stems 2–3 dm. high, hirsutulous and hispid; leaves obovate, broadly oblanceolate, or oblong; heads corymbose, peduncled, about 8 mm. high; disk 1 cm. or less wide; bracts sparingly strigose-hirsute. Hills: Neb.—Wyo.—N.M.—Tex. *Plain—Son.* Jl–S.

12. **C. víscida** (A. Gray) Greene. Stem 1–3 dm. high, resinous-granuliferous and hispidulous; lower leaves spatulate and petioled, the upper obovate, elliptic or ovate, sessile, 1–2 cm. long; heads few or solitary; involucres 8 mm. high and 10 mm. wide; bracts resinous-granuliferous and hispidulous. *C.*

THISTLE FAMILY 787

villosa viscida A. Gray. Mountains and dry ledges: Colo.—Kans.—Tex.—Ariz. *Plain—Mont.* My–N.

13. C. asprélla Greene. Stem erect, 2–4 dm. high, hispidulous and sparingly hirsute, simple up to the inflorescence; head corymbose; involucres 7–8 mm. high and broad. *C. compacta* Greene. *C. arida* A. Nels. Dry hills and mountains: Mont.—Neb.—Kans.—N.M.—Ariz. *Plain—Submont.—Subalp.* Jl–S.

14. C. híspida (Hook.) DC. Stems 2–3 dm. high, hirsute and resinous granuliferous; lower leaves oblanceolate and petioled, the upper lanceolate, sessile, 1–3 cm. long; heads corymbose and peduncled; involucres 7–8 mm. high and 8–12 mm. wide. *C. villosa hispida* A. Gray. Sandy river valleys: N.D,—Sask.—N.M.—Ariz.—Calif.—B.C. *Plain—Submont.* Je–Au.

15. C. stenophýlla (A. Gray) Greene. Perennial, with a cespitose caudex; stem 2–4 dm. high, very hispid, stiff; leaves 2–3 cm. long, spreading, early deciduous, hispid and resinous-granuliferous, the hairs with pustulate bases; heads subtended by a few narrow leaves; involucre 8 mm. high and 1 cm. broad; bracts granuliferous and sparingly hirsute. *C. villosa stenophylla* A. Gray. Dry plains: Kans.—Colo.—Tex—Ark. *St. Plains.* Au–S.

11. HETEROTHÈCA Cass.

Annual, biennial or perennial herbs. Leaves alternate, dentate or entire. Heads corymbosely cymose, terminating the branches, usually radiate. Involucre hemispheric; bracts imbricate, in several series, narrow. Receptacle flat, alveolate. Ray-flowers pistillate and fertile; ligules yellow. Disk-flowers hermaphrodite, or sometimes staminate; corollas yellow. Style-branches flat, with lanceolate or triangular appendage. Achenes pubescent, those of the ray-flowers thick, turgid or triangular, those of the disk flattened. Pappus of the former wanting, or coroniform, of the latter double, the outer of short bristles or scales, the inner of numerous long bristles.

1. H. subaxilláris (Lam.) Britton & Rusby. An annual or biennial; stem 3–10 dm. high; leaves oval or oblong, the lower 5–7 cm. long, petioled, the upper sessile, clasping; heads several; involucre 6–10 mm. long; bracts linear; ligules 10–25. *H. Lamarckii* Cass. Dry soil: Del.—Fla.—Ariz.—Kans.; Mex. *Austral.—Son.* Jl–S.

12. CHRYSOTHÁMNUS Nutt. RABBIT BRUSH.

Shrubs or undershrubs, with narrow leaves. Inflorescence mostly paniculate-cymose. Heads discoid. Involucres turbinate; bracts coreaceous or chartaceous, in distinct, usually 5, vertical rows, without herbaceous tips. Flowers all alike, hermaphrodite and fertile, 5–15 in the head; corollas yellow, slender, 5-toothed. Style-branches with exserted subulate or setaceous-filiform appendages. Achenes slender, terete or slightly angled. Pappus white, of numerous slender bristles. *Bigelovia* T. & G., in part.

Bracts of the involucre acuminate.
  Achenes glabrous or nearly so; bracts not foliaceous.
    Margins of the leaves glabrous.   1. *C. pulchellus.*
    Margins of the leaves scabrous-ciliolate.   2. *C. Baileyi.*
  Achenes pubescent; outer bracts more or less foliaceous. 3. *C. Howardii.*
Bracts merely acute not foliaceous; achenes strigose.
  Bracts and mature leaves perfectly glabrous.   4. *C. graveolens.*
  Bracts erose-ciliate on the margins, tomentose on the back.
    Leaves 1 mm. wide or less, spreading.   5. *C. plattensis.*
    Leaves 1–2.5 mm. wide, ascending.   6. *C. frigidus.*

1. C. pulchéllus (A. Gray) Greene. A shrub, 6–10 dm. high; branches glabrous, striate; leaves narrowly linear, 1–2 cm. long, glabrous; involucre 10–15 mm. high; bracts 5-ranked in 5 or 6 series, lanceolate; achenes sparingly

pubescent or glabrous. *Bigelovia pulchella* A. Gray. Dry hills: Kans.—Tex.
—N.M.—Utah; northern Mex. *St. Plains—Son.* Au–S.

2. C. Baileyi Woot. & Standl. A low shrub, 2–5 dm. high, densely
branched; branches glabrous; leaves linear, glabrous except the scabrous-
ciliolate margins, about 2 cm. long, 1–2 mm. wide; involucre 10–12 mm. high;
bracts 5-ranked, in about 5 series, lanceolate, cuspidate; achenes glabrous.
Arid regions: Kans.—N.M.—Tex. *St. Plains.* Jl–S.

3. C. Howárdii (Parry) Greene. A low shrub,
1–3 dm. high; branches white-tomentose, ascending;
leaves linear-filiform, 3–5 cm. long; involucres 10–12
mm. high; bracts lanceolate, attenuate, tomentose
and arachnoid-hairy; corollas scarcely 1 cm. long;
achenes strigose. *B. Howardii* A. Gray. Dry hills:
Neb.—Wyo.—Colo. Jl–S.

4. C. gravèolens (Nutt.) Greene. *Fig. 548.* A
shrub, 1–1.5 m. high; branches green, pannose only
when young; leaves linear, 4–7 cm. long, 1–2 mm.
wide, slightly pannose when young; heads numerous
in a compound corymb; involucre about 8 mm. high;
bracts lanceolate, in 4 series; achenes sparingly strigose. *B. graveolens* A. Gray. *C. nauseosus* Britton [B]. Cañons and bad-lands and dry hills: N.D.
—Neb.—N.M.—Utah—Mont. *Plain—Mont.* Au–O.   *f.548*

5. C. platténsis Greene. Low undershrub, 2–3 dm. high; branches more
or less pannose-tomentose; leaves linear-filiform, tomentose, 3–5 cm. long;
bracts in about 3 series, lanceolate, acute, glabrous except the erose-ciliate
margins; achenes hirsute-strigose. Plains and hills: Mont.—N.D.—S.D.—Colo.
Plain—Submont. Jl–O.

6. C. frígidus Greene. An undershrub, 2–6 dm. high; branches white-
tomentose; leaves narrowly linear, 3–5 cm. long; bracts in 2 or 3 series, lanceolate, the outer acute, the inner obtuse; achenes strigose. *C. nauseosus typicus*
Hall & Clements. Plains: N.D.—Sask.—Colo.—N.D.—Mont. *Plain—Mont.*
Jl–S.

13. OONÓPSIS Greene.

Perennial herbs, with woody caudex or crown, and erect leafy stems.
Leaves linear or lanceolate, entire. Heads in terminal cymes or solitary, radiate or discoid. Involucres campanulate or hemispheric; bracts well imbricate,
flat, cuspidate-acuminate. Ray-flowers, if present, pistillate and fertile; ligules
yellow. Disk-flowers hermaphrodite and fertile; corollas yellow, cylindric,
with 5 short teeth. Tips of the style-branches ovate, shorter than the stigmatic
parts. Achenes glabrous or strigose, many-nerved. Pappus of rather few
coarse bristles.

Plant more or less villous; leaves flat; heads radiate.   1. *O. argillacea.*
Plant glabrous; leaves involute; heads discoid.   2. *O. Engelmanni.*

1. O. argillàcea A. Nels. *Fig. 549.* Stems numerous, 5–8 cm. long,
sparingly villous; leaves entire, narrowly linear, acute; involucres slightly
villous, turbinate; bracts oblong-ovate or lanceolate, acuminate; rays about 10;
ligules 5–6 mm. long. Clayey slopes: Wyo.—S.D. *Submont.* Je–Jl.

2 O. Engelmánni (A. Gray) Greene. A perennial with a crepitose
woody caudex; stems 1–2 dm. high, glabrous; leaves narrowly linear, 3–5 cm.
long, 2–3 mm. wide, rigid; heads in a congested cyme; involucre about 12 mm.

high, turbinate-campanulate; bracts ovate or ellip-
tic, short-cuspidate; achenes puberulent. *Bigelovia
Engelmanni* A. Gray. Dry plains: Colo.—Kans.—
S.D. *Plain.* Au O.

14. PRIONÓPSIS Nutt.

Annual or biennial caulescent glabrous herbs.
Leaves alternate, sessile, spinulose-toothed. Heads
radiate, showy. Involucre broadly hemispheric;
bracts well imbricate, in several series, the outer
spreading. Receptacle naked. Ray-flowers numerous, pistillate; ligules yellow. Disk-flowers numerous, hermaphrodite and fertile; corolla yellow.
Style-branches with linear-lanceolate appendages.
Achenes glabrous, those of the rays broader.
Pappus of a few firm capillary bristles, the outer
shorter, all deciduous.

f. 547.

1. P. ciliáta Nutt. Annual; stem 8–15 dm. high, sparingly branched;
leaves numerous, oval, ovate or oblong, saliently dentate; heads few, corymbose; involucre about 1 cm. high, 2–2.5 cm. broad; bracts linear-lanceolate,
acuminate, squarrose; ligules 1–1.5 cm. long. *Aplopappus ciliatus* DC. [G].
River banks and hills: Mo.—Colo.—Tex. *Texan.* Au–S.

15. SIDERÁNTHUS Nutt. IRON PLANT

Perennial or annual herbs, rarely suffrutescent at the base. Leaves alternate, with spinulose-tipped lobes or teeth. Heads radiate, or in one species
discoid. Involucre campanulate or hemispheric; bracts well imbricate, in several series, appressed or with loose tips, green. Receptacle flat, naked, usually
pitted. Ray-flowers pistillate, fertile, or wanting; ligules yellow. Disk-flowers
hermaphrodite, fertile; corolla yellow; tube scarcely dilated into a distinct
throat. Anthers obtuse at the base. Style-branches flattened, with lanceolate appendages. Achenes obtuse, pubescent, mostly 8–10-nerved. Pappus
of 1–3 series of many unequal capillary bristles, persistent.

Leaves spinescent-toothed, not pinnatifid.
    Heads discoid; plant perennial, with a woody caudex.   1. *S. grindelioides.*
    Heads radiate; plant annual.   2. *S. annuus.*
Leaves pinnatifid.
    Plant more or less floccose, not at all glandular.   3. *S. spinulosus.*
    Plant glabrous or nearly so.   4. *S. glaberrimus.*

1. S. grindelioídes (Nutt.) Britton. A low cespitose perennial; stems
erect, 1–3 dm. high, somewhat puberulent; leaves thick, obovate or oblanceolate, spinulose-serrate, 1–3 cm. long, crisp-hirsute or strigose; involucres hemispheric, 6–7 mm. high, 1 cm. broad; bracts in 4 or 5 series, linear-lanceolate,
puberulent; achenes somewhat flattened. *Aplopap-
pus Nuttallii* T. & G. Dry plains and hills: Sask.—
S.D.—Neb.—Ariz.—Nev.—Alta. *Plain—Submont.—
Son.* Jl–Au.

2. S. ánnuus Rydb. An annual; stem branched
above, more or less glandular-pubescent; leaves
oblanceolate, spinulose-dentate, 3–6 cm. long; heads
corymbose; involucres hemispheric, about 8 mm.
high, viscid-puberulent; bracts linear-lanceolate,
acuminate, in 4 or 5 series; ligules light yellow, 5–6
mm. long; achenes obtusely angled. *S. rubiginosus*
Britt., in part [B]. Sandy places: Neb.—Colo.—Tex.
Plain—St. Plains. Au–S.

3. S. spinulósus (Pursh) Sweet. *Fig. 550.* A
perennial; stem 1–4 dm. high, branched; leaves 1–3

f. 550.

cm. long, pinnatifid; segments linear or oblong, toothed or pinnately parted; involucres 6–8 mm. high; bracts linear-lanceolate, spinulose-tipped, villous-canescent; ligules 6–7 mm. long, bright yellow; achenes somewhat flattened. *Aplopappus spinulosus* DC. [G]. Plains and prairies: Man.—Tex.—Ariz.—Alta; northern Mex. *Prairies—Plain—Son.* Je–S.

4. **S. glabérrimus** Rydb. A perennial; stems several, 1–3 dm. high; leaves rigid, about 2 cm. long, bluish-green, glabrous, pinnatifid; segments oblong, spinulose-tipped; involucre depressed-hemispheric, 8–15 mm. broad; bracts linear-lanceolate, acute or spinulose-tipped; ligules 4–5 mm. long; achenes somewhat flattened. Dry plains: N.D.—Okla.—Colo.—Wyo. *Plain.* Je–S.

16. STENÒTUS Nutt.

Low cespitose perennials, with woody caudex and narrow, entire, more or less coriaceous leaves, which are mostly basal. Involucre hemispheric, or nearly so; bracts imbricate in a few rows, appressed, thin, lanceolate to oval, not herbaceous. Receptacle naked, alveolar. Ray-flowers pistillate and fertile; ligules yellow. Disk-flowers hermaphrodite and fertile; corollas tubular, some-what enlarged upwards, deeply 5-toothed. Anthers obtuse at the base. Style-branches stigmatic their whole length, with subulate appendages. Achenes grayish- or white-villous. Pappus of numerous white soft capillary scabrous bristles.

Bracts lanceolate, acute.   1. *S. acaulis.*
Bracts oval to oblong, obtuse.   2. *S. armerioides.*

1. **S. acaùlis** Nutt. A cespitose, tufted, subacaulescent perennial; leaves mostly basal, oblanceolate or narrowly spatulate, 3-nerved, scabrous-puberulent; flowering stem scapiform, 2–5 cm. high; head solitary, 1 cm. high; ligules 8 mm. long. *Aplopappus acaulis* A. Gray. Dry hills and bad-lands: Sask.—S.D.—Colo.—Calif.—Wash. *Plain—Submont.* My–Je.

2. **S. armerioìdes** Nutt. A perennial, with a cespitose caudex; flower-ing stems 5–15 cm. high, with a few small stem-leaves, glabrous or slightly puberulent; basal leaves linear or linear-oblanceolate, 3-nerved, glabrous or nearly so, 3–7 cm. long; heads 10–13 mm. high; ligules 10–12 mm. long. *Aplopappus armerioides* A. Gray. Dry hills and bad-lands: Man.—Neb.—N.M.—Wyo.—Sask. *Plain—Submont.* Je–Jl.

17. PYRROCÒMA Hook.

Perennial herbs, with thick taproots. Leaves alternate, sometimes with spinulose teeth. Heads terminal, or in the axils of the upper leaves, radiate. Involucres hemispheric; bracts more or less foliaceous; receptacle flat, alveolar. Ray-flowers numerous, fertile or sterile; ligules yellow, sometimes not longer than the disk. Disk-flowers yellow, perfect; corollas cylindric, slightly if at all dilated upwards, with short, erect teeth. Appendages of the style-branches subulate-linear, longer than the stigmatic portion. Achenes linear, 3-angled, striate, usually glabrous. Pappus reddish brown or yellow, of numerous rigid bristles.

1. **P. lanceolàta** (Hook.) Greene. Stems 2–4 dm. high, glabrous or nearly so; basal leaf-blades lanceolate, usually more or less dentate, glabrous; stem-leaves narrowly lanceolate, sessile; heads several, corymbose; involucres 15–20 mm. broad; bracts lanceolate, long-acute; ligules about 10 mm. long. *A. lance-olatus* T. & G. Meadows and flats: Sask.—Neb.—Wyo.—B.C. *Plain.* Jl–Au.

18. **ISOPÁPPUS** T. & G.

More or less rough-hairy annual or biennial herbs. Leaves alternate, narrow, entire or toothed, 1-nerved. Involucre campanulate; bracts imbricate in 2 or 3 series, not herbaceous. Receptacle alveolate. Ray-flowers pistillate, fertile, 5–8; ligules yellow. Disk-flowers 10–20, hermaphrodite, fertile. Anthers not sagittate at the base. Style-branches with narrow, hirsute appendages. Achenes terete, narrowed below, silky-villous. Pappus of a single series of rough capillary bristles.

1. **I. válidus** Rydb. *Fig. 555.* Annual; stem 3–6 dm. high, hispid and densely glandular-hairy, branched above, very leafy; leaves oblanceolate, the lower petioled, 3–10 dm. long, sharply serrate with lanceolate teeth directed forward; leaves of the inflorescence entire, lance-subulate, attenuate; inflorescence corymbiform, the branches densely glandular; involucres hemispheric, about 1 cm. high; bracts narrowly lance-linear, in 5–6 series, white-margined; ligules 12–18, linear, 7–8 mm. long; achenes narrowly obconic, strigose, 2 mm. long; pappus reddish brown, 2 mm. long. Sandy hills and valleys: Kans.—Tex. Jl O.

19. **SOLIDÀGO** L. Goldenrod.

Perennial caulescent herbs, mostly with rootstocks, rarely shrubby at the base. Leaves alternate, often toothed. Heads small, paniculate or thyrsoid, radiate. Involucre campanulate, turbinate, or subcylindric; bracts in several series, ribless. Receptacle small, alveolate. Ray-flowers pistillate, fertile; ligules short, yellow, or in one species white. Disk-flowers hermaphrodite, fertile; corollas usually yellow, tubular, 5-lobed. Anthers obtuse at the base. Style-branches flattened; appendages lanceolate. Achenes terete, usually ribbed. Pappus of many rough capillary bristles, in 1 or 2 series.

Bracts of the involucres appressed, without green tips.
  Heads in small axillary clusters; leaves pinnately veined, the upper not much reduced.   1. Axillares.
  Heads in a terminal thyrsus or panicle, sometimes also in clusters in the axils of the reduced upper leaves.
    Branches of the panicle or thyrsus erect, not secund, nor the tops nodding; leaves pinnately veined.
      Plant pubescent with long hairs; heads 3–5 mm. high, in a narrow, raceme-like thyrsus, often leafy at the base.   2. Bicolores.
      Plant glabrous or puberulent, not long-hairy; heads 5–12 mm. high in a more ample panicle.   3. Speciosae.
    Branches of the thyrsus or panicle recurved, spreading, distinctly secund, or if the inflorescence is more simple, its top more or less nodding.
      Leaves pinnately veined, rarely 3-ribbed.   4. Junceae.
      Leaves, at least the lower ones, distinctly 3-ribbed.
        Leaves, at least the lower ones, oblanceolate, spatulate, or obovate, the upper usually much smaller.
          Plant glabrous or nearly so.   5. Glaberrimae.
          Plant canescent-puberulent.   6. Nemorales.
        Leaves all lanceolate; stem equally leafy throughout; plants sobolif-erous.   7. Serotinae.
Bracts of the involucres, at least the outer ones, with squarrose-spreading tips; heads in a dense thyrsus.   8. Petiolares.

1. Axillares.
Stem angled; leaves with winged petioles; blades broadly oval or ovate.
                        1. *S. flexicaulis.*
Stem terete; leaves sessile; blades lanceolate or oblong-lanceolate.   2. *S. caesia.*

2. Bicolores.
Ligules white; midrib of the bracts conspicuously dilated above.   3. *S. bicolor.*

Ligules yellow; midrib of the bracts slightly, if at all, dilated above.   4. *S. hispida.*

3. SPECIOSAE.

Bracts acute.
Stem and leaves scabrous-puberulent.   5. *S. Lindheimeriana.*
Stem and leaves glabrous.   14. *S. glaberrima.*
Bracts obtuse; stem at least below, and leaves except the scabrous-ciliolate margins, glabrous or nearly so.
Lower leaves narrowly oblanceolate, the upper linear-lanceolate; branches of the inflorescence nearly erect.   6. *S. uliginosa.*
Lower leaves broadly oblanceolate; the upper broadly lanceolate, elliptic, or ovate; branches of the inflorescence ascending-spreading.
Inflorescence narrow and dense; plant pale.   7. *S. pallida.*
Inflorescence more open, with elongate branches; plant not pale.
Lower leaves broadly oblanceolate, dentate, the upper broadly lanceolate or ovate.   8. *S. speciosa.*
Lower leaves lanceolate, the upper oblong or oblong-lanceolate, all entire.   9. *S. rigidiuscula.*

4. JUNCEAE.

Stem hirsute.   10. *S. rugosa.*
Stem glabrous, except in the inflorescence.
Branches of the inflorescence more or less pubescent, spreading; involucre cylindro-turbinate, few-flowered.
Inner bracts 3 mm. long, obtuse.   11. *S. ulmifolia.*
Inner bracts 4 mm. long, acute.   12. *S. microphylla.*
Branches of the inflorecence glabrous, numerous, strongly ascending or suberect at the base, curving outwards.   13. *S. juncea.*

5. GLABERRIMAE.

Stem-leaves elongate, linear-lanceolate.
Basal leaves oblanceolate, with ovate teeth; secondary fascicled leaves rarely present in the axils of the primary leaves.
Panicle open, with secund branches; bracts oblong-lanceolate.   14. *S. glaberrima.*
Panicle narrow; branches ascending, scarcely secund; bracts linear-lanceolate.   15. *S. missouriensis.*
Basal leaves narrowly linear-oblanceolate, with lanceolate, saliently tipped teeth; secondary fascicled leaves usually present in the axils of the primary leaves.   16. *S. moritura.*
Stem-leaves short, lanceolate.   17. *S. Hapemaniana.*

6. NEMORALES.

Bracts oblong or linear, obtuse; stem-leaves scarcely 3-ribbed; plant not soboliferous.
Basal leaves narrowly oblanceolate, 1–2.5 cm. broad.
Basal leaves short-petioled, the petioles much shorter than the blades.   18. *S. pulcherrima.*
Basal leaves long-petioled, the petioles often nearly as long as the blades.   19. *S. longipetiolata.*
Basal leaves broadly oblanceolate or spatulate, 2–4 cm. broad.   20. *S. nemoralis.*
Bracts lanceolate or ovate, acute; stem-leaves distinctly 3-ribbed; plant soboliferous.
Stem-leaves oblanceolate or elliptic; bracts lanceolate.   21. *S. trinervata.*
Stem-leaves obovate, thick; bracts ovate.   22. *S. mollis.*

7. SEROTINAE.

Stem glabrous up to the inflorescence.
Leaves glabrous, or glabrous except the margins and the veins beneath; involucres 4–5 mm. high.   23. *S. serotina.*
Leaves finely pubescent beneath; involucres about 3 mm. high.   24. *S. Somesii.*
Stem more or less pubescent.
Involucres 3.5–5 mm. high.
Leaves rather thin, softly pubescent, not strongly veiny.
Leaves less than 1 cm. wide, broadest at the middle; plant 3–7 dm. high.
Leaves usually more than 1 cm. wide, broadest below the middle; plant 5–20 dm. high.   25. *S. Lunellii.*

Pubescence not longer on the ribs on the lower
surface of the leaves ; branches of the in-
florescence spreading.   26. *S. altissima.*
Pubescence longer on the ribs on the lower
surface of the leaves ; branches of the inflorescence long, strongly ascending and
usually leafy below, recurved-spreading
only at the end.   27. *S. procera.*
Leaves rather firm, scabrous above, hirsutulous be-
neath and strongly veined.   28. *S. dumetorum.*
Involucres 2–3 mm. high.
Plant green, sparingly pubescent ; leaves narrowly
lanceolate or linear-lanceolate.   29. *S. canadensis.*
Plant yellowish- or grayish-green, densely puberu-
lent or pubescent.
Lower surface of the leaves puberulent with short
straight hairs.   30. *S. gilvocanescens.*
Lower surface of the leaves softly pubescent with
crisp hairs.   31. *S. satanica.*
8. PETIOLARES.
Plant green ; leaves rather thin, dull.   32. *S. petiolaris.*
Plant canescent ; leaves firm, shining.   33. *S. Wardii.*

1. S. flexicaulis L. A perennial ; stem often zigzag, glabrous, angled,
3 10 dm. high ; leaf-blades broadly ovate or oval, sharply serrate, acuminate,
thin, 5–15 cm. long ; heads in very short axillary clusters, about 6 mm. high ;
bracts obtuse or acutish ; ligules 3 or 4 ; achenes hirsute. *S. latifolia* L. [G].
Rich woods and shaded banks: N.B.—Ga.— Kans.—N.D. *Canad.—Allegh.* Jl–S.

2. S. caèsia L. A perennial ; stem glabrous, often glaucous, bluish or
purple, 3–10 dm. high ; leaves lanceolate or oblong-lanceolate, serrate, acuminate, glabrous, 5–12 cm. long ; heads in very short axillary clusters ; about 5
mm. high ; bracts linear, obtuse, glabrous ; achenes pubescent. Woods and
thickets: Me.—Fla.—Tex.—Minn. *E. Temp.* Au–O.

3. S. bicolor L. A perennial ; stem 2–12 dm. high, more or less pubescent with long white hairs ; lower leaves long-petioled, the upper sessile ; blades
obovate or oblanceolate, 5–10 cm. long, dentate, hirsute on both sides ; inflorescence either elongate and narrow, with short branches of clustered heads,
leafy below, or divided into several similar, elongate branches ; involucres 3–5
mm. high ; bracts obtuse, yellowish, with a green midrib ; achenes glabrous.
Dry soil: N.B.—Man.—Mo.—Ga. *Canad.—Allegh.* Au–O.

4. S. híspida Muhl. A perennial ; stem 4–10 dm. high, short-hirsute ;
lower leaves petioled, the upper sessile ; blades of the former obovate, oblanceolate or oval, obtuse, crenate-serrate, 5–10 cm. long ; those of the latter oblanceolate or oblong, acute ; inflorescence as in the preceding, but usually more open ;
involucres 5–6 mm. high ; bracts obtuse, straw-colored, with green midrib ;
achenes glabrous. *S. bicolor concolor* T. & G. Dry soil: Newf.—Man.—Neb.
—Ark.—Ga. *Canad.—Allegh.* Au–O.

5. S. Lindheimeriàna Scheele. A perennial, with a stoloniferous rootstock ; stem cinereous-puberulent, 3–8 dm. high ; leaves broadly lanceolate, elliptic, or oval, acute, sessile, scabrous-puberulent, somewhat glossy ; heads in a
dense, narrow thyrsus ; involucres campanulate, 5.5–7 mm. high ; achenes glabrous. Limestone bluffs and rocky woods: Kans.—Tex.; northern Mex. *Texan.*
—Son. Au–S.

6. S. uliginòsa Nutt. A perennial, with an elongate rootstock ; stem
simple, 3–10 dm. high, glabrous, usually purplish ; leaves glabrous, except the
scabrous-ciliolate margins, the lower petioled, 1–2 dm. long, narrowly oblanceolate, serrate, the upper sessile, smaller, linear-lanceolate, entire ; inflorescence
narrow, with appressed branches ; involucre 4–5 mm. high ; bracts linear oblong,
obtuse ; achenes glabrous. Swamps: Newf.—Man.—Minn.—Pa.—S.C. *Canad.*
—Allegh. Jl–S.

7. **S. pállida** (Porter) Rydb. A perennial, with a short erect rootstock; stem 4–8 dm. high, glabrous; leaves obovate-spatulate to lanceolate, 5–15 cm. long, acute, entire, pale; heads very numerous in a dense, elongate-conic panicle; involucres about 5 mm. high; bracts oblong or linear-oblong, obtuse, glabrous; achenes glabrous or nearly so. *S. speciosa pallida* Porter [B]. Hills: Sask.—Minn.—Neb.—Colo. *Plain—Submont.* Au–S.

8. **S. speciòsa** Nutt. A perennial, with a rootstock; stem 5–20 dm. high, glabrous below, scabrous above; leaves thick, smooth, except the scabrous-ciliolate margins, the lower long-petioled, 1.5–3 dm. long, 5–10 cm. wide, dentate, the upper much smaller and sessile; heads numerous, in an elongated panicle; involucre 4.5–5.5 mm. high; bracts oblong; ligules 5, large; achenes glabrous. *S. conferta* Mill. (?). Rich soil: N.S.—Minn.—Kans.—Ark.—N.C. Au–O.

9. **S. rigidiúscula** (T. & G.) Porter. A perennial, with a thick rootstock; stem 6–10 dm. high, glabrous, or sparingly hispidulous above; leaves thick, glabrous except the strongly ciliolate margins, the lower ones 5–10 dm. long, 1–2.5 cm. wide, petioled, the upper sessile, gradually smaller; involucre 4.5–5 mm. high; bracts oblong; achenes glabrous. *S. speciosa angustata* T. & G. [G]. *S. Chandonnetii* Steele (?). Open ground: Ohio—Minn.—S.D.—Tex.—Ala. *Prairie—Ozark—Texan.* Au–O.

10. **S. rugòsa** Mill. A perennial, with a rootstock; stem stout, 3–20 dm. high, branched above; leaf-blades oval to lance-elliptic, acute, serrate, usually pubescent on both sides, 2–5 cm. long, 1–3.5 cm. wide, the lower petioled; inflorescence broad with recurved branches; heads 3–4 mm. high; bracts linear, obtuse. Dry soil: Newf.—Man.—Tex.—Fla. *E. Temp.* Jl–N.

11. **S. ulmifòlia** Muhl. A perennial, with a stoloniferous rootstock; stem 5–15 dm. high, branched above; branches puberulent; leaves thin, lanceolate or ovate, serrulate, acuminate, usually sparingly pubescent with long hairs beneath, the lower 7–12 cm. long, with margined petioles; the upper sessile; inflorescence with a few long spreading secund branches; ligules small, deep yellow. Woods and copses: N.S.—Minn.—Kans.—Tex.—Ga. *E. Temp.* Au–S.

12. **S. microphýlla** Engelm. A perennial, with a stoloniferous rootstock; stem 8–13 dm. high, glabrous, striate; leaves somewhat pubescent, thick, the lower spatulate, 6–10 cm. long, shallowly serrate, wing-petioled, the upper elliptic to oblong-ovate, acute or acuminate, slightly toothed or entire; inflorescence with long spreading branches; achenes pubescent. Dry soil: Kans.—Tex.—Ark. *Ozark.* Au–S.

13. **S. júncea** Ait. A perennial, with a horizontal rootstock; stem 5–12 dm. high, glabrous; leaves glabrous, serrate, the lower petioled, 1.5–4 dm. long, elliptic or broadly oblanceolate; the upper sessile, linear-lanceolate or oblong; 3–6 cm. long; inflorescence a flat-topped panicle with spreading branches; involucre 3–5 mm. high; bracts pale, oblong, blunt or acutish; achenes glabrous or nearly so; ligules 5–12, small. Copses and banks: N.B.—Sask.—Mo.—N.C. *Canad.—Allegh.* Je–N.

14. **S. glabérrima** Martens. A perennial, with a horizontal rootstock; stem 2–4 dm. high, glabrous up to the inflorescence; lower leaves narrowly oblanceolate, 3–15 cm. long, serrate or entire, glabrous, ciliolate on the margins; inflorescence flat-topped or round-topped, with curved, spreading branches; involucre about 5 mm. high; achenes sparingly hirsutulous. *S. missouriensis* A. Gray; not Nutt. [B, G]. *S. inornata* Lunell, a form with more open inflorescence. Plains and hills: Man.—Mich.—Mo.—Tex.—Ariz.—Ida. *Prairie—Plain—Mont.* Jl–S.

15. **S. missouriénsis** Nutt. A perennial, with a horizontal rootstock; stem slender, 2–4 dm. high, glabrous; leaves linear-lanceolate, more or less

triple-ribbed, glabrous, scabrous-ciliolate on the margins, 5–15 cm. long, acute, the lower petioled and sometimes serrate, the upper sessile and entire; panicle narrow, with ascending branches, scarcely secund; involucres 4–5 mm. high; bracts linear, acute, glabrous. *S. Tolmieana* A. Gray. Mountains and hills: Wash.—Ore.—Colo.—S.D.—Alta. *Submont.—Mont.* Jl–Au.

16. **S. moritùra** Steele. A perennial, with a running rootstock; stem assurgent below, glabrous throughout, 3–6 dm. high; leaves numerous, all narrowly linear-oblanceolate or linear-lanceolate, the lower 10–15 cm. long, the upper smaller, scabrous-ciliolate on the margins; inflorescence short, ovoid or pyramidal, with strongly recurved-spreading branches; involucres 4–5 mm. high; ligules 8–9; achenes strigose. *S. missouriensis fasciculata* Holz. Prairies: Ill.—Iowa—Kans.—Mo. *Prairie.* Jl–S.

17. **S. Hapemaniàna** Rydb. A perennial, with a horizontal rootstock; stem about 3 dm. high, glabrous, leafy; lower leaves oblanceolate, short-petioled, 4–7 cm. long, scabrous-ciliolate on the margins, otherwise glabrous, serrate, acute, the upper sessile, gradually becoming smaller; inflorescence a short corymbiform panicle; branches less than 4 cm. long, recurved-spreading, secund; involucre 5 mm. high; bracts imbricate, linear-lanceolate, glabrous, acute; achenes glabrous. Sand-hills: Kearney County, Neb. Jl.

18. **S. pulchérrima** A. Nels. A perennial, with a cespitose rootstock; stems 3–8 dm. high, striate, cinereous-puberulent; lower leaves narrowly oblanceolate, entire, 6–12 cm. long; upper leaves oblong or linear, all cinereous-puberulent; inflorescence narrow, often leafy; branches short and recurved-spreading; involucres 5–6 mm. high; achenes puberulent. *S. diffusa* A. Nels. Dry plains: Sask.—Minn.—Neb.—Colo.—Ariz.—Mont. *Prairie—Mont.* Jl–S.

19. **S. longipetiolàta** Mackenzie & Bush. A tufted perennial, with thick rootstock; stem 3–6 dm. high, finely puberulent; leaves 5–18 dm. long, 3–11 mm. wide, linear-oblanceolate, long-petioled, crenate, the upper smaller, entire, sessile, all finely puberulent, but green; inflorescence one-sided, usually short, either with short branches and a more elongate recurved top, or with several longer, recurved branches; involucre 4–5 mm. high; bracts oblong, thickened at the end, the inner ones often acutish; ligules 8–12; achenes puberulent. Rocky woods and barrens: Ill.—Minn.—Kans.—Mo. *Prairie.*

20. **S. nemoràlis** Ait. A perennial, with a thick rootstock, often forming clumps; stem 2–7 dm. high, canescent-puberulent; lower leaves broadly oblanceolate or spatulate, long-petioled, crenate, 5–15 cm. long, 3-ribbed, canescent-puberulent, the upper linear or linear-oblanceolate, reduced, entire, 1–3 cm. long; inflorescence one-sided, sometimes simple, with short branches and a racemiform, recurved top, sometimes more compound, with several long racemiform branches; involucre 3.5–5 mm. high; bracts yellowish, linear-oblong, obtuse; ligules 5–9. Dry open places, rarely in woods: N.S.—Man.—Ark.—Fla. *E. Temp.* Jl–N.

21 **S. trinervàta** Greene. A perennial, with cespitose rootstock; stem decumbent at the base, cinereous-puberulent, 3–6 dm. high; leaves oblanceolate, 5–10 cm. long, puberulent, scabrous-ciliolate on the margins, entire or nearly so; involucre about 6 mm. high; bracts lanceolate, acute; achenes hispidulous. Plains and hills: S.D.—N.M.—Ariz.—Utah. Jl–S.

22. **S. móllis** Bartl. A perennial, with a horizontal rootstock; stem 2–4 dm. high, cinereous-puberulent, strict, leafy; leaves obovate, oval, or broadly spatulate, or the upper elliptic, canescent, the lower more or less serrate; panicle dense and usually leafy; involucres 5–6.5 mm. high; bracts ovate or oblong, acute; achenes sparingly hirsutulous. *S. nemoralis incana* A. Gray. Dry plains: Man.—Tex.—Colo.—Mont. *Plain—Mont.* Jl–S.

23. S. serótina Ait. *Fig. 551.* A perennial,
with a stout rootstock; stem 5–20 dm. high, glabrous
up to the inflorescence; leaves lanceolate, 5–15 cm.
long, sessile, usually serrate, glabrous or nearly so;
inflorescence pyramidal, dense, with spreading,
curved branches; bracts linear or linear-lanceolate,
obtuse or acutish; achenes hirsutulous. *S. Pitcheri*
Nutt. Meadows, valleys, and banks: N.B.—Ga.—
Ark.—Colo.—B.C. *Temp.* Jl–S.

f.551.

24. S. Somèsii Rydb. A perennial, with a
rootstock; stem about 1 m. high, terete, purplish,
glabrous up to the inflorescence, leafy; leaves lanceolate or the lower oblanceolate, 5–10 cm. long, serrate, acuminate, glabrate above, appressed short-
hairy beneath, hispid-ciliolate on the margins; inflorescence elongate, leafy; branches short and
spreading, hispidulous; involucre about 3 mm. high; bracts linear, acutish;
ligules about 10; achenes sparingly hirsute. Valleys: Iowa. Au.

25. S. Lunéllii Rydb. A perennial, with a rootstock; stem 3–7 dm.
high, finely and softly puberulent, simple up to the inflorescence; leaves
3-ribbed, lance-linear or the lower linear-oblanceolate, finely puberulent, 4–10
cm. long, 5–10 mm. wide, acute; inflorescence short, with many recurved-
spreading, secund branches; involucres 3.5–4 mm. high; bracts narrowly linear-
lanceolate, puberulent; ligules about 15; achenes glabrous. Dry soil: N.D.—
Sask. Jl–S.

26. S. altíssima L. A perennial, with a rootstock; stem densely puberulent, 7–20 dm. high, grooved; leaves thickish, lanceolate, serrate or sometimes
entire, minutely puberulent above, short-pilose beneath, 5–15 cm. long, acute
or acuminate; inflorescence broadly pyramidal; involucre usually 3.5–4 mm.
high; bracts linear-lanceolate, acute; achenes sparingly hirsute. *S. canadensis
scabriuscula* Porter [B]. Rich open ground: Me.—Minn.—Kans.—Ga. *Canad.
—Allegh.* Au–O.

27. S. procèra Ait. A perennial, with a stoloniferous rootstock; stem
7–15 dm. high, mostly pilose; leaves lanceolate, entire or nearly so, puberulent
above, short-pubescent on the surface and densely pilose on the veins beneath,
5–8 cm. long; inflorescence large, often 2 dm. long and 1.5 dm. broad, rhomboidal; involucres 3.5–4 mm. high; bracts linear-lanceolate, acute; achenes
sparingly hirsute. *S. canadensis procera* T. & G. [B]. *S. perornata* Lunell.
S. altissima procera Fernald [G]. Rich ground: Mich.—N.D.—Kans.—Tenn.
—W.Va. *Allegh.* Au–N.

28. S. dumetòrum Lunell. A perennial, with a horizontal rootstock;
stem 5–10 dm. high, densely puberulent; leaves lanceolate, acute, thick, more or
less serrate, strongly veined, revolute and scabrous-ciliate on the margins,
scabrous above, densely hispidulous-puberulent beneath; inflorescence short-
conic, with short branches; involucres about 4 mm. high; bracts linear, acute;
achenes sparingly hispidulous. *S. scabriuscula* Rydb., in part. Dry soil: Sask.
—Mo.—Tex.—Colo.—Mont. *Prairie—Plain.* Jl–S.

29. S. canadénsis L. A perennial, with a horizontal rootstock; stem
6–15 dm. high, puberulent; leaves sessile, 5–10 cm. long, rarely 1.5 cm. wide,
narrowly lanceolate, acuminate, serrate, sparingly puberulent on both sides, or
glabrous above; inflorescence a broadly pyramidal panicle; bracts narrowly
linear, acute or obtusish; achenes sparingly hairy. Among bushes: Labr.—
W.Va.—Colo.—N.D.—Man. *Boreal.—Mont.* Jl–O.

30. S. gilvocanéscens (Rydb.) Smyth. A perennial, with a horizontal
rootstock; stem slender, 3–6 dm. high, puberulent; leaves narrowly lanceolate,

3–6 dm. long, sparingly serrate or entire, puberulent-canescent; inflorescence small and dense; bracts narrowly linear, acute; achenes sparingly hairy. *S. canadensis gilvocanescens* Rydb. [B, G]. Sandy soil: Man.—Kans.—Utah—Mont. *Prairie—Submont.* Jl–Au.

31. S. satánica Lunell. A perennial, with a rootstock; stem 5–8 dm. high, minutely puberulent, or scabrous above, glabrate below; leaves lanceolate, minutely appressed-pubescent or scabrous above, densely and softly cinereous-pubescent beneath, prominently 3-ribbed, serrate or the upper entire, 4–8 cm. long; inflorescence small, ovoid, with many small heads crowded on the branches, involucres 2.5–3 mm. high; bracts linear-lanceolate. Woods: Devils Lake, N.D.

32. S petiolàris Ait. A perennial, with a woody rootstock; stem 3–10 dm. high, puberulent; leaves oblong to oval or oblanceolate, mucronate, entire or nearly so, or the lower somewhat serrate, 1 7 cm long, roughish-puberulent, hispid-ciliolate on the margins; heads 0–7 mm high, in a narrow terminal thyrsus, leafy below; bracts lanceolate or lance-subulate, pubescent, the outer only squarrose. Dry woods: N.C.—Ill.—Kans.—Tex.—Fla. *Austral.* S–O.

33. S. Wárdii Britton. A perennial, with a thick, woody rootstock; stem 4–10 dm. high, puberulent; leaves lanceolate or ovate lanceolate, entire or with a few teeth, firm, 5–9 cm. long, pale, usually shining, scabrous-puberulent; panicle, narrow, leafy; heads about 8 mm. high; bracts lanceolate, acute, canescent, squarrose. *S. petiolaris Wardii* Fernald [G]. Plains and prairies: Mo.—Ark.—Tex.—w Neb. *Prairie—Plain—Texan.* Au–O.

20. EUTHÀMIA Nutt. BUSHY GOLDENROD.

Erect perennial herbs, with long rootstocks and narrow entire punctate leaves. Heads small, numerous in small clusters, forming convex or flat-topped corymbiform cymes, radiate. Involucres campanulate or turbinate; bracts imbricate, appressed, somewhat glutinous. Receptacles flat, fimbrillate. Ray-flowers pistillate, fertile; ligules small. Disk-flowers hermaphrodite, fertile. Anthers obtuse at the base. Style-branches with lanceolate appendages. Achenes broadened at the top, pubescent.

Inflorescence round-topped : inner bracts decidedly acute.	1. *E. occidentalis.*
Inflorescence flat-topped : inner bracts mostly obtuse.	
Leaves distinctly 3–5-ribbed ; ligules 10–20.	
Outer bracts ovate, acute, the inner oblong, straw-colored, scarcely green-tipped ; leaves broadly linear, scabrous on the margins and ribs.	2. *E. camporum.*
Outer bracts lanceolate or oblong, the inner linear or lance-linear, with more or less thickened green tips.	
Leaves and branches of the inflorescence hispidulous-pubescent.	3. *E. Nuttallii.*
Leaves glabrous, except the scabrous margins and ribs ; branches of the inflorescence also minutely scabrous.	4. *E. glutinosa.*
Leaves 1-ribbed or indistinctly 3-ribbed ; ligules 5–10.	
Leaves linear-lanceolate, not viscid, 4–6 mm. wide.	5. *E. leptocephala.*
Leaves linear, strongly punctate and glutinous.	6. *E. gymnospermoides.*

1. E. occidentális Nutt. Stem 5–20 dm. high, glabrous, striate; leaves linear, triple-veined, glabrous, obsoletely scabrous on the margins; bracts narrowly lanceolate, thin, ciliolate; achenes sparingly hairy. *Solidago occidentalis* Nutt. Moist ground: Alta.—S.D.—N.M.—Calif.—B.C. *Plain.* Je–N.

2. E. campòrum Greene. *Fig. 552.* Stem 3–5 dm. high, puberulent-scabrous on the angles above; leaves linear, triple-veined, but the lateral veins often glabrous or nearly so on both sides, scabrous-puberulent on the margins; outer bracts ovate, acutish, the inner oblong-linear, more or less green-tipped; achenes strigose. Closely related to *E. graminifolia* (L.) Nutt., of the East. Low ground: Man.—Kans.—Colo.—Alta. *Plain.* Jl–O.

3. **E. Nuttállii** Greene. Stem 3–5 dm. high, angled, scabrous-hispidulous in the inflorescence; leaves narrowly linear, 3-ribbed, 5–10 cm. long, 3–5 mm. wide, scabrous on both sides, especially on the margins and the ribs beneath; heads 4–5 mm. high, the bracts narrower and thicker than in the preceding; achenes strigose. *Solidago lanceolata* Am. auth., in part. *S. graminifolia Nuttallii* Fernald [G]. Meadows: N.S. — Va. — Ala. — Neb. — Wis. Au–O.

4. **E. glutinòsa** Rydb. Stem 3–6 dm. high, glabrous or nearly so, much branched above; leaves narrowly linear, 3–6 dm. long, 3–4 mm. wide, densely punctate and viscid, scabrous on the margins; involucres 6–7 mm. high, turbinate; bracts in about 5 series, the lowest lanceolate, acute, the inner oblong, obtusish, viscid, with a thickened green tip; ligules oblong, nearly 2 mm. long. Hills and prairies: Mo.—Kans.—Neb.—Minn.(?). S–O.

f. 552.

5. **E. leptocéphala** (T. & G.) Greene. Stem simple to near the summit, 3–6 dm. high, glabrous, angled; leaves linear-lanceolate, glabrous, except the scabrous margins, 4–6 cm. long, somewhat punctate; heads in small clusters, sessile or short-peduncled; involucre elongate-turbinate, 5–6 mm. high; bracts straw-colored, linear, obtuse. *S. leptocephala* T. & G. [G]. Moist sandy soil: Miss.—Mo.—Neb.—Tex. Au–O.

6. **E. gymnospermoìdes** Greene. Stem 3–7 dm. high, glabrous, angled, branched from below the middle; leaves linear, attenuate, 4–7 cm. long, glabrous except the scabrous margins; heads in small clusters; involucre turbinate, 5–6 mm. high; bracts straw-colored, with green viscid tips, linear to oblong, obtuse. *S. gymnospermoides* (Greene) Fern. [G]. Dry soil: Kans.—Miss.—Tex. Au–O.

21. OLIGONEÙRON Small. GOLDENROD.

Stout perennials with rootstocks. Leaves alternate, entire or crenate, the lower petioled. Heads radiate, in compound, corymbiform cymes. Involucre campanulate; bracts imbricate, in several series, thickish, with several parallel ribs. Receptacle pitted. Ray-flowers pistillate, fertile; ligules yellow. Disk-flowers hermaphrodite and fertile; corolla yellow, dilated above, 5-lobed. Anthers obtuse at the base. Style-branches with oblong appendages. Achenes few–several-nerved. Pappus of many capillary bristles.

Leaves broad and flat, not 3-ribbed, canescent.
  Plant green, very rough; achenes glabrous throughout; inflorescence ample.   1. *O. rigidum.*
  Plant yellowish-cinereous; inflorescence small, dense.
    Pubescence rather harsh, short and spreading; achenes hairy above.   2. *O. canescens.*
    Pubescence soft and dense, more appressed; achenes glabrous.   3. *O. bombycinum.*
Leaves conduplicate, 3-ribbed, nearly glabrous.   4. *O. Riddellii.*

1. **O. rígidum** (L.) Small. Rootstock thick, woody; stem densely rough-pubescent, 3–15 dm. high, very leafy; leaves thick, rigid, rough on both sides, the basal ones petioled, oblanceolate, 1–3 dm. long, crenate, the upper sessile, elliptic to oval, mostly entire; cyme mostly round-topped; involucre 6–8 mm. high; bracts oblong, rounded at the apex; ligules 6–10. *Solidago rigida* L. [G, B]. Dry ground: Ont.—Mass.—Ga.—Ark.—Kans.—Sask. *Canad.—Allegh.* Au–S.

2. O. canéscens Rydb. Rootstock stout, woody; stem 5–7 dm. high, stout, canescent; basal leaves long-petioled, oval, blades 5–7 cm. long, thick, canescent on both sides, crenulate; upper stem-leaves sessile, oval; inflorescence congested; involucre nearly 1 cm. high; bracts oblong, rounded at the apex, densely canescent. Prairies and river valleys: Sask.—Neb.—N.M.—Wash.— Alta. *Plain—Submont.—W. Temp.* Au–O.

3. O. bombýcinum Lunell. Stems several from a black fibrous-covered base, 2.5–3 dm. high, softly white-pubescent, very leafy; leaves oblong, thick, entire with soft velvety pubescence on both sides, the upper clasping; head clustered in a short oblong spike; involucres 6–8 mm. high; bracts oblong, puberulent. Moist soil: Butte, N.D. *Prairie.* S.

4. O. Riddéllii (Frank) Rydberg. Rootstock thick; stem glabrous or slightly pubescent above, 3–10 dm. high; leaves numerous, the upper linear-lanceolate, 1–15 dm. long, glabrous, except the scabrous margins, clasping, the basal ones long-petioled, narrowly oblanceolate, 2–3 dm. long, 8–20 mm. broad; involucre campanulate, 5 mm. high; bracts broadly oblong, obtuse, few-ribbed; ligules 7–9, narrow; achenes glabrous, 5-nerved. *Solidago Riddellii* Frank. [G, B]. Moist prairies: Ont.—Ohio—Mo.—Iowa—Minn. Au–S.

22. BÉLLIS L. DAISY.

Annual, biennial, or perennial herbs. Leaves alternate, or mostly basal. Heads solitary at the ends of the branches or of a leafless scape, radiate. Involucre hemispheric; bracts herbaceous, in 1 or 2 series, nearly equal. Receptacle convex or conic, naked. Ray-flowers pistillate, fertile; ligules white. Disk-flowers hermaphrodite and perfect; corollas yellow. Anthers obtuse at the base. Style-branches flattened, with short, triangular appendages. Achenes flattened. Pappus none or a ring of minute bristles.

1. B. integrifòlia Michx. A slender, diffusely branched, caulescent annual or biennial; stem 1–4 dm. high; lower leaves spatulate, 2–7 cm. long, with a winged petiole, upper leaves oblong to linear; heads at the ends of the branches; involucre 5–6 mm. high, 8–10 mm. broad; ligules white, 5–8 mm. long; bracts scarious-margined. Moist ground: Ky.—Tenn.—Tex. Kans. *Ozark.* My–Jl.

23. APHANOSTÈPHUS DC.

Annual or biennial, caulescent herbs. Leaves alternate, entire, toothed or pinnately lobed. Heads solitary at the ends of the branches, radiate. Involucre hemispheric; bracts narrow, imbricate, in a few series, scarious-margined. Receptacle convex or conic. Ray-flowers pistillate, fertile; ligules white or purplish. Disk-flowers hermaphrodite and fertile; corollas yellow. Anthers obtuse at the base. Style-branches with obtuse appendages. Achenes many-ribbed. Pappus an entire, toothed or ciliate crown.

1. A. skirróbasis (DC.) Trel. Stem erect, much branched, angled, pubescent, 2–6 dm. high; lower leaves spatulate or oblanceolate, obtuse, 2–10 cm. long, coarsely toothed or lobed, the upper linear or nearly so, entire; involucre fully 1 cm. broad; ligules white, 1 cm. long, narrow. *A. arkansanus* A. Gray. Dry soil: Kans.—Tex.—N.M. *Texan.* My–Au.

24. BOLTÒNIA L'Hér.

Perennial caulescent herbs. Leaves alternate, glabrous, narrow, entire, sessile. Heads numerous, corymbose-paniculate, radiate. Involucre broadly campanulate to hemispheric; bracts in few series. Receptacle convex or conic, pitted. Ray-flowers numerous, pistillate and fertile. Disk-flowers hermaphrodite and fertile. Anthers obtuse at the base. Style-branches flat, with short appendages. Achenes broadest above the middle, with winged or thickened margins. Pappus of a series of short squamellae and often 2–4 slender bristles.

Bracts acute, broadest below the middle.
Bracts obtuse or mucronate, broadest above the middle.

1. *B. asteroides.*
2. *B. latisquama.*

1. B. asteroìdes L'Hér. *Fig. 554.* A stoloniferous perennial; stem 1–2 m. high; leaves linear or linear-oblanceolate, 5–7 cm. long, acute, sessile; involucres 2.5–3 mm. high, 6–12 mm. broad; ligules white or pink. Low ground: N.J.—Man.—N.D.—La.—Fla. *E. Temp.* Jl–S.

2. B. latisquàma A. Gray. A stoloniferous perennial; stem much branched, 1–2 m. high; leaves linear, sessile, acute, 2–10 cm. long; involucres about 4 mm. high; ligules light violet-blue. River valleys: Minn.—N.D.—Kans.—Ark. Au–S.

25. TOWNSÉNDIA Hook.

Tufted scapose or caulescent perennials or biennials, often with a cespitose caudex. Leaves alternate, often crowded, entire, often broadest above the middle. Involucres hemispheric; bracts imbricate in several series, herbaceous, with more or less scarious margins; receptacle broad, flat alveolate. Ray-flowers pistillate, perfect; ligules, white, rose, or purple. Disk-flowers mostly perfect. Anthers obtuse and entire at the base. Style-branches with lanceolate appendages. Ray-achenes usually 3-angled, and disk-achenes flattened. Pappus of a single series of scales or awns, or that of the ray-flowers sometimes reduced to a crown of small scales.

Bracts acuminate; more or less caulescent biennials, with spreading branches from the base.
1. *T. grandiflora.*
Bracts acute or obtuse; acaulescent perennials, with the heads sessile among the leaves.
  Plant glabrate, cinereous only when young; leaves oblanceolate or spatulate.
2. *T. exscapa.*
  Plant permanently cinereous; leaves linear or linear-oblanceolate.
3. *T. sericea.*

1. T. grandiflòra Nutt. *Fig. 553.* A biennial, with a taproot; stem 4–16 cm. high, branched at the base, densely strigose; leaves narrowly linear-oblanceolate, strigose; involucres 10–15 mm. high, 15–20 mm. broad; bracts strigose, broadly lanceolate, acuminate, fimbriate; ligules rose-purple or pinkish; achenes hirsutulous; pappus of the ray-flowers reduced to a short crown. Plains and hills: S.D.—Okla.—Colo.—Wyo. Je–Au.

2. T. exscàpa (Richards.) Porter. Acaulescent cespitose perennial; leaves oblanceolate, mostly acute; involucre 12–15 mm. high, 1.2–2 cm. broad; bracts narrowly lanceolate, in 4 or 5 series; pappus of the ray- and disk-achenes of many bristles, about twice as long as the sparingly hirsute achenes. *T. intermedia* Rydb. Dry soil: Neb.—Kans.—Ariz.—Wyo. Alt. 1000–1500 m. Ap–My.

3. T. serìcea Hook. A cespitose, acaulescent perennial; leaves sericeous-strigose, 2–5 cm. long; involucres 1 cm. high, 1–3 cm. broad; bracts narrowly lanceolate, in 4 or 5 series; achenes hirsute; pappus of the ray-flowers of many bristles, often shorter than those of the disk-flowers, the latter twice as long as the achenes. *T. mensana* M. E. Jones. Plains: Man.—S.D.—Tex.—N.M.—Alta. Alt. 1000–2500 m. Ap–Je.

26. BRACHYÁCTIS Ledeb.

Leafy-stemmed annuals with entire leaves. Heads small, radiate, numerous, paniculate. Involucre campanulate; bracts in 2 or 3 series, linear or oblong, almost equal in length, the outer foliaceous and loose. Ray-flowers pistillate, fertile, in several series, more numerous than the disk-flowers; corollas white, shorter than the pappus, with short or no ligules. Disk-flowers hermaphrodite and fertile, with slender corollas. Anthers obtuse at the base. Style-branches with lanceolate appendages. Achenes narrow, not compressed, 2- or 3-nerved, appressed-pubescent. Pappus a single series of nearly white bristles.

Bracts oblong or oblong-linear, obtuse.   1. *B. frondosa.*
Bracts narrowly linear, acute.   2. *B. angusta.*

1. B. frondòsa (Nutt.) A. Gray. An annual; stem 2–5 dm. high, glabrous or nearly so; leaves linear or lance linear, 3–8 cm. long, glabrous, ciliate on the margins; involucre about 8 mm. high, fully 1 cm. broad; bracts in 3 distinct series; ligules 2 mm. long, pinkish. *Aster frondosus* Nutt. [G]. Saline soil and along streams: Wyo.—Colo.—Calif.—Ore. P. E. I. *W. Temp.* Au–O.

2. B. angústa (Lindl.) Britton. An annual; stem erect, branched, glabrous, 2–6 dm. high; leaves narrowly linear, 3–10 cm. long, glabrous, slightly ciliolate on the margins; involucres 6–8 mm. high, 8–10 mm. broad; bracts in 2 series, almost equal; ligule rudimentary or none. *Aster angustus* (Lindl.) T. & G. [G, B]. Saline soil: Que.—Man.—Mo.—Utah—Alta. *Canad.—Submont.* Jl–S.

27. XYLORRHÌZA Nutt.

Perennial caulescent herbs, with thick woody taproots and short caudices. Leaves many, spinulose tipped, entire or spinulose-toothed. Heads large, solitary at the ends of the branches, radiate. Involucres hemispheric; bracts mostly in 2 or 3 series, acuminate, herbaceous, carinate below. Ray-flowers pistillate; ligules showy, white or pinkish. Disk-flowers perfect. Appendages of the style-branches lanceolate, acute. Achenes oblong, compressed. Pappus fulvous, simple, of rather few coarse bristles.

1. X. glabriúscula Nutt. *Fig. 556.* Stems 1–2 dm. high, glabrous; leaves linear-oblanceolate, 2–5 cm. long, glabrous; involucres 8–10 mm. high, 10–15 mm. wide; bracts linear-lanceolate, acuminate, glabrous, erose on the margins; ligules rose-colored, 10–12 mm. long. Alkali plains: Wyo.—S.D. *Plain.* Je–Jl.

f. 556.

28. UNÀMIA Greene. SNEEZEWORT ASTER.

Perennial caulescent herbs, with horizontal rootstocks. Leaves alternate, narrow, entire. Involucre turbinate-campanulate; bracts narrow, not foliaceous, more or less keeled. Heads small, corymbose, radiate. Ray-flowers pistillate, fertile; ligules white or ochroleucous. Disk-flowers hermaphrodite and perfect; corollas yellow. Style-branches with lance-subulate appendages. Achenes 4- or 5-angled, not compressed, glabrous. Pappus of white rigid bristles, somewhat clavate at the end.

Ligules white: involucral bracts acute or the inner obtusish.   1. *U. alba.*
Ligules ochroleucous; outer involucral bracts obtuse or acutish,
    the inner ones rounded or at least obtuse at the apex.   2. *U. lutescens.*

1. U. álba (Nutt.) Rydb. Stem 2–5 dm. high, smooth or scabrous; lower leaves linear-oblanceolate, 3–5-ribbed, 5–15 cm. long, firm, the upper linear;

involucre 5 mm. high, 5–8 mm. broad; outer bracts lance-subulate, the inner ovate-lanceolate; ligules 5–6 mm. long. *Inula alba* Nutt. *Aster ptarmicoides* T. & G. [B, G]. Rocky banks and bluffs: Vt.—N.Y.—Colo.—Sask. *Boreal.— Submont.* Jl–S.

2. **U. lutéscens** (Lindl.) Rydb. Stem 2–4 dm. high, scabrous, at least above; leaves similar to those of preceding species; involucre 5–6 mm. high, 6–8 mm. broad; outer bracts lanceolate, the inner oblong or elliptic; ligules about 8 mm. long. *Diplopappus lutescens* Lindl. *Aster ptarmicoides lutescens* A. Gray [B, G]. Banks and plains: Ill.—Wis.—Sask.—Man. *Prairie—Plain.* Au–S.

29. ÁSTER L. ASTER.

Usually perennial, seldom annual, leafy herbs. Leaves alternate, petioled, sessile or clasping, entire or toothed. Heads usually several or many, radiate. Involucre hemispheric, campanulate, or turbinate; bracts imbricate in several series, at least with herbaceous tips, not carinate. Receptacle flat or convex, alveolate. Ray-flowers pistillate, fertile; ligules elongate, blue, purple, pink, violet, or white. Disk-flowers hermaphrodite and fertile; corolla with a narrow cylindric tube, a campanulate throat, and 5 short teeth, red, brown, and purple. Anthers obtuse at the base. Style-branches flattened, with subulate or lanceolate, or ovate-acute appendages. Achenes more or less flattened, nerved or ribbed, pubescent or glabrous. Pappus of numerous capillary bristles, mostly in a single series.

Lower leaf-blades cordate or at least ovate, long-petioled, usually more or less serrate.
  Upper leaves not with cordate-clasping winged petioles.
    Bracts and branches of the inflorescence glandular-pubescent; ligules violet or purple.   1. MACROPHYLLI.
    Bracts and branches of the inflorescence not glandular.
      Ligules white; achenes narrow.   2. DIVARICATI.
      Ligules blue or purplish; achenes decidedly flattened.
        Bracts of the involucre recurved-spreading, pubescent.
                                                                     3. ANOMALI.
        Bracts of the involucre appressed, glabrous (except in *A. Finkii*).
                                                                     4. SAGITTIFOLII.
  Upper leaves with winged petioles, which are cordate-clasping at the base.
                                                                     5. UNDULATI.
Lower leaf-blades neither cordate at the base nor long-petioled.
  Involucral bracts and peduncles glandular.
    Leaves oval, ovate, or the lower obovate, sharply and coarsely serrate; bracts broadly lanceolate, thick, well imbricate.   6. CONSPICUI.
    Leaves lanceolate or linear, entire or slightly toothed; bracts linear-lanceolate, almost equal in length.
      Lower leaves petioled; bracts lanceolate, firm, ascending; stem-leaves linear.   7. PAUCIFLORI.
      Lower leaves as well as the upper sessile.
        Bracts narrowly linear, almost equal in length, attenuate.
                                                                     8. MODESTI.
        Bracts thicker, oblanceolate, well imbricate; leaves firm.
          Stem-leaves sessile, slightly, if at all, clasping.   9. OBLONGIFOLII.
          Stem-leaves deeply cordate-clasping.   12. PATENTES.
  Involucral bracts and peduncles not glandular.
    Bracts bristle-tipped, pubescent on the back.   10. MULTIFLORI.
    Bracts not bristle-tipped.
      Outer bracts neither foliaceous nor exceeding the inner.
        Bracts more or less pubescent on the back.
          Leaves silvery-white.   11. SERICEI.
          Leaves not silvery.
            Upper leaves cordate-clasping and auricled at the base.
                                                                    12. PATENTES.
            Upper leaves obovate to lanceolate, neither cordate-clasping nor auricled at the base.
              Bracts lanceolate, wholly foliaceous, often tinged with purple.
                                                                    13. RADULINI.
              Bracts oblanceolate or linear, with a rhombic green tip.
                                                                    14. GRISEI.
      Bracts glabrous or merely ciliolate on the margins.

Involucre densely elongate-turbinate; bracts green, imbricate in 5 or 6 series, rounded on the back.   15. TURBINELLI.
Involucre hemispheric, campanulate or broadly turbinate; bracts not rounded on the back.
  Plant perfectly glabrous or nearly so.
    Bracts linear-subulate, without distinct green tips; leaves linear, not clasping; plant annual.   16. EXILES.
    Bracts whitish-coreaceous below, with distinct subrhombic green tips; upper leaves clasping; plant perennial.
      17. LAEVES.
  Plant at least with pubescent decurrent lines on the upper part of the stem and branches.
    Bracts linear, oblong or lanceolate, all acute.
      Leaves not clasping.   18. SALICIFOLII.
      Leaves tapering to the clasping base.   21. PUNICEI.
    Bracts oblanceolate, the outer ones obtuse.
      19. ADSCENDENTES.
Outer bracts foliaceous, equaling or exceeding the inner.
  Bracts pubescent on the back; stem densely pubescent all over.
    20. NEBRASKENSES.
  Bracts glabrous or merely with ciliolate margins.
    Disk 8 mm. high or more; inflorescence pyramidal.
      21. PUNICEI.
    Disk 5-7 mm. high; inflorescence round-topped.
      Leaves lanceolate or oblong, clasping.   22. PHYLLODES.
      Leaves linear or lance-linear, not clasping.   23. FULCRATI.

1. MACROPHYLLI.

One species.   1. A. macrophyllus.

2. DIVARICATI.

One species.   2. A. divaricatus.

3. ANOMALI.

One species.   3. A. anomalus.

4. SAGITTIFOLII.

Leaves nearly all distinctly serrate, thin; ligules pale (except in A. hirtellus).
  Bracts distinctly pubescent; leaf-blades deeply cordate, densely pubescent beneath.   4. A. Finkii.
  Bracts glabrous or the margins ciliolate.
    Stem and leaves densely and finely short-pubescent throughout.   5. A. Drummondii.
    Stem not finely pubescent throughout, usually hairy only along decurrent lines; leaves, if densely pubescent, with rather coarse hairs.
      Involucre 5-10 mm. high; bracts acute or acuminate.
        Heads numerous, paniculate; involucre 6-8 mm. high; lower stem-leaves distinctly cordate.
          Inflorescence racemose-paniculate.
            Lower stem-leaves lance-cordate, crenate, gradually acute, sparingly hairy or glabrate on both sides.   6. A. sagittifolius.
            Lower stem-leaves ovate-cordate, serrate, rather densely short-pubescent beneath, scabrous above.   7. A. hirtellus.
          8. A. Saundersii.
          Inflorescence corymbose-paniculate.
        Heads fewer; involucre about 1 cm. high; lower stem-leaves mostly ovate.
          Petioles, midribs, and stem glabrous or short-hairy.   9. A. Lindleyanus.
          Petioles, midribs, and usually also the stem with long white hairs.   10. A. Wilsonii.
      Involucre 4-6 mm. high; bracts obtuse or merely acutish.   11. A. cordifolius.
Leaves, at least the upper ones, entire or nearly so, thick and firm; ligules dark.
  Leaves glabrous or nearly so above.   12. A. Shortii.
  Leaves rough-pubescent on both sides.   13. A. azureus.

5. UNDULATI.

One species.   14. A. undulatus.

6. CONSPICUI.

One species.   15. A. conspicuus.

7. PAUCIFLORI.

One species.   16. *A. pauciflorus.*

8. MODESTI.

Stem not hispid; heads distinctly peduncled; leaves thin,
slightly, if at all, auricled at the base.   17. *A. major.*
Stem short-hispid; heads crowded, often subsessile, sub-
tended by bract-like leaves; leaves thick, strongly auri-
cled at the base.   18. *A. Novae-Angliae.*

9. OBLONGIFOLII.

Stem-leaves sessile, slightly, if at all, clasping.
 Leaves pubescent, mostly spreading or reflexed.
  Leaves linear-oblong or linear-lanceolate, hirsutu-
   lous, not rigid; plant divaricately branched.   19. *A. oblongifolius.*
  Leaves oblong, very scabrous-hispidulous, plant fas-
   tigiately branched.   20. *A. Kumleini.*
 Leaves glabrous, except the hispid-ciliolate margins.   21. *A. Fendleri.*
Stem-leaves deeply cordate-clasping.   31. *A. patens.*

10. MULTIFLORI.

Pubescence of the stem and branches spreading or reflexed,
 hirsutulous.
 Ligules white.
  Heads 6–8 mm. high; bracts spatulate.   22. *A. crassulus.*
  Heads less than 5 mm. high; bracts oblanceolate.   23. *A. exiguus.*
 Ligules sky-blue or pink; bracts linear-oblanceolate.   24. *A. amethystinus.*
Pubescence of the stem and branches appressed or ascend-
 ing.
 Bracts very unequal in length, well imbricated, the
  outer shorter.
  Bracts narrowly oblanceolate, especially the inner,
   acutish; heads less than 5 mm. high.
   Ligules white.
    Stem much branched; heads racemosely dis-
     posed on the more or less diverging
     branches.   25. *A. ericoides.*
    Stem simple below; heads few at the ends of
     the erect branches.   26. *A. stricticaulis.*
                                                                27. *A. Batesii.*
   Ligules sky-blue.
  Bracts broadly oblanceolate or the outer spatulate,
   very obtuse.   28. *A. polycephalus.*
 Bracts almost equal in length, or the outer sometimes
  longer; heads few or solitary at the ends of the
  branches.   29. *A. commutatus.*

11. SERICEI.

One species.   30. *A. sericeus.*

12. PATENTES.

Bracts broad, oblong; branches divaricate.
 Bracts with spreading tips, slightly glandular-puberulent. 31. *A. patens.*
 Bracts appressed, merely strigose.   32. *A. patentissimus.*
Bracts narrow, linear; branches ascending-spreading.   33. *A. tenuicaulis.*

13. RADULINI.

One species   34. *A. meritus.*

14. GRISEI.

One species.   35. *A. Woldeni.*

15. TURBINELLI.

One species.   36. *A. turbinellus.*

16. EXILES.

One species.   37. *A. exilis.*

17. LAEVES.

Green tips of the bracts broadly rhombic; leaves of the
 branches much reduced.   38. *A. laevis.*
Green tips of the bracts narrowly rhombic or rhombic-
 oblanceolate; leaves of the branches gradually but not
 conspicuously reduced.   39. *A. Geyeri.*

18. SALICIFOLII.

Heads on strictly secund branches.
 Stem glabrate or slightly pubescent.

Cauline leaves linear or lance-linear, minutely hack-
toothed or entire.   40. *A. vimineus.*
Cauline leaves oblong or lanceolate, serrate.   41. *A. lateriflorus.*
Stem villous-hirsute.   42. *A. hirsuticaulis.*
Heads scattered, not secund.
Heads small; involucre 4–5 mm. high, rarely 6 mm.
high, 4–8 mm. broad.
Plant puberulent or pubescent all over.
Inflorescence racemose-paniculate; ligules white.   43. *A. missouriensis.*
Inflorescence corymbiform-paniculate; ligules dark
purple.   35. *A. Woldeni.*
Plants glabrous or nearly so, except the upper part
of the stem.
Stem-leaves narrowly linear-lanceolate to subu-
late, entire.
Involucre hemispheric, 5–6 mm. high, 7–8 mm.
broad.   44. *A. glabellus.*
Involucre somewhat turbinate, acute at the
base, about 4 mm. high and 4–5 mm. broad   45. *A. parviceps.*
Stem-leaves narrowly lanceolate, often toothed.
Inflorescence open, with long leafy branches.   46. *A. Tradescanti.*
Inflorescence narrow, with short branches.   47. *A. Jacobaeus.*
Heads larger; involucre 5–8 mm. high and 8–10 mm.
broad.
Stem-leaves lanceolate, mostly somewhat toothed.
Ligules mostly bluish or purplish; leaves thickish,
firm.
Bracts with acute tips.   48. *A. salicifolius.*
Bracts attenuate at the tip.   51. *A. fluviatilis.*
Ligules mostly white; leaves thinner.
Leaves narrowly lanceolate, blunt-toothed or
entire.   49. *A. paniculatus.*
Leaves broadly lanceolate, sharply toothed, at
least above the middle.   50. *A. acutidens.*
Stem-leaves linear, entire, mostly somewhat clasping.
Bracts in 3 or 4 series, the outer shorter.
Ligules purple.   51. *A. fluviatilis.*
Ligules white.
Heads usually many, in an elongate pan-
icle.
Disk about 1 cm. high and broad.
Bracts not with white midrib;
leaves narrowly linear.   52. *A. Oosterhoutii.*
Bracts with white midrib; leaves
lanceolate or oblong-lanceolate.   53. *A. laetevirens.*
Disk 6–8 mm. high and broad.   54. *A. longulus.*
Heads fewer, in a corymbiform panicle.   55. *A. junciformis.*
Bracts in 2 series, almost equal.   56. *A. longifolius.*

19. ADSCENDENTES.

Lower leaves narrowly linear-oblanceolate.   57. *A. Nuttallii.*
Lower leaves oblanceolate.   58. *A. adscendens.*

20. NEBRASKENSES.

One species.   59. *A. nebraskensis.*

21. PUNICEI.

Stem-leaves abruptly contracted into a winged petiole, which
is dilated at the base.   60. *A. prenanthoides.*
Stem-leaves cordate-clasping.
Stem more or less hispid, often purplish; leaves dark
green above.
Outer bracts narrowly linear, caudate-attenuate, minutely ciliolate; inflorescence ample.   61. *A. puniceus.*
Outer bracts broadly linear, merely acute, hispid-ciliate; inflorescence narrow.   62. *A. Forwoodii.*
Stem not hispid, light green, rather softly pubescent
along decurrent lines; bracts linear, acute, glabrous, scarcely ciliate; leaves light green.
Leaves scabrous-hispidulous above; those of the
branches lanceolate; bracts mostly green, gradu-
ally becoming paler at the base.   63. *A. lucidulus.*
Leaves bluish green, glabrous, except the ciliolate
margins, those of the branches more or less
cordate-sagittate; bracts with well marked
oblanceolate green tips.
Inflorescence open, much branched; leaves firm,
the upper ones broad at the base.   64. *A. laeviformis.*

Inflorescence narrow, with few heads; leaves thin,
the upper ones broadest near the middle, nar-
rower at the base.   65. *A. clivorum.*

22. PHYLLODES.

Bracts broadly oblanceolate; upper leaves slightly auricled.   66. *A. phyllodes.*
Bracts linear-oblanceolate; auricles of the upper leaves
large, rounded, extending around the stem.   67. *A. oticus.*

23. FULCRATI.

One species.   68. *A. Mearnsii.*

1. A. macrophýllus L. A perennial, with a branched rootstock; stem
4–9 dm. high, angled, often red; basal leaves several, long-petioled; blades
broad, cordate, coarsely toothed, acuminate, 7–15 cm. long, rough above; lower
stem-leaves similar, but smaller, the upper oblong, with short winged petioles,
or the uppermost sessile; inflorescence corymbiform; disk 10–15 mm. broad;
ligules 12–16, mostly lavender, 10–14 mm. long. Shaded places: N.B.—N.C.—
Minn. *Canad.—Allegh.* Au–S.

2. A. divaricàtus L. A perennial, with a tufted rootstock; stem 4–6
dm. high, flexuose, terete, glabrous; basal leaves long-petioled; blades thin,
smoothish, elongate-cordate, sharply dentate, 5–10 cm. long, acuminate; stem-
leaves similar, but short-petioled, those of the inflorescence sessile, ovate, nearly
entire; inflorescence corymbiform; involucre 5 mm. high, about 8 mm. broad;
bracts lanceolate, ciliate on the margins; ligules 6–9, nearly 1 cm. long. Open
woods and thickets: Me.—Ga.—Tenn.—Man. *Canad.—Allegh.* Au–O.

3. A. anómalus Engelm. A perennial, with a thick rootstock; stem 3–9
dm. high, rough; basal and lower stem-leaves long-petioled; blades ovate-cordate, entire or repand, rough on both sides, acuminate, 7–10 cm. long, 2–5 cm.
wide, the upper ones lanceolate to linear, with short winged petioles; involucre
turbinate-hemispheric, about 6 mm. high and 10 mm. broad; bracts lanceolate,
attenuate; ligules 30–45, violet, 10–12 mm. long. Limestone cliffs: Ill.—Iowa
—Kans.—Ark. *Ozark.* S–O.

4. A. Fínkii Rydb. A perennial, with a rootstock; stem 3–4 dm. high,
finely and densely short-pilose, paniculately branched above; leaves, except
those of the inflorescence, petioled, the petioles nearly equaling the blades in
length; the blades deeply cordate, 5–7 cm. long, 2.5–4 cm. wide, coarsely serrate, abruptly short-acuminate, short-pubescent above, rather densely pilose
beneath, the basal auricles rounded; panicle dense; involucre 5–6 mm. high;
bracts linear, acute, the oblanceolate green tips strigose; ligules apparently
pale purple or rose, 6 mm. long. Woods: Iowa. *Allegh.* S.

5. A. Drummóndii Lindl. A perennial, with a thick rootstock; stem
6–15 dm. high, finely canescent; blades of the lower leaves thick, ovate-lanceolate, cordate at the base, sharply toothed, 5–10 cm. long, acuminate, rough
above, canescent-puberulent beneath; upper stem-leaves lanceolate, cordate or
rounded at the base, with winged petioles, those of the branches sessile; involucre 6–7 mm. high, 8–10 mm. broad; bracts linear-lanceolate, with oblanceolate green tips; ligules 8–15, purplish blue, 6–8 mm. long. Prairies and
borders of woods: Ohio—Tenn.—Ark.—Tex.—Minn. *Prairie—Texan.* Au–O.

6. A. sagittifòlius Willd. A perennial, with a thick rootstock; stem
slender, 6–15 dm. high, glabrous or sparingly pubescent above, paniculately
branched; blades of the lower leaves thin, glabrous above, glabrate or slightly
roughened, pubescent beneath, ovate-lanceolate or lanceolate, cordate or sagittate at the base, sharply serrate, acuminate, 7–15 cm. long, the petioles narrowly if at all margined; upper stem-leaves lanceolate, sessile or with short
winged petioles, serrate or entire, those of the inflorescence linear-subulate;
heads many, racemose-paniculate; involucre turbinate; bracts linear-subulate,
attenuate, green-tipped, loose; ligules 10–15, light blue or purple, 6–8 mm.
long. Dry soil: N.B.—N.C.—Kans.—N.D. Au–O.

7. **A. hirtéllus** Lindl. A perennial, with a rootstock; stem 5–10 dm. high, glabrous below, with decurrent pubescent lines and paniculately branched above; petioles 2–7 cm. long, often slightly winged; leaf-blades ovate in outline, more or less cordate at the base, coarsely serrate at the middle, acuminate and entire towards the apex, scabrous-hispidulous above, more or less pilose beneath; heads numerous; involucre about 6 mm. high; bracts linear-lanceolate, acute, the green tips oblanceolate or slightly rhombic, glabrous; ligules 10–15, purplish, 5–6 mm. long. Rich woods: Wis.—Minn.—Mo. *Allegh.* S–O.

8. **A. Saundérsii** Burgess. A perennial, with a branching rootstock; stem 3–6 dm. high, glabrous except the puberulent branches of the inflorescence; petioles of the basal leaves marginless, those of the lower stem-leaves winged; blades thin, ovate-cordate, 6–10 cm. long; those of the upper leaves lanceolate, often somewhat clasping; involucre turbinate, about 6 mm. high and 10 mm. broad; ligules bluish purple, 6 mm. long. Woodlands and along streams: N.D.—S.D.—Iowa. *Prairie.* Jl–Au.

9. **A. Lindleyànus** T. & G. A perennial, with a horizontal rootstock; stem 3–10 dm. high, crisp-hairy in lines above; blades cordate, 5–15 cm. long, rather firm; middle leaves with ciliolate winged petioles, with clasping bases, the uppermost leaves lanceolate, sessile; involucre 7–8 mm. high, about 1 cm. broad; bracts linear-subulate, with a green oblanceolate midrib; ligules bluish purple, 10–12 mm. long; disk-corollas reddish purple. River banks: Lab.—N.H.—Ohio—Wyo.—Yukon. *Boreal.—Mont.*

10. **A. Wilsònii** Rydb. A perennial, with a horizontal rootstock; stem 3–6 dm. high, often purplish, more or less pubescent with long white hairs; lower leaf-blades 5–10 cm. long, usually more or less hirsute on both sides, glabrate, serrate, acuminate, the upper lanceolate, sessile, subentire; involucre 7–8 mm. high, scarcely 1 cm. broad; bracts subulate, attenuate, with green oblanceolate midribs; ligules 8–10 mm. long, bluish purple. River valleys: Mackenzie, B.C.—w Ont. *W. Boreal.* Au–S.

11. **A. cordifòlius** L. A perennial, with a thick rootstock; stem 3–15 dm. high, glabrous or nearly so, much branched; blades of the lower leaves thin, broadly cordate, 5–12 cm. long, acuminate, sharply serrate, more or less pubescent; upper leaves ovate or lanceolate, sessile or short-petioled; involucre turbinate, about 6 mm. high; bracts green-tipped; ligules 10–20, blue or violet, 6–8 mm. long. Woods and thickets: N.B.—Ga.—Mo.—Man. *Canad.—Allegh.* Au–O.

12. **A. Shórtii** Hook. A perennial, with a thick rootstock; stem 6–12 dm. high, more or less rough, paniculately branched; blades of the lower leaves thick, lanceolate-cordate, entire, glabrous above, finely pubescent beneath, 5–15 cm. long, the upper ones lanceolate, tapering below into winged petioles, those of the branches very small; involucre about 6 mm. high and 8 mm. broad; ligules 10–15, violet, 10–12 mm. long. Banks and open woods: Pa.—Ga.—Tenn.—Iowa. *Allegh.* Au–O.

13. **A. azùreus** Lindl. A perennial, with a thick rootstock; stem 8–12 dm. high, rough, branched above; blades of the lower leaves ovate- or lanceolate-cordate, thick, entire or slightly crenate, mostly acute, 5–15 cm. long, upper leaves lanceolate to linear, sessile or short-petioled; heads numerous; involucre 6–7 mm. high, about 10 mm. broad, turbinate; bracts with rhombic green tips; ligules 10–20, bright blue, 6–8 mm. long. Prairies and copses: N.Y.—Ga.—Tex.—Minn. *Allegh.* Au–O.

14. **A. undulàtus** L. A perennial, with a branched rootstock; stem stiff, 3–10 dm. high, rough-pubescent, divaricately branched above; basal leaves long-petioled, orbicular or ovate, soft-downy; lower cauline leaves ovate-cordate, 5–12 cm. long, serrate, the petioles with cordate bases; upper stem-leaves ovate

or lanceolate, with broadly winged petioles with cordate-clasping bases, those of the branches lanceolate or subulate, reduced; involucre turbinate, 5 mm. high, 8 mm. broad; bracts lanceolate, with rhombic-lanceolate pubescent green tips; ligules 8–15, pale violet, 6–10 mm. long. Dry soil: Ont.—Fla.—Ala.—Ark.—Minn. *E. Temp.* S–O.

15. **A. conspícuus** Lindl. A perennial, with a horizontal rootstock; stem 4–6 dm. high, scabrous-puberulent; leaves 1–1.5 cm. long, acute or short-acuminate, scabrous above, puberulent beneath; inflorescence corymbiform; involucre broadly campanulate, 1 cm. high, 1–1.5 cm. broad; ligules violet, 1 cm. long. Hillsides, open woods: Sask.—S.D.—Wyo.—Wash.—B.C. *W. Temp.—Mont.* Jl–S.

16. **A. pauciflòrus** Nutt. A perennial, with a slender creeping rootstock; stem 1.5–5 dm. high, branched, glabrous; leaves somewhat fleshy, entire, glabrous, the basal ones petioled, linear-oblanceolate, 3–10 cm. long, the upper linear, sessile; inflorescence flat-topped; heads few; involucre 7–8 mm. high, bracts in 2 or 3 series, linear-lanceolate, acute; ligules blue or white, 5 mm. long. *A. carnosus* M. E. Jones. Saline soil: Man.—Tex.—Ariz.—Utah. *Prairie—Plain—Son.* My–Au.

17. **A. màjor** (Hook.) Porter. A perennial, with a horizontal rootstock; stem 5–10 dm. high, more or less long-villous; leaves 5–10 cm. long, lanceolate, acuminate or attenuate, thin, entire or minutely denticulate; inflorescence flat-topped; involucre 1 cm. high, 12–15 mm. broad; ligules bluish purple, fully 1 cm. long. *A. modestus* Lindley [G]. Moist woods: w Ont.—Ore.—Alaska. *W. Boreal.* Au–S.

18. **A. Nòvae-Ángliae** L. *Fig. 557.* A perennial, with a horizontal rootstock; stem 4–25 dm. high, hirsute; leaves lanceolate or oblong, 3–12 cm. long, entire, acute, firm, hispidulous on both sides; inflorescence leafy, much branched, flat-topped; involucre 8–10 mm. high, about 15 mm. broad; ligules reddish purple or rose-colored, nearly 1 cm. long. Low ground: Que.—S.C.—Colo.—Sask. *Canad.—Allegh.* Au–O.

19. **A. oblongifòlius** Nutt. A perennial, with a long rootstock; stem 3–7 dm. high, hirsute; leaves divaricate, sessile, often slightly clasping, entire, hirsutulous on both sides, 3–5 cm. long, 4–8 mm. wide, those of the branches smaller; involucres hemispheric, 6 mm. high, 10 mm. broad; bracts linear or linear-oblong, spreading; ligules 20–30, violet-purple or pink, 6–10 mm. long. Prairies: Pa.—Va.—Tenn.—Tex.—Minn. *Allegh.—Prairie.* Au–O.

20. **A. Kumleìni** Fries. A perennial, with a horizontal branched rootstock; stem 2–5 dm. high, branched above, often yellowish or straw-colored, scabrous-puberulent; leaves rigid, oblong, sessile, 2–3 cm. long, those of the branches reduced and more spreading; involucre 6–7 mm. high, 8–10 mm. broad; bracts in 3 or 4 series, linear-oblanceolate, squarrose; ligules bluish violet, 7–8 mm. long. *A. oblongifolius rigidulus* A. Gray [G]. Plains: Wis.—Mo.—Tex.—Colo.—S.D. *Prairie—Plain.* Au–O.

21. **A. Féndleri** A. Gray. A perennial, with a cespitose rootstock or caudex; stem 1–3 dm. high, stiff, sparingly hirsutulous; leaves linear, 1-nerved, 2–3 cm. long; heads few; involucre turbinate, 7–8 mm. high, fully 1 cm. broad; bracts acute, lanceolate, in 3 or 4 series; ligules violet, 8 mm. long. Plains and sand-hills: Kans.—Tex.—N.M.—Colo. *Plain—Submont.* S.

22. A. cràssulus Rydb. A perennial, with a horizontal stoloniferous rootstock; stem erect or ascending, branched above, 4–7 dm. high; stem-leaves linear or oblong-linear, 3–6 cm. long, 2–3 mm. wide, sessile and slightly clasping, those of the branches only 3–5 mm. long; heads numerous, borne at the ends of short leafy branches; involucre 5–8 mm. high, 8–10 mm. broad; ligules numerous, white, 5–7 mm. long. *A. hebecladus* A. Nels.; not DC. Plains: N.D.—Colo.—Calif.—Ida. *Plain—Mont.* Jl–O.

23. A. exíguus (Fern.) Rydb. A perennial, with a creeping rootstock; stem erect, branched and very leafy above, with divergent branches; leaves linear or oblong, sessile, 0.5–3 cm. long, entire, densely hirsutulous; heads numerous, racemosely arranged, at the ends of short leafy branches; involucre about 4 mm. high and broad; ligules 3–5 mm. long. *A. multiflorus exiguus* Forn. [(+]. Prairies and plains: Vt.—Pa.—Tex.—Ariz.—Wash.; also Mex. *Temp.—Submont.* Au–O.

24. A. amethýstinus Nutt. A perennial, with a branched rootstock; stem strict, 5 dm. high or more, densely short hirsute, with ascending branches; leaves linear, 1–4 cm. long, 3–5 mm. wide, sessile and somewhat clasping, hirsutulous on both sides, acute and minutely spinulose-tipped; heads subracemose on leafy branches; involucre turbinate, 4–5 mm. high, 6–8 mm. broad; bracts narrowly oblanceolate, minutely spinulose-pointed, squarrose, hispidulous and slightly glandular on the back; ligules about 20, sky-blue or pink, 5 mm. long. Open places: Me.—Pa.—Mo.—Iowa. *Canad.*

25. A. ericoìdes L. A perennial, with a rootstock; stem much branched, 3–6 dm. high; branches mostly ascending-spreading; leaves linear, 1–5 cm. long, sparingly hirsute-strigose, or glabrate above; heads mostly racemosely disposed, each at the end of a short leafy branch; involucre about 4 mm. high and broad; ligules white, 3–4 mm. long. *A. multiflorus* Ait. [G, B, R]. Prairies and dry ground: Me.—Ga.—Mex.—Colo.—Mont. *E. Temp.—Submont.* Au–O.

26. A. stricticaùlis (T. & G.) Rydb. Stem 3–6 dm. high, slender; leaves narrowly linear, 1–5 cm. long, 1–2 mm. wide, sparingly strigose or glabrate; involucres 3–4 mm. high and nearly as broad; bracts in 3 series, narrowly oblanceolate, almost glabrous, rather thin; ligules white, 3 mm. long. Meadows and river valleys: Sask.—Neb.—Wash.—Alta. *Plain.* Au–S.

27. A. Batèsii Rydb. A perennial, with a branched rootstock; stem 4–6 dm. high, terete, with numerous ascending branches; leaves linear, sessile, 1–3 cm. long, 1–2 mm. wide, hispidulous-strigose on both sides, hispidulous-ciliolate on the margins, spinulose-tipped; involucres turbinate, about 4 mm. high, 6–7 mm. broad; bracts oblanceolate, squarrose, with spinulose pubescent green tips; ligules about 15, sky-blue, 3–4 mm. long; disk-flowers flesh-colored. (Mistaken for *A. amethystinus* Nutt.; perhaps a hybrid, *A. ericoìdes* ✕ *Kumleini*). Prairies: Neb

28. A. polycéphalus Rydb. A perennial, with a thick cespitose rootstock; stem with many ascending branches, 3–8 dm. high; leaves linear, 2–5 cm. long, hispidulous-strigose, or in age glabrate; heads numerous, racemosely arranged; involucres 5–7 mm. high and as broad; rays white, 4–6 mm. long. Plains and hills: Alta.—Minn.—Tex.—Ariz. *Plain—Submont.—Son.* Jl–S.

29. A. commutàtus T. & G. A perennial, with a branched rootstock; stem branched above, with usually spreading branches, coarsely strigose; leaves linear, 1–4 cm. long, densely hispid-strigose; involucre 6–8 mm. high and fully as wide; heads solitary at the ends of leafy branches; bracts nearly of the same length, the outer passing into leaves of the branches; ligules white, 4–5 mm. long. *A. adsurgens* Greene. Plains and river banks: Man.—Kans.—Colo.—Alta. *Prairie—Submont.* Au.

30. A. serìceus Vent. A perennial, with a short rootstock; stem 3–6 dm. high, corymbosely branched above, glabrous, straw-colored; lower leaves oblanceolate, tapering into a winged petiole; stem-leaves sessile, with a broad base, oblong, entire, mucronate, densely silvery-silky; involucre turbinate, 8–10 mm. high, 10–12 mm. broad; bracts oblong to lanceolate, acute, silvery-canescent; ligules 15–26, reddish violet, becoming blue, 12–16 mm. long. Prairies: Ill.—Tenn.—Tex.—N.D.—Man. *Prairie—Texan.* Au–S.

31. A. pàtens Ait. A perennial, with a long creeping rootstock; stem 3–9 dm. high, rough; leaves sessile, auriculate-clasping, ovate-oblong or oval, scabrous on both sides, rigid, entire, 2.5–7.5 cm. long, those of the branches much smaller, heads solitary at the ends of the branches; involucre 8–10 mm. high, 10–12 mm. broad; bracts acute, with spreading tips; ligules 20–30, bluish or violet, 8–12 mm. long. Open places: Mass.—Fla.—Tex.—Minn. (?) *E. Temp.* Au–O.

32. A. patentìssimus Lindl. A perennial, with a rootstock; stem 3–5 dm. high, very rough, much branched; leaves sessile, clasping, ovate to triangular-oblong, acute, hirtellous on both sides, 1–3 cm. long, 3–10 mm. wide, those of the branches lanceolate, 1–5 mm. long; involucre decidedly turbinate, about 10 mm. high, 10–12 mm. broad; bracts in 5 or 6 series, broadly lanceolate, canescent; ligules dark violet, 8–10 mm. long. Dry woods: Mo.—Kans.—Tex.—Miss. *Ozark.* Au–S.

33. A. tenuicaùlis (Mohr) Burgess. A perennial, with a creeping rootstock; stem 3–6 dm. high, rough; leaves sessile, clasping, ovate, scabrous on both sides, 1–4 cm. long, entire; those of the slender branches lanceolate or subulate, minute; involucre 6–8 mm. high, 8–10 mm. broad; bracts attenuate; ligules 15–20, blue or violet, 8 mm. long. *A. patens gracilis* Hook. [G]. Dry soil: Ga.—Kans.—Tex. *Austral.*

34. A. mérìtus A. Nels. A perennial, with a cespitose rootstock; stem ascending, 2–4 dm. high; leaves somewhat crenate-serrate or some entire, glabrous above, puberulent beneath, 3–7 cm. long; inflorescence corymbiform; involucre turbinate-campanulate, 8 mm. high, 10 mm. broad; bracts in 3 or 4 series, acute; ligules purple or violet, about 15 mm. long. Mountains: Mack.—S.D.—Wyo.—B.C. *W. Boreal.—Mont.* Jl–Au.

35. A. Wòldeni Rydb. A perennial; stem 6 dm. high or more, densely villous-pilose, at least above, corymbosely branched; lower leaves 10–15 cm. long, oblanceolate, tapering below into a winged petiole, crenate-serrate above the middle, rather densely hispidulous on both sides; upper leaves much smaller, entire, oblong, sessile or slightly clasping; involucre turbinate, 5 mm. high and fully as broad; bracts well imbricate in 4 or 5 series, lance-oblong, acute, the green tips more or less rhombic; ligules 20–25, purple, 5–6 mm. long. Prairies: Iowa. S.

36. A. turbinéllus Lindl. A perennial, with a thick rootstock; stem 5–10 dm. high, paniculately branched, glabrous or somewhat short-pubescent; leaves firm, lanceolate, 5–7 cm. long, entire, ciliate, acuminate or acute, the basal ones petioled, the upper sessile; involucre more than 1 cm. high; ligules 10–20, violet, 6–10 mm. long. Prairies: Ill.—La.—Kans.—Neb. *Prairie.* S–O.

37. A. éxilis Ell. A glabrous annual; stem 3–18 dm. high, paniculately branched; leaves linear, sessile, 2–10 cm. long, entire, or the lowest oblong and petioled, those of the branches small and subulate; involucre turbinate, 5–6 mm. high, 6–8 mm. broad; bracts linear-subulate, imbricate in 3 or 4 series; ligules purplish, 4 mm. long. Wet, especially saline soil: S.C.—Fla.—Tex.—Kans. *Austral.* Au–N.

38. A. laèvis L. A perennial, with a thick rootstock; stem 3–12 dm. high, rigid, glabrous, glaucous; basal leaves with winged petioles; blades ovate or lanceolate, more or less serrate, glabrous and glaucous; upper stem-leaves ovate or subcordate, auriculate-clasping; involucre 8–9 mm. high, about 1 cm. broad; ligules blue. Open woodlands: Ont.—La.—N.M.—Alta. *E. Temp.—Mont.* Au–O.

39. A. Gèyeri (A. Gray) Howell. Stem 5–10 dm. high, glabrous; lower leaves 1–2 dm. long, with winged petioles; blades oblanceolate, entire or sparingly serrate, glabrous, glaucous; involucres about 8 mm. high and 1 cm. wide; ligules blue, 8–10 mm. long; achenes glabrous. Valleys: Alta.—S.D.—Colo.—Wash. *Submont.—Mont.* Jl–S.

40. A. vimíneus Lam. A perennial, with a thick rootstock; stem 6–15 dm. high, glabrous, divergently branched; leaves 7–12 cm. long, acuminate, those of the branches smaller; heads numerous, short-peduncled; involucre broadly turbinate, 3–5 mm. high, 5–6 mm. broad; bracts linear, acute, green-tipped, appressed; ligules many, white or rose-colored in fading, 4 mm. long. Moist soil: Ont.—Mass. Fla.—Ark.—Kans.—Man. *E. Temp.* Jl–O.

41. A. lateriflòrus (L.) Britton. A perennial, with a short rootstock; stem 3–15 dm. high, puberulent or glabrous; leaves 5–12 cm. long, acuminate, those of the branches smaller, linear-oblong; heads numerous, short-peduncled or sessile; involucre broadly turbinate, 5–6 mm. high, 6–8 mm. broad; bracts in 4 series, linear-oblong, obtuse or acute, with short green tips; ligules white or pale purple, 2–3 mm. long. Woods: N.S.—N.C.—Tex.—Minn. *E. Temp.* Au–O.

42. A. hirsuticaùlis Lindl. Perennial, stem erect, 4–10 dm. high, villous throughout, or glabrate towards the base, paniculately branched; leaves thin, glabrous on both sides or pilose on the veins beneath, lanceolate or linear-lanceolate, entire or serrulate, with distant, ascending teeth, the lower 5–15 cm. long, 4–20 mm. wide; heads very numerous, subsessile on racemiform branches; involucre 4–5 mm. high; bracts in 2 or 3 series, linear, acute, the green tips oblanceolate; ligules about 20, white, 4 mm. long. *A. lateriflorus hirsuticaulis* Porter [G]. Woods: Me.—Minn.—Ala. N.C. *Allegh.* S–O.

43. A. missouriénsis Britton. A perennial, with a creeping rootstock, stem 5–10 dm. high, villous-hirsute; leaves thin, oblanceolate, acute, 5–10 cm. long, more or less serrate, villous-puberulent, or finely pubescent beneath; heads paniculate; involucres fully 6 mm. high, about 8 mm. broad; bracts linear, acute; ligules white, 5–6 mm. long. Moist places: Mo.—Minn.—Kans.—Tex. *Prairie—Ozark.*

44. A. glabéllus Nees. A perennial, with a thick branched rootstock; stem 3–9 dm. high, glabrous or nearly so, or in var. *pilosus* somewhat villous-hirsute, much branched and bushy; leaves firm, the basal ones spatulate, obtuse, dentate, petioled; stem-leaves linear, acute, 2–7 cm. long, those of the branches linear-subulate, numerous; bracts linear-lanceolate or linear-lanceolate, acute or short-acuminate, in 3 series, with subrhombic tips; ligules 15–25, white or tinged with purple, 5–6 mm. long. *A. ericoides* Am. auth. Dry soil: Me.—Fla.—Ky.—Miss.—Minn. Au–O.

45. A. párviceps (Burgess) Mackenzie & Bush. A perennial, with a short rootstock; stem 3–5 dm. high, sparingly hirsute or pilose, branched above; basal leaves spatulate, petioled; stem-leaves linear or lance-linear, 2–7 cm. long, often with leafy branchlets in their axils; leaves of the branches subulate, much reduced; bracts linear-subulate in 3 or 4 series, acute; ligules 10–12, white, about 4 mm. long. *A. ericoides parviceps* Burgess [B]. *A. depauperatus pariceps* Fernald [G]. Prairies and open woods: Ill.—Mo.—Iowa. *Prairie.* u–S.

46. A. Tradescánti L. A perennial, with a rootstock; stem 5–15 dm. high, paniculately branched; stem-leaves numerous, 7–15 cm. long, acuminate, thin; heads very numerous; involucre hemispheric or somewhat turbinate, 4–6 mm. high; bracts in 4 or 5 series, linear, acute, with oblanceolate green tips; ligules many, white or nearly so, 4–6 mm. long. Swamps, fields and copses: Ont.—Fla.—Mo.—Minn. *E. Temp.* Au–O.

47. A. Jacobaèus Lunell. Stem slender, 4–8 dm. high, glabrous up to the inflorescence; leaves rather firm, lanceolate, acuminate, mostly denticulate, 2–7 cm. long; inflorescence narrow, raceme-like; heads subsessile or short-peduncled; involucre 5–6 mm. high; bracts in 3 series, acute, green-tipped; ligules white or pale violet-purple. Low meadows: N.D. Au–S.

48. A. salicifòlius Lam. A perennial, with a rootstock; stem 5–20 dm. high, paniculately branched; leaves lanceolate, firm, glabrous or scabrous, 6–10 cm. long, the lower more or less dentate; involucre 6–8 mm. high, 8–10 mm. broad; bracts with narrowly oblanceolate green tips, or the outer wholly green; ligules white or violet, 6–8 mm. long. *A chelonicus* Lunell. *A. lautus prionoides* Lunell. Wooded banks: Me.—Fla.—Tex.—Colo.—Sask. *E. Temp.— Mont.* Au–O.

49. A. paniculàtus Lam. A perennial, with a rootstock; stems 5–20 dm. high, paniculately branched; leaves oblong-lanceolate to narrowly linear-lanceolate, mostly attentuate, thin, glabrous, sharply denticulate, or the upper entire; involucre 5–6 mm. high, 8–10 mm. broad; bracts with narrowly linear-oblanceolate green tips; ligules white, or rarely pink or pale lilac, 7–8 mm. long. Wooded banks: N.B.—Va.—Colo.—Sask. *Canad.—Prairie—Mont.* Au–O.

50. A. acùtidens Smyth. A perennial; stem glabrous below, with pubescent decurrent lines above, 1 m. high or more, paniculately branched above; stem-leaves broadly oblanceolate, sessile, acuminate, 8–15 cm. long, 2–3 cm. wide, usually coarsely dentate above the middle, rather thick, glabrous on both sides, scabrous-hirsutulous along the margins, those of the branches smaller and entire; heads very numerous; involucre 7 mm. high, 1 cm. broad or more; bracts linear, acute, the green tips narrowly oblanceolate; ligules 20–30, white, about 8 mm. long. *A. paniculatus acutidens* Burgess. Low ground: Minn.—Kans.—Va.—Wis. *Allegh.* Au–S.

51. A. fluviátilis Osterhout. A perennial, with a creeping rootstock; stem slender, 6–10 dm. high, leafy and branched; leaves lance-linear, sessile, slightly clasping, entire or with fine teeth directed forward, the larger 1 dm. long; inflorescence paniculate; involucre 6–7 mm. high, nearly 1 cm. broad; bracts in 4 series, oblanceolate, acute, with oblanceolate green tips; ligules purple, 7–8 mm. long. *A. caerulescens* Auth.; not DC. Along streams: Colo.—Iowa. *Prairie—Plain.* Au–S.

52. A. Osterhoùtii Rydb. Stem branched and leafy, 5–10 dm. high; leaves 5–12 cm. long, 8 mm. wide or less, those of the branches much smaller; involucres 7–8 mm. high; bracts linear to oblanceolate, apiculate; ligules white or rarely pinkish, nearly 1 cm. long. *A. durus* Lunell (?). Along ditches and streams: Colo.—N.D.—Sask.—Alta. *Plain.* Au–S.

53. A. laetévirens Greene. Stem erect, 3–6 dm. high, red or purplish; leaves lanceolate or oblong-lanceolate, thin, light green, somewhat glaucescent; involucres 8 mm. high and about 1 cm. wide; ligules 8–10 mm. long, white or pinkish. Along streams: N.D.—Colo.—Ida. *Submont.* Au–S.

54. A. lóngulus Sheldon. A perennial, with a rootstock; stem strict, branched above, 6–15 dm. high, hispid in lines; leaves narrowly linear, 3–10 cm. long, sessile, glabrous on both sides, or scabrous above; inflorescence paniculate; involucre 4–5 mm. high, 6–7 mm. broad; bracts acute, the outer ones wholly green; ligules 6 mm. long, white or pale lilac. Swampy ground: Minn.—Neb.—Colo.—B.C. *Prairie—Plain.* Jl–S.

55. A. junciférmis Rydb. A perennial, with a slender horizontal rootstock; stem slender, 3–5 dm. high, simple below; leaves narrowly linear, 4–8 cm. long, 2–4 mm. wide; involucre about 6 mm. high and 1 cm. broad; inner bracts linear, acute, with oblanceolate green tips, the outer oblong or oblanceolate, obtuse, often almost wholly green; ligules white, 6–8 mm. long. Swamps: Wis.—Colo.—Wash.—B.C. *W. Boreal.*

56. A. longifòlius Lam. A perennial, with a long creeping rootstock; stem glabrous or slightly pubescent, branched above; leaves linear-lanceolate, entire or nearly so, acuminate, clasping at the base, 7–20 cm. long, 5–15 mm. wide; involucres hemispheric, 8–10 mm. high; bracts narrow, green, acute; ligules many, violet or purple, 10–12 mm. long. Swamps: Lab.—Mass.— Minn.—Sask. *E. Boreal.* Au–S.

57. A. Nuttállii T. & G. A perennial, with a cespitose rootstock; stem 2–6 dm. high; leaves glaucous, glabrous; stem-leaves narrowly linear; inflorescence paniculate; involucre 5–6 mm. high, less than 1 cm. broad; ligules bluish purple, 6–8 mm. long. *A. orthophyllus* Greene. Mountains and hills: Neb.—Colo.—Utah. *Submont.—Mont.* Jl–Au.

58. A. adscéndens Lindl. A perennial, with a cespitose rootstock; stem ascending or decumbent at the base, 2–8 dm. high; leaves firm, the lower petioled, 6–12 cm. long, the upper linear and sessile; inflorescence paniculate; involucres 5–6 mm. high, nearly 1 cm. broad; ligules bluish-purple, 5–8 mm. long. Valleys· Sask.— Colo. –Nev,—Alta. *Submont.–Mont.* Jl–S.

59. A. nebraskénsis Britton. A perennial, with a rootstock; stem erect, stiff, hirsutulous, rather simple, 4–7 dm. high; leaves thick, ascending, oblong-lanceolate, entire, sessile, acute, 3–7 cm. long, 8–12 mm. wide; heads few; involucre hemispheric, 6 mm. high; bracts green, oblong, acute; ligules purple, about 12 mm. long. Lake shores: Neb. *Prairie.* Au–S.

60. A. prenanthoìdes Muhl. A perennial, with a creeping rootstock; stem 3–6 dm. high, pubescent in lines, flexuose, branched; leaves thin, 7–15 cm. long, ovate or lanceolate, sharply and coarsely serrate, scabrous above, glabrous beneath, caudate-acuminate; heads numerous; involucre hemispheric, about 7 mm. high and 10 mm. broad; bracts linear, acute, green; ligules 20–30, violet, 8–12 mm, long. Moist ground: Mass.—Tenn.—Iowa—Minn. *Allegh.* Au- O.

61. A. puníceus L. A perennial, with a thick rootstock; stem 5–25 dm. high, stout, purplish, branched above, hispid; leaves lanceolate or oblong-lanceolate, 7–15 cm. long, acuminate, sharply serrate, or the upper entire, rough above, pubescent on the veins beneath; heads numerous; involucre hemispheric, nearly 1 cm. high, 1–1.5 cm. broad; bracts linear or oblong, attenuate, green, loose; ligules 20–40, light violet or pale purple, 10–14 mm. long. Swamps: Newf.—Ga. –Ala.—N.D.—Man. *Canad.—Allegh.* Jl–N.

62. A. Forwoòdii S. Wats. A perennial, with a rootstock; stem 4–6 dm. high, rough-hirsute, strict; leaves sessile, obovate, oblanceolate or lanceolate, narrowed to an auriculate base, hirsute on the veins beneath, 5–10 cm. long; serrate, or the upper entire; inflorescence corymbiform; disk 10–12 mm. high, 12–15 mm. broad; bracts linear, short-acuminate, equaling the disk; ligules 10–12 mm. long. Mountains: S.D.—N.D. *Plain—Submont.* Au.

63. A. lucídulus (A. Gray) Rydb. A perennial; stem 1–2 m. high, more or less angled, paniculately branched, glabrous below, pubescent in lines above; stem-leaves oblanceolate or lanceolate, more or less clasping at the base, acuminate, sharply serrate, 8–10 cm. long, 1.5–2.5 cm. wide, scabrous-hispidulous above, especially near the margins, those of the branches smaller and entire; heads very numerous; disk 8 mm. high and 15 mm. wide; bracts linear, acute, the green tips narrowly oblanceolate; ligules about 30, purple, 1 cm. long. *A. lucidus* Wender.; not *A. puniceus lucidulus* A. Gray [G]. Copses: Me.—Minn. —Mo. *Canad.* Au–O.

64. A. laevifórmis Rydb. Perennial, with a rootstock; stem 6–10 dm. high, glabrous up to the inflorescence, its branches slightly pilose along decurrent lines; stem-leaves ovate, sessile and clasping, slightly serrate along the middle, 6–10 cm. long, 2–4 cm. wide, glabrous except the ciliolate margins, acute, the upper ones more cordate-sagittate, entire; heads numerous, in a large panicle; disk 7–8 mm. high, 1 cm. wide; bracts imbricate, the green tips rhombic-oblanceolate; ligules about 20, bright blue, 8 mm. long. Copses and wet places: Minn. *Prairie.* Au.

65. A. clivòrum Lunell. Stem slender, 4–5 dm. high, simple; leaves large, up to 12 cm. long; the lower oblanceolate, acuminate-denticulate, the upper oblong-lanceolate, all clasping at the narrowed base; involucre 6–7 mm. high; bracts in 3 series, linear, acute, with lanceolate green tips and white-margined towards the base; rays bluish, 6 mm. long. Foothills: Turtle Mountains, N.D. S.

66. A. phyllòdes Rydb. A perennial, with a horizontal rootstock; stem 5–10 dm. high, leafy, glabrous below, pubescent in lines above; leaves thin, the lower usually serrate, with small distant teeth, glabrous on both sides, ciliolate on the margins; disk about 8 mm. high and 12 mm. broad; outer bracts oblanceolate, acute, foliaceous, usually slightly exceeding the disk; ligules violet, 8–10 mm. long. Meadows: Neb.—Colo. *Prairie—Plain.* Au.

67. A. òticus Rydb. A perennial; stem 1 m. high or more, paniculately branched above, glabrous, except the pubescent decurrent lines in the inflorescence; lower leaves lance-linear, about 1 dm. long and 1 cm. wide, attenuate, slightly denticulate or entire, clasping, with large rounded auricles, scabrous-puberulent above, glabrous beneath, those of the branches more lance-oblong, acute, 4 cm. long or less; inflorescence much branched; involucre 5–6 mm. high; bracts linear, acute, the outer equaling the inner; ligules about 20, pale purple, 6–7 mm. long. Copses: Minn. Au.

68. A. Meárnsii Rydb. A perennial, with a rootstock; stem 3–10 dm. high, leafy, branched, glabrous below, sparingly pubescent above; leaves 5–10 cm. long, sessile, entire, glabrous beneath, scabrous or glabrate above; disk about 8 mm. high and barely 1 cm. broad; bracts oblanceolate, acute, some longer than the disk; ligules about 8 mm. long, white or pinkish. River banks: Sask.—S.D.—Wyo.—B.C. *W. Boreal.—Submont.* Jl–O.

30. MACHAERANTHÈRA Nees. Tansy of Viscid Aster.

Annual, biennial, or short-lived perennial, leafy, caulescent herbs, with taproots, never with rootstocks. Leaves alternate, pinnatifid, serrate, or rarely entire, the teeth, lobes, and apex bristle-tipped. Heads corymbose or paniculate, radiate. Involucre hemispheric or turbinate; bracts with narrow, thick, herbaceous tips, imbricate in several series. Receptacle pitted, the pits with toothed or lacerate edges. Ray-flowers numerous, pistillate and fertile; ligules violet or purple. Disk-flowers hermaphrodite and fertile; corollas yellow, turning red or brown. Anthers rounded at the base. Style-branches with subulate or lanceolate appendages. Achenes elongate-turbinate, pubescent or glabrous. Pappus of numerous firm rough bristles.

Leaves once or twice pinnatifid ; plant annual ; achenes terete.   1. *M. tanacetifolia.*
Leaves merely spinulose-toothed ; plant biennial or perennial ;
    achenes somewhat compressed.
    Bracts linear-subulate, their green tips longer than the
        straw-colored lower portion ; inflorescence narrow, ra-
        cemiform ; heads subsessile.   2. *M. sessiliflora.*
    Bracts linear, with lanceolate or rhombic tips, which are
        usually shorter than the straw-colored lower por-
        tion ; inflorescence open ; heads peduncled.
        Leaves cinereous.   3. *M. canescens.*
        Leaves glabrous or nearly so.   4. *M. ramosa.*

1. **M. tanacetifòlia** (H.B.K.) Nees. Annual; stem branched, 1–4 dm. high, glandular-pubescent; leaves 1–4 cm. long, oblanceolate in outline, their divisions oblanceolate or linear, entire or toothed; involucre 1 cm. high, 15–20 mm. broad; bracts linear, attenuate; ligules 12–20 mm. long, blue-purple. *Aster tanacetifolius* H.B.K.; *M. coronopifolia* (Nutt.) A. Nels. Plains and hills, in sandy soil: Tex.—S.D.—Mont.—Ariz.; also Mex. *Plain—Submont.* Je–O.

2. **M. sessiliflòra** (Nutt.) Greene. Biennial; stem leafy, 3–5 dm. high; leaves linear, saliently toothed, 3–6 cm. long, scabrous-puberulent and nearly glabrous; involucre about 1.5 cm. broad and 1 cm. high; bracts subulate, canescent, scarcely glandular; ligules rose-purple. Hills and plains: S.D.—Neb.—m N.M.—Wyo. *Plain—Submont.* Au–S.

3. **M. canéscens** (Pursh) A. Gray. *Fig. 558.* Biennial; stem profusely branched, 1–4 dm. high, canescent; lower leaves spatulate or oblanceolate, slightly toothed; stem leaves linear, usually entire; involucre 7–8 mm. high, about 1 cm. broad; bracts linear-lanceolate, acute, canescent; ligules 8 mm. long, dark bluish purple. *Aster canescens* Pursh. *M. magna* A. Nels. Arid places and sandy banks: Sask.—N.D.—Colo.—Utah—B.C. *Plain—Submont.* Jl–S.

4. **M. ramòsa** A. Nels. Biennial, branched at the base; stem ascending or spreading, 2–4 dm. high; leaves green, nearly glabrous, linear-oblanceolate or linear, the earlier sharply dentate; involucres hemispheric, about 7 mm. high and 1 cm. wide; bracts linear, acute or short-acuminate, glandular-puberulent; ligules purple, 8 mm. long. Mountains and plains: Wyo.—Colo.—Neb.—S.D. *Submont.—Mont.* Jl–S.

31. DOELLINGÙRIA Nees.

Perennial caulescent herbs, with rootstocks. Leaves alternate, entire, ciliate. Heads in corymbiform cymes, radiate. Involucre campanulate; bracts imbricate in 2–4 series, narrow, appressed, without herbaceous tips. Receptacle pitted. Ray-flowers few, pistillate; ligules white. Disk-flowers perfect, hermaphrodite; corollas with a slender tube, expanded into the 5-toothed limb. Anthers obtuse at the base. Style-branches with subulate to ovate appendages. Achenes slightly broadened above. Pappus double, the outer of short bristles or squamellae, the inner of capillary bristles, usually clavellate at the apex.

Stem and leaves glabrous, except the upper part of the former and the margins and veins of the latter. 1. *D. umbellata.*
Stem and leaves decidedly rough-pubescent. 2. *D. pubens.*

1. **D. umbellàta** (Mill.) Nees. Rootstock woody, stem 6–25 dm. high; leaves narrowly elliptic to linear or lance-elliptic, 7–15 cm. long, acuminate; involucre turbinate, 4–5 mm. high; bracts puberulent at the tips, the outer lanceolate, the inner linear; ligules 4–6 mm. long, white. *Aster umbellatus* Mill. [G]. Thickets and in moist soil: Newf.—Ga.—Ark.—S.D.—Sask. *Canad.—Allegh.* Jl–O.

2. **D. pùbens** (A. Gray) Rydb. Rootstock woody; stem 5–10 dm. high, puberulent; leaves lanceolate or ovate-lanceolate, 5–15 cm. long, entire, acute, green and scabrous above, pale and tomentulose beneath; involucre about 5 mm. high; bracts linear-lanceolate, slightly tomentose; ligules 4 mm. long,

white. *A. umbellatus pubens* A. Gray [G]. *D. umbellata pubens* Britton [B]. Meadows and open woods: Upper Mich.—Wis.—Neb.—Sask. *Prairie.* Jl–S.

32. IONÁCTIS Greene.

Low branching perennial, often somewhat woody at the base. Leaves alternate, numerous, narrow, 1-nerved. Heads radiate, at the ends of the branches. Involucre more or less turbinate; bracts coriaceous, imbricate in several series, appressed. Ray-flowers pistillate, perfect; ligules violet or purple. Disk-flowers hermaphrodite and fertile. Style-branches with subulate appendages. Achenes villous. Pappus double, the inner of long capillary bristles, the outer much shorter.

1. **I. linariifòlia** (L.) Greene. A cespitose perennial; stems stiff, scabrous-puberulent, corymbosely branched above, 2–6 dm. high; leaves linear or nearly so, rigid, entire, ciliate on the margins, those of the branches reduced and appressed; involucre about 1 cm. high, 1–1.5 cm. broad; bracts keeled on the back; ligules violet, 8–10 mm. long. *Aster linariifolius* L. [G]. Dry rocky soil: Newf.—Fla.—Tex.—Minn. (?). *E. Temp.* Jl–O.

33. LEUCELÈNE Greene.

Suffrutescent low perennials, with diffuse branches. Leaves alternate, linear or subulate, or the lower spatulate or oblanceolate. Heads solitary at the ends of the branches, radiate. Involucre turbinate; bracts imbricate in several series, herbaceous, with narrow scarious margins. Ray-flower pistillate and fertile; ligules white, turning reddish. Disk-flowers hermaphrodite and fertile; corollas white, tubular-funnelform. Style-branches with obtuse appendages. Anthers obtuse at the base. Achenes long and slender. Pappus of long slender white bristles.

Leaves of the branches 6–12 mm. long.	1. *L. alsinoides.*
Leaves of the branches 2–5 mm. long.	2. *L. ericoides.*

1. **L. alsinoìdes** Greene. A perennial, with a cespitose caudex; stem about 1 dm. high, strigose; lower leaves linear-oblanceolate, about 1 cm. long, the upper subulate, appressed, all hispid-ciliate; involucre 5–6 mm. high and broad; bracts strigose, slightly glandular-puberulent; ligules 5 mm. long. Dry hills: Tex.—Kans.—Colo.—N.M. *Plain—Submont.* Ap–Je.

2. **L. ericoìdes** (Torr.) Greene. Like the preceding, but the leaves of the branches much shorter; involucre barely 5 mm. high and broad; bracts less imbricate, strigose. *Aster ericaefolius* Rothr. Plains: Kans.—Tex.—Colo. *Plain.* Je–S.

34. ERÍGERON L. FLEABANE.

Annual or perennial herbs. Leaves alternate, entire, toothed, or lobed. Heads solitary or few, corymbose or paniculate, peduncled, radiate or rarely discoid. Involucre hemispheric or saucer-shaped; bracts in 1 or 2, rarely 3, series, subequal or slightly imbricate. Receptacle flat, naked. Ray-flowers pistillate and fertile; ligules white, pink, purple, rarely ochroleucous or yellow, or wanting. Disk-flowers hermaphrodite and fertile; corolla yellow. Anthers obtuse at the base. Style-branches flattened, mostly with obtuse appendages. Achenes flattened, usually 2-nerved, sometimes terete or angled. Pappus in a single series of capillary bristles, sometimes with an additional outer series of shorter bristles.

Bracts of the involucre in 1 or 2 series of almost equal length, not thickened on the back.
Ligules inconspicuous, erect or ascending, usually involute and incurved, often inside them a series of rayless pistillate flowers; leaves entire.
1. ACRES.
Ligules conspicuous, spreading, flat; no rayless pistillate flowers inside.

Plants without runners.
  Perennials, with rootstocks or woody caudices.
    Leaves dissected into linear or oblong divisions.   2. COMPOSITI.
    Leaves entire or merely toothed.
      Stem densely cespitose from a thick taproot.
        Stem low, less than 2 dm. high, usually bearing 1 head; leaves
          filiform, mostly basal.   3. RADICATI.
        Stem leafy.
          Head solitary, terminal; leaves numerous, linear.
                              12. HYSSOPIFOLII.
          Heads several, corymbose; the lower leaves oblanceolate.
                              4. PUMILI.
      Stems tall, usually solitary at the ends of often-branched rootstocks.
        Upper stem-leaves ample, ovate or lanceolate; leaves usually
          3-ribbed.   5. MACRANTHI.
        Upper stem-leaves linear-lanceolate, reduced; none of the leaves
          3-ribbed.   6. GLABELLI.
  Annuals or biennials, or perennials by means of offsets.
    Stem-leaves not cordate-clasping; annuals or biennials.
      Stem simple, with a few large heads; disk about 1 cm. broad or
        more.   7. ASPERI.
      Stem much branched, leafy, with numerous small heads; disk 6–9
        mm. broad.   8. ANNUI.
    Stem-leaves broad, clasping; usually perennials by means of offsets.
                          9. PHILADELPHICI.
  Plants at first with scapiform peduncles, later producing lateral runners.
                        10. FLAGELLARES.
Bracts of the involucres in 3 or 4 series, more or less imbricate and the outer suc-
  cessively shorter, thickened on the back; cespitose perennials.
                        11. CAESPITOSI.

1. ACRES.

Bracts of the involucre linear, abruptly acute, never glandular; inflorescence strictly
  racemose.
    Low, 1–2 dm. high, usually branched at the base; stem-leaves sessile.

                                               1. E. minor.

    Taller, 3–6 dm. high, simple; lower stem-leaves petioled;
      individual peduncles elongate.   2. E. lonchophyllus.
Bracts of the involucre linear-subulate, long-attenuate, more
  or less glandular-puberulent; inflorescence corymbiform
  or paniculate.
    Plant tall, 3–6 dm. high; bracts merely glandular-puberu-
      lent, with a few hairs at the base.   3. E. droebachiensis.
    Plant low, 1–3 dm. high; bracts both hirsute and glandu-
      lar-puberulent throughout.   4. E. jucundus.

2. COMPOSITI.

Leaves twice or thrice ternate.   5. E. compositus.
Leaves ternately or quinately cleft at the apex.   6. E. trifidus.

3. RADICATI.

One species.   7. E. nematophyllus.

4. PUMILI.

One species.   8. E. pumilus.

5. MACRANTHI.

Stem and leaves glabrous, the latter with ciliate margins;
  bracts glandular-puberulent and with scattered hairs.   9. E. speciosus.
Stem and leaves hairy; bracts hirsute.   10. E. subtrinervis.

6. GLABELLI.

One species.   11. E. glabellus.

7. ASPERI.

Pubescence of the stem and leaves ascending, kinky.
  Lower leaves distinctly toothed, oblanceolate, almost gla-
    brous; ligules about 1 cm. long.   12. E. tardus.
  Lower leaves entire, linear or linear-oblanceolate, distinctly
    pubescent.
    Leaves narrowly linear-oblanceolate, 1–2 dm. long; lig-
      ules about 1 cm. long.   13. E. anodontus.
    Leaves oblanceolate, 3–10 cm. long; ligules 6–8 mm.
      long.   14. E. asper.
Pubescence of the stem and leaves spreading.
  Lower leaves entire or rarely with a few teeth.
    Pubescence short and rather sparse, spreading.   15. E. oligodontus.
    Pubescence long, shaggy, usually reflexed.   16. E. Drummondii.
  Lower leaves distinctly toothed.
    Lower leaves narrowly oblanceolate, 5–15 cm. long,

sharply serrate with ascending teeth.	17. *E. oxydontus.*
Lower leaves spatulate or oblanceolate, 3–10 cm. long, crenate-serrate with mucronate teeth.	18. *E. abruptorum.*

8. ANNUI.

Stem and leaves strigose, or sparingly long-hirsute or glabrous. Upper stem-leaves linear, entire.	19. *E. ramosus.*
Upper leaves ovate to lanceolate, toothed.	20. *E. annuus.*
Stem and leaves densely pubescent, with short spreading leaves.	
Annuals ; pappus scant and simple.	21. *E. Bellidiastrum.*
Biennials ; pappus double, the outer of short subulate squamellae.	22. *E. divergens.*

9. PHILADELPHICI.

Upper stem-leaves few and reduced ; ligules about 50, rather broad, lilac or pink.	23. *E. pulchellus.*
Upper stem-leaves many, well developed ; ligules more than 100, narrow.	
Basal leaves crenate or dentate, with broad rounded teeth ; upper stem-leaves lanceolate, acute.	24. *E. philadelphicus.*
Basal leaves dentate, with sharp, triangular salient teeth ; upper stem-leaves triangular-lanceolate, acuminate.	25. *E. purpureus.*

10. FLAGELLARES.

One species.	26. *E. flagellaris.*

11. CAESPITOSI.

Ligules bluish, purple or white.	
Achenes terete or nearly so ; leaves not triple-nerved.	27. *E. canus.*
Achenes flattened.	
Leaves more or less distinctly triple-nerved.	
Stem erect, about 3 dm. high, usually with several heads ; stem-leaves linear.	28. *E. subcanescens.*
Stem decumbent at the base, 1–2 dm. high, with 1–3 heads ; stem-leaves oblong.	29. *E. caespitosus.*
Leaves not triple-nerved.	
Ligules blue or purple, about 1 mm. wide.	30. *E. laetevirens.*
Ligules white, 1.5–2 mm. wide.	31. *E. montanensis.*
Ligules ochroleucous or yellow.	32. *E. peuciphyllus.*

12. HYSSOPIFOLII.

One species.	33. *E. hyssopifolius.*

1. E. minor (Hook.) Rydb. Biennial; stem more or less hirsute; basal leaves spatulate or oblanceolate, 3–6 cm. long, ciliate or glabrate; stem-leaves linear, 3–10 cm. long, acute; peduncles erect, 1–5 cm. long; involucre 6–8 mm. high, 10–15 mm. broad; bracts hirsute; ligules about 2 mm. long, rose-colored or white. *E. glabratus minor* Hook.; *E. armeriaefolius* A. Gray; not Turcz. Damp places: Man.—S.D.—Colo.—Utah—B.C. *Plain—Mont.* Jl–S.

2. E. lonchophyllus Hook. Biennial; stem more or less hirsute; lower leaves petioled, narrowly oblanceolate, 6–15 cm. long, hirsute-ciliate or glabrate; peduncles usually elongate, 2–8 cm. long; involucre about 8 mm. high and 15 mm. broad; bracts hirsute; ligules 2–3 mm. long, erect, white. *E. racemosus* Nutt. Wet places: Sask.—N.D.—Colo.—Nev.—Mont. *Plain—Mont.* Jl–Au.

3. E. droebachiénsis Müll. Biennial; stem simple, glabrous or sparingly hirsute; basal leaves 5–10 cm. long, ciliate; stem-leaves narrowly oblanceolate to linear-lanceolate; involucre about 6 mm. high and 1 cm. broad; ligules about 3 mm. long, white. *E. acris droebachiensis* Blytt [B]. *E. acris asteroides* (Andrz.) DC. [G]. Woods: N.B.—Sask.—Colo.—Alaska; also in Eur. *Boreal.—Mont.* Jl–Au.

4. E. jucúndus Greene. Biennial or usually perennial; stems often several, puberulent and sparingly hirsute; basal leaves spatulate or oblanceolate, petioled, 3–5 cm. long, sparingly hirsute; stem-leaves linear-lanceolate or oblong; heads solitary or few; involucre 6 mm. high, 8–10 mm. broad; bracts often purple-tinged; ligules erect, pink, 2–2.5 mm. long. *E. acris debilis* A. Gray [B]. *E. debilis* Rydb. Wet places: Que.—Man.—Colo.—Utah—B.C. *Boreal.—Subalp.* Jl–Au.

5. E. compósitus Pursh. Perennial, with a cespitose caudex; stem scapi·form, 3–20 cm. high; leaves basal, crowded, usually twice ternate (or the earlier simply ternate), 2–8 cm. long, with linear or spatulate divisions, more or less hirsute and glandular, or almost glabrous (var. *nudus* Rydb.); heads solitary; involucre about 7 mm. high, 10–18 mm. broad; bracts linear-subulate, hirsute and glandular-puberulent; ligules 5–8 mm. long, white or purplish, or wanting (var. *discoideus* A. Gray). *E. multifidus* Rydb. Mountains and rocky places: Greenl.—Sask.—S.D.—Colo.—Calif.—Alaska. *Arct.—Mont.—Alp.* Jl-Au.

6. E. trífidus Hook. A subacaulescent cespitose perennial; peduncles 5–10 cm. high; leaves basal, crowded, 2–5 cm. long, the earlier 3-cleft, the later ones often with the lateral divisions again 2-cleft; divisions oblanceolate or spatulate; involucre 6–7 mm. high, 10–15 mm. broad; bracts linear-subulate, hirsute and glandular-puberulent; ligules about 5 mm. long, white or pinkish. *E. Gormani* Greene. Stony crests: Greenl.—Alta.—S.D.—Colo.— Yukon. *Arct.—Mont.—Alp.* My–Jl.

7. E. nematophýllus Rydb. Stems several, strigose, about 5 cm. high, few-leaved; leaves linear-filiform, 2–4 cm. long, less than 1 mm. wide, strigose; heads solitary; involucres 4–5 mm. high, 8–10 mm. broad; bracts linear, acute, hirsute-strigose; ligules pinkish or white, 4–5 mm. long; outer pappus of short bristles. *E. Eatonii* Coult. & Nels., in part; not A. Gray. Rocky hills: S.D.—Colo.—Wyo. *Submont.* My–Jl.

8. E. púmilus Nutt. Perennial, with a short caudex; stems erect, 1–3 dm. high, hirsute; lower leaves narrowly oblanceolate, 2–10 cm. long, hirsute, the upper linear; heads several; involucre 7–8 mm. high, 12–18 mm. broad; bracts narrowly linear, acute, hirsute; ligules 50–80, white, 7–10 mm. long; pappus double, the outer of short bristles. Dry plains: Man.—Kans.—Utah—B.C. *Plain—Submont.* My–Jl.

9. E. speciòsus DC. Stem 3–5 dm. high; basal leaves linear-oblanceolate, ciliate, 5–10 cm. long, petioled; upper leaves sessile, narrowly lanceolate; heads corymbose; involucres 7–8 mm. high, 12–18 mm. broad; ligules blue or violet, numerous, 12–18 mm. long. Mountains: Alta.—S.D.—Colo.—Utah—Ore.—B.C. *Submont.—Mont.* Jl–S.

10. E. subtrinérvis Rydb. *Fig. 559.* Perennial, with a short rootstock; stem 3–6 dm. high, hirsute with short hairs; lower leaves oblanceolate, 5–10 cm. long, hirsutulous on both sides; inflorescence corymbiform; involucre 7–8 mm. high, 12–20 mm. broad; bracts hirsute; ligules numerous, violet, rose, or purplish, rarely white, about 1 cm. long. Wooded mountain-sides: S.D.—N.M.—Utah—Wash. *Submont.—Mont.* Jl–Au.

559.

11. E. glabéllus Nutt. Perennial, with a cespitose rootstock; stem decumbent at the base, 1.5–4 dm. high, glabrous or sparingly hirsute above; basal leaves oblanceolate, 5–10 cm. long, glabrous; stem-leaves narrowly linear-lanceolate, the upper much reduced; heads 1–3; involucre 6–7 mm. high, 1–1.5 cm. broad; bracts linear, acuminate; ligules purple. *E. rubicundus* Greene, a low monocephalous form. Hills and mountains: Man.—Wis.—N.M.—Mack. *Plain—Subalp.* Jl–Au.

12. E. tárdus Lunell. Stem 3–4.5 dm. high, striate, sparingly appressed-pubescent; leaves at length glabrate, the lower ones wing-petioled, oblanceolate,

1–1.5 dm. long, the upper linear and entire; bracts 7–8 mm. long, linear, strigose; ligules white, about 1 cm. long. Roadsides: N.D. Je.

13. E. anodóntus Lunell. Stem simple, 3–4.5 dm. high, sparingly pubescent with ascending hairs; lower leaves narrowly linear-oblanceolate, 1–2 dm. long, 1 cm. broad or less, entire, sparingly pubescent, the upper linear, 4–7 cm. long; bracts 7–8 mm. long, linear, strigose; ligules white, about 1 cm. long. Dry pastures: N.D. Je.

14. E. ásper Nutt. Biennial, with a short taproot; stem strict, erect, 2–3 dm. high, shortly hirsute-strigose; lower leaves linear-oblanceolate, 3–10 cm. long, entire or minutely denticulate, short-pubescent, upper leaves linear-lanceolate, acuminate; heads 1–4; involucre 5–6 mm. high, about 1 cm. broad, hirsute; bracts linear; ligules white, 6–8 mm. long. *E. anicularum* Greene. *E. multicolor* Lunell. *E. subcostatus* Lunell, a large form. Prairies and hills: Man.—Minn.—Sask.—S.D. *Plain.* Je–Jl.

15. E. oligodóntus Lunell. Stem 2.5–4 dm. high, copiously pubescent with short spreading hairs, lower leaves narrowly oblanceolate, 5–15 cm. long, sparingly denticulate or entire, pubescent with short spreading hairs; upper leaves linear, entire; bracts 6–7 mm. long, linear, hirsute; ligules white, 8–10 mm. long. *E. procerus* Lunell, a large form. Meadows: Man.—Minn.—N.D. Je.

16. E. Drummóndii Greene. Stem 2–4 dm. high, strict; lower leaves oblanceolate, short-petioled, densely hirsute, entire or denticulate; upper leaves lance-linear; heads 1–4, with long peduncles; involucres 6–7 mm. high, 10–15 mm. broad; ligules numerous, very narrow, about 1 cm. long, pink or lilac. *E. glabellus pubescens* Hook. Hillsides and mountains: Alta.—Mont.—B.C.—Yukon. *Plain.* Je–Au.

17. E. oxydóntus Lunell. Stem simple, 3–3.5 dm. high, pubescent, with short spreading hairs; lower leaves narrowly linear-oblanceolate, 5–15 cm. long, serrate with 3–5 large cuspidate teeth on each margin; upper leaves linear and entire; bracts about 7 mm. long, linear, hirsute; ligules white, 6–8 mm. long. Wet soil: N.D. Je.

18. E. abruptòrum Lunell. Stem simple, 2–3 dm. high, pubescent, with short spreading hairs; lower leaves spatulate or broadly oblanceolate, 3–10 cm. long, hirsute, crenate-serrate, with mucronate teeth; upper leaves linear, entire; bracts 5–6 mm. long, linear, hirsute; ligules white, about 6 mm. long. Bluffs: N.D. Je.

19. E. ramòsus (Walt.) B.S.P. Annual; stem 3–6 dm. high, corymbosely branched, strigose; basal leaves petioled, 5–10 cm. long, spatulate or oblanceolate; stem-leaves linear or nearly so; heads numerous; involucre hirsute, 3–4 mm. high, 6–8 mm. broad; bracts linear, acute; ligules numerous, white, about 5 mm. long; inner pappus of few bristles or in the rays wanting. *E. strigosus* Muhl. *E. obscurus* Lunell, an undeveloped form. Dry places: N.S.—Fla.—Calif.—B.C. *Temp.* Je–S.

20. E. ánnuus (L.) Pers. Annual or biennial; stem sparingly hirsute or glabrous, 3–12 dm. high, branched above; leaves 4–10 cm. long, the lower wing-petioled, the upper dentate; heads corymbose-paniculate; involucre 3–4 mm. high; bracts linear or linear-oblanceolate; ligules many, pink or violet, rarely white, 5–8 mm. long. Fields and meadows: N.S.—Ga.—Mo.—Kans.—Minn. *Canad.—Allegh.* Je–O.

21. E. Bellidiástrum Nutt. Annual; stem corymbosely branched above, grayish-hirsutulous; leaves linear-spatulate or linear-oblanceolate, obtuse, entire, densely hirsutulous; heads numerous; involucres 3–4 mm. high, 7–8 mm. broad; bracts linear, acute, hirsute; ligules light purple or white, 5–6 mm. long. Low ground: S.D.—Kans.—Ariz.—Wyo. *Plain—Mont.* Jl–O.

22. **E. divérgens** T. & G. Biennial or rarely perennial, with a taproot; branched at the base; stems several, 2–4 dm. high, densely hirsute with short hairs; basal leaves spatulate or oblanceolate, petioled, 2–6 cm. long, entire or rarely few-lobed; stem-leaves linear, sessile; heads numerous; involucre 4–5 mm. high, 8–10 mm. broad, hirsute; bracts linear, hirsute; ligules numerous, bluish purple or lilac, about 5 mm. long. Low plains and river banks: Mont.—S.D.—Tex.—Calif.—Wash. *Plain—Mont.—W. Temp.* My–Au.

23. **E. pulchéllus** Michx. Perennial, with stolons and offsets; stem villous or hirsute, 1–15 dm. high; leaves mostly basal, spatulate or elliptic, 3–12 cm. long, entire or toothed, villous; stem-leaves lanceolate or oblanceolate; heads 1–5; involucre 6–8 mm. high; bracts linear, acuminate; ligules 10–15 mm. long. Open woods: N.S.—Minn.—La.—Fla. *E. Temp.* Ap–Je.

24. **E. philadélphicus** L. Perennial, with stolons and offsets; stem 3–10 dm. high, branched above, soft-hirsute; basal leaves oblanceolate, 5–15 cm. long, scarcely petioled, softly pubescent; lower stem leaves oblong, the upper lanceolate, acute, half-clasping; involucre 4–5 mm. high, 8–12 mm. broad, sparingly hirsute; bracts broadly linear, acute; ligules white or pinkish, 5–6 mm. long. Wet fields, meadows and woodlands: Lab.—Fla.—Calif.—B.C. *Temp.* My–Au.

25. **E. purpùreus** Ait. Perennial, with stolons and offsets; stem 3–10 dm. high, sparingly hirsute or glabrate above; basal leaves 6–15 cm. long, oblanceolate, acute, sparingly pubescent; upper leaves triangular lanceolate or ovate, with a broad cordate base, acuminate; involucre 5 mm. high, 10–12 mm. broad; bracts broadly linear; ligules rose-colored or reddish purple, 8–10 mm. long. Wet meadows: Man.—Mont.—Wash.—Mack. *W. Boreal.* Je–Jl.

26. **E. flagellaris** A. Gray. Branched at the base; stem decumbent, usually rooting at the ends and producing new plantlets; basal leaves spatulate or oblanceolate, 2–4 cm. long, strigose; stem-leaves linear; peduncles naked, 4–10 cm. long; involucre 4 mm. high and 8 mm. broad, hirsute; bracts linear, acute; ligules 5 mm. long, white or purplish. *E. stolonifer* Greene; *E. Mac-Dougalii* Heller. Banks of streams and meadows: S.D.—N.M.—Ariz.—Mont.; also Mex. *Plain—Subalp.—Son.* Ap–Au.

27. **E. cànus** A. Gray. Perennial, with a cespitose caudex; stem decumbent at the base, 1–2 dm. high, strigose-canescent; basal leaves linear-oblanceolate, 1–4 cm. long, canescent-strigose; stem-leaves narrowly linear; heads solitary; involucre 5–6 mm. high, 10–12 mm. broad; bracts lanceolate-linear, acute; ligules purplish or white, 6 mm. long. Dry plains: S.D.—Neb.—N.M.—Wyo. *Plain—Mont.* Jl–Au.

28. **E. subcanéscens** Rydb. Perennial, with a cespitose caudex; stem erect, canescent-hirsutulous; basal leaves linear-oblanceolate, mostly obtuse, 6–10 cm. long, triple-nerved, densely canescent; involucre 5–6 mm. high, 10–12 mm. broad; bracts linear, acute, canescent-hirsutulous; ligules white, rarely pinkish, about 8 mm. long. Stony slopes and plains: Sask.—Neb.—Colo.—B.C. *Plain.* Jl–Au.

29. **E. caespitòsus** Nutt. Perennial, with a cepitose caudex; stem 1–2 dm. high, canescent; basal leaves oblanceolate, 5–10 cm. long, less distinctly triple-nerved, obtuse, densely hirsutulous-canescent; involucre 6–7 mm. high, 10–15 mm. wide; bracts linear, hirsute-canescent, acute; ligules 40–50, white, 7–10 mm. long. Dry hills and plains: Man.—Neb.—Utah—Yukon. *Plain—Submont.* Jl–Au.

30. **E. laetévirens** Rydb. Stems several, 1.5–2 dm. high, slender, erect or ascending, silky-strigose; basal leaves very narrowly linear-oblanceolate, acute, 5–10 cm. long, sparingly strigose; stem-leaves narrowly linear; heads solitary; involucre 6 mm. high, 12–15 mm. broad, grayish villous-hirsute; ligules numer-

822 CARDUACEAE

ous, blue or purple, 7–8 mm. long. Mountain slopes: Wyo.—S.D. *Submont.—Mont.* Jl–Au.

31. E. montanénsis Rydb. Stem 1–2 dm. high, softly strigose; basal leaves narrowly linear to linear-oblanceolate, sparingly strigose, 5–10 cm. long; stem-leaves narrowly linear; heads solitary; involucre about 7 mm. high and 12–15 mm. broad; bracts linear, acuminate, densely hirsute-villous; ligules white or pinkish, 5–7 mm. long, 1.5–2 mm. wide. *E. Tweedyanus* Canby & Rose. *Wyomingia Tweedyana* A. Nels. Dry hills and plains: Mont.—Wyo.—S.D. *Plain—Mont.* My–Au.

32. E. peuciphýllus A. Gray. Perennial, with a cespitose caudex; stems 1–2 dm. high, cinereous-strigose, erect; leaves filiform, cinereous, 2–5 cm. long; heads usually solitary; involucre 4–5 mm. high, 8–10 mm. broad; ligules 5–6 mm. long. *E. filifolius* Piper; not Nutt. Dry hills: B.C.—Calif.—Ida.—Sask. My–Jl.

33. E. hyssopifòlius Michx. Perennial, with a branched caudex or rootstock; stem 1–2 dm. high, sparingly pilose, with ascending hairs, or glabrate; heads solitary at the ends of the branches; peduncles 3–10 cm. long; leaves narrowly linear, more or less spreading, 1–2 cm. long, 1–2 mm. wide, 3-ribbed; involucre 5 mm. high, broadly turbinate; leaves linear, acute, strigose; ligules 20–30, white, or tinged with purple, 6–7 mm. long. Calcareous rocks: Labr.—Me.—Mich.—Man.—Mack. *Boreal.* Je–Jl.

35. LÉPTILON Raf. Horseweed, Canada Fleabane.

Annual or biennial caulescent herbs. Leaves alternate, narrow. Heads small, panicled. Involucre campanulate; bracts in 2 or 3 series. Receptacle naked. Ray-flowers few, pistillate, fertile, with short white or purplish ligules. Disk-flowers hermaphrodite and fertile, usually with 4-lobed corollas. Anthers obtuse at the base. Style-branches flat, with short appendages. Achenes flattened, often pubescent. Pappus of many capillary brittle bristles.

Stem strict; lower leaves spatulate, toothed; ligules white. 1. *L. canadense.*
Stem divaricately branched; leaves all linear, entire; ligules purplish. 2. *L. divaricatum.*

1. L. canadénse (L.) Britton. Annual; stem erect, 2–30 dm. high, hirsute or glabrate; lower leaves 2–10 cm. long, hirsute or at least ciliate; upper leaves linear, sessile; heads numerous in an elongate panicle; involucre 3–4 mm. high and as broad; ligules numerous, white. *Erigeron canadensis* L. [G]. Waste places and dry soil: Lab.—Fla.—Calif.—B.C.; also Mex., W.Ind., and introduced in Eur. *Temp.—Trop.* Je–N.

2. L. divaricàtum (Michx.) Raf. Annual; stem diffusely branched, 0.5–3 dm. high, pubescent; leaves linear or subulate, entire, 1–2.5 cm. long; involucre about 2 mm. high, 3–4 mm. broad; ligules purple, very short. *Erigeron divaricatus* Michx. [G]. Sandy soil and waste places: Ind.—Tenn.—Tex.—S.D. Je–O.

36. BÁCCHARIS L. Groundsel Tree.

Perennial caulescent herbs or shrubs, usually with glabrous resinous foliage. Leaves alternate, often leathery, entire or toothed. Heads discoid, dioecious, corymbose or paniculate. Involucre campanulate; bracts imbricate in several series, those of the pistillate flowers more numerous. Receptacle flat, naked, pitted. Corollas mostly yellow; those of the staminate flowers tubular, those of the pistillate ones filiform. Anthers entire and obtuse at the base. Pappus of filiform bristles; those of the pistillate heads usually elongate in fruit, very fine; those of the staminate heads shorter, scabrous and often tortuose.

Pappus bristles in several series; plant 3–6 dm. high, herbaceous, with a woody base.
                                                                1. *B. Wrightii*.
Pappus bristles in a single series; plant 1–5 m. high, shrubby.
    Leaves oblong or lance-oblong, mostly obtuse.   2. *B. salicina*.
    Leaves linear or nearly so, mostly acute.   3. *B. neglecta*.

f 560

1. B. Wrightii A. Gray. A herbaceous perennial; stem diffusely branched, glabrous and angled; lower leaves linear, 2–4 cm. long, the upper smaller, linear-subulate; involucre of the staminate heads about 7 mm. high, 10–12 mm. broad; bracts in 3 series, lanceolate, acute, erose-ciliate; involucre of the pistillate heads 8–10 mm. high, 12–15 mm. broad; bracts in 5 or 6 series, with more evident scarious margins; achenes oblong, 8–10-ribbed, puberulent, cross-rugose, pappus brownish, 10–12 mm. long. Saline soil: Kans.—Colo.—Ariz.; also Mex. *St. Plains.—Son.* Ap–Je.

2. B. salicina T. & G. *Fig. 560.* A shrub, 1–5 m. high; branches green, glabrous, angled; leaves oblong or oblong-lanceolate, denticulate or entire, 3–5 cm. long; heads in small clusters; involucre of the staminate heads 4–5 mm. high and slightly broader; bracts in 4 or 5 series, ovate, yellowish green, erose-ciliate; involucre of the pistillate heads 6–8 mm. high and broad; inner bracts lanceolate; achenes 10 ribbed, glabrous; pappus white, soft, 10–12 mm. long. Saline soil: Kans.—Utah—N.M.—Tex. *St. Plains — Son.* Je–Jl.

3. B. neglecta Britton. A glabrous shrub, about 1 m. high, the branches paniculate, slender, ascending; leaves faintly 3-nerved, nearly sessile, 2–7 cm. long, 2–6 mm. wide; heads in small, short-peduncled clusters; involucres campanulate, 4 mm. high, the outer bracts ovate, acute or somewhat obtuse, the inner lanceolate, acuminate; pistillate pappus-bristles capillary, dull white. Saline soil: Neb.—Tex.—Mex. *Texan.* Jl–S.

Tribe 4. **GNAPHALIEAE.** Heads discoid. Involucres mostly scarious. Pistillate flowers with filiform corollas, the hermaphrodite ones with tubular corollas with entire styles, and sterile. Plants polygamo-dioecious or monoecious. Anthers sagittate at the base. Pappus of capillary bristles or none.

37. PLÙCHEA Cass. Marsh Fleabane.

Somewhat glandular, odorous herbs. Leaves alternate. Heads many-flowered, apparently discoid. Involucre imbricate. Receptacle flat, naked. Central flowers few, hermaphrodite but sterile, the corolla tubular, 5-cleft, the other flowers pistillate, fertile with filiform corollas. Achenes 4–5 angled; pappus of a single series of scabrous bristles.

1. P. camphoràta (L.) DC. Annual; stem 3–15 dm. high; leaves short-petioled, ovate, lanceolate or oblong, 6–12 cm. long, serrate or entire, reticulately-veined; heads in a flat-topped corymb, 6–8 mm. high; bracts ovate to lanceolate, puberulent; achenes pubescent. Salt marshes, mostly along the coast: Mass.—Fla.—Tex.; Kans.; Mex. Au–O.

38. DIAPÈRIA Nutt.

Low caulescent annuals. Leaves alternate, floccose. Heads globose, glomerate, discoid. Involucre woolly; bracts few, more or less scarious, in several series. Receptacle convex, chaffy. Paleae of the pistillate flowers barely concave, scarious; those of the central flowers similar or with woolly tips. Marginal flowers pistillate, fertile, with filiform corollas and 2-cleft style. Central flowers 2–5, hermaphrodite and usually sterile, with undivided style. Achenes obcompressed, smooth or very minutely papillose. Pappus wanting. *Evax* DC., in part.

1. D. prolifera Nutt. Annual; stem erect, simple or with ascending branches from the base, loosely tomentose; leaves spatulate, 1 cm. long or less; heads in terminal glomerules, subtended by a rosette of leaves, from which develop later 1–4 branches bearing similar terminal glomerules; involucre cylindric or oblong; paleae subtending the fertile flowers chartaceous-scarious, naked; those subtending the staminate flowers herbaceous and woolly-tipped. *Evax prolifera* Nutt. *Filago prolifera* Britton [B]. Dry ground: S.D.—Ark. —Tex.—Ariz. *Plain—Son.* My–Jl.

39. ANTENNÀRIA Gaertner. CAT'S-PAWS, PUSSY-TOES, EVERLASTING, LADIES' TOBACCO.

More or less woolly-tomentose perennial herbs, usually stoloniferous or soboliferous. Leaves alternate, the basal ones in the stoloniferous species forming rosettes. Plants dioecious, the staminate plants usually rarer; heads discoid, usually corymbose. Involucres turbinate or hemispheric; bracts imbricate in several series, those of the staminate heads usually broader. Receptacle flat, not chaffy. Pistillate flowers fertile, their corollas filiform, truncate; style 2-cleft; pappus-bristles capillary, united at the base and falling off together. Staminate or rather hermaphrodite flowers with tubular, 5-lobed corollas, sterile, their styles and ovaries rudimentary; pappus-bristles usually somewhat clavate at the apex; anthers caudate at the base. Achenes terete or nearly so.

Pappus of the staminate plants with clavate or scarious, dilated tips.
  Plant surculose-proliferous, with leafy stolons.
    Leaves of the rosettes small, 1-ribbed or indistinctly 3-ribbed, usually less
      than 3 cm. long.   1. DIOICAE.
    Leaves of the rosettes larger, distinctly 3-ribbed, 3–12 cm. long.
                                       2. PLANTAGINIFOLIAE.
  Plant not surculose-proliferous, with short erect stolons; leaves linear, densely
    villous-tomentose.   3. PULCHERRIMAE.
Pappus of the staminate plants not clavate; achenes puberulent; plants low and
  densely cespitose.   4. DIMORPHAE.

1. DIOICAE.
Leaves of the rosettes 5–15 mm. long, rhombic-spatulate; tomentum very fine and
  silky, appressed.   1. A. microphylla.
Leaves of the rosettes spatulate or oblanceolate, rounded at
  the apex; 1.5–4 cm. long; tomentum looser, at least
  above.
  Rosette-leaves spatulate, abruptly contracted into a peti-
    ole-like base.
    Upper surface of the leaves dull, permanently floc-
      cose, or tardily becoming glabrate in age.
      Pistillate plant 2–4 dm. high.
        Tips of the bracts rose-colored or pink.   2. A. rosea.
        Tips of the bracts white or nearly so.
          Leaves white-tomentose on both sides;
            stem-leaves ample, oblanceolate.
          Pistillate heads 7–8 mm. high; outer
            bracts obtuse or rounded at the apex.   3. A. oxyphylla.
          Pistillate heads fully 10 mm. high;
            bracts all acute or the outer acutish.   10. A. aureola.
        Leaves greener and tardily becoming gla-
         brate above; stem-leaves small, lin-
         ear.
         Stolons short, ascending, rather leafy.   4. A. neodioica.
         Stolons well developed, scaly, with well
          developed leaves only at the apex.   5. A. petaloidea.
    Pistillate plants rarely more than 1.5 dm. high.
      Involucres 5–8 mm. high.
        Leaves equally white-tomentose on both
         sides; bracts of both pistillate and

staminate plants oblong, with a con-
  spicuous brown or green spot at the
  base.   6. *A. sedoides.*
Leaves glabrate above ; bracts of the pistil-
  late plants linear.
  Plant with very short stolons.   7. *A. parvula.*
  Plant with slender, elongate stolons.   8. *A. Lunellii.*
Involucres 8–10 mm. high.
  Outer bracts of the pistillate plant rounded
    at the apex, the inner acute ; plant usu-
    ally less than 1 dm. high.   9. *A. aprica.*
  Outer bracts of the pistillate plant acute
    or obtuse, the inner attenuate ; plant
    usually about 2 dm. high.   10. *A. aureola.*
Upper surface of the leaves bright green and gla-
  brous, even when young ; pistillate plant 2–4 dm.
  high.   11. *A. canadensis.*
Rosette-leaves oblanceolate or obovate, gradually taper-
  ing to the sessile base ; leaves glabrate above.
  Stolons short and leafy ; pistillate plant low, about
    1 dm. high.
    Involucres of the pistillate plant 7–8 mm., of the
      staminate plant 5 6 mm. high ; rosette-leaves
      oblanceolate.   12. *A. longifolia.*
    Involucres of the pistillate plant 8–10 mm. high ;
      those of the staminate ones 6–7 mm. high ;
      rosette-leaves obovate.   13. *A. campestris.*
  Stolons long, their leaves, except those of the ter-
    minal rosette, narrow and scale-like ; plant
    usually 2–3 dm. high.
    Branches of the inflorescence very short ; heads
      glomerate or rarely racemosely arranged ; in-
      volucres 7–9 mm. high.   14. *A. neglecta.*
    Branches of the inflorescence longer ; heads corym-
      bose ; involucre 9–10 mm. high.   15. *A. chelonica.*

2. PLANTAGINIFOLIAE.

Upper surface of the leaves dark green and glabrous from
  the beginning ; stem, stolons, and stem-leaves more or
  less glandular.   16. *A. Parlinii.*
Upper surface of the leaves dull, permanently tomentose, or
  tardily glabrate in age ; plant not glandular, except
  the upper part of the stem slightly so in a form of
  A. fallax.
  Heads small ; involucre of the pistillate plants 6–8 mm.
    high.   17. *A. plantaginifolia.*
  Heads larger ; involucre of the pistillate plants 8–10 mm.
    high.
    Leaves of the rosettes rhombic-obovate ; blade broad-
      est at the middle, usually acutish.   18. *A. fallax.*
    Leaves of the rosettes spatulate or obovate, rounded
      at the apex ; blade broadest above the middle.
      Rosette-leaves 3–7 cm. long, loosely floccose above. 19. *A. occidentalis.*
      Rosette-leaves 2–4 cm. long, finely and densely
        white silky-tomentose on both sides, or gla-
        brate above.   20. *A. obovata.*

3. PULCHERRIMAE.

One species.   21. *A. pulcherrima.*

4. DIMORPHAE.

One species.   22. *A. dimorpha.*

1. **A. microphýlla** Rydb. Stem slender, strict, 2–3 dm. high ; stem-
leaves linear-oblong ; heads 5–30, conglomerate, 5–6 mm. high ; involucre slightly
glandular ; bracts of the pistillate heads linear-oblong to linear-lanceolate,
greenish or straw-colored, the inner acute ; those of the staminate heads oval
or elliptic, rounded at the apex. *A. parvifolia* Greene ; not Nutt. *A. solstitialis*
Lunell, the staminate plant, which is smaller. Dry hills and plains : Man.—
N.D.—Neb.—N.M.—B.C. Plain—Subalp. Je–Jl.

2. **A. rósea** (D. C. Eat.) Greene. Stems of both plants 2–4 dm. high,
slender ; leaves narrowly spatulate or oblanceolate, acute, 1.5–2 cm. long, white,
with close tomentum ; heads short-peduncled in rounded clusters ; involucre

about 5 mm. high; bracts of the pistillate plants linear to oblong, obtuse, usually rose-colored; those of the rare staminate plant broadly elliptic, pale pink. *A. parvifolia* Rydb.; not Nutt.; *A. dioica rosea* D. C. Eat.; *A. sordida* Greene. Meadows: Alta.—S.D.—Colo.—Calif.—Yukon. *Plain—Mont.—W. Temp.* Je–Au.

3. **A. oxyphýlla** Greene. Stem of the pistillate plant 3–4 dm. high, that of the staminate plant 2–3 dm. high; leaves of the rosettes spatulate-obovate or broadly oblanceolate, densely tomentose beneath, less so above; heads many in a rounded corymb; involucre 6–7 mm. high; bracts of the pistillate heads with dull white tips, the outer ovate, obtusish, the inner lanceolate or linear-lanceolate, acute. Hills and mountains: S.D.—Neb.—Wyo.—Mont. *Submont.—Mont.* My–Jl.

4. **A. neodioìca** Greene. Stem of the pistillate plant 1–4 dm. high, that of the staminate plant about 1 dm. high; leaves of the rosettes obovate-spatulate or spatulate, 1–4 mm. long, 5–15 mm. broad; heads corymbose; involucre of the pistillate heads 6–8 mm. high, its bracts linear, brownish at the base, the inner acute, the outer obtuse; that of the staminate heads 5 mm. high, its bracts oblong, obtuse; style pale. Open woods: Newf.—Va.—Ind.—Minn.—S.D. *Canad.* My–Jl.

5. **A. petaloìdea** Fern. Stem 2–4 dm. high; stolons rather short, scaly, leafy at the end; basal leaves and those of the offsets spatulate-obovate to oblanceolate, abruptly contracted into the petiole-like base; heads corymbose, about 1 cm. broad, few on the fertile plant, densely clustered on the sterile; bracts lanceolate, brownish, and with long petal-like, white-scarious obtuse tips. Fields and open woods: Que.—N.Y.—Mich.—Minn.—Ont. *Canad.* My–Jl.

6. **A. sedoìdes** Greene. Stem about 1 dm. high; leaves of the rosettes spatulate, less than 1 cm. long, densely white-tomentose on both sides; stem-leaves linear or linear-spatulate, small; heads 4–6, conglomerate, sessile; involucres 5–6 mm. high, sometimes more or less viscid; tips of the bracts dirty white, obtuse or acutish. *A. arida viscidula* E. Nels.; *A. viscidula* Rydb. Mountains: Alta.—Colo.—Man. *Plain—Mont.* Jl.

7. **A. párvula** Greene. Stem 2.5–7 cm. high; leaves of the rosettes 1–1.5 cm. long, oval, sometimes suborbicular, white-tomentose on both sides; heads few, subsessile, large for the plant. Foothills: S.D.

8. **A. Lunéllii** Greene. Stems of the pistillate plants 5–15 cm. high, those of the staminate ones scarcely 5 cm. high; leaves of the rosettes obovate-spatulate, 2 cm. long or less, less than 1 cm. wide; stem-leaves small, linear, acute; pistillate heads few, conglomerate or at length racemose; involucres about 6 mm. high; bracts dark brown at the base, linear, obtuse; staminate heads clustered; involucre 5 mm high; bracts oblong, obtuse. Prairies: N.D. My–Je.

9. **A. áprica** Greene. *Fig. 561.* Stem in both the staminate and pistillate plants low, less than 1.5 dm. high; leaves of the rosettes spatulate or cuneate-oblanceolate, 1–2 cm. long, densely tomentose on both sides; stem-leaves 1 cm. long, linear, acute; heads 3–6, sessile or nearly so, conglomerate; involucre of the pistillate plants 7–9 mm. high, usually with brown spots and white or rarely pinkish tips, oblong, obtuse or acutish; involucre of the staminate heads 6–7 mm. high; bracts oval or elliptic, obtuse. *A. pumila* Greene (?). *A. modesta* Greene. Plains, hills and mountains: Man.—Minn.—Neb.—N.M.—Utah—B.C. *Plain—Subalp.* My–Jl.

f. 561

10. **A. auròola** Lunell. Stem of the pistillate plant 1–3 dm. high, floccose; leaves of the rosettes 2–3 cm. long, 5–8 mm. wide, cuneate-obovate, densely and persistently silky-tomentose on both sides; stem-leaves 1.5–2.5 cm. long, the lower oblong, the upper narrowly lanceolate; heads 1–10, campanulate; involucre 8–10 mm. long; bracts in 4 series, lanceolate to linear-lanceolate, the inner attenuate, with a brown spot and long white scarious tips. Dry soil: N.D. *Prairie.* Jl.

11. **A. canadénsis** Greene. Plant forming broad mats; stems of the pistillate plants 3–5 dm. high, slender, those of the staminate plants about 1 dm. high; leaves of the rosettes spatulate, 2–3 cm. long, 0.5–1 cm. broad; stemleaves linear, acute; heads corymbose; involucre of the pistillate plants 8–10 mm. high; its bracts lance-linear, acute, greenish or brownish at the base; involucre of the staminate plants 5 mm. high, its bracts broadly oblong, obtuse; style pale, drying brownish. Dry open places. Nowf.—N.Y.—Mich. – Man. *E. Boreal.* My–Jl.

12. **A. longifòlia** Greene. Stems 1–1.5 dm. high; leaves of the rosettes oblanceolate, acute or obtuse, glabrate above, 2–5 cm. long, less than 1 cm. wide, without distinct petioles; heads of the pistillate plants in small glomerules; involucre 7–8 mm. high; bracts brownish at the base, the inner linear and acute, the outer linear-spatulate and obtuse; involucre of the staminate plants 5–6 mm. high; bracts oblong, obtuse or notched. Open woods and prairies: Mo.—Iowa—Neb.—Okla. *Prairie.*

13. **A. campéstris** Rydb. Stems low, about 1 dm. high or less; leaves of the rosettes obovate-cuneate, 2–4 cm. long and about 1 cm. wide, glabrate above, tomentose beneath; stem-leaves small, linear, acute; heads 3–6, sessile, conglomerate; involucre of the pistillate plant 8–9 mm. high, outer bracts oblong, obtuse, the inner ones linear-lanceolate, acute or acuminate; staminate involucre 6–7 mm. high; bracts elliptic or obovate, rounded at the apex. *A. nebraskensis* Greene. *A. angustiarum* Lunell. Plains: Man.—Mich.– Mo.– Kans.—Sask.—B.C. *Prairie—Plain.* My–Je.

14. **A. negléecta** Greene. Stems of the pistillate plants 1–4 dm. high, those of the staminate plants 1 dm. high or less; leaves of the rosettes oblanceolate or cuneate-spatulate, 3–6 cm. long, 0.5–1.5 cm. wide, glabrate above in age; heads at first crowded, later sometimes becoming more racemose; involucre of the pistillate plants 7–9 mm. high, its bracts linear, brownish at the base, the inner ones acute; involucre of the staminate plants about 6 mm. high; bracts broadly oblong, obtuse. Pastures and open woods: N.B.—Va.—Kans.—Minn. *Canad.—Allegh.* Ap–Je.

15. **A. chelónica** Lunell. Stems of the pistillate plants 1–3 dm. high, those of the staminate plants scarcely 1 dm. high; leaves of the rosettes oblanceolate, 2–4 cm. long, 0.5–1.5 cm. wide; stem-leaves linear; heads of the pistillate plants corymbose; involucre nearly 10 mm. high; bracts linear, brownish at the base, the inner acute; heads of the staminate plants in capitate clusters; involucre 8–9 mm. high; bracts broad, obtuse. Woodlands: Turtle Mountains, N.D. Je.

16. **A. Parlínii** Fernald. Stems of the pistillate plants 3–5 dm. high, those of the staminate plants about 2 dm. high; leaves of the rosettes oblong-spatulate or obovate-spatulate, obtuse or acutish, 3–8 cm. long, 1–3.5 cm. wide; heads corymbose; involucre of the pistillate plants 8–10 mm. high, its bracts linear, acute, that of the staminate plants 5–6 mm. high, its bracts oblong, obtuse. *A. arnoglossa* Greene. Rich soil, open woods: Me.—D.C.– Iowa. *Canad.– Allegh.* My–Je.

17. **A. plantaginifòlia** Hook. Stems of the pistillate plants 2–5 dm. high, those of the staminate plants 1–2 dm. high; leaves of the rosettes ovate-

spatulate or spatulate, acute, 3–7 cm. long, 1–3 cm. wide; heads corymbose on short peduncles; involucre of the pistillate plants 6–8 mm. high, that of the staminate ones 5–6 mm. high; bracts of the former linear, acute, greenish, or purplish below, those of the latter oblong, obtuse; styles usually purplish. Dry soil: Me.—Fla.—Tex.—Minn.—Man. *E. Temp.* Ap–Je.

18. **A. fállax** Greene. Stems of the pistillate plants 2–4 dm. high, those of the staminate ones 1–2 dm. high, in one form (*A. ambigens* (Greene) Fernald) slightly glandular; leaves of the rosettes 4–10 cm. long, 2–5 cm. wide, rhombic-obovate or obovate-spatulate; heads in crowded corymbs; involucre of the pistillate plants 8–10 mm. high, that of the staminate ones about 6 mm. high; bracts of the former linear, acute, light brownish green at the base, those of the latter oblong, obtuse; styles mostly pale. Open woods: Me.—D.C. —Kans.—Minn. *Canad.—Allegh.* Ap–Je.

19. **A. occidentàlis** Greene. Stems of the pistillate plants 2–4 dm. high, those of the staminate ones 1–2 dm. high; leaves of the rosettes 3–7 cm. long, 1–2.5 cm. wide; stem-leaves linear-lanceolate; heads corymbose; involucre of the pistillate plants 8–9 mm. high, that of the staminate ones 5–6 mm. high; bracts of the former lance-linear, acute, with green or brown bases, those of the latter oblong, obtuse; styles usually purplish. Rich soil: Que.—N.H.—Iowa —Minn.—Man. (?). *Canad.* My–Je.

20. **A. obováta** E. Nels. Stems 2–3 dm. high; leaves of the rosettes about 3 cm. long, more or less distinctly 3-ribbed, obovate-cuneate, permanently tomentose on both sides; stem-leaves small, oblong-linear; heads 3–7, corymbose; involucre about 8 mm. high; bracts of the pistillate plants oblong to linear-lanceolate, with purplish-brown spots below, acute or acuminate, or the outermost obtusish. Hills and mountains: Man.—S.D.—Colo. *Submont.— Mont.* Je–Jl.

21. **A. pulchérrima** (Hook.) Greene. Perennial with a rootstock; stem 3–5 dm. high; basal leaves oblanceolate, 8–12 cm. long, loosely tomentose, 3-ribbed; stem-leaves narrower; heads several, corymbose; involucre of the pistillate plants nearly 1 cm. high, turbinate; bracts in 6 or 7 series, the outer broadly ovate, obtusish, the inner lanceolate, acuminate, with brown bases and dirty-white or brownish tips; staminate involucre 5–6 mm. high; bracts with rounded brownish or dirty white tips. Hillsides and mountains: Que. (Anticosti)—Ont. (James Bay)—Sask.—Colo.—Wash.—Yukon. *Boreal.—Mont.* Je–Au.

22. **A. dimórpha** (Nutt.) T. & G. Pulvinate-cespitose perennial; stems 1–3 cm. high; leaves narrowly oblanceolate, appressed-tomentose on both sides, 1–2 cm. long; heads solitary; pistillate involucre 1–1.5 cm. high, turbinate, tomentose only at the base; outer bracts ovate, acute, the inner narrowly lanceolate, attenuate; staminate involucre hemispheric, 6–7 mm. high; bracts oblong, obtuse, darker brown. Dry plains: Mont.—Neb.—Colo.—Nev.—B.C. *Plain—Submont.* Ap–Jl.

40. **ANÁPHALIS** DC. Pearly Everlasting.

Tomentose, caulescent, perennial herbs, with rootstocks. Leaves alternate, narrow, entire, sessile. Heads discoid, polygamo-dioecious. Involucre hemispheric; bracts imbricate, in several series, mostly pearly white. Receptacle convex, not chaffy. Pistillate heads usually with a few hermaphrodite flowers in the center; their corollas filiform; styles 2-cleft. Hermaphrodite flowers with tubular, 5-toothed corollas and undivided styles, sterile. Pappus-bristles capillary, falling off separately.

Leaves narrowly linear-lanceolate, taper-pointed, dark green above; inflorescence
    usually ample.   1. *A. margaritacea.*
Leaves narrowly lanceolate, acute, often yellowish green; inflor-
    escence usually congested.   2. *A. subalpina.*

1. A. margaritàcea (L.) B. & H. *Fig. 562.*
Perennial, with a creeping rootstock; stem erect,
3–10 dm. high; leaves 6–15 cm. long, 4–6 mm. wide,
tomentose beneath, floccose and usually glabrate
above, often revolute-margined; involucre fully 7
mm. high; bracts obtuse. Open woods and clearings:
Newf.—N.C.—Kans.—Man.; Eur. *Canad.—Allegh.*
Jl–Au.

2. A. subalpìna (A. Gray) Rydb. Perennial,
with a creeping rootstock; stem 3–4 dm. high; leaves
comparatively thin, 3–8 cm. long, 5–15 mm. wide,
densely white-tomentose beneath, greener and sometimes glabrate above; involucres 6–7 mm. high;
bracts broadly ovate, mostly acute. *A. margaritacea
subalpina* A. Gray. Open mountain woods: S.D.—
Colo.—Utah—B.C. *Submont.—Subalp.* Je–Au.

f 562

41. GNAPHÀLIUM L. Everlasting, Cudweed.

Annual, biennial, or perennial, caulescent herbs, with woolly and sometimes
glandular foliage. Leaves alternate, entire, narrow. Heads discoid, glomerate
or paniculate in our species. Involucre campanulate or hemispheric; bracts
scarious, imbricate in several series; receptacle not chaffy, usually pitted. Marginal flowers pistillate, fertile, in several series, with filiform corollas and
2-cleft styles. Central flowers hermaphrodite, few, with tubular corollas and
also 2-cleft styles. Anthers sagittate at the base. Achenes terete or slightly
flattened; pappus of capillary bristles, distinct or united at the base, deciduous.

Bristles of the pappus not united at the base, falling off separately.
Heads not leafy-bracted; involucre well imbricate, its bracts scarious, white or
tinged with brownish, rose, or yellow; plants mostly tall.
Leaves not decurrent; plant not glandular.   1. *G. obtusifolium.*
Leaves distinctly decurrent, as well as the stem decidedly
glandular.   2. *G. Macounii.*
Heads leafy-bracted; involucre little imbricate, its bracts
brown or greenish; plants low.
Plant loosely floccose; leaves broad, spatulate, oblong, or
oblanceolate.   3. *G. palustre.*
Plant appressed-tomentose; leaves, except the lowest ones,
narrowly oblanceolate or linear.
Stem diffusely branched; glomerules crowded, cymosely
disposed.
Upper leaves narrowly oblanceolate.   4. *G. uliginosum.*
Upper leaves narrowly linear.   5. *G. exilifolium.*
Stem mostly simple, erect; glomerules spicately dis-
posed.   6. *G. Grayi.*
Bristles of the pappus united in a ring at the base and falling
off together; leaves oblanceolate or spatulate.   7. *G. purpureum.*

1. G. obtusifòlium L. Annual; stem simple below, branched at the summit, 3–10 dm. high; leaves linear-lanceolate, sessile, glabrous above, 2–7 cm.
long, 4–8 mm. wide; heads numerous; involucre 6 mm. high; bracts white,
tinged with brown, glabrous or nearly so, obtuse. *G. polycephalum* Michx. [G].
Open places: N.S.—Fla.—Tex.—Man. *E. Temp.* Au–S.

2. G. Macoûnii Greene. Biennial, with a taproot; stem 5–8 dm. high,
corymbosely branched above, glandular-pubescent, loosely tomentose above;
leaves lanceolate to linear, soon green and glandular above, white-tomentose
beneath, 5–10 cm. long, adnate-decurrent; involucre glabrous, 5 mm. high;
bracts ovate or the inner lanceolate, acute, white or straw-colored, turning
rusty. *G. decurrens* Ives [G]; not L. *G. Ivesii* Nels. & Macbr. Open ground:
N.S.—Pa.—N.M.—Ariz.—B.C. *Temp.—Mont.* Jl–S.

3. **G. palústre** Nutt. Low spreading annual; stem at first erect, later diffusely branched and spreading, 5–20 cm. high; leaves spatulate or the uppermost oblong or lanceolate, 1–2 cm. long; heads in small, leafy-bracted clusters; involucre woolly below, 3–4 mm. high; bracts oblong to linear or the inner with white tips. Wet places: Mont.—S.D.—Neb.—N.M.—Calif.—B.C. *W. Temp.—Submont.* My–Au.

4. **G. uliginòsum** L. Annual; stem at first simple, erect, later diffusely branched and spreading; leaves narrowly oblanceolate; heads in small leafy-bracted clusters; involucre floccose at the base, 3–4 mm. high; bracts oblong to linear, mostly acute, brown. Wet places: Newf.—Va.—Colo.—Utah—Ore. —B.C.; also northern Eur. *Boreal.—Mont.* Jl–Au.

5. **G. exilifòlium** A. Nels. Low annual; stem branching at the base; branches decumbent or assurgent, 8–12 cm. long, sparingly floccose; leaves 2–4 cm. long; involucre about 3 mm. high; bracts lanceolate, acute, brown below, white-tipped. *G. angustifolium* A. Nels.; not Lam. Wet places: Colo.—Wyo. —S.D. *Submont.—Mont.* Jl–S.

6. **G. Gràyi** Nels. & Macbr. Strict and simple annual; stem 1–3 dm. high, appressed-tomentose; leaves all linear, appressed-tomentose, 1–4 cm. long; involucre about 3 mm. long; bracts oblong to linear, wholly brown, or the innermost with lighter tips, obtuse. *G. strictum* A. Gray; not Moench. Wet places: Wyo.—S.D.—Neb.—N.M.—Ariz. *Submont.—Mont.* Au–S.

7. **G. purpùreum** L. Perennial, with a short rootstock; stem 1–4 dm. high, floccose; leaves spatulate to oblanceolate, 2–8 cm. long, glabrate above; glomerules of the heads arranged in a narrow spike-like panicle; involucre about 4 mm. high; bracts yellowish brown or purplish, acute. Wet soil: Me.— Fla.—Tex.—Kans.—Minn. *E. Temp.* Je–Jl.

Tribe 5. **INULEAE.** Heads radiate. Involucres imbricate and herbaceous. Receptacle naked. Style-branches of the hermaphrodite and fertile flowers linear, rounded at the apex. Pappus of filiform bristles.

42. ÍNULA L. ELECAMPANE.

Perennial leafy herbs. Leaves alternate, sometimes mainly basal, with toothed blades, the lower usually petioled. Heads radiate, usually showy. Involucre hemispheric or campanulate; bracts in several series, often foliaceous. Receptacle flat or convex, pitted, not chaffy. Ray-flowers pistillate, fertile; ligules yellow. Disk-flowers hermaphrodite; corollas tubular. Anthers sagittate at the base, with caudate auricles. Style-branches narrow, those of the disk-flowers obtuse. Achenes 5-ribbed. Pappus of several capillary bristles.

1. **I. Helènium** L. A stout perennial, with a thick root; stem 1–2 m. high, branched; lower leaves oblong or elliptic, denticulate, 2–4 dm. long; stem-leaves oblong to ovate, acute or acuminate, tomentose beneath, sessile or clasping; involucre 1–1.5 cm. high; ligules numerous, narrowly linear, 2–3 cm. long. Waste places: N.S.—N.C.—Mo.—Minn.; escaped from cult.

Tribe 6. **ADENOCAULEAE.** Heads discoid but heterogamous. Both hermaphrodite and pistillate flowers with tubular 4 or 5-toothed corollas. Receptacle naked, small. Pappus none.

43. ADENOCAÙLON Hook.

Perennial herbs, with rootstocks. Leaves alternate, with winged petioles and dilated blades. Inflorescence paniculate, the peduncles beset with stalked glands. Heads several-flowered, discoid. Involucre turbinate, with a few herbaceous thin bracts. Marginal flowers pistillate, 4-lobed or somewhat 2-lipped. Central flowers hermaphrodite, but sterile; corollas broadly funnelform, deeply 4- or 5-cleft, all white. Anthers sagittate and the auricles minutely caudate. Achenes obovoid or clavate, very obtuse, beset with stipitate glands above. Pappus none.

1. **A. bícolor** Hook. *Fig. 563.* Perennial, with a creeping rootstock; stem erect, 3–10 dm. high, more or less floccose; leaf-blades deltoid-cordate, 5–10 cm. long, sinuately dentate or somewhat lobed, soon green and glabrate above, densely white-tomentose beneath; involucre 2.5–3 mm. high; bracts 4 or 5, ovate, acute, reflexed in fruit; achenes 4–6, clavate, 7–8 mm. long. Damp woods: B.C.—Calif.— Mont. Mich.— w Ont. Je–Au.

Tribe 7. **HELIANTHEAE.** Heads radiate or rarely discoid. Marginal flowers if present pistillate and usually ligulate. Disk-flowers hermaphrodite, with tubular 4- or 5-lobed corollas. Receptacle paleaceous. Anthers sometimes sagittate at the base, but never caudate. Style-branches of the disk-flowers usually with hairy, subulate, lanceolate or conic appendages. Pappus various or none, but not of capillary bristles.

44. MELAMPÓDIUM L.

Annual or perennial herbs, or shrubs. Leaves opposite, often narrow, entire, toothed, or pinnatifid. Heads radiate. Involucre double, the outer of 4 or 5 partly united flat bracts, the inner of a series of concave bracts, each embracing an achene and deciduous with it. Receptacle convex or conic. Ray-flowers in a single series, pistillate; ligules white or yellow. Disk-flowers hermaphrodite but sterile. Anthers entire at the base. Achenes broadened upwards, somewhat incurved. Pappus wanting.

1. **M. leucánthum** T. & G. *Fig. 564.* Perennial, with a taproot and a short caudex; stems several, branched below, 1–3 dm. high, cinereous-strigose; leaves linear or the lower oblanceolate, 3–5 cm. long, entire or sinuately lobed; involucre about 5 mm. high and fully 1 cm. broad, cinereous; ligules white, about 1 cm. long and 5 mm. wide. Dry plains: Kan.—Tex.—Ariz.—Colo. Mr–O.

45. POLÝMNIA L. LEAF-CUP.

Perennial leafy-stemmed herbs, or rarely shrubby plants. Leaves mostly opposite, petioled, toothed, angled, or lobed. Heads radiate. Involucre double, the outer of 5 loose bracts, the inner of 5 or more thin bracts, subtending the marginal achenes. Receptacle flat, chaffy. Ray-flowers pistillate, fertile, the corollas with short-pubescent tubes; ligules yellow or white. Disk-flowers hermaphrodite but sterile. Anthers 2-toothed at the base. Achenes short, swollen, glabrous. Pappus wanting.

Leaves palmately lobed; ligules yellow; achenes striate.   1. *P. Uvedalia.*
Leaves pinnately lobed; ligules white, very small; achenes 3-ribbed.   2. *P. canadensis.*

1. P. Uvedàlia L. *Fig. 565.* Stem 1–3 m. high, pubescent to nearly glabrous; leaf-blades 3–5-lobed or angled, 1–5 dm. broad, decurrent on the petioles; outer bracts of the involucre ovate or elliptic, 9–14 cm. long, ciliate; ray-flowers 10–14; ligules linear-oblong to elliptic, 1.5–2 cm. long; disk yellow; achenes black, obovoid, 5–6 mm. long, flattened. Moist thickets: N.Y.—Mich.—Kans. (?)—Tex.—Fla. *E. Temp.* Jl–Au.

2. P. canadénsis L. Stem 5–10 dm. high, viscid-hirsute; leaf-blades oblong to ovate, 1–2 dm. long, pinnately lobed, the lobes toothed or sinuate; head drooping; outer bracts pubescent, 6–7 mm. long; ligules few, inconspicuous, suborbicular, 1–2 mm. long, slightly 3-lobed; achenes about 3 mm. long, 3-angled. Thickets: Ont.—Minn.—Kans.—Ga. *Canad.—Allegh.* Je–S.

f. 565.

46. SÍLPHIUM L. Compass-plant, Prairie Dock, Rosin-weed, Cup-plant.

Coarse perennial herbs, often with resinous sap. Leaves alternate or opposite, from entire to bipinnatifid. Heads radiate, showy. Involucre campanulate or hemispheric, many-flowered; bracts broad, in a few series. Receptacle flat, with numerous paleae subtending the disk-flowers. Ray-flowers pistillate, fertile, in 2 or 3 series; ligules yellow or rarely white. Disk-flowers hermaphrodite, but sterile. Anthers 2-toothed or entire at the base. Achenes flat and broad, 2-winged, often emarginate at the apex. Pappus wanting, or of 2 small awns.

Leaves not connate-clasping at the base; stem terete or slightly 4-angled.
  Leaves pinnately parted.   1. *S. laciniatum.*
  Leaves merely toothed or entire.
    Stem scape-like, leafless or nearly so, except at the base.   2. *S. terebinthinaceum.*
    Stem leafy; leaves mostly opposite.
      Involucral bracts pubescent and ciliate; leaves mostly scabrous-hispidulous on both sides.   3. *S. integrifolium.*
      Involucral bracts glabrous, except the hispidulous-ciliolate margins; leaves smooth or nearly so beneath, thick.   4. *S. speciosum.*
Leaves connate-clasping, perfoliate; stem square in cross-section.   5. *S. perfoliatum.*

1. S. laciniàtum L. *Fig. 566.* Stem 1–3.5 m. high, coarsely hispid; leaves mostly basal, turning vertically, pointing north and south, 1–4 dm. long, divided into lanceolate or linear, entire or pinnatifid segments; involucre 2–2.5 cm. high; bracts lanceolate or ovate; ligules yellow, 3–5 cm. long. Compass-plant. Prairies: Mich. — N.D. — Tex. — Ala. *Prairie—Texan.* Jl–S.

2. S. terebinthinàceum Jacq. Stem 1–3 m. high; basal leaves long-petioled; blades ovate to oblong, cordate at the base, 1–6 dm. long, dentate; involucre 1–5 cm. high; bracts rounded and obtuse; ligules 15–25, yellow, 2–3 cm. long. Prairie Dock. Prairies and oak-openings: Ont.—Ga.—La.—Iowa—Minn. *Canad.—Allegh.—Prairie.* Jl–S.

f. 566.

3. S. integrifòlium Michx. Stem 8–15 dm. high, rough-pubescent; leaves mostly sessile, ovate or lanceolate, entire or den-

tate, 7–12 cm. long, scabrous-hispidulous on both sides; involucre 1–1.5 cm.
high; bracts triangular-ovate to lanceolate, acute or the inner obtuse; ligules
15–23, yellow, 2–3 cm. long. ROSIN-WEED. Prairies: Mich.—Miss.—Tex.—Neb.
—Minn. *Prairie—Texan.* Au–S.

4. S. speciòsum Nutt. Stem 6–12 dm. high, glabrous, sulcate, straw-
colored; leaves opposite, sessile, entire or serrulate, or the lower with short,
winged petioles and sometimes coarsely serrate, mostly ovate, 1–2 dm. long,
scabrous above, smooth beneath; involucre 12–15 mm. high; outer bracts deltoid or lanceolate, the inner ones broadly ovate or subreniform; ligules 15–25,
bright yellow, 2–3 cm. long. *S. integrifolium laeve* T. & G. Dry prairies: Ill.
—Neb.—Tex.—Ark. *Prairie—Ozark—Tex.* Au–S.

5. S. perfoliàtum L. Stem 1–2.5 m. high, glabrous; leaves connate-per-
foliate, ovate or lanceolate, 2–6 dm. long, dentate; involucre 1–2 cm. high and
fully as broad; bracts ovate to lanceolate or oval, or the inner spatulate; ligules 15–25, bright yellow, 2.5–3 cm. long. CUP-PLANT. Prairies: Ont.—Ga.—
La.—Neb—Minn. *Canad.—Allegh.—Prairie.* Jl–S.

47. BERLANDIÈRA DC.

Perennial, caulescent or scapose herbs. Leaves alternate, sometimes basal,
toothed or pinnatifid. Heads radiate. Involucre hemispheric or depressed,
many-flowered; bracts in 3 series, the inner becoming thin and reticulate. Receptacle flat, chaffy. Ray-flowers few, pistillate, fertile; ligules yellow. Disk
flowers hermaphrodite, but sterile. Anthers entire or minutely toothed at the
base. Style branches in the disk-flowers united. Achenes broad, flat, wingless,
more or less adnate to the adjacent supporting bract. Pappus obsolete, caducous, or of 2 caducous awns.

Leaves toothed; heads corymbose.   1. *B. tomana.*
Leaves lyrate-pinnatifid; heads solitary at the ends of the branches.   2. *P. lyrata.*

1. B. texàna DC. Plant simple; stem 5–12 dm. high, villous-tomentose;
leaves ovate or triangular-lanceolate, 5–15 cm. long, crenate, cordate at the base,
tomentose beneath; involucre depressed, 3–5 cm. broad; bracts spatulate to
oblong, ciliolate; ligules 2 cm. long. Hillsides: Mo.—La.—Tex.—Kans. *Ozark.*
Jl–Au.

2. B. lyràta Benth. Plant usually tufted, cinereous; stems 1–4 dm.
high, striate, puberulent; leaves lanceolate in outline, pinnatifid, 5–15 cm.
long, cinereous, especially beneath, the segments oblong to oval, crenate; involucre hemispheric, 2 cm. broad; bracts canescent, the outer ovate or obovate,
the inner broadly obovate-cuneate; ligules striate, oblong or oblong-cuneate,
1–1.5 cm. long; achenes obovate, 5–6 mm. long. Dry plains and hills: Ark.—
Kans.—Ariz.—Tex. *St. Plains—Son.* Ap–Au.

48. ENGELMÁNNIA T. & G.

Perennial herbs, with alternate pinnatifid leaves. Heads radiate. Involucres hemispheric, or campanulate; bracts in 2 or 3 series, the outer narrow,
the inner broad. Receptacle flat, chaffy. Ray-flowers 8–10, pistillate, fertile;
ligules yellow. Disk-flowers hermaphrodite, but sterile, partly embraced by the
paleae. Anthers 2-toothed at the base. Achenes flat, broader upwards, 1-ribbed
on each side. Pappus an irregular crown.

1. E. pinnatífida T. & G. Stem 3–7 dm. high, hirsute or hispid; leaves
5–15 cm. long, pinnatifid, the upper sessile; outer bracts linear, the middle ones
suborbicular, with linear tips, and the inner oval or obovate, ciliate; ligules
8–10, golden yellow, 1 cm. long. Plains and hillsides: La.—Kans.—Colo.—Ariz.
St. Plains—Son.

49. PARTHÈNIUM L. FEVERFEW.

Perennial herbs or shrubs, rarely annuals. Leaves alternate, toothed, pinnatifid, or dissected. Heads radiate, but ligules inconspicuous. Involucre campanulate, hemispheric, or shallower; bracts in 2 or 3 series, appressed, obtuse. Receptacle convex or conic, chaffy. Ray-flowers usually 5, pistillate and fertile; ligules short and broad, white or nearly so. Disk-flowers hermaphrodite but sterile, embraced by the paleae. Achenes flat, margined, keeled on the inner face, tipped by the persistent corollas. Pappus of 2 or 3 scales or awns.

Stem glabrous below, puberulent above, from a thickened rootstock.
1. *P. integrifolium.*
Stem low, hispid, from a slender stoloniferous rootstock.
2. *P. hispidum.*

1. P. integrifòlium L. Perennial herb; stem 4–12 dm. high, corymbose above; lower leaves long-petioled, the blades oblong or lanceolate, crenately toothed or cut-lobed below the middle, 5–15 cm. long; upper leaves ovate or oval, sessile or partly clasping, once to thrice crenate; heads numerous; involucre globular, 4–6 mm. high; bracts oblong to suborbicular; ligules 2 mm. long, funnelform, retuse. Dry soil: Md.—Minn.—Tex.—Ga. *Allegh.—Austral.* My–S.

2. P. hispidum Raf. *Fig. 567.* Perennial herb, stem 6 dm. high or less, hispid; leaves similar to those of the preceding, irregularly crenate, often more or less lyrate, hispid on both sides; heads few, somewhat larger, the outer bracts broader. *P. repens* Eggert. [G]. Barrens: Mo.—Kans.—Tex. *Ozark.* Ap–Jl.

50. ZÍNNIA L.

Annual or perennial herbs or shrubby plants. Leaves opposite, commonly narrow, entire or sparingly toothed. Heads radiate. Involucres campanulate to nearly cylindric; bracts in 3–5 series, firm, appressed, rather dry. Receptacle conic to nearly cylindric, chaffy. Ray-flowers pistillate, fertile; their ligules red, purple, or yellow, or variegated, in ours yellow, persistent and becoming papery. Disk-flowers hermaphrodite, fertile, enveloped in the paleae. Achenes various, those of the rays 3-angled, those of the disk-flowers flattened. Pappus of one to several teeth or awns, or lacking. *Crassina* Scepin.

1. Z. grandiflòra Nutt. Perennial, with a woody root and short caudex; stem 1–2 dm. high, branched, puberulent; leaves linear, more or less distinctly 3-nerved, impressed-punctate, scabrous-hispidulous, 2–3 cm. long; heads solitary, short-peduncled; involucres 6–9 mm. high; ligules yellow, turning whitish, 12–15 mm. long; disk-corollas dark brick-red. *C. grandiflora* (Nutt.) Kuntze. Plains: Tex.—Kans.—Colo.—Ariz. *St. Plains—Son.—Submont.* My–S.

51. HELIÓPSIS Pers. OX-EYE.

Perennial or rarely annual, leafy-stemmed herbs. Leaves opposite, petioled, toothed. Heads radiate, corymbose or solitary. Involucres hemispheric or campanulate; bracts nearly equal, in 2 or 3 series, the outer herbaceous. Receptacle conic or convex, chaffy. Ray-flowers pistillate, fertile, with oblong yellow ligules, more or less persistent on the fruit. Disk-flowers numerous, hermaphrodite, fertile, enveloped by paleae. Stigmas with short conic hirsute appendages. Achenes 3- or 4-angled, truncate. Pappus none or reduced to a short, 1–4-toothed annular border.

Leaves very scabrous, thick, abruptly contracted into the petiole.

1. *H. scabra.*

Leaves glabrous or nearly so, at least beneath, thin, dark green, with a cuneate base, gradually tapering into the petiole.

2. *H. helianthoides.*

F. 568.

1. H. scàbra Dunal. *Fig. 568.* Perennial, with a short crown and a fascicle of fibrous roots, stem 5–15 dm. high, more or less scabrous-hispidulous; leaf-blades ovate or subcordate, 5–10 cm. long, scabrous-hispidulous, strongly veined, triple-nerved, coarsely dentate; involucres 8–12 mm. high, 15–25 mm. broad, canescent; bracts oblong, obtuse, in 2 series, the outer usually longer; ligules 15–25 mm. long. Dry soil and river banks: Me.—N.Y.—Mo.—N.M.—B.C. *Plain—Submont.*

2. H. helianthoides (L.) Sweet Perennial; stem 3–15 dm. high; leaf-blades ovate or lanceolate, thinnish, 8–20 cm. long, sharply serrate, cuneate at the base; involucre 8–10 mm. high, 18–20 mm. broad; bracts oblong-lanceolate, ciliate, the outer with spreading or reflexed tips; ligules bright yellow, 2–3.5 cm. long; achenes truncate or obscurely 2–4-toothed at the apex. Thickets or open woods: Ont.—Minn.—Ala.—Fla. *E. Temp.* Jl S.

52. ECLÍPTA L. YERBA DE TAJO.

Annual, leafy-stemmed herbs. Leaves opposite, narrow, entire-margined or toothed. Heads radiate, but inconspicuous. Involucre hemispheric or campanulate; bracts in two series, the outer usually somewhat longer than the inner. Receptacle flat or somewhat convex, chaffy. Ray-flowers pistillate, fertile; ligules small, white. Disk-flowers perfect, fertile; style branches with obtuse or triangular tips. Ray-achenes triangular; disk-achenes flattened. Pappus of several short teeth or wanting. *Verbesina* L., in part.

1. E. álba (L.) Hassk. Stem 2–10 dm. long, finely pubescent, leaves sessile, elliptic to lanceolate or linear, 3–10 cm. long, acute or short-acuminate, remotely toothed or undulate, involucre broadly campanulate, 5–6 mm. broad; bracts united to about the middle, ovate, acute, ciliolate, ligules 1.5 mm. long; achenes cuneate, 2.5 mm. long, rugose. *V. alba* L. *E. erecta* L. Waste places: Mass.—Minn.—Tex.—Fla.; Mex.; naturalized in the North.

53. RUDBÉCKIA L. CONE-FLOWER, GOLDEN GLOW, NIGGER-HEADS.

Perennial, or rarely annual or biennial, caulescent herbs. Leaves alternate, entire, toothed, or pinnatifid, petioled or sessile. Heads radiate or discoid, many-flowered. Involucres hemispheric; bracts imbricate in 2 or more series, more or less foliaceous, loose and spreading. Receptacle convex, conic, or in fruit cylindric, chaffy; paleae not spinescent. Ray-flowers neutral; ligules yellow, or partly or rarely wholly brown-purple or crimson, or none. Disk-flowers hermaphrodite, fertile; corollas with a short but manifest tube, 5-lobed. Style-branches with blunt or subulate pubescent tips. Achenes 4-angled, obtuse or truncate. Pappus a 4-toothed crown, a low border, or wanting.

Style-branches with short, obtuse tips; pappus present; leaves broad or lobed.
  Paleae of the receptacle obtuse or truncate and pubescent at the tips.
    Leaves, at least the lower ones, deeply lobed
      Plant glabrous or sparingly pubescent; disk greenish yellow.
        Crown of the achenes with 4 distinct acute teeth; leaves usually scabrous above; disk 1–1.5 cm. broad.   1. *R. laciniata.*
        Crown of the achenes with 4 less distinct rounded teeth; leaves glabrous; disk 1.5–2 cm. broad.   2. *R. ampla.*

Plant cinereous-pubescent ; disk dull brown.
Leaves entire or merely toothed.
Paleae of the receptacle aristate or subulate ; lower leaves 3-parted.
Style-branches with slender subulate tips ; pappus wanting ; leaves mostly narrow, not lobed.
Leaf-blades oblong to elliptic.
Leaf-blades linear or narrowly lance-linear.

3. *R. subtomentosa.*
4. *R. grandiflora.*
5. *R. triloba.*
6. *R. hirta.*
7. *R. sericea.*

S.569.

1. R. laciniàta L. *Fig. 569.* Perennial, with a rootstock; stem 5–20 dm. high, glabrous or nearly so; lower leaf-blades pinnately 3–7-foliolate or deeply divided, with toothed or laciniate lanceolate divisions, pubescent above and on the margins; upper stem-leaves with shorter petioles, 3–5-cleft, the uppermost 3-lobed or merely toothed; involucral bracts oblong or lanceolate, reflexed, unequal; ligules 2–4 cm. long, soon drooping ; achenes 4-angular; pappus of 4 short teeth. Moist thickets: Que.—Man.—N.D.—Kans.—Fla. *E. Temp.* Jl–S.

2. R. ámpla A. Nels. Perennial, with a root-stock and thick crown; stem 1–2 m. high, glabrous, striate, branched above; basal leaves large, long-petioled, pinnately 3-divided; divisions ovate or lanceolate, 2–3-cleft, and coarsely serrate, glabrous and slightly paler beneath, somewhat hairy above; upper stem-leaves 3-parted or entire, toothed, acuminate; involucre about 1 cm. high and 3 cm. wide; bracts very unequal, oblong or ovate-oblong, mostly obtuse, reflexed in age, 8–12 mm. long; ligules 3–5 cm. long, 8–15 mm. wide; disk ovoid in fruit; pappus coroniform, sometimes 4-toothed. Wet places: Sask.—S.D.—N.M.—Ariz.—Ida. *Plain* —*Mont.* Jl–Au.

3. R. subtomentòsa Pursh. Perennial; stem 5–15 dm. high, cinereous-pubescent; lower leaves long-petioled, the blades 3-lobed or 3-parted, the terminal lobe elliptic to lanceolate, acuminate, serrate, the lateral ones smaller and narrower; upper stem-leaves mostly undivided; involucre 1 cm. high and 2–2.5 cm. broad, the bracts linear or nearly so, acuminate; ligules yellow, 2–3 cm. long; disk hemispheric or ovoid, brown-purple; pappus-squamellae crenately toothed. Prairies: Ill.—Kans.—Tex.—La. *Prairie—Texan.* Jl–S.

4. R. grandiflòra C. C. Gmelin. Perennial; stem 5–10 dm. high, scabrous or hispid throughout; leaves mostly near the base, scabrous on both sides, elliptic-ovate to lanceolate, acute or acuminate, serrate or dentate, the blades 6–15 cm. long; heads solitary or few; involucres 1.5–2.5 cm. broad; bracts linear, acuminate; ligules 3–3.5 cm. long, drooping; disk ovoid; pappus crenate or toothed. Dry prairies: Mo.—Kans. (?)—Okla.—Tex.—La. *Ozark—Texan.* Jl–S.

5. R. tríloba L. Perennial; stem 5–15 dm. high, hirsute or hispid; basal leaves long-petioled, the blades ovate to oblong, serrate, cordate at the base; blades of the lower stem-leaves or some of them palmately 3-lobed or 3-parted, 3–10 cm. long, those of the upper undivided, ovate to lanceolate, acuminate, narrowed at the base or wing-petioled; heads several; involucre 5–15 mm. broad; bracts linear, 7–11 mm. long, at last reflexed; disk subglobose; ligules 8–12, yellow or with orange base, 1.5–2.5 cm. long; pappus a minute crown. Moist soil: N.J.—Mich.—Kans.—Tex.—Ga. *Allegh.* Je–O.

6. R. hírta L. Perennial; stem erect, usually straw-colored, 3–5 dm. high, usually purple-dotted, hirsute; lower leaves long-petioled; blades oblong-lanceolate or oblanceolate, 5–10 cm. long, entire or remotely denticulate, obtuse, 3-ribbed, hirsute; upper stem-leaves linear and subsessile; peduncles 5–15 cm.

long; involucres 15–18 mm. high, 3–3.5 cm. broad; bracts oblong-linear, hispid; ligules yellow, 1.5–3 cm. long. *E. flava* T. V. Moore. *E. longipes* T. V. Moore. Hillsides and plains: Que.—Fla.—Colo.—B.C. *E. Temp.—Plain—Mont.* Jl–Au.

7. E. serícea T. V. Moore. Perennial; stem 3–12 dm. high, somewhat hispid; leaves linear or nearly so, 6–15 cm. long, remotely toothed or entire, strigose; involucre 12–20 mm. broad; bracts linear to lanceolate, 1–2 cm. long; ligules 10–15, yellow, 2–4 cm. long; disk ovoid-conic; paleae ciliate at the tip; pappus wanting. Woods and fields: Mo.—S.D.—Ala.—Ga. Jl–S.

54. ECHINÀCEA Moench. PURPLE CONE-FLOWER.

Perennial caulescent herbs. Leaves mostly alternate, entire or toothed. Heads radiate, solitary or few, mostly long-peduncled. Involucres rather flat; bracts in 2–4 series, narrow, herbaceous, squarrose. Receptacle low-conic, chaffy. Ray-flowers several, neutral, with rudimentary styles; ligules spreading or drooping, purple or rose, seldom yellow or white. Disk-flowers hermaphrodite, fertile. Paleae awned, surpassing the flowers, persistent. Achenes acutely 4-angled. Styles with acute or obtuse, hispid appendages. Pappus a crown produced into a small tooth at each angle. *Brauneria* Neck.

Lower leaves ovate, 5-ribbed, mostly dentate; awns of the pappus-squamellae as long as the body.   1. *E. purpurea.*
Lower leaves linear-lanceolate, 3-ribbed; awns of the pappus-squamellae shorter than the body.
    Ligules drooping, 4–8 cm. long; leaves hirsute, not pustulate.   2. *E. pallida.*
    Ligules spreading, or slightly reflexed, 2–3 cm. long; leaves hispid, the hairs with pustulate bases.   3. *E. angustifolia.*

1. E. purpúrea (L.) Moench. Perennial, with a horizontal or inclined root; stem scabrous, 6–12 dm. high; leaf-blades ovate to lanceolate, 5–12 cm. long, serrate or dentate, cuneate or cordate at the base; disk 2–2.5 cm. high; involucral bracts linear or lance-linear, finely hispid; ligules 12–20, purple, rarely crimson or whitish, 2.5–5 cm. long. *B. purpurea* (L.) Britton. Open woods and fields: Pa.—Mich.—Iowa—La.—N.C. *Allegh.* Jl–O.

2. E. pállida (Nutt.) Britton. Perennial, with a vertical fusiform root; stem more or less hispid, 5–10 dm. high; leaf-blades linear or narrowly elliptic, 5–10 cm. long, entire; disk 1.5–2 cm. high; involucral bracts lanceolate, hispid, with lax tips; ligules 10–15, rose-colored, 2-cleft at the apex. Dry soil: Mich.—Minn.—Tex. Ala. *Allegh.—Austral.* My–Jl.

J.570.

3. E. angustifòlia DC. *Fig. 570.* Perennial plants, with a taproot; stem 3–6 dm. high, erect, hispid; lower leaves petioled, the upper subsessile; blades lanceolate to nearly linear, 3-ribbed, scabrous-hispidulous; involucres about 1 cm. high, 2–3 cm. wide; bracts lanceolate, hispid; ligules about 2–2.5 cm. long, light purple, spreading. Prairies: Sask.—Man.—Tex.—Colo.—Mont. *Prairie—Plain.* Je–Au.

55. DRACÓPSIS Cass.

Annual glaucous caulescent herbs. Leaves alternate, thickish, entire or slightly serrate, clasping, 1-ribbed. Heads radiate, many-flowered. Involucral bracts few, foliaceous. Receptacle slender. Ray-flowers neutral, the ligules yellow, often brownish-purple at the base. Disk-corollas brownish. Style-branches with small pubescent tips. Achenes terete, striate, minutely transversely wrinkled. Pappus wanting.

1. **D. amplexicaùlis** Vahl. Stem 3–7 dm. high; stem-leaves spatulate or oblong to ovate, sessile and clasping, 4–10 cm. long; bracts lanceolate or lance-linear, 6–10 mm. long; ligules 5–9, 1–2.5 cm. long; disk oblong-cylindric at maturity; achenes 2 mm. long. *Rudbeckia amplexicaulis* Vahl [G]. Moist places: Ga.—La.—Tex.—Kans. *Austral.*

56. LÉPACHYS Raf. Cone-flower.

Annual, biennial, or perennial caulescent herbs. Leaves alternate, pinnatifid with narrow divisions. Heads radiate, long-peduncled. Involucres many-flowered, rather flat. Ray-flowers neutral; ligules yellow, or partly brown at the base, rarely wholly brown-purple, spreading or drooping. Disk-flowers hermaphrodite, usually enveloped by the paleae; corollas gray or yellowish, almost without a tube. Style-branches with blunt or lanceolate tips. Achenes flattened, broad-margined or winged, deciduous with the scale. Pappus of 1 or 2 teeth, or wanting. *Ratibida* Raf., hyponym.

Style-branches with short obtuse tips; ligules 3 cm. long or less; leaf-segments linear. Disk in fruit oblong, about 1 cm. long; pappus of 1 or 2 awn-like teeth, without intermediate squamellae.   1. *L. Tagetes.*
Disk in fruit cylindraceous, 2–4 cm. long; pappus with a series of squamellae.   2. *L. columnifera.*
Style-branches with lance-subulate tips; ligules usually 3–5 cm. long; leaf-segments lanceolate.   3. *L. pinnata.*

1. **L. Tagètes** (James) A. Gray. Perennial, with taproot; stems several, branched, 2–4 dm. high, grayish-strigose; leaves pinnately divided with 3–7 linear divisions, hispidulous; heads rather short-peduncled; bracts reflexed, oblong, 3–5 mm. long; ligules 7–10 mm. long; receptacle about 1 cm. high in fruit. *R. Tagetes* Barnh. [B, R]. Plains: Tex.—Kans.—Colo.—Ariz. *Plain—Submont.* Jl–Au.

2. **L. columnífera** (Nutt.) Rydb. Perennial, with a taproot; stems often several, 3–6 dm. high, hispidulous-strigose, angled; leaves pinnately divided, hispidulous; divisions of the stem-leaves 5–9, oblong to linear, entire or rarely 2- or 3-cleft; heads long-peduncled; involucral bracts oblong, 5–8 mm. long, soon spreading; ligules yellow or in the variety *pulcherrima* partly or wholly brown-purple, 1–3 cm. long; receptacle in fruit 2–4 cm. long. *L. columnaris* (Sims) T. & G. [G]. *R. columnifera* Woot. & Standl. [B, R]. Plains: Sask.—Tenn.—Tex.—Ariz.—B.C. *Prairie—Plain—Submont.* My–S.

3. **L. pinnàta** (Vent.) T. & G. Perennial; stem branched, 6–15 dm. high, strigillose or scabrous; leaves pinnately 3–7-divided or -parted, the uppermost small and entire; bracts linear or oblong, reflexed; ligules 4–10, yellow, drooping; inner margin of the achenes with a short tooth. *R. pinnata* Barnh. [B, R]. Prairies: N.Y.—Man.—S.D.—Tex.—Fla. *E. Temp.* Je–S.

57. GALINSÒGA R. & P.

Annual leafy-stemmed herbs. Leaves opposite, petioled, toothed. Heads radiate, small. Involucre campanulate or hemispheric; bracts imbricate in 2 series, thin. Receptacle conic, chaffy. Ray-flowers few, pistillate, fertile; ligules short, white. Disk-flowers hermaphrodite, fertile. Anthers sagittate at the base. Style-branches with acute appendages. Achenes 4 or 5-angled, or those of the rays slightly flattened. Pappus of the ray-flowers of several bristles or wanting, that of the disk-flowers of lacerate or fimbriate squamellae.

Pappus of the disk-flowers aristate, fimbriate, that of the ray-flowers of linear scales; stem and peduncles with spreading hairs.   1. *G. ciliata.*
Pappus of the disk-flowers not aristate, that of the ray-flowers wanting; stem and peduncles with appressed hairs.   2. *G. parviflora.*

1. **G. ciliàta** (Raf.) Blake. Stem 1–5 dm. high, pilose, especially near the nodes, and on the peduncles; leaves pilose, the blades ovate, 2–5 cm. long, coarsely serrate; involucre 2.5 mm. high, bracts ovate, ciliolate; ligules 4 or 5,

white, 1.5 mm. long. Waste places: Que.—Fla.—Tex.—Colo.—Minn.; introd. from Mex.—Cent.Am. Jl–S.

2. **G. parviflòra** Cav. Stem 1–7 dm. high, branched; leaf-blades glabrous, ovate or ovate-lanceolate, 2–5 cm. long, dentate or undulate; involucre 2–2.5 mm. high; bracts ovate; ligules 4 or 5, only 1–1.5 mm. long; pappus-squamellae in the disk-flowers 8–15. Waste places: Mass.—Ga.—Kans.—Mex. —Ore.; nat. from S.Am.

58. MARSHÁLLIA Schreb.

Perennial caulescent herbs. Leaves alternate, simple, entire, 3-ribbed. Heads discoid, long-peduncled. Involucre campanulate or hemispheric; bracts narrow, nearly equal, in 1 or 2 series, foliaceous. Receptacle convex or conic, chaffy, the paleae narrow, rigid. Flowers hermaphrodite and fertile; corollas white, pink, or purple, with slender tubes. Anthers sagittate at the base. Style-branches elongate, truncate. Achenes obpyramidal, 5-angled. Pappus of 5 or 6 acute or acuminate squamellae.

1. **M. caespitòsa** Nutt. Stem 2–4 dm. high, tufted; leaves mainly at the base of the stem, spatulate or linear, 2–10 cm. long, entire; peduncles pubescent under the head; bracts narrowly linear-lanceolate, 7–9 mm. long; flowers many, the corolla-tube 6–7.5 mm. long, tomentose; achenes 2–2.5 mm. long, pubescent on the angles. Calcareous soil: Ark.—Kans.—Tex.—La. *Ozark.* My–Je.

59. BALSAMORRHÌZA Hook. BALSAM-ROOT.

Low perennials, with almost scapose stems, numerous petioled basal leaves, and a thick edible root, its bark exuding a terebinthine balsam. Involucre mostly hemispherical; bracts in several series, more or less foliaceous, especially the outer ones. Receptacle almost flat, beset with concave paleae, which loosely embrace the disk-flowers. Ray-flowers present, fertile; ligules yellow. Disk-flowers hermaphrodite, fertile. Anthers not caudate. Style-branches with filiform or slender subulate hispid appendages. Achenes mostly glabrous, those of the disk-flowers quadrangular with intermediate nerves, those of the ray-flowers flattened. Pappus none.

f.576.

1. **B. sagittàta** (Pursh) Nutt. *Fig. 576.* Leaves mostly basal, long-petioled; blades from cordate to hastate or sagittate, mostly entire, white-tomentose on both sides, 1–2 dm. long; peduncles 3–5 dm. high, tomentose; involucres floccose, about 2 cm. high and 2.5 cm. broad; bracts lanceolate or linear-lanceolate, usually all appressed; ligules 2–3 cm. long, 8–10 mm. broad, oblong. Hillsides: Sask.—S.D.—Colo.—Calif.—B.C. *Submont.—Mont.* Ap.

60. HELIÁNTHUS L. SUNFLOWER, GROUND ARTICHOKE.

Erect, mostly branched annuals or perennials, with simple leaves and large, peduncled, corymbose or solitary heads. Involucre hemispherical or depressed; its bracts imbricated in several series. Receptacle flat or convex, rarely conic, with paleae subtending the disk-flowers. Ray-flowers neutral; ligules yellow, spreading, mostly entire. Disk-flowers perfect, fertile; their corollas tubular, yellow, brown, or purple. Anthers entire or minutely 2-toothed at the base. Style-branches with hirsute appendages. Achenes more or less 4-angled and somewhat compressed. Pappus of 2 scales or awns, sometimes with a few additional smaller intermediate ones, deciduous.

Annuals ; disk purple or dark brown.   1. ANNUI.
Perennials.
  Disk dark brown or purple.
    Leaf-blades linear.   2. SALICIFOLII.
    Leaf-blades subrhombic to ovate.   3. SCABERRIMI.
  Disk yellow or light brown.
    Stem scape-like, the leaves mostly at the base.   4. OCCIDENTALES.
    Stem not scape-like, the leaves evenly distributed along the stem.
      Leaf-blades rounded or obtuse at the base, at least the upper subsessile or
        sessile.   5. DIVARICATI.
      Leaf-blades tapering at the base (rounded in *H. chartaceus*), distinctly
        petioled.   6. TUBEROSI.

1. ANNUI.

Bracts hispid-ciliate, ovate or obovate, acuminate.
  Lower leaf-blades at least ovate or cordate, distinctly toothed.
    Heads nodding, the bracts loose, long-acuminate; lower leaf-blades coarsely
      toothed, with large rounded basal lobes and the base between the lobes
      conspicuous, cuneate.   1. *H. annuus.*
    Heads not nodding or slightly so; bracts appressed.
      abruptly acuminate; lower leaf-blades not coarsely
      toothed, mostly ovate, if cordate not conspicuously
      cuneate between the basal lobes.   2. *H. lenticularis.*
  Leaf-blades lanceolate or narrowly deltoid, minutely
    toothed or entire.   3. *H. aridus.*
Bracts not ciliate, canescently strigose, lanceolate.   4. *H. petiolaris.*

2. SALICIFOLII.

One species.   5. *H. salicifolius.*

3. SCABERRIMI.

Leaves mostly near the base ; the blades abruptly contracted
  into the winged petioles.   6. *H. atrorubens.*
Leaves scattered along the stem, the blades of the lower
  ones gradually narrowed into the petioles.
    Stem-leaves lanceolate ; bracts acuminate.   7. *H. rigidus.*
    Stem-leaves rhombic-ovate to rhombic-lanceolate.   8. *H. subrhomboideus.*

4. OCCIDENTALES.

One species.   9. *H. occidentalis.*

5. DIVARICATI.

Stem glabrous or scabrous ; leaves 3-ribbed from the base.
  Petioles less than 1 cm. long.   10. *H. divaricatus.*
  Petioles 2–3 cm. long.   17. *H. chartaceus.*
Stem hirsute.
  Leaves sessile, 3-ribbed above the base.   11. *H. mollis.*
  Leaves, at least the lower, short-petioled, 3-ribbed from
    the base.   12. *H. hirsutus.*

6. TUBEROSI.

Leaf-blades ovate.
  Leaves comparatively long-petioled, the petioles one
    quarter to one half as long as the blades.
    Bracts of the involucre ovate-lanceolate, appressed.   13. *H. laetiflorus.*
    Bracts of the involucre lanceolate, acuminate, spread-
      ing.
      Stem more or less hirsute or hispid.
        Leaves grayish green, densely scabrous-his-
          pidulous above, hispid beneath ; stem very
          hispid ; bracts with spreading tips.   14. *H. Besseyi.*
        Leaves green ; bracts appressed or ascending ;
          stem sparingly hispid or hirsute.
          Leaf-blades tapering at the base, often de-
            current on the petiole, pubescent
            throughout beneath.
            Leaves sparingly rough-hairy beneath.   15. *H. tuberosus.*
            Leaves densely soft-hairy beneath.   16. *H. subcanescens.*
          Leaf-blades rounded at the base, sparingly
            hairy beneath.   17. *H. chartaceus.*
        Stem glabrous or puberulent.   27. *H. decapetalus.*
  Leaves short-petioled, the petioles less than one fifth as
    long as the blades.   18. *H. nitidus.*
Leaf-blades lanceolate or lance-linear.
  Stem more or less scabrous or hispid.
    Leaves on both sides, as well as the stem, very sca-
      brous.
      Leaf-blades narrowly lanceolate, mostly pinnately
        veined.   19. *H. Maximiliani.*
      Leaf-blades ovate-lanceolate, triple-ribbed.   20. *H. Rydbergii.*

Leaves scabrous above, usually hirsute beneath; stem more or less hirsute.

Upper leaves mostly alternate and indistinctly triple-ribbed. 21. *H. giganteus.*

Upper leaves, as well as the lower, opposite, distinctly triple-ribbed.

Leaf-blades tapering at the base.

Leaves broadly lanceolate, coarsely toothed; bracts with loose lax tips. 22. *H. trachelifolius.*

Leaves narrowly lanceolate, denticulate; bracts appressed. 24. *H. subtuberosus.*

Leaf-blades, at least the lower ones, rounded at the base. 12. *H. hirsutus.*

Stem, except the upper portion, glabrous or nearly so, and glaucous; leaves petioled.

Leaves all or nearly all opposite.

Bracts rather few, broadly lanceolate, scarcely exceeding the disk, more or less spreading; leaves paler beneath 23. *H. strumosus.*

Bracts numerous, linear-lanceolate, longer than the disk, appressed; leaves not paler beneath. 24. *H. subtuberosus.*

Leaves, at least the upper, alternate; bracts numerous, linear-lanceolate, longer than the disk, loose.

Leaves leathery, narrowly lanceolate.

Lower leaves coarsely toothed; bracts hirsute, ciliate. 25. *H. grosseserratus.*

Leaves all denticulate; bracts ciliolate below the middle. 26. *H. fascicularis.*

Leaves membranous, broadly lanceolate. 27. *H. decapetalus.*

1. H. ánnuus L. *Fig. 571.* Stem 1–4.5 m. high, hispid or scabrous; leaves mostly alternate, the lower long petioled; blades 3-ribbed, 1–3 dm. long, scabrous on both sides or pubescent beneath; involucre 2–4 cm. high, 1–2 dm. broad, nodding; achenes usually dark, spotted, often 1 cm. long. Waste places: Me.—Minn.—Iowa—Mo.—S.C.; escaped from cultivation, perhaps a cultivated form of the next. Jl–S.

2. H. lenticuláris Dougl. Stem 1–2 m. high; leaves mostly alternate, the lower long-petioled; blades ovate, 1–2 dm. long, sometimes cordate at the base, dentate, hispidulous-scabrous, the upper more lanceolate, subentire; involucres 1.5–2 cm. high, 4–5 cm. broad; achenes grayish-strigose, about 6 mm. long. Plains and alluvial soil, also in waste places and cultivated ground: Sask.—Tex.—Calif.—Wash. *Plain—Submont.* Je–S.

3. H. áridus Rydb. Stem 3–8 dm. high, more or less hispid; leaves all petioled; blades lanceolate, 4–7 cm. long, entire or crenate, acute, hispid-scabrous on both sides; disk 1.5–2 cm. wide; ligules oblong to oval, 15–20 mm. long, 6–8 mm. wide; achenes cuneate, almost black, finely strigose, about 5 mm. long. Arid soil: Sask.—Neb.—N.M.—Ariz.—B.C. *Plain—Submont.* Je–S.

4. H. petioláris Nutt. Stem 3–10 dm. high, hirsute-strigose; leaves petioled; blades lanceolate or ovate-lanceolate, mostly cuneate at the base, 5–8 cm. long, hispidulous-scabrous; involucres about 1 cm. high, 2–3 cm. broad; ligules golden yellow, 1.5–2 cm. long; achenes about 5 mm. long, strigose. Dry plains and waste places: Man.—Mo.—Tex.—Calif.—B.C. *Plain—Mont.* Je–S.

5. H. salicifólius A. Dietr. Stem smooth and glabrous, 5–30 dm. high, very leafy; leaves mostly alternate, narrowly linear, 2–4 dm. long, 4–8 mm. wide, scabrous-puberulent; involucres fully 1 cm. high and about 2 cm. wide; bracts linear-subulate, long-attenuate, ciliolate on the margins; rays 15–18 mm. long. *H. orgyalis* DC. [G]. Dry plains: Mo.—Tex.—e Colo. *Plain.* Au–S.

6. **H. atrórubens** L. Stem 6–18 dm. high, hirsute or hispid; leaves opposite, mostly near the base; blades ovate, oval, or lanceolate, 5–20 cm. long, more or less toothed, cuneate to truncate at the base; involucre 1 cm. high, 1.5 cm. broad; bracts oblong, obtuse or mucronate; ligules 1.5–2 cm. long. Open woods: Va.—Minn.—La.—Fla. *Allegh.—Carol.* Au–O.

7. **H. rígidus** (Cass.) Desf. Stem rigid, scabrous, 5–25 dm. high, usually tinged with purple; leaves few, opposite; blades leathery, those of the lower leaves elliptic, 1–2 dm. long, those of the upper lanceolate, scabrous, 3-ribbed, serrulate or serrate; involucre 2–2.5 cm. broad; bracts ovate to lanceolate; disk brown-purple; ligules 2.5–4 cm. long. *H. scaberrimus* Ell. [G]. Prairies: Mich.—Man.—Tex.—Ga. *Allegh.—Austral.* Au–S.

8. **H. subrhomboídeus** Rydb. Perennial, with a rootstock; stem 3–6 dm. high, terete, tinged with red, sparingly hirsute; leaves opposite; blades firm, very scabrous, triple-veined, serrulate, the lower ovate or obovate-spatulate, the upper rhombic-ovate or rhombic-lanceolate, 5–10 cm. long; heads solitary, rarely 2 or 3; involucres 10–12 mm. high and 1.5–2 cm. broad; bracts in 4 or 5 series, oblong, acutish, white-ciliolate; ligules about 1.5 cm. long. Plains and prairies: Ill.—Man.—Ark.—N.M.—Alta. Jl–S. Perhaps variant of no. 7.

9. **H. occidentális** Riddell. Stem scabrous, 6–15 dm. high, sparingly branched above; leaves few, opposite; blades oblong to ovate, 5–15 cm. long, entire or denticulate, the upper ones reduced; involucres 7–9 mm. high, 1 cm. broad; bracts ovate-lanceolate, acuminate, nearly glabrous; ligules 11–17, yellow, 1.5–2 cm. long; disk yellow. Dry soil: Ohio—Minn.—Mo.—Fla. *Allegh.— Austral.* Au–S.

10. **H. divaricàtus** L. Perennial, with a rootstock; stem 3–10 dm. high, simple, smooth below, hispidulous-scabrous above; leaves opposite, subsessile, divaricate, lanceolate or elongate-deltoid, 5–15 cm. long, gradually attenuate above, rounded at the base, 3-ribbed, scabrous on both sides; involucres about 1 cm. high and 2 cm. broad; bracts narrowly lanceolate, acuminate, hirsute-ciliate on the margins, hirsutulous on the back; ligules 8–12, about 2 cm. long. Dry ground: Que.—Fla.—La.—Sask. *E. Temp.—Plain.* Jl–S.

11. **H. móllis** Lam. Stem 6–10 dm. high, canescent; leaves opposite; blades ovate, 5–15 cm. long, acute or acuminate, entire or serrate, sessile or clasping, canescent; bracts oblong to lanceolate, acuminate, the outer spreading; disk 2–3 cm. broad; ligules 13–26, 2–3 cm. long. Dry barren grounds: N.Y.—Iowa—Kans.—Tex.—Ga. *Allegh.—Austral.* Au–S.

12. **H. hirsùtus** Raf. Stem 5–12 dm. high, hispid or hirsute; leaves opposite; blades firm, lanceolate, 5–15 cm. long, acuminate, scabrous above, hispid beneath, serrate or subentire; involucre 10–12 mm. high, 1.5–2 cm. broad; bracts lanceolate, caudate-attenuate, hispidulous; ligules 12–15, 2.5–3.5 cm. long. Dry soil: Pa.—S.D.—Kans.—Tex.—Ga. *Allegh.—Austral.*

13. **H. laetiflòrus** Pers. Stem scabrous, 1–3 m. high; leaves short-petioled; blades ovate-lanceolate, 3-ribbed, rough on both sides, serrate, long-acuminate, 1–2.5 dm. long, the upper often alternate; involucres 1 cm. high, 1.5–2 cm. broad; bracts ovate-lanceolate, ciliate, not exceeding the disk; ligules 15–25, 2 cm. long. Prairies and barrens: Pa.—Minn.—Iowa—Mo. *Allegh.* Au–S.

14. **H. Bésseyi** Bates. Perennial, producing tubers; stem 5–10 dm. high, scabrous-hispidulous; leaves mostly opposite, the petioles 1–2 dm. long, the blades ovate, 6–10 cm. long, 3–5 cm. wide, thick, grayish green, densely hispidulous-scabrous above, with pustulate-based hairs, densely hirsute beneath when young, more scabrous in age; bracts lanceolate, somewhat squarrose, strigose, hirsute-ciliate below the middle; involucre 8–10 mm. high, 1.5–2 cm. broad; disk yellow; ligules 15–18 mm. long. *H. nebraskensis* Cockerell. Dry places: N.D.—Okla. *Plain.* Au–S.

15. H. tuberòsus L. Perennial, with tubers; stem 1–3 m. high, branching at the summit, more or less hirsute; lower leaves usually opposite, the upper alternate, petioled; blades ovate or subcordate, acuminate, firm, 3-ribbed, scabrous above, pubescent beneath, 5–30 cm. long, usually dentate; involucres about 1.5 cm. high, 2–3 cm. broad; bracts lanceolate, attenuate, hirsute at least on the margins; ligules 12–20, 2.5–3.5 cm. long. Alluvial soil: N.S.— Ga.—Ark. —Neb.—Sask. *E. Temp.—Plain.* Au–O.

16. H. subcanéscens A. Gray. Perennial, with tubers; stem 1 m. high or more, roughly hispidulous; leaves mostly opposite, or the upper alternate, the petioles 3–7 cm. long, the blades ovate, triple-ribbed, 1–2.5 dm. long, 5–11 cm. wide, coarsely serrate, hispidulous above, the hairs with pustulate bases, densely and softly short-pubescent beneath; heads several; bracts broadly lanceolate, attenuate, ascending, strigose on the back, hirsute-ciliolate throughout; disk 1.5–2 cm. broad, light brown; ligules 12–15, entire, 3–4 cm. long, 5–8 mm. wide. Bottom-land and copses: Wis.—Mo.—Kans.—Man. *Prairie.* Au–S.

17. H. chartàceus E. E. Wats. Perennial; stem 1 m. high or more, sub-rate, sparingly scabrous-hispidulous; leaves mostly opposite, the petioles 2–3 cm. long, the blades ovate deltoid, 1–2 cm. long, 3–7 cm. wide, long-acuminate at the apex, mostly rounded at the base, finely serrate, scabrous-hispidulous above, the hairs with pustulate bases, sparingly hispid beneath; heads few; bracts lanceolate, acuminate, ascending, hirsute on the back, hispid-ciliate below; disk light yellowish brown, 15 mm. broad; ligules 12–15, 15–20 mm. long, 4 mm. wide. Banks and open woods: Wis.—Mo.—Neb.—Minn. *Prairie.* Au–S.

18. H. nítidus Lunell. Perennial; stem 6–10 dm. high, glabrous or slightly scabrous below, hispidulous in the inflorescence; leaves opposite, rarely verticillate in 3 's, or the upper alternate; petioles less than 1 cm. long, ciliolate, the blades ovate or oval, 4–10 cm. long, thick, 3-ribbed, scabrous above, strigose or scabrous beneath; involucre about 1 cm. high, 2 cm. broad; bracts linear lanceolate, attenuate, squarrose; ligules 13–20, about 1.5 cm. long. Ravines and prairies: N.D. *Prairie.* Au–S.

19. H. Maximiliàni Schrad. Perennial, with a thick rootstock; stem 5–30 dm. high, scabrous-hispidulous; leaves mostly alternate and subsessile, lanceolate to linear, entire or denticulate, 5–15 cm. long, 1-ribbed, very scabrous on both sides; heads many in a narrow panicle; involucres about 1.5 cm. high and 3 cm. wide; bracts lanceolate-subulate, long-attenuate, strigose-canescent or somewhat hispidulous; ligules golden yellow, 15–30 mm. long. Prairies, plains, and river banks: Man.—Mo.—Tex.—Wyo.—B.C. *Prairie—Plain— Submont.*

20. H. Rydbérgii Britton. Perennial, with fusiform tuberous roots; stem 7–15 dm. high, scabrous; leaf-blades ovate-lanceolate or the upper lanceolate, thick, pale green, scabrous, 3 ribbed, sparingly serrate, 7–10 cm. long; involucre 12–15 m. high, about 2 cm. broad; bracts lanceolate, acuminate, ciliate; ligules 2.5 cm. long. (?) *H. apricus* Lunell (a more entire-leaved form). Sandhills and arid places: N.D.—Neb. *Prairie—Plain.*

21. H. gigantèus L. Perennial, with a rootstock, some of the roots often becoming fusiform; stem 1–3 m. high, more or less hirsute; leaves lanceolate or oblong-lanceolate, 5–15 cm. long, scabrous above, hirsutulous beneath, short-petioled or subsessile, serrulate or denticulate, rarely entire; heads in an open panicle; involucres 12–15 mm. high and about 3 cm. broad; bracts linear-lanceolate, attenuate, hirsute-ciliate; ligules pale yellow, 1.5–2 cm. long. Low ground: Me.—Fla.—La.—Colo.—B.C. *E. Temp.—Plain.* Au–S.

22. H. trachelifòlius Willd. Stem 7–20 dm. high, pubescent or rough by the persistent hair-bases; leaves opposite; blades lanceolate, 8–20 cm. long, sharply serrate, acuminate, revolute-margined, scabrous above, pubescent beneath; involucres 1 cm. high, 1.5 cm. broad; bracts lanceolate to linear-lanceolate, caudate-acuminate, with lax tips; ligules 9–20, 3–4 cm. long. Dry soil: Pa.—Minn.—Tex.—Ga. *Allegh.—Austral.* Au–S.

23. **H. strumòsus** L. Stem 5–20 dm. high, smooth and glabrous up to the inflorescence, often glaucous; leaves opposite; blades ovate to lanceolate, thin, 8–20 cm. long, acuminate, serrate or entire, whitish beneath; involucre 1.3 cm. high, 1.5–2 cm. broad; bracts lanceolate, attenuate, ciliate, with spreading tips; ligules 11–15, bright yellow, 2.5–4 cm. long. Woods and banks: Me.—Minn.—Ark.—Ga. *Canad.—Allegh.* Jl–S.

24. **H. subtuberòsus** Bourgeau. Perennial, with tubers; stem 3–6 dm. high, glabrous or nearly so; leaves short-petioled, opposite; blades narrowly lanceolate, acute at both ends, finely serrate, scabrous-hispidulous above, hirsutulous beneath, 5–8 cm. long, involucres about 1.5 cm. high, 2.5–3 cm. wide; bracts linear-lanceolate or subulate, attenuate, hirsute, the outer spreading; ligules about 3 cm. long; disk yellow. *H. borealis* E. E. Wats. Valleys and plains: Sask.—Minn.—Wyo.—Mont. *Plain.* Jl–Au.

25. **H. grosseserràtus** Martens. Perennial; stem 1–3 m. high, glabrous and often purplish; leaves alternate, or some opposite, short-petioled; blades elongate-lanceolate, 1–2.5 dm. long, gradually acuminate, sharply serrate, slightly scabrous above, paler and short-pubescent beneath; heads in an open panicle; involucre about 1.5 cm. high, 2.5–3 cm. wide; bracts linear-subulate, attenuate; ligules golden yellow, 2–3 cm. long. Plains and prairies: N.Y.—Pa.—Tex.—N.M.—Utah—Sask. *E. Temp.—Plain.* Au–S.

26. **H. faciculàris** Greene. Perennial, with a rootstock and fusiform thickened roots; stem glabrous and somewhat glaucous, 5–10 dm. high; leaves alternate or the lower opposite, 7–15 cm. long, short-petioled; blades lanceolate or linear-lanceolate, scabrous on both sides; heads in an open panicle; involucres about 1.5 cm. high and 3 cm. broad; bracts linear-subulate, attenuate; ligules 14–18, deep yellow, 2.5–3 cm. long; disk yellowish brown. *H. giganteus utahensis* D. C. Eat. *H. utahensis* A. Nels. (?) *H. coloradensis* Cockerell. Closely related to *H. subtuberosus* and *H. Nuttallii.* Mountain valleys: Sask.—S.D.—N.M.—Ariz.—Alta. *Plain—Submont.* Au–S.

27. **H. decapétalus** L. Stem 5–15 dm. high, glabrous below the inflorescence; lower leaves opposite; blades ovate to oblong-lanceolate, 8–20 cm. long, acuminate, coarsely toothed, cuneate or truncate at the base, scabrous above; involucre 1.3 cm. high, 1.5 cm. broad; bracts linear to linear-lanceolate, 1.5–2 cm. long, ciliate; ligules 2.5–3 cm. long. Thickets: Que.—Minn.—Neb.—Ga. *Allegh.* Au–S.

61. **HELIANTHÉLLA** T. & G.

Caulescent perennials with taproots. Leaves alternate or opposite, more or less distinctly triple-ribbed. Involucres hemispheric or flat; bracts more or less imbricate, often foliaceous; receptacle chaffy. Ray-flowers neutral, showy; ligules yellow. Disk-flowers hermaphrodite and fertile; tube of the corollas half as long as the throat; lobes short, ovate, puberulent. Achenes flat, cuneate-obovate, emarginate or obcordate. Pappus of several squamellae between the awns or paleaceous teeth. Appendages of the style-branches obtuse, short, spatulate or oblong.

1. **H. quinquenérvis** (Hook.) A. Gray. Stem 5–15 dm. high, glabrous or sparingly hirsute; leaves mostly opposite; blades ovate-lanceolate or elliptic-lanceolate, acuminate, 1–2.5 dm. long, usually with two pairs of the lateral veins prominent, sparingly hirsute; heads long-peduncled, nodding; involucres about 2 cm. high and 4–5 cm. broad; bracts lanceolate, acuminate, hirsute-ciliate; ligules 15–20, pale yellow, 2.5–3 cm. long. Mountains, along streams: S.D.—N.M.—Utah—Ida.—Mont. *Submont.—Mont.* Jl–Au.

62. ACTINÓMERIS Nutt.

Caulescent perennial herbs. Leaves alternate or opposite, with simple toothed blades, usually decurront on the stem. Heads small, with few narrow, recurved or spreading bracts. Receptacle chaffy, convex or conic, becoming globose. Ray-flowers neutral or wanting; ligules yellow or white. Disk-flowers perfect, fertile, enclosed in the paleae. Achenes flattened, winged; pappus of 2 spreading squamellae, sometimes with 2 or 3 smaller awns or scales. *Ridan* Adans.

1. A. alternifòlia (L.) DC. Stem 5–25 dm. high, winged, corymbosely branched above; leaves oblong or elliptic, 1–3 dm. long, acuminate at both ends, serrate, finely pubescent; involucre 1–1.5 cm. broad, the bracts linear or lance-linear, acute; ligules 2–8, yellow, irregular; achenes obovate, 5–6 mm. long. *R. alternifolius* (L.) Britton [BB]. *Verbesina alternifolia* (L.) Britton [B]. Rich soil and in thickets: N.Y.—Neb.—Kans.—La.—Fla. *Allegh.—Austral.* Au–S.

63. VERBESÌNA L. CROWNBEARD.

Caulescent perennial herbs, rarely shrubby plants. Leaves alternate or opposite, entire or toothed, decurrent on the stem. Heads radiate or discoid. Involucre hemispheric or campanulate; bracts imbricate in few series. Receptacle convex or conic, chaffy. Ray-flowers neutral or fertile; ligules yellow or white, inconspicuous, or wanting. Disk-flowers fertile, embraced by the paleae. Achenes of the ray-flowers 3-angled, those of the disk-flowers flattened, winged or (in ours) wingless. Pappus of 1–2 awns, sometimes accompanied by as many small scales. *Phaothusa* Gaertner.

Involucre campanulate, 4–6 mm. broad.
Involucre hemispheric, 12–15 mm. broad.

1. *V. virginica.*
2. *V. helianthoides.*

1. V. virgínica L. Stem 5–18 dm. high, tomentose or puberulent, strongly winged; leaves alternate, membranous, lanceolate to oval or ovate, 5–20 cm. long, acute or acuminate, serrate or entire, narrowed into the winged petioles; involucre 6 mm high, the bracts linear, erect; ligules white, oval, 4–7 mm. long; achenes oblong, 3 mm. long, ciliate; pappus-awns as long. *P. virginica* (L.) Britton. Dry soil: Pa.—Kans.—Tex.—Fla. *Allegh.—Austral.* Au–S.

2. V. helianthoìdes Michx. *Fig. 572.* Stem 5–10 dm. high, finely pubescent, winged; leaves alternate, lanceolate to elliptic, 4–12 cm. long, acute or acuminate, serrate, hirsute beneath, sessile and decurrent; heads solitary or clustered; involucre 8–10 mm. high, the bracts linear or lance-linear; ligules yellow, 2–3 cm. long; achenes 4–5 mm. long, *f.572.* winged, pubescent, with 2 small pappus-awns. *P. helianthoides* (Michx.) Britton. Woods and prairies: Ohio—Iowa—Kans.—Tex.—Ga. *Allegh.—Austral.* Je–Jl.

64. XIMENÈSIA Cav.

Annual caulescent herbs, ours canescent. Leaves alternate or sometimes opposite, toothed or lobed. Heads solitary or few, radiate, peduncled. Involucres rather flat; bracts narrow, spreading, foliaceous, nearly equal. Ray-flowers pistillate, fertile; ligules yellow. Disk-flowers numerous, hermaphrodite, fertile. Anthers somewhat sagittate at the base. Style-branches with slender pubescent appendages. Achenes flat, winged. Pappus of short awns, without intermediate scales.

1. **X. exauriculàta** (Robins. & Greenm.) Rydb. Annual, with a taproot; stem 3–6 dm. high, white-tomentose; leaf-blades ovate or rhombic-lanceolate, saliently dentate, minutely strigose but green above, densely white-strigose beneath, 3–7 cm. long; heads few, paniculate; involucres scarcely 1 cm. high, 1.5 cm. broad; ligules orange, 12–15 mm. long. *Verbesina encelioides exauriculata* Robins. & Greenm. [G]. Mountain valleys: Mont.—Kans.—Tex.—Ariz. *Plain—Son.—Mont.* Je–O.

65. COREÓPSIS L. TICKSEED.

Annual or perennial herbs. Leaves various. Heads on long peduncles, radiate. Involucres campanulate, their bracts in two series, all more or less united at the base, the outer ones usually narrow and foliaceous, the inner broad, variously colored, in ours orange or brown, scarious or with scarious margins. Receptacle flat or slightly convex. Achenes flat or more or less convex on the back. Pappus of two fimbriate squamellae, two awns or minute teeth, or none.

Style-branches with acute or cuspidate tips ; perennials.
  Leaves simple or pinnately lobed ; paleae of the receptacle broad at the base,
    attenuate at the apex ; style-tips cuspidate.
    Stem leafy only at the base, with long naked peduncle-like branches.
      Plant glabrous, except the ciliate leaf-bases.   1. *C. lanceolata.*
      Plant pubescent.   2. *C. crassifolia.*
    Stem uniformly leafy to near the top ; peduncles
      short.   3. *C. grandiflora.*
  Leaves palmately lobed or divided ; paleae of the recep-
    tacle linear or clavate ; style-tips conic.
    Leaves petioled.   4. *C. tripteris.*
    Leaves sessile.
      Leaves 3-cleft to about the middle, 3-ribbed below.   5. *C. palmata.*
      Leaves divided to the base into linear or filiform
        divisions.   6. *C. verticillata.*
Style-branches with truncate or obtuse tips ; annuals.
  Achenes wingless.   7. *C. tinctoria.*
  Achenes wing-margined.
    Lower leaves, as well as the upper ones, with linear
      divisions ; ligules more than 1 cm. long, uniformly
      yellow or orange.   8. *C. Atkinsoniana.*
    Lower leaves with oval or oblong divisions ; ligules
      less than 1 cm. long, with a dark spot at the base.   9. *C. cardaminifolia.*

1. **C. lanceolàta** L. Perennial, with a short rootstock; stem 2–5 dm. high, glabrous; leaves opposite, the lower spatulate or oblanceolate, petioled, the upper linear-lanceolate or linear, obtuse; involucres 8 mm. high and 12–15 mm. broad; outer bracts lanceolate, only slightly shorter than the ovate-lanceolate inner ones; ligules orange, 12–25 mm. long, coarsely toothed; achenes broadly wing-margined; pappus obsolete. Rich soil: Ont.—Fla.—La.—N.M.—Colo. *E. Temp.—Plain.* My–Au.

2. **C. crassifòlia** Ait. Perennial, with a rootstock; stem ascending, 2–4 dm. high, leafy below; leaves opposite, the lower ones simple, petioled, entire, oblong to obovate-spatulate, the upper sessile, oblong, acute; involucre 1–1.5 cm. broad; outer bracts lance-ovate, 6–8 mm. long, the inner broadly ovate, about 1 cm. long; ligules obovate or cuneate, 3-lobed, the middle lobe notched; achenes oval, winged. *C. lanceolata villosa* Michx. [G]. Dry soil: Ill.—Kans.—La.—Fla.—S.C. *Austral.* My–Au.

3. **C. grandiflòra** Hogg. Perennial; stem erect, 3–6 dm. high; leaves opposite, the lower ones usually entire, spatulate or lanceolate, the upper divided into linear or lance-linear divisions; involucre about 2 cm. broad; outer bracts lanceolate, 6–9 mm. long, the inner ovate, slightly longer; ligules obovate or cuneate, 3-lobed, the middle lobe toothed, 1.5–2 cm. long; achenes orbicular, winged. Dry soil: Mo.—Kans.—Tex.—N.M. (?)—Ga. *Austral.* My–Au.

4. **C. trípteris** L. Perennial; stem glabrous or nearly so, erect, 1–3 m. high, leafy, branched above; leaves opposite, petioled, 3-divided, the segments lanceolate or elliptic, the middle one often again parted, glabrous or pubescent; uppermost leaves simple and entire; outer bracts linear, obtuse, 2–3 mm. long; inner bracts ovate, about 5 mm. long; ligules elliptic, entire or notched; achenes elliptic-oblong, narrowly winged. Moist thickets: Pa.—Fla.—La.—Kans.—Minn. *Allegh.—Austral.* Je–S.

5. **C. palmàta** Nutt. Perennial; stem glabrous, erect, 5–10 dm. high, angled; leaves opposite, 3-cleft to the middle, or the uppermost simple, the divisions linear-oblong; involucre 1 cm. broad; outer bracts linear, rigid, obtuse, 7–9 mm. long, the inner oblong-obovate, slightly longer; ligules oblong-obovate, irregularly notched; achenes elliptic-oblong, narrowly winged. Ill.—Minn.—Neb. Tex.—La. *Prairie Texan.* Je–Jl.

6. **C. verticillàta** L. Perennial; stem glabrous, 5–10 dm. high, striate; leaves opposite, palmately 3-divided, the divisions once or twice pinnately parted into linear-filiform divisions; involucre 6–8 mm. broad; outer bracts linear, 5–6 mm. long, obtuse, the inner oblong, slightly longer; ligules oblong, entire or slightly notched; achenes oblong obovate, narrowly winged. Dry soil: Ont.—Minn.—Kans.—N.C. *Canad.—Allegh.* Je–S.

f. 573.

7. **C. tinctòria** Nutt. *Fig. 573.* Annual, with a taproot; stem 1–10 dm. high, glabrous; lower leaves bipinnately, the upper ones pinnately divided into linear divisions, glabrous; involucres about 7 mm. high, 10–15 mm. wide; outer bracts oblong, in fruit reflexed, inner bracts ovate or lance-ovate, acute; ligules 8–18 mm long, orange, with or without purple base. Low ground: Man.—Wis.—Tex. Ariz.—e B.C. *Temp.—Plain.* Je–Au.

8. **C. Atkinsoniàna** Dougl. Annual or perennial (?), with a taproot; stem 4–10 dm. high, glabrous; leaves opposite, pinnately or bipinnately divided into linear divisions, glabrous; involucres about 6 mm. high, 15 mm. broad; outer bracts oblong, obtuse, the inner broadly ovate; ligules orange, 12–15 mm. long. River banks: B.C.—N.D.—Ore. *W. Temp.—Plain.* Jl–Au.

9. **C. cardaminifòlia** (DC.) T. & G. Annual; stem glabrous, striate, 2–5 dm. high; leaves opposite, short, the lower once or twice pinnatifid into elliptic-oblong to linear-lanceolate divisions, the upper few-lobed or entire, linear; involucre about 1 cm. broad; outer bracts lanceolate, 3–4 mm. long, the inner ovate, about twice as long; ligules obovate, 3-lobed; achenes elliptic oblong, the wings one third as broad as the body. Low ground: Kans.—Ariz.—Tex.—La.; Mex. *Texan—Son.* My–O.

66. **BÌDENS** L. BEGGAR-TICKS, BUR-MARIGOLD.

Caulescent herbs, ours all annual. Leaves mostly opposite, simple or pinnatifid. Heads numerous, radiate or discoid. Involucres campanulate or turbinate, of two series of bracts, distinct or united only at the very base, the outer foliaceous, the inner scarious and appressed. Receptacle flat, chaffy. Ray-flowers neutral; ligules yellow or white, or none. Disk-flowers hermaphrodite and fertile. Achenes flat, 4-angled, or rarely almost terete, beakless. Pappus of 2–4 barbed awns or teeth, rarely none.

Achenes flattened, broadest above the middle; ligules yellow or none.
   Ligules inconspicuous or none.
      Leaves usually undivided; pappus-awns 3 or 4.

Disk-corollas pale yellowish green, 4-lobed; outer involucral bracts foil-
aceous, broad.
　Outer bracts erect or nearly so; lower leaves petioled.
　　　　　　　　　　　　　　　　　　　　　　　　　　1. *B. comosa.*
　　Outer bracts spreading; leaves all sessile.　　　2. *B. acuta.*
Disk-corollas orange; involucral bracts short, or
　some of the outer ones foliaceous, but narrow.　3. *B. connata.*
　Leaves 3–5-divided; pappus-awns usually 2.
　　Leaf-segments narrow, linear or lanceolate in out-
　　　line.　　　　　　　　　　　　　　　　　　　　6. *B. Sandbergii.*
　　Leaf-segments broadly lanceolate, toothed.
　　　Outer bracts 8–16, ciliate, the inner constricted
　　　　at the tip; corolla pale.
　　　　Bracts (except the ciliolate margins) and the
　　　　　leaves glabrous or nearly so.　　　　　　4. *B. vulgata.*
　　　　Bracts, leaves, and upper part of the stem
　　　　　pubescent with curved hairs.　　　　　　5. *B. puberula.*
　　　Outer bracts 4–8, nearly glabrous, the inner not
　　　　constricted; corollas orange.
　　　　Leaves very thin; heads 5–7 mm. high;
　　　　　awns 3–4.5 mm. long.　　　　　　　　　7. *B. discoidea.*
　　　　Leaves firmer; heads 1 cm. high or more;
　　　　　awns 4–6 mm. long.　　　　　　　　　8. *B. frondosa.*
　Ligules usually conspicuous.
　　Leaves undivided; pappus-awns 2–4, retrorsely barbed.
　　　Paleae with orange tips; 3-nerved.
　　　　Ligules 2–3 cm. long; outer bracts rarely sur-
　　　　　passing the inner ones; leaves oblanceolate,
　　　　　scarcely clasping.　　　　　　　　　　9. *B. laevis.*
　　　　Ligules 1.5–2 cm. long; outer bracts usually as
　　　　　long as the ligules or longer, reflexed; leaves
　　　　　distinctly clasping.　　　　　　　　　10. *B. glaucescens.*
　　　Paleae light yellow or pale throughout.
　　　　Outer bracts not more than twice as long as the
　　　　　ligules.
　　　　　Plant stout and tall; leaves broadly elliptic-
　　　　　　oblanceolate, more or less clasping;
　　　　　　ligules usually 1–1.5 cm. long.
　　　　　　Stem glabrous; outer bracts linear, not
　　　　　　　exceeding the ligules, obtuse, appressed.　11. *B. elliptica.*
　　　　　　Stem more or less pubescent with fleshy
　　　　　　　hairs; outer bracts lance-linear, acut-
　　　　　　　ish, loose.　　　　　　　　　　　12. *B. filamentosa.*
　　　　　Plant slender or low; leaves narrowly linear-
　　　　　　oblanceolate, scarcely clasping; ligules
　　　　　　less than 1 cm. long.
　　　　　　Outer bracts spreading, seldom longer
　　　　　　　than the inner; ray-achenes greenish,
　　　　　　　striate.　　　　　　　　　　　　　13. *B. prionophylla.*
　　　　　　Outer bracts reflexed, shorter than the
　　　　　　　inner; achenes blackish.　　　　　　14. *B. gracilenta.*
　　　　Outer bracts 3–5 times as long as the ligules.　15. *B. leptopoda.*
　Leaves pinnately divided; pappus-awns 2 or none.
　　Achenes in fruit broadly obovate.
　　　Involucre nearly glabrous; pappus-awns very
　　　　slender or wanting.　　　　　　　　　　　16. *B. aristosa.*
　　　Involucre hispid; achenes bidenticulate.　　　17. *B. involucrata.*
　　Achenes in fruit narrowly cuneate, short-awned.　18. *B. trichosperma.*
Achenes spindle-shaped, tapering above; ligules nearly white.　19. *B. bipinnata.*

1. B. comòsa (A. Gray) Wiegand. Annual; stem 3–8 m. high, glabrous;
leaves simple, 5–10 cm. long, narrowly elliptic or lance-elliptic, acuminate, serrate; involucres about 12 mm. high and 12–18 mm. broad; outer bracts 6–8,
linear or linear-lanceolate, mucronate, entire, 2–5 times as long as the disk,
nearly erect; achenes olive-green or brownish, smooth; awns 3. Wet ground:
Mass.—Va.—Colo.—Neb. *Allegh.—Plain.* Au–O.

2. B. acùta (Wiegand) Britton. Annual; stem 2–10 dm. high, glabrous;
leaves simple, lanceolate to oblong-lanceolate, sessile, sharply serrate, 4–8 cm.
long, sparingly ciliate; involucre 1.5–3 cm. broad, broader than high; outer
bracts foliaceous, longer than the disk, ciliate; inner bracts narrowly deltoid,
acute; achenes cuneate, sparingly retrorsely hairy; awns 3, retrorsely barbed,
the longer more than half as long as the achene. *B. comosa acuta* Wiegand
[G]. *B. riparia* Greene. Wet soil: Mo.—N.D.—Kans. *Prairie.* Au–S.

3. **B. connàta** Muhl. Annual; stem 3–14 dm. high, glabrous; leaves undivided, lanceolate to elliptic, rarely some of the lower 3-parted, coarsely saliently toothed; outer bracts 4 or 5, linear-spatulate, glabrous, rarely exceeding the disk; ligules, if present, half as long as the disk; achenes 4–6 mm. long, narrow, thick, black, with yellowish warts; awns 4, slender, with erect and retrorse barbs. Swamps: N.H.—Man.—Neb.—Mo.—N.C. *Canad.—Alleyh.* Au–O.

4. **B. vulgàta** Greene. Annual; stem 5–10 dm. high, glabrous; leaves pinnately 3–5-foliolate, glabrous or nearly so; leaflets lanceolate, acute, serrate, short-stalked; involucres 10–12 mm. high, 15–25 mm. broad; outer bracts 8–16, linear or linear-oblanceolate, ciliate, entire, rarely exceeding the disk; ligules usually present, equaling the disk; achenes brown or olivaceous, smooth or merely papillose, Wet ground: Ont.—N.C.—Calif.—B.C. *Temp.—Plain.* Au–S.

5. **B. pubérula** (Wieg.) Rydb. Annual; stem 3–10 dm. high, glabrous below, pubescent above; leaves pinnately 3–5-foliolate, usually decidedly pubescent on both sides; lanceolate, short-petioluled, serrate, acuminate, 4–8 cm. long; outer bracts 10–13, foliaceous, some at least exceeding the disk, hispid-ciliate on the margins, hirsute on the back, oblanceolate; disk about 1 cm. high; disk-corollas yellow; achenes olive-brown, slightly pubescent, flat; awns 2, 3–4 mm. long. *B. frondosa puberula* Wieg. Marshes and fields: Wis.—Mo.—Sask. *Prairie.* Au–S.

6. **B. Sandbérgii** Rydb. Annual; stem 3–6 dm. high, glabrous, usually purplish, branched; leaves pinnately divided, 7–10 cm. long, the divisions narrowly lanceolate, acuminate, toothed; outer bracts 3–5, linear or linear-oblanceolate, 1–1.5 cm. long; heads discoid, 10–15 mm. broad; paleae of the disk and lobes of the corollas orange, the tube of the latter yellow; achenes 4–6 mm. long, narrow; awns 3, strongly retrorsely barbed. *B. connata pinnata* S. Wats. [G]; not *B. pinnata* Noronha. Wet places: Hennepin and Ramsey counties, Minn. Jl. Perhaps a hybrid of *B. connata* and *B. bipinnata.*

7. **B. discoídea** (T. & G.) Britt. Annual; stem diffusely branched, 3–6 dm. high, glabrous, slender; leaves ternate, slender-petioled; leaflets rhombic to obovate-lanceolate, acuminate, coarsely serrate, 2.5–7 cm. long; outer bracts usually 4, oblanceolate, twice as long as the disk; ray-flowers usually wanting; achenes flat, narrowly cuneate, strigose, 4 mm. long; pappus-awns 2, short. Swamps: Mass.—Va.—La.—Minn. *Allegh.* Jl–S.

8. **B. frondòsa** L. Annual; stem 3–10 dm. high; leaves pinnately 3–5-foliolate, glabrous or nearly so, scabrous on the margins; leaflets 3–8 cm. long, lanceolate or oblong-lanceolate, sharply serrate, acuminate; involucres 10–15 mm. high and broad; outer bracts 4–8, spatulate-linear, entire, ciliate; ligules usually present, golden yellow, equaling the disk; achenes black, tuberculate. Wet places: N.B.—Fla.—Utah—B.C. *Temp.—Plain.* Jl–O.

9. **B. laèvis** (L.) B.S.P. Annual; stem 5–10 dm. high, glabrous; leaves undivided, 7–15 cm. long, sessile; heads sometimes nodding; outer bracts 7–8, rarely longer than the disk; ligules 2–3 cm. long, golden yellow; disk-corollas orange, 5-lobed; achenes cuneate, not tubercled; awns 2–4, retrorsely barbed. *B. chrysanthemoides* Michx. Swamps, mostly near the coast: Mass.—Wis.—Ga.; reported from Kans. and Iowa. Au–O.

10. **B. glaucéscens** Greene. *Fig. 574.* Annual; stem 3–7 dm. high, glabrous; leaves linear-lanceolate, 5–15 cm. long, serrate, sessile and clasping, somewhat paler beneath; heads nodding, 1 cm. high, 2–3 cm. wide; ligules often 1 cm. long; awns 4. *B. cernua* of western reports. In water and wet grounds: Minn.—Kans.—Utah—Ida.—Alta. *Plain—Submont.* Jl–O.

f. 574.

11. B. elliptica (Wieg.) Gleason. Annual; stem glabrous, 5–15 dm. high; leaves 1–2 dm. long, 2–4 cm. wide, elliptic-oblanceolate, tapering and clasping below, closely serrate-dentate; heads usually nodding; outer bracts usually longer than the inner, but shorter than the ligules; paleae mostly 3-striate, pale; achenes cuneate, retrorsely hispid on the 3 or 4 ribs, olive-brown when ripe; awns 3 or 4, ascending or erect. *B. cernua elliptica* Wieg. Swamps: Mass.—Pa.—Kans.—Minn. *Allegh.* Au–S.

12. B. filamentòsa Rydb. Stem 5–15 dm. high, usually more or less pubescent, with fleshy hairs; leaves narrowly lanceolate, 10–15 cm. long, 1.5–3 cm. wide, sharply serrate, with ascending teeth, acuminate, tapering to the clasping base, scabrous on the margins; heads cymose; outer bracts foliaceous, linear-lanceolate, acute or obtuse, loose, hispidulous-ciliate, often equaling the rays; disk fully 1 cm. high and about 2 cm. broad; ligules bright yellow; inner paleae pale, 3–4-striate; achenes darkly olive-brown, retrorsely hispidulous on the 4 angles; awns 4, ascending. Marshes: Que.—Iowa—Neb.—N.D. Au–S. Confused with *B. cernua radians* of Europe, which has more coarsely toothed leaves and the paleae with more striae, 5–7, even on the inner paleae.

13. B. prionophýlla Greene. Annual; stem slender, 3–5 dm. high, more or less pubescent; leaves linear or linear-lanceolate, sharply serrate, 5–10 cm. long, sessile; heads.nodding, about 1 cm. high and 2 cm. broad; achenes greenish; awns 3 or 4, very unequal. Water and wet places: Ont.—N.Y.—S.D.— (?) Colo.—Mont. *Prairie—Plain.* Au–S.

14. B. gracilénta Greene. Annual; stem slender, 3–5 dm. high, purplish, glabrous; leaves narrowly lanceolate, 5–10 cm. long, remotely serrulate; heads nodding; outer bracts shorter than the inner, reflexed; ligules 5–8 mm. long, often wanting; disk 1–1.5 cm. wide; achenes cuneate, 3–4-ribbed, retrorsely hispidulous, almost black; awns 3 or 4, strongly retrorsely barbed. Swamps: Mich.—Iowa—Minn. *Prairie.* Au–S.

15. B. leptópoda Greene. Stem slender, 6–10 dm. high, sparingly pubescent or glabrous; leaves linear-lanceolate, distinctly serrate, 4–10 cm. long, 4–18 mm. wide; heads long-peduncled; outer bracts linear or lance-linear, 2–5 cm. wide; ligules elliptic, 8–10 mm. long; achenes narrowly cuneate, thickly 4-angled, the angles retrorsely hispidulous, pustulate. Swamps: Que.—Iowa— Man. *Prairie.*

16. B. aristòsa (Michx.) Britton. Annual or biennial, 3–10 dm. high; leaves pinnately 5–7-divided, the segments lanceolate, acuminate, serrate, incised or pinnatifid, slightly pubescent beneath; heads numerous; outer bracts 8–10, linear-spatulate, not surpassing the inner; ligules 6–10, golden-yellow; ovaries ciliate and awnless; achenes obovoid, flat, strigose and ciliate; awns 2, slender, as long as the achenes, the barbs erect or reflexed. *Coreopsis aristosa* Michx. Swamps: Mich.—Iowa—Kans.—Tex.—La. *Prairie—Texan.* Au–O.

17. B. involucràta (Nutt.) Britton. Annual or biennial; stem 3–10 dm. high, branched, puberulent; leaves pinnately divided, the segments narrow, linear-lanceolate, long-acuminate, incised or pinnatifid; outer bracts 12–20, linear, acutish, hispid on the backs and margins, longer than the inner; ligules golden-yellow; disk-corollas 5-lobed; achenes obovate, flat, strigose-ciliate, bidenticulate. *Coreopsis involucrata* Nutt. Swamps: Ill.—Neb.—Tex.—Ark.; Del. *Prairie—Texan.* Jl–S.

18. B. trichospérma (Michx.) Britton. Annual or biennial; stem 5–15 dm. high, glabrous; leaves pinnatifid, the segments narrowly lanceolate, serrate or incised, acute or acuminate; outer bracts linear-spatulate, equaling the disk;

ligules bright yellow, 12–25 mm. long; disk-corollas deep yellow, 5-lobed; achenes slightly strigose-ciliate above; awns 2, as long as the breadth of the achene, the barbs erect. Swamps: Mass.—Minn.—Iowa—Ky.—Ga. *Allegh.* Au–O.

19. **B. bipinnàta** L. Annual; stem 3–7 dm. high, 4-angled, glabrous; leaves bipinnatifid, the primary divisions deltoid in outline, the ultimate ones lanceolate, incised or lobed; outer bracts 7–10, linear, shorter than the inner; ligules few, about equaling the disk, entire; disk-corollas 5-lobed; achenes in fruit longer than the involucre, the outer ones shorter than the inner; awns 3 or 4, retrorsely barbed. Damp soil: R.I.—S.D.—Ariz.—Fla. *Allegh.— Austral.* Jl–O.

67. THELESPÉRMA Less.

Glabrous perennial or annual, with finely dissected or rarely linear entire leaves and long-peduncled heads. Involucre hemispherical or campanulate, of two distinct series of bracts, the outer narrow, short, often spreading, the inner broad, united to about the middle into a cup, their free lobes scarious-margined. Receptacle flat, with a broad, 2-nerved, white scarious palea subtending each disk-flower. Ray-flowers present, neutral, or none; ligules, if present, yellow, mostly toothed. Disk-flowers perfect, fertile, corolla-tube slender; limb 5-toothed. Anthers obtuse and entire at the base. Style-tips acute. Achenes oblong or linear, terete or slightly compressed. Pappus of 2 retrorsely hispid awns, or sometimes none.

Heads radiate; annual or biennial.
  Leaf-segments linear-filiform, 1 mm. or less wide.   1. *T. trifidum.*
  Leaf-segments linear, over 1 mm. wide, plant perennial.   2. *T. intermedium.*
Heads discoid; perennials, with rootstock or woody caudex.   3. *T. gracile.*

1. **T. trifidum** (Poir.) Britton. A slender annual or biennial; stem branched, 3–9 dm. high, leafy; leaves bipinnately divided, the inner united not higher than the middle, with broad scarious margins; ligules 12–15 mm. long; awns of the pappus rather short, not longer than the width of the achenes. In dry soil: S.D.—Tex.—N.M.—e Colo. *Plain—Submont.* Je–Au.

2. **T. intermèdium** Rydb. *Fig. 575.* A glabrous bushy biennial or perhaps perennial; stems several, 2–6 dm. high; leaves once or twice pinnately divided; involucre about 1 cm. broad; outer bracts very narrowly linear-lanceolate, about half as long as the inner, these united to about the middle, scarious-margined; ligules 10–12 mm. long; teeth of the pappus longer than the width of the achenes. On plains in dry or sandy soil: Neb.—Wyo.—Colo. *Prairie—Plain—Submont.* Je–Au.

f.575.

3. **T. grácile** (Torr.) A. Gray. Stem 3–9 dm. high, branched and leafy; leaves as in the preceding, the upper reduced and often simple; heads 12–15 mm. broad; outer bracts oblong, obtuse, one fourth as long as the inner, the latter united to the middle or beyond; awns of the pappus longer than the width of the achenes. On plains and prairies: Neb.—Tex.—Ariz.—Wyo. *Prairie— Plain—Mont.* My–Au.

68. MEGALODÓNTA Greene. WATER MARIGOLD.

Immersed aquatic perennials. Leaves opposite, the immersed ones crowded, many times dissected into filiform divisions, the few emersed ones lanceolate, dentate, slightly connate. Heads single, radiate; involucre hemispheric; bracts in two series, oval or oblong, the outer shorter than the inner. Receptacle chaffy. Ray-flowers neutral, the ligules yellow, obovate or oblong, obtuse. Disk-flowers hermaphrodite and fertile. Achenes thick, cylindric, truncate at each end, smooth. Pappus of 3–6 slender divergent awns, barbed at the apex.

1. **M. Béckii** (Torr.) Greene. Stem 5–25 dm. long; immersed leaves 3–5 cm. long, the emersed ones 1–4 cm. long; ligules 6–10, golden-yellow; achenes 1–1.5 cm. long; awns 12–25 mm. long. *Bidens Beckii* Torr. [G, B]. Ponds and slow streams: Me.—Que.—Man.—Mo.—N.J. *Canad.—Allegh.* Au–S.

69. **MÀDIA** Molina.

Glandular-viscid, heavy-scented annuals. Leaves alternate, entire, narrow. Heads radiate. Involucre campanulate; bracts in a single series, strongly inflexed on the margins enclosing the ray-achenes; paleae in a single series enclosing the disk-flowers as an inner involucre inside the rays. Ray-flowers pistillate, fertile; ligules inconspicuous, cuneate, 3-lobed. Disk-flowers hermaphrodite and fertile. Achenes angled, those of the ray-flowers laterally flattened, very oblique. Pappus wanting.

1. **M. glomeràta** Hook. Stem 3–5 dm. high, leafy, hirsute, glandular in the inflorescence; leaves linear, pilose and hirsute-ciliate; heads more or less glomerate; involucres about 6 mm. high and 4 mm. wide; rays 2–5, sometimes none; achenes of the rays somewhat curved, 1-nerved on each face, those of the disk 4- or 5-angled. Open places in the mountains: Sask.—Minn.—Iowa—Colo.—Calif.—B.C. *W. Temp.—Mont.* Jl–S.

Tribe 8. **HELENIEAE.** Heads radiate or sometimes discoid. Marginal flowers, if present, ligulate. Disk-flower with tubular 4–5-lobed corollas. Receptacle naked, *i.e.*, without squamellae, only in *Gaillardia* bristly. Bracts of the involucre herbaceous or membranous but not scarious. Style-branches of the disk-flowers usually with hairy appendages, rarely truncate. Pappus paleaceous, or aristiform, but not capillary. Plants without oil glands.

70. **PSILÓSTROPHE** DC. Paper Flower.

Tomentose or woolly perennial (all ours) or rarely biennial herbs, or low shrubs. Leaves alternate, entire or lobed. Involucre cylindric or campanulate; bracts in a single series, equal, connivent, sometimes with 1–4 scarious ones within; receptacle naked. Ray-flowers few, pistillate and fertile; ligules broad, 3-lobed, yellow, becoming papery, persistent. Disk-flowers hermaphrodite and fertile; corollas yellow. Style-branches truncate at the apex. Achenes linear, striate. Pappus of lacerate or ciliate squamellae. *Riddellia* Nutt.

1. **P. villòsa** Rydb. *Fig. 581.* Stem 1.5–6 dm. high, branched above, loosely white-woolly; lower leaves oblanceolate, entire, dentate or rarely pinnatifid, 5–10 cm. long, the upper small, oblanceolate, sessile; heads several, short-peduncled; ligules lemon-yellow, 4 mm. long, 2- or 3-toothed; achenes glabrous; squamellae linear-lanceolate. Sandy soil: Kans.—Tex. *Texan.* Je–S.

f.581.

71. **HYMENOPÁPPUS** L'Hér.

More or less tomentose-floccose perennials or biennials, with mostly once- or twice-pinnate leaves. Heads corymbose or rarely solitary, discoid. Involucre hemispheric, or nearly so; bracts 6–12, in 1 or 2 series, subequal, appressed, more or less colored and scarious-margined. Receptacle naked. Disk-flowers all perfect and fertile; corollas yellow. Styles with short conic appendages. Achenes mostly obpyramidal, 4- or 5-angled, the faces 1–3 nerved. Pappus of 10–20 hyaline obtuse scales without costa, or none.

Corolla-lobes lanceolate, about equaling the short-campanulate throat; biennials.
  Bracts broadly obovate or ovate; achenes hirsutulous at least on the angles.
    Bracts with white tips.   1. *H. scabiosaeus.*
    Bracts with sulphur-yellow tips.   2. *H. sulphureus.*
  Bracts oblong or oblong obovate, with greenish white tips;
    achenes merely puberulent.   3. *H. corymbosus.*
Corolla-lobes shorter than the throat.
  Pappus more than 1 mm. long, not hidden by the hairs of
    the achenes; plant biennial.   4. *H. tenuifolius.*
  Pappus less than 1 mm. long, hidden by the hairs of the
    achenes, or sometimes none; perennial.   5. *H. filifolius.*

1. **H. scabiosaeus** L'Hér. Biennial; stem 3–7 dm. high, often purplish, glabrate or sparingly floccose, corymbosely branched; stem leaves once or twice pinnatifid with linear or oblong divisions, 5–15 cm. long, green and glabrate above, grayish-floccose beneath; heads 10–15 mm. broad; bracts obovate, bright white; corollas ochroleucous; achenes short-hirsute; squamellae less than 1 mm. long. *H. carolinensis* (Lam.) Porter [B, G]. Prairies and plains: S.C.—Kans.—Tex.—Fla. *Austral.* My–Jl.

2. **H. sulphureus** Rydb. Biennial; stem 3–6 dm. high, white-tomentose, branched above; lower leaves pinnately divided into oblong or linear lobes, white-tomentose beneath, glabrate above; stem leaves pinnatifid into linear divisions; heads 10–12 mm. broad; bracts obovate, with sulphur-yellow tips; corollas ochroleucous; achenes hirsutulous on the angles. Prairies and plains: Kans.—Tex. *Texan.* My–Je.

3. **H. corymbosus** T. & G. Biennial; stem 3–7 dm. high, glabrate below, glandular and puberulent above; leaves bipinnatifid with narrowly linear divisions, floccose when young, glabrate in age; heads numerous, 10–12 mm. broad; bracts oblong-obovate, with greenish white tips; corolla yellowish; achenes puberulent, strongly muricate and striate; squamellae minute. Plains and prairies: Mo.—Kans.—Tex.—Ark. *Ozark.* Ap–Je.

4. **H. tenuifolius** Pursh. *Fig. 577.* Biennial or short-lived perennial, slightly tomentose, in age often glabrate, 3–6 dm. high, branched and leafy; leaves, except the uppermost, once- to thrice-pinnate, with linear divisions, 5–30 mm. long; heads numerous, 8–12 mm. broad; corollas dirty white or cream-colored; achenes densely pubescent. Plains and dry prairies: N.D.—Wyo.—N.M.—Tex.—Ark. *Plain—St. Plains.* Je–S.

f.577.

5. **H. filifolius** Hook. Perennial; stem 2–3.5 dm. high, sparingly branched; leaves similar to those of *H. tenuifolius;* heads few, 10–15 mm. broad; bracts obovate-oblong, with broad scarious margins, tomentose; corollas bright yellow; throat about 1.5 mm. long; achenes densely villous, the long hairs hiding the small pappus; squamellae less than 1 mm. long. On plains and prairies: Alta.—Sask.—Kans.—N.M.—Nev.—Wash. *Plain—Submont.* Je–S.

72. OTHÀKE Raf.

Erect glandular annual herbs. Leaves alternate, mostly entire. Heads corymbose or paniculate, radiate or discoid. Inflorescence campanulate to obconic; bracts narrow, in 1 or 2 series, herbaceous, usually colored, at least at the tips. Receptacle small, flat, naked. Ray-flowers, when present, pistillate, fertile; ligules purple or rose-colored, broad, 3-cleft. Disk-flowers fertile; corollas pink or rose, 5-divided almost to the tube. Style-branches filiform, pubescent throughout. Achenes linear to narrowly obpyramidal, 4-angled. Pappus of 6–12 lanceolate, strongly ribbed squamellae.

Heads radiate; leaves lanceolate to linear-lanceolate.	1. O. sphacelata.
Heads discoid; leaves narrowly linear.	2. O. macrolepis.

1. **O. sphacelàta** (Nutt.) Rydb. Stem stout, branched, hispid, glandular-pubescent and viscid above, 2–6 dm. high; leaves petioled, lanceolate or linear-lanceolate, scabrous on both sides, indistinctly 3-veined; involucre 10–15 mm. high; ligules 8–10, rose-purple, 1–2 cm. long; pappus-scales of the disk-flowers 6–8, lanceolate, more than half as long as the achenes. *Polypteris Hookeriana* A. Gray, in part; not *Palafoxia Hookeriana* T. & G. Sandy plains: Tex.— Neb.—Colo.—N.M.; n Mex. *Plain—Son.—Submont.*

2. **O. macrólepis** Rydb. Stem 3–4 dm. high, strigose-puberulent, and glandular on the upper parts; leaves strigose-puberulent on both sides, 3–5 cm. long, 2–3 mm. wide; involucres 9–10 mm. high and about as wide; bracts 8–12, linear-oblanceolate, abruptly acute; disk-corollas 14–15 mm. long; pappus-squamellae 6–8, lanceolate, caudate-acuminate, 5–6 mm. long. Plains: Colo.— Kans. *Plain.* Au.

73. PICRADENIÓPSIS Rydb.

Low perennial canescent herbs, more or less woody at the base. Leaves opposite, pedately 3–5-divided, with linear divisions, or the upper entire. Heads radiate, corymbose. Involucre campanulate; bracts subequal, in two series. 3-nerved, the outer more or less keeled. Ray-flowers few, pistillate and fertile; ligules yellow, short. Disk-flowers many, hermaphrodite and fertile; corollas yellow. Achenes elongate-obpyramidal, glandular or hispidulous. Pappus of a crown of 8 ovate or lanceolate squamellae.

1. **P. oppositifòlia** (Nutt.) Rydb. Stem 1–2 dm. high, branched below, strigose-puberulent; leaves 1–5 cm. long, impressed-punctate and strigose-puberulent; involucres 5–7 mm. high and 7–10 mm. broad; ligules 3–4 mm. long; achenes glandular. *Bahia oppositifolia* A. Gray. Alkaline soil and dry plains: N.D.—Tex.—Ariz.—Mont. *Plain—Submont.* Je–S.

74. CHAENÁCTIS DC. Morning Brides.

Annual, biennial, or perennial herbs. Leaves alternate, usually pinnately dissected. Heads discoid, but the marginal flowers often with enlarged corollas. Involucre campanulate or turbinate; bracts herbaceous in 2 series; receptacle alveolate, naked or in some species bristly. Flowers hermaphrodite and fertile; corollas yellow, white, or flesh-colored, the throat cylindric or in some species funnelform in the marginal flowers. Achenes linear, terete or oval in cross-section, or obscurely 4-angled, pubescent. Pappus of 4–16 hyaline squamellae.

1. **C. Douglásii** (Hook.) H. & A. Stem 2–5 dm. high, sparingly floccose, in age glabrate, glandular-puberulent in the inflorescence; leaves bipinnatifid with numerous short, crowded, obtuse lobes, more or less floccose; heads paniculate; involucres 10–13 mm. high, 15–20 mm. broad; bracts linear, obtuse, glandular-puberulent; pappus-squamellae oblong or linear, obtuse, about 5 mm. long. Sandy and rocky places: Alta.—S.D.—N.M.—Calif.—B.C. *Plain—Mont.* Je–Jl.

75. TETRANEÙRIS Greene.

Perennial (all ours) or annual, caulescent or scapose, more or less villous or silky herbs. Leaves usually entire, rarely some of them lobed, alternate, either clustered at the base or scattered along the stem, or both. Heads radiate, rarely discoid, on slender peduncles. Involucre hemispheric, or nearly so; bracts several in two subequal series; receptacle convex, naked. Ray-flowers 10–20, pistillate and fertile; ligules yellow, oblong, broad 3- (rarely 4-) toothed; nerves 4. Disk-flowers many, hermaphrodite and fertile; corollas with 5 short pubescent teeth. Anther-tips broad, triangular or ovate. Style-branches dilated-truncate, somewhat penicillate at the tips. Achenes obpyramidal, 5-angled, about 3 times as long as broad, densely hirsute. Pappus of about 5 hyaline squamellae, with a strong midrib, which sometimes is produced into an awn.

Leafy-stemmed annuals. 1. T. linearifolia.
Scapose perennials.
  Bases of the leaves not wider than the blades.
    Leaves silvery-silky; ligules 6–8 mm. long. 2. T. acaulis.
    Leaves greenish, sparingly silky; ligules 8–10 mm. long. 3. T. simplex.
  Bases of the leaves 2–4 times as wide as the linear leaf-
    blades; branches of the caudex or rootstock elongate. 4. T. fastigiata.

1. **T. linearifòlia** (Hook.) Greene. Slender, branched annual; stem 2–3 dm. high; lower leaves linear-oblanceolate, 2–4 cm. long; upper leaves narrowly linear; peduncles 3–10 cm. long; involucre 5–7 mm. broad, silky-villous; bracts oblong; ligules 5–6 mm. long, 3 mm. wide; achenes densely hairy; pappus-squamellae lanceolate or ovate, awned. Dry plains: Kans.—Tex.—Mex. *Texan.*

2. **T. acaùlis** (Pursh) Greene. *Fig. 578.* Leaves crowded, narrowly linear-oblanceolate, 2–8 cm. long, 2–6 mm. wide, mostly acutish; involucres 7–8 mm. high, 10–15 mm. broad, hemispheric, silky-villous; bracts linear to oblong, or spatulate, obtuse; ligules 6–8 mm. long, yellow with orange veins; pappus-squamellae ovate, abruptly aristate. *Gaillardia acaulis* Pursh. *Actinella acaulis* Nutt. *T. incana* A. Nels. *T. eradiata* A. Nels., a rayless form. Dry hills and plains: N.D.—Tex.—N.M.—Ida.—Mont. *Plain—Submont.*

3. **T. símplex** A. Nels. Leaves green, sparingly appressed-pubescent, glabrate in age, ascending or erect, 2–5 mm. wide, acute; involucres 12–18 mm. broad, 8–10 mm. high, silky-villous; ligules 8–10 mm. long, 4–7 mm. wide, golden yellow, with orange veins; pappus-squamellae lance-ovate, abruptly aristate. *Actinea simplex* A. Nels. Dry hills: N.D.—Colo.—Wyo. *Plain.* Je–Jl.

4. **T. fastigiàta** Greene. Leaves basal, very narrowly linear-oblanceolate, 2–4 cm. long, 1–2 mm. wide, strongly punctate, glabrous or with a few long hairs near the base; scape 5–10 cm. long, striate, punctate and sparingly strigose; involucre about 5 mm. high and 10 mm. wide, hemispheric; bracts elliptic, obtuse; ligules light yellow, about 10 mm. long and 5 mm. wide; pappus-squamellae ovate, acuminate. *T. stenophylla* Rydb. [B]. *Actinella fastigiata* A. Nels. Dry plains: Kans.—Colo.—Tex. *St. Plains.* My–Jl.

76. HYMENÓXYS Cass. Colorado Rubber-plant.

Caulescent annuals, biennials, or perennials. Leaves alternate, commonly pinnatifid into narrowly linear divisions. Heads (in ours) radiate. Involucre of two series of bracts, 5–16 in each series, the outer usually narrower and more or less united at the base, rather firm, entire, acute, the inner broader, either similar, or else more or less erose on the margin and abruptly acuminate or mucronate. Ray-flowers 5–16, pistillate and perfect; ligules usually conspicuous, 3-lobed. Disk-flowers many, hermaphrodite, perfect. Style-branches obtuse or truncate, penicillate. Achenes obpyramidal, 5-angled, densely hirsute with ascending hairs. Pappus of 5 hyaline squamellae, with faint or indistinct midrib, usually acute or acuminate. *Picradenia* Hook.

Perennials, with a multicipital caudex; inner bracts fimbriate or erose, abruptly acute, acuminate, or mucronate, unlike the outer ones; disk-corollas not expanded at the mouth; outer bracts united to the middle; free portions ovate or ovate-lanceolate.
1. *H. Richardsonii.*

Annuals, with numerous heads; inner bracts not very different from the outer, neither abruptly acute nor mucronate, nor evidently erose; throat of the disk-corollas somewhat funnelform.
2. *H. odorata.*

1. H. Richardsònii (Hook.) Cockerell. Stem 1–3 dm. high, glabrous or nearly so; leaves 5–10 cm. long, pinnatifid, with narrowly linear divisions 2–4 cm. long and 1–2 mm. wide; involucres about 8 mm. high and 10–15 mm. broad; ligules bright yellow, 7–10 mm. long. *Picradenia Richardsonii* Hook. *Actinella Richardsonii* Nutt. *Hymenopappus ligulaeflorus* A. Nels. *P. ligulaeflora* A. Nels. Dry plains: N.D.—Sask.—Mont.—Alta. *Plain.* Je.

2. H. odoràta DC. Stem branched near the base, 1–3 dm. high, sparingly pilose or glabrous; leaves pinnatifid, with linear-filiform divisions; involucres about 4 mm. high, 6–8 mm. broad; ligules cuneate, 3–5 mm. long; pappus-scales ovate, aristate. *Philozera multiflora* Buckley. *Picradenia multiflora* Greene. *Actinella odorata* A. Gray. Dry plains: Tex.—Kans.—Colo.—Ariz. *St. Plains —Son.* Je–Au.

77. FLAVÈRIA Juss.

Glabrous or puberulent herbs. Leaves opposite, narrow, entire or dentate, sessile, often more or less connate. Heads numerous, narrow, in dense cymes or glomerules. Involucre of 1–8 subequal bracts, and sometimes 1 or 2 smaller outer ones. Receptacle small, naked or setose. Ray-flowers solitary, pistillate and fertile, or wanting; ligules entire, emarginate, or 3-dentate, yellow, inconspicuous. Disk-flowers 1–15, hermaphrodite and fertile. Style of the hermaphrodite flowers 2-parted, reflexed, obtuse. Achenes oblong, somewhat compressed, 10-ribbed; pappus in ours wanting.

1. F. campéstris J. R. Johnston. Annual; stem 2–6 dm. high, glabrous; leaves lanceolate to linear, serrulate to entire, more or less distinctly 3-ribbed; heads in close, subsessile clusters at the ends of the branches, 2–5-flowered; involucres cylindric, about 5 mm. long; bracts mostly 3, linear-oblong, glabrous; ligule elliptic, 1.5–2 mm. long, exceeding the disk. *F. angustifolia* A. Gray; not Pers. Alkali grounds: Tex.—Colo.—N.M.; Mex. *Son.* Jl–Au.

78. HELÈNIUM L. SNEEZEWEED.

Annual or perennial leafy herbs. Leaves alternate, narrow, impressed-punctate, more or less decurrent, producing wings on the stem. Heads usually radiate. Involucre rotate; bracts narrow, linear or subulate in 2 or 3 series, subequal or the inner shorter, at first spreading, usually soon reflexed. Receptacle from convex to ovoid or conic, naked or sometimes with a few bracts between the ray- and disk-flowers. Ray-flowers usually present, in most species pistillate and fertile, in the rest neutral and sterile; ligules large, conspicuous, 3 or 4-toothed or -lobed, cuneate in outline. Disk-flowers numerous, hermaphrodite and fertile; corollas yellow, brown, or purplish. Achenes truncately obpyramidal, 4- or 5-angled, with as many intermediate ribs, pubescent on the angles and ribs. Pappus of 4–8 thin scarious squamellae, with or without midribs.

Ray-flowers pistillate.
   Annuals; leaves narrowly linear-filiform, not decurrent.   1. *H. tenuifolium.*
   Perennials; leaves broader, more or less decurrent.
      Pappus about one fourth as long as the glandular-granu-
        liferous disk-corollas, only slightly exceeding the
        corolla-tube; plant glabrate or minutely puberulent.
        Ligules 1.5–3 cm. long; leaves long-acuminate, sharply
         toothed.   2. *H. altissimum.*
        Ligules 1–1.5 cm. long; leaves acute or short-acumi-
         nate, not sharply toothed.   3. *H. latifolium.*
      Pappus fully half as long as the puberulent disk-corollas,
        twice as long as the tube; plant densely pubescent.
        Ligules 6–10 mm. long; bracts only slightly exceeding
         the disk.   4. *H. montanum.*
        Ligules more than 10 mm. long; bracts much exceed-
         ing the disk.   5. *H. macranthum.*
Ray-flowers neutral.   6. *H. polyphyllum.*

1. **H. tenuifolium** Nutt. Annual; stem branched above, 2–4 dm. high, glabrous; leaves glabrous, linear-filiform, 1–2 cm long, 1–2 mm. wide; heads corymbose; involucre about 1 cm. broad; bracts glabrous, thin, long-attenuate; ligules 5–10, cuneate, 5–10 mm. long; disk globose; achenes hispid on the angles; pappus-squamellae ovate, awned. Moist or wet places: Va.—Kans.—Tex.—Fla.; Cuba. *Austral.—Trop.*

2. **H. altissimum** Link. Perennial; stem 1 m. high or more, minutely puberulent, winged; leaves narrowly oblanceolate, 1–2 dm. long, 1.5–5 cm. wide, long-acuminate at the apex, cuneate at the base, sharply serrate, minutely puberulent; bracts subulate; ligules 1.5–3 cm. long, 6–10 mm. wide; disk globose, 1.5–2 cm. broad, yellow; achenes long-hirsute; squamellae lanceolate, fimbriate. Low ground: Ill.—Minn.—Mo.—Kans. *Prairie.* Jl–S.

3. **H. latifolium** Mill. Perennial; stem 3–10 dm. high, glabrous or nearly so; leaves sessile, oblanceolate, elliptic, or lanceolate, 3–15 cm. long, serrulate or the upper entire, finely puberulent or in age glabrate, punctate; bracts subulate, 5–7 mm. long; ligules 10–15 mm. long, 5–10 mm. wide; disk globose, 10–13 mm. wide; achenes hispid on the ribs; squamellae lanceolate, erose-fimbriate, acuminate. Moist places: Conn.—Minn.—Miss.—Fla. *E. Temp.* Jl–S.

4. **H. montanum** Nutt. *Fig. 579.* Perennial; stem 2–10 dm. high, more or less puberulent; leaves lanceolate, decurrent, 5–10 cm. long, denticulate or subentire; heads numerous, paniculate-cymose; involucres about 8 mm. high, 10–15 mm. broad; ray-flowers fertile; ligules 6–10 mm. long, cuneate, puberulent beneath, golden yellow; pappus-scales lanceolate, acuminate. Meadows: Mont.—Kan.—N.M.—Wash.—B.C. *Prairie—Plain—Submont.* Jl–S.

f. 579.

5. **H. macranthum** Rydb. Perennial; stem 3–12 dm. high; leaves broadly lanceolate or ovate-lanceolate, dentate or entire, thinner than in the preceding; heads with longer naked peduncles; involucres over 1 cm. high, 15–20 mm. broad; ligules 12–20 mm. long, orange. *H. grandiflorum* Nutt.; not Gilib. *H. autumnale grandiflorum* A. Gray. Wet places: Mack.—Sask.—Wyo.—Ore.—B.C. *W. Temp.—Plain—Submont.* Jl–S.

6. **H. polyphyllum** Small. Perennial; stem 3–8 dm. high, broadly winged, puberulent; basal leaves oblanceolate, the upper linear-lanceolate, 2–10 cm. long, acute, entire or undulate, puberulent; heads numerous, corymbose; receptacle ovoid; ligules yellow, 6–15 mm. long; disk ovoid-globose, purple, 8–10 mm. broad; achenes hispid on the angles. Moist places and woods: Ill.—Kans.—Ark.—Ga. *Austral.* Jl–O.

79. GAILLÀRDIA Foug. BLANKET-FLOWER.

Annual, biennial, or perennial herbs, either scapose or caulescent. Leaves alternate or basal, entire, toothed, or pinnatifid, often in the same species. Heads radiate or discoid, peduncled. Involucre saucer-shaped or rotate; bracts in 2 or 3 series, ovate, oblong, or lanceolate, at least the upper half strongly reflexed. Receptacle convex to subglobose, alveolate, usually more or less fimbrillate, the fimbrillae from soft and short-conic, to elongate-subulate, stiff, and awn-like. Ray-flowers usually neutral, rarely pistillate and fertile, often wanting; ligules, if present, broad, cuneate or flabelliform, deeply 3-cleft, yellow or purple, or both. Disk-flowers hermaphrodite and fertile. Anthers auricled at the base. Style-branches appendaged; appendages from glabrous and short to hispidulous and filiform. Achenes broadly obpyramidal, about twice as long as broad, wholly or partly covered by long, stiff, ascending hairs.

Fimbrillae of the receptacle reduced to soft teeth ; lobes of the disk-corollas caudate, with a broad base.
    Heads long-peduncled ; branches slender.   1. *G. lanceolata.*
    Heads very short-peduncled ; branches stout.   2. *G. fastigiata.*
Fimbrillae of the receptacle spinescent ; lobes of the disk-corollas subulate, covered with moniliform hairs.
    Plant perennial; ligules usually wholly yellow.   3. *G. aristata.*
    Plant annual ; ligules partly or wholly purple.
        Leaves densely short-hairy beneath, the lower inclined to be round-lobed, the upper acute ; squamellae narrowly lanceolate, gradually attenuate into the awn.   4. *G. pulchella.*
        Leaves sparingly long-hairy beneath or glabrate, the lower acutely lobed, the upper acuminate ; squamellae ovate or ovate-lanceolate, abruptly acuminate into the awn.   5. *G. Drummondii.*

1. G. lanceolàta Michx. Annual or biennial; stem 3–7 dm. high, cinereous-puberulent; lower leaves oblanceolate or spatulate, 5–10 cm. long, dentate or sinuate, hirsute-puberulent; stem-leaves oblanceolate, oblong, or linear, entire or remotely serrate, 2–8 cm. long; peduncles 5–20 cm. long; bracts oblong or lanceolate, canescent; ray-flowers neutral; ligules yellow, cuneate, 1–2 cm. long, 3-cleft; disk purple; pappus-squamellae narrowly lanceolate, awned. Sandy soil: S.C.—Kans.—Tex.—Fla. *Austral.* Ap–S.

2. G. fastigiàta Greene. Biennial; stem 3–4 dm. high, canescent-puberulent; lower leaves spatulate, undulate or toothed, 5–8 cm. long, scabrous-puberulent; upper stem-leaves oblanceolate or linear-oblong; peduncles 1–4 cm. long; bracts lanceolate, canescent; ligules 10–12 mm. long, yellow, or copper-red at the base; disk purple, 15–20 mm. broad; corollas 6 mm. long; pappus squamellae lanceolate, awned. Waste or sandy places: Kans.—Tex. *Texan.* Je–O.

3. G. aristàta Pursh. Stem 2–6 dm. high, simple, leafy, more or less hirsute and puberulent; leaves oblanceolate in outline, the lower petioled, the upper sessile, grayish-puberulent, entire, dentate, lobed, or pinnatifid; involucres 2–3 cm. broad; bracts lance-subulate, long-attenuate, surpassing the disk, hirsute-villous; disk-corollas purple; ligules wholly yellow, or purplish at the base, 1.5–3 cm. long; pappus-squamellae lanceolate, long-aristate. *G. bracteosa* Standl. Hills and plains: Man.—S.D. —Kans.—Ore.—B.C. *W. Temp.—Plain—Submont.* Je–S.

4. G. pulchélla Foug. *Fig. 580.* Stem branched, leafy, 2–4 dm. high; puberulent; lower leaves oblanceolate or spatulate in outline, petioled, puberulent, entire, dentate or sinuately pinnatifid, the upper leaves sessile, lanceolate, acute or acuminate; involucres 15–20 mm. broad;

f.580.

bracts lanceolate, acuminate, hirsute; disk-flowers purple; ligules 10–18 mm. long, purple below with yellow tips, or sometimes wholly purple. Plains: La.— Mo.—Neb.—Colo.—Ariz.; n Mex. *Son.—St. Plains—Plain.* Ap–Au.

5. **G. Drummóndii** (Hook.) DC. Stem 3–6 dm. high, puberulent and with moniliform hairs; lower leaves oblanceolate, 5–10 cm. long, with triangular lobes; bracts linear-lanceolate, long-attenuate, hirsute and ciliate; ligules 15–18 mm. long, deeply 3-cleft, usually purple with yellowish tips. Plains and open spaces: Tex.—n Mex.; cult. and escaped in Kans. *Son.* Ap–Au.

Tribe 9. **TAGETEAE.** Heads radiate or discoid. Involucre of a single series of bracts, dotted or striped with oil glands. Leaves also gland-dotted. Marginal flowers, if present, pistillate, ligulate, and fertile. Disk-flowers hermaphrodite, with tubular, 4- or 5-lobed corollas. Receptacle naked. Style-branches with or without appendages. Pappus various.

80. **BOÉBERA** Willd. FETID MARIGOLD.

Annual or perennial herbs. Leaves opposite or the upper alternate, pinnately or bipinnately dissected. Heads usually radiate. Involucre hemi-spheric, turbinate or broadly campanulate, with a distinct calyculum; principal bracts 8–10, broad, thin and somewhat scarious on the margins, with 3–7 conspicuous glands; receptacle puberulent. Ray-flowers pistillate and fertile; ligules oblong to rounded-oval, yellow or orange. Disk-flowers many. Style branches with a short, conic or obtuse appendage. Achenes hirsute, more or less angled, somewhat compressed, cuneate-oblong. Pappus of 8–15 squamellae, each dissected into 5–10 bristles. *Dysodia* Cav.; not Lour.

f. 582.

1. **B. pappòsa** (Vent.) Rydb. *Fig. 582.* Ill-scented annual; stem leafy, branched, 1–3 dm. high; leaves pinnately or bipinnately divided into linear lobes; involucres 6–8 mm. high and as broad; outer bracts linear, green, the inner 8–10, oblong, scarious-tipped, brownish green or purplish-tinged, with conspicuous oblong glands; ligules few, inconspicuous. *Dyssodia chrysanthemoides* Lag. *D. papposa* (Vent.) Hitchc. [G]. Prairies, river valleys, roadsides, and waste places: Ohio—La.—Ariz.—Mont.; Mex. *Prairie—Son.— Submont.* Jl–S.

81. **THYMOPHÝLLA** Lag. TINY TIM.

Annual or perennial herbs. Leaves opposite or alternate, or both, pinnately dissected into narrow lobes. Heads usually radiate, peduncled. Involucre turbinate, campanulate, or hemispheric, with or without a few small accessory bracts below; principal bracts in one or two more or less distinct series, equal in length and more or less united, each with 1–5 glands towards the tip. Receptacle naked. Ray-flowers pistillate and fertile, rarely wanting. Disk-flowers hermaphrodite, fertile. Style-tips obtuse or truncate. Achenes clavate, 4- or 5-angled. Pappus of normally 10 (rarely 11–20) squamellae most often in two series. *Hymenatherum* Cass. *Lowellia* A. Gray.

1. **T. aùrea** (A. Gray) Greene. Glabrous annual; stem branched from below, leafy, 1–1.5 dm. high; leaves alternate, pinnately parted into 7–9 linear-filiform divisions; involucres 4–5 mm. high, 5–6 mm. broad; bracts oblanceolate, with conspicuous glands; ligules about 12, oblong, 5–6 mm. long; pappus of 6–8 quadrate or oblong, erose-truncate scales. *Lowellia aurea* A. Gray. *Hymenatherum aureum* A. Gray. *Dyssodia aurea* A. Nels. Dry plains: Colo.— Kan.—Tex.—N.M. *Son.* Je–O.

82. PÉCTIS L. LEMON-SCENT.

Rather low, branching, mostly aromatic or strong-scented, annual or perennial herbs. Leaves opposite, glandular-dotted, mostly entire, usually with several pairs of marginal bristles near the base. Heads usually small, solitary or cymose, radiate. Involucre from cylindric or oblong to campanulate or turbinate; bracts 3–12, free, in a single series, without calyculum, glandular-dotted, rounded-carinate at least below; receptacle naked. Ray-flowers few, pistillate and fertile; ligules yellow or tinged with red or purple. Disk-flowers rather few, hermaphrodite and fertile; corollas yellow. Style hispidulous, the short branches obtuse and without appendages. Achenes linear, terete or somewhat angled, pubescent or glabrate. Pappus various, of few or many squamellae, awns, or bristles, or rarely reduced to a mere crown.

f.583.

1. **P. angustifòlia** Torr. *Fig. 583.* Glabrous annual, lemon-scented; stem branched, 5–20 cm. high; leaves narrowly linear; involucres 4–5 mm. high, 2–3 mm. broad; bracts linear, about 8; ligules oblong, about 3 mm. long. *P. papposa sessilis* M. E. Jones. ''Sand-draws'' and sandy hillsides: Tex.—Neb.—Colo.—Ariz.; Mex. *Prairie—Son.—Mont.* Ap–O.

Tribe 10. **ANTHEMIDEAE.** Heads radiate or discoid. Bracts of the involucre imbricate, wholly or partly scarious. Marginal flowers pistillate, ligulate or filiform-tubular, or wanting. Disk-flowers hermaphrodite, tubular. Receptacle chaffy or naked. Anthers not caudate. Style-branches in the disk-flowers usually truncate. Pappus squamellate, a small crown, or wanting.

83. ACHILLÈA (Vaillant) L. YARROW, MILFOIL.

Perennial caulescent, usually villous herbs. Leaves alternate, from serrate to tripinnatifid. Heads several, usually radiate. Involucre campanulate to hemispheric; bracts imbricate in 3 or 4 series, the outer usually much shorter. Receptacle conic or convex, chaffy; paleae membranous. Ray-flowers few, 5–12, pistillate and fertile; ligules short and broad, in most species white or sometimes pink or purple. Disk-flowers 15–75, hermaphrodite and fertile; corollas yellowish white or straw-colored. Anthers with ovate obtuse tips. Style-branches in the pistillate flowers oblong, obtuse, in the hermaphrodite ones with truncate, fimbriate tips. Achenes oblong or obovate, obcompressed, callous-margined, glabrous. Pappus wanting.

Leaves pinnatifid to tripinnatifid.
  Bracts with dark brown or black margins.   1. *A. subalpina.*
  Bracts with light brown or straw-colored margins.
    Ligules 2.5–4 mm. long; leaf-segments crowded, ascending.   2. *A. lanulosa.*
    Ligules 1.5–3 mm. long; leaf-segments more or less
        spreading, not crowded.
      Rachis merely margined; ligules 1.5–2.5 mm. long.   3. *A. occidentalis.*
      Rachis distinctly winged; ligules 2.5–3 mm. long.
        Leaves conspicuously punctate, their ultimate seg-
          ments strongly callous-tipped; ligules purplish.   4. *A. asplenifolia.*
        Leaves not conspicuously punctate, their ultimate
          segments not callous-tipped; ligules white,
          rarely pink.   5. *A. Millefolium.*
Leaves serrate-incised, not pinnatifid.   6. *A. multiflora.*

1. **A. subalpìna** Greene. Stem slender, 1–3
dm. high, villous with long hairs; leaves 3–10 cm.
long, linear in outline, more or less long-villous; segments short and narrow, linear, much crowded;
heads in a dense crowded corymb; bracts elliptic,
with dark brown or blackish margins, the innermost
usually acute, the rest obtuse; ligules 2–3 mm. long
and broad. *A. lanulosa alpicola* Rydb. *A. alpicola*
Rydb. High mountains: Alta.—Man.—N.M.—Nev.
—D.C. *Boreal. —Subalp.—Alp.* Jl–Au.

2. **A. lanulòsa** Nutt. *Fig. 584.* Stem 3–6 dm.
high, copiously villous with long silky hairs; leaves
5–10 dm. long, linear in outline, densely long-villous,
bipinnatifid; segments (primary and secondary)
short and directed forward, the secondary linear-
lanceolate; bracts elliptic, obtuse, with green mid-
rib and straw-colored (rarely brownish) margins. Plains and mountain valleys: Man.—Kans.—N.M.—Calif.—B.C.; adv. eastward to Ont.; n Mex. *W.
Temp.—Plain—Subalp.* My–S.

3. **A. occidontàlis** Raf. Stem 3–6 dm. high, usually rather copiously
villous with long silky hairs; leaves lanceolate in outline, 5–10 cm. long, twice
or thrice pinnatifid; ultimate segments narrow, linear; involucres about 4 mm.
high and 3 mm. broad; bracts elliptic, obtuse, with a green midrib and straw-
colored margins. Roadsides and prairies: Pa.—S.C.—Ark.—Neb.—S.D.; adv.
in Colo. *Prairie.* My–Au.

4. **A. asplenifòliu** Vent. Stem 3–6 dm. high, sparingly silky-villous;
lower leaves lanceolate in outline, the upper linear, glabrous and punctate
above, sparingly villous beneath; primary divisions divaricate, involucre 4–5
mm. high, 3–4.5 mm. broad; outer bracts ovate, obtuse, the inner oblong, acute;
margins light brown; ligules mostly 5, nearly orbicular, purple or dark rose,
2.5 mm. long. Roadsides: N.S.—N.C.—S.D.; adv. from Eur. Jl–S.

5. **A. Millefòlium** L. Stem 2–6 dm. high, somewhat villous, with short
hairs; leaves oblong in outline, 5–10 cm. long, 8–20 mm. broad, finely villous
or glabrate, twice pinnatifid; primary segments more or less spreading, more
or less decurrent into the wing-margins; secondary segments short, lanceolate,
flat, spinulose-tipped; involucres 4–5 mm. high, 3–4 mm. broad; bracts lance-
elliptic or elliptic, obtuse, with brown (rarely nearly black) margins; ligules
2–3 mm. long and broad. Shores, hillsides, and roadsides: Newf.—Va.—Colo.
—B.C.; nat. or adv. from Eur. Jo–S.

6. **A. multiflòra** Hook. Stem 3–10 dm. high, villous; leaves linear, 5–10
cm. long, 5–8 mm. broad, somewhat pubescent or glabrate, pectinately cleft into
lanceolate, serrulate lobes; involucres 4–5 mm. high, 4–6 mm. broad; bracts
more or less villous, elliptic with brown margins; ligules 10–12, suborbicular,
1–1.5 mm. long. Valleys: Mack.—Sask.—Alaska. *W. Boreal.* Jl–Au.

84. CÒTA J. Gay. YELLOW CHAMOMILE.

Annual or perennial herbs. Leaves alternate, pinnatifid or bipinnatifid.
Involucre saucer-shaped; bracts in 3 series, rather firm. Receptacle convex,
chaffy; palea membranous, cuspidate. Ray-flowers pistillate and fertile, with
white or yellow ligules. Disk-flowers hermaphrodite and fertile. Achenes obcompressed, somewhat triquetrous, the two faces 10-nerved. Pappus a membranous entire crown.

1. **C. tinctòria** (L.) J. Gay. A leafy perennial; stem 3–6 dm. high, hirsute-villous; leaves 2–5 cm. long, pinnatifid with a winged rachis and oblong

incised divisions; bracts villous; ligules 20–30, yellow, 12–15 mm. long. *Anthemis tinctoria* L. Waste places and fields: Me.—N.J.—Iowa; nat. from Eur.

85. ÁNTHEMIS (Mich.) L. CORN CHAMOMILE.

Annual or perennial herbs. Leaves alternate, mostly pinnatifid or bipinnatifid. Heads solitary at the ends of the branches, usually radiate. Involucre saucer-shaped; bracts in about 3 series of nearly the same length, usually thin; receptacle conic to hemispheric, chaffy; paleae membranous with a distinct midrib. Ray-flowers pistillate and fertile; ligules white. Disk-flowers hermaphrodite and fertile; corollas yellow. Anthers with ovate obtuse tips. Style-branches truncate and fimbriate at the apex. Achenes subcylindric or somewhat obovoid, not angled, with filiform ribs. Pappus wanting, or a minute crown.

1. **A. arvénsis** L. A leafy annual; stem 2–5 dm. high, branched, hirsute-villous; leaves 3–5 cm. long, bipinnatifid, with linear-lanceolate cuspidate lobes, hirsute-villous; involucre 4–5 mm. high, 7–12 mm. broad; bracts villous-hirsute; receptacle conic; paleae linear-lanceolate, abruptly cuspidate; ligules white, 7–12 mm. long, 3–4 mm. wide; achenes subcylindric, 10-ribbed; pappus a minute border. Fields and waste places: Me.—Ga.—Ore.—B.C.; adv. or nat. from Eur.

86. MARÙTA Cass. MAYWEED, DOG FENNEL, DOG CHAMOMILE.

Leafy annuals. Leaves alternate, bipinnatifid, with narrow divisions. Heads radiate, solitary at the ends of the branches. Involucre saucer-shaped; bracts in 2 series, subequal. Receptacle conic, chaffy towards the apex; paleae subulate, stiff, rather persistent. Ray-flowers 10–15, neutral; ligules white. Disk-flowers hermaphrodite and fertile; corolla-tube cylindric, longer than the funnelform throat; lobes 5, ovate, spreading. Anthers with ovate obtuse tips. Style included, with short branches, truncate and fimbriate at the apex. Achenes nearly cylindric, 10-ribbed, glandular-tubercled. Pappus wanting.

1. **M. Cótula** (L.) DC. *Fig. 585.* Annual, with fetid odor; stem glabrous or slightly pubescent above, 3–6 dm. high; leaves 3–5 cm. long, twice or thrice pinnatifid, with narrow linear-filiform divisions, glabrate or somewhat hairy; involucre 8–12 mm. broad; bracts oblong, obtuse, somewhat pubescent; ligules 10–18, white, about 1 cm. long, at length reflexed. *Anthemis Cotula* L. [G]. Fields and waste places: Newf.—Fla.—Calif.—B.C.—Yukon; nat. from Eur. *Temp.—Plain—Submont.* Jl–S.

f.585.

87. CHAMOMÍLLA (Hall.) Gilib. CHAMOMILE, PINEAPPLE-WEED, GREEN DOG-FENNEL.

Mostly annual glabrous herbs. Leaves alternate, once to thrice pinnatifid into narrow divisions. Heads solitary or corymbose, radiate or discoid. Involucre saucer-shaped to hemispheric; bracts in 2–4 series, somewhat imbricate, obtuse, scarious-margined; receptacle conic, hemispheric, or subglobose, naked. Ray-flowers, if present, pistillate and fertile; ligules white. Disk-flowers numerous, hermaphrodite and fertile; corollas yellow; anthers with ovate obtuse tips; style-branches short, with truncate tips. Achenes usually asymmetric, with 3–5 ribs on the inner half, the back being nerveless. Pappus of a more or less developed crown or margin. *Matricaria* L., in part.

Heads discoid.   1. *C. suaveolens.*
Heads radiate.
    Bracts with white or light brown margins.   2. *C. inodora.*
    Bracts with dark brown or almost black margins.   3. *C. Hookeri.*

1. C. suavèolens (Pursh) Rydb. Annual; stem glabrous, very leafy, much branched, 1–4 dm. high; leaves glabrous, 2 or 3 times pinnatifid, with linear acute lobes; involucres 7–10 mm. broad; bracts glabrous, oval to oblong, with broad scarious margins; ligules none; disk-corollas yellow, 4-lobed; receptacle conic; achenes oblong, slightly angular; pappus an obscure crown. *M discoidea* DC. *M. matricarioides* (Less.) Porter. *M. suaveolens* Buchenau [G]. Moist and sandy places, roadsides, and waste places: Alaska—Calif.—Ariz.—N.D.; adv. or nat. eastward to Newf.—N.J.—N.Y., and in Eur. *W. Temp.—Plain—Submont.* Mr–S.

2. C. inodòra (L.) Gilib. Annual; stem 2–4 dm. high, glabrous; leaves pinnatifid, 3–6 cm. long, with linear-filiform divisions, glabrous or nearly so; involucre saucer-shaped, 10–15 mm. broad; bracts in 2–3 subequal series, linear-oblong, with white or light brown scarious margins; ligules present, white, 7–10 mm. long; achenes dark brown, with 3 strong ribs on the inner side, rugose on the back. *Matricaria inodora* L. Waste places: Newf.—Pa.—Minn.—Ont.; Ore.—Ida.; Colo.; adv. or nat. from Eur.

3. C. Hŏŏkŏrı (Schultz-Bip.) Rydb. Annual; stem 1–4 dm. high, grooved, glabrous; leaves 3–7 cm. long, glabrous, bipinnatifid with filiform divisions; heads solitary or few; bracts in 2 or 3 subequal series, oblong, obtuse; ligules 15–25, white, 8–10 mm. long; achenes brown, with 3 ribs on the inner side. *Matricaria grandiflora* Britton. *M. inodora grandiflora* Ostenf. Coasts: Greenl.—Man.—Alaska; Eur. *Arctic –Subarct.* Jl–Au.

88. LEUCÁNTHEMUM (Tourn.) Mill. Ox-eye Daisy.

Perennial herbs with rootstocks. Leaves alternate, dentate or entire or in some exotic species pinnatifid. Heads radiate, rarely discoid, solitary, rather long-peduncled at the ends of the stem. Involucre saucer-shaped; bracts many, rather narrow, in 2–4 series, somewhat imbricate. Receptacle flat or convex, naked. Ray-flowers in a single series, pistillate and fertile; ligules usually well-developed, white. Disk corollas yellow. Anthers with ovate tips. Style of the ray-flowers slightly exserted, with short oblong branches, that of the disk-flowers included and with short branches, truncate and fimbriate at the apex. Achenes all cylindric, 10-ribbed or 10-angled. Pappus wanting.

f.586.

1. L. vulgàre Lam. *Fig. 586.* Stem 3–10 dm. high, glabrous; basal leaves petioled, obovate or spatulate, coarsely dentate or incised; stem-leaves sessile, narrowly oblanceolate to nearly linear, dentate; involucres 12–15 mm. broad; bracts lanceolate, margined by a narrow brown band and a scarious margin; ligules white, 12–15 mm. long. *Chrysanthemum Leucanthemum* L. [G]. *L. Leucanthemum* (L.) Rydb. Meadows, roadsides and pastures: Lab.—Fla.—Utah—B.C.; nat. from Eur. *Plan—Submont.* My–N.

89. TANACÈTUM (Tourn.) L. Tansy.

Strongly aromatic, leafy, mostly perennial herbs, with rootstocks. Leaves alternate, once to thrice pinnatifid, conspicuously punctate. Heads corymbose, usually several or many, radiate, but the rays often inconspicuous and not exceeding the disk. Involucre hemispheric or broader; bracts in 2 or 3 series, usually narrow, not very unequal in length; receptacle convex, naked. Ray-flowers pistillate and fertile; ligules yellow, varying from erect, 3-lobed, concave and scarcely exceeding the disk to spreading, flat, and well developed. Disk-flowers many, hermaphrodite and fertile; corollas yellow. Style in the ray-flowers slightly exserted, in the disk-flowers included. Anthers with obtuse tips. Achenes subcylindric, those of the ray-flowers mostly 3-angled, those of the disk-flowers 5-angled. Pappus-squamellae united into a short crown.

1. **T. vulgàre** L. Stout perennial; stem 5–10 dm. high, glabrous or nearly so; leaves bipinnatifid, with oblong-acute serrate divisions, glabrous or nearly so; heads numerous in a corymbiform panicle; involucre 6–10 mm. broad; flowers yellow; marginal flowers with short oblique 3-toothed limbs. Roadsides and waste places: N.S.—N.C.—Miss.—S.D.—Man.; Calif.—Wash.; escaped from cultivation; native of Eurasia. Jl–S.

90. BALSÁMITA Desf. Costmary, Mint Geranium.

Aromatic perennial, herbs. Leaves alternate, conspicuously punctate. Heads corymbose, radiate or discoid. Involucre hemispheric; bracts in 4 or 5 series, imbricate. Receptacle convex. Ray-flowers, if present, pistillate and fertile, in ours wanting. Disk-flowers hermaphrodite and fertile. Achenes cylindric-oblong, 10-nerved; pappus a denticulate border.

1. **B. màjor** Desf. A leafy perennial; stems 5–12 dm. high, grooved, puberulent; lower leaves 1–2 dm. long, puberulent or pilose, obovate or oblanceolate, crenate or dentate, rounded at the apex; upper leaves sessile, crenate, often incised or lobed at the base; heads discoid; outer bracts green, the inner with scarious erose tips. *Chrysanthemum Balsamita* L. [B, G]. Waste places and around dwellings: Me.—N.Y.—Ind.—S.D.; escaped from cultivation; nat. of Eur. and Orient.

91. ARTEMÍSIA L. Wormwood, Mugwort, Cudweed, Sage-brush.

Perennial or annual herbs, or shrubs, usually bitter-aromatic. Leaves alternate. Heads small, most commonly panicled and nodding when young, apparently discoid; the marginal flowers without ligules or wanting. Involucre campanulate to hemispheric; bracts in 2–4 series, at least the inner ones more or less scarious. Receptacle naked or hairy, convex or conic. Marginal flowers (functionally ray-flowers) pistillate and fertile or wanting; corollas cylindric or subcylindric and somewhat tapering upwards, 2–3-, rarely 4-toothed, often somewhat oblique; style more or less exserted, 2-cleft; style-branches linear-filiform and subterete to oblong and somewhat flattened. Disk-flowers hermaphrodite, fertile or sterile; corollas campanulate, funnelform or trumpet-shaped, 5-toothed, regular; anthers with lanceolate or subulate tips; style either 2-cleft and the branches more or less recurved, truncate at the apex, with an erose or fimbriate apex, or else entire and surmounted by an erose or fimbriate cup; achenes ellipsoid, terete, usually glabrous, without pappus.

Disk-flowers sterile, their styles undivided or with short erect branches; receptacle naked.
 Annual or perennial herbs; style of the disk-flowers undivided, ending in a cup-shaped penicillate appendix. 1. Dracunculoides.
 Low shrubs or undershrubs; style of the disk-flowers usually more or less 2-cleft, each branch erect with truncate penicillate or erose apex. 2. Filifoliae.
Disk-flowers fertile, their styles 2-cleft; branches more or less recurved.
 Marginal pistillate flowers present.
  Receptacle hairy. 3. Frigidae.
  Receptacle naked.
   Annual or biennial herbs with bipinnatifid leaves. 4. Annuae.
   Perennials.
    Leaves not dissected into narrowly linear more or less divergent divisions, or these, if narrow, very few, directed forward, and entire.
     Leaves greener and glabrate above, at least in age (except in *A. Brittonii*). 5. Vulgares.
     Leaves white-tomentose on both sides. 6. Gnaphaloides.
    Leaves once or twice dissected into linear or filiform, more or less spreading divisions.
     Leaves once pinnatifid; plant herbaceous; style of the ray-flowers long-exserted. 7. Wrightianae.

Leaves twice pinnatifid; plant suffruticose; style of the ray-flowers
short-exserted.   8. PONTICAE.
Marginal pistillate flowers wanting; receptacle naked; shrubs 3–40 dm. high;
heads in terminal panicles; leaves entire, usually 3-toothed.
     9. TRIDENTATAE.

1. DRACUNCULOIDES.

Leaves all entire or the lower trifid.	
Leaves densely pubescent when young.	1. *A. glauca.*
Leaves glabrous.	
Branches of the inflorescence strict.	2. *A. dracunculoides.*
Branches of the inflorescence drooping at the apex.	3. *A. cernua.*
Leaves, at least the lower, pinnatifid or bipinnatifid.	
Heads very small, 2–3 mm. broad, numerous in large leafy panicles; plants mostly tall, 3–20 dm. high.	
Biennials; outer bracts acute.	
Plant glabrous or nearly so.	4. *A. caudata.*
Plant pubescent, especially when young.	5. *A. Forwoodii.*
Perennials, with a rootstock or caudex; outer bracts obtuse.	6. *A. camporum.*
Heads larger, 4–5 mm. broad, in narrow spike-like panicles; plants low, rarely 4 dm. high.	7. *A. Bourgeauana.*

2 FILIFOLIAE.

One species.	8. *A. filifolia.*

3. FRIGIDAE.

Plant 5–15 dm. high, erect; leaves twice or thrice pinnatifid.	9. *A. Absinthium.*
Plant 2–5 dm. high, often decumbent at base; leaves twice ternate.	10. *A. frigida.*

4. ANNUAE.

Heads 2–3 mm. high, 2–4 mm. broad, in dense axillary spikes.	11. *A. biennis.*
Heads 1.5 mm. high, 1.5–2.0 mm. broad, in lax racemes ending the branches.	12. *A. annua.*

5. VULGARES.

Leaf-segments again divided or lobed; heads very numerous; involucre campanulate, 4 mm. high, 10–30-flowered; leaf-segments obovate in outline.	13. *A. vulgaris.*
Leaves or their segments entire or nearly so.	
Heads crowded, racemose, comparatively large, 25–50-flowered; involucre hemispheric; leaves lanceolate, thin, deeply lobed; involucre glabrous or nearly so.	14. *A. elatior.*
Heads very numerous in dense compound panicles; involucre campanulate or ellipsoid.	
Upper leaves lanceolate or oblanceolate, usually more than 4 mm. wide.	
Leaves finely serrate throughout.	15. *A. serrata.*
Leaves not finely serrate.	
Involucre 4–5 mm. high; leaves large, 5–20 cm. long, lanceolate or ovate-lanceolate in outline, with very few divisions, directed forward.	16. *A. Herriotii.*
Involucre 3–4 mm. high; leaves shorter; lobes of the lower leaves often spreading.	
Leaves of a lanceolate type; involucre campanulate.	17. *A. ludoviciana.*
Leaves of a cuneate type; involucre hemispheric.	18. *A. Brittonii.*
Upper leaves linear or narrowly linear-oblanceolate, less than 4 mm. wide, at least some of them with long, narrow, alternate, salient, falcate lobes.	
Involucre about 3 mm. high and 2 mm. broad; leaf-segments mostly opposite, not falcate.	19. *A. Lindheimeriana.*
Involucres about 4 mm. high and broad; leaf-segments alternate, falcate.	20. *A. falcata.*

6. GNAPHALOIDES.

Leaves all entire or the basal ones merely toothed.	
Involucre 3–4 mm. high, 2–3 mm. broad; heads less than 25-flowered.	
Corollas dark brown or purplish; leaves usually less tomentose above, the lower serrate towards the apex.	21. *A. gnaphalodes.*

Corollas light brown or yellow; leaves equally white-
tomentose on both sides.
Involucre 2–2.5 mm. broad; leaves very narrow,
linear or lance-linear, acute.   22. *A. pabularis.*
Involucre 2–3 mm. broad; leaves broader, if
lance-linear, more or less acuminate.
    Upper leaves ovate to lanceolate, acute; heads
densely crowded and usually erect.   23. *A. Purshiana.*
    Upper leaves lanceolate to lance-linear, acu-
minate; heads less crowded, mostly spread-
ing. (Entire-leaved forms of)   25. *A. diversifolia.*
Involucre 4–5 mm. high, 4–7 mm. broad; heads usually
25–40-flowered.   24. *A. longifolia.*
Leaves, at least the lower ones, more or less lobed or divided, with simple long and narrow lobes.
    Lower leaves with long and narrow lobes, directed
forward.   25. *A. diversifolia.*
    Lower leaves with short and broad, often salient lobes. 18. *A. Brittonii.*

7. WRIGHTIANAE.

One species.   26. *A. Carruthii.*

8. PONTICAE.

Leaves green above.   27. *A. Abrotanum.*
Leaves white on both sides.   28. *A. pontica.*

9. TRIDENTATAE.

Involucre 4–5 mm. high, 3–5 mm. broad; leaves mostly
entire.   29. *A. cana.*
Involucre 2–4 mm. high, 2–2.5 mm. broad; leaves, at least
some of them, 3-toothed at the apex.   30. *A. tridentata.*

1. **A. glaùca** Pallas. Perennial; stem 3–6 dm. high, silky-pubescent;
leaves entire, linear, 2–5 dm. long, 2–5 mm. wide, or with a few linear divisions,
silky-canescent; heads small, numerous in a large panicle; involucres about 13
mm. broad; bracts elliptic, pubescent, scarious-margined; flowers yellow.
Prairies and banks: Man.—Neb.—Colo.—Alta.; Siberia. *Prairie—Plain.* Jl–O.

2. **A. dracunculoìdes** Pursh. Perennial; stem 5–15 dm. high, glabrous;
leaves narrowly linear, or some 3-cleft with similar divisions, glabrous, 3–7 dm.
long, 1–3 mm. wide; heads very numerous, in a compound panicle, nodding;
involucres broad, glabrous, yellowish green; bracts elliptic, obtuse; ray-flowers
10–20. Prairies and plains: Man.—Mo.—Tex.—Calif.—B.C. *W. Temp.—
Prairie—Plain—Submont.*

3. **A. cérnua** Nutt. Perennial, with a taproot; stems erect, branched,
glabrous, often tinged with red, the branches drooping at the ends; lower
leaves pinnately divided, with linear-filiform divisions, glabrous; upper leaves
simple; peduncles slender, 3–6 mm. long; involucre hemispheric, 2 mm. high;
bracts yellowish with a green rib, obtuse; ray-flowers about 15. Waste places:
Iowa—Kans.—Mo. *Prairie.* Jl–S.

4. **A. caudàta** Michx. Biennial; stem glabrous, 5–20 dm. high; leaves
1–3-pinnately divided into linear-filiform divisions, glabrous or nearly so, the
lower petioled, 7–15 cm. long; heads very numerous in a narrow panicle, nodding; involucres 2–3 mm. broad; bracts glabrous, elliptic, obtuse, yellowish
green; flowers yellow. Sandy soil: N.B.—Fla.—Tex.—Colo.—Sask. *E. Temp.*
Jl–S.

5. **A. Forwoòdii** S. Wats. Biennial; stem 4–7 dm. high, somewhat pubescent when young; basal leaves 5–10 cm. long, petioled, bi- or tripinnatifid into
narrowly linear divisions, loosely silky-canescent, at least when young; heads
numerous in a dense panicle, mostly erect, heterogamous; involucre subglobose,
2–2.5 mm. high, 2.5–3 mm. broad; bracts in 3 or 4 series, the outermost lanceolate or ovate, acute, half as long as the innermost ones; ray-flowers 15–20.
A. caudata calvens Lunell. Plains: Ont.—Mich.—Neb.—Sask. *Prairie—Plain.*

6. **A. campòrum** Rydb. Perennial; stem 3–5 dm. high, striate, often
purplish; basal leaves clustered, 4–10 cm. long, petioled, twice or thrice pinna-

tifid into linear or oblong divisions, more or less silky-canescent; upper stem-leaves sessile, pinnatifid, with narrowly linear divisions; heads numerous in a narrow leafy panicle, nodding; involucre hemispheric, 2.5–3 mm. high and fully as broad; bracts in 3 or 4 series, scarious-margined, the outermost ovate, obtuse, a little more than half as long as the innermost ones; ray-flowers 15–20. *A. canadensis* Nutt.; not Michx. *A. Forwoodii* Rydb. (Fl. Colo.); not S. Wats. Plains and hills. Ont.—Sask.—Neb.—Ariz.—Yukon. *Prairie—Plain—Submont.*

7. **A. Bourgeaüàna** Rydb. Perennial; stem 3–4 dm. high, rather densely silky-villous, more or less tinged with red; basal leaves clustered, 4 6 dm. long, petioled, sericeous-canescent on both sides, twice pinnatifid, with oblanceolate divisions; stem-leaves mostly sessile, rather small, with linear divisions, or trifid or entire; heads very numerous, in dense leafy panicles; involucre hemispheric, about 4 mm. high and 5 mm. broad; bracts yellowish, silky-villous, scarious-margined, the outermost about half as long as the innermost, ovate, acute; ray-flowers about 20. Plains: Sask.—N.D. *Plain.*

8. **A. filifòlia** Torr. Undershrub, 3–10 dm. high; branches erect, minutely tomentulose, leafy; leaves all slender, entire and filiform, or some 3-parted with filiform divisions, minutely tomentulose, more or less fascicled; heads very small and very numerous, in narrow leafy panicles; involucres 1 1.5 mm. broad, tomentulose, 3–5-flowered; sterile central flowers 1–3. *A. plattensis* Nutt. Plains: Neb.—Tex.—Ariz.—Nev.; Mex. *Son.—Plain.* Ap–O.

9. **A. Absínthium** L. Shrubby; stem 5–15 dm high, finely canescent; leaves 5–12 cm. long, once to thrice pinnately divided into oblong or lanceolate, obtuse divisions, finely canescent, especially beneath; heads numerous in a large panicle, with racemiform branches, nodding; involucres 4–5 mm. broad, canescent; outer bracts linear, the inner oval, scarious-margined. Waste places and around dwellings: Newf.—N.C.—Mont.—Man.; escaped, native of Eur. Au–S.

10. **A. frígida** Willd. Perennial, with a cespitose, suffrutescent caudex; stem 2–5 dm. high, finely canescent; leaves twice pinnatifid, with linear-filiform divisions, silvery white, in age often turning brownish; heads numerous, racemosely disposed, nodding; involucres 4–6 mm. broad, canescent or tomentose; bracts oblong or lanceolate; corollas yellow, glabrous. Dry plains and hills: Ont.—Wis.—Tex.—Ariz.—Ida.—Alaska; Asia. *Plain –Mont.* Jl–O.

11. **A. biénnis** Willd. Biennial; stem 3–10 dm. high, glabrous; leaves pinnately or bipinnately parted into lanceolate, laciniate-dentate divisions, glabrous; heads small, numerous, in a dense leafy panicle; involucres 2–3 mm. wide, glabrous; bracts elliptic, dark green, scarious-margined; corollas yellow, glabrous. Wet places: N.S.—N.J.—Calif.—B.C.—Mack. *Prairie—Plain—Mont.* Au–D.

12. **A. ánnua** L. Annual; stem tall, 3–10 dm. high, striate, glabrous; leaves 2–7 cm. long, pinnately or bipinnately divided into lance-oblong, laciniate divisions; heads numerous in small racemes, nodding; involucre hemispheric, 1.5 mm. high, 1.5–2.5 mm. broad; bracts 8, glabrous, the outer oblong, the inner elliptic, with green back and scarious margins; ray-flowers and disk-flowers each 5 or 6, yellow. Waste places: Prince Edward Island –Va.—Ark. —Iowa; Calif.; nat. from the Old World.

13. **A. vulgàris** L. A stout perennial, with a rootstock; stem 5–15 dm. high, sparingly tomentose when young; lower and middle leaves ovate in outline, pinnately or bipinnately divided to near the midrib, green and glabrate above, white-tomentose beneath, 5–10 cm. long; primary divisions 5–7, obovate, oblong, or oblanceolate, acute, the ultimate lanceolate, often few-toothed; upper leaves pinnatifid, with lanceolate or linear divisions, or simple; heads very numerous in a leafy panicle; involucre about 4 mm. high, 3–4 mm. broad, more or less tomentose; bracts ovate to oblong, obtuse; ray-flowers 6–12. Waste places: Newf.—

Man.—Wis.—Ala.—Ga.; B.C.; escaped from cultivation; nat. of the Old World.

14. A. elàtior (T. & G.) Rydb. Perennial, with a creeping rootstock; stem 5–10 dm. high, glabrous or slightly floccose; leaves glabrous or nearly so above, tomentose beneath, acuminate, 5–10 cm. long, the lower pinnatifid, with lanceolate, long-acuminate lobes, the upper trifid or entire; heads numerous, in a panicle, nodding; involucres about 5 mm. broad, shining, nearly glabrous; bracts with yellowish or brownish scarious margins, oval; corollas yellow or brownish. *A. Tilesii elatior* T. & G. Valleys: Mack.—Man.—Mont.—Wash.—Alaska. *Subarct.—Submont.—Subalp.* Jl–S.

15. A. serràta Nutt. Perennial; stem 1–3 m. high; leaves numerous, lanceolate or lance-linear, acuminate, closely serrate with small teeth, green above, white-tomentose beneath, 7–15 cm. long; heads numerous, 3 mm. high, 2 mm. broad; bracts 10, slightly floccose; outer bracts ovate, acute, the inner elliptic, obtuse; ray-flowers 6–8; disk-flowers 10–12. Alluvial soil: Ill.—Mo.—Kans.—N.D.; intr. east to Pa. *Prairie.* Jl–S.

16. A. Herriótii Rydb. Tall perennial, with a rootstock; stem simple, 6–10 dm. high, finely tomentose; leaves entire or sparingly and sharply toothed, 5–20 cm. long, glabrate and green above, densely white-tomentose beneath; heads very numerous, erect, in a narrow dense panicle; involucre 4–5 mm. high, 2.5–3 mm. broad; bracts tomentose, the outer ovate-lanceolate, acute; inner bracts elliptic, mostly obtuse; ray-flowers 6–8. Banks: Alta.—Sask.—Minn.—S.D. *Plain.*

17. A. ludoviciàna Nutt. *Fig. 587.* Perennial, with a cespitose rootstock; stem usually branched, more or less tomentose, 3–6 dm. high; lower leaves oblanceolate, 3–7-lobed above the middle, 3–8 cm. long, soon glabrate and dark green above, densely white-tomentose beneath; lobes lanceolate and falcate; upper leaves lanceolate and entire; heads very numerous and small, in a compound open panicle; involucres 3–4 mm. high, 2–3 mm. broad, usually densely tomentose; bracts elliptic; corollas light brown. Prairies, cañons, and mountain-sides: Minn.—Mo.—Tex.—Ariz.—Utah. *Prairie—Plain—Submont.* Au–S.

f.587.

18. A. Brittònii Rydb. Perennial, with a horizontal rootstock; stem rather stout, 3–6 dm. high, white-floccose; lower leaves thick, cuneate or oblong-oblanceolate, 3–5-lobed, mostly above the middle, densely white-tomentose on both sides, the lobes lanceolate; upper leaves ovate-lanceolate, entire; panicles narrow; involucres 3–4 mm. high, about 2 mm. broad; bracts ovate and densely floccose; flowers about 15, light brown or yellow. Hills and plains: Wyo.—Ore.—Ariz.—N.M.—Kans.; n Mex. *Submont.* Au–O.

19. A. Lindheimeriàna Scheele. Perennial; stem 3–10 dm. high, striate, sparingly floccose; lower leaves usually pinnatifid, with narrow, linear divisions, 4–5 cm. long, floccose and glabrate above, white-tomentose beneath; upper stem-leaves entire, linear-lanceolate or linear; heads numerous, narrow, nodding, 3 mm. high, 2 mm. broad; bracts 10–12, densely tomentose, ovate to elliptic; ray-flowers 5–7. Prairies: Mo.—Kans.—Tex.—Mex. *Texan.* Au–O.

20. A. falcàta Rydb. Tall perennial, with a rootstock; stem 6–10 dm. high, loosely floccose; leaves linear or narrowly linear-lanceolate, 10–15 cm. long, green and glabrate above, white-tomentose beneath, the upper simple,

falcate, the lower usually with several alternate, spreading and strongly falcate, narrow lobes; panicle very large; involucre about 4 mm. high and fully
as broad, densely floccose; outer bracts ovate, acute, the inner oval and obtuse;
ray-flowers 5–7. Banks: S.D.—Sask. *Plain.* Au–S.

21. A. gnaphalòdes Nutt. Perennial, with a slender, horizontal, cespitose rootstock; stem 3–6 dm. high, white-tomentose; lower leaves oblanceolate,
entire or serrate, 2–10 cm. long, tomentose on both sides; upper leaves lanceolate or linear, entire; heads very numerous, in rather dense panicles; involucres about 4 mm. high, 2–3 mm. wide, about 20-flowered. Prairies and river
banks: Ont.—Mo.—Tex.—Colo—Alta.; adv. eastward to N.H. and N.Y.
Prairie—Plain—Submont.

22. A. pabulàris (A. Nels.) Rydb. Perennial, with a creeping rootstock,
stem 2–4 dm. high, slender, white-tomentose; leaves linear or narrowly linear-
lanceolate, the lower often toothed, densely tomentose, inclined to become
yellowish, 3–5 cm. long; heads small, in a narrow spike-like inflorescence; involucres tomentose, about 3 mm. high and 2 mm. broad, about 15-flowered.
A. rhizomata pabularis A. Nels. Saline soil: w Minn.—Iowa—Neb.—Colo. –
Ida. *Prairie—Plain.* Jl–Au.

23. A. Purshiàna Besser. Perennial, with a horizontal rootstock; stem
3–6 dm. high, tomentose; leaves ovate to lanceolate, entire or the lower somewhat toothed, white-tomentose on both sides; heads small, in a dense, contracted panicle; involucres about 3 mm. high, 2 mm. broad, tomentose; flowers
about 15. Plains: Sask.—Neb.—Calif.—B.C. *Plain—Submont.* Au–O.

24. A. longifòlia Nutt. Perennial, with a suffruticose base or woody
caudex; stem 3–12 dm. high, silvery-tomentose; leaves linear or lance-linear,
tomentose on both sides, in age greenish above, with revolute margins; heads
erect, in a narrow raceme-like panicle; involucres 5–6 mm. broad, tomentose;
bracts ovate or oval; flowers about 30, yellow. *A. natronensis* A. Nels., a broad-
leaved form. Plains in alkaline soil: Man.—Colo.—Ida.—Wash. *Plain.* Jl–Au.

25. A. diversifòlia Rydb. White-tomentose perennial, with horizontal
rootstock; stem simple, leafy, 5–10 dm. high; leaves densely tomentose on both
sides, subsessile, 5–10 cm. long, the lower usually pinnately cleft into 3–5
narrowly lanceolate-acuminate lobes, which are directed forward, the upper or
all entire, linear-lanceolate, passing into the bracts of the inflorescence; inflorescence a narrow panicle, 1.5–3 dm. long; heads numerous; involucres 3–4
mm. high and 3 mm. broad; bracts oblong or ovate-oblong, scarious-margined,
densely villous-tomentose; flowers light yellow. Valleys: Ont. (Hudson Bay)—
Kans.—Colo.—Calif.—B.C. *Plain—Submont.* Jl–S.

26. A. Carrùthii Wood. Perennial, with a short cespitose caudex; stem
3–5 dm. high, white-tomentose, very leafy; leaves pinnatifid, with narrowly
linear divisions, with revolute margins, densely white tomentose on both sides,
the uppermost often entire, linear-filiform; heads very numerous in a dense
panicle; involucre about 2.5 mm. high, 2 mm. broad, densely tomentose; outer
bracts lanceolate, the inner oval, broadly scarious-margined. *A. kansana* Britton [G]. *A. coloradensis* Osterhout, a form with broader segments. Dry plains
and prairies: Mo.—Colo.—Utah—Tex. *Prairie—Son.—Mont.* Au–S.

27. A. Abrótanum L. A much branched shrub, 5–10 dm. high; lower
leaves petioled, 4–6 cm. long, twice pinnately dissected into linear-filiform divisions, with revolute margins, slightly tomentose beneath; heads numerous,
nodding in leafy panicles with racemiform branches; involucre hemispheric,
about 3 mm. high and 5 mm. broad; bracts canescent, the outer linear-lanceolate, the rest broadly oval, scarious; ray-flowers about 10; disk-flowers 15–20.
Around dwellings: N.B.—Mass.—N.C.—Colo.—Sask.; escaped from cultivation, native of s Eur. and the Orient.

28. **A. póntica** L. A suffruticose perennial, with a creeping rootstock; stem 5 dm. high, canescent-tomentulose; leaves 1–3 cm. long, bipinnatifid with linear divisions, grayish-tomentulose; heads numerous, nodding in dense panicles; involucres 3 mm. high, 5 mm. broad, ray-flowers 10–15, disk-flowers 25–30. Waste places: Me.—Pa.—Iowa—Man.; escaped from cultivation.; nat. of Eur.

29. **A. càna** Pursh. Shrub, 3–10 dm. high; branches strict, canescent; leaves linear or lance-linear, acute at both ends, 1–4 cm. long, silvery-canescent on both sides, or the lower with 2–3 acute teeth or lobes at the apex; heads glomerate, in a leafy panicle; involucres 5–9-flowered; outer bracts linear or subulate, the inner oval, all more or less tomentose; flowers yellowish. *A. columbiana* Nutt. Plains and hills: Sask.—Minn.—Colo.—Utah—Ore.—Mont. *Plain —Mont.* Au–S.

30. **A. tridentàta** Nutt. Shrub, 1–4 m. high, much branched; bark of the old stems shreddy; branches erect, silvery-canescent; leaves cuneate, 1–2 cm. long, 3-toothed (rarely 4–7-toothed) at the apex, or the upper linear-cuneate and entire, silvery-canescent; heads paniculate, numerous; involucres well imbricate, about 4 mm. long and 2 mm. wide, 5–8-flowered; bracts oval-elliptic, at least the outer densely tomentose; flowers yellow or brownish. Dry plains and hills: N.D.—Neb.—Colo.—Calif.—B.C. *Plain—Mont.* Jl–O.

Tribe 11. **SENECIONEAE.** Heads radiate or discoid. Involucres of a series of equal bracts, often supported by a few small ones forming a calyculum. Receptacle naked. Marginal flowers, if present, pistillate and ligulate, rarely with filiform corollas. Disk-flowers hermaphrodite, with tubular 4–5-toothed corollas. Anthers sometimes sagittate at the base but not caudate. Style-branches of the disk-flowers usually truncate. Pappus of capillary bristles.

92. **TUSSILÀGO** L. Coltsfoot.

Acaulescent perennial herbs, with creeping rootstocks. Leaves long-petioled, appearing after the flowers, with orbicular-cordate sinuately dentate blades, white-tomentose beneath. Heads solitary, on a scaly scape, radiate. Involucre broadly turbinate; bracts in a single series. Receptacle plane, naked, alveolate. Ray-flowers pistillate, fertile, in several series; ligules yellow. Disk-flowers hermaphrodite, but sterile. Anthers slightly sagittate at the base. Styles of the ray-flowers with linear-lanceolate branches, those of the disk-flowers undivided. Achenes 5–10-ribbed. Pappus of numerous capillary bristles.

1. **T. Fárfara** L. Scape 0.5–4.5 dm. high, with lanceolate scales; petioles 5–20 cm. long; leaf-blades 5–15 cm. broad, with a deep sinus; disk of the head about 2 cm. broad; ligules 6–7 mm. long, one third mm. wide or less. Roadsides and clayey banks: N.S.—Minn.—N.J.; nat. from Eur. Ap–Je.

93. **PETASÌTES** L. Sweet Coltsfoot, Butterbur.

Perennial herbs, with thick creeping rootstocks. Leaves basal, long-petioled, with ample reniform, cordate, triangular, or sagittate blades, usually tomentose beneath, generally appearing later than the scaly-bracted flowering stems. Heads racemose or corymbose, subdioecious. Involucre campanulate, bracts in one series, herbaceous. Flowers of the fertile plants all or most of them pistillate, fertile; corolla irregularly 2–5-toothed and cylindric, or else ligulate. Flowers of the substerile plant mostly hermaphrodite, but sterile, with a few pistillate ones at the margin; corolla of the sterile flowers tubular and 5-toothed; style 2-lobed, but ovary sterile. Achenes narrow, 5–10-ribbed. Pappus of soft white bristles.

Leaf-blades sagittate or cordate.
  Leaves repand-denticulate, with numerous teeth, pinnately veined.
    1. *P. sagittatus.*
    Leaves cleft one third to one half to the midrib; blades decidedly broader than long; two or more pairs of lateral veins rising from the base.
    2. *P. vitifolius.*
  Leaf-blades round-reniform, pedately veined and lobed.
    3. *P. palmatus.*

1. **P. sagittatus** (Pursh) A. Gray. *Fig. 588.* Leaf-blades cordate, ovate-cordate to deltoid-cordate, 1–3 dm. long, 1–2 dm. wide, with a deep basal sinus, glabrate above, white-tomentose beneath; scape 2–3 dm. high; heads corymbose; involucres obconic, 8–9 mm. high, glandular-pubescent as well as tomentose; ligules of the substerile heads 5–7 mm. long, 1–1.5 mm. wide, those of the subfertile heads smaller, 3–5 mm. long, less than 1 mm. wide. *Tussilago sagittata* Pursh. *Petasites dentata* Blankinship. Wet grounds: Lab.– Man —S,D. —Colo.– Alaska. *Boreal.—Plain— Mont.* My–Jl.

2. **P. vitifolius** Greene. Leaf-blades broadly cordate or deltoid-reniform, 5–12 cm. long, 8–20 cm. broad, with two or more lateral veins rising from the base, usually 5-lobed, the basal lobes usually deeply 2-cleft and the terminal one 3–5-lobed; lobes coarsely dentate; scape 2–3 dm. high; involucres over 1 cm. high, glandular and floccose; ligules of the substerile heads about 7 mm. long and 1.5 mm. wide. *P. trigonophyllus* Greene [G]. Wet places: Man.—Alta. *F Boreal.* Je–Jl.

3. **P. palmatus** (Ait.) A. Gray. Leaf-blades reniform in outline, 5–15 cm. long, 10–30 cm. wide, sparingly villous tomentulose beneath, pedately veined and lobed; lobes oblanceolate, coarsely few-toothed with broadly triangular, mucronate teeth; scape stout, 1.5–6 dm. high; heads corymbose; involucres about 1 cm. high, somewhat glandular-pubescent, tomentulose at the base; ligules of the substerile heads about 5 mm. long, 1 mm. wide, those of the fertile heads 2.5–3 mm. long, 0.5 mm. wide. Boggy places: Newf.—N.Y.— Minn.—Alta. *Plain—Boreal.* Je–Au.

94. **HAPLOÉSTES** A. Gray.

Suffrutescent perennials. Leaves opposite, linear-filiform, somewhat fleshy. Heads cymose, radiate. Involucres campanulate; bracts 4 or 5, broadly oval, strongly overlapping. Receptacle flat, naked. Ray-flowers few, pistillate, and fertile. Disk-flowers hermaphrodite and fertile. Achenes linear, terete, glabrous, ribbed. Pappus of a single series of rigid, scabrous, white bristles.

1. **H. Gréggii** A. Gray. Stem 3–6 dm. high, branched, glabrous, leafy; heads cymose; involucres hemispheric, about 4 mm. high; bracts glabrous, obovate, shorter than the disk; flowers yellow; ligules 3–5, oblong, about 3 mm. long. Saline soil: Tex.—Kans.—Colo.—N.M.; n Mex. *St. Plains—Son.* Je–S.

95. **ÁRNICA** L. ARNICA.

Perennial caulescent herbs, ours with rootstocks. Leaves mostly opposite, entire or toothed. Heads usually radiate, rarely discoid, several or solitary. Involucre campanulate or turbinate; bracts equal, in 1 or 2 series. Receptacle flat, naked, pubescent, or fimbrillate. Ray-flowers pistillate and fertile; ligules yellow. Disk-flowers many, hermaphrodite and fertile; corolla tubular, yellow. Anthers sometimes sagittate. Achenes narrow, 5–10-ribbed, pubescent or glabrous. Pappus a single series of capillary, scabrous bristles.

Lower stem-leaves linear to lanceolate or oblanceolate, the upper ones entire, acu-
minate, and reduced.
Involucre turbinate ; base of the stem without a tuft of brown fibers.
Lower stem-leaves short-petioled or subsessile, entire or merely denticulate.
Leaves and stem white-villous when young ; bracts purple-tinged.
Bracts 15 or less, narrowly lanceolate, densely long-villous at the base.
                                                   1. *A. alpina.*
Bracts broadly lanceolate ; densely short-villous
throughout.   2. *A. Lowii.*
Leaves and lower part of the stem glabrous or nearly
so ; bracts green.   3. *A. Rydbergii.*
Lower stem-leaves long-petioled, dentate.   4. *A. arnoglossa.*
Involucre hemispheric ; base of the stem with a tuft of brown
fibers.   5. *A. fulgens.*
Lower stem-leaves cordate or broadly ovate, long-petioled, the
upper ones mostly similar, ample.   6. *A. cordifolia.*

1. **A. alpìna** (L.) Olin. Stem 1–3 dm. high, loosely villous; basal leaves
with margined petioles, 4–10 cm. long; blades lanceolate, 3-nerved, acuminate;
stem-leaves 1 or 2 pairs, sessile, lanceolate, with long acumination; heads usually solitary; involucre turbinate, about 1 cm. high, loosely white-villous;
densely so at the base. *A. angustifolia* Vahl. Arctic-alpine situations:
Greenl.—Lab.—Man.—B.C.—Alaska; Wyo.; n Eur. *Arctic—Subarctic.—Alp.*
Au.

2. **A. Lòwii** Holm. Stem 2–3 dm. high, grayish-pubescent, with short
hairs, lower leaves narrowly oblanceolate, strongly 3-ribbed, densely villous on
both sides, entire-margined, attenuate, 6–10 cm. long; upper stem-leaves linear-
lanceolate, sessile; heads 1–3; involucre broadly turbinate, 12–13 mm. high;
ligules 15, pale yellow, 15 mm. long. Coast: Ont.—Man. *Subarct.*

3. **A. Rydbérgii** Greene. Stem 2–4 dm. high, sparingly villous; basal
leaves oblanceolate, entire, petioled; stem-leaves 3–4 pairs, glabrate or sparingly
puberulent, villous-ciliate on the margins, 3–8 cm. long, remotely dentate,
strongly 3-nerved, the lower short-petioled and spatulate or oblanceolate, the
upper lanceolate and sessile; heads 1–3; involucres turbinate, 10–12 mm. high,
villous; bracts lanceolate, acute; ligules usually orange; achenes densely silky-
pilose. *A. caespitosa* A. Nels. *A. aurantiaca* Greene. Hills and mountains:
Alta.—S.D.—Colo.—Wash. *Mont.—Subalp.*

4. **A. arnoglóssa** Greene. Stem 3–5 dm. high,
rather slender, glandular-puberulent and with scattered hairs; lower leaf-blades lanceolate, acute, wing-
petioled, remotely serrate, strongly 5-ribbed, firm,
sparsely scabrous, pale beneath; upper stem-leaves
lanceolate, sessile; heads 1–5, usually 3; involucres
turbinate, 9–10 mm. high, glandular-puberulent;
achenes slightly hirsute. *A. chionopappa* Fernald
[G]. Hills: Que.—N.B.—S.D.—Wyo.—Alta. *E.
Boreal.—Submont.—Mont.* Je–Au.

5. **A. fúlgens** Pursh. *Fig. 589.* Stem 2–6 dm.
high, finely villous, somewhat glandular above; leaves
mostly basal, broadly oblong to nearly linear, 3–5-
ribbed, 5–10 cm. long, densely short-pilose, entire,
tapering into winged petioles; stem-leaves usually
2 pairs, linear, sessile, the upper small; heads usu-
ally solitary, involucres hemispheric, 12–15 mm. high, densely villous; bracts
18–20, lanceolate or linear-lanceolate, acute; ligules orange, 1.5 cm. long;
achenes hirsute. *A. pedunculata* Rydb. *A. monocephala* Rydb., a low broad-
leaved form. Hills and meadows: Sask.—N.D.—Colo.—Calif.—Wash. *Prairie
—Submont.—Subalp.* Je–Au.

6. **A. cordifòlia** Hook. Stem 2–6 dm. high, somewhat viscid; basal leaf-
blades broadly cordate, usually coarsely toothed, 3–10 cm. long, puberulent,

and viscid-villous on the veins; stem-leaves 2–4 pairs, all except the uppermost petioled and cordate or rarely ovate; involucres campanulate-turbinate, 15–18 mm. high; bracts oblanceolate, acuminate. *A. abortiva* Greene, a small-leaved form. Wooded hills: Alta.—S.D.—Colo.—Calif.—B.C. *Submont.—Subalp. Je–Au.*

96. MESADÈNIA Raf. INDIAN PLANTAIN.

Perennial caulescent herbs, with a clump of fibrous-fleshy roots. Leaves alternate, strongly ribbed, the lower ones petioled, the upper ones sessile, often reduced. Heads discoid, in corymbiform panicles. Involucre nearly cylindric, bracts 5, rarely 6. Receptacles small, usually raised in the center into a conic or pyramidal point. Flowers 5 or 6, hermaphrodite and fertile; corollas dirty white, ochroleucous, or purplish. Anthers entire at the base. Style-branches short, truncate or obtuse at the enlarged apex, with an apical hair-brush. Achenes oblong, terete, somewhat 10-ribbed, tapering towards the base. Pappus of a single series of white, barbellate or scabrous, capillary bristles, on a small disk, deciduous. *Cacalia* L., in part.

Basal leaf-blades reniform or broadly cordate, the lateral ribs united at the base; involucral bracts neither winged nor keeled.
  Stem decidedly furrowed and angled.   1. *M. Muhlenbergii.*
  Stem terete, not furrowed.   2. *M. atriplicifolia.*
Basal leaf-blades lance-elliptic to oval or ovate, the lateral
  ribs free; involucral bracts keeled or winged on the back.   3. *M. tuberosa.*

1. **M. Muhlenbérgii** (Schultz-Bip.) Rydb. Stem 1–2 m. high; basal leaf-blades reniform, 1–2.5 dm. wide, green on both sides, slightly lobed, sinuately dentate; upper leaf-blades suborbicular, coarsely dentate, cuneate at the base; involucre 1 cm high, 4 mm. broad; bracts linear-oblong, acute, scarious-margined; achenes black, 4 mm. long. *Cacalia reniformis* Muhl. [G]; not Lam. *M. reniformis* Raf. [B] Woods: N.J.—Minn.—Ala. *On. Allegh.* Jl–S.

2. **M. atriplicifòlia** (L.) Raf. Stem 1–2 m. high, glabrous, more or less purplish; basal leaf-blades 1–1.5 dm. wide, reniform or broadly cordate, coarsely sinuate-toothed, glaucous beneath; upper stem-leaves ovate or rhombic, 3–7-lobed, or those of the inflorescence lanceolate and often entire; involucre 1 cm. high and 4–5 mm. broad; bracts lance-oblong, scarious-margined; achenes 5 mm. long, obtusely 10-angled. *C. atriplicifolia* L. [G]. Woods: Ont.—Minn.—Kans.—Fla. *E. Temp.* Jl–S.

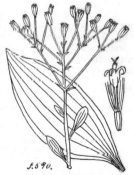

3. **M. tuberòsa** (Nutt.) Britton. *Fig. 590.* Stem 5–15 dm. high, striate, glabrous; basal leaf-blades firm, from lance-elliptic to oval or ovate, acute at each end or obtuse at the apex, 5–9-ribbed, entire-margined or denticulate; upper stem-leaves much reduced, subsessile; involucre 1 cm. long, 5 mm.

1590.

broad; bracts linear-oblong, acute; achenes dark brown, 5 mm. long, bluntly 10-ribbed. *C. tuberosa* Nutt. [G]. Wet prairies: Ont.—Minn.—Tex.—Ala. *E. Temp.* Je–Au.

97. SYNÓSMA Raf.

Perennial, caulescent, glabrous herbs. Leaves alternate, short-petioled, often triangular or hastate. Heads discoid, corymbose. Involucre nearly cylindric, with a calyculum; bracts several in a single series, those of the calyculum few, subulate; receptacle flat, naked. Flowers all perfect, fertile; corollas 5-lobed, dirty white or ochroleucous, or pinkish. Style-branches obtuse or truncate, unappendaged. Achenes narrow, subcylindric, 5–10-ribbed. Pappus of numerous soft white capillary bristles.

1. **S. suavèolens** (L.) Britton. Stem 5–15 dm. high, glabrous, striate; leaves numerous, hastate, 8–20 cm. long, acute or acuminate, sharply dentate; heads 4–6 mm. broad, 20–30-flowered; bracts 12–15, linear, 7–11 mm. long. *Cacalia suaveolens* L. [G]. Woods: Conn.—Minn.—Iowa—Tenn.—Fla. *Allegh.* —*Austral.* Au–O.

98. ERECHTÌTES Raf. Fire-weed, Pilewort.

Annual or perennial caulescent herbs. Leaves alternate, from entire to pinnatifid. Heads racemose or paniculate, discoid, but the marginal flowers pistillate. Involucre cylindric or cylindro-campanulate, usually subtended by a calyculum; principal bracts narrow, equal, in a single series; those of the calyculum subulate, usually spreading. Receptacle flat, naked. Corolla of the marginal flowers slender, usually 3–5-toothed, ampliate and oblique at the throat. Disk-flowers hermaphrodite and fertile; tube of the corolla long and slender; limb 5-toothed. Style-branches elongate, truncate or obtuse at the apex. Achenes cylindric, 5-angled, 10–20-striate. Pappus of numerous soft, capillary bristles.

1. **E. hieracifòlia** (L.) Raf. Annual; stem 3–20 dm. high, sparingly hirsute or glabrous, striate; leaves thin, the lower short-petioled, 3–15 cm. long, spatulate or oblanceolate, dentate or incised, the upper sessile, lanceolate, incised, or lobed and toothed; heads 12–15 mm. high, 5–7 mm. thick; bracts linear, abruptly acute, olive-green; corollas ochroleucous; achenes 10-ribbed; pappus white. Thickets and waste places: Newf.—Minn.—Tex.—Fla. *E. Temp.* Jl–S.

99. SENÈCIO L. Groundsel, Ragwort, Squaw-weed.

Annual or perennial herbs. Leaves alternate, entire, toothed or pinnatifid. Heads several or many, rarely solitary, radiate or discoid. Involucre from hemispheric or campanulate to cylindric; bracts many, strictly in one series with a few smaller, forming a calyculum at the base. Receptacle flat or slightly convex, naked, often pitted. Ray-flowers, if present, pistillate and fertile, ligulate. Disk-flowers more numerous, hermaphrodite and fertile; corollas yellow, tubular. Anthers obtuse or rarely slightly sagittate at the base. Style-branches in the disk-flowers spreading and recurved, truncate. Achenes terete or slightly flattened, 5–10-ribbed. Pappus copious, of capillary bristles.

Perennials, with a more or less developed rootstock or caudex, if less developed
  bearing a tuft of fibrous-fleshy roots.
  Plants equally leafy throughout.
    Leaves or their divisions not narrowly linear or filiform.
      Leaves merely toothed or entire-margined, not pinnatifid.
          1. Triangulares.
      Leaves pinnatifid.   2. Eremophili.
    Leaves or their divisions linear to filiform ; plants usually suffruticose at the
      base.   3. Filifolii.
  Plants with the upper stem-leaves more or less reduced.
    Rootstock well developed, horizontal or ascending, woody.
      Basal leaves entire, white-tomentose.   4. Cani.
      Basal leaves, at least most of them, toothed to pinnatifid.
        Leaves and stems more or less floccose, tardily becoming glabrate.
          5. Tomentosi.
        Leaves and stem glabrous or slightly floccose when young.
          6. Aurei.
    Rootstock very short, erect, of short duration, with numerous fleshy-fibrous
      roots.   7. Integrifolii.
Annuals, with taproots.
  Plants simple, floccose ; heads radiate.   8. Palustres.
  Plants branched, viscid ; heads discoid.   9. Vulgares.

1. Triangulares.

Heads radiate.   1. *S. crassulus.*
Heads discoid.   2. *S. rapifolius.*

2. EREMOPHILI.

Involucre 6–7 mm. high. 5–6 mm. broad.   3. *S. ambrosioides.*
Involucre 7–10 mm. high, 6–8 mm. broad.   4. *S. eremophilus.*

3. FILIFOLII.

Leaves, except those of the branches, pinnatifid   5. *S. Riddellii.*
Leaves entire, or the lower rarely with a single pair of filiform
  lobes.   6. *S. spartioides.*

4. CANI.

Heads 7–8 mm. high ; stem 1–2 dm. high ; stem leaves mostly
  entire.   7. *S Purshianus.*
Heads 10 mm. high or more ; stem 3–5 dm. high ; stem-leaves
  often pinnatifid.   8. *S. canus.*

5. TOMENTOSI.

Some of the basal leaves merely dentate ; heads radiate.
  Leaf-bladed thin, glabrous.   18. *S. pauperculus.*
  Leaf-blades rather thick, floccose at least when young.   9. *S. plattensis.*
All leaves pinnatifid, with toothed segments ; heads discoid.   10. *S. lanatifolius.*

6. AUREI.

Heads radiate.
  Leaves thick, more or less fleshy ; plant low, less than 2
    dm. high.
    Cyme dense ; upper leaves usually pinnatifid with nar-
     row lobes.   11. *S. densus.*
    Cyme more open ; stem-leaves entire, crenate or dentate.   12. *S. oblanceolatus.*
  Leaves thin ; plant usually more than 2 dm. high ; cyme
    open.
    Basal leaf-blades cordate or truncate or rarely rounded
     at the base.
     Basal leaf-blades suborbicular in outline, as broad
      as long or broader, usually deeply cordate at the
      base.   13. *S. aureus.*
     Basal leaf-blades oval or rounded-ovate, longer than
      broad, shallowly cordate to rounded at the
      base.
      Plant glabrate ; leaves very thin.   14. *S. pseudaureus.*
      Plant floccose when young ; leaves firmer.   15. *S. semicordatus.*
    Basal leaf-blades tapering at the base.
     Basal leaf-blades rounded-obovate to suborbicular.   16. *S. rotundus.*
     Basal leaf-blades oblanceolate or oblong.
      Leaves subglaucous.   17. *S. Willingii.*
      Leaves not glaucous.   18. *S. pauperculus.*
Heads discoid.   19. *S. pauciflorus.*

7. INTEGRIFOLII.

Tall bog-plants, 5–20 dm. high ; basal leaves long-petioled.   20. *S. hydrophilus.*
Meadow and wood plants, 3–10 dm. high ; leaves short-petioled.
  Leaves sharply dentate.   1. *S. crassulus.*
  Leaves entire-margined or denticulate.
    Bracts of the involucre linear-lanceolate, long-attenuate.   21. *S. integerrimus.*
    Bracts of the involucre linear or oblong, abruptly acute.
     Bracts about half as long as the well-developed disk.   22. *S. lugens.*
     Bracts nearly as long as the disk.   23. *S. columbianus.*

8. PALUSTRES.

One species.   24. *S. palustris.*

9. VULGARES.

One species.   25. *S. vulgaris.*

1. S. crássulus A. Gray. Stem 2–5 dm. high, glabrous ; basal leaves 8–15
cm. long ; blades obovate or lanceolate, acutish or obtuse, dentate or denticulate,
tapering into a winged petiole ; upper stem-leaves oblong, ovate or lanceolate,
sessile and clasping at the base ; heads corymbose ; involucre campanulate, 12–15
mm. high and 10–12 mm. broad ; bracts oblong, acute, very thick and fleshy, the
inner with membranous margins, slightly shorter than the disk, the calyculate
ones linear-subulate, about one-third as long ; ligules bright yellow, about 1 cm.
long ; achenes glabrous. Subalpine meadows: S.D.—N.M.—Utah—Ida. *Submont.—Alp.* Je–Au.

2. S. rapifólius Nutt. Stem 2–5 dm. high, terete or slightly obtuse-
angled ; leaves rather thin, sharply dentate or incised ; blades of the lower

leaves obovate or broadly oblanceolate, obtuse or acutish, 3–10 cm. long, tapering into wing-margined and auricled petioles, half-clasping at the base; upper leaves subsessile, oval, oblong, or lanceolate, clasping and auricled at the base; heads numerous, panicled, 6–8 mm. high and about 5 mm. broad; bracts oblong, acute, thick on the back, membranous-margined, about two thirds as long as the brownish yellow disk; ligules none; achenes glabrous. *S. lactucinus* Greene. Rocky wooded hillsides and mountains: Ida.—Colo.—S.D. *Submont.* Jl–Au.

3. **S. ambrosioìdes** Rydb. Stem glabrous, leafy, 4–10 dm. high; leaves lanceolate or oblanceolate in outline, pinnatifid to near the midrib or the lower incised, mostly short-petioled, the lobes lanceolate, coarsely dentate or incised; heads corymbose-paniculate; bracts connate, linear, acute, with black tips, the calyculate ones subulate, 4–5 mm. long; ligules light yellow, 5–6 mm. long; achenes minutely scabrous-puberulent on the angles. Mountains: Wyo.—S.D.— N.M.—Utah. *Mont.—Subalp.* Jl–Au.

4. **S. eremóphilus** Richards. Stem stout, 5–10 dm. high; leaves deeply pinnatifid, 1–2 dm. long, the lower petioled, the upper sessile; lobes lanceolate, usually toothed or cleft; heads many, corymbose, 10–12 mm. high; bracts linear-lanceolate, costate, black-tipped; ligules 6–8 mm. long; achenes papillose-scabrous on the angles. Wet ground: Man.—S.D.—Wyo.—Alta.—Mack. *Plain —Submont.* Au–S.

5. **S. Riddéllii** T. & G. Stem 3–10 dm. high, glabrous, leafy; lower leaves pinnatifid, with linear or linear-filiform divisions, glabrous; heads corymbose, 10–12 mm. high; involucres campanulate, glabrous; bracts linear, acute, somewhat carinate; ligules about 12, light yellow, 9–12 mm. long; achenes somewhat canescent. *S. Fremontii* (T. & G.) Rydb. [B]; not T. & G. Plains: Tex.—Neb.—Wyo.—N.M. *Plain—Submont.* Je–O.

6. **S. spartioìdes** T. & G. Stem 3–6 dm. high, leafy throughout, often several from one root; leaves glabrous, narrowly linear, entire; heads corymbose; involucre nearly cylindric; bracts about 12, carinate, linear, acute, the calyculate ones subulate, minute; ligules few, 8–10 mm. long; achenes silky-canescent. Banks and plains: Tex.—S.D.—Wyo.—Ariz. *Plain—Mont.* Jl–O.

7. **S. Purshiànus** Nutt. Stems 1–2 dm. high; basal leaves broadly spatulate to linear-oblanceolate, entire, 3–5 cm. long, obtuse; cyme small, corymbiform; heads 2–10; involucre campanulate, 7–9 mm. high; bracts linear-lanceolate, acute; ligules bright yellow, about 6 mm. long; achenes glabrous. *S. laramiensis* A. Nels. Dry hills and mountain sides: Sask.—S.D.—Tex.—Utah— B.C. *Plain—Alp.* Au.

8. **S. cànus** Hook. Stems 3–5 dm. high; basal leaves petioled, entire, 4–15 cm. long, obovate or spatulate, thick, obtuse; stem-leaves oblong or lanceolate in outline, more or less lyrate-pinnatifid or lobed, sessile, auricled; cyme compound-corymbose; heads generally many, 10–12 mm. high; bracts narrowly linear-lanceolate; ligules bright yellow, 8–10 mm. long; achenes glabrous. Dry hills: Man.—Neb.— Colo.—Calif.—B.C. *Plain—Submont.* My–Au.

9. **S. platténsis** Nutt. *Fig. 591.* Stem 2–5 dm. high, loosely floccose, very tardily glabrate; basal leaves petioled, 3–10 cm. long; blades ovate or elliptic, floccose when young, glabrate in age; lower stem-leaves petioled, lyrate-pinnatifid, with oblong lobes, the upper sessile, reduced, linear-lanceolate in outline, pinnatifid or toothed; heads corymbose; involucres campanulate, 6–7 mm. high, tomentose at the base; bracts lance-linear, acute, the calyculate ones minute, subulate; ligules about 8 mm. long, orange;

f.591.

achenes hispidulous on the angles. *S. pseudotomentosus* Mack. & Bush. Prairies and river valleys: w Ont.—Mo.—Tex.—Colo.—e Mont.—Sask. *Prairie —Plain.* My–Je.

10. **S. lanatifòlius** Osterhout. Stems several, 1–2 dm. high, usually branched, very leafy to the inflorescence; leaves linear, pectinate-pinnatifid, the divisions crenate-toothed with inrolled edges, especially those of the lower leaves; both stem and leaves pannose-canescent; heads numerous and crowded, about 8 mm. high; involucres campanulate, glabrate; bracts about 7, the calyculate ones minute; flowers 15; rays none; achenes glabrous. Hills and mountains: Colo.—Neb. *Plain—Mont.*

11. **S. dónsus** Greene. Stem 1.5–3 dm. high, tomentose at the base and at the axils of the leaves; basal leaves petioled, 4–7 cm. long; blades oblance-olate or linear-oblanceolate, serrate, thick, slightly floccose when young; lower stem-leaves similar but laciniate or pinnately lobed; upper stem-leaves reduced, more or less pinnatifid; heads in a small dense corymb; involucres 6–7 mm. high, hemispheric, glabrous; bracts linear or oblong, abruptly acuminate, scari-ous-margined; ligules 6–7 mm. long. *S. aureus compactus* A. Gray. *S. compactus* Rydb. [B]; not Kirk. *S. condensatus* Rydb.; not Greene. Plains: Man. —Neb.—e Colo.—N.D. *Plain.* My–Je.

12. **S. oblanceolàtus** Rydb. Stem stout, about 2 dm. high, floccose at the base of the leaves, basal leaves oblanceolate, thick and fleshy, 4–6 cm. long, obtuse, entire, crenate or dentate, with entire tapering bases, petioled; stem-leaves similar, or the upper reduced and linear; heads about 8 mm. high; bracts linear, acute, yellowish green, the calyculate ones minute and few; ligules 5–6 mm. long. *S. suavis* Lunell. Dry plains: N.D.—Tex.—N.M.—Utah—Wyo. *Plain—Mont.* My–Jl.

13. **S. aùreus** L. Stem 3–7 dm. high, corymbose above; basal leaves long-petioled; blades ovate or suborbicular, cordate or truncate at the base, crenate, 2–12 cm. long, rounded at the apex, glabrous; stem-leaves pinnatifid, more or less lyrate, the segments often cuneate and toothed; heads slender-peduncled, 6–7 mm. high; bracts linear; ligules deep yellow, 5.5–7.5 mm. long; achenes glabrous, 3 mm. long. Swamps and meadows: Newf.—Man.—S.D.—Tex.—Fla. *E. Temp.* My–Jl.

14. **S. pseudaùreus** Rydb. Stem 3–8 dm. high; basal leaves long-peti oled; blades thin, broadly ovate or subcordate, serrate, 3–7 cm. long; stem-leaves more or less laciniate or lyrate-pinnatifid, the upper clasping; heads 8–10 mm. high; involucres hemispheric; bracts lance-linear, acute; ligules bright yellow, about 8 mm. long; achenes glabrous. *S. Burkei* Greenm. Wet mead-ows: Mack.—Man.—Minn.—N.M.—Calif.—B.C. *Submont.—Alp.* Je–S.

15. **S. semicordàtus** Mack. & Bush. Plant sparingly floccose when young, in age the tomentum usually remaining at the bases of the petioles and invo-lucres; stem 3–5 dm. high; basal leaves long-petioled; blades from rounded-oval to elliptic, subcordate, rounded, or truncate at the base, 2–7 cm. long, 1–4 cm. wide, crenate; stem-leaves oblong or lanceolate, clasping, cleft or lacini-ate; heads corymbose; involucre 6–7 mm. high, 10–14 mm. wide; ligules 8–12, 5–6 mm. long; achenes glabrous. Low prairies: Ill.—Iowa—Kans.—Tex.—Mo. *Prairie—Texan.* My–Je.

16. **S. rotúndus** (Britton) Small. Stem 2–5 dm. high, corymbose above; leaves mostly near the base, with short petioles; blades rounded-obovate to sub-orbicular, rounded at the apex, coarsely toothed, 2–7 cm. long and almost as broad; stem-leaves small, lyrate-pinnatifid; involucre 4–5 mm. high; bracts narrowly linear; ligules light yellow, 8–10 mm. long; achenes 3 mm. long, glabrous. *S. obovatus rotundus* Britt. [G]. Banks and moist places: Ohio.—Kans.—Tex.—La.—Ky. *Allegh.—Austral.* Ap–Je.

17. S. Willíngii Greenm. Stem erect, 2.5–3 dm. high, glabrous; leaves oblong-lanceolate, 3–12 cm. long, crenate-serrate to pinnatifid into oblong entire or subentire lobes, floccose-tomentulose along the veins when young; heads 8–10 mm. high; involucre campanulate, sparingly floccose; bracts linear-lanceolate; ligules yellow; achenes glabrous. Gravelly soil: Man.—Alta. *Plain.*

18. S. paupérculus Michx. Stem often tomentose at the base and the nodes, 2–6 dm. high; basal leaves petioled; blades oblanceolate, oblong, or oblong-spatulate, 2–7 cm. long, crenate, sometimes with salient teeth at the base; stem-leaves once or twice pinnatifid; involucre 4–5 mm. high; bracts narrowly linear; ligules 4–5 mm. long; achenes often hispidulous on the margins. *S. Balsamitae* Muhl. [B, G]. Stony soil: Lab.—Minn.—Neb.—Tex.—N.C. Canad.—*Allegh.* My–Jl.

19. S. pauciflòrus Pursh. Stem glabrous, 3–5 dm. high, slender; basal leaves petioled; blades oval, ovate or subcordate, crenate or rarely serrate, 1–3 cm. long; stem-leaves lyrate or pinnatifid, the upper reduced; heads about 8 mm. high; involucre campanulate; bracts narrowly linear, acute, usually purple-tinged; achenes glabrous. *S. discoideus* (Hook.) Britton [B]. Low ground: Lab.—Que.—Mich.—Wyo. —Calif.—B.C.—Alaska. *Boreal.*—*Mont.*—*Subalp.* Je–Au.

20. S. hydróphilus Nutt. Stem 5–20 dm. high, striate, hollow, strict and simple; basal leaves generally long-petioled, 1–3 dm. long, thick; oblanceolate and entire; stem-leaves subsessile, lanceolate, oblong or linear, the upper much reduced and bract-like; heads numerous; involucres cylindro-campanulate, about 1 cm. high and 5–8 mm. broad; bracts oblong, thin, membranous-margined, acute, more or less black-tipped, shorter than the disk; ligules few, light yellow, about 5 mm. long; achenes glabrous. Swampy places: S.D.—Colo.— Calif.—B.C. *Plain*—*Submont.* Je–Au.

21. S. integérrimus Nutt. Stem 3–10 dm. high, striate, leafy, branched above; basal leaves oblong-lanceolate, entire or sinuate, petioled, acute, somewhat fleshy; stem-leaves linear-lanceolate, sessile, sometimes auricled at the base; heads numerous, 12 mm. or in fruit even 16 mm. high and 10–15 mm. broad; involucres campanulate; bracts narrowly lanceolate, long-acuminate, thick and fleshy, somewhat shorter than the disk; ligule dull yellow, about 1 cm. long; achenes glabrous. Wet meadows: Man.—Iowa—Colo.—Sask. *Plain* —*Submont.* Je–Au.

22. S. lùgens Richards. Stem slender, terete, about 3 dm. high; basal leaves 5–10 cm. long, narrowly oblanceolate, sinuate-dentate; upper stem-leaves much reduced, linear or linear-lanceolate; cyme corymbiform; heads 5–7, campanulate, about 1 cm. high, 8–12 mm. broad; bracts rather thick, linear-oblong or lanceolate, acute, conspicuously black-tipped; rays about 8 mm. long; achenes glabrous. Wet places: Subarctic America—Man.—Ida.—Alaska. *Boreal.*— *Subarct.*—*Subalp.* Je–Jl.

23. S. columbiànus Greene. Stem 3–6 dm. high, floccose when young; leaves sinuate-dentate, the lower petioled, oblanceolate or spatulate, the uppermost linear-lanceolate, much reduced, sessile; inflorescence a short corymb; involucres campanulate, about 1 cm. high; ligules 5–8 mm. long; achenes glabrous. *S. lugens* A. Gray, in part. *S. atriapiculatus* Rydb. Valleys: Sask.— Minn.—Colo.—Ida.—Alaska. *Plain*—*Submont.* Je–Au.

24. S. palústris (L.) Hook. Stem arachnoid-villous when young, 1.5–5 dm. high, rather fleshy, strongly striate; lower leaves with winged petioles; blades lanceolate in outline, sinuately dentate to laciniate-pinnatifid, the upper linear, more or less clasping; cyme crowded, corymbiform; involucres hemispherical, about 8 mm. high, not calyculate; bracts narrowly linear-lanceolate, acuminate; ligules broad and short, oblong, light yellow, 4–5 mm. long; achenes glabrous. Wet ground: Lab.—Ia.—Sask.—Alaska and the Arctic coast; Eur. *Subboreal.—Subarctic.* My–Jl.

25. S. vulgáris L. Annual; stem erect, 1–4 dm. high, more or less pilose or glabrate; leaves sinuate-pinnatifid, glabrous or slightly arachnoid, with oblong to rounded incised divisions, the upper auriculate-clasping; heads cym-ose; involucres campanulate-cylindric, 7–9 mm. high, glabrous; bracts linear-subulate, black-tipped, the calyculate ones minute; rays none; achenes canes-cent-puberulent. Waste places: Lab.—Newf. N.C.—Calif.—Alaska; nat. from Eur. F–Au.

Tribe 12. **CYNAREAE.** Heads discoid. Involucre imbricate in many series. Receptacle not paleaceous but often fimbrillate or setose. Flowers all alike, hermaphrodite and with tubular lobed corollas, or in *Centaurea* the marginal flowers usually neutral, with enlarged more or less funnel-form corollas. Anthers caudate at the base. Style-branches without ap-pendages, often united, with a hairy ring below.

100. ARCTIUM L. Burdock

Coarse biennial herbs. Leaves alternate, petioled, with broad cordate or sagittate blades. Heads discoid. Involucres globular; bracts imbricate, in many series, subulate, with hooked tips. Receptacle flat, setose. Flowers hermaphrodite and fertile. Corolla purple or white; tube slender. Anthers sagittate at the base. Achenes oblong, somewhat compressed and 3-angled, truncate. Pappus of numerous short and rigid bristles, falling off separately.

Involucre less than 2 cm. broad; inner bracts not exceeding the flowers. .... 1. *A. minus.*
Involucre 2.5 cm. broad or more; inner bracts at least equaling the flowers. .... 2. *A. Lappa.*

1. A. minus Schk. Stem stout, 5–20 dm. high; puberulent, branched; leaves petioled; blades 2–4 cm. long, cordate, repand, grayish-tomentulose beneath; heads in a leafy panicle, often with racemiform branches; involucres 12–20 mm. high, glabrous; the inner bracts purple-tipped; corollas rose-purple. *Lappa minor* DC. Waste places: N.S.—Ga.—Ala.—Colo.—Ida.—B.C.; nat. from Eur. Au–S.

2. A. Láppa L. Stem 1–2 m. high; leaf-blades thin, broadly ovate-cor-date, the lower 4–5 cm. long, grayish-tomentose beneath; heads corymbose; involucre glabrous or nearly so; bracts longer and more spreading or reflexed; corollas rose-purple. Waste places: N.B.—N.Y.—Neb.—S.D.—Minn.; Mont.; nat. or adv. from Eur. Jl–O.

101. ÉCHINOPS L. Globe Thistle.

Tall thistle-like herbs. Leaves alternate, spinose, pinnatifid or dentate. Heads 1-flowered, conglomerate into a spherical head, with a small common outer involucre of reflexed bracts. Involucre proper cylindric, of several spines-cent bracts. Corolla-tube cylindric, 4-angled; pappus coroniform or paleaceous.

1. E. sphaerocéphalus L. Stem 1–2 m. high; leaves oblanceolate, 2–4 dm. long, pinnately lobed with triangular, lobed and spinose divisions, rough and greenish above, white-tomentose beneath; head-like flower-cluster 2–3 cm. in diameter, globose; flowers white or bluish. Around dwellings: Mass.—Va.—Iowa; escaped from cultivation; nat. of Eur. Jl–Au.

102. CÍRSIUM (Tourn.) Hill. Thistle.

Stout perennial or biennial herbs. Leaves alternate, with spinosely tipped lobes or teeth, often decurrent on the stem. Heads discoid. Involucre globose, campanulate or urceolate, of many imbricate bracts in several series. Receptacle flat, bristly. Flowers hermaphrodite and fertile, or the plant rarely dioecious. Corollas all alike, tubular, deeply 5-cleft. Anthers sagittate at the base, the two lobes produced into longer or shorter tails. Styles filiform or sometimes thickened, or with a ring at the base of the stigmatic portion. Achenes obovate or oblong, somewhat flattened. Pappus of one series of slender plumose bristles, united at the base and falling off together.

Bracts of the involucre conspicuously arachnoid-pubescent, all spine-tipped.
    1. *C. lanceolatum.*
Bracts slightly, if at all, arachnoid-pubescent: if somewhat so, the inner ones innocuous.
  Flowers all perfect.
    Inner bracts with loose-twisted innocuous tips; none with a glutinous dorsal ridge, or the outer ones with a narrow one in *C. Hillii;* heads often solitary; flowers rose-purple.
      Leaves not at all tomentose, but somewhat arachnoid-pubescent when young.
        Outer bracts without a glutinous ridge, not floccose.
        2. *C. Drummondii.*
        Outer bracts with a narrow glutinous ridge near the tip, arachnoid-floccose when young.
        3. *C. Hillii.*
      Leaves more or less tomentose beneath; heads clustered; flowers white or nearly so.
      4. *C. foliosum.*
    Inner bracts without loose-twisted tips, except in *C. plattense;* bracts with conspicuous agglutinous dorsal ridge.
      Leaves white-tomentose beneath; bracts bristle-pointed.
        Flowers ochroleucous.
          Inner bracts more or less dilated and crisp at the tip; stem-leaves deeply pinnately lobed.
          5. *C. plattense.*
          Inner bracts neither dilated nor crisp at the tip.
            Stem-leaves deeply lobed, with rather stout spines.
            6. *C. Nelsonii.*
            Stem-leaves sinuate-dentate.
            7. *C. nebraskense.*
        Flowers rose or purple.
          Leaves, at least the upper ones, entire or slightly lobed, tomentose on both sides.
          8. *C. oblanceolatum.*
          Leaves pinnately divided or deeply lobed.
            Leaves pale or gray and more or less floccose above when young.
              Involucre less than 4 cm. broad.
                Leaves deeply pinnatifid, with linear-lanceolate lobes.
                9. *C. Flodmanii.*
                Leaves with triangular or triangular-lanceolate lobes.
                10. *C. undulatum.*
              Involucres 4–7 cm. broad.
                Spines of the bracts rarely more than 5 mm. long; leaves broad, not conspicuously decurrent.
                11. *C. megacephalum.*
                Spines of the middle bracts fully 1 cm. long; leaves narrower, conspicuously decurrent.
                12. *C. ochrocentrum.*
            Leaves deep green and hispid above.
              Leaves entire or sinuately lobed, only the lowermost rarely pinnatifid.
                Involucres 4–6 cm. broad, 3–3.5 cm. high.
                13. *C. iowense.*
                Involucre 3–4 cm. broad, 2.5–3 cm. high.
                14. *C. altissimum.*
              Leaves deeply pinnatifid, the divisions often again lobed.
              15. *C. discolor.*
         Leaves green on both sides; bracts mostly without prickle-points; involucre 2–3 cm. broad, about 2 cm. high.
          16. *C. muticum.*
  Flowers dioecious.
    Leaves deeply pinnately lobed.
      17. *C. arvense.*
    Leaves entire, bristly ciliate or the lower slightly lobed.
      18. *C. setosum.*

1. C. lanceolàtum (L.) Hill. Stem 1–1.5 m. high, more or less villous when young; leaves deeply pinnatifid, with lanceolate lobes and long stout spines, grayish-tomentose or glabrate beneath; heads scattered at the ends of the branches, 4–5 cm. high; bracts lanceolate, attenuate, with long yellowish spines; flowers rose-purple. *Carduus lanceolatus* L. [B]. Waste places and roadsides: N.S.—Ga.—Calif.—B.C.; nat. from Eur. Jl–S.

2. C. Drummóndii T. & G. Stem low, 1–3 dm. high, slightly arachnoid-hairy; leaves oblanceolate in outline, green, somewhat arachnoid, but not at all tomentose, with triangular lobes and weak spines; heads 4–6 cm. high; bracts thin, brownish, the outer ovate, tipped with bristle-like spines; corollas rose-purple. *Carduus Drummondii* Coville. Hills and mountains; Man.—S.D.—B.C. Submont.—Mont. Jl–Au.

3. C. Hillii (Canby) Fernald. Stem 3–6 dm. high, somewhat arachnoid-hairy, from a hollow fusiform root; leaves green, somewhat arachnoid, the lower oblanceolate, the upper oblong in outline, pinnatifid with deltoid, dentate and short-spinose lobes; heads solitary or few; bracts numerous, the outer with short and weak spine-tips and a narrow but conspicuous glutinuous ridge near the apex, the inner unarmed, with a slightly dilated, lanceolate, crisp and twisted tip; flowers red-purple. *Cnicus Hillii* Canby. *Carduus Hillii* Porter [B]. Fields: Ont.—Man.—Iowa—Pa. Je–Jl.

4. C. foliósum (Hook.) DC. Stem 2–6 dm. high, more or less arachnoid-hairy; leaves light green, but more or less arachnoid above, tomentose beneath, from deeply pinnatifid with oblong lobes to nearly entire, with rather weak yellow spines; heads clustered at the end of the branches, leafy-bracted, 3–5 cm. high; outer bracts ovate, glabrous, with short weak spines; corollas white or nearly so. *Carduus foliosus* Hook. *Cirsium scariosum* Rydb. (Fl. Mont. & Fl. Colo.). Mountain valleys: Sask.—S.D.—Colo.—Utah–D.C. Submont.—Mont.

5. C. plattènse (Rydb.) Cockerell. Stem 3–7 dm. high, white-tomentose; leaves deeply pinnatifid into oblong divisions, 3–5 cm. long, tipped with slender short yellow spines, densely white-tomentose beneath, more sparingly floccose above; heads 4–5 cm. high; bracts very numerous, yellow, with dark thick glutinous backs, all except the innermost broad, ovate, lanceolate, with short yellow spreading spines. *Carduus plattensis* Rydb. [B]. Sandy places: Neb.—Colo. *Plain.* Je–Jl.

6. C. Nelsònii (Pammel) Rydb. Stem 6–10 dm. high, tomentose; leaves deeply pinnatifid, densely white-tomentose beneath, loosely floccose above; lobes with yellow spines; heads scattered, hemispheric, 3–3.5 cm. high, 3–4 cm. wide; bracts very numerous, all except the innermost lanceolate and tipped with short, flattened, yellow, spreading spines. *Cnicus Nelsonii* Pammel. Plains: Wyo.—S.D. *Plain.*

7. C. nebraskénse (Britton) Lunell. Stem 3–5 dm. high, tomentose; leaves linear-oblong or lanceolate, white-tomentose beneath, loosely floccose above, entire or irregularly toothed; heads solitary, 3–3.5 cm. high, 3–4 cm. broad; bracts numerous, lanceolate, with short, slender, spreading spines: inner bracts linear-lanceolate, tapering into linear-lanceolate spreading tips. *Carduus nebraskensis* Britton [B]. Magnesia cliffs: N.D.—w Neb.—Wyo. *Plain.* Jl.

8. C. oblanceolàtum (Rydb.) K. Schum. Stem rather slender, 3–6 dm. high, densely white-tomentose; lower leaves about 1 dm. long, thick, merely spinulose-dentate or rarely with a few triangular lobes, loosely floccose above, densely white-tomentose beneath; spines slender, 2–5 mm. long; upper leaves lanceolate, sessile and half-clasping; heads 1–3, campanulate, 3–3.5 cm. high, 1.5–2.5 cm. broad; bracts slightly floccose at first, with slender divergent spines 3–5 mm. long, the innermost unarmed; flowers rose-colored. *Carduus oblanceolatus* Rydb. Mountains: Colo.—Wyo.—Utah—Mont. Jl–S.

9. **C. Flodmánii** (Rydb.) Arthur. Stem comparatively slender, 5–10 dm. high, white-floccose, from creeping rootstocks; leaves deeply pinnatifid (or those of the seedlings entire) with narrowly lanceolate lobes, tipped with short prickles, densely tomentose beneath, floccose above; heads 3–4 cm. high and nearly as broad; bracts more or less floccose when young, the outer ovate and with short ascending or spreading spines; the innermost linear-lanceolate, attenuate to a weak tip; corollas rose or rose-purple. *Carduus canescens* Pammel mainly, *Cirsium canescens* A. Gray [G]; not Nutt.; *Carduus Flodmanii* Rydb. [B]; *Carduus filipendulus* A. Nels.; not Rydb. Meadows and river valleys: Sask.—Man.—Ia.—Colo.—Mont. *Plain—Submont.*

10. **C. undulàtum** (Nutt.) Spreng. Stem 3–6 dm. high, white-tomentose; leaves pinnately lobed or parted, with ovate or triangular lobes and moderately strong prickles, white-tomentose beneath, floccose or rarely in age glabrous above; heads solitary or few, scattered, 3–4 cm. high and about as broad; bracts numerous except the innermost, ovate to ovate-lanceolate, with short divergent spines, the innermost lance-linear, attenuate into a weak point; corollas rose or rose-purple, rarely white. *Carduus undulatus* Nutt. [B]. Dry plains: Man.—Mich.—Tex.—Ariz.—B.C. *Prairie—Plain—Submont.* Jl–S.

11. **C. megacéphalum** (A. Gray) Cockerell. Stem 3–7 dm. high, tomentose; leaves broadly oblanceolate or lanceolate, pinnatifid, with short triangular lobes, moderately stout-spined, densely tomentose beneath, floccose, but in age becoming green above; heads peduncled, solitary or few, 5–7 cm. high and as broad; bracts numerous, all except the innermost with short and slender divergent spines and conspicuous dorsal glutinous ridges, the inner tapering into slender tips; corollas rose or purple, rarely white. *Cnicus undulatus megacephalus* A. Gray [B]. Prairies and plains: Man.—N.D.—Mo.—Tex.—Colo.—Ida. *Prairie—Plain—Submont.* Je–S.

C. Flodmanii × megacephalum. Intermediate between the two parents, leaves much broader and with broader lobes, heads larger and bracts more glutinous than in *C. Flodmanii*. Colo.

C. megacephalum × ochrocentrum. Leaves broad, resembling those of *C. ochrocentrum*, but not decurrent, and involucral bracts with short spines as in *C. megacephalum*. Neb.—Kans.

12. **C. ochrocéntrum** A. Gray. Stem stout, 3–15 dm. high, tomentose; leaves pinnatifid, with crowded, triangular segments and long, yellow spines, the upper leaves usually conspicuously decurrent, white-tomentose beneath, floccose above, greener but scarcely glabrate in age; heads 4–6 cm. high; bracts numerous, lanceolate, with conspicuous broad glutinous ridges, the inner bracts attenuate; spines yellow, stout; corollas purple, rose, or rarely white. *Carduus ochrocentrus* Greene [B]. Plains: Neb.—Tex.—Ariz. *Son.—Plain—Prairie.* Je–Au.

13. **C. iowénse** (Pammel) Fern. Stem 1–1.5 m. high, striate, hirsute to nearly glabrous; lower leaves oblanceolate, pinnately lobed with triangular spine-tipped lobes, green and rough above, white floccose beneath, the upper more entire, lanceolate; heads 4–5 cm. high; bracts with a prominent dorsal ridge, the outer short, with weak spinose spreading tips, the inner unarmed; flowers purple. *Cnicus iowensis* Pammel. Ledges: Iowa—Kans.—S.D.

14. **C. altíssimum** (L.) Spreng. Stem 1–3 m. high, somewhat pubescent, branched; leaves 1–3 dm. long, oblong or elliptic, the lower sometimes lobed, the upper merely toothed with spine-tipped teeth, sparingly hispid above, densely white-tomentose beneath; heads many, bracts numerous, the outer with a short glutinous ridge and a short slender spreading spine; the inner with a serrulate tip; flowers rose-purple. *Carduus altissimus* L. [B]. Thickets and fields, especially in alluvial soil: Mass.—N.D.—Tex.—Fla. *Allegh.—Austral.* Au–S.

15. C. díscolor (Muhl.) Spreng. *Fig. 592.*
Stem 1–2 m. high, furrowed, hirsutulous; basal
leaves 3–4 dm. long, deeply pinnatifid into linear-
lanceolate, lobed and spino-tipped divisions, white-
tomentose beneath; upper leaves much smaller; heads
many; involucre 2–3 cm. high; bracts numerous, the
outer with a glutinous dorsal ridge and weak spreading tip, the inner with an entire spineless tip; flowers rose-purple. *Carduus discolor* Nutt. [B]. Hills
and fields: N.B.—Man.—Neb.– Ga. *Canad.—Allegh.*

f.592.

16. C. mùticum Michx. Stem 1–2.5 m. high,
branched above, glabrate; leaves numerous, oblong
in outline, 2–4 dm. long, once or twice pinnatifid,
with lanceolate, weakly spine-tipped lobes, slightly
hairy above, somewhat arachnoid beneath; involucre
woolly at the base; bracts numerous, the outer obtuse
and mucronate, the inner attenuate. *Carduus muticus* Pers. [B]. Thickets:
Newf.—Sask.—N.D.—Tex.—Fla. *E. Temp.* Jl S.

17. C. arvénse (L.) Scop. Perennial, with a horizontal rootstock, dioecious; stems striate, 3–10 dm. high, branched above, hirsutulous or glabrous;
leaves sessile, slightly clasping, or the lower petioled, green on both sides, glabrous or slightly pubescent, pinnatifid with oblong, lanceolate or triangular
lobes; heads numerous, corymbose, about 2 cm. high; involucres of the staminate heads hemispheric, those of the pistillate ones campanulate; corollas purple
or rarely white. *Cnicus arvensis* L.; *Carduus arvensis* Robs. [B]. Fields, roadsides, and waste places: Newf.—Va.—Utah—B.C.; nat. from Eur. Je–S.

18. C. setòsum (Willd.) Bieb. Like the last, but the leaves flat, the
lower slightly lobed, the upper entire or nearly so, spinulose-ciliate on the
margins. *C. arvense integrifolium* Wimmer & Grab. [G]. Fields: Que.—Md.—
Neb.—N.D.; nat. from se Eur. Jo–S.

103. CÁRDUUS L. Thistle.

Stout annual or biennial herbs. Leaves alternate, strongly decurrent on
the stem, pinnately lobed or toothed, the lobes and teeth spinulose-tipped.
Heads discoid. Involucres mostly hemispheric, with many more or less foliaceous imbricate bracts, in several series. Receptacle flat, bristly. Flowers
hermaphrodite and fertile. Corollas all alike, tubular, deeply 5-cleft. Anthers
sagittate at the base, more or less caudate. Styles filiform. Pappus of many
merely scabrous bristles.

Heads large, 3–4 cm. broad, nodding on long peduncles.   1. *C. nutans.*
Heads small, 1–2.5 cm. broad, usually clustered, short-peduncled,
  not nodding.   2. *C. acanthoides.*

1. C. nùtans L. Biennial; stem 5–10 dm. high; leaves lanceolate in outline, deeply pinnatifid, with deltoid, laciniate, spinose divisions, glabrate, 7–15
cm. long; bracts lanceolate, foliaceous, with the prominent rib prolonged into
a spine; flowers purple. Waste places: N.B.—D.C.—Iowa; adv. from Eurasia.
Je–S.

2. C. acanthoides L. Annual or biennial; stem 5–10 dm. high; leaves
1–3 dm. long, pinnatifid, with broadly ovate divisions, which are again lobed
and spinulose; bracts linear, the outer spreading; flowers purple. Waste
places: N.S.—N.J.—Iowa—Neb.; adv. from Eur. Jl–Au.

104. ONOPÓRDUM (Vaill.) L. Cotton Thistle, Scotch Thistle.

Coarse biennial or perennial herbs. Leaves alternate, lobed, spinescent, conspicuously decurrent on the stem. Heads discoid. Involucre globose; bracts imbricate in many series, tipped with long spines. Receptacle flat, honeycombed, not bristly. Flowers all hermaphrodite and fertile. Corolla mostly purplish; tube slender; throat dilated, and limb 5-cleft. Filaments pilose; anthers sagittate at the base. Achenes 4-angled, oblong. Pappus of many plumose or barbellate bristles, in several series, united at the base.

1. **O. Acánthium** L. Biennial; stem 1–3 m. high, white-tomentose, wing-angled by the decurrent leaves; leaf-blades oblong, densely white-tomentose, with triangular lobes, very spiny, the lower often 3 dm. long; heads 3–5 cm. broad; bracts floccose and arachnoid-hairy, ending in short spines; flowers pale purple. Waste places: N.S.—N.J.—Mich.; Utah—Colo.; native of Eurasia, cultivated and escaped. Jl–S.

105. CÁRTHAMUS L. SAF-FLOWER, FALSE SAFFRON.

Tall thistle-like herbs. Leaves alternate, spinosely toothed or lobed. Heads discoid; flowers mostly hermaphrodite or rarely the marginal ones pistillate. Involucres campanulate or subglobose; bracts in several series, dry, the outer, like the leaves, spinulose at the apex and margins. Corollas yellowish, white or purple, slender, deeply 5-cleft. Anthers sagittate at the base. Achenes glabrous, obovoid. Pappus paleaceous in several series.

1. **C. tinctòrius** L. Stem 3–10 dm. high, glabrous; leaves ovate, clasping, spiny-toothed, glabrous; involucre ovoid; bracts leaf-like, imbricate; flowers orange. Waste places: Iowa—Kans.; cultivated for the dyestuff extracted from the corollas, and escaped.

106. CENTAÚREA L. BLUEBOTTLE, CORNFLOWER, BACHELOR'S BUTTON, STAR THISTLE.

Perennial or annual herbs. Leaves alternate, entire or lobed. Heads discoid. Involucres ovoid or globose, or urn-shaped; bracts imbricate in several series, with appressed bases and spreading, lacerate, erose or spine-tipped appendages. Receptacle flat, densely bristly. Flowers all hermaphrodite and fertile, or the marginal ones usually neutral, and with funnelform, dilated corollas. Corolla of the fertile flowers purple, pink, yellow, or blue, with slender tube and 5-cleft limb. Anthers sagittate at the base. Style-branches short and obtuse. Achenes compressed or obtusely 4-angled, obliquely attached, and crowned by a disk or margin. Pappus of several series of bristles or paleae, rarely wanting.

Involucral bracts with entire, pectinate, or fimbriate, broad appendages, not spinescent; corollas not yellow.
  Involucre 1–1.5 cm. high and broad.
    Appendages of the bracts entire, merely denticulate; corollas rose purple.
                                                    1. *C. Picris.*
    Appendages of the bracts lacerate or fimbriate.
      Marginal corollas rose, narrowly funnelform, with linear lobes.
        Bracts not contracted at the base of the colored appendages; leaf-segments narrow.   2. *C. maculosa.*
        Bracts contracted at the base of the colored appendages.
          Appendages dark colored, pectinately dissected.   3. *C. nigrescens.*
          Appendages light colored, merely lacerate.   4. *C. Jacea.*
      Marginal corollas usually blue, funnelform, with lanceolate lobes.   5. *C. Cyanus.*
  Involucre 2–3 cm. high, 3–5 cm. broad; marginal corollas flesh-colored, with long linear lobes.   6. *C. americana.*
Involucral bracts armed with spines; corollas yellow.
  Spines of the bracts 12–18 mm. long, straw-colored, spreading or reflexed.   7. *C. solstitialis.*
  Spines of the bracts 3–9 mm. long, spreading or ascending.
    Stem-leaves decurrent; corollas yellow.   8. *C. melitensis.*
    Stem-leaves sessile; corollas bright white.   9. *C. diffusa.*

1. **C. Pícris** Pall. Stem 3–5 dm. high, finely puberulent, striate; lower leaves pinnately lobed, 1 cm. long, the upper entire, glabrate in age, linear or lance-linear, those of the branches 1–2 cm. long; involucre 1–1.5 cm. broad, and about as high; bracts broadly obovate, with membranous scarious tips, the outer rounded, the inner acute at the apex; corollas purplish. Waste places: Mo.—Minn.—N.D.—Ida.—Utah; adv. from south Russia and the Orient. Je–Au.

2. **C. maculòsa** Lam. Annual or biennial; stem loosely floccose or glabrate, branched; leaves pinnatifid into linear or linear-oblong divisions, 7–10 cm. long, or the uppermost smaller and entire; heads peduncled; involucre ovoid-campanulate, bracts with black, fimbriate appendages, or the inner ones longer and erose; corollas purple or rose, rarely white, the marginal ones narrowly funnelform. Waste ground: Mass.—Kans.—S.D.—Minn.; adv. from Eur. Jl–Au.

3. **C. nigréscens** Willd. Perennial; stem 5–10 dm. high, sulcate, more or less pubescent, much branched; lower leaves oblanceolate in outline, petioled, 2–3 dm. long, with a few oblong lobes, or merely toothed, or entire, sparingly pubescent; upper stem leaves oblong, oblanceolate, or lanceolate, obtuse, toothed or entire; involucre 1.5–2 cm. high, about 2 cm. broad; bracts lanceolate; appendages broadly ovate, black or dark brown, pectinately cleft; corollas rose-purple, the marginal narrowly funnelform, with linear lobes. *C. decipiens* Thuill. *C. Jacea lacera* Koch. [G]. Around dwellings and waste places: N.S.—N.Y.—Neb.; nat. from Eur. Jl–O.

4. **C. Jàcea** L. Stem branched, 3–6 dm. high; leaves lanceolate or oblanceolate, slightly denticulate, the lower petioled, the upper sessile, resembling those of the preceding; heads subglobose; outer bracts pale, ovate, the inner linear and brown, the appendages brown, entire or slightly lacerate, corollas rose-purple, the marginal ones with long linear lobes. Waste places: Mass.—N.J.—Iowa; adv. from Eur. Je–S.

5. **C. Cȳanus** L. Annual; stem 3–7 dm. high, with slender, ascending branches, more or less floccose; leaves linear or linear lanceolate, entire or the lowest dentate or somewhat pinnatifid; heads long-peduncled; involucres round-urceolate, about 15 mm. high; bracts greenish-yellow, with dark brown fimbriate margins and tips; marginal corollas funnelform, blue, varying to rose or white. Waste places and around dwellings: Que.—Va.—Calif.—B.C.; native of Eur.; escaped from cultivation. Je–S.

6. **C. americàna** Nutt. *Fig. 593.* Stem 5–18 dm. high, sulcate, glabrous; leaves oblanceolate, oblong or linear-lanceolate, 2–8 cm. long, entire or denticulate; involucre hemispheric; bracts numerous; marginal corollas 2–2.5 cm. long; achenes 5–6 mm. long, hairy, with an oblique basal scar, pappus longer, of unequal bristles. Prairies and plains: Mo.—Kans.—Ariz.—Tex.—La.; Mex. *Ozark—Texan —Son.* My–Au.

f.593

7. **C. solstitiàlis** L. Annual; stem much branched, 3–6 dm. high, canescent; leaves decurrent on the margins of the stem, canescent, the lower pinnatifid with oblong or ovate divisions, the upper entire and linear; involucres round-urceolate, about 1 cm. high; bracts yellowish, ending in 3–5 spines; the central one of the middle bracts very stout, 1–2 cm. long; corollas yellow. Waste places: Mass.—N.Y.—ne Kans.—Iowa—Ont.; Calif.—Utah; adv. from the Mediterranean region. Jl–S.

8. C. meliténsis L.   Annual; stem branched, 3–5 dm. high, angled and winged, grayish-pubescent; basal leaves lyrate-dissected, with rounded or obtuse lobes, 5–12 cm. long; lower stem-leaves pinnately lobed, the upper entire, lanceolate, decurrent; heads subsessile, leafy-bracted at the base; corollas yellow; marginal neutral flowers wanting.   Ballast and waste places: Mass.—Ga. Mo.—Minn.; Wash.—Calif.—Ariz.; nat. from Eur.

9. C. diffùsa Lam.   Stem densely branched, sharply angled, thinly grayish-pubescent; basal leaves soon evanescent, bipinnately parted; lower stem-leaves with a few pinnate lobes, the upper narrowly oblong, sessile, entire, 1 cm. long, grayish-pubescent; heads numerous, sessile, oblong-cylindric; involucres minutely glandular-puberulent.   Sioux County, Iowa; adv. from se Eur.

Family 168.   **CICHORIACEAE.**   Chicory Family.

Herbs with a bitter or milky sap.   Leaves mostly alternate, sometimes all basal.   Flowers all alike, hermaphrodite and fertile, in involucrate heads.   Involucres of 1–several series of bracts, sometimes subtended by a series of smaller bracts (calyculum).   Receptacle flat or nearly so.   Corollas all gamopetalous, split on one side part way down, the upper portion modified into a strap-shaped ligule.   Stamens 5; anthers united into a tube around the pistil; sacs auricled or sagittate at the base, appendaged at the apex.   Style 2-cleft; branches filiform, naked, stigmatic only towards the base.

A. Pappus none.   1. SERINIA.
B. Pappus present.
  I. Pappus of plumose bristles, often more or less paleaceous at the base.
    Outer involucral bracts loose and spreading; achenes cross-wrinkled.
                                           2. PICRIS.
    Bracts all appressed; achenes not cross-wrinkled.
      Achenes truncate at the apex, not beaked.   3. PTILORIA.
      Achenes with long beaks.
        Receptacle not chaffy; leaves grass-like and heads solitary.
                                   4. TRAGOPOGON.
        Receptacle chaffy; leaves lyrate, heads corymbose. 5. HYPOCHAERIS.
  II. Pappus not plumose.
    a. Pappus consisting, at least partly, of squamellae or these reduced and united into a crown.
      1. Involucres simple and naked, i. e., without smaller calyculate ones below; pappus of both squamellae and bristles; flowers yellow.
        Bracts of the involucre 5–8, erect in fruit; pappus bristles and scales each 5.   6. CYMBIA.
        Bracts of the involucre 9–18, reflexed in fruit, not keeled; pappus bristles usually more numerous than the scales.
        Annuals; pappus-scales 5, obovate or rounded.   7. KRIGIA.
        Perennials; pappus-scales 10–15, linear or oblong.
                                     8. CYNTHIA.
      2. Involucres double, either imbricate or with smaller calyculate ones below.
        Flowers blue; pappus crown-like of small numerous blunt squamellae in 2 or more series; tall perennials.   9. CICHORIUM.
        Flowers yellow; pappus of large squamellae; low annuals, acaulescent or nearly so.
          Squamellae of the pappus 5, cleft at the apex with an awn in the notch; involucres calyculate.   18. AGOSERIS.
          Squamellae of the pappus 20–30, very narrow, linear-lanceolate, tapering into a bristle-like apex; bracts nearly equal, in two series.   17. NOTHOCALAIS.
    b. Pappus of capillary bristles, not plumose, slightly, if at all, broadened below.
      1. Achenes not flattened.
        a. Pappus-bristles promptly deciduous, mainly together, only a few of the stouter ones in some species remaining.
          Achenes more or less narrowed into a beak, 10-striate; bracts not scarious.   15. YOUNGIA.
          Achenes not beaked, columnar, 5–15-ribbed, truncate at the apex; bracts only slightly scarious-margined.   10. MALACOTHRIX.

b. Pappus persistent, tardily falling off separately, or together only by the breaking off of the beak.
  Beak of the achenes none or a mere attenuation.
    Flowers rose or purplish.
      Stems rush-like and striate; leaves narrowly linear-lanceolate or reduced; achenes tapering at the summit.
        11. LYGODESMIA.
      Stems not rush-like; leaves ample; achenes tapering to the base, terete or 4 or 5-angled.   12. NABALUS.
    Flowers yellow or white.
      Heads several, rarely solitary; stem leafy.
        Achenes tapering upwards; pappus white.
          Achenes not dilated into a pappiferous disk.
            Bracts in fruit more or less thickened at the base or on the midrib.   13. CREPIS.
            Bracts not thickened on the back.
              14. HETEROPLEURA.
            Achenes contracted into a more or less distinct beak, enlarged at the apex into a pappiferous disk.
              15. YOUNGIA.
        Achenes not tapering upwards; pappus in ours sordid or reddish; bracts not thickened.
          Leaves simple and entire-margined or denticulate; inflorescence in ours more or less corymbiform; bracts narrow, green.   16. HIERACIUM.
          Leaves divided or lobed, inflorescence thyrsoid-paniculate; bracts broad and colored.
            12. NABALUS.
      Heads solitary on a leafless scape.   17. NOTHOCALAIS.
  Beaks of the achenes distinct and slender; plants scapiferous.
    Achenes not muricate-spinulose; involucres more or less imbricate.   18 AGOSERIS.
    Achenes muricate-spinulose, at least near the apex; involucres of a single series of principal bracts and several or numerous calyculate ones below.
      Style-branches slender, filiform; pappus without woolly ring at the base,   19. TARAXACUM.
      Style-branches short, obtuse; pappus base surrounded by a soft-villous ring.   20 PYRRHOPAPPUS.
2. Achenes flattened; leafy-stemmed plants with paniculate heads.
    Achenes narrowed at the top or beaked; pappus-bristles falling separately; involucres cylindraceous.   21. LACTUCA.
    Achenes truncate at the top; pappus-bristles falling off more or less in connection; involucres hemispherical or campanulate.
      22. SONCHUS.

1. SERÍNIA Raf.

Annual herbs. Leaves alternate or opposite, at least some of them clasping, entire or pinnatifid. Heads small; bracts usually 5, membranous, concave, all alike. Receptacle flat, naked. Flowers few. Disk-flowers pistillate and fertile; ligules yellow or orange, truncate or toothed. Disk-flowers hermaphrodite, fortile. Anthers sagittate at the base. Style-branches slender. Achenes obovoid, 8–10-ribbed. Pappus none or obsolete. *Apogon* Ell.

1. S. oppositifòlia (Raf.) Kuntze. Stem glabrous or glandular on the peduncles, under the heads, 0.5–3 dm. high; leaves spatulate to linear, 2–12 cm. long, entire to sinuate-pinnatifid; involucre 4 mm. high; flowers pure yellow, slightly surpassing the bracts; achenes 1.5 mm. long, ribbed and minutely wrinkled. *A. humilis* Ell. Sandy soil: S.C.—Kans.—Tex.—Fla. *Austral.* Mr–My.

2. PÍCRIS L.

Coarse rough-bristly annuals or biennials, with yellow flowers, terminating leafy stems. Heads many-flowered. Outer bracts loose or spreading. Achenes terete, with 5–10 rugose ribs, not beaked. Pappus of 1 or 2 rows of plumose bristles.

1. P. echioìdes L. Stem rather tall, corymbosely branched, the bristles somewhat barbed; leaves lanceolate or broader, clasping, spinescent; the outer

bracts ovate, subcordate, spinescent, the inner narrow, becoming thickened below; pappus densely plumose. Waste places; rarely adv. from Eur. Jl–S.

3. PTILÒRIA Raf.

Rush-like slender plants, with the upper leaves reduced and bract-like. Heads small, 3–20-flowered. Involucres cylindric, rarely campanulate. Bracts few, in a single series, with a few small calyculate ones at the base. Flowers pink or flesh-colored. Achenes prismatic, truncate at both ends, strongly angled, glabrous, often rugose, the base broad and hollowed. Pappus of 10–20 bristles, somewhat paleaceous at the base, plumose at least above the middle. *Stephanomeria* Nutt.

Pappus plumose to the base; involucres 7–10 mm. high; leaves, at least the lower
 ones, broad, oblanceolate in outline, and runcinate.   **1.** *P. ramosa.*
Pappus merely scabrous, or hirsutulous at the base; leaves runcinate, but narrow.
                      2. *P. pauciflora.*

1. P. ramòsa Rydb. Stem branched at the base, with more or less spreading branches, striate, puberulent below, 2–3 dm. high; lower leaves lanceolate, runcinate-divided; upper leaves linear and entire; involucres 8–10 mm. high, usually 5-flowered; bracts proper usually 5, lance-linear, usually acutish; achenes strongly rugose; pappus white, plumose to the base. (?) *S. runcinata* Nutt. Dry hills: Neb.—Colo.—Mont. *Plain—Submont.* Je–S.

2. P. pauciflòra (Torr.) Raf. Stem 2–4 dm. high, with stiff ascending branches; leaves linear or nearly so, the lower runcinate-pinnatifid with narrow divisions, the upper entire; involucres 9–10 mm. high; bracts about 5, linear, obtuse; achenes rugose; pappus tawny; bristles naked at the base. (?) *S. runcinata* Nutt. Dry plains: Kans.—Tex.—Ariz.—Nev. *St. Plains—Son.* Je–S.

4. TRAGOPÒGON (Tourn.) L. SALSIFY, OYSTER-PLANT.

Tall perennial herbs, with fleshy taproots and long alternate grass-like leaves clasping at the base. Involucres nearly cylindric; bracts in a single series, united at the base. Flowers many, yellow or purplish; ligules truncate and 5-toothed at the apex. Achenes muricate, 5–10-ribbed, produced at the apex into a distinct beak. Pappus of plumose bristles connate at the base.

Flowers yellow; involucral bracts equaling or shorter than the flowers.
                    1. *T. pratensis.*
Flowers purple; involucral bracts much longer than the flowers. **2.** *T. porrifolius.*

1. T. praténsis L. Stem 3–6 dm. high, glabrous; leaves linear-lanceolate, long-attenuate; peduncles scarcely fistulose, little if at all thickened below the heads; bracts linear-lanceolate, attenuate, 2.5–3 cm. long; outer achenes muricate. Fields and waste places: N.B.—N.J.—Utah—Mont.—Man.; adv. or nat. from Eur. Je–Au.

2. T. porrifòlius L. Stem 6–10 dm. high, glabrous; leaves clasping, with lanceolate bases, long-attenuate; peduncles distinctly fistulose, gradually thickened below the head; bracts 8–10, linear-lanceolate, long-attenuate, 3–5 cm. long; outermost achenes strongly muricate. Fields and waste places: Ont.—N.C.—N.M.—Utah—Calif.—B.C.; escaped from cultivation. Je–Au.

  T. porrifolius × pratensis. Corolla pale lilac; involucral bracts 8–10, mostly 8.
With the parents. Ont.—N.J.—Colo.—Mont.

5. HYPOCHAÈRIS L. CAT'S-EAR, GOSMORE.

Subscapose annuals or perennials. Leaves mostly in basal clusters or rosettes, often lyrate. Involucres campanulate; bracts rather few, but imbricate and the outer successively shorter. Receptacle chaffy; paleae thin and narrow. Flowers yellow. Achenes fusiform or oblong, 10-ribbed, tapering at the top into a more or less distinct beak. Pappus of capillary plumose bristles, or some of the outer bristles shorter and not plumose.

1. **H. radiàta** L. Perennial; stem 5–20 dm. high, corymbosely branched, almost leafless; basal leaves lyrate-pinnatifid, with the lateral lobes oblong, obtuse, hirsute; involucre about 25 mm. high; ligules 7–10 mm. long. Waste places and fields: N.Y.—N.J.—Colo.; Calif.—B.C.; adv. or nat. from Eurasia. My–O.

6. CÝMBIA (T. & G.) Standley.

Annual acaulescent herbs. Scapes tufted. Leaves in a basal rosette, entire to pinnatifid. Heads solitary on the scapes, broad. Bracts 5–8, in 2 subequal series, erect in fruit, thin, keeled. Receptacle flat, naked. Ray-flowers few; ligules yellow. Disk-flowers hermaphrodite, fertile; anthers sagittate at the base. Style-branches slender, obtuse. Achenes ribbed, transversely wrinkled. Pappus of 5 broad obovate squamellae and 5 alternating bristles twice as long as the squamellae, about equaling the achenes.

1. **C. occidentàlis** (Nutt.) Standley. Leaves obovate to oblong, 1–4 cm. long, entire to lyrately lobed or pinnatifid; scape slender, glandular, 0.5–2 dm. high; involucre 4–5 mm. high; achenes 1.5 mm. long. *Krigia occidentalis* Nutt. [G]. *Adopogon occidentale* Kuntze [B]. Prairies: Mo.—Kans.—Tex. *Ozark—Texan.* Mr–My.

7. KRIGIA Schreber.

Annual caulescent herbs. Scapes several, each with a single head. Leaves clustered at the base, entire, coarsely toothed or pinnatifid. Involucres campanulate; bracts 9–18, linear or lanceolate, in 2 subequal series, reflexed in fruit, not keeled. Receptacle flat, naked. Ray-flowers fertile; ligules yellow; flowers hermaphrodite and fertile. Anthers sagittate at the base. Achenes narrowly turbinate, 5-angled. Pappus of 5 rounded or obovate squamellae and 5 or 10 bristles several times as long.

f. 594.

1. **K. virgínica** (L.) Willd. *Fig. 594.* Leaves linear in outline, 2–12 cm. long, the earlier ones entire, the later ones pinnatifid; scapes erect, 3–40 cm. long, glandular-hirsute under the head; involucre 4–5 mm. high; achenes 1.5 mm. long; pappus-bristles twice as long. *Adopogon carolinianum* (Walt.) Britton [B]. Dry soil: Me.—Minn.—Tex.—Fla. Ap–Au.

8. CÝNTHIA D. Don.

Perennial herbs with pale green foliage. Leaves mainly basal. Involucre campanulate; bracts in 1 or 2 subequal series, without calyculum. Receptacle naked. Flowers yellow. Achenes oblong or somewhat turbinate, 15–20-ribbed, beakless. Pappus of 10–15 outer short squamellae and 15–20 capillary inner bristles. *Adopogon* Neck.

Plant with a few stem-leaves and several heads.	1. *C. virginica.*
Plant with a leafless scape and a single head.	2. *C. Dandelion.*

1. **C. virgínica** (L.) D. Don. Stem glabrous, 3–6 dm. high, 1- or 2-leaved, branched above; basal leaves spatulate, wing-petioled, runcinately lobed, toothed or entire, 5–10 cm. long; cauline leaves partly clasping, with a broad base; bracts 10–15, linear-lanceolate, acutish, glabrous, 8–10 mm. long; flowers orange. *Krigia amplexicaulis* Nutt. [G]. *Adopogon virginicum* (L.) Kuntze [B]. Moist woods: Ont.—Ga.—Colo.—Man. *Allegh.—Plain—Submont.*

2. C. Dandèlion (L.) DC. Perennial, with tubers; scape erect, 0.5–5 dm. high; leaves clustered, spatulate to linear, denticulate to pinnatifid, 5–15 cm. long; bracts linear, glabrous, 9–14 mm. long; flowers yellow; achenes 2 mm. long, glandular-ribbed, pappus-bristles twice as long. *Krigia Dandelion* Nutt. [G]. *Adopogon Dandelion* Kuntze [B]. Moist soil: Md.—Kans.—Tex.—Fla. *Allegh.—Austral.* Mr–Jl.

9. CICHÓRIUM (Tourn.) L. CHICORY.

Herbs with stiff branching stems and mostly basal leaves, the stem-leaves reduced and bract-like. Involucres cylindraceous; bracts in 2 series, the outer somewhat spreading, the inner erect and partly enfolding the achenes at the base. Flowers blue, rose, or pink. Achenes 5-angled, truncate and beakless at the apex. Pappus of 1–3 series of short blunt squamellae.

1. C. Íntybus L. *Fig. 595.* Perennial, with a taproot; stems slightly hispid, 3–10 dm. high; basal leaves spreading, runcinate-pinnatifid, spatulate in outline, 7–15 cm. long; stem-leaves smaller, lanceolate or oblong, lobed or entire, clasping and auricled at the base; heads numerous, 1–4 together in sessile clusters; bracts proper about 8, glandular-ciliate. *f. 595.* Roadsides, fields, and waste places: N.S.—N.C.—Calif.—B.C.; nat. from Eur. Jl–S.

10. MALACÒTHRIX DC. DESERT DANDELION.

Annual or perennial herbs, commonly with a basal cluster of leaves. Involucre campanulate or turbinate; bracts narrow, acute or acuminate, imbricate. Receptacle naked or with a few deciduous bristles. Flowers yellow, white, or pinkish. Achenes short, truncate at the apex, with a denticulate border, 10–15-ribbed. Pappus of numerous capillary, white-scabrous bristles, more or less united at the base and falling off together or with 1–8 stronger outer ones more persistent.

1. M. sonchoìdes (Nutt.) T. & G. Annual; stem branched at the base with ascending branches; leaves oblanceolate or obovate in outline, lyrate-pinnatifid, with oblong, dentate lobes; heads paniculate; involucres 8–10 mm. high; bracts in 3 series, the outer ovate-lanceolate, the inner linear-lanceolate; achenes cylindric, 15-ribbed, with 4 or 5 ribs slightly stronger, not winged; outer persistent bristles usually none. Plains: Neb.—Kans.—s Calif.—Nev.—Ida. *Plain—Son.* My–Je.

11. LYGODÉSMIA D. Don. WILD ASPARAGUS, SKELETON-WEED,
PRAIRIE PINK.

Rush-like herbs, with narrow alternate leaves. Involucre cylindric; bracts 5–12, linear, equal, with a few calyculate ones at the base. Flowers 5–12, pink or rose-colored. Achenes slender, 4–8-ribbed, slightly tapering towards the summit. Pappus of numerous soft bristles, falling off separately.

Perennials, with deep-seated rootstocks.   1. *L. juncea.*
Annuals.   2. *L. rostrata.*

1. L. júncea (Pursh) D. Don. Stem 2–4 dm. high, much branched from near the base; lower leaves linear, 2–10 cm. long, those of the branches reduced and subulate; involucres usually 5-flowered; calyculate bracts lanceolate, 1–2 mm. long; bracts proper 5, narrowly linear; achenes 5–6 mm. long; pappus sordid. Plains and prairies: Man.—Mo.—N.M.—Nev.—Alta. *Prairie—Plain — Submont.* Je–Au.

2. **L. rostràta** A. Gray. Stem erect, 3–10 dm. high, with strongly ascending branches; leaves narrowly linear, obscurely 3-nerved, 5–20 cm. long; heads numerous, ending short scaly branches; calyculate bracts lanceolate, 2–3 mm. long; bracts proper 8–9, narrowly linear; flowers as many; achenes fusiform, 5–8-striate; pappus dull white. Cañons and sandy plains: Sask.—Man.—Kans. —Colo. *Plain.* Jl–S.

12. NÀBALUS Cass. RATTLESNAKE-ROOT.

Perennial caulescent herbs. Heads small, clustered, and usually nodding. Involucre narrow, cylindric to campanulate, bracts narrow, in 1 or 2 subequal series, subtended by a few small calyculate ones. Receptacle flat, naked. Flowers pale, cream-colored or purplish. Achenes oblong or columnar, terete or 4- or 5-angled, often 10-ribbed, sometimes slightly tapering towards the base. Pappus of numerous capillary, rather rigid, pale or brown bristles. *Prenanthes* L., in part.

Involucre glabrous, 8–10-flowered; basal leaves with cordate, hastate, or deltoid
   blades.
    Involucres 2 mm. thick, 5–7-flowered; pappus straw-colored.   1. *N. altissimus.*
    Involucre 3–6 mm. thick, 8–16-flowered; pappus cinnamon-
      colored.   2. *N. albus.*
Involucre pubescent.
   Basal leaves oblanceolate or spatulate; inflorescence thyr-
    soid; involucre 12–16-flowered.
     Stem glabrous; upper leaves clasping   3. *N. racemosus.*
     Stem hirsute; upper leaves not clasping.   4. *N. aspor.*
   Basal leaves with deltoid blades; inflorescence corymbose-
    paniculate; involucre 20–25-flowered.   5. *N. crepidineus.*

1. **N. altíssimus** (L.) Hook. Stem erect, glabrous or nearly so, 1–2 m. high; lower leaf blades 5–15 cm. long, light green, thin, hastately 3–5-lobed, on slender petioles, the upper ones hastate, ovate to lanceolate, dentate, rounded or cordate, or the uppermost cuneate at the base; involucre 10–12 mm. high, 2 mm. thick; flowers greenish or yellowish; achenes slender, 4–5 mm. long. *Prenanthes altissima* L. [G]. Woods and hillsides: Newf.—Man.—Tenn.— Ga. *Canad.—Allegh.* Jl–O.

2. **N. álbus** (L.) Hook. Stem 5–15 dm. high, glabrous, often tinged with purple; leaves petioled; blades deltoid-hastate, denticulate, dentate, lobed, or divided, or the upper lanceolate and entire, thin, glabrous, paler beneath; heads numerous, paniculate, pendulous, about 12 mm. high and 6 mm. broad; principal bracts about 8, purplish, linear; corollas greenish or white; pappus reddish-brown. *P. alba* L. [G]. Open woods: Me.—Ga. —Ky.—Minn.—Sask. *Canad.—Allegh.* Au–S.

3. **N. racemòsus** (Michx.) DC. *Fig. 596.* Stem 5–10 dm. high, glabrous or pubescent above; basal leaves glabrous, entire or dentate, with winged petioles; stem-leaves sessile, oblong or elliptic, partly clasping; heads in a long, spike-like thyrsus; involucres campanulate, 12–14-flowered, 10–12 mm. high; bracts narrowly linear, obtuse, hirsute; flowers rose or purplish; pappus fulvous. *P. racemosa* Michx. [G]. Moist ground: N.B.—N.J.—Colo.—Mont.— Alta. *Canad.—Plain—Mont.* Au–S.

f.596.

4. N. ásper (Michx.) T. & G. Stem 1–2 dm. high, scabro-pubescent; lower leaf-blades oblanceolate or oblong, the upper lanceolate, 4–20 cm. long, irregularly toothed; involucres 12–14 mm. high; bracts linear or the outer subulate, hirsute; flowers cream-colored, the ligules 1 cm. long; achenes sordid-brown to straw-colored. *P. aspera* Michx. [G]. Dry prairies: Ohio—Minn. —Kans.—La. *Allegh.—Prairie.* Au–S.

5. N. crepidíneus (Michx.) DC. Stem 1–2 m. high; lower leaf-blades hastate, the upper ovate-hastate to oblong, 1–3 dm. long, tapering at the base into winged petioles, coarsely dentate; involucre about 12 mm. high; bracts linear, strigose-hirsute; flowers ochroleucous; achenes 4–5 mm. long, 12–15-ribbed; pappus dirty-brown. *P. crepidinea* Michx. [G]. Thickets: N.Y.—Minn.—Kans.—Tenn. *Allegh.* Au–O.

13. CRÈPIS L.

Annual, biennial, or perennial (all ours), mostly caulescent herbs. Involucre cylindric or campanulate; bracts in a single series, equal, with a more or less thickened midrib; small outer calyculum present or absent. Flowers yellow. Achenes 10–30-ribbed, narrowed towards the summit. Pappus of numerous white and soft bristles.

Plant neither canescent nor furfurascent; heads hemispherical to turbinate.
  Stem leafy throughout; introduced annuals or biennials.
    Stem-leaves lanceolate, with auriculate-clasping bases; achenes smooth.
                                                                          1. *C. capillaris.*
    Stem-leaves linear, merely sessile; achenes scabrous on
      the ribs.   2. *C. tectorum.*
  Stem mostly leafy at the base. the stem-leaves few and re-
    duced, native perennials.
    Involucres glabrous or tomentulose when young; neither
      hirsute nor glandular.   3. *C. glauca.*
    Involucres and peduncles hirsute or glandular or both.
      Leaf-blades oblanceolate or oblong to linear-oblance-
        late.
        Leaves long-petioled; petioles half as long to
          fully as long as the blades, not winged.
          Blades of the basal leaves broadly oblance-
            olate; stem usually with 1 or 2 leaves;
            involucres over 1 cm. high.   4. *C. petiolata.*
          Blades of the basal leaves narrowly linear-
            lanceolate or linear-oblanceolate; stem
            scapiform; involucres less than 1 cm.
            high.   5. *C. glaucella.*
        Leaves subsessile or with short-winged petioles.
          Leaves glabrous or nearly so; involucres
            with a few short hairs.   6. *C. perplexans.*
          Leaves hairy.
            Involucres sparingly short-hirsute; pe-
              duncles and stem usually glabrous;
              involucres less than 1 cm. high.   7. *C. runcinata.*
            Involucres, peduncles, and usually also
              the stem copiously glandular-hirsute;
              involucre over 1 cm. high.   8. *C. platyphylla.*
      Leaf-blades obovate to elliptic; involucres and pe-
        duncles densely glandular-hispid with long
        hairs; involucres over 1 cm. high; leaves hairy,
        1–2 dm. long.   9. *C. riparia.*
Plant more or less canescent or furfurascent, especially the
  leaves; heads mostly narrow, often almost cylindrical.
  Involucres cylindric; principal bracts of the involucres 5–8
    (rarely 9–14).   10. *C. intermedia.*
  Involucres campanulate; principal bracts of the involucres
    9–18.   11. *C. occidentalis.*

1. C. capillàris (L.) Wallr. Stem 3–6 dm. long, ascending; lower leaves spatulate or oblanceolate in outline, laciniate-pinnatifid to dentate; involucre 6–7 mm. high; principal bracts lanceolate, the calyculate ones appressed;

achenes 10-ribbed, slightly fusiform. *C. virens* L. [B]. Fields and waste places: Conn.—N.D.—Pa.; adv. from Europe. Jl–S.

2. **C. tectòrum** L. Stem slender, 2-4 dm. high, branched from the base; leaves narrow, the lower runcinate or dentate, 1–1.5 dm. long, the upper linear, entire or lobed, often revolute-margined; involucre 6–10 mm. high, pubescent; principal bracts lanceolate, the calyculate ones linear, spreading; achenes fusiform, attenuate above. Fields: Conn.–N.D.—Neb.—N.J.; adv. from Eur. Jl–S.

3. **C. glaùca** (Nutt.) T. & G. Stem scape-like, 3–6 dm. high, glabrous; leaves oblanceolate in outline, tapering below into short petioles, more or less runcinate-pinnatifid, with triangular or lanceolate lobes or sometimes entire, glabrous and somewhat glaucous; involucres turbinate, about 1 cm. high; bracts narrowly linear-lanceolate, acute, glabrous. *C. lancifolia* Greene. Valleys: Sask.—Colo.– Nev.—Alta. *Plain—Mont.* Je–Au.

4. **C. petiolàta** Rydb. Stem with 1–3 leaves, glabrous and often purplish below, more or less glandular-hirsute above; basal leaf-blades usually obtuse, sinuate-dentate or entire, about 1 dm. long, glabrous and glaucous; stem-leaves oblanceolate and usually sessile; involucre about 12 mm. high, turbinate-campanulate, with glandular black hairs; bracts linear-lanceolate, acuminate. Mountains: Minn.—S.D.– Colo.—Utah—Wyo. *Prairie—Submont.—Mont.* Jl–Au.

5. **C. glaucólla** Rydb. Stem slender, about 3 dm. high, glabrous and shining; basal leaves glabrous and shining, somewhat glaucous, thin, 5–10 cm. long; blades oblanceolate, acute at the apex, remotely sinuate-dentate with divaricate or retrorse short teeth or entire; stem-leaves 1 or 2, much reduced, 1–2 cm. long, linear or nearly so; involucre turbinate, sparingly hirsute; bracts linear-lanceolate, acuminate. *C. dacotica* Lunell. Wet meadows: Sask.– S.D.—Colo.—Mont. *Submont.—Mont.* Jl–S.

6. **C. perpléxans** Rydb. Stem scape-like; leaves about 1 dm. long, glaucous and glabrous or rarely with a few hairs on the midrib below and on the margin of the narrow base, oblanceolate, runcinate-toothed; involucres turbinate-campanulate, about 1 cm. high, more or less glandular-hairy, with yellowish hairs; bracts linear-lanceolate, acuminate; calyculum about one third as long. Valleys: Man.—Neb.—Colo.—Alta. *Plain—Mont.* Je–Au.

7. **C. runcinàta** (James) T. & G. *Fig. 597.* Stem scapiform, 3–5 dm. high, slightly pubescent or glabrous; leaves mostly basal, oblanceolate or spatulate, tapering below, dentate or entire or runcinate, rather densely glandular-hirsute; involucres turbinate, glandular-hirsute, with yellowish hairs; bracts narrowly lance-linear. Valleys: Man.—N.D.—Colo.—Alta. Je–Au.

8. **C. platyphýlla** Greene. Stem scape-like, 3–6 dm. high, glandular-hirsute in the inflorescence, otherwise glabrous; leaves basal, hirsute oblanceolate, dentate or somewhat runcinate, 10–15 cm. long; involucres broadly turbinate, 12–15 mm. high, densely glandular-hirsute; bracts lance-linear. *C. runcinata hispidulosa* Howell. (?) *C. subcarnosa* Greene. Meadows and bogs: Mont.—Wyo.—Utah—Neb.— Ore. *Plain—Submont.* Je–Au.

9. **C. ripària** A. Nels. Stems 2–4 dm. high, subscapose, minutely and sparsely pubescent below, glandular-hirsute in the inflorescence; leaves mostly basal, usually rounded at the apex, coarsely and irregularly dentate or entire,

somewhat runcinate towards the base, 1–2 dm. long, sparingly hirsute on the veins; involucres broadly turbinate, 12–15 mm. high, densely glandular with dark hairs; principal bracts linear. (?) *C. aculeolata* Greene. River-banks: Neb.—Colo.—Utah—Wyo. *Plain—Submont.* Jl–Au.

10. C. intermèdia A. Gray. Stem 3–5 dm. high, canescent-tomentulose; basal leaves broadly lanceolate, acuminate, canescent-tomentulose, usually divided only halfway to the midrib into lanceolate-acuminate lobes, which are usually directed downward; stem-leaves subsessile, the uppermost linear and entire; involucres subcylindric, tomentose, 12–14 mm. high; bracts proper about 7, linear-lanceolate, more or less scarious-margined. Hillsides: Sask.—Colo.— Calif.—B.C. *Submont.* My–Au.

11. C. occidentàlis Nutt. Stem stout, 1–3 dm. high, canescent-tomentose, usually also glandular-hispid in the inflorescence; basal leaves petioled; blades 1–2 dm. long, broadly lanceolate or oblanceolate, runcinate-lobed or cleft, usually about halfway to the midrib, canescent-tomentose, acute; lobes triangular or lanceolate, often toothed; involucres campanulate, 12–15 mm. high; bracts proper 9–18, tomentose. Plains and hillsides: Sask.—S.D.—Colo.—Ariz.—Calif. —B.C. *Plain—Submont.* My–Au.

14. HETEROPLEÙRA Schultz-Bip.

Perennial leafy-stemmed herbs, of the habit of *Hieracium.* Involucre campanulate; bracts in 1 or 2 series, subequal, narrow, with a few calyculate ones at the base, not thickened on the back. Flowers light yellow. Achenes tapering from the base to the summit, 10-ribbed, the alternating ribs much stronger. Pappus of numerous dirty white soft bristles.

1. H. Féndleri (Schultz-Bip.) Rydb. Stem scape-like, 2–3 dm. high, sparingly setose-hirsute; basal leaves spatulate or obovate, with short wing-margined petioles, setose-hirsute, the hairs long with pustulate bases; stem-leaves few, the upper or all reduced, narrowly linear, and bract-like; heads rather few, racemiform-paniculate, long-peduncled, nodding in anthesis; involucres campanulate, 12–14 mm. high; bracts proper lance-linear, puberulent or glabrate, with or without a few long hairs; achenes reddish or blackish. *Heteropleura ambigua* (A. Gray) Schultz-Bip. *Hieracium Fendleri* Schultz-Bip. *H. nigrocollinum* S. Wats. Woods: S.D.—N.M.—Ariz. *Submont.—Mont.* Je–Au.

15. YOÚNGIA Cass.

Low or depressed perennials, with many basal leaves. Heads several, 8–15-flowered. Involucres cylindric; bracts linear, obtuse, in single series, not thickened on the back, with 3 or 4 short calyculate ones at the base. Flowers yellow. Achenes narrow, 10-striate, tapering above and there enlarged into a disk, bearing the pappus. Pappus of numerous white capillary deciduous bristles.

1. Y. nàna (Richards.) Rydb. Depressed perennial, with creeping rootstock; stem 1–5 cm. high; leaves mostly basal, petioled, 2–5 cm. long; blades obovate or spatulate, entire, repand-dentate or lyrate; involucres 8–10 mm. high; bracts proper linear, obtuse, with the backs thickened at the base; achenes cylindric, about 5 mm. long, 0.5 mm. thick. *Crepis nana* Richards. *Y. pygmaea* Ledeb. High mountains: Lab.—Alta.—Utah—Ida.—B.C.; Asia. *Mont.—Subalp.* Jl–Au.

16. HIERÀCIUM (Tourn.) L. HAWKWEED.

Perennial, mostly caulescent herbs, with entire or dentate leaves and paniculate heads. Involucre cylindric or campanulate; bracts in 1–3 series, subequal or somewhat imbricate, not thickened on the back, with a few small calyculate ones at the base. Flowers yellow, seldom white or orange. Achenes cylindric or fusiform, not tapering towards the apex, truncate, 10–15-ribbed. Pappus of 1 or 2 series of sordid or brownish, fragile capillary bristles.

Rootstock slender ; stolons present ; flowers orange.   1. *H. aurantiacum.*
Rootstock short, stout, not stoloniferous ; flowers yellow.
  Heads 8–20 mm. in diameter.
    Stem leafy towards the base ; upper stem-leaves few and
      reduced ; leaves of the inflorescence not foliaceous.
      Inflorescence corymbiform ; achenes columnar.
        Leaves narrowly oblanceolate, glaucous.   2. *H. florentinum.*
        Leaves, at least the lower, obovate or elliptic.   3. *H. venosum.*
    Inflorescence cylindro - paniculate or racemiform ;
      achenes fusiform.
      Leaves and lower part of the stem slightly pilose ;
        basal leaves rounded or obtuse at the apex.   4. *H. Gronovii.*
      Leaves and stem densely shaggy, with coarse
        brownish or white hairs ; basal leaves acute. 5. *H. longipilum.*
    Stem leafy up to the inflorescence ; leaves of the inflores-
      cense foliaceous.   6. *H. scabrum.*
  Heads 2.5–4.5 cm. in diameter ; stem leafy.
    Leaves ovate to lanceolate, the upper rounded or subcor-
      date at the base ; stem more or less hirsute.   7. *H. canadense.*
    Leaves lance-linear, narrowed at the base ; stem pubern-
      lent.   8. *H. scabriusculum.*

1. **H. aurantiacum** L. Stolons numerous and slender ; scape 2–6 dm. high, 1–2-bracted ; leaves oblanceolate, 6–15 cm. long, long-hirsute ; heads 1.5–2.5 cm. broad. Fields and meadows: Que.—Iowa—Pa.; nat. from Eur. Je–Jl.

2. **H. florentinum** All. Scape 2–8 dm. high, sparingly hirsute ; basal leaves oblanceolate or spatulate, sparingly setose, 5–10 cm. long ; heads corymbose, 8–12 mm. broad ; bracts linear, pilose and somewhat glandular ; flowers yellow. Meadows and fields: Que.—N.Y.—Iowa; nat. from Eur. My–Jl.

3. **H. venosum** L. Scape 2–7 dm. high, naked or with 1 or 2 leaves, glabrous ; leaves entire, or glandular-denticulate, thin, glabrous, often purple-veined and mottled above, glaucous beneath, 3–10 cm. long, the lower obovate or spatulate ; heads 1–1.5 cm. broad ; peduncle slightly glandular above. Dry woods and sandy places: Me.—Man.—Neb.—Ky.— Ga. *Canad.—Allegh.* My–S.

4. **H. Gronovii** L. Stem 3–12 dm. high, leafy below the middle, villous at the base ; basal leaves oblong or obovate, 5–15 cm. long, setose above, minutely pubescent with branched hairs beneath ; pedicels slightly glandular ; heads 1–1.8 cm. broad, 15–20-flowered. Sandy soil: Mass.—Kans.—Tex.— Fla. *E. Temp.* Jl–O.

5. **H. longipilum** Torr. Stem 4–12 dm. high, long-hairy ; basal leaves spatulate or oblanceolate, acute, 1–3 dm. long ; stem-leaves much reduced, lanceolate ; pedicels copiously glandular-pilose ; heads 1.5–2 cm. broad, 20–30-flowered. Open woods and prairies: Ont.—Minn.—Tex.—Ill. *Allegh.—Ozark.* Jl–S.

6. **H. scabrum** Michx. Stem 3–12 dm. high, rough-hairy ; leaves elliptic to obovate-spatulate, 3–12 cm. long, subentire, hairy on both sides, paler beneath ; peduncle and its branches white-tomentose and glandular ; head 12–14 mm. broad, 40–50-flowered. Dry woods and pastures: N.S.—Minn.—Kans. —Ga. *Canad.—Allegh.* Jl–S.

7. **H. canadense** Michx. Stem stout, 2–12 dm. high ; leaves ovate to lanceolate, coarsely toothed, acute, firm, 3–7 cm. long, 1–2 cm. wide ; heads corymbose, 2.5 cm. broad ; involucre 12 mm. high. Dry woods and thickets: N.S.— N.J.—Ill.—S.D.— Man. *Canad.—Allegh.* Jl–S.

8. **H. scabriusculum** Schwein. *Fig. 598.* Stem erect, rather simple and strict, leafy ; lower leaves oblanceolate, the rest lanceolate, sessile, 3–8 cm.

f.598.

long, distantly and acutely dentate, rather firm, paler beneath, scabrous-puberulent; heads corymbose; involucres rounded-campanulate, 10–12 mm. high; bracts glabrous or nearly so, usually blackish. *H. macranthum* Nutt. *H. umbellatum* Am. auth. [B]. Open woods, hills, and valleys: Man.—Minn.—Wis.—Ore.—B.C. *Prairie—Plain—Submont.*

17. NOTHOCÁLAIS (A. Gray) Greene.

Acaulescent perennial of the habit of *Agoseris*, with narrow, entire or pinnatifid leaves. Involucres oblong-campanulate; bracts in 2 or 3 series, nearly equal, without calyculum. Receptacle naked. Flowers yellow. Achenes attenuate-fusiform, beaked. Pappus of 20–30 narrow, linear-lanceolate, in one species nearly bristle-like, squamellae tapering into an awn.

1. **N. cuspidàta** (Pursh) Greene. Leaves somewhat glaucescent, linear lanceolate, caudately attenuate, more or less wavy, villous-tomentulose on the margins when young; scape 6–30 cm. long; involucres turbinate, about 2 cm. high; bracts lanceolate or linear-lanceolate, gradually acuminate, somewhat tinged with purple; pappus of 40–50 bristles. *Troximon cuspidatum* Pursh. Prairies and plains: Wis.—Mo.—Colo.—S.D. *Prairie—Plain—Submont.* Ap–Je.

18. AGÓSERIS Raf. GOAT CHICORY.

Perennial herbs, with strong taproots, or annuals, mostly acaulescent. Involucre campanulate to nearly cylindric; bracts imbricate in a few series, the outer broader and shorter. Flowers yellow, orange, or purplish. Achenes fusiform or oblong, 10-ribbed, narrowed above into a beak. Pappus of numerous capillary white bristles. *Troximon* Nutt.

Bracts villous-ciliate, at least on the margins.
　Outer bracts much broader than the inner, often obtusish; plant generally low
　　and leaves short, obtuse.　　　　　　　　　　　　　　　　　1. *A. pumila.*
　Outer bracts usually not much broader than the inner;
　　plant 3–5 dm. high; leaves long and acute.　　　　　　　2. *A. scorzoneraefolia.*
Bracts glabrous; involucres, if at all hairy, tomentose only
　at the very base.
　Leaves linear-oblanceolate.　　　　　　　　　　　　　　　3. *A. glauca.*
　Leaves narrowly linear.　　　　　　　　　　　　　　　　　4. *A. parviflora.*

1. **A. pùmila** (Nutt.) Rydb. Leaves 5–15 cm. long, usually broadly oblanceolate, entire or merely dentate, glabrous and somewhat glaucous; scape 1–2 dm. (rarely 3 dm.) high, stout, more or less villous under the heads; involucres campanulate, about 2 cm. high, 2–2.5 cm. broad; bracts villous-ciliate on the margins and usually also on the back; outer bracts ovate to oblong, often obtuse or rounded at the apex; flowers light yellow, with purplish veins, turning pinkish in age. *T. pumilum* Nutt. Hills: Mont.—N.D.—Colo. *Submont.* —*Alp.* Jl–Au.

2. **A. scorzoneraefòlia** (Schrad.) Greene. Leaves oblanceolate or linear-oblanceolate, scarcely glaucous, glabrous, 1–3 dm. long, usually acute, entire, or rarely dentate; scape villous under the head; involucres 2–3 cm. high, 2–3.5 cm. wide; bracts lanceolate or the inner linear-lanceolate, acute, villous-ciliate at least on the margins; flowers light yellow, with purplish veins, turning pinkish in age. *T. glaucum dasycephalum* T. & G. Hillsides: Alta.—S.D.—Colo.—Nev. —Ore.—B.C. *Plain—Submont.* Jl–Au.

3. **A. glaùca** (Nutt.) Greene. Leaves 1–2 dm. long, glaucous, not two-ranked, rarely pinnatifid; scape 2–4 dm. high, glabrous; involucres turbinate or turbinate-campanulate; bracts lanceolate or linear-lanceolate, acute; corollas light yellow, turning pinkish. (?) *A. vicinalis* Greene. *T. glaucum* Nutt. Prairies and meadows: Man.—S.D.—Colo.—Utah—Wash.—B.C. *Plain—Mont.* Je–Au.

4. **A. parviflòra** (Nutt.) Greene. Leaves 1–1.5 dm. long, 2–5 mm. wide; more or less distinctly 2-ranked; scapes decumbent at the base, 1–2 dm. high,

glabrous; involucres turbinate, about 15 mm. high; bracts linear-lanceolate; corollas yellow, turning pink in drying. *T. parviflorum* Nutt. Meadows: Man. —N.D.—Colo.—Utah—Wash. *Plain—Submont.* Je–Au.

19. TARÁXACUM (Haller) Ludw. DANDELION.

Acaulescent perennial herbs, with pinnatifid fleshy taproots, toothed or rarely entire leaves, and the heads solitary on naked hollow scapes. Involucre campanulate; bracts equal, subtended by a well-developed calyculum of shorter bracts, the calyculate bracts in several series. Receptacle naked. Flowers yellow. Achenes mostly fusiform, 4- or 5-angled, 8–10-ribbed, retrorsely spinulose above, contracted into a slender beak. Pappus of numerous persistent bristles spreading in fruit. *Leontodon* L., in part.

Achenes bright red; terminal lobe of the leaves small.   1. *T. erythrospermum.*
Achenes greenish or brownish yellow; bracts numerous;
    leaves broad and the terminal lobe large.   2. *T. officinale.*

1. **T. erythrospermum** Andrz. Leaves 1–2 dm long, oblanceolate in outline, deeply runcinate-pinnatifid to near the midrib; lobes with a triangular base, more or less caudate lanceolate; scape 1–3 dm. high; involucres about 15 mm. high; inner bracts 12–20, linear, with a lanceolate, slightly scarious margined base; achenes 3 mm. long, bright red; ridges spinulose above and muricate below to the very base. *L. erythrospermum* Eichw. Roadsides and around dwellings: Me.—W.Va.– Kans. –Wyo.—Alta.; nat. from Eur. My–Au.

2. **T. officinale** Weber. Leaves 1–3 dm. long, oblanceolate in outline, usually deeply runcinate, with triangular, more or less cut-toothed lobes, the terminal division large, deltoid or deltoid-ovate; scape 1–3 dm. high; involucres about 15 mm. high; inner bracts linear-lanceolate, 15–25, with narrow scarious margins, rarely corniculate; achenes greenish or brownish yellow, the ridges spinuloso above, then muricate and nearly smooth at the base. *L. Taraxacum* L. [BB, R]. *T. Taraxacum* Karst. [B]. Around dwellings, fields, and roadsides: Lab.—S.C.—Calif.–Alaska; nat. from Eur. *Temp.—Mont.* My–N.

20. PYRRHOPAPPUS DC. FALSE DANDELION.

Biennial or perennial herbs. Leaves toothed, pinnatifid or entire, sometimes basal. Heads long-peduncled. Involucres double, inner bracts several, narrow, slightly united, the outer ones much smaller. Flowers yellow, the ligules toothed. Anthers sagittate at the base. Style-branches stout, obtuse. Achenes fusiform, beaked, 5-ribbed, pubescent or scabrous. Pappus of brown capillary bristles, and a ring of short hairs at the base. *Sitilias* Raf.

Plant leafy-stemmed, glabrous or nearly so.   1. *P. carolinianus.*
Plant scapose, with a single head, hirsute or pubescent.   2. *P. grandiflorus.*

1. **P. carolianus** (Walt.) DC. Stem erect, 1–1.5 dm. high, branched; leaves deeply pinnatifid to nearly entire, inner bracts 12–25 mm. long, narrowly linear, the outer narrower, one-third to one-half as long; achenes 5 mm. long, reddish. *S. caroliniana* Raf. [B]. Dry soil: Del.—Kans.—Tex.—Fla. *Allegh.—Austral.* Ap–Jl.

2. **P. grandiflorus** Nutt. Perennial, with a tuberous root; scape 1–3 dm. high, erect or ascending; leaves basal, pinnatifid, 3–20 cm. long; inner bracts narrowly linear, the outer less than half as long; achenes 4 mm. long, brown. *P. scaposus* DC. *Sitilias grandiflora* Greene [B]. Prairies: Kans.— Tex.—Ark. *Texan.* Ap–Je.

21. LACTÙCA (Tourn.) L. LETTUCE.

Tall leafy-stemmed herbs, with paniculate heads. Involucres cylindric, or in fruit conic; bracts imbricate, in 3 or more series. Achenes obcompressed,

1–5-nerved on the faces, contracted into a beak dilated at the apex. Pappus of numerous capillary bristles, which fall off separately.

Achenes with a slender beak; pappus white.
  Outer bracts (calyculum) not more than half as long as the bracts proper; flowers yellow, rarely tinged with blue.
    Heads 6–8-flowered; achenes several-nerved, not rugose; leaves spinulose on the ribs.
      Leaves sinuate-dentate; achenes dark.   1. *L. virosa.*
      Leaves sinuately pinnatifid; achenes light colored.   2. *L. Scariola.*
    Heads 12–20-flowered; achenes 1–3-nerved, transversely rugose.
      Leaves glabrous.
        Lower leaves sinuate-pinnatifid, thin, pale beneath; body of the achenes equaling the beak.   3. *L. canadensis.*
        Lower leaves, as well as the upper, entire, thick, sometimes spinulose on the margins; body of the achenes longer than the beak.   4. *L. sagittifolia.*
      Leaves hirsute or setose on the midrib beneath.
        Stem leafy below the middle; leaves not spinulose-toothed; bracts of the inflorescence minute.   5. *L. hirsuta.*
        Stem leafy throughout; leaves spinulose-toothed; bracts of the inflorescence somewhat foliaceous.
          Flowers yellow.   6. *L. ludoviciana.*
          Flowers blue.   7. *L. campestris.*
  Bracts in 5 or 6 series, gradually increasing upwards, a distinct calyculum therefore not evident; flowers blue.   8. *L. pulchella.*
Achenes beakless or with a broad neck.
  Pappus white.
    Leaves toothed; achenes without a neck.   9. *L. villosa.*
    Leaves pinnatifid; achenes with a short neck.   10. *L. floridana.*
  Pappus brown.   11. *L. spicata.*

1. **L. viròsa** L. Biennial, with a branched root; stem erect, 5–20 dm. high, more or less hispid towards the base; leaves numerous, oblong or oblanceolate, obtuse or mucronate, horizontal, 1–3 dm. long, spinulose-denticulate, sessile and clasping; panicle loosely branched; involucres 10–12 mm. high, 4–6 mm. broad; bracts in about 4 series, lanceolate, thick; flowers light yellow, turning bluish; achenes with broader margins than those of the next. *L. integrata* A. Nels. Waste places and fields: Me.—Ga.—Calif.—B.C.; nat. or adv. from Eur. Je–S.

2. **L. Scariola** L. Biennial, with a branched root; stem 3–10 dm. high, sparingly prickly-bristly below; leaves pinnatifid or lobed, rarely merely sinuate, tending to turn edgewise into a vertical position; involucres ovoid-cylindric, about 10 mm. high, 4 mm. broad; bracts imbricate in about 4 series, lanceolate or the inner linear-lanceolate; flowers yellow, turning bluish; achenes elliptic-oblanceolate, slightly margined, 5-nerved on each side. Fields and waste places: Mass.—Tenn.—Calif.—B.C.; nat. from Eur. Jl–Au.

3. **L. canadénsis** L. Biennial; stem 1–3 m. high, often mottled, glabrous or nearly so; basal leaves mostly spatulate or oblong, dentate or pinnatifid; stem-leaves 1–3 dm. long, sinuately pinnatifid; involucres 10–12 mm. high; outer bracts lanceolate, inner linear; flowers yellow; achenes oval-oblong, almost black, transversely rugose, 3-nerved. Moist places: N.S.—Fla.—Colo.—Alta. *E. Temp.—Plain—Submont.*

4. **L. sagittifòlia** Ell. Biennial; stem 1–3 m. high; leaves obovate or oblanceolate, or the upper lanceolate, 1–3 dm. long, irregularly dentate, narrowed below, but partly clasping; involucre 1–2 cm. high; inner bracts linear or lance-linear; flowers yellow or reddish; achenes 5 mm. long, strongly ribbed, not margined. Banks: N.B.—Minn.—Kans.—Ga. *Canad.—Allegh.* Jl–S.

5. **L. hirsùta** Muhl. Biennial, more or less hirsute; stem 6–12 dm. high; leaves 1–2 dm. long, sinuate-pinnatifid, hirsute at the base; involucre 12–15

mm. high, often purplish; bracts lanceolate to linear, the inner scarious-margined; flowers white to purplish-yellow; achenes 3–5 mm. long, shorter than the beak. Dry soil: Me.—Minn.—Tex.—Ala. *Canad.—Allegh.* Jl–S.

6. L. ludoviciàna (Nutt.) DC. *Fig. 599.* Biennial; stem 5–15 dm. high, glabrous; leaves oblong-oblanceolate in outline, sinuately lobed or pinnatifid, with rounded, ovate or lanceolate, spinulose-denticulate lobes; involucres 15–20 mm. high, outer bracts ovate, the inner linear-lanceolate, scarious-margined; flowers yellow; achenes brown or black, 4 mm. long, broadly oval or obovate, 3-ribbed, obscurely transversely rugose. River-banks and wet places: Man.—Mo.—Tex.—Colo.—Mont. *Prairie—Submont.* Jl–S.

f.599.

7. L. campéstris Greene. Very similar to *L. ludoviciana,* but flowers blue. River banks and wet places: Minn.—Okla.—Wash. *Prairie—Submont.* Jl–S.

8. L. pulchélla (Pursh) DC. Perennial, with a deep rootstock; stem 3–10 dm. high, glabrous, leaves linear-lanceolate, lanceolate, or oblong, acute, entire, dentate or even some of them pinnatifid, 5–20 cm. long, somewhat glaucous; panicle usually narrow; involucres 16–20 mm. high; flowers blue; achenes oblong-lanceolate, about 4 mm. long, rather strongly 3-nerved on the faces, gradually tapering into the short beak. *L. sylvatica* A. Nels. Wet meadows: Sask.—Mo.—N.M.—Calif.—B.C. *Prairie—Submont.* My Au.

9. L. villòsa Jacq. Annual; stem slender, 1–2 dm. high, glabrous; leaves oval to oblong-lanceolate, 8–20 cm. long, acuminate, narrowed into the winged petiole; involucre 10–13 mm. high; bracts linear-lanceolate, purplish; flowers blue; achenes 4 mm. long, thick-margined, 3-ribbed on each face. Banks and thickets: N.Y.—Neb.—Ky.—Fla. *Allegh.—Austral.* Jl–S.

10. L. floridàna (L.) Gaertner. Annual; stem stout, 1–3 m. high, often mottled; leaves lyrate-pinnatifid, 3.5–4.5 dm. long; involucre 10–12 mm. high; bracts deep green or purplish; flowers blue; achenes 6–7 mm. long, narrowed into a neck. Hillsides and open ground: N.Y.—Minn.—Kans.—La.—Fla. *Allegh.—Austral.* Jl–S.

11. L. spicàta (Lam.) Hitchc. Annual or biennial; stem glabrous, 5–35 dm. high; leaves deeply pinnatifid, more or less hispid on the veins beneath; lobes somewhat recurved; involucres campanulate, about 1 cm. high; bracts imbricate in about 4 series, lanceolate; flowers blue to nearly white; achenes elliptic, 5-nerved. *L. leucophaea* (Willd.) A. Gray. Moist ground: Newf.—N.C.—Colo.—Ida.—Man. *Canad.—Allegh.—Mont.* Jl–O.

22. SÓNCHUS (Tourn.) L. Sow Thistle.

Leafy-stemmed, mostly glaucous herbs, with cymose or umbellate heads. Involucre campanulate; bracts few, thin, with many shorter ones at the base. Flowers yellow. Achenes obcompressed, ribbed, not beaked. Pappus of numerous white, soft, fine bristles, mainly falling off together.

Perennial; achenes slightly compressed; heads about 2 cm. high.
Involucre glandular-pubescent.   1. *S. arvensis.*
Involucre glabrous.   2. *S. uliginosus.*
Annual; achenes strongly compressed; involucres usually glabrous; heads about 15 mm. high.
Auricles of the leaves acute; achenes transversely wrinkled.   3. *S. oleraceus.*
Auricles of the leaves rounded; achenes not transversely wrinkled.   4. *S. asper.*

1. **S. arvénsis** L. Stem 5–12 dm. high, leafy below; lower leaves run-cinate-pinnatifid, spinulose-dentate, with short petioles, glabrous; upper leaves lanceolate, sessile, clasping, undivided or pinnatifid; bracts linear-lanceolate, in about 3 series; achenes oblong, slightly flattened, with thick ribs, transversely rugose. Fields and roadsides: Newf.—N.J.—Utah—Ida.—B.C.; nat. from Eur. Jl–O.

2. **S. uliginòsus** Bieb. Perennial, with a creeping rootstock; stem 4–10 dm. high, leafy, glabrous, sulcate; lower leaves more or less runcinate, lobed or merely toothed, oblanceolate in outline, 1.5–2 dm. long, spinulose-dentate, glabrous, tapering towards the base; upper stem-leaves lanceolate, clasping, undivided, with rounded basal auricles; involucre 15–20 mm. broad; bracts linear-lanceolate, glabrous, in about 4 series; achenes oblong, turgid, 2.5 mm. long, with thick, transversely rugose ribs. Fields: N.J.—Pa.—Minn.—N.D.; nat. from Eur. Au–S.

3. **S. oleràceus** L. Stem 5–30 dm. high, glabrous; lower leaves petioled, lyrate-pinnatifid, 1–2.5 dm. long; lobes lanceolate; spinulose-dentate, or the terminal one large and triangular or triangular-hastate; upper leaves similar but auriculate-clasping; involucral bracts linear-lanceolate, imbricate in about 4 series; achenes 2.5 mm. long, oblong, 3-ribbed. Fields and waste places: N.S.—Fla.—Calif.—Wash.; Mex., C.Am., and S.Am.; nat. from Eur. My–N.

4. **S. ásper** (L.) All. *Fig. 600.* Stem 5–30 dm. high, glabrous; lower leaves obovate or spatulate-petioled, the upper oblong or lanceolate, auriculate-clasping, undivided to deeply runcinate-pinnatifid, with broad lobes, spinulose-denticulate; bracts linear-lanceolate, imbricate in about 4 series, the outer with the midribs more or less thickened in age; achenes oblong, 3-ribbed, about 3 mm. long. Waste places and fields: N.S.—Fla.—Calif.—B.C.; W.Ind. and Mex.; nat. from Eur. My–N.

Appendices and Index

SUMMARY

Families	Genera	Species
Ophioglossaceae	2	9
Osmundaceae	1	3
Polypodiaceae	17	56
Marsileaceae	1	1
Salviniaceae	1	1
Equisetaceae	1	13
Isoëtaceae	1	4
Lycopodiaceae	1	11
Selaginellaceae	1	3
PTERIDOPHYTA	26	101
Pinaceae	7	12
Juniperaceae	3	6
Taxaceae	1	1
GYMNOSPERMAE	11	19
Typhaceae	1	2
Sparganiaceae	1	10
Zannichelliaceae	3	33
Najadaceae	1	3
Scheuchzeriaceae	2	3
Alismaceae	5	17
Elodeaceae	2	4
Hydrocharitaceae	1	1
Poaceae	89	388
Cyperaceae	14	256
Araceae	5	5
Lemnaceae	3	9
Xyridaceae	1	1
Eriocaulaceae	1	1
Commelinaceae	2	13
Pontederiaceae	3	4
Melanthiaceae	6	11
Juncaceae	2	38
Alliaceae	4	14
Liliaceae	8	15
Convallariaceae	9	16
Dracaenaceae	1	2
Calochortaceae	1	2
Trilliaceae	2	10
Smilacaceae	2	9
Amaryllidaceae	3	3
Dioscoreaceae	1	1
Iridaceae	4	11
Orchidaceae	26	54
MONOCOTYLEDONES	203	936
Saururaceae	1	1
Salicaceae	2	50
Myricaceae	2	2
Juglandaceae	2	8
Betulaceae	2	14
Corylaceae	3	4
Fagaceae	2	17
Ulmaceae	2	11
Moraceae	2	4
Cannabinaceae	2	3
Urticaceae	5	12
Polygonaceae	11	81
Chenopodiaceae	16	62
Amaranthaceae	5	16
Nyctaginaceae	3	12
Phytolaccaceae	1	1
Tetragoniaceae	2	2
Portulacaceae	5	13
Corrigiolaceae	3	8

Families	Genera	Species
Alsinaceae	9	34
Caryophyllaceae	9	23
Ceratophyllaceae	1	1
Nymphaeaceae	2	8
Nelumbonaceae	1	1
Cabombaceae	1	1
Ranunculaceae	22	85
Podophyllaceae	3	3
Berberidaceae	2	3
Menispermaceae	3	3
Lauraceae	2	2
Anonaceae	1	1
Papaveraceae	4	8
Fumariaceae	4	11
Brassicaceae	45	136
Capparidaceae	4	6
Droseraceae	1	4
Sarraceniaceae	1	1
Podostemonaceae	1	1
Crassulaceae	2	6
Penthoraceae	1	1
Parnassiaceae	1	4
Saxifragaceae	12	22
Hydrangeaceae	2	3
Grossulariaceae	4	16
Plantanaceae	1	1
Hamamelidaceae	1	1
Rosaceae	27	142
Malaceae	6	36
Amygdalaceae	1	19
Mimosaceae	4	6
Caesalpiniaceae	6	9
Krameriaceae	1	1
Fabaceae	54	222
Geraniaceae	2	10
Oxalidaceae	3	7
Linaceae	2	9
Balsaminaceae	1	4
Limnanthaceae	1	1
Zygophyllaceae	2	3
Rutaceae	2	2
Simarubaceae	1	1
Polygalaceae	1	8
Euphorbiaceae	13	49
Callitrichaceae	1	3
Celastraceae	2	2
Aquifoliaceae	2	3
Empetraceae	1	1
Anacardiaceae	2	16
Staphyleaceae	1	1
Hippocastanaceae	1	3
Aceraceae	2	8
Sapindaceae	2	2
Rhamnaceae	2	10
Vitaceae	5	14
Tiliaceae	1	1
Malvaceae	11	24
Hypericaceae	4	17
Elatinaceae	2	4
Tamaricaceae	1	1
Cistaceae	3	7
Violaceae	3	37
Loasaceae	3	6
Passifloraceae	1	2
Cactaceae	4	15
Thymelaeaceae	1	1
Elaeagnaceae	2	4
Lythraceae	6	8
Melastomataceae	1	2

Families	Genera	Species	Families	Genera	Species
Onagraceae	20	78	Martyniaceae	1	1
Haloragidaceae	3	7	Bignoniaceae	2	2
Araliaceae	2	5	Acanthaceae	3	4
Ammiaceae	44	74	Phrymaceae	1	1
Cornaceae	3	11	Plantaginaceae	2	16
Pyrolaceae	5	12	Rubiaceae	6	25
Monotropaceae	3	3	Caprifoliaceae	9	33
Ericaceae	12	16	Adoxaceae	1	1
Vacciniaceae	8	19	Santalaceae	1	4
Primulaceae	11	30	Loranthaceae	1	2
Ebenaceae	1	1	Aristolochiaceae	2	5
Sapotaceae	1	2	Cucurbitaceae	4	4
Oleaceae	4	9	Campanulaceae	3	9
Gentianaceae	9	27	Lobeliaceae	1	10
Menyanthaceae	2	2	Valerianaceae	2	8
Apocynaceae	0	12	Dipsaceae	2	2
Asclepiadaceae	6	30	Ambrosiaceae	5	25
Convolvulaceae	5	19	Carduaceae	106	542
Cuscutaceae	1	13	Cichoriaceae	22	65
Polemoniaceae	8	33			
Hydrophyllaceae	4	12	DICOTYLEDONES	826	2932
Heliotropiaceae	4	5			
Boraginaceae	16	60	Pteridophyta	26	101
Verbenaceae	2	12	Gymnospermae	11	19
Lamiaceae	25	90	Angiospermae—		
Solanaceae	12	44	Monocotyledones	203	936
Scrophulariaceae	29	107	Dicotyledones	826	2932
Lentibulariaceae	4	7			
Orobanchaceae	3	5	Total	1066	3988

ABBREVIATIONS OF AUTHORS' NAMES*

Adans.   Adanson, Michel, 1727–1806.
Agardh.   Agardh, Carl Adolf, 1785–1859.
Ait.   Aiton, William, 1731–1793.
Ait. f.   Aiton, William Townsend, 1766–1849.
All.   Allioni, Carlo, 1728–1804.
Ames.   Ames, Oakes, 1874–
Anders.   Andersson, Nils Johan, 1821–1880.
Andr.   Andrews, Henry C., 17——18—.
Andrz.   Andrzejowski, Anton Lukianowicz, 1784–1868.
Angstr.   Angström, Johan, 1813–1879.
Antoine.   Antoine, Franz, 1815–1886.
Ard.   Arduino, Pietro, 1728–1805.
Arn.   Arnott, George Arnott Walker, 1799–1868.
Arth.   Arthur, Joseph Charles, 1850–
Aschers.   Ascherson, Paul Friedrich August, 1834–1913.
Aschers. & Graebn.   Ascherson, Paul Friedrich August, 1834–1913; Graebner, Karl Otto Robert Peter Paul, 1871–
Ashe.   Ashe, William Willard, 1872–

B.S.P.   Britton, Nathaniel Lord, 1859–; Sterns, Emerson Ellick, 1846–1926; Poggenburg, Justus Ferdinand, 1840–1893.
Bailey.   Bailey, Liberty Hyde, 1858–
Balbis.   Balbis, Giovanni Battista, 1765–1831.
Baldw.   Baldwin, William 1779–1819.
Balf.   Balfour, John Hutton, 1808–1884.
Ball.   Ball, Carleton Roy, 1873–
Banks.   Banks, Joseph, 1743–1820.
Barneoud.   Barneoud, François Marius, 1821–18—.
Barnh.   Barnhart, John Hendley, 1871–
Barratt.   Barratt, Joseph, 1796–1882.
Bart.   Barton, Benjamin Smith, 1766–1815.
Bart., W.   Barton, William Paul Crillon. 1786–1856.
Bartl.   Bartling, Friedrich Gottlieb, 1798–1875.
Batchelder.   Batchelder, Frederick William, 1838–1911.
Bates.   Bates, John Mallery, 1846–1930.
Bauhin, C.   Bauhin, Gaspard (Caspar), 1560–1624.
Baxter.   Baxter, William, 1787–1871.
Beal.   Beal, William James, 1833–1924.
Beauv.   Palisot de Beauvois, Ambroise Marie François Joseph, 1752–1820.
Bebb.   Bebb, Michael Schuck, 1833–1895.
Beck.   Beck, Lewis Caleb, 1798–1853.
Beeby.   Beeby, William Haddon, 1849–1910.
Benn.   Bennett, Arthur, 1843–1929.
Benth.   Bentham, George, 1800–1884.
Benth. & Hook.   Bentham, George, 1800–1884; Hooker, Joseph Dalton, 1817–1911.
Berch. & Presl.   Berchtold, Friedrich von, 1781–1876; Presl, Jan Swatopluk, 1791–1849.
Berg.   Bergius, Peter Jonas, 1730–1790.
Bernh.   Bernhardi, Johann Jacob, 1774–1850.
Bertero.   Bertero, Carlo Giuseppe, 1789–1831.
Besser.   Besser, Wilibald Swibert Joseph Gottlieb, 1784–1842.
Best.   Best, George Newton, 1846–1926.
Beurl.   Beurling, Pehr Johan, 1800–1866.
Beyer.   Beyer, Rudolf, 1852–
Beyr.   Beyrich, Heinrich Karl, 1796–1834.
Bickn.   Bicknell, Eugene Pintard, 1859–1925.
Bieb.   Marschall von Bieberstein, Friedrich August, 1768–1826.
Bigel.   Bigelow, Jacob, 1787–1879.
Bisch.   Bischoff, Gottlieb Wilhelm, 1797–1854.
Bissell.   Bissell, Charles Humphrey, 1857–1925.
Björnstr.   Björnström, Fredrik Johan, 1833–1889.
Blake.   Blake, Sidney Fay, 1892–
Blanchard.   Blanchard, William Henry, 1850–1922.
Blankinship.   Blankinship, Joseph William, 1862–
Blume.   Blume, Carl Ludwig von, 1796–1862.
Blytt.   Blytt, Mathias Numsen, 1789–1862.
Boehm.   Boehmer, George Rudolf, 1723–1803.
Boiss.   Boissier, Pierre Edmond, 1810–1885.
Bolton.   Bolton, James, 17——1799.
Booth.   Booth, William Beattie, 1804–1874.
Boott.   Boott, Francis, 1792–1863.
Boott, W.   Boott, William, 1805–1887.

* Prepared by Dr. J. H. Barnhart.

Borkh.	Borkhausen, Moriz Balthasar, 1760–1806.
Bory.	Bory de Saint-Vincent, Jean Baptiste Georges Marcellin, 1778–1846.
Bosc.	Bosc, Louis Augustin Guillaume, 1759–1828.
Bosch.	Bosch, Roelof Benjamin van den, 1810–1862.
Bourgeau.	Bourgeau, Eugène, 1813–1877.
Boynton.	Boynton, Frank Ellis, 1859–
Br., A.	Braun, Alexander Carl Heinrich, 1805–1877.
Br., P.	Browne, Patrick, 1720–1790.
Br., R.	Brown, Robert, 1773–1858.
Brack.	Brackenridge, William Dunlop, 1810–1893.
Brainerd.	Brainerd, Ezra, 1844–1924.
Brand, A.	Brand, August, 1863–
Brewer.	Brewer, William Henry, 1828–1910.
Brign.	De Brignoli di Brunnhoff, Giovanni, 1774–1857.
Briq.	Briquet, John Isaac, 1870–1931.
Britt.	Britton, Nathaniel Lord, 1859–
Britt. & Br	Britton, Nathaniel Lord, 1859–; Brown, Addison, 1830–1913.
Britt. & Rose.	Britton, Nathaniel Lord, 1859–; Rose, Joseph Nelson, 1862–1928.
Britt. & Rusby.	Britton, Nathaniel Lord, 1859–, Rusby, Henry Hurd, 1855–
Brot.	Brotero, Felix d'Avellar, 1744–1828.
Brown, S.	Brown, Stewardson, 1867–1921.
Buch.	Buchenau, Franz Georg Philipp, 1831–1906.
Buckl.	Buckley, Samuel Botsford, 1809–1884.
Bunge.	Bunge, Alexander von, 1803–1890.
Burgess.	Burgess, Edward Sandford, 1855–1928.
Burgsd.	Burgsdorf, Friedrich August Ludwig von, 1747–1802.
Burm.	Burman, Johannes, 1706–1779.
Bush.	Bush, Benjamin Franklin, 1858–
Butler.	Butler, Bertram Theodore, 1872–
Butters & St. John.	Butters, Frederic King, 1878–; St. John, Harold, 1892–
Canby.	Canby, William Marriott, 1831–1904.
Canby & Rose.	Canby, William Marriott, 1831–1904; Rose, Joseph Nelson, 1862–1928.
Carey.	Carey, John, 1797–1880.
Carr.	Carrière, Elie Abel, 1818–1896.
Casp.	Caspary, Johann Xaver Robert, 1818–1887.
Cass.	Cassini, Alexandre Henri Gabriel, 1781–1832.
Cassidy.	Cassidy, James, 18——1889.
Catesby.	Catesby, Mark, 1679–1749.
Cav.	Cavanilles, Antonio José, 1745–1804.
Celak.	Celakovsky, Ladislav Josef, 1834–1902.
Cerv.	Cervantes, Vicente, 1759–1829.
Chab.	Chabert, Alfred, 1836–1916.
Chaix.	Chaix, Dominique, 1730–1799.
Cham.	Chamisso, Ludolf Adalbert von, 1781–1838.
Cham. & Schlecht.	Chamisso, Ludolf Adalbert von, 1781–1838; Schlechtendal, Diederich Franz Leonhard von, 1794–1866.
Chapm.	Chapman, Alvan Wentworth, 1809–1899.
Chase.	Chase, Mary Agnes (Merrill), 1869–
Chatelain.	Chatelain, Jean Jacques, 17——17—.
Chev.	Chevallier, François Fulgis, 1796–1840.
Choisy.	Choisy, Jacques Denis, 1799–1859.
Chr., C.	Christensen, Carl Frederik Albert, 1872–
Clarke.	Clarke, Charles Baron, 1832–1906.
Clayton.	Clayton, John, 1685–1773.
Cockerell.	Cockerell, Theodore Dru Alison, 1866–
Cohn.	Cohn, Ferdinand Julius, 1828–1898.
Coleman.	Coleman, Nathan, 1825–1887.
Coss. & Germ.	Cosson, Ernest Saint-Charles, 1819–1889; Germain de Saint-Pierre, Jacques Nicolas Ernest, 1815–1882.
Cosson.	Cosson, Ernest Saint-Charles, 1819–1889.
Coult.	Coulter, John Merle, 1851–1928.
Coult. & Evans.	Coulter, John Merle, 1851–1928; Evans, Walter Harrison, 1863–
Coult. & Fish.	Coulter, John Merle, 1851–1928; Fisher, Elmon McLean, 1861–
Coult. & Nels.	Coulter, John Merle, 1851–1928; Nelson, Aven, 1859–
Coult. & Rose.	Coulter, John Merle, 1851–1928; Rose, Joseph Nelson, 1862–1928.
Cov.	Coville, Frederick Vernon, 1867–
Cov. & Britt.	Coville, Frederick Vernon, 1867–; Britton, Nathaniel Lord, 1859–
Crantz.	Crantz, Heinrich Johann Nepomuk von, 1722–1797.
Crép.	Crépin, François, 1830–1903.
Curtis.	Curtis, William, 1746–1799.
Curtis, M. A.	Curtis, Moses Ashley, 1808–1872.
Cyrill.	Cirillo (Cyrillus), Domenico Maria Leone, 1739–1799.
Daniels.	Daniels, Francis Potter, 1869–
Darby.	Darby, John, 1804–1877.

Darl.	Darlington, William, 1782–1863.
DC.	Candolle, Augustin Pyramus de, 1778–1841.
DC., A.	Candolle, Alphonse Louis Pierre Pyramus de, 1806–1893.
Decne.	Decaisne, Joseph, 1807–1882.
Desf.	Desfontaines, Réné Louiche, 1750–1833.
DesMoul.	DesMoulins, Charles Robert Alexandre, 1798–1875.
Desv.	Desvaux, Nicaise Auguste, 1784–1856.
Dewey.	Dewey, Chester, 1784–1867.
Dickson.	Dickson, James, 1738–1822.
Diels.	Diels, Friedrich Ludwig Emil, 1874–
Dietr.	Dietrich, Friedrich Gottlieb, 1768–1850.
Dill.	Dillenius, John James, 1684–1747.
Dode.	Dode, Louis Albert, 1875–
Domin.	Domin, Karel, 1882–
Don, D.	Don, David, 1799–1841.
Don, G.	Don, George, 1798–1856.
Donn.	Donn, James, 1758–1813.
Dougl.	Douglas, David, 1798–1834.
Drejer.	Drejer, Salomon Thomas Nicolai, 1813–1842.
Druce.	Druce, George Claridge, 1851–
Drude.	Drude, Carl Georg Oscar, 1852–
Duby.	Duby, Jean Etienne, 1798–1885.
Duchesne.	Duchesne, Antoine Nicolas, 1747–1827.
Dufr.	Dufresne, Pierre, 1786–1836.
Duham.	Duhamel du Monceau, Henri Louis, 1700–1781.
Dulac.	Dulac, Joseph, 18— –189–.
Dum.	Dumortier, Barthélemy Charles Joseph, 1797–1878.
Dum.-Cours.	Dumont de Courset, Georges Louis Marie, 1746–1824.
Dunal.	Dunal, Michel Félix, 1789–1856.
Durand.	Durand, Elie Magloire, 1794–1873.
Durand, T.	Durand, Théophile Alexis, 1855–1912.
Durand & Jacks.	Durand, Théophile Alexis, 1855–1912; Jackson, Benjamin Daydon, 1846–1927.
Durieu.	Durieu de Maisonneuve, Michel Charles, 1797–1878.
DuRoi.	DuRoi, Johann Philipp, 1741–1785.
Eastw.	Eastwood, Alice, 1859–
Eat.	Eaton, Amos, 1776–1842.
Eat., A. A.	Eaton, Alvah Augustus, 1865–1908.
Eat., D. C.	Eaton, Daniel Cady, 1834–1895.
Eat. & Wright.	Eaton, Amos, 1776–1842; Wright, John, 1811–1846.
Eggert.	Eggert, Heinrich Karl Daniel, 1841–1904.
Ehrh.	Ehrhart, Friedrich, 1742–1795.
Eichw.	Eichwald, Karl Eduard von, 1794–1876.
Ell.	Elliott, Stephen, 1771–1830.
Ellis.	Ellis, John, 1710–1776.
Endl.	Endlicher, Stephan Ladislaus, 1804–1849.
Engelm.	Engelmann, George, 1809-1884.
Engelm. & Gray.	Engelmann, George, 1809–1884; Gray, Asa, 1810–1888.
Eschsch.	Eschscholtz, Johann Friedrich Gustav von, 1793–1831.
Evans, W. H.	Evans, Walter Harrison, 1863–
Fabr.	Fabricius, Philipp Conrad, 1714–1774.
Farw.	Farwell, Oliver Atkins, 1867–
Fassett.	Fassett, Norman Carter, 1900–
Fedde.	Fedde, Friedrich Karl Georg, 1873–
Fée.	Fée, Antoine Laurent Apollinaire, 1789–1874.
Fenzl.	Fenzl, Eduard, 1808–1879.
Fern.	Fernald, Merritt Lyndon, 1873–
Fern. & Brack.	Fernald, Merritt Lyndon, 1873– ; Brackett, Amelia Ellen, 1896–1926.
Fern. & Eames.	Fernald, Merritt Lyndon, 1873– ; Eames, Arthur Johnson, 1881–
Fern. & Rydb.	Fernald, Merritt Lyndon, 1873– ; Rydberg, Per Axel, 1860–1931.
Fern. & Wieg.	Fernald, Merritt Lyndon, 1873– ; Wiegand, Karl McKay, 1873–
Fisch.	Fischer, Friedrich Ernst Ludwig von, 1782–1854.
Fisch. & Avé-Lall.	Fischer, Friedrich Ernst Ludwig von, 1782–1854; Avé-Lallement, Julius Leopold Eduard, 1803–1867.
Fisch. & Mey.	Fischer, Friedrich Ernst Ludwig von, 1782–1854; Meyer, Carl Anton von, 1795–1855.
Fisch. & Trautv.	Fischer, Friedrich Ernst Ludwig von, 1782–1854; Trautvetter, Ernst Rudolf von, 1809–1889.
Flügge.	Flügge, Johann, 1775–1816.
Focke.	Focke, Wilhelm Olbers, 1834–1922.
Forbes.	Forbes, James, 1773–1861.
Forsk.	Forskål, Pehr, 1732–1763.
Forst., T. F.	Forster, Thomas Furley, 1761–1825.
Foug.	Fougeroux de Bondaroy, Auguste Denis, 1732–1789.
Fourn.	Fournier, Eugène Pierre Nicolas, 1834–1884.

Fourr.   **Fourreau**, Pierre Jules, 1844–1871.
Frank.   **Frank**, Joseph C., 1782–1835.
Fraser.   **Fraser**, John, 1750–1811.
Fresen.   **Fresenius**, Johann Baptist Georg Wolfgang, 1808–1866.
Fries.   **Fries**, Elias Magnus, 1794–1878.
Fritsch.   **Fritsch**, Karl, 1864–
Fröl.   **Frölich**, Joseph Aloys von, 1766–1841.

Gaertn.   **Gaertner**, Joseph, 1732–1791
Gand.   **Gandoger**, Michel, 1850–1926.
Garcke.   **Garcke**, Friedrich August, 1819–1904.
Gates, R. R.   **Gates**, Reginald Ruggles, 1882–
Gatt.   **Gattinger**, Augustin, 1825–1903.
Gaud.   **Gaudichaud-Beaupré**, Charles, 1789–1854.
Gaudin.   **Gaudin**, Jean Francois Gottlieb Philippe, 1766–1833.
Gay, J.   **Gay**, Jacques Etienne, 1786–1864.
Gesn.   **Gesner**, Conrad, 1510–1565.
Geyer.   **Geyer**, Carl Andreas, 1809–1853.
Gilib.   **Gilibert**, Jean Emmanuel, 1741–1814.
Glatf.   **Glatfelter**, Noah Miller, 1837–1911.
Gleason   **Gleason**, Henry Allan, 1882–
Glox.   **Gloxin**, Benjamin Peter, 17––17––.
Gmel.   **Gmelin**, Johann Georg, 1709–1775.
Gmel., C. C.   **Gmelin**, Carl Christian, 1762–1837.
Gmel., J. F.   **Gmelin**, Johann Friedrich, 1748–1804.
Gmel., S. G.   **Gmelin**, Samuel Gottlieb, 1745–1774.
Godr.   **Godron**, Dominique Alexandre, 1807–1880.
Goldie.   **Goldie**, John, 1793–1886.
Good.   **Goodenough**, Samuel, 1743–1827.
Goodding.   **Goodding**, Leslie Newton, 1880–
Gordon.   **Gordon**, George, 1806–1879.
Gouan.   **Gouan**, Antoine, 1733–1821.
Graebner.   **Graebner**, Karl Otto Robert Peter Paul, 1871–
Grah.   **Graham**, Robert, 1786–1845.
Gray, A.   **Gray**, Asa, 1810–1888.
Gray, S. F.   **Gray**, Samuel Frederick, 1766–1830.
Greene.   **Greene**, Edward Lee, 1842–1915.
Greenm.   **Greenman**, Jesse More, 1867
Grev. & Hook.   **Greville**, Robert Kaye, 1794–1866; **Hooker**, William Jackson, 1785–1865.
Griseb.   **Grisebach**, August Heinrich Rudolf, 1814–1879.
Gron.   **Gronovius**, Jan Fredrik, 1690–1762.
Guers.   **Guersent**, L. B., 1776–1848.
Gunn.   **Gunnerus**, Johan Ernst, 1718–1773.
Guss.   **Gussone**, Giovanni, 1787–1866.

H. & A.   **Hooker**, William Jackson, 1785–1865; **Arnott**, George Arnott Walker, 1799–1868.
H. B. K.   **Humboldt**, Friedrich Heinrich Alexander von, 1769–1859; **Bonpland**, Aimé Jacques Alexandre, 1773–1858; **Kunth**, Carl Sigismund, 1788–1850.
Hack.   **Hackel**, Eduard, 1850–1926.
Hall, E.   **Hall**, Elihu, 1822–1882.
Hall & Clements.   **Hall**, Harvey Monroe, 1874–; **Clements**, Frederic Edward, 1874–
Haller.   **Haller**, Albrecht von, 1708–1777.
Hamilt.   **Hamilton**, William, 1783–1856.
Hamilt., A.   **Hamilton**, Arthur.
Hartm.   **Hartman**, Carl Johan, 1790–1849.
Harv. & Gray.   **Harvey**, William Henry, 1811–1866; **Gray**, Asa, 1810–1888.
Haussk.   **Haussknecht**, Heinrich Carl, 1838–1903.
Haw.   **Haworth**, Adrian Hardy, 1768–1833.
Hayne.   **Hayne**, Friedrich Gottlob, 1763–1832.
Heist.   **Heister**, Lorenz, 1683–1758.
Heller.   **Heller**, Amos Arthur, 1867–
Herb.   **Herbert**, William, 1778–1847.
Hieron.   **Hieronymus**, Georg Hans Emmo Wolfgang, 1846–1921.
Hill.   **Hill**, John, 1716–1775.
Hill, E. J.   **Hill**, Ellsworth Jerome, 1833–1917.
Hitchc.   **Hitchcock**, Albert Spear, 1865–
Hitchc., E.   **Hitchcock**, Edward, 1793–1864.
Hitchc. & Chase.   **Hitchcock**, Albert Spear, 1865–; **Chase**, Mary Agnes (Merrill), 1869–
Hoffm.   **Hoffmann**, Georg Franz, 1761–1826.
Hoffmgg.   **Hoffmannsegg**, Johann Centurius von, 1766–1849.
Hogg.   **Hogg**, Thomas, 1777–1855.
Holl. & Britt.   **Hollick**, Charles Arthur, 1857–; **Britton**, Nathaniel Lord, 1859–
Holm.   **Holm**, Herman Theodor, 1854–
Holz.   **Holzinger**, John Michael, 1853–1929.
Hook.   **Hooker**, William Jackson, 1785–1865.

Hook. & Grev.	Hooker, William Jackson, 1785–1865; Greville, Robert Kaye, 1794–1866.
Horkel.	Horkel, Johann, 1769–1846.
Hornem.	Hornemann, Jens Wilken, 1770–1841.
Host.	Host, Nicolaus Thomas, 1761–1834.
Houba.	Houba, Julien, 1843–
House.	House, Homer Doliver, 1878–
Houst.	Houstoun, William, 1695–1733.
Howe.	Howe, Elliot Calvin, 1828–1899.
Howell.	Howell, Thomas, 1842–1912.
Hubbard.	Hubbard, Frederick Tracy, 1875–
Huds.	Hudson, William, 1730–1793.
Humb. & Bonpl.	Humboldt, Friedrich Heinrich Alexander von, 1769–1859; Bonpland, Aimé Jacques Alexandre, 1773–1858.
Huth.	Huth, Ernst, 1845–1897.
Ives.	Ives, Eli, 1779–1861.
Jacq.	Jacquin, Nikolaus Joseph von, 1727–1817.
James.	James, Edwin, 1797–1861.
Jedwabnick.	Jedwabnick, Elisabeth.
Jennings.	Jennings, Otto Emery, 1877–
Johnston, J. R.	Johnston, John Robert, 1880–
Jones, M. E.	Jones, Marcus Eugene, 1852–
Juss.	Jussieu, Antoine Laurent de, 1748–1836.
Juss., A.	Jussieu, Adrien Henri Laurent de, 1797–1853.
Kalm.	Kalm, Pehr, 1716–1779.
Karst.	Karsten, Gustav Karl Wilhelm Hermann, 1817–1908.
Kearney.	Kearney, Thomas Henry, 1874–
Kellogg.	Kellogg, Albert, 1813–1887.
Ker.	Ker, John Bellenden, 1764–1842.
Kirchner.	Kirchner, Emil Otto Oskar von, 1851–1925.
Kirk.	Kirk, Thomas, 1828–1898.
Kit.	Kitaibel, Paul, 1757–1817.
Kl. & Garcke.	Klotzsch, Johann Friedrich, 1805–1860; Garcke, Friedrich August, 1819–1904.
Klatt.	Klatt, Friedrich Wilhelm, 1825–1897.
Klotzsch.	Klotzsch, Johann Friedrich, 1805–1860.
Knerr.	Knerr, Ellsworth Brownell, 1861–
Knight.	Knight, Ora Willis, 1874–1913.
Koch.	Koch, Wilhelm Daniel Joseph, 1771–1849.
Koch, E.	Koch, Karl Heinrich Emil, 1809–1879.
Koehne.	Koehne, Bernhard Adalbert Emil, 1848–1918.
Koeler.	Koeler, Georg Ludwig, 1764–1807.
Kostel.	Kosteletzky, Vincenz Franz, 1801–1887.
Krock.	Krocker, Anton Johann, 1743–1823.
Krok.	Krok, Thorgny Ossian Bolivar Napoleon, 1834–1921.
Kühlewein.	Kühlewein, Paul Edward, 1798–1870.
Kükenth.	Kükenthal, Georg, 1864–
Kuhn.	Kuhn, Maximilian Friedrich Adalbert, 1842–1894.
Kunth.	Kunth, Carl Sigismund, 1788–1850.
Kuntze.	Kuntze, Carl Ernst Otto, 1843–1907.
Kunze.	Kunze, Gustav, 1793–1851.
Kusnez.	Kusnezow, Nicolai Ivanovič, 1864–
L.	Linnaeus, Carl, 1707–1778.
L. f.	Linné, Carl von, 1741–1783.
Labill.	Labillardière, Jacques Julien Houtton de, 1755–1834.
Laest.	Laestadius, Lars Levi, 1800–1861.
Lag.	Lagasca y Segura, Mariano, 1776–1839.
Lam.	Lamarck, Jean Baptiste Antoine Pierre Monnet de, 1744–1829.
Lamb.	Lambert, Aylmer Bourke, 1761–1842.
Lange.	Lange, Johan Martin Christian, 1818–1898.
Leavenw.	Leavenworth, Melines Conklin, 1796–1862.
LeConte.	LeConte, John Eatton, 1784–1860.
Ledeb.	Ledebour, Carl Friedrich von, 1785–1851.
Leggett.	Leggett, William Henry, 1816–1882.
Lehm.	Lehmann, Johann Georg Christian, 1792–1860.
Leiberg.	Leiberg, John Bernhard, 1853–1913.
Lej.	Lejeune, Alexander Louis Simon, 1779–1858.
Lemmon.	Lemmon, John Gill, 1832–1908.
Less.	Lessing, Christian Friedrich, 1809–1862.
Lév.	Léveillé, Augustin Abel Hector, 1863–1918.
Leyss.	Leysser, Friedrich Wilhelm von, 1731–1815.
L'Hér.	L'Héritier de Brutelle, Charles Louis, 1746–1800.
Lightf.	Lightfoot, John, 1735–1788.
Liljebl.	Liljeblad, Samuel, 1761–1815.
Lindb.	Lindberg, Sextus Otto, 1835–1889.

Lindl.	Lindley, John, 1799–1865.
Link.	Link, Johann Heinrich Friedrich, 1767–1851.
Lloyd & Underw.	Lloyd, Francis Ernest, 1868– ; Underwood, Lucien Marcus, 1853–1907.
Lodd.	Loddiges, Conrad, 1738–1826.
Loefl.	Loefling, Pehr, 1729–1756.
Loisel.	Loiseleur-Deslongchamps, Jean Louis Auguste, 1774–1849.
Loud.	Loudon, John Claudius, 1783–1843.
Lour.	Loureiro, João de, 1710–1791.
Ludw.	Ludwig, Christian Gottlieb, 1709–1773.
Lunell.	Lunell, Joel, 1851–1920.
Mack. & Bush.	Mackenzie, Kenneth Kent, 1877– ; Bush, Benjamin Franklin, 1858–
Mackenzie.	Mackenzie, Kenneth Kent, 1877–
MacMill.	MacMillan, Conway, 1867–1929.
Marsh.	Marshall, Humphry, 1722–1801.
Mart.	Martius, Carl Friedrich Philipp von, 1794–1868.
Mart. & Gal.	Martens, Martin, 1797–1863 ; Galeotti, Henri Guillaume, 1814–1858.
Martens.	Martens, Martin, 1797–1863.
Maxim.	Maximowicz, Karl Johann, 1827–1891.
Maxon.	Maxon, William Ralph, 1877–
Medic.	Medicus, Friedrich Casimir, 1736–1808.
Meinsh.	Meinshausen, Karl Friedrich, 1810–1890.
Meissn.	Meissner, Carl Friedrich, 1800–1874.
Merr.	Merrill, Elmer Drew, 1876–
Mert. & Koch.	Mertens, Franz Carl, 1764–1831 ; Koch, Wilhelm Daniel Joseph, 1771–1849.
Mett.	Mettenius, George Heinrich, 1823–1866.
Meyer, C. A.	Meyer, Carl Anton von, 1795–1855.
Meyer, E.	Meyer, Ernst Heinrich Friedrich, 1791–1858.
Meyer, G. F. W.	Meyer, Georg Friedrich Wilhelm, 1782–1856.
Mich.	Micheli, Pier' Antonio, 1679–1757.
Michx.	Michaux, André, 1746–1802.
Michx. F.	Michaux, François André, 1770–1855.
Mieg.	Mieg, Achilles, 1731–1799.
Milde.	Milde, Carl August Julius, 1824–1871.
Mill.	Miller, Philip, 1691–1771.
Mill. & Standl.	Miller, Gerrit Smith, 1869– ; Standley, Paul Carpenter, 1884–
Millsp.	Millspaugh, Charles Frederick, 1854–1923.
Millsp. & Chase.	Millspaugh, Charles Frederick, 1854–1923 ; Chase, Mary Agnes (Merrill), 1869–
Millsp. & Sherff.	Millspaugh, Charles Frederick, 1854–1923 ; Sherff, Earl Edward, 1886–
Mirb.	Mirbel, Charles François Brisseau, 1776–1854.
Mitchell.	Mitchell, John, 1680?–1768.
Moc.	Mocino, José Mariano, 1757–1820.
Moench.	Moench, Conrad, 1744–1805.
Mohr, C.	Mohr, Charles Theodore, 1824–1901.
Molina.	Molina, Juan Ignacio, 1737–1829.
Moore.	Moore, Thomas, 1821–1887.
Moore, T. V.	Moore, Thomas Verner, 1877–
Moq.	Moquin-Tandon, Christian Horace Bénédict Alfred, 1804–1863.
Moric.	Moricand, Moïse Etienne, 1779–1854.
Morong.	Morong, Thomas, 1827–1894.
Müll.	Müller, Otto Fridrich, 1730–1784.
Muell.-Arg.	Müller, Jean, 1828–1896. ("Mueller-Argoviensis.")
Muenchh.	Muenchhausen, Otto von, 1716–1774.
Muhl.	Muhlenberg, Gotthilf Henry Ernest, 1753–1815.
Munro.	Munro, William, 1818–1880.
Murr.	Murray, Johan Andreas, 1740–1791.
Mutis.	Mutis, José Celestino, 1732–1808.
Nash.	Nash, George Valentine, 1864–1921.
Necker.	Necker, Noel Joseph de, 1730–1793.
Nees.	Nees von Esenbeck, Christian Gottfried Daniel, 1776–1858.
Nees & Arn.	Nees von Esenbeck, Christian Gottfried Daniel, 1776–1858 ; Arnott, George Arnott Walker, 1799–1868.
Nees & Eberm.	Nees von Esenbeck, Theodor Friedrich Ludwig, 1787–1837 ; Ebermaier, Karl Heinrich, 1802–1870.
Nees & Meyen.	Nees von Esenbeck, Christian Gottfried Daniel, 1776–1858 ; Meyen, Franz Julius Ferdinand, 1804–1840.
Nels., A.	Nelson, Aven, 1859–
Nels., E.	Nelson, Elias Emanuel, 1876–
Nels. & Macbr.	Nelson, Aven, 1859– ; Macbride, J. Francis, 1892–
Niedzu.	Niedenzu, Franz Josef, 1857–
Nieuwl.	Nieuwland, Julius Aloysius Arthur, 1878–
Noronha.	Noronha, Fernando, 17——–1787.

Norton.	Norton, John Bitting Smith, 1872–
Nutt.	Nuttall, Thomas, 1786–1859.
Nym.	Nyman, Carl Fredrik, 1820–1893.
Oakes.	Oakes, William, 1799–1848.
Oakes & Tuckerm.	Oakes, William, 1799–1848; Tuckerman, Edward, 1817–1886.
Oeder.	Oeder, Georg Christian von, 1728–1791.
Olin.	Olin, Emil Hjalmar Fredrik, 1869–1915.
Olney.	Olney, Stephen Thayer, 1812–1878.
Opiz.	Opiz, Philipp Maximilian, 1787–1858.
Ort.	Gómez Ortega, Casimiro, 1740–1818.
Ostenf.	Ostenfeld, Carl Emil Hansen, 1873–1931.
Osterh.	Osterhout, George Everett, 1858–
Paine.	Paine, John Alsop, 1840–1912.
Pall.	Pallas, Peter Simon, 1741–1811.
Pammel.	Pammel, Louis Hermann, 1862–1931.
Parl.	Parlatore, Filippo, 1816–1877.
Parry.	Parry, Charles Christopher, 1823–1890.
Pax.	Pax, Ferdinand Albin, 1858–
Paxt.	Paxton, Joseph, 1801–1865.
Payson.	Payson, Edwin Blake, 1893–1927.
Peck.	Peck, Charles Horton, 1833–1917.
Pennell.	Pennell, Francis Whittier, 1886–
Pers.	Persoon, Christiaan Hendrik, 1761–1836.
Peterm.	Petermann, Wilhelm Ludwig, 1806–1855.
Philippi.	Philippi, Rudolf Amandus, 1808–1904.
Pickering.	Pickering, Charles, 1805–1878.
Piper.	Piper, Charles Vancouver, 1867–1926.
Planch.	Planchon, Jules Émile, 1823–1888.
Plum.	Plumier, Charles, 1646–1704.
Poir.	Poiret, Jean Louis Marie, 1755–1834.
Poll.	Pollich, Johann Adam, 1740–1780.
Pollard.	Pollard, Charles Louis, 1872–
Porter.	Porter, Thomas Conrad, 1822–1901.
Porter & Coulter.	Porter, Thomas Conrad, 1822–1901; Coulter, John Merle, 1851–1928.
Prantl.	Prantl, Karl Anton Eugen, 1849–1893.
Prescott.	Prescott, John D., 17——1837.
Presl.	Presl, Karel Bořiwog, 1794–1852.
Presl, J. & C.	Presl, Jan Swatopluk, 1791–1849; Presl, Karel Bořiwog, 1794–1852.
Prince.	Prince, William Robert, 1795–1869.
Prov.	Provancher, Léon, 1820–1892.
Purpus.	Purpus, Joseph Anton, 1860–
Pursh.	Pursh, Frederick Traugott, 1774–1820.
Pylaie.	La Pylaie, Auguste Jean Marie Bachelot de, 1786–1856.
R. & P.	Ruiz Lopez, Hipólito, 1754–1815; Pavon, José, 175––1844.
R. & S.	Roemer, Johann Jacob, 1763–1819; Schultes, Joseph August, 1773–1831.
Raf.	Rafinesque, Constantine Samuel, 1783–1840.
Raim.	Raimann, Rudolf, 1863–1896.
Ramaley.	Ramaley, Francis, 1870–
Ray.	Ray, John, 1627–1705.
Red.	Redouté, Pierre Joseph, 1759–1840.
Regel.	Regel, Eduard August von, 1815–1892.
Rehder.	Rehder, Alfred, 1863–
Reich.	Reichard, Johann Jacob, 1743–1782.
Reichenb.	Reichenbach, Heinrich Gottlieb Ludwig, 1793–1879.
Retz.	Retzius, Anders Jahan, 1742–1821.
Rich.	Richard, Louis Claude Marie, 1754–1821.
Richards.	Richardson, John, 1787–1865.
Richter.	Richter, Karl, 1855–1891.
Ricker.	Ricker, Percy Leroy, 1878–
Riddell.	Riddell, John Leonard, 1807–1865.
Riv.	Rivinus, Augustus Quirinus. 1652–1723.
Robbins.	Robbins, James Watson, 1801–1879.
Robins. & Fern.	Robinson, Benjamin Lincoln, 1864–; Fernald, Merritt Lyndon, 1873–
Robins. & Greenm.	Robinson, Benjamin Lincoln, 1864–; Greenman, Jesse More, 1867–
Robinson, B. L.	Robinson, Benjamin Lincoln, 1864–
Robs.	Robson, Stephen, 1741–1779.
Roehl.	Röhling, Johann Christoph, 1757–1813.
Roemer.	Roemer, Johann Jacob, 1763–1819.
Roemer, M.	Roemer, M. J.
Rose.	Rose, Joseph Nelson, 1862–1928.
Rosendahl.	Rosendahl, Carl Otto, 1875–

Rostk.	Rostkovius, Friedrich William Theophil, 1770–1848.
Rostk. & Schmidt.	Rostkovius, Friedrich William Theophil, 1770–1848; Schmidt, Wilhelm Ludwig Ewald, 1804–1843.
Roth.	Roth, Albrecht Wilhelm, 1757–1834.
Rothr.	Rothrock, Joseph Trimble, 1839–1922.
Rottb.	Rottböll, Christen Friis, 1727–1797.
Rowlee.	Rowlee, Willard Winfield, 1861–1923.
Roxb.	Roxburgh, William, 1751–1815.
Royen.	Royen, Adrian van, 1704–1779.
Royle.	Royle, John Forbes, 1799–1858.
Rudge.	Rudge, Edward, 1763–1846.
Rupp.	Ruppius, Heinrich Bernhard, 1688–1719.
Rupr.	Ruprecht, Franz Josef, 1814–1870.
Rusby.	Rusby, Henry Hurd, 1855–
Rydb.	Rydberg, Per Axel, 1860–1931.
Salisb.	Salisbury, Richard Anthony, 1761–1829.
Sarg.	Sargent, Charles Sprague, 1841–1927.
Scepin.	Scepin, Constantin, 1727–17—.
Schaffner, J. H.	Schaffner, John Henry, 1866–
Scheele.	Scheele, Georg Heinrich Adolf, 1808–1864.
Scheutz.	Scheutz, Nils Johan Wilhelm, 1836–1889.
Schk.	Schkuhr, Christian, 1741–1811.
Schlecht.	Schlechtendal, Diederich Franz Leonhard von, 1794–1866.
Schleich.	Schleicher, Johann Christoph, 1768–1834.
Schleiden	Schleiden, Matthias Jacob, 1804–1881.
Schmidel.	Schmidel, Casimir Christoph, 1718–1792.
Schmidt, F. W.	Schmidt, Franz Wilibald, 1764–1796.
Schneid., C. K.	Schneider, Camillo Karl, 1876–
Schott.	Schott, Heinrich Wilhelm, 1704–1865.
Schrader.	Schrader, Heinrich Adolph, 1767–1836.
Schrank.	Schrank, Franz von Paula von, 1717–1835.
Schreb.	Schreber, Johann Christian Daniel von, 1739–1810.
Schuette.	Schuette, Joachim Heinrich, 1821–1908.
Schult.	Schuller, Josef August, 1773–1821
Schultz, F.	Schultz, Friedrich Wilhelm, 1804–1876
Schultz-Bip.	Schultz, Carl Heinrich, 1805–1867, ("Schultz Bipontinus.")
Schulz, O. E.	Schulz, Otto Eugen, 1874–
Schum.	Schumacher, Heinrich Christian Friedrich, 1757–1830
Schum., K.	Schumann, Karl Moritz, 1851–1904.
Schur.	Schur, Philipp Johann Ferdinand, 1799–1878.
Schw.	Schweinitz, Lewis David von, 1780–1834.
Schw. & Torr.	Schweinitz, Lewis David von, 1780–1834; Torrey, John, 1796–1873.
Schweinf.	Schweinfurth, Georg August, 1836–1925.
Scop.	Scopoli, Johann Anton, 1723–1788.
Scribn.	Scribner, Frank Lamson-, 1851–
Scribn. & Ball.	Scribner, Frank Lamson-, 1851–; Ball, Carleton Roy, 1873–
Scribn. & Merr.	Scribner, Frank Lamson-, 1851–; Merrill, Elmer Drew, 1876–
Scribn. & Rydb.	Scribner, Frank Lamson-, 1851–; Rydberg, Per Axel, 1860–1931.
Scribn. & Smith.	Scribner, Frank Lamson-, 1851–; Smith, Jared Gage, 1866–
Scribn. & Williams.	Scribner, Frank Lamson-, 1851–; Williams, Thomas Albert, 1865–1900.
Seubert.	Seubert, Moritz, 1818–1878.
Shafer.	Shafer, John Adolph, 1863–1918.
Sharp, S. S.	Sharp, Seymour Sereno, 1893–
Shear.	Shear, Cornelius Lott, 1865–
Sheldon.	Sheldon, Edmund Perry, 1869–
Short & Peter.	Short, Charles Wilkins, 1794–1863; Peter, Robert, 1805–1894.
Shull.	Shull, George Harrison, 1874–
Shuttlew.	Shuttleworth, Robert James, 1810–1874.
Sibth.	Sibthorp, John, 1758–1796.
Sibth. & Sm.	Sibthorp, John, 1758–1796; Smith, James Edward, 1759–1828.
Sieb. & Zucc.	Siebold, Philipp Franz von, 1796–1866; Zuccarini, Joseph Gerhard, 1797–1848.
Siebold.	Siebold, Philipp Franz von, 1796–1866.
Sims.	Sims, John, 1749–1831.
Slosson.	Slosson, Margaret, 187––
Small.	Small, John Kunkel, 1869–
Smith.	Smith, James Edward, 1759–1828.
Smith, J. G.	Smith, Jared Gage, 1866–
Smith & Rydb.	Smith, Jared Gage, 1866–; Rydberg, Per Axel, 1860–1931.
Smyth.	Smyth, Bernard Bryan, 1843–1913.
Soland.	Solander, Daniel Carl, 1733–1782.
Somes.	Somes, Melvin Philip, 18–––1928.
Spach.	Spach, Edouard, 1801–1879.
Spenner.	Spenner, Fridolin Carl Leopold, 1798–1841.
Spreng.	Sprengel, Curt Polycarp Joachim, 1766–1833.
Spring.	Spring, Anton Friedrich, 1814–1872.

Standl.	Standley, Paul Carpenter, 1884–
Stapf.	Stapf, Otto, 1857–
Steele.	Steele, Edward Strieby, 1850–
Stephan.	Stephan, Christian Friedrich, 1757–1814.
Steud.	Steudel, Ernst Gottlieb, 1783–1856.
St. Hil.	Saint-Hilaire, Auguste François César Prouvençal de, 1779–1853.
Stokes.	Stokes, Jonathan, 1755–1831.
Stuntz.	Stuntz, Stephen Conrad, 1875–
Sudw.	Sudworth, George Bishop, 1864–1927.
Suksd.	Suksdorf, Wilhelm Nikolaus, 1850–
Sulliv.	Sullivant, William Starling, 1803–1873.
Sw.	Swartz, Olof Peter, 1760–1818.
Swallen.	Swallen, Jason Richard, 1903–
Sweet.	Sweet, Robert, 1783–1835.
Swezey.	Swezey, Goodwin Deloss, 1851–
T. & G.	Torrey, John, 1796–1873; Gray, Asa, 1810–1888.
Tausch.	Tausch, Ignaz Friedrich, 1793–1848.
Tenore.	Tenore, Michele, 1780–1861.
Thal.	Thal, Johann, 1542–1583.
Thellung.	Thellung, Albert, 1881–1928.
Thomas.	Thomas, David, 1776–1859.
Thompson.	Thompson, Charles Henry, 1870–1931.
Thouars.	Aubert du Petit-Thouars, Aubert, 1758–1831.
Thuill.	Thuillier, Jean Louis, 1757–1822.
Thunb.	Thunberg, Carl Peter, 1743–1828.
Thurb.	Thurber, George, 1821–1890.
Tidest.	Tidestrom, Ivar, 1865–
Todaro.	Todaro, Agostino. 1818–1892.
Torr.	Torrey, John, 1796–1873.
Torr. & Frém.	Torrey, John, 1796–1873; Frémont, John Charles, 1813–1890.
Torr. & Hook.	Torrey, John, 1796–1873; Hooker, William Jackson, 1785–1865.
Tourn.	Tournefort, Joseph Pitton de, 1656–1708.
Tratt.	Trattinnick, Leopold, 1764–1849.
Trel.	Trelease, William, 1857–
Trev.	Treviranus, Ludolf Christian, 1779–1864.
Trevisan.	Trevisan de Saint-Léon, Vittore Benedetto Antonio, 1818–1897.
Trew.	Trew, Christoph Jakob, 1695–1769.
Trin.	Trinius, Carl Bernhard von, 1778–1844.
Trin. & Rupr.	Trinius, Carl Bernhard von, 1778–1844; Ruprecht, Franz Josef, 1814–1870.
Tuckerm.	Tuckerman, Edward, 1817–1886.
Turcz.	Turczaninow, Nicolaus, 1796–1864.
Uline & Bray.	Uline, Edwin Burton, 1867–; Bray, William L, 1865–
Underw.	Underwood, Lucien Marcus 1853–1907.
Urban & Gilg.	Urban, Ignatz, 1848–1931; Gilg, Ernst Friedrich, 1867–
Vahl.	Vahl, Martin. 1749–1804.
Vail.	Vail, Anna Murray, 1863–
Vaill.	Vaillant, Sébastien, 1669–1722.
Vasey.	Vasey, George, 1822–1893.
Vasey & Holz.	Vasey, George, 1822–1893; Holzinger, John Michael, 1853–1929.
Vasey & Scribn.	Vasey, George, 1822–1893; Scribner, Frank Lamson-, 1851–
Vent.	Ventenat, Étienne Pierre, 1757–1808.
Vilm., E.	Vilmorin, Elisa (Bailly) de, 18—–1868.
Voss.	Voss, Andreas, 1857–1924.
Wahl.	Wahlenberg, Göran, 1780–1851.
Waldst. & Kit.	Waldstein, Franz de Paula Adam von, 1759–1823; Kitaibel, Paul, 1757–1817.
Wallr.	Wallroth, Carl Friedrich Wilhelm, 1792–1857.
Walp.	Walpers, Wilhelm Gerhard, 1816–1853.
Walt.	Walter, Thomas, 1740–1789.
Wang.	Wangenheim, Friedrich Adam Julius von, 1749–1800.
Warder.	Warder, John Aston, 1812–1883.
Wats., E. E.	Watson, Elba Emanuel, 1871–
Wats., S.	Watson, Sereno, 1826–1892.
Wats. & Coult.	Watson, Sereno, 1826–1892; Coulter, John Merle, 1851–1928.
Watt.	Watt, David Allan Poe, 1830–
Waugh.	Waugh, Frank Albert, 1869–
Webb & Berth.	Webb, Philip Barker, 1793–1854; Berthelot, Sabin, 1794–1880.
Webb & Moq.	Webb, Philip Barker, 1793–1854; Moquin-Tandon, Christian Horace Bénédict Alfred, 1804–1863.
Weber.	Weber, George Heinrich, 1752–1828.
Wedd.	Weddell, Hugh Algernon, 1819–1877.
Weigel.	Weigel, Christian Ehrenfried von, 1748–1831.
Weinm.	Weinmann, Johann Anton, 1782–1858.

Wender.	Wenderoth, Georg Wilhelm Franz, 1774–1861.
Wendl.	Wendland, Heinrich Ludolph, 1791–1869.
Wheeler.	Wheeler, William Archie, 1876–
White.	White, Theodore Greely, 1872–1901.
Wibel.	Wibel, August Wilhelm Eberhard Christoph, 1775–1814.
Wieg.	Wiegand, Karl McKay, 1873–
Wight, W.	Wight, William Franklin, 1874–
Willd.	Willdenow, Carl Ludwig, 1765–1812.
Williams.	Williams, Thomas Albert, 1865–1900.
Wimmer.	Wimmer, Christian Friedrich Heinrich, 1803–1868.
Wimmer & Grab.	Wimmer, Christian Friedrich Heinrich, 1803–1868; Grabowski, Heinrich Emanuel, 1792–1842.
Witasek.	Witasek, Johanna, 18— –1910.
With.	Withering, William, 1741–1799.
Wolfgang.	Wolfgang, Johann Friedrich, 1776–1859.
Wood.	Wood, Alphonso, 1810–1881.
Woodson.	Woodson, Robert Everard, 1904–
Woot.	Wooton, Elmer Ottis, 1865–
Woot. & Standl.	Wooton, Elmer Ottis, 1865– ; Standley, Paul Carpenter, 1884–
Wormsk.	Wormskjold, Morten, 1783–1845.
Wright, S. H.	Wright, Samuel Hart, 1825–1905.
Wulfen.	Wulfen, Franz Xaver von, 1728–1805.
Wylie.	Wylie, Robert Bradford, 1870–
York.	York, Harlan Harvey, 1877–
Zea.	Zea, Francisco Antonio, 1770–1822.
Zinn.	Zinn, Johann Gottfried, 1727–1759.
Zuccagni.	Zuccagni, Attilio, 1754–1807.

GLOSSARY

Abnormal. Differing from the usual structure.

Abortion. Imperfect development or non-development of an organ.

Abortive. Imperfectly formed, rudimentary, or barren.

Abruptly pinnate. Pinnate without an odd leaflet at the end.

Acaulescent. Stemless or apparently so.

Accrescent. Growing larger after flowering.

Accumbent (cotyledon). Having the edges against the radicle or hypocotyl.

Acerose. Needle-shaped, as the leaves of pines.

Achene. A small, dry and hard, 1-celled, 1-seeded, indehiscent fruit.

Acicular. Slenderly needle-shaped.

Aculeate. Prickly; beset with prickles.

Acuminate. Tapering at the end.

Acute. Sharp-pointed, but less so than acuminate.

Adnate. An organ adhering to a different one; united, as the inferior ovary with the calyx-tube. *Adnate anther,* one attached for its whole length to the filament.

Adsurgent = *assurgent.*

Adventitious. Out of the usual place.

Adventive. Not indigenous, but apparently becoming naturalized.

Aerial. Growing in or pertaining to the air; hence above the ground or the water.

Aestivation. The arrangement of the parts in a flower bud.

Alate. Winged.

Albumen. See Endosperm.

Alliaceous. Onion-like, in aspect or odor, or taste.

Alternate. Not opposite to each other.

Alveolar. Containing sockets or pits.

Alveolate. Honeycombed; having angular depressions separated by thin partitions.

Ament. A catkin, or peculiar scaly unisexual spike.

Amentaceous. Catkin-like, or catkin-bearing.

Amphitropous (ovule or seed). Half-inverted and straight, with the hilum lateral.

Amplexicaul. Clasping the stem.

Anastomosing. Connecting so as to form a well-defined network.

Anatropous (ovule). Inverted and straight, with the micropyle next the hilum.

Androecium. The whole set of stamens.

Androgynous. Having both staminate and pistillate flowers in the same inflorescence, or in *Carex* in the same spikelet, the former above the latter.

-androus (in composition). A suffix referring to stamens.

Angiospermous. Having the seeds borne within a pericarp.

Annual. Of only one year's duration. *Winter annual,* a plant from autumn-sown seed which blooms and fruits in the following spring.

Annular. In the form of a ring.

Annulate. Furnished with a ring or annulus.

Annulus. A ring, like that of the spore-case of most ferns.

Anterior (in the flower). The side towards the bract (external).

Anther. The part of the stamen which contains the pollen.

Antherid. The male organ of reproduction in ferns and mosses.

Antheriferous. Anther-bearing.

Anthesis. The time of expansion of a flower.

Apetalous. Having no petals.

Aphyllopodic. Without leaves at the base.

Aphyllous. Destitute of leaves, at least of green leaves.

Apical. Situated at the apex or tip.

Apiculate. With a minute point.

Appressed. Lying close and flat.

Approximate. Near together.

Arachnoid. Cobwebby; of slender entangled hairs.

Arborescent. Tree-like, in size or shape.

Archegone, or *Archegonium* (plural *archegonia*). The female organ in mosses and ferns.

Arcuate. Bent or curved like a bow.

Areola (*-ae*). A little, usually angular, space on the surface.

Areolate. Marked out into little areas; reticulate.

Aril. A fleshy organ growing about the hilum.

Arilliform. Resembling an aril.

Arillate. Having an aril.

Aristate. Tipped by an awn or bristle.

Aristulate. Diminutive of aristate.

Articulate. Jointed; having a node or joint.

Ascending. Growing obliquely upward, or upcurved.

Asexual. Without sex.

Assurgent. Ascending.

Attenuate. Slenderly tapering; becoming very narrow.

Auricle. An ear-shaped appendage.

Auriculate. Furnished with auricles.

Awl-shaped. Sharp-pointed from a broader base.

Awn. A slender bristle-like organ.

Axial = *Axile.*

Axil. The upper angle formed by a leaf or branch with the stem.

Axile. In the axis of an organ.

Axillary. Borne at or pertaining to an axil.

Axis. The central line of any organ or support of a group of organs; a stem, etc.

Baccate. Berry-like.

Banner. Upper petal of the papilionaceous flower; vexillum or standard.

Barbed. Furnished with rigid points or short bristles, usually reflexed like the barb of a fish-hook.

Barbellate. Finely barbed.

Barbulate. Finely bearded.

Basal, Basilar. At or pertaining to the base.

Basifixed. Attached by the base.

Bast. The fibrous portion of the inner bark.

Beaked. Ending in a beak or prolonged tip.

Bearded. With long or stiff hairs of any sort; awns of grasses are sometimes called a beard.

Berry. A fruit with pericarp wholly pulpy.

Bi- or *Bis-.* A Latin prefix signifying two, twice, or doubly.

Biconvex. Convex on both sides; lens-shaped.

Bidentate. Having two teeth.

Bidentulate. Diminutive of bidentate.

Biennial. Of two years' duration.

Bifid. Two-cleft.

Bilabiate. Two-lipped.

Bilocular. Two-celled.

Binate. Two together.

Bipinnate (leaf). Twice pinnate.

Bipinnatifid. Twice pinnatifid.

Biserial, Biseriate. Occupying two rows, one within the other.

Bisexual. Having both stamens and pistils.

Biternate. Twice ternate (principal divisions 3, each with 3 leaflets).

Bivalvular. Two-valved.

Bladdery. Thin and inflated.

Blade. The flat expanded part of a leaf.

Bract. A leaf, usually small, subtending a flower or flower-cluster, or a sporange.

Bracteate. With bracts.

Bracteolate. Having bractlets.

Bracteose. With numerous or conspicuous bracts.

Bractlet. A secondary bract, borne on a pedicel, or immediately beneath a flower; sometimes applied to minute bracts.

Bristle. A stiff hair or any similar outgrowth.

Bulb. A subterranean leaf-bud with fleshy scales or coats.

Bulbiferous. Bearing bulbs.

Bulblet. A small bulb, especially one borne upon the stem.

Bulbose, Bulbous. Bulb-like in form.

Caducous. Dropping off very early.

Calcarate. Produced into or having a spur.

Callus. An extension of the inner scale of a grass spikelet; a protuberance.

Calyculate. Having bracts around the calyx imitating an outer calyx.

Calyptrate. Furnished with a calyptra, or coming off as a lid or extinguisher.

Calyx. The outer of two series of floral leaves.

Campanulate. Bell-shaped.

Campylotropous (ovule or seed). So curved as to bring the apex and base nearly together.

Canaliculate. Channeled; longitudinally grooved.

Cancellate. Reticulated, with the meshes sunken.

Canescent. With gray or hoary fine pubescence.

Capillary. Hair-like in form; as fine as hair or slender bristles.

Capitate. Shaped like a head; collected into a head or dense cluster.

Capitellate. Diminutive of *capitate.*

Capitulum. A little head.

Capsular. Belonging to or of the nature of a capsule.

Capsule. A dry dehiscent fruit, composed of more than one carpel.

Carinate. Keeled; with a longitudinal ridge.

Carpel. The modified leaf forming the ovary, or a part of a compound ovary.

Carpophore. The slender prolongation of the floral axis which in the Umbelliferae supports the pendulous ripe carpels.

Cartilaginous. Of the texture of cartilage; firm and tough.

Caruncle. An appendage to a seed at the hilum.

Carunculate. With a caruncle.

Caryopsis. The grain; fruit of grasses, with a thin pericarp adherent to the seed.

Catkin. A scaly deciduous spike of flowers, an ament.

Caudate. With a slender tail-like appendage.

Caudex. The persistent base of an otherwise annual herbaceous stem.

Caudicle. Stalk of a pollen-mass in the Orchid and Milkweed families.

Caulescent. Having a manifest stem.

Cauline. Pertaining to the stem.

Cell. A cavity of an anther or ovary.

Chaff. Thin dry scales.

Chaffy. Furnished with chaff, or of the texture of chaff.

Channeled. Deeply grooved longitudinally, like a gutter.

Chartaceous. Papery in texture.

Chlorophyll. Green coloring matter of plants.

Choripetalous. Applied to a corolla whose petals are distinct.

Chorisepalous. Applied to a calyx whose sepals are distinct.

Ciliate (foliar organs). Beset on the margin with a fringe.

Ciliolate. Minutely ciliate.

Cinereous. Ash-grayish; the color of ashes.

Circinate. Coiled downward from the apex.

Circumscissile. Transversely dehiscent, the top falling away as a lid.

Clavate. Club-shaped.

Claw. The narrow or stalk-like base of some petals.

Cleft. Cut about half-way to the mid-vein.

Cleistogamous. Fertilized in the bud, without the opening of the flower.

Coalescence. The union of parts or organs of the same kind.

Cochleate. Coiled or shaped like a snail shell.

Cohesion. The union of one organ with another of the same kind.

Columella. A term applied to the persistent axis of the capsule.

Columnar. Like a column.

Coma. Tuft of hairs at the ends of some seeds.

Commissure. The surface by which one carpel joins another, as in the Umbelliferae.

Composite. A plant belonging to Carduaceae, Cichoriaceae, or Ambrosiaceae (constituting the old Compositae).

Concave. With the surface curved in.
Conduplicate. Folded lengthwise.
Conglomerate. Densely clustered.
Coniferous. Cone-bearing.
Connate. Similar organs more or less united.
Connective. The end of the filament between the anther-sacs.
Connivent. Converging.
Convolute. Rolled up longitudinally.
Coralloid. Resembling coral.
Cordate. Heart-shaped with the point upward.
Coriaceous. Leathery in texture.
Corm. The enlarged fleshy base of a stem, bulb-like but solid.
Corniculate. Furnished with a small horn or spur.
Corolla. The inner of two series of floral leaves.
Coroniform. Shaped like a crown.
Corrugate. Wrinkled or in folds.
Corymb. A convex or flat-topped flower-cluster of the racemose type, with pedicels or rays arising from different points on the axis.
Corymbose. Borne in corymbs; corymb-like.
Costa. A rib; the midrib of a leaf, etc.
Costate. Ribbed.
Cotyledon. A rudimentary leaf of the embryo.
Crateriform. In the shape of a saucer or cup, hemispherical or more shallow.
Creeping (stems). Growing flat on or beneath the ground and rooting.
Cremocarp. A fruit consisting of two or more indehiscent, inferior, one-seeded carpels, separating at maturity from each other and from the axis.
Crenate. With rounded teeth.
Crenulate. Diminutive of crenate.
Crested, Cristate. Bearing any elevated appendage like a crest.
Crinite. Bearded with long hairs, etc.
Crown. An inner appendage to a petal, or to the throat of a corolla.
Crucifer. A member of Brassicaceae, or Mustard Family, from the cross-like corolla.
Crustaceous. Hard and brittle in texture; crust-like.
Cucullate. Hooded, or resembling a hood.
Culm. The stem of grasses and sedges.
Cuneate, Cuneiform. Wedge-shaped.
Cupulate. Cup-shaped.
Cusp. A sharp stiff point.
Cuspidate. Sharp-pointed; ending in a cusp.
Cyme. A convex or flat flower-cluster of the determinate type, the central flowers first unfolding.
Cymose. Arranged in cymes; cyme-like.
Cymule. A small cyme.

Deciduous. Falling away at the close of the growing period.
Decompound. More than once compound or divided.
Decumbent. Reclining, but with the summit ascending.
Decurrent (leaf). Extending down the stem below the insertion.
Decussate. Alternating in pairs at right angles, or in threes.

Deflexed. Turned abruptly downward.
Dehiscence. The opening of an ovary, anther-sac, or sporange to emit the contents.
Dehiscent. Opening to emit the contents.
Dentate. Toothed, especially with outwardly projecting teeth.
Denticulate. Diminutive of dentate.
Dentiform. Tooth-like.
Depressed. Somewhat flattened from above.
Dextrorse. Turned to the right.
Di- (as a prefix in compounds). Two or twice.
Diadelphous (stamens). United by filaments in two sets.
Diandrous. Having two stamens.
Dicarpellary. Composed of two carpels.
Dichotomous. Forking regularly by pairs.
Dicotyledonous. Having two cotyledons.
Didymous. Twin-like.
Didynamous. With two stamens longer than the other two.
Diffuse. Loosely spreading.
Digitate. Diverging, like the spread fingers.
Digynous (flower). Having two pistils.
Dimerous. In two parts. Referring to a flower constructed on the numerical plan of two.
Dimidiate. Appearing as if cut in half.
Dimorphism (in flowers). Possessing two forms of flowers, one with short styles and long stamens, the other with long styles and short stamens.
Dimorphous. Of two forms.
Dioecious. Bearing staminate flowers or antherids on one plant, and pistillate flowers or archegones on another of the same species.
Disciform, Disk-shaped. Flat and circular, like a disk or quoit.
Discoid. Heads of Compositae composed only of tubular flowers; rayless; like a disk.
Disk. A development of the receptacle at or around the base of the pistil. In Compositae, the tubular flowers of the head as distinct from the ray.
Dissected. Cut or divided into numerous segments.
Dissepiment. A partition-wall of an ovary or fruit.
Distichous. In two vertical ranks.
Distinct. Separate; not united; evident.
Divaricate. Widely divergent.
Divided. Cleft to the base or to the midrib.
Dorsal. On the back; pertaining to the back.
Dorsiventral. In the plane running through the axis from above to below; contrary to lateral.
Drupaceous. Drupe-like.
Drupe. A simple fruit, usually indehiscent, with fleshy exocarp and bony endocarp.
Drupelet. Diminutive of drupe.

E- or *Ex-* (a prefix in compounds). Destitute of.
Echinate. Prickly.

Ellipsoid. A solid body, elliptic in section.

Elliptic. With the outline of an ellipse; usually narrowly oval.

Emarginate. Notched at the apex.

Embryo. A rudimentary plant in the seed.

Emersed. Raised out of water.

Endocarp. The inner layer of the pericarp.

Endogenous. Forming new tissue within.

Endogens. Monocotyledons.

Endosperm. The substance surrounding the embryo of a seed; albumen.

Ensiform. Sword-shaped, as the leaves of *Iris*.

Entire. Without divisions, lobes, or teeth.

Ephemeral. Continuing for only a day or less.

Epicarp. The outer layer of the pericarp.

Epigynous. Upon the ovary.

Epiphyte. A plant that grows upon another plant, but does not derive its sustenance from it.

Epiphytic. Growing on other plants, but not parasitic.

Equitant. Astride, used of conduplicate leaves which enfold each other in two ranks, as in *Iris*.

Erose. Irregularly margined, as if gnawed.

Estipulate. Without stipules.

Evanescent. Early disappearing.

Evergreen. Bearing green leaves throughout the year.

Excurrent. With a tip projecting beyond the main part of the organ.

Exfoliating. Peeling off in layers.

Exocarp. The outer layer of the pericarp.

Exogenous. Forming new tissue outside the older.

Exogens. Dicotyledons.

Exserted. Prolonged past surrounding organs.

Extravaginal innovation. Where the new shoot breaks through the basal sheath and produces a horizontal stolon.

Extrorse. Facing outward.

Falcate. Scythe-shaped.

Farinaceous. Mealy in texture.

Farinose. Covered with a mealy powder.

Fascicle. A dense cluster.

Fastigiate. Stems or branches which are nearly erect and close together.

Faveolate, Favose. Honeycombed; same as alveolate.

Fenestrate. With window-like markings.

Fenestration. Transparent spots or openings.

-ferous (in composition). Bearing.

Ferruginous. Color of iron rust.

Fertile. Bearing spores, or bearing seed.

Fibrillate, Fibrillose. Furnished with or abounding in fine fibres.

Fibrovascular. Composed of woody fibres and ducts.

Filament. The stalk of a stamen; also any slender thread-shaped appendage.

Filamentose, Filamentous. Bearing or formed of slender threads.

Filiform. Thread-like.

Fimbriate. Fringed.

Fimbrillate. Minutely fringed.

Fistular, Fistulose. Cylindrical and hollow, as the leaves of an onion.

Flabellate, or *Flabelliform.* Fan-shaped; applied to leaves, etc.

Flaccid. Without rigidity; lax and weak.

Flagellate. Producing slender runners or prolongations.

Flagelliferous. Bearing flagella.

Flagellum (-a). A slender runner or prolongation.

Flexuous. Zigzag; bending alternately in opposite directions.

Floccose. With loose tufts of wool-like hairs.

Floret. A small flower, usually one of a dense cluster.

Foliaceous. Leaf-like in texture or appearance.

-foliate. Having leaves.

Foliolate. With separate leaflets.

-foliolate. Having leaflets.

Foliose. Leafy.

Follicle. A fruit consisting of a single carpel, dehiscing by the ventral suture.

Follicular. Like a follicle.

Fornicate. Arched over, as the corona of some Boraginaceae, closing the throat of the corolla.

Fornix (ces). Small arching crest in the throat of the corolla.

Foveate, Foveolate. More or less pitted.

Free. Not adnate to other organs.

Frond. The leaf of ferns and some other cryptogams; in Lemnaceae, the thallus-like stem which functions as foliage.

Frutescent, Fruticose. More or less shrub-like.

Fugacious. Soon falling off or perishing.

Funiculus. The stalk of an ovule or seed.

Funnelform, Funnel-shaped. Expanding gradually upwards, like a funnel.

Furfuraceous. Resembling bran; scurfy.

Furfurescent. Becoming scurfy.

Fuscous. Grayish brown.

Fusiform. Spindle-shaped; swollen in the middle and narrowing toward each end.

Galea. A hood-like part of a perianth or corolla; upper lip of a two-lipped corolla.

Galeate. Helmet-shaped; having a galea.

Gamopetalous. With united petals; same as *monopetalous* and *sympetalous.*

Gamosepalous. With united sepals; same as monosepalous.

Geminate. Like twins.

Geniculate. Bent abruptly, like a knee.

Gibbous. Enlarged or swollen on one side.

Glabrate. Becoming glabrous with age, or almost glabrous.

Glabrous. Devoid of hairs.

Gland. A secreting cell, or group of cells.

Glandular. With glands, or gland-like.

Glaucous. Covered or whitened with a bloom.

Globose. Spherical or nearly so.

Glochidiate. Barbed at the tip.

Glomerate. In a compact cluster.

Glomerule. A dense capitate cyme.

Glumaceous. Resembling glumes.

Glume. The scaly bract of the spikelets of grasses and sedges.

Glutinous. Covered with a sticky exudation.

Granuliferous. Bearing or covered with small granules.

Gregarious. Growing in groups or colonies.

Gymnospermous. Bearing naked seeds, without an ovary.

Gynaecandrous. Having staminate and pistillate flowers in the same spikelet, the latter above the former.

Gynandrous. Having the stamens and pistils more or less united.

Gynobase. A prolongation or enlargement of the receptacle supporting the ovary.

Gynoecium. The whole set of pistils.

Gynophore. An elongation of the receptacle bearing the pistil ; a stalk of the pod made up from a part of the receptacle ; compare stipe.

Habit. The general aspect of a plant, or its mode of growth.

Habitat. The situation in which a plant grows in a wild state.

Halberd-shaped. The same as hastate.

Hamate. Crooked, hooked.

Hastate. Halberd-shaped ; like sagittate, but with the basal lobes diverging.

Head. A dense cluster of sessile or nearly sessile flowers on a very short axis or receptacle.

Herb. A plant with no persistent woody stem above ground.

Herbaceous. Leaf-like in texture and color ; pertaining to an herb.

Hermaphrodite (flower). Having both stamens and pistils ; same as perfect.

Heterogamous. Producing more than one kind of flowers.

Heteromorphous. Having flowers of different forms as regards the size or relative position of the essential organs.

Hexa-, in compounds, means six.

Hexamerous. Consisting of six parts or members. Applied to a flower that is constructed on the numerical plan of six.

Hilum. The scar or area of attachment of a seed or ovule.

Hirsute. With rather coarse stiff hairs.

Hirtellous. Minutely hirsute.

Hispid. With bristly stiff hairs.

Hispidulous. Diminutive of hispid.

Hoary. Grayish white ; see *canescent*.

Homogamous. A head or cluster with flowers all of one kind.

Hyaline. Thin and translucent.

Hybrid. A cross between two species.

Hydrophilous. Water-loving.

Hydrophyte. A water-plant.

Hypanthium. A calyx-like enlargement of the flower-axis or receptacle, often surrounding or enclosing the pistils and bearing the calyx and corolla and often the stamens on its margin.

Hypocotyl. The rudimentary stem of the embryo ; also termed *radicle.*

Hypogynous. Inserted under the pistil.

Imbricate. Overlapping (as shingles on a roof).

Immersed. Growing wholly under water.

Imperfect flowers. Wanting either stamens or pistils.

Incised. Cut sharply and irregularly, more or less deeply.

Incision. A cut ; a narrow opening between two lobes.

Included. Not at all protruded from the surrounding envelope.

Incumbent (embryo). Cotyledons with the back of one of them against the hypocotyl.

Indehiscent. Not splitting open.

Indigenous. Native to the country.

Induplicate. Valvate aestivation in which the margins of the leaves are inflexed or folded inward.

Indurated. Hardened.

Indusiate. With an indusium.

Indusium. The proper (often shield-shaped) covering of the sorus or fruit-dot in Ferns.

Inequilateral. Unequal-sided.

Inferior. Lower or below ; outer or anterior. *Inferior ovary,* one that is adnate to the hypanthium.

Inflated. Turgid and bladdery.

Inflexed. Bent inwards.

Inflorescence. The flowering part of plants ; its mode of arrangement.

Infra- (in compound words). Below, being below.

Infra-axillary. Inserted some distance below the axils.

Innocuous. Harmless, hence unarmed or spineless.

Innovation. An offshoot from the stem.

Inserted. Attached to or growing out of.

Insertion. The place or the mode of attachment of an organ to its support.

Inter- (in composition). Between.

Internerve. Space between the nerves.

Internode. Portion of a stem or branch between two nodes.

Interval. Space between ridges.

Intramarginal. Within and near the margin.

Intravaginal innovation. Where the new shoot starts inside a basal sheath and continues to grow, remaining between it and the stem.

Introrse. Facing inward.

Involucel. A secondary involucre.

Involucrate. With an involucre, or like one.

Involucre. A whorl of bracts subtending a flower or flower-cluster.

Involute. Rolled inwardly.

Irregular. A flower in which one or more of the organs of the same series are unlike the rest.

Keel. A projecting ridge on a surface, like the keel of a boat ; the two anterior petals of a papilionaceous corolla.

Labellum. The odd (lower) petal of orchids.

Labiate. Lipped; belonging to the Labiatae or LAMIACEAE.

Lacerate. Irregularly cleft, as if torn.

Lacinia (-ae). Lobe.

Laciniate. Cut into narrow lobes or segments.

Lamina. A plate or blade; the blade of a leaf or a petal.

Lanate. Woolly; clothed with long and soft entangled hairs.

Lanceolate. Considerably longer than broad, tapering upward from the middle or below; lance-shaped.

Lanose. Densely lanate.

Lanuginous. Cottony or woolly.

Lateral. Belonging to or borne on the side.

Latex. The milky sap of certain plants.

Leaflet. One of the divisions of a compound leaf.

Legume. A simple dry fruit, usually dehiscent along both sutures

Leguminosae. Plants of the families Fabaceae, Caesalpiniaceae, and Mimosaceae.

Leguminous. Pertaining to a legume or to the Leguminosae.

Lemma. The lower of the two bracts inclosing the flower in the grasses.

Lenticels. Small oval dots which appear upon the branches of cork-forming Dicotyledons during the first year's growth, and which, by further growth during the early part of the second year, rupture the epidermis.

Lenticular. Lens-shaped.

Lepidote. Beset with small scurfy scales.

Liana. A woody vine.

Ligneous. Woody, or having a woody texture.

Ligulate. Provided with or resembling a ligule.

Ligule. A strap-shaped corolla, as in the ray-flowers of Compositae; a thin scarious projection from the summit of the sheath in grasses.

Limb. The expanded part of a petal, sepal, or gamopetalous corolla.

Linear. Long and narrow, with parallel margins.

Lip. The principal lobe of a bilabiate corolla or calyx; the odd and peculiar petal in the orchis family; the labellum.

-locular (in composition). Having cells.

Loculicidal (dehiscence). Splitting down through the middle of the back of each cell.

Lodicules. Minute hyaline scales subtending the flower in grasses.

Loment. A jointed legume, usually constricted between the seeds.

Lunate. Crescent-shaped.

Lunulate. Diminutive of lunate.

Lurid. Dirty brown.

Lutescent. Yellowish, or becoming yellow.

Lyrate. Pinnatifid, with the terminal lobe or segment considerably larger than the others.

Macrosporange. Sporange containing macrospores.

Macrospore. The larger of two kinds of spores borne by a plant, usually giving rise to a female prothallium.

Mammillate. Furnished with nipple-shaped processes.

Marcescent. Withering, but remaining attached.

Medullary. Pertaining to the pith or medulla.

Membranaceous, Membranous. Thin and rather soft and more or less translucent.

-merous (in composition). A suffix referring to parts, as 2-merous, having two parts of each kind.

Mesa. Dry tableland.

Mesocarp. Middle layer of a pericarp.

Micropyle. Orifice of the ovule, and corresponding point on the seed.

Microsporange. Sporange containing microspores.

Microspore. The smaller of two kinds of spores borne by a plant, usually giving rise to a male prothallium; pollen-grain.

Midvein, Midrib. The central vein or rib of a leaf or other organ.

Monadelphous. Stamens united by their filaments into one set.

Moniliform. Like a string of beads.

Monocephalous. Bearing only one head.

Monocotyledonous (embryo). Having only one cotyledon.

Monoecious. Bearing stamens and pistils on the same plant, but in different flowers.

Mucro. A short and small abrupt tip.

Mucronate. Provided with a mucro.

Mucronulate. Diminutive of mucronate.

Multicellular. Consisting of many cells.

Multifid. Cleft into many lobes or segments

Multilocular. Possessing many loculi or cavities.

Muricate. Roughened with short hard processes.

Muriculate. Very finely muricate.

Muticous. Pointless, or blunt.

Napiform. Turnip-shaped.

Naturalized. Not indigenous to the region, but so firmly establishel as to have become part of the flora.

Nectariferous. Nectar-bearing; having a nectary.

Nectary An organ which secretes nectar.

Nerve. A simple or unbranched vein or slender rib.

Nigrescent. Becoming black or blackish.

Node. The place upon a stem which normally bears a leaf or whorl of leaves.

Nodose. Furnished with knots or nodes.

Nodulose. Diminutive of nodose.

Nut. An indehiscent one-seeded fruit with a hard or bony pericarp.

Nutlet. Diminutive of nut.

Ob-, as a prefix, signifies inversion, as follows:

Obcompressed. Flattened the opposite of the usual way.

Obconic. Conical, but with the point of attachment at the apex.

Obcordate. Inversely heart shaped.

Oblanceolate. Inverse of lanceolate.

Oblong. Longer than broad, with the sides nearly parallel or somewhat curving.

Obovate. Inversely ovate.

Obovoid. Inversely ovoid.

Obpyramidal. Like an inverted pyramid. *i.e.*, pyramidal with the base uppermost.

Obsolete. Imperfectly developed or rudimentary.

Obtuse. Blunt, or rounded.

Ochroleucous. Yellowish white.

Ocrea. A sheathing stipule.

Ocreate. Having sheathing stipules.

Offset. Short branch next the ground which takes root.

Operculate. With an operculum.

Operculum. A lid.

Orbicular. Approximately circular in outline.

Orthotropous (ovule or seed). Erect, with the orifice or micropyle at the apex.

Oval. Broadly elliptic.

Ovary. The part of the pistil that contains the ovules.

Ovate. In outline like a longitudinal section of a hen's egg.

Ovoid. Shaped like a hen's egg.

Ovule. The macrosporangium of flowering plants, becoming the seed after fertilization.

Ovuliferous. Bearing ovules.

Palate. A rounded projection of the lower lip of a personate corolla, closing the throat.

Palea (plural *paleae*). Chaff; the chaff or bracts on the receptacle of many Compositae.

Paleaceous. Chaffy.

Palet. The upper thin chaffy or hyaline bract which with the lemma encloses the flowers in Grasses.

Pallid. Pale.

Palmate. Diverging radiately like the fingers.

Palmately. In a palmate manner.

Panduriform. Fiddle-shaped.

Panicle. A compound flower-cluster of the racemose type.

Panicled, Paniculate. Borne in a panicle; resembling a panicle.

Pannose. Of the appearance or texture of felt.

Papilionaceous (corolla). Having a standard, wings, and keel, as in the peculiar corolla of the Pea Family.

Papilla. A little nipple-shaped protuberance.

Papillate, Papillose. Covered with papillae.

Pappiferous. Pappus-bearing.

Pappus. The modified calyx-limb in Compositae, forming a crown of very various character at the summit of the achene.

Papyraceous. Having a papery texture.

Parasitic. Growing on and deriving nourishment from another plant.

Parietal. Borne on or pertaining to the wall or inner surface of a capsule.

Parted. Cleft nearly, but not quite to the base.

Pectinate. Pinnatifid with narrow, closely set segments; comb-like.

Pedate. Palmately divided or parted, with the lateral segments 2-cleft.

Pedicel. The stalk of a single flower in a flower-cluster.

Pedicellate, Pedicelled. Furnished with a pedicel.

Peduncle. A primary flower-stalk, supporting either a cluster or a solitary flower.

Peduncled, Pedunculate. Furnished with a peduncle.

Peltate. Shield-shaped and flat, with a stalk on its lower surface.

Pendulous. More or less hanging or declined. *Pendulous ovule,* one that hangs from the side of the cell.

Penicillate. With a tuft of hairs or hair-like branches.

Penta- (in compounds). Five.

Pentagonal. Five-angled.

Pepo. A melon-like fruit.

Perennial. Lasting from year to year.

Perfect (*flower*). Having both stamens and pistils.

Perianth. The floral envelopes of the flower, especially used when calyx and corolla cannot be distinguished.

Pericarp. The wall of the fruit, or seed-vessel.

Perigynium (-*a*). The more or less inflated sac-like organ surrounding the pistil in *Carex.*

Perigynous. Borne around the ovary.

Persistent. Long-continuous, as a calyx upon the fruit, leaves through winter, etc.

Personate. Masked; bilabiate, and the throat closed by a prominent palate.

Petal. One of the leaves of the corolla.

Petaloid. Petal-like; resembling or colored like petals.

Petiolate. Having a petiole.

Petiole. The leaf-stalk.

Petioled. Furnished with a petiole.

Petiolulate. With a petiolule (leaflet).

Petiolule. The stalk of a leaflet.

Phaenogamous. Having flowers with stamens and pistils and producing seeds.

Phyllode, Phyllodium (-*a*). A somewhat dilated petiole having the form of and serving as a leaf-blade.

Phyllodic. With a leafy base.

-phyllous (in composition). A suffix referring to leaves; as *gamophyllous,* with united leaves; *diphyllous,* with two leaves, etc.

Pilose. Hairy, with soft hairs.

Pinna (pl. *pinnae*). One of the primary divisions of a pinnate or compoundly pinnate frond or leaf.

Pinnate (leaf). Compound, with the leaflets arranged on each side of a common petiole.

Pinnatifid. Pinnately cleft.

Pinnatisect. Pinnately divided.

Pinnule. A division of a pinna.

Pistil. The central organ of a flower containing the macrosporanges (ovules).

Pistillate. Provided with pistils, and, in its more proper sense, without stamens.

Pitted. Marked with small depressions or pits.

Placenta. An ovule-bearing surface.

Plane. With a flat, not curved surface.

Plano-convex. Plane on one side and convex on the other.

Plicate. Folded into plaits, like a fan.

Plumose. Having fine hairs on each side, like the plume of a feather.

Plumule. The rudimentary terminal bud of the embryo.

Pod. Any dry and dehiscent fruit.

Pointed. Acuminate.

Pollen. Pollen grain. See Microspore.

Pollinia. The pollen-masses of the Orchid and Milkweed Families.

Polliniferous. Bearing pollen.

Poly-, in compounds means many.

Polyadelphous. Applied to stamens which are united by their filaments into many sets.

Polygamous. Bearing both perfect and imperfect flowers.

Polymorphous. Of several forms.

Polypetalous. Possessing many petals. Applied by the older botanists to flowers having the petals distinct or ununited.

Polysepalous. With the sepals distinct.

Pome. The fleshy fruit of the Apple Family.

Posterior. On the side towards the axis ; see anterior.

Prickles. Sharp elevations of the bark, and coming off with it, as in the rose.

Prismatic. Of the shape of a prism, angular, with flat sides, and of nearly uniform size throughout.

Procumbent. Trailing on the ground.

Proliferous. Bearing offshoots ; a shoot, a branch, a rosette, or a flower producing a shoot ending in a similar organ.

Prostrate. Lying flat on the ground.

Prothallium. The sexual generation of Pteridophyta.

Pruinose. Frosted ; covered with a powder like hoar-frost.

Pseudo- (in combinations). Falsely.

Pteridophytes. Fern-plants ; ferns and their allies.

Puberulent. Minutely pubescent.

Pubescent. Covered with hairs.

Pulverulent. Dusted ; covered apparently with fine powder.

Pulvinate. Cushioned, or shaped like a cushion.

Punctate. Dotted with depressions or with translucent internal glands or colored dots.

Puncticulate. Minutely punctate.

Pungent. Terminating in a rigid sharp point ; acrid.

Pustular, Pustulate. With blister-like elevations.

Pustule. Blister or blister-like process.

Putamen. The bony part of a stone-fruit.

Pyriform. Pear-shaped.

Pyxis. A capsule whose dehiscence is circumscissile, or which opens by a circular, horizontal line, so that the upper part comes off like a lid.

Quadrate. Nearly square in form.

Raceme. An elongated indeterminate flower-cluster with each flower pedicelled.

Racemose. In racemes, or resembling a raceme.

Rachilla. The axis of the spikelet in grasses.

Rachis. The axis of a compound leaf, or of a spike or raceme.

Radially. Spreading from a common center.

Radiant. With the marginal flowers enlarged and ray-like.

Radiate. Bearing ray-flowers ; spreading from or arranged around a common center.

Radical. Belonging to the root, or apparently coming from the root.

Radicle. The rudimentary stem of the embryo ; hypocotyl.

Rameal. Belonging to a branch.

Ramification. Branching.

Ray. One of the peduncles or branches of an umbel ; the flat marginal flowers in Compositae.

Receptacle. The end of the flower stalk, bearing the floral organs ; or, in Compositae, bearing the flowers ; also in some ferns, an axis bearing sporanges.

Reclined. Turned or curved downwards.

Recurved. Curved backwards.

Reflexed. Bent backward abruptly.

Regular. Having the members of each part alike in size and shape.

Reniform. Kidney-shaped.

Repand. With a somewhat wavy margin.

Replicate. The form of vernation in which the apex of the leaf is bent backward toward the base.

Replum. The septum of certain pods that persists after the valves have fallen away.

Resiniferous. Producing resin.

Reticulate. In the form of network.

Retrorse. Directed back or downward.

Retuse. With a shallow notch at a rounded apex.

Revolute. Rolled backward.

Rhacheola. The prolongation of the secondary axis of the spikelets of sedges, within the perigynium ; compare *rachilla.*

Rhachis. See Rachis.

Rhizome. A rootstock.

Rhombic, Rhomboidal. Somewhat lozenge-shaped ; obliquely four-sided.

Rib. A primary or prominent vein of a leaf.

Ringent. Gaping, as the mouth of an open bilabiate corolla.

Rostellum. Beak of the style in Orchids.

Rostrate. Bearing a beak or a prolonged appendage.

Rosulate. In the form of a rosette.

Rosuliferous. Bearing rosettes.

Rotate (corolla). Wheel-shaped ; flat and circular in outline.

Rotund. Rounded in outline.

Rudiment. A very partially developed organ ; a vestige.

Rudimentary. Imperfectly developed, or in an early stage of development.

Rufous. Reddish brown.

Rugose. Wrinkled.

Runcinate. Sharply pinnatifid or incised, the lobes or segments turned backward.

Runner. A filiform or very slender stolon.

Sac. A pouch, especially the cavity of an anther.

Saccate. Like a sac or pouch ; furnished with a sac.

Sagittate. Like an arrow-head, with the lobes turned downward.

Salver-shaped (corolla). Having a slender tube abruptly expanded into a flat limb.

Samara. An indehiscent winged fruit.

Saprophyte. A plant which grows on dead organic matter.

Scabrous. Rough to the touch.

Scale. A minute, rudimentary, or vestigial leaf.

Scape. A peduncle rising from the ground, naked or without proper foliage.

Scapose. Bearing or resembling a scape.

Scarious. Thin, dry, and translucent, not green.

Scorpioid. Coiled up in the bud, or in the beginning of growth, unrolling in expanding.

Scrobiculate. Possessing minute or shallow depressions.

Secund. Borne along one side of an axis.

Segment. One of the parts of a leaf or other like organ that is cleft or divided.

Semi- (in compounds). Half.

Sepal. One of the leaves of a calyx.

Septate. Divided in compartments by cross-partitions.

Septicidal (capsule). Dehiscing through the partitions and between the cells.

Septum (plural *septa*). A partition, as of a pod, etc.

Sericeous. Silky; clothed with satiny pubescence.

Serrate. Having teeth pointing forward.

Serrulate. Finely serrate.

Sessile. Without a stalk.

Seta (*-ae*). A bristle, or a slender body resembling a bristle.

Setaceous. Bristle-like.

Setose. Bristly.

Setulose. Having minute bristles.

Sheath. A tubular envelope, as the lower part of the leaf in grasses.

Sheathing. Enclosing as by a sheath.

Silicle. A short silique.

Silique. The peculiar pod of Brassicaceae.

Sinuate. Strongly wavy.

Sinus. The cleft or recess between two lobes.

Smooth. Without roughness.

Sobol, Sobole. A creeping underground stem.

Soboliferous. Bearing underground mostly horizontal branches.

Sorus (pl. *sori*). A heap or cluster, applied to the fruit-dots of ferns.

Spadiceous. Like or pertaining to a spadix.

Spadix. A fleshy spike of flowers.

Spathaceous. Resembling a spathe; furnished with a spathe.

Spathe, A bract, usually more or less concave, subtending a spadix.

Spatulate. Shaped like a spatula; spoon-shaped.

Spermatozoid. A motile ciliated male reproductive cell.

Spicate. Arranged in or resembling a spike.

Spiciform. Spike-like.

Spike. An elongate flower-cluster, with sessile or nearly sessile flowers.

Spikelet. Diminutive of spike; especially applied to flower-clusters of grasses and sedges.

Spine. A sharp woody or rigid outgrowth from the stem.

Spinescent. Tipped by or degenerating into spines or thorns.

Spinose. Thorny; with spines or structures similar to spines.

Spiricle. Delicate coiled thread on the surface of seeds and achenes.

Spirilliferous. Bearing or having spiricles.

Sporange, Sporangium (*-a*). A sac containing spores.

Spore. The reproductive organ in Cryptogams which roughly corresponds to a seed but possesses no embryo.

Sporocarp. The fruit-cases of certain Cryptogams containing sporangia or spores.

Sporophyll. A spore-bearing leaf.

Sporophyte. The sexual generation of plants.

Spreading. Diverging nearly at right angles; nearly prostrate.

Spur. A hollow projection.

Squamella (*-ae*). A scale-like member of the pappus, borne on the achenes of some Composites.

Squamiform. Resembling a scale.

Squarrose. With spreading or projecting parts.

Stamen. The organ of a flower which bears the microspores (pollen-grains).

Staminate. Possessing stamens. Applied to flowers which have stamens but no pistils.

Staminiferous. Bearing stamens.

Staminodium. A sterile stamen, or other organ in the position of a stamen.

Standard. The upper, usually broad, petal of a papilionaceous corolla.

Stellate. Star-like.

Sterigmata. The projections from twigs, bearing the leaves, in some genera of Pinaceae.

Sterile. Without spores, or without seed.

Stigma. That part of a pistil through which fertilization by the pollen is effected.

Stigmatic. Belonging to or characteristic of the stigma.

Stipe. The stalk-like lower portion of a pistil; the leaf-stalk of a fern.

Stipitate. Provided with a stipe.

Stipular. Belonging to stipules.

Stipulate. Having stipules.

Stipules. The appendages on each side of the base of certain leaves.

Stolon. A basal branch rooting at the nodes.

Stoloniferous. Producing or bearing stolons.

Stoma (pl. *stomata*). An orifice in the epidermis of a leaf communicating with internal air-cavities.

Stramineous. Straw-colored.

Striate. Marked with slender longitudinal grooves or channels.

Strict. Very straight and upright.

Strigillose. Diminutive of strigose.

Strigose. With appressed stiff hairs.

Strobilaceous. Like a pine-cone.

Strobile. An inflorescence marked by imbricated bracts or scales, as in the pine-cone.

Strophiolate. With a strophiole.

Strophiole. An appendage to a seed at the hilum.

Style. The usually attenuated portion of the pistil connecting the stigma and ovary.

Stylopodium. A disk-like expansion at the base of a style, as in Umbelliferae.

Sub- (in compound words). Somewhat, almost, in a subordinate grade, of inferior rank, beneath.

Submarginal. Near the margin; situated under the margin.

Subulate. Awl-shaped.

Succulent. Soft and juicy.

Sucker. A shoot from subterranean branches.

Suffrutescent. Slightly or obscurely shrubby.

Suffruticose. Very low and woody; diminutively shrubby.

Sulcate. Grooved longitudinally.

Superior (ovary). Free from the calyx or hypanthium.

Supra- (in compound words). Above, being above.

Supra-axillary. Inserted some distance above the axils.

Surculose. Producing shoots from the rootstock.

Suspended (ovule). Hanging from the apex of the cell.

Suture. A line of splitting or opening.

Symmetrical. Applied to a flower with the different series of its parts of equal numbers.

Sympetalous. With united petals.

Syngenesious. With stamens united by their anthers.

Synonym. A superseded or unused name.

Taproot. A stout vertical root which continues the main axis of the plant.

Tawny. Dull yellowish, with a tinge of brown.

Tendril. A thread-shaped process used for climbing.

Terete. Circular in cross-section.

Ternary. Consisting of three.

Ternate. Divided into three segments, or arranged in threes.

Tessellate. Checkered.

Testa. Outer coat or covering of seed.

Tetra- (in compounds). Four.

Tetradynamous. Applied to stamens when there are six in the flower, four of them longer than the other two.

Tetragonal. Four-angled.

Tetramerous. Applied to flowers constructed on the numerical plan of four.

Thalloid. Resembling a thallus.

Thallus. A plant body without differentiation of stem, leaf, and root.

Throat. The orifice of a gamopetalous corolla or calyx; the part between the tube proper and the limb.

Thyrsoid. Like a thyrsus.

Thyrsus. A congested cyme.

Tomentose. Covered with tomentum.

Tomentulose. Diminutive of tomentose.

Tomentum. Densely matted woolly hairs.

Torose. Cylindrical with contraction at intervals.

Torsion. Twisting of an organ.

Tortuous. Twisted or bent.

Torulose. Diminutive of torose.

Torus. The receptacle of a flower.

Transverse. Across; in a right and left direction.

Tri (in composition). Three or thrice.

Triandrous. Having three stamens.

Trichotomous. Three-forked.

Tridentate. Three-toothed.

Trifoliolate. Having three leaflets.

Trigonous. Three-angled.

Triquetrous. Having three salient angles, the sides concave or channeled.

Truncate. Ending abruptly, as if cut off transversely.

Tuber. A thickened and short subterranean branch, having numerous buds.

Tubercle. The persistent base of the style in some Cyperaceae; a small projection.

Tuberculate. With rounded projections.

Tuberiferous. Bearing tubers.

Tuberous. Resembling a tuber.

Tumid. Swollen.

Tunicate. Coated; invested with layers, as an onion.

Turbinate. Top-shaped.

Turion. A scaly shoot from a subterranean bud.

Turioniferous. Bearing turions or suckers like the shoots of Asparagus.

Twining. Ascending by coiling around a support.

Umbel. A determinate, usually convex flower-cluster, with all the pedicels arising from the same point.

Umbellate. Borne in umbels; resembling an umbel.

Umbellet. A secondary umbel.

Umbellifer. A member of Ammiaceae, or Carrot Family.

Umbonate. Bearing a stout projection in the center; bossed.

Unarmed. Destitute of spines, prickles, and the like.

Uncinate. Hooked, or in form like a hook.

Undulate. With a wavy surface; repand.

Unguiculate. Contracted at base into a claw.

Uni- (in compounds). One.

Unicellular. One-celled.

Unifoliolate. Applied to a compound leaf that has but one leaflet, as the leaves of the Orange and Lemon.

Unilocular. Possessing one locule or cell.

Uniserial. Arranged in one series. Applied to parts that are arranged in one horizontal whorl.

Unisexual. Having only one kind of sex-organs; applied also to flowers having only stamens or pistils.

Urceolate. Urn-shaped.

Utricle. A small, thin-walled, one-seeded fruit.

Utricular. Like a small bladder.

Vaginate. Surrounded by a sheath.

Valvate. Meeting by the margins in the bud, not overlapping; dehiscent by valves.

Valve. One of the pieces into which a dehiscent pod, or any similar body, splits.

Veinlet. A smaller branch of a vein.

Veins. Threads of fibro-vascular tissue in a leaf or other organ, especially those which branch (as distinguished from nerves).

Velum. A fold of the inner side of the leaf-base in *Isoetes*.

Velutinous. Velvety; with dense fine pubescence.

Venation. The veining of leaves, etc.

Ventral, Ventrally. Being on the side next the axis.

Ventricose. Swelling unequally, or inflated on one side.

Venulose. Finely veiny.

Vernation. The arrangement of leaves in the bud.

Verrucose. Warty; covered with protuberances.

Versatile. An anther attached at or near its middle to the filament.

Verticil. A whorl.

Verticillaster. A pair of opposite cymes that occur in the axils of the leaves of Mints.

Verticillate. Whorled.

Vesicle. A small bladder or air-cavity.

Vespertine. Belonging to the evening; applied to flowers that open at nightfall.

Vexillum. The standard of a papilionaceous flower.

Villosulous. Diminutive of villous.

Villous. Bearing long and soft hairs.

Virgate. Wand-shaped; slender, straight and erect.

Viscid. Glutinous; sticky.

Whorl. A group of three or more similar organs, radiating from a node. Verticil.

Wing. Any membranous expansion.

Woolly. Clothed with long and entangled soft hairs.

Zoospore. Literally, an animal spore; a vegetable spore endowed with the power of locomotion, and therefore appearing like an animal.

Zygomorphous, Zygomorphic (flowers). Divisible into similar halves in only one plane.

INDEX

[This comprehensive index covers both volumes of the work. Volume One contains pages 1 through 503 (in part); Volume Two contains pages 503 through the end of the text.]

The Latin names of genera and species are printed in ordinary Roman font, the synonyms are indicated by *italic*, the common or English names are in SMALL CAPITALS, and the Latin family names are in large CAPITALS.

A CATALOGUE OF SELECTED DOVER BOOKS
IN ALL FIELDS OF INTEREST

A CATALOGUE OF SELECTED DOVER BOOKS
IN ALL FIELDS OF INTEREST

AMERICA'S OLD MASTERS, James T. Flexner. Four men emerged unexpectedly from provincial 18th century America to leadership in European art: Benjamin West, J. S. Copley, C. R. Peale, Gilbert Stuart. Brilliant coverage of lives and contributions. Revised, 1967 edition. 69 plates. 365pp. of text.

21806-6 Paperbound $3.00

FIRST FLOWERS OF OUR WILDERNESS: AMERICAN PAINTING, THE COLONIAL PERIOD, James T. Flexner. Painters, and regional painting traditions from earliest Colonial times up to the emergence of Copley, West and Peale Sr., Foster, Gustavus Hesselius, Feke, John Smibert and many anonymous painters in the primitive manner. Engaging presentation, with 162 illustrations. xxii + 368pp.

22180-6 Paperbound $3.50

THE LIGHT OF DISTANT SKIES: AMERICAN PAINTING, 1760-1835, James T. Flexner. The great generation of early American painters goes to Europe to learn and to teach: West, Copley, Gilbert Stuart and others. Allston, Trumbull, Morse; also contemporary American painters—primitives, derivatives, academics—who remained in America. 102 illustrations. xiii + 306pp. 22179-2 Paperbound $3.00

A HISTORY OF THE RISE AND PROGRESS OF THE ARTS OF DESIGN IN THE UNITED STATES, William Dunlap. Much the richest mine of information on early American painters, sculptors, architects, engravers, miniaturists, etc. The only source of information for scores of artists, the major primary source for many others. Unabridged reprint of rare original 1834 edition, with new introduction by James T. Flexner, and 394 new illustrations. Edited by Rita Weiss. 6⅝ x 9⅝.

21695-0, 21696-9, 21697-7 Three volumes, Paperbound $13.50

EPOCHS OF CHINESE AND JAPANESE ART, Ernest F. Fenollosa. From primitive Chinese art to the 20th century, thorough history, explanation of every important art period and form, including Japanese woodcuts; main stress on China and Japan, but Tibet, Korea also included. Still unexcelled for its detailed, rich coverage of cultural background, aesthetic elements, diffusion studies, particularly of the historical period. 2nd, 1913 edition. 242 illustrations. lii + 439pp. of text.

20364-6, 20365-4 Two volumes, Paperbound $6.00

THE GENTLE ART OF MAKING ENEMIES, James A. M. Whistler. Greatest wit of his day deflates Oscar Wilde, Ruskin, Swinburne; strikes back at inane critics, exhibitions, art journalism; aesthetics of impressionist revolution in most striking form. Highly readable classic by great painter. Reproduction of edition designed by Whistler. Introduction by Alfred Werner. xxxvi + 334pp.

21875-9 Paperbound $2.50

How to Know the Wild Flowers, Mrs. William Starr Dana. This is the classical book of American wildflowers (of the Eastern and Central United States), used by hundreds of thousands. Covers over 500 species, arranged in extremely easy to use color and season groups. Full descriptions, much plant lore. This Dover edition is the fullest ever compiled, with tables of nomenclature changes. 174 full-page plates by M. Satterlee. xii + 418pp. 20332-8 Paperbound $2.50

Our Plant Friends and Foes, William Atherton DuPuy. History, economic importance, essential botanical information and peculiarities of 25 common forms of plant life are provided in this book in an entertaining and charming style. Covers food plants (potatoes, apples, beans, wheat, almonds, bananas, etc.), flowers (lily, tulip, etc.), trees (pine, oak, elm, etc.), weeds, poisonous mushrooms and vines, gourds, citrus fruits, cotton, the cactus family, and much more. 108 illustrations. xiv + 290pp. 22272-1 Paperbound $2.00

How to Know the Ferns, Frances T. Parsons. Classic survey of Eastern and Central ferns, arranged according to clear, simple identification key. Excellent introduction to greatly neglected nature area. 57 illustrations and 42 plates. xvi + 215pp. 20740-4 Paperbound $1.75

Manual of the Trees of North America, Charles S. Sargent. America's foremost dendrologist provides the definitive coverage of North American trees and tree-like shrubs. 717 species fully described and illustrated' exact distribution, down to township, full botanical description; economic importance; description of subspecies and races; habitat, growth data; similar material. Necessary to every serious student of tree-life. Nomenclature revised to present. Over 100 locating keys. 783 illustrations. lii + 934pp. 20277-1, 20278-X Two volumes. Paperbound $6.00

Our Northern Shrubs, Harriet L. Keeler. Fine non-technical reference work identifying more than 225 important shrubs of Eastern and Central United States and Canada. Full text covering botanical description, habitat, plant lore, is paralleled with 205 full-page photographs of flowering or fruiting plants. Nomenclature revised by Edward G. Voss. One of few works concerned with shrubs. 205 plates, 35 drawings. xxviii + 521pp. 21989-5 Paperbound $3.75

The Mushroom Handbook, Louis C. C. Krieger. Still the best popular handbook: full descriptions of 259 species, cross references to another 200. Extremely thorough text enables you to identify, know all about any mushroom you are likely to meet in eastern and central U. S. A.: habitat, luminescence, poisonous qualities, use, folklore, etc. 32 color plates show over 50 mushrooms, also 126 other illustrations. Finding keys. vii + 560pp. 21861-9 Paperbound $3.95

Handbook of Birds of Eastern North America, Frank M. Chapman. Still much the best single-volume guide to the birds of Eastern and Central United States. Very full coverage of 675 species, with descriptions, life habits, distribution, similar data. All descriptions keyed to two-page color chart. With this single volume the average birdwatcher needs no other books. 1931 revised edition. 195 illustrations. xxxvi + 581pp. 21489-3 Paperbound $3.25

CATALOGUE OF DOVER BOOKS

AMERICAN FOOD AND GAME FISHES, David S. Jordan and Barton W. Evermann. Definitive source of information, detailed and accurate enough to enable the sportsman and nature lover to identify conclusively some 1,000 species and sub-species of North American fish, sought for food or sport. Coverage of range, physiology, habits, life history, food value. Best methods of capture, interest to the angler, advice on bait, fly-fishing, etc. 338 drawings and photographs. 1 + 574pp. 6⅝ x 9⅜.

22383-1 Paperbound $4.50

THE FROG BOOK, Mary C. Dickerson. Complete with extensive finding keys, over 300 photographs, and an introduction to the general biology of frogs and toads, this is the classic non-technical study of Northeastern and Central species. 58 species; 290 photographs and 16 color plates. xvii + 253pp.

21973-9 Paperbound $4.00

THE MOTH BOOK: A GUIDE TO THE MOTHS OF NORTH AMERICA, William J. Holland. Classical study, eagerly sought after and used for the past 60 years. Clear identification manual to more than 2,000 different moths, largest manual in existence. General information about moths, capturing, mounting, classifying, etc., followed by species by species descriptions. 263 illustrations plus 48 color plates show almost every species, full size. 1968 edition, preface, nomenclature changes by A. E. Brower. xxiv + 479pp. of text. 6½ x 9¼.

21948-8 Paperbound $5.00

THE SEA-BEACH AT EBB-TIDE, Augusta Foote Arnold. Interested amateur can identify hundreds of marine plants and animals on coasts of North America; marine algae; seaweeds; squids; hermit crabs; horse shoe crabs; shrimps; corals; sea anemones; etc. Species descriptions cover: structure; food; reproductive cycle; size; shape; color; habitat; etc. Over 600 drawings. 85 plates. xii + 490pp.

21949-6 Paperbound $3.50

COMMON BIRD SONGS, Donald J. Borror. 33⅓ 12-inch record presents songs of 60 important birds of the eastern United States. A thorough, serious record which provides several examples for each bird, showing different types of song, individual variations, etc. Inestimable identification aid for birdwatcher. 32-page booklet gives text about birds and songs, with illustration for each bird.

21829-5 Record, book, album. Monaural. $2.75

FADS AND FALLACIES IN THE NAME OF SCIENCE, Martin Gardner. Fair, witty appraisal of cranks and quacks of science: Atlantis, Lemuria, hollow earth, flat earth, Velikovsky, orgone energy, Dianetics, flying saucers, Bridey Murphy, food fads, medical fads, perpetual motion, etc. Formerly "In the Name of Science." x + 363pp.

20394-8 Paperbound $2.00

HOAXES, Curtis D. MacDougall. Exhaustive, unbelievably rich account of great hoaxes: Locke's moon hoax, Shakespearean forgeries, sea serpents, Loch Ness monster, Cardiff giant, John Wilkes Booth's mummy, Disumbrationist school of art, dozens more; also journalism, psychology of hoaxing. 54 illustrations. xi + 338pp.

20465-0 Paperbound $2.75

MATHEMATICAL PUZZLES FOR BEGINNERS AND ENTHUSIASTS, Geoffrey Mott-Smith. 189 puzzles from easy to difficult—involving arithmetic, logic, algebra, properties of digits, probability, etc.—for enjoyment and mental stimulus. Explanation of mathematical principles behind the puzzles. 135 illustrations. viii + 248pp.
20198-8 Paperbound $1.25

PAPER FOLDING FOR BEGINNERS, William D. Murray and Francis J. Rigney. Easiest book on the market, clearest instructions on making interesting, beautiful origami. Sail boats, cups, roosters, frogs that move legs, bonbon boxes, standing birds, etc. 40 projects; more than 275 diagrams and photographs. 94pp.
20713-7 Paperbound $1.00

TRICKS AND GAMES ON THE POOL TABLE, Fred Herrmann. 79 tricks and games—some solitaires, some for two or more players, some competitive games—to entertain you between formal games. Mystifying shots and throws, unusual caroms, tricks involving such props as cork, coins, a hat, etc. Formerly *Fun on the Pool Table*. 77 figures, 95pp.
21814-7 Paperbound $1.00

HAND SHADOWS TO BE THROWN UPON THE WALL: A SERIES OF NOVEL AND AMUSING FIGURES FORMED BY THE HAND, Henry Bursill. Delightful picturebook from great-grandfather's day shows how to make 18 different hand shadows: a bird that flies, duck that quacks, dog that wags his tail, camel, goose, deer, boy, turtle, etc. Only book of its sort. vi + 33pp. 6½ x 9¼. 21779-1 Paperbound $1.00

WHITTLING AND WOODCARVING, E. J. Tangerman. 18th printing of best book on market. "If you can cut a potato you can carve" toys and puzzles, chains, chessmen, caricatures, masks, frames, woodcut blocks, surface patterns, much more. Information on tools, woods, techniques. Also goes into serious wood sculpture from Middle Ages to present, East and West. 464 photos, figures. x + 293pp.
20965-2 Paperbound $2.00

HISTORY OF PHILOSOPHY, Julián Marías. Possibly the clearest, most easily followed, best planned, most useful one-volume history of philosophy on the market; neither skimpy nor overfull. Full details on system of every major philosopher and dozens of less important thinkers from pre-Socratics up to Existentialism and later. Strong on many European figures usually omitted. Has gone through dozens of editions in Europe. 1966 edition, translated by Stanley Appelbaum and Clarence Strowbridge. xviii + 505pp.
21739-6 Paperbound $3.00

YOGA: A SCIENTIFIC EVALUATION, Kovoor T. Behanan. Scientific but non-technical study of physiological results of yoga exercises; done under auspices of Yale U. Relations to Indian thought, to psychoanalysis, etc. 16 photos. xxiii + 270pp.
20505-3 Paperbound $2.50

Prices subject to change without notice.
Available at your book dealer or write for free catalogue to Dept. GI, Dover Publications, Inc., 180 Varick St., N. Y., N. Y. 10014. Dover publishes more than 150 books each year on science, elementary and advanced mathematics, biology, music, art, literary history, social sciences and other areas.

8641